AF768322

Springer Series in

MATERIALS SCIENCE 116

Springer Series in
MATERIALS SCIENCE

Editors: R. Hull R. M. Osgood, Jr. J. Parisi H. Warlimont

The Springer Series in Materials Science covers the complete spectrum of materials physics, including fundamental principles, physical properties, materials theory and design. Recognizing the increasing importance of materials science in future device technologies, the book titles in this series reflect the state-of-the-art in understanding and controlling the structure and properties of all important classes of materials.

99 **Self-Organized Morphology in Nanostructured Materials**
Editors: K. Al-Shamery and J. Parisi

100 **Self Healing Materials**
An Alternative Approach to 20 Centuries of Materials Science
Editor: S. van der Zwaag

101 **New Organic Nanostructures for Next Generation Devices**
Editors: K. Al-Shamery, H.-G. Rubahn, and H. Sitter

102 **Photonic Crystal Fibers**
Properties and Applications
By F. Poli, A. Cucinotta, and S. Selleri

103 **Polarons in Advanced Materials**
Editor: A.S. Alexandrov

104 **Transparent Conductive Zinc Oxide**
Basics and Applications in Thin Film Solar Cells
Editors: K. Ellmer, A. Klein, and B. Rech

105 **Dilute III-V Nitride Semiconductors and Material Systems**
Physics and Technology
Editor: A. Erol

106 **Into The Nano Era**
Moore's Law Beyond Planar Silicon CMOS
Editor: H.R. Huff

107 **Organic Semiconductors in Sensor Applications**
Editors: D.A. Bernards, R.M. Ownes, and G.G. Malliaras

108 **Evolution of Thin-Film Morphology**
Modeling and Simulations
By M. Pelliccione and T.-M. Lu

109 **Reactive Sputter Deposition**
Editors: D. Depla and S. Mahieu

110 **The Physics of Organic Superconductors and Conductors**
Editor: A. Lebed

111 **Molecular Catalysts for Energy Conversion**
Editors: T. Okada and M. Kaneko

112 **Atomistic and Continuum Modeling of Nanocrystalline Materials**
Deformation Mechanisms and Scale Transition
By M. Cherkaoui and L. Capolungo

113 **Crystallography and the World of Symmetry**
By S.K. Chatterjee

114 **Piezoelectricity**
Evolution and Future of a Technology
Editors: W. Heywang, K. Lubitz, and W. Wersing

115 **Lithium Niobate**
Defects, Photorefraction and Ferroelectric Switching
By T. Volk and M. Wöhlecke

116 **Einstein Relation in Compound Semiconductors and Their Nanostructures**
By K.P. Ghatak, S. Bhattacharya, and D. De

117 **From Bulk to Nano**
The Many Sides of Magnetism
By C.G. Stefanita

Volumes 50–98 are listed at the end of the book.

Kamakhya Prasad Ghatak
Sitangshu Bhattacharya
Debashis De

Einstein Relation in Compound Semiconductors and Their Nanostructures

With 253 Figures

Springer

Professor Dr. Kamakhya Prasad Ghatak
University of Calcutta, Department of Electronic Science
Acharya Prafulla Chandra Rd. 92, 700 009 Kolkata, India
E-mail: kamakhyaghatak@yahoo.co.in

Dr. Sitangshu Bhattacharya
Nanoscale Device Research Laboratory
Center for Electronics Design Technology
Indian Institute of Science, Bangalore-560012, India
E-mail: isbsin@yahoo.co.in

Dr. Debashis De
West Bengal University of Technology, Department of Computer Sciences and Engineering
700 064 Kolkata, India
E-mail: debashis.de@wbut.ac.in

Series Editors:

Professor Robert Hull
University of Virginia
Dept. of Materials Science and Engineering
Thornton Hall
Charlottesville, VA 22903-2442, USA

Professor R. M. Osgood, Jr.
Microelectronics Science Laboratory
Department of Electrical Engineering
Columbia University
Seeley W. Mudd Building
New York, NY 10027, USA

Professor Jürgen Parisi
Universität Oldenburg, Fachbereich Physik
Abt. Energie- und Halbleiterforschung
Carl-von-Ossietzky-Strasse 9–11
26129 Oldenburg, Germany

Professor Hans Warlimont
Institut für Festkörper-
und Werkstofforschung,
Helmholtzstrasse 20
01069 Dresden, Germany

Springer Series in Materials Science ISSN 0933-033X

ISBN 978-3-540-79556-8 e-ISBN 978-3-540-79557-5

Library of Congress Control Number: 2008931052

© Springer-Verlag Berlin Heidelberg 2009

This work is subject to copyright. All rights are reserved, whether the whole or part of the material is concerned, specifically the rights of translation, reprinting, reuse of illustrations, recitation, broadcasting, reproduction on microfilm or in any other way, and storage in data banks. Duplication of this publication or parts thereof is permitted only under the provisions of the German Copyright Law of September 9, 1965, in its current version, and permission for use must always be obtained from Springer-Verlag. Violations are liable to prosecution under the German Copyright Law.

The use of general descriptive names, registered names, trademarks, etc. in this publication does not imply, even in the absence of a specific statement, that such names are exempt from the relevant protective laws and regulations and therefore free for general use.

Typesetting: Data prepared by SPi using a Springer TEX macro package
Cover concept: eStudio Calamar Steinen
Cover production: WMX Design GmbH, Heidelberg

SPIN: 12116942 57/3180/SPi
Printed on acid-free paper

9 8 7 6 5 4 3 2 1

springer.com

Preface

In recent years, with the advent of fine line lithographical methods, molecular beam epitaxy, organometallic vapour phase epitaxy and other experimental techniques, low dimensional structures having quantum confinement in one, two and three dimensions (such as inversion layers, ultrathin films, nipi's, quantum well superlattices, quantum wires, quantum wire superlattices, and quantum dots together with quantum confined structures aided by various other fields) have attracted much attention, not only for their potential in uncovering new phenomena in nanoscience, but also for their interesting applications in the realm of quantum effect devices. In ultrathin films, due to the reduction of symmetry in the wave–vector space, the motion of the carriers in the direction normal to the film becomes quantized leading to the quantum size effect. Such systems find extensive applications in quantum well lasers, field effect transistors, high speed digital networks and also in other low dimensional systems. In quantum wires, the carriers are quantized in two transverse directions and only one-dimensional motion of the carriers is allowed. The transport properties of charge carriers in quantum wires, which may be studied by utilizing the similarities with optical and microwave waveguides, are currently being investigated. Knowledge regarding these quantized structures may be gained from original research contributions in scientific journals, proceedings of international conferences and various review articles. It may be noted that the available books on semiconductor science and technology cannot cover even an entire chapter, excluding a few pages on the Einstein relation for the diffusivity to mobility ratio of the carriers in semiconductors (DMR). The DMR is more accurate than any one of the individual relations for the diffusivity (D) or the mobility (μ) of the charge carriers, which are two widely used quantities of carrier transport in semiconductors and their nanostructures.

It is worth remarking that the performance of the electron devices at the device terminals and the speed of operation of modern switching transistors are significantly influenced by the degree of carrier degeneracy present in these devices. The simplest way of analyzing such devices, taking into account the degeneracy of the bands, is to use the appropriate Einstein relation to express the performances at the device terminals and the switching speed in terms of carrier concentration (S.N. Mohammad, *J. Phys.* ***C***, **13**, 2685 (1980)). It is well known from the fundamental works of Landsberg (P.T. Landsberg, *Proc. R. Soc.* ***A***, **213**, 226, (1952); *Eur. J. Phys*, **2**, 213, (1981)) that the Einstein relation for degenerate materials is essentially determined by their energy band structures. It has, therefore, different values in different materials having various band structures and varies with electron concentration, the magnitude of the reciprocal quantizing magnetic field, the quantizing electric field as in inversion layers, ultrathin films, quantum wires and with the superlattice period as in quantum confined semiconductor superlattices having various carrier energy spectra.

This book is partially based on our on-going researches on the Einstein relation from 1980 and an attempt has been made to present a cross section of the Einstein relation for a wide range of materials with varying carrier energy spectra, under various physical conditions.

In Chap. 1, after a brief introduction, the basic formulation of the Einstein relation for multiband semiconductors and suggestion of an experimental method for determining the Einstein relation in degenerate materials having arbitrary dispersion laws are presented. From this suggestion, one can also experimentally determine another two seemingly different but important quantities of quantum effect devices namely, the Debye screening length and the carrier contribution to the elastic constants. In Chap. 2, the Einstein relation in bulk specimens of tetragonal materials (taking n-Cd_3As_2 and n-$CdGeAs_2$ as examples) is formulated on the basis of a generalized electron dispersion law introducing the anisotropies of the effective electron masses and the spin orbit splitting constants respectively together with the inclusion of the crystal field splitting within the framework of the $\boldsymbol{k.p}$ formalism. The theoretical formulation is in good agreement with the suggested experimental method of determining the Einstein relation in degenerate materials having arbitrary dispersion laws. The results of III–V (e.g. InAs, InSb, GaAs, etc.), ternary (e.g. $Hg_{1-x}Cd_xTe$), quaternary (e.g. $In_{1-x}Ga_xAs_{1-y}P_y$ lattice matched to InP) compounds form a special case of our generalized analysis under certain limiting conditions. The Einstein relation in II–VI, IV–VI, stressed Kane type semiconductors together with bismuth are also investigated by using the appropriate energy band structures for these materials. The importance of these materials in the emergent fields of opto- and nanoelectronics is also described in Chap. 2.

The effects of quantizing magnetic fields on the band structures of compound semiconductors are more striking than those of the parabolic one and are easily observed in experiments. A number of interesting physical features originate from the significant changes in the basic energy wave vector relation of the carriers caused by the magnetic field. Valuable information could also be obtained from experiments under magnetic quantization regarding the important physical properties such as Fermi energy and effective masses of the carriers, which affect almost all the transport properties of the electron devices. Besides, the influence of cross-field configuration is of fundamental importance to an understanding of the various physical properties of various materials having different carrier dispersion relations. In Chap. 3, we study the Einstein relation in compound semiconductors under magnetic quantization. Chapter 4 covers the influence of crossed electric and quantizing magnetic fields on the Einstein relation in compound semiconductors. Chapter 5 covers the study of the Einstein relation in ultrathin films of the materials mentioned.

Since Iijima's discovery (S. Iijima, *Nature* **354**, 56 (1991)), carbon nanotubes (CNTs) have been recognized as fascinating materials with nanometer dimensions, uncovering new phenomena in different areas of nanoscience and technology. The remarkable physical properties of these quantum materials make them ideal candidates to reveal new phenomena in nanoelectronics. Chapter 6 contains the study of the Einstein relation in quantum wires of compound semiconductors, together with carbon nanotubes.

In recent years, there has been considerable interest in the study of the inversion layers which are formed at the surfaces of semiconductors in metal–oxide–semiconductor field-effect transistors (MOSFET) under the influence of a sufficiently strong electric field applied perpendicular to the surface by means of a large gate bias. In such layers, the carriers form a two dimensional gas and are free to move parallel to the surface while their motion is quantized in the perpendicular to it leading to the formation of electric subbands. In Chap. 7, the Einstein relation in inversion layers on compound semiconductors has been investigated.

The semiconductor superlattices find wide applications in many important device structures such as avalanche photodiode, photodetectors, electro-optic modulators, etc. Chapter 8 covers the study of the Einstein relation in nipi structures. In Chap. 9, the Einstein relation has been investigated under magnetic quantization in III-V, II-VI, IV-VI, HgTe/CdTe superlattices with graded interfaces. In the same chapter, the Einstein relation under magnetic quantization for effective mass superlattices has also been investigated. It also covers the study of quantum wire superlattices of the materials mentioned. Chapter 10 presents an initiation regarding the influence of light on the Einstein relation in optoelectronic materials and their quantized structures which is itself in the stage of infancy.

In the whole field of semiconductor science and technology, the heavily doped materials occupy a singular position. Very little is known regarding the dispersion relations of the carriers of heavily doped compound semiconductors and their nanostructures. Chapter 11 attempts to touch this enormous field of active research with respect to Einstein relation for heavily doped materials in a nutshell, which is itself a sea. The book ends with Chap. 12, which contains the conclusion and the scope for future research.

As there is no existing book devoted totally to the Einstein relation for compound semiconductors and their nanostructures to the best of our knowledge, we hope that the present book will be a useful reference source for the present and the next generation of readers and researchers of solid state electronics in general. In spite of our joint efforts, the production of error free first edition of any book from every point of view enjoys the domain of impossibility theorem. Various expressions and a few chapters of this book have been appearing for the first time in printed form. The positive suggestions of the readers for the development of the book will be highly appreciated.

In this book, from Chap. 2 to the end, we have presented 116 open and 60 allied research problems in this beautiful topic, as we believe that a proper identification of an open research problem is one of the biggest problems in research. The problems presented here are an integral part of this book and will be useful for readers to initiate their own contributions to the Einstein relation. This aspect is also important for PhD aspirants and researchers. We strongly contemplate that the readers with a mathematical bent of mind would invariably yearn for investigating all the systems from Chapters 2 to 12 and the related research problems by removing all the mathematical approximations and establishing the appropriate respective uniqueness conditions. Each chapter except the last one ends with a table containing the main results.

It is well known that the studies in carrier transport of modern semiconductor devices are based on the Boltzmann transport equation which can, in turn, be solved if and only if the dispersion relations of the carriers of the different materials are known. In this book, we have investigated various dispersion relations of different quantized structures and the corresponding electron statistics to study the Einstein relation. Thus, in this book, the alert readers will find information regarding quantum-confined low-dimensional materials having different band structures. Although the name of the book is extremely specific, from the content one can infer that it will be useful in graduate courses on semiconductor physics and devices in many Universities. Besides, as a collateral study, we have presented the detailed analysis of the effective electron mass for the said systems, the importance of which is already well known, since the inception of semiconductor science. Last but not the least, we do hope that our humble effort will kindle the desire of anyone engaged in materials research and device development, either in academics or in industries, to delve deeper into this fascinating topic.

Acknowledgments

Acknowledgment by Kamakhya Prasad Ghatak

I am grateful to A.N. Chakravarti, my Ph.D thesis advisor, for introducing an engineering graduate to the classics of Landau Liftsitz 30 years ago, and with whom I spent countless hours delving into the sea of semiconductor physics. I am also indebted to D. Raychaudhuri for transforming a network theorist into a quantum mechanic. I realize that three renowned books on semiconductor science, in general, and more than 200 research papers of

B.R. Nag, still fire my imagination. I would like to thank P.T. Landsberg, D. Bimberg, W.L. Freeman, B. Podor, H.L. Hartnagel, V.S. Letokhov, H.L. Hwang, F.D. Boer, P.K. Bose, P.K. Basu, A. Saha, S. Roy, R. Maity, R. Bhowmik, S.K. Dasgupta, M. Mitra, D. Chattopadhyay, S.N. Biswas and S.K. Biswas for several important interactions.

I am particularly indebted to K. Mukherjee, A.K. Roy, S.S. Baral, S.K. Roy, R.K. Poddar, N. Guhochoudhury, S.K. Sen, S. Pahari and D.K. Basu, who acted as mentors in the difficult moments of my academic career. I thank my department colleagues and the members of my research team for their help. P.K. Sarkar of the semiconductor device laboratory has always helped me. I am grateful to S. Sanyal for her help and academic advice. I also acknowledge the present Head of the Department, S.N. Sarkar, for creating an environment for the advancement of learning, which is the logo of the University of Calcutta, and helping me to win an award in research and development from the All India Council for Technical Education, India, under which the writing of many chapters of this book became a reality. Besides, this book has been completed under the grant (8023/BOR/RID/RPS-95/2007-08) as sanctioned by the said Council in their research promotion scheme 2008 of the Council.

Acknowledgment by Sitangshu Bhattacharya

I am indebted to H.S. Jamadagni and S. Mahapatra at the Centre for Electronics Design and Technology (CEDT), Indian Institute of Science, Bangalore, for their constructive guidance in spite of a tremendous research load and to my colleagues at CEDT, for their constant academic help. I am also grateful to my sister, Ms. S. Bhattacharya and my friend Ms. A. Chakraborty for their constant inspiration and encouragement for performing research work even in my tough times, which, in turn, forms the foundation of this twelve-storied book project. I am grateful to my teacher K.P. Ghatak, with whom I work constantly to understand the mysteries of quantum effect devices.

Acknowledgment by Debashis De

I am grateful to K.P. Ghatak, B.R. Nag, A.K. Sen, P.K. Roy, A.R. Thakur, S. Sengupta, A.K. Roy, D. Bhattacharya, J.D. Sharma, P. Chakraborty, D. Lockwood, N. Kolbun and A.N. Greene. I am highly indebted to my brother S. De for his constant inspiration and support. I must not allow a special thank you to my better half Mrs. S. De, since in accordance with Sanatan Hindu Dharma, the fusion of marriage has transformed us to form a single entity, where the individuality is being lost. I am grateful to the All India Council of Technical Education, for granting me the said project jointly in their research promotion scheme 2008 under which this book has been completed.

Joint Acknowledgments

The accuracy of the presentation owes a lot to the cheerful professionalism of Dr. C. Ascheron, Senior Editor, Physics Springer Verlag, Ms. A. Duhm, Associate Editor Physics, Springer and Mrs. E. Suer, assistant to Dr. Ascheron. Any shortcomings that remain are our own responsibility.

Kolkata, India
June 2008

K.P. GHATAK
S. BHATTACHARYA
D. DE

Contents

List of Symbols

α	Band nonparabolicity parameter
a	The lattice constant
a_0, b_0	The widths of the barrier and the well for superlattice structures
A_0	The amplitude of the light wave
$\overrightarrow{A}$	The vector potential
$A(E, n_z)$	The area of the constant energy 2D wave vector space for ultrathin films
B	Quantizing magnetic field
B_2	The momentum matrix element
b	Bandwidth
c	Velocity of light
C_1	Conduction band deformation potential
C_2	A constant which describes the strain interaction between the conduction and valance bands
ΔC_{44}	Second order elastic constant
ΔC_{456}	Third order elastic constant
Δ	Crystal field splitting constant
Δ_0	Interface width
$\Delta\left(\frac{1}{B}\right)$	Period of SdH oscillation
d_0	Superlattice period
D	Diffusion constant
$\frac{D}{\mu}$	Einstein relation/diffusivity-mobility ratio in semiconductors
$D_0(E)$	Density-of-states (DOS) function
$D_{\mathrm{B}}(E)$	Density-of-states function in magnetic quantization
$D_{\mathrm{B}}(E, \lambda)$	Density-of-states function under the presence of light waves
d_x, d_y, d_z	Nanothickness along the x, y and z-directions
$\Delta_{\parallel}$	Spin–orbit splitting constant parallel to the C-axis
$\Delta_{\perp}$	Spin–orbit splitting constant perpendicular to the C-axis
Δ	Isotropic spin–orbit splitting constant

d^3k	Differential volume of the k space
$\in$	Energy as measured from the center of the band gap
ε	Trace of the strain tensor
ε_0	Permittivity of free space
ε_∞	Semiconductor permittivity in the high frequency limit
ε_{sc}	Semiconductor permittivity
ΔE_g	Increased band gap
$\lvert e \rvert$	Magnitude of electron charge
E	Total energy of the carrier
E_0, ζ_0	Electric field
E_g	Band gap
E_i	Energy of the carrier in the ith band.
E_{ki}	Kinetic energy of the carrier in the ith band
E_F	Fermi energy
$\bar{E}_{FB}$	Fermi energy in the presence of cross-fields configuration
$\bar{E}_{F0}$	Fermi energy in the electric quantum limit
$\bar{E}_0$	Energy of the electric sub-band in electric quantum limit
E_{FB}	Fermi energy in the presence of magnetic quantization
E_n	Landau subband energy
E_{Fs}	Fermi energy in the presence of size quantization
E_{Fis}, E_{Fiw}	Fermi energy under the strong and weak electric field limit
$\bar{E}_{Fs}, \bar{E}_{Fw}$	Fermi energy in the n-channel inversion layer under the strong and weak electric field quantum limit
$\bar{E}_{0s}, \bar{E}_{0w}$	Subband energy under the strong and weak electric field quantum limit
$\bar{E}_{Fn}$	Fermi energy for nipis
E_{FSL}	Fermi energy in superlattices
$\vec{\varepsilon}_s$	Polarization vector
E_{FQWSL}	Fermi energy in quantum wire superlattices with graded interfaces
E_{F_L}	Fermi energy in the presence of light waves
$E_{F_{BL}}$	Fermi energy under quantizing magnetic field in the presence of light waves
E_{F2DL}	2D Fermi energy in the presence of light waves
E_{F1DL}	1D Fermi energy in the presence of light waves
E_{g_0}	Un-perturbed band-gap
Erfc	Complementary error function
Erf	Error function
E_{F_h}	Fermi energy of heavily doped materials
$\bar{E}_{hd}$	Electron energy within the band gap
F_s	Surface electric field
$F(V)$	Gaussian distribution of the impurity potential
$F_j(\eta)$	One parameter Fermi–Dirac integral of order j
f_0	Equilibrium Fermi–Dirac distribution function of the total carriers

f_{0i}	Equilibrium Fermi–Dirac distribution function of the carriers in the ith band
g_{v}	Valley degeneracy
G	Thermoelectric power under classically large magnetic field
G_0	Deformation potential constant
g^*	Magnitude of the band edge g-factor
h	Planck's constant
$\hat{H}$	Hamiltonian
$\hat{H}'$	Perturbed Hamiltonian
$H(E - E_n)$	Heaviside step function
$\widehat{i}$, $\widehat{j}$ and $\widehat{k}$	Orthogonal triads
i	Imaginary unit
I	Light intensity
j_{ci}	Conduction current contributed by the carriers of the ith band
k	Magnitude of the wave vector of the carrier
k_{B}	Boltzmann's constant
λ	Wavelength of the light
$\bar{\lambda}_0$	Splitting of the two spin-states by the spin orbit coupling and the crystalline field
$\bar{l}$, $\bar{m}$, $\bar{n}$	Matrix elements of the strain perturbation operator
L_x, L_z	Sample length along x and z directions
L_0	Superlattices period length
L_{D}	Debye screening length
m_1	Effective carrier masses at the band-edge along x direction
m_2	Effective carrier masses at the band-edge along y direction
m_3	The effective carrier masses at the band-edge along z direction
m_2'	Effective-mass tensor component at the top of the valence band (for electrons) or at the bottom of the conduction band (for holes)
m_i^*	Effective mass of the ith charge carrier in the ith band
$m_{\|}^*$	Longitudinal effective electron masses at the edge of the conduction band
$m_{\perp}^*$	Transverse effective electron masses at the edge of the conduction band
m^*	Isotropic effective electron masses at the edge of the conduction band
$m_{\perp,1}^*$, $m_{\|,1}^*$	Transverse and longitudinal effective electron masses at the edge of the conduction band for the first material in superlattice
m_{r}	Reduced mass
m_0, m	Free electron mass
m_{v}	Effective mass of the heavy hole at the top of the valance band in the absence of any field
m, n	Carbon nanotubes chiral indices
n	Landau quantum number
n_x, n_y, n_z	Size quantum numbers along the x, y and z-directions

n_{1D}, n_{2D}	1D and 2D carrier concentration
n_{2Ds}, n_{2Dw}	2D surface electron concentration under strong and weak electric field
$\bar{n}_{2Ds}$, $\bar{n}_{2Dw}$	Surface electron concentration under the strong and weak electric field quantum limit
n_i	Miniband index for nipi structures
$N_{nipi}(E)$	Density-of-states function for nipi structures
$N_{2DT}(E)$	2D Density-of-states function
$N_{2D}(E, \lambda)$	2D density-of-states function in the presence of light waves
$N_{1D}(E, \lambda)$	1D density-of-states function in the presence of light waves
n_0	Total electron concentration
$\bar{n}_0$	Electron concentration in the electric quantum limit
n_i	Carrier concentration in the ith band
P	Isotropic momentum matrix element
P_n	Available noise power
$P_{\|\|}$	Momentum matrix elements parallel to the direction of crystal axis
$P_{\perp}$	Momentum matrix elements perpendicular to the direction of crystal axis
$\vec{r}$	Position vector
S_i	Zeros of the airy function
$\vec{s}_0$	Momentum vector of the incident photon
t	Time scale
t_c	Tight binding parameter
T	Absolute temperature
$\tau_i(E)$	Relaxation time of the carriers in the ith band
$u_1(\vec{k}, \vec{r}), u_2(\vec{k}, \vec{r})$	Doubly degenerate wave functions
$V(E)$	Volume of k space
V_0	Potential barrier encountered by the electron
$V(\vec{r})$	Crystal potential
x, y	Alloy compositions
z_t	Classical turning point
μ_i	Mobility of the carriers in the ith band
μ	Average mobility of the carriers
$\zeta(2r)$	Zeta function of order $2r$
$\Gamma(j+1)$	Complete Gamma function
η	Normalized Fermi energy
η_g	Impurity scattering potential
ω_0	Cyclotron resonance frequency
θ	Angle
μ_0	Bohr magnetron,
ω	Angular frequency of light wave
$\uparrow', \downarrow'$	Spin up and down function

1

Basics of the Einstein Relation

1.1 Introduction

It is well known that the Einstein relation for the diffusivity-mobility ratio (DMR) of the carriers in semiconductors occupies a central position in the whole field of semiconductor science and technology [1] since the diffusion constant (a quantity very useful for device analysis where exact experimental determination is rather difficult) can be obtained from this ratio by knowing the experimental values of the mobility. The classical value of the DMR is equal to $(k_{\mathrm{B}}T \,/\, |e|)$, ($k_{\mathrm{B}}, T$, and $|e|$ are Boltzmann's constant, temperature and the magnitude of the carrier charge, respectively). This relation in this form was first introduced to study the diffusion of gas particles and is known as the Einstein relation [2,3]. Therefore, it appears that the DMR increases linearly with increasing T and is independent of electron concentration. This relation holds for both types of charge carriers only under non-degenerate carrier concentration although its validity has been suggested erroneously for degenerate materials [4]. Landsberg first pointed out that the DMR for semiconductors having degenerate electron concentration are essentially determined by their energy band structures [5, 6]. This relation is useful for semiconductor homostructures [7, 8], semiconductor–semiconductor heterostructures [9, 10], metals–semiconductor heterostructures [11–19] and insulator–semiconductor heterostructures [20–23]. The nature of the variations of the DMR under different physical conditions has been studied in the literature [1–3, 5, 6, 24–50] and some of the significant features, which have emerged from these studies, are:

(a) The ratio increases monotonically with increasing electron concentration in bulk materials and the nature of these variations are significantly influenced by the energy band structures of different materials;
(b) The ratio increases with the increasing quantizing electric field as in inversion layers;

(c) The ratio oscillates with the inverse quantizing magnetic field under magnetic quantization due to the Shubnikov de Hass effect;
(d) The ratio shows composite oscillations with the various controlled quantities of semiconductor superlattices.
(e) In ultrathin films, quantum wires and field assisted systems, the value of the DMR changes appreciably with the external variables depending on the nature of quantum confinements of different materials.

Before the in depth study of the aforementioned cases, the basic formulation of the DMR for multi-band non-parabolic degenerate materials has been presented in Sect. 1.2. Besides, the suggested experimental method of determining the DMR for materials having arbitrary dispersion laws has also been included in Sect. 1.3.

1.2 Generalized Formulation of the Einstein Relation for Multi-Band Semiconductors

The carrier energy spectrum in the ith band in multi-band semiconducting materials can be expressed as [31]

$$E = \left(\frac{\hbar^2 k^2}{2m_i^*(E)}\right) + E_i = E_{\mathrm{ki}} + E_i, \tag{1.1}$$

where E is the total energy of the carrier as measured from the edge of the band in the vertically upward direction, $\hbar = h / 2\pi, h$ is Planck constant, k is the magnitude of the wave vector of the carrier, $m_i^*(E)$ is the effective mass of the charge carrier, E_i is the energy of the carrier in the ith band in the z-direction and E_{ki} is the kinetic energy of the carrier in the ith band.

The carrier concentration n_i in the ith band can be written as

$$n_i(E_{\mathrm{Fi}}) = (4\pi^3)^{-1} \int f_{0i}\, \mathrm{d}^3 k, \tag{1.2}$$

where $E_{\mathrm{Fi}} = E_{\mathrm{F}} - E_i, E_{\mathrm{F}}$ is the Fermi energy, f_{0i} the Fermi–Dirac equilibrium distribution function of the carriers in the ith band can, in turn, be expressed as

$$f_{0i} = \left[1 + \exp\left[(k_{\mathrm{B}}T)^{-1}(E_{\mathrm{ki}} + E_i - E_{\mathrm{F}})\right]\right]^{-1}, \tag{1.3}$$

and $\mathrm{d}^3 k$ is the differential volume of k space.

The solution of the Boltzmann transport equation under relaxation time approximation leads to the expression of the conduction current j_{ci} contributed by the carriers in the ith band in the presence of an electric field ζ_0 in the z-direction as given by [31]

$$j_{\mathrm{ci}} = -\left(4\pi^3\right)^{-1}\left(\zeta_0 e^2 / \hbar^2\right) \int \left(\nabla_{kz} E\right)^2 \tau_i(E)\left(\partial f_{0i} / \partial E_{\mathrm{ki}}\right) \mathrm{d}^3 k = |e|\left(n_i \mu_i \zeta_0\right), \tag{1.4}$$

where μ_i is the mobility and $\tau_{\mathrm{i}}(E)$ is the relaxation time. For scattering mechanisms, for which the relaxation time approximation is invalid, (1.4) remains invariant where $\tau_{\mathrm{i}}(E)$ is being replaced by $\phi_i(E)$. The perturbation in the distribution function can be written as

$$f_i \equiv f_{0i} - \left[(\nabla_{k_z} E)\left(\frac{\partial f_{0i}}{\partial E_{\mathrm{ki}}}\right)\phi_i(E)\right]\left(\frac{e\zeta_0}{\hbar}\right),$$

The current density due to conduction mechanism can be expressed as

$$J_c = |e| \sum_i n_i \mu_i \zeta_0 = |e| \, \mu n_0 \zeta_0,$$

where μ is the average mobility of the carriers and n_0 is the total electron concentration defined by $n_0 = \sum_i n_i$.

It may be noted that the diffusion current density will also exist when the carrier concentration varies with the position and consequently the concentration gradient is being created. Let us assume that it varies along the z-direction, under these conditions, both E_{F} and E_i will in general be functions of z. The application of the same process leads to the expression of the diffusion current density contributed by the carriers in the ith band as

$$j_{\mathrm{Di}} = -\left(4\pi^3\right)^{-1}\left(\frac{e}{\hbar^2}\right)\int (\nabla_{k_z} E)^2 \tau_i(E)\left(\frac{\partial f_{0i}}{\partial z}\right) \mathrm{d}^3 k, \tag{1.5}$$

We note that

$$\frac{\partial f_{0i}}{\partial z} = \frac{\partial f_{0i}}{\partial E_{\mathrm{Fi}}}\frac{\partial E_{\mathrm{Fi}}}{\partial z} = -\frac{\partial f_{0i}}{\partial E_{\mathrm{ki}}}\frac{\partial E_{\mathrm{Fi}}}{\partial z},$$

and

$$\frac{\partial n}{\partial z} = \frac{\partial}{\partial z}\sum_i n_i\left(E_{\mathrm{Fi}}\right) = \frac{\partial E_{\mathrm{Fi}}}{\partial z}\beta_i, \tag{1.6}$$

where

$$\beta_i = \sum_j \frac{\partial n_j\left(E_{\mathrm{Fi}}\right)}{\partial E_{\mathrm{Fj}}}\frac{\partial E_{\mathrm{Fj}}}{\partial E_{\mathrm{Fi}}} \tag{1.7}$$

in which j stands for the jth band.

Using (1.5), (1.6) and (1.7), one can write

$$J_{\mathrm{Di}} = \frac{1}{4\pi^3}\frac{e}{\hbar^2}\frac{\partial n_0}{\partial z}\int (\nabla_{k_z} E)^2 \tau_i(E)\frac{\partial f_{0i}}{\partial E_{ki}}\beta_i^{-1}\,\mathrm{d}^3 k = -\frac{e}{|e|}\frac{\partial n_0}{\partial z} n_i \mu_i \beta_i^{-1}. \tag{1.8}$$

Hence the total diffusion current is given by

$$j_D = \sum_i j_{\mathrm{Di}} = -\left(\frac{e}{|e|}\right)\left(\frac{\partial n_0}{\partial z}\right)\sum_i n_i \mu_i (\beta_i)^{-1} = -De\left(\frac{\partial n_0}{\partial z}\right), \tag{1.9a}$$

where D is the diffusion constant.

Thus, we get [31]

$$D=\frac{1}{|e|}\left[\sum_i n_i\mu_i\left(\beta_i\right)^{-1}\right] \tag{1.9b}$$

and

$$\frac{D}{\mu}=\left[\frac{1}{|e|}\right]n_0\frac{\sum\limits_i n_i\mu_i\beta_i^{-1}}{\sum\limits_i n_i\mu_i} \tag{1.10}$$

When E_is are z invariant, (1.10) assumes the well known form as [31]

$$\frac{D}{\mu}=\left(\frac{n_0}{|e|}\right)/\left(\frac{\mathrm{d}n_0}{\mathrm{d}E_F}\right). \tag{1.11}$$

The electric quantum limit as in inversion layers and nipi structures refers to the lowest electric sub-band and for this particular case $i=j=0$. Therefore, (1.10) can be written as

$$\frac{D}{\mu}=\left(\frac{\bar{n}_0}{|e|}\right)/\left(\frac{\mathrm{d}\bar{n}_0}{\mathrm{d}\left(\bar{E}_{F0}-\bar{E}_0\right)}\right), \tag{1.12}$$

where $\bar{n}_0, \bar{E}_{F0}$ and $\bar{E}_0$ are the electron concentration, the energy of the electric sub-band and the Fermi energy in the electric quantum limit.

It should be noted that (1.11) is valid for different kinds of multi-band materials and low dimensional systems if the contribution of the charge density to the internal potential is small except for inversion layers and nipi structures. For these cases (1.10) should be used for the evaluation of DMR. For inversion layers and nipis under the electric quantum limit and for heavily doped semiconductors, (1.12) may be used.

1.3 Suggestions for the Experimental Determination of the Einstein Relation in Semiconductors Having Arbitrary Dispersion Laws

(a) It is well-known that the thermoelectric power of the carriers in semiconductors in the presence of a classically large magnetic field is independent of scattering mechanisms and is determined only by their energy band spectra [51]. The magnitude of the thermoelectric power G can be written as [51]

$$G=\frac{1}{|e|\,Tn_0}\int\limits_{-\infty}^{\infty}\left(E-E_{\mathrm{F}}\right)R\left(E\right)\left[-\frac{\partial f_0}{\partial E}\right]\mathrm{d}E, \tag{1.13}$$

where $R\,(E)$ is the total number of states. Equation (1.13) can be written under the condition of carrier degeneracy [52,53] as

$$G = \left(\frac{\pi^2 k_B{}^2 T}{3\,|e|\,n_0}\right)\left(\frac{\partial n_0}{\partial E_F}\right). \tag{1.14}$$

The use of (1.11) and (1.14) leads to the result [52]

$$\frac{D}{\mu} = \left(\frac{\pi^2 k_B{}^2 T}{3\,|e|^2\,G}\right). \tag{1.15}$$

Thus, the DMR for degenerate materials can be determined by knowing the experimental values of G.

The suggestion for the experimental determination of the DMR for degenerate semiconductors having arbitrary dispersion laws as given by (1.15) does not contain any energy band constants. For a fixed temperature, the DMR varies inversely as G. Only the experimental values of G for any material as a function of electron concentration will generate the experimental values of the DMR for that range of n_0 for that system. Since G decreases with increasing n_0, from (1.15) one can infer that the DMR will increase with increase in n_0. This statement is the compatibility test so far as the suggestion for the experimental determination of DMR for degenerate materials is concerned.
(b) For inversion layers and the nipi structures, under the condition of electric quantum limit, (1.13) assumes the form

$$G = \left(\frac{\pi^2 k_B{}^2 T}{3\,|e|\,\bar{n}_0}\right)\left[\frac{\mathrm{d}\bar{n}_0}{\mathrm{d}\left(\bar{E}_{F0} - \bar{E}_0\right)}\right]. \tag{1.16}$$

Using (1.16) and (1.12) one can again obtain the same (1.15). For quantum wires and heterostructures with small charge densities, the relation between D/μ and G is thus given by (1.15). Equation (1.15) is also valid under magnetic quantization and also for cross-field configuration. Thus, (1.15) is independent of the dimensions of quantum confinement. We should note that the present analysis is not valid for totally k-space quantized systems such as quantum dots, magneto-inversion and accumulation layers, magneto size quantization, magneto nipis, quantum dot Superlattices and quantum well Superlattices under magnetic quantization. Under the said conditions, the electron motion is possible in the broadened levels. The experimental results of G for degenerate materials will provide an experimental check on the DMR and also a technique for probing the band structure of degenerate compounds having arbitrary dispersion laws.
(c) In accordance with Nag and Chakravarti [32]

$$\frac{D}{\mu} = P_n\,|e|\,b, \tag{1.17}$$

where P_n is the available noise power in the band width b. We wish to remark that (1.17) is valid only for semiconductors having non-degenerate electron

concentration, whereas the compound small gap semiconductors are degenerate in general.
(d) In this context, it may be noted that the results of this section find the following two important applications in the realm of quantum effect devices:
(1) It is well known that the Debye screening length (DSL) of the carriers in the semiconductors is a fundamental quantity, characterizing the screening of the Coulomb field of the ionized impurity centers by the free carriers. It affects many special features of the modern semiconductor devices, the carrier mobilities under different mechanisms of scattering, and the carrier plasmas in semiconductors [53–55]. The DSL (L_{D})can, in general, be written as [54–56]

$$L_{\mathrm{D}} = \left(\frac{|e|^2}{\varepsilon_{\mathrm{sc}}} \frac{\partial n_0}{\partial E_{\mathrm{F}}} \right)^{-1/2}, \tag{1.18}$$

where $\varepsilon_{\mathrm{sc}}$ is the semiconductor permittivity.

Using (1.18) and (1.14), one obtains

$$L_{\mathrm{D}} = \left(3\,|e|^3\, n_0 G \,/\, \varepsilon_{\mathrm{sc}} \pi^2 k_{\mathrm{B}}^2 T \right)^{-1/2}. \tag{1.19}$$

Therefore, we can experimentally determine L_D by knowing the experimental curve of G vs. n_0 at a fixed temperature.
(2) The knowledge of the carrier contribution to the elastic constants are useful in studying the mechanical properties of the materials and has been investigated in the literature [57–60]. The electronic contribution to the second- and third-order elastic constants can be written as [57–60]

$$\Delta C_{44} = -\frac{G_0^2}{9} \frac{\partial n_0}{\partial E_{\mathrm{F}}}, \tag{1.20}$$

and

$$\Delta C_{456} = \frac{G_0^3}{27} \frac{\partial^2 n_0}{\partial E_{\mathrm{F}}^2}, \tag{1.21}$$

where G_0 is the deformation potential constant. Thus, using (1.14), (1.20) and (1.21), we can write

$$\Delta C_{44} = \left[-n_0 G_0^2\, |e|\, G \,/\, (3\pi^2 k_{\mathrm{B}}^2 T) \right], \tag{1.22}$$

and

$$\Delta C_{456} = \left(n_0\, |e|\, G_0^3 G^2 \,/\, (3\pi^4 k_{\mathrm{B}}^3 T) \right) \left(1 + \frac{n_0}{G} \frac{\partial G}{\partial n_0} \right). \tag{1.23}$$

Thus, again the experimental graph of G vs. n_0 allows us to determine the electronic contribution to the elastic constants for materials having arbitrary spectras.

1.4 Summary

Section 1.2 of this chapter presents the expression of the Einstein relation together with the special practical cases. The formulation of the Einstein relation requires the relation between the electron concentration and the Fermi energy, which, in turn, is determined by the respective energy band structure. For various materials the electron dispersion relations are different and consequently all the subsequent formulations change radically introducing new information. The dispersion relation for bulk materials gets modified under magnetic quantization, in inversion layers, ultrathin films, quantum wires, and with various types of semiconductor superlattices. The electron energy spectrum also changes in a fundamental way for heavily doped semiconductors and also in the presence of external photo-excitation, respectively. We shall study these aspects in the incoming chapters. The experimental determination of DMR has been investigated in Sect. 1.3 for materials having arbitrary band structures and this suggestion is dimension independent. Besides, the experimental methods for determining the Debye screening length and the

Table 1.1. Main results of Chap. 1

(a) The generalized expression for the DMR can be written as

$$\frac{D}{\mu} = \left(\frac{1}{|e|}\right) n_0 \frac{\sum\limits_i n_i \mu_i \beta_i^{-1}}{\sum\limits_i n_i \mu_i}, \quad (1.10)$$

For E_i's independent of z, (1.10) gets simplified to the well-known form as

$$\frac{D}{\mu} = \left(\frac{n_0}{|e|}\right) \Big/ \left(\frac{\mathrm{d}n_0}{\mathrm{d}E_{\mathrm{F}}}\right). \quad (1.11)$$

For inversion layers and nipis under electric quantum limit, (1.10) transforms into the form

$$\frac{D}{\mu} = \left(\frac{\bar{n}_0}{|e|}\right) \Big/ \left(\frac{\mathrm{d}\bar{n}_0}{\mathrm{d}\left(\bar{E}_{\mathrm{F}0} - \bar{E}_0\right)}\right). \quad (1.12)$$

(b) The DMR, the screening length and the carrier contribution to the elastic constants can be experimentally determined by knowing the experimental curve of the thermoelectric power under large magnetic field vs. the carrier concentration as given by the following, respectively.

$$\frac{D}{\mu} = \left(\frac{\pi^2 k_{\mathrm{B}}^2 T}{3|e|^2 G}\right), \quad (1.15)$$

$$L_{\mathrm{D}} = \left(3|e|^3 n_0 G \,/\, \varepsilon_{\mathrm{sc}} \pi^2 k_{\mathrm{B}}^2 T\right)^{-1/2}, \quad (1.19)$$

$$\Delta C_{44} = [-n_0 G_0^2 |e| G \,/\, (3\pi^2 k_{\mathrm{B}}^2 T)], \quad (1.22)$$

$$\Delta C_{456} = \left(n_0 |e| G_0^3 G^2 \,/\, (3\pi^4 k_{\mathrm{B}}^3 T)\right) \left(1 + \frac{n_0}{G}\frac{\partial G}{\partial n_0}\right), \quad (1.23)$$

carrier contribution to the elastic constants have also been suggested in this context. As a condensed presentation, the main results have been presented in Table 1.1.

References

1. H. Kroemer, IEEE Trans. Electron Devices **25**, 850 (1978)
2. A. Einstein, Ann. der Physik **17**, 549 (1905); W. Nernst, Z. Phys. Chem. Leipzig **2**, 613 (1888); J.S. Townsend, Trans. R. Soc. **193A**, 129 (1900)
3. C. Wagner, Z. Physik. Chem. **B21**, 24 (1933); C. Herring, M.H. Nichols, Rev. Mod. Phys. **21**, 185 (1949); P.T. Landsberg, *Thermodynamics and Statistical Mechanics.* (Oxford University Press, Oxford, 1978); *In Recombination in Semiconductors.* (Cambridge University Press, UK, 1991)
4. R.W. Lade, Proc. IEEE **52**, 743 (1965)
5. P.T. Landsberg, Proc. R. Soc. **A 213**, 226 (1952); Proc. Phys. Soc. **A 62**, 806 (1949)
6. P.T. Landsberg, Eur. J. Phys **2**, 213 (1981)
7. C.H. Wang, A. Neugroschel, IEEE Electron. Dev. Lett. **ED-11**, 576 (1990)
8. I.-Y. Leu, A. Neugroschel, IEEE Trans. Electron. Dev. **ED-40**, 1872 (1993)
9. F. Stengel, S.N. Mohammad, H. Morkoç, J. Appl. Phys. **80**, 3031 (1996)
10. H.-J. Pan, W.-C. Wang, K.-B. Thai, C.-C. Cheng, K.-H. Yu, K.-W. Lin,C.-Z. Wu, W.-C. Liu, Semiconduct. Sci. Technol. **15**, 1101 (2000)
11. S.N. Mohammad, J. Appl. Phys. **95**, 4856 (2004)
12. V.K. Arora, Appl. Phys. Lett. **80**, 3763 (2002)
13. S.N. Mohammad, J. Appl. Phys. **95**, 7940 (2004)
14. S.N. Mohammad, Philos. Mag. **84**, 2559 (2004)
15. S.N. Mohammad, J. Appl. Phys. **97**, 063703 (2005)
16. K. Suzue, S.N. Mohammad, Z.F. Fan, W. Kim, O. Aktas, A.E. Botchkarev, H. Morkoç, J. Appl. Phys. **80**, 4467 (1996)
17. S.N. Mohammad, Z.F. Fan, W. Kim, O. Aktas, A.E. Botchkarev, A. Salvador, H. Morkoç, Electron. Lett. **32**, 598 (1996)
18. Z. Fan, S.N. Mohammad, W. Kim, O. Aktas, A.E. Botchkarev, K. Suzue, H. Morkoç, J. Electron. Mater. **25**, 1703 (1996)
19. C. Lu, H. Chen, X. Lv, X. Xia, S.N. Mohammad, J. Appl. Phys. **91**, 9216 (2002)
20. S.G. Dmitriev, Yu.V. Markin, Semiconductors **34**, 931 (2000)
21. M. Tao, D. Park, S.N. Mohammad, D. Li, A.E. Botchkerav, H. Morkoç, Philos. Mag. B **73**, 723 (1996)
22. D.G. Park, M. Tao, D. Li, A.E. Botchkarev, Z. Fan, S.N. Mohammad, H. Morkoç, J. Vac. Sci. Technol. B **14**, 2674 (1996)
23. Z. Chen, D.G. Park, S.N. Mohammad, H. Morkoç, Appl. Phys. Lett. **69**, 230 (1996)
24. P.T. Landsberg, J. Appl. Phys. **56**, 1696 (1984); P.T. Landsberg, A.G. Guy, Phys. Rev. B **28**, 1187 (1983); P.T. Landsberg, Phys. Rev. B **33**, 8321 (1986); S.A. Hope, G. Feat, P.T. Landsberg, J. Appl. Phys. **61**, 4909 (1987); Y. Roichman, N. Tessler, Appl. Phys. Lett. **80**, 1948 (2002); J.M.H. Peters, Eur. J. Phys. **3**, 19 (1982); H. Van Cong, S. Brunet, S. Charar, Phys. Stat. Solidi B **109**, K1 (1982); H. Van Cong, Phys. Stat. Solidi A **56**, 395 (1979); H. Van Cong, Solid State Electron. **24**, 495 (1981)

25. S.N. Mohammad, S.T.H. Abidi, J. Appl. Phys. **61**, 4909 (1987); S.N. Mohammad, S.T.H. Abidi, Solid State Electron. **27**, 1153 (1985); S.N. Mohammad, S.T.H. Abidi, J. Appl. Phys. **56**, 3341 (1984); M.A. Sobhan, S.N. Mohammad, J. Appl. Phys. **58**, 2634 (1985); S.N. Mohammad, A.V. Bemis, IEEE Trans. Electron. Dev. **ED-39**, 282 (1992); S.N. Mohammad, R.L. Carter, Philos. Mag. B **72**, 13 (1995); S.N. Mohammad, Solid State Electron. **46**, 203 (2002); S.N. Mohammad, J. Chen, J.-I. Chyi, H. Morkoç, Appl. Phys. Lett. **56**, 937 (1990)
26. P.T. Landsberg, S.A. Hope, Solid State Electron. **20**, 421 (1977); S.A. Hope, G. Feat, P.T. Landsberg, J. Phys. A. Math. Gen. **14**, 2377 (1981)
27. W. Elsäber, E.O. Göbel, Electron. Lett. **19**, 335 (1983); R. Hilfer, A. Blumen, Phys. Rev. A **37**, 578 (1988); T.G. Castner, Phys. Rev. B **55**, 4003 (1997); E. Barkai, V.N. Fleurov, Phys. Rev. E. **58**, 1296 (1998); Y.Y. Roichman, N. Tessler, Appl. Phys. Lett. **80**, 1948 (2002); T.H. Nguyen, S.K. O'Leary, Appl. Phys. Lett. **83**, 1998 (2003); T.H. Nguyen, S.K. O'Leary, J. Appl. Phys. **98**, 076102 (2005); C.G. Rodrigues, Á.R. Vasconcellos, R. Luzzi, J. Appl. Phys. **99**, 073701 (2006)
28. R.K. Jain, Phys. Stat. Sol. (a) **42**, K221 (1977); B.A. Aronzon, E.Z. Meilikhov, Phys. Stat. Sol. (a)**19**, 313 (1973)
29. A.N. Chakravarti, B.R. Nag, Int. J. Electron. **37**, 281 (1974); P.N. Butcher, A.N. Chakravarti, S. Swaminathan, Phys. Stat. Sol. (a) **25**, K47 (1974)
30. B.R. Nag, A.N. Chakravarti, Solid State Electron. **18**, 109 (1975); B.R. Nag, A.N. Chakravarti, Phys. Stat. Sol. (a) **22**, K153 (1974)
31. B.R. Nag, A.N. Chakravarti, P.K. Basu, Phys. Stat. Sol. (a) **68**, K75 (1981)
32. B.R. Nag, A.N. Chakravarti, Phys. Stat Sol. (a) **67**, K113 (1981)
33. A.N. Chakravarti, D.P. Parui, Phys. Lett. **40A**, 113 (1972); A.N. Chakravarti, D.P. Parui, Phys. Lett. **43A**, 60 (1973); A.N. Chakravarti, D.P. Parui, Phys. Stat. Sol. (a) **14**, K23 (1972); A.N. Chakravarti, D.P. Parui, Phys. Stat. Sol. (a) **14**, K55 (1972); A.N. Chakravarti, D.P. Parui, Canad. J. Phys. **51**, 451 (1973); D. Mukherjee, A.N. Chakravarti, B.R. Nag, Phys. Stat. Sol (a) **26**, K27 (1974); S. Ghosh, A.N. Chakravarti, Phys. Stat. Sol. (b), **147**, 355 (1988)
34. S. Choudhury, D. De, S. Mukherjee, A. Neogi, A. Sinha, M. Pal, S. K. Biswas, S. Pahari, S. Bhattacharya, K.P. Ghatak, J. Comp. Theo. Nanosci. **5**, 375 (2008); S. Mukherjee, S.N. Mitra, P.K. Bose, A.R. Ghatak, A. Neogi, J.P. Banerjee, A. Sinha, M. Pal, S. Bhattacharya, K.P. Ghatak, J. Comp. Theo. Nanosci. **4**, 550 (2007); A. Khan, A. Das, Appl. Phys. A **89**, 695 (2007)
35. J.P. Bouchaud, A. Georges, Phys. Rep. **195**, 127 (1996); V. Blickle, T. Speck, C. Lutz, U. Seifert, C. Bechinger, Phys. Rev. Lett. 98, 210601 (2007); Y. Kang, E. Jean, C.M. Fortmonn, Appl. Phys. Lett. **88**, 112110 (2006); F. Neumann, Y.A. Genenko, H.V. Seggern, J. Appl. Phys. **99**, 013704 (2006); J. van de Lagemaat, Phys. Rev. B. **73**, 235319 (2005); Q. Gu, E.A. Schiff, S. Grneber, F. Wang, R. Schwarz, Phys. Rev. Lett. **76**, 3196 (1996); M.Y. Azbel, Phys. Rev. B. **46**, 15004 (1992)
36. A.H. Marshak, Solid State Electron. **30**, 1089 (1987); A.H. Marshak, C.M.V. Vliet, Proc. IEEE **72**, 148 (1984); C.M.V. Vliet, A. van der Zeil, Solid State Electron. **20**, 931 (1977)
37. A.N. Chakravarti, K.P. Ghatak, A. Dhar, K.K. Ghosh, S. Ghosh, Appl. Phys. **A26**, 169 (1981)
38. B.R. Nag, *Electron Transport in Compound Semiconductors.* (Springer-Verlag, Germany, 1980); O. Madelung, *Semiconductors: Data Handbook*, 3rd edn.

(Springer, Berlin, 2004); M. Krieehbaum, P. Kocevar, H. Pascher, G. Bauer, *IEEE QE* **24**, 1727 (1988)
39. K.P. Ghatak, S.N. Biswas, Nanostruct. Mater. **2**, 91 (1993); K.P. Ghatak, S.N. Biswas, J. Appl. Phys. **70**, 4309 (1991); K.P. Ghatak, B. Mitra, M. Mondal, Ann. der Physik **48**, 283 (1991); B. Mitra, K.P. Ghatak, Physica Scripta **42**, 103 (1990); B. Mitra, K.P. Ghatak, in Phys. Lett. **135A**, 397 (1989); M. Mondal, K.P. Ghatak, Ann. der Physik **46**, 502 (1989); K.P. Ghatak, N. Chattopadhyay, M. Mondal, J. Appl. Phys. **63**, 4536 (1988); J. Low Temp. Phys. **73**, 321 (1988)
40. K.P. Ghatak, A. Ghoshal, S.N. Biswas, Nouvo Cimento **15D**, 39 (1993)
41. K.P. Ghatak, D. Bhattacharyya, Phys. Lett. **A 184**, 366 (1994)
42. K.P. Ghatak, D. Bhattacharyya, Physica Scripta **52**, 343 (1995); M. Mondal, K.P. Ghatak, J. Mag. Mag. Mater. **62**, 115 (1986)
43. K.P. Ghatak, B. Nag, D. Bhattacharyya, J. Low Temp. Phys. **14**, 1 (1995)
44. K.P. Ghatak, M. Mondal, Thin Solid Films **148**, 219 (1987)
45. K.P. Ghatak, A.K. Choudhury, S. Ghosh, A.N. Chakravarti, Appl. Phys. **23**, 241 (1980)
46. K.P. Ghatak, Influence of band structure on some quantum processes in tetragonal semiconductors, D. Eng. Thesis, Jadavpur University, Kolkata, India, 1991
47. K.P. Ghatak, N. Chattopadhyay, M. Mondal, Appl. Phys. A **44**, 305 (1987)
48. S.N. Biswas, K.P. Ghatak, in *Proceedings of the Society of Photo-optical and Instrumentation Engineers (SPIE)*, Quantum Well and Superlattice Physics, USA, 1987, vol. 792, p. 239; K.P. Ghatak, M. Mondal, S. Bhattacharyya, SPIE, USA, 1990, vol. 1284, p. 113; K.P. Ghatak, S. Bhattacharyya, M. Mondal, SPIE, USA, 1990, vol. 1307, p. 588; K.P. Ghatak, B. De, Defect engineering in semiconductor growth, processing and device technology materials research society proceedings (MRS) Spring meeting, 1992, vol. 262, p. 911; K.P. Ghatak, B. De, The Proceedings of Wide Band Gap Semiconductors Symposium, MRS, 1992, p. 377; S. Bhattacharya, K.P. Ghatak, S.N. Biswas, SPIE, 836, Optoelectronic materials, Devices, Packaging Interconnects, 1988, pp. 72–87
49. M. Mondal, K.P. Ghatak, J. Phys. C (Sol. State.) **20**, 1671 (1987); M. Mondal, S.N. Banik, K.P. Ghatak, Canad. J. Phys. **67**, 72 (1989)
50. K.P. Ghatak, M. Mondal, J. Appl. Phys. **70**, 1277 (1992)
51. W. Zawadzki, Adv. Phys. **23**, 435 (1974); B.M. Askerov, N.F. Gashimzade, M.M. Panakhov, Sov. Phys. Solid State **29**, 465 (1987); S.P. Zelenin, A.S. Kondratev, A.I. Kuchma, Sov. Phys. Semicond. **16**, 355 (1982); A.V. Germaneko, G.M. Minkov, Phys. Stat. Sol. (b) **184**, 9 (1994); W. Zawadzki, in *Two-Dimensional Systems, Hetrostructures and Superlattices*, vol. 53, ed. by G. Bower, F. Kuches, H. Haiunich, Springer Ser. Solid-State Sci. (Springer, Berlin, 1984), p. 79; B. Mitra, K.P. Ghatak, Phys. Lett. **141**, 81 (1989); L.J. Singh, S. Choudhury, D. Baruah, S.K. Biswas, S. Pahari, K.P. Ghatak, Physica B **368**, 188 (2005); K.P. Ghatak, B. De, B. Nag, P.K. Chakraborty, Nonlin. Opt. Quant. Opt. **16**, 212 (1996); K.P. Ghatak, M. Mitra, B. Goswami, B. Nag, Nonlin. Opt. **16**, 167 (1996); K.P. Ghatak, B. Nag, Nanostruct. Mater. **2**, 14 (1995); K.P. Ghatak, B. Mitra, Nouvo Cimento **15D**, 97 (1993)
52. S.K. Biswas, A.R. Ghatak, A. Neogi, A. Sharma, S. Bhattacharya, K.P. Ghatak, Physica E **36**, 163 (2007); K.P. Ghatak, S. Bhattacharya, S. Bhowmik, R. Benedictus, S. Choudhury, J. Appl. Phys. **103**, 1 (2008); M. Mondal, A. Ghoshal, K.P. Ghatak, Nouvo Cimento D **14D**, 63 (1992); K.P. Ghatak, Nouvo Cimento D **13D**, 1321 (1992); K.P. Ghatak, M. Mondal, J. Appl. Phys.

70, 299 (1991); K.P. Ghatak, S.N. Biswas, J. Low Temp. Phys. **78**, 219 (1990); K.P. Ghatak, M. Mondal, J. Appl. Phys. **65**, 3480 (1989); J. Appl. Phys. **66**, 3056 (1989)

53. K.P. Ghatak, B. De, M. Modal, S.N. Biswas, Epitaxial Hetrostructures, MRS Symposium Proceedings Spring Meeting **198**, 327 (1990); K.P. Ghatak, S.N. Biswas, Long Wave Length Semiconductor Devices, Materials and Processes MRS Symposium Proceedings, MRS Spring Meeting **216**, 465 (1990); K.P. Ghatak, B. De, Modern Perspective on Thermoelectrics and Related Materials, MRS Symposium Proceedings, Spring Meeting **234**, 55 (1991); K.P. Ghatak, B. De, Modern Perspective on Thermoelectrics and Related Materials, MRS Symposium Proceedings, Spring Meeting **234**, 59 (1991); K.P. Ghatak, Proceedings of SPIE, USA, Fiber Optic and Laser Sensors IX **1584**, 435 (1992)
54. R.B. Dingle, Phil Mag **46**, 813 (1955); D. Redfield, M.A. Afromowitz, Phil. Mag. **18,** 831 (1969); H.C. Cassey, F. Stern, J. Appl. Phys. **47**, 631 (1976); M. Mondal, K.P. Ghatak, Physics Lett. **102A**, 54 (1984); P.K. Chakraborty, G.C. Datta, K.P. Ghatak, Physica Scripta **68**, 368 (2003); B. Mitra, D.K. Basu, B. Nag, K.P. Ghatak, Nonlinear Opt. **17**, 171 (1997); K.P. Ghatak, S. Bhattacharya, J. Appl. Phys. **102**, 073704 (2007)
55. K.P. Ghatak, S. Bhattacharya, H. Saikia, D. Baruah, A. Saikia, K.M. Singh, A. Ali, S.N. Mitra, P.K. Bose, A. Sinha, J. Comput. Theor. Nanosci. **3**, 727 (2006)
56. E.O. Kane, Solid State Electron. **8**, 3 (1985); T. Ando, A.H. Fowler, F. Stern, Rev. Modern Phys. **54**, 437 (1982); P.K. Basu, *Optical Processes in Semiconductors.* (Oxford University Press, 2001); A.N. Chakravarti, D. Mukherjee, Phys. Lett. **53A**, 403 (1975); A.N. Chakravarti, S. Swaminathan, Phys. Stat. Sol. (a) **23**, K191 (1974); A.N. Chakravarti, Phys. Stat. Sol. (a) **25**, K 105 (1974)
57. A.K. Sreedhar, S.C. Gupta, Phys. Rev. B **5**, 3160 (1972); R.W. Keyes, IBM. J. Res. Develop. **5**, 266 (1961); R.W. Keyes, Solid State Phys. **20**, 37 (1967)
58. S. Bhattacharya, S. Chowdhury, K.P. Ghatak, J. Comput. Theor. Nanosci. **3**, 423 (2006); S. Choudhury, L.J. Singh, K.P. Ghatak, Physica B **365**, 5 (2005); L.J. Singh, S. Choudhary, A. Mallik, K.P. Ghatak, J. Comput. Theor. Nanosci. **2**, 287 (2005); K. P. Ghatak, J.Y. Siddiqui, B. Nag, Phys. Lett. A **282**, 428 document (2001); K.P. Ghatak, J.P. Banerjee, B. Nag, J. Appl. Phys. **83**, 1420 (1998); B. Nag, K.P. Ghatak, Nonlin. Opt. **19**, 1 (1998); K.P. Ghatak, B. Nag, Nanostruct. Mater. **10**, 923 (1998); B. Nag, K.P. Ghatak, J. Phys. Chem. Sol. **58**, 427 (1997)
59. K.P. Ghatak, D.K. Basu, B. Nag, J. Phys. Chem. Sol. **58**, 133 (1997); K.P. Ghatak, J.P. Banerjee, B. Goswami, B. Nag, Nonlin. Opt. Quant. Opt. **16**, 241 (1996); K.P. Ghatak, J.P. Banerjee, D. Bhattacharyya, B. Nag, Nanotechnology **7**, 110 (1996); K. P. Ghatak, J.P. Banerjee, M. Mitra, B. Nag, Nonlin. Opt. **17**, 193 (1996); B. Nag, K. P. Ghatak, Phys. Scr. **54**, 657 (1996); K.P. Ghatak, B. Mitra, Phys. Scr. **46**, 182 (1992); K.P. Ghatak, Int. J. Electron. **71**, 239 (1991)
60. K.P. Ghatak, B. De, S.N. Biswas, M. Mondal, Mechanical Behavior of Materials and Structures in Microelectronics, MRS Symposium Proceedings, Spring Meeting **2216**,191 (1991); K.P. Ghatak, B. De, MRS Symposium Proceedings **226**, 191 (1991); K.P. Ghatak, B. Nag, G. Majumdar, Proceedings of MRS **379**, 109 (1995); D. Baruah, S. Choudhury, K.M. Singh, K.P. Ghatak, J. Phys. Conf. Series **61**, 80 (2007)

2

The Einstein Relation in Bulk Specimens of Compound Semiconductors

2.1 Investigation on Tetragonal Materials

2.1.1 Introduction

It is well known that the $A^2_{III}B^5_{II}$ and the ternary chalcopyrite compounds are called tetragonal materials due to their tetragonal crystal structures [1]. These materials find extensive use in non-linear optical elements [2], photodetectors [3] and light emitting diodes [4]. Rowe and Shay [5] showed that the quasi-cubic model [6] can be used to explain the observed splitting and symmetry properties of the band structure at the zone center of **k** space of the aforementioned materials. The s-like conduction band is singly degenerate and the p-like valence bands are triply degenerate. The latter splits into three subbands because of the spin–orbit and the crystal field interactions. The largest contribution to the crystal field splitting is from the non-cubic potential [7]. The experimental results on the absorption constants, the effective mass, and the optical third order susceptibility indicate that the fact that the conduction band in the same materials corresponds to a single ellipsoidal revolution at the zone center in k-space [1, 8]. Introducing the crystal potential in the Hamiltonian, Bodnar [9] derived the electron dispersion relation in the same material under the assumption of an isotropic spin–orbit splitting constant. It would, therefore, be of much interest to investigate the DMR in these compounds by including the anisotropies of the spin–orbit splitting constant and, the effective electron mass together with the inclusion of crystal field splitting, within the framework of **k.p** formalism since, these are the important physical features of such materials [1].

In what follows, in Sect. 2.1.2 on the theoretical background the expressions for the electron concentration and the DMR for tetragonal compounds have been derived on the basis of the generalized dispersion relation. In Sect. 2.1.3, it has been shown that the corresponding results for III–V, ternary and quaternary materials form special cases of our generalized analysis. The expressions for n_0 and DMR for semiconductors whose energy band structures are defined by the two-band model of Kane and that of parabolic energy bands have further been formulated under certain constraints. For the purpose of numerical computations, n-Cd_3As_2 and n-$CdGeAs_2$ have been used as examples of $A^2_{III}B^5_{II}$ and the ternary chalcopyrite compounds, which are being extensively used in Hall pick-ups, thermal detectors, and non-linear optics [3]. In addition, the DMR has also been numerically investigated by taking n-InAs and n-InSb as examples of III–V semiconductors, n-$Hg_{1-x}Cd_xTe$ as an example of ternary compounds and n-$In_{1-x}Ga_xAs_yP_{1-y}$ lattice matched to InP as example of quaternary materials in accordance with the three and the two band models of Kane together with parabolic energy bands, respectively, for the purpose of relative comparison. The importance of the aforementioned materials in electronics has been discussed in Sect. 2.1.3. Section 2.1.4 contains the results and discussions.

2.1.2 Theoretical Background

The form of **k.p** matrix for tetragonal semiconductors can be expressed, extending Bodnar's [9] relation, as

$$H = \begin{bmatrix} H_1 & H_2 \\ H_2{}^+ & H_1 \end{bmatrix}, \tag{2.1}$$

where $H_1 \equiv \begin{bmatrix} E_g & 0 & P_{\|}k_z & 0 \\ 0 & (-2\Delta_{\|}/3) & (\sqrt{2}\Delta_{\perp}/3) & 0 \\ P_{\|}k_z & (\sqrt{2}\Delta_{\perp}/3) & -(\delta+\frac{1}{3}\Delta_{\|}) & 0 \\ 0 & 0 & 0 & 0 \end{bmatrix}$ and $H_2 \equiv \begin{bmatrix} 0 & -f_{,+} & 0 & f_{,-} \\ f_{,+} & 0 & 0 & 0 \\ 0 & 0 & 0 & 0 \\ f_{,+} & 0 & 0 & 0 \end{bmatrix}$,

in which E_g is the band gap, $P_{\|}$ and $P_{\perp}$ are the momentum matrix elements parallel and perpendicular to the direction of crystal axis respectively, δ is the crystal field splitting constant, $\Delta_{\|}$ and $\Delta_{\perp}$ are the spin–orbit splitting constants parallel and perpendicular to the C-axis respectively, $f_{,\pm} \equiv \left(P_{\perp}/\sqrt{2}\right)(k_x \pm ik_y)$ and $i = \sqrt{-1}$. Thus, neglecting the contribution of the higher bands and the free electron term, the diagonalization of the above matrix leads to the dispersion relation of the conduction electrons in bulk specimens of tetragonal compounds [1] as

$$\psi_1(E) = \psi_2(E)\,k_s^2 + \psi_3(E)\,k_z^2, \tag{2.2}$$

where

$$\psi_1(E) \equiv E(E+E_g)\left[(E+E_g)\left(E+E_g+\Delta_{\|}\right)+\delta\left(E+E_g+\frac{2}{3}\Delta_{\|}\right) + \frac{2}{9}\left(\Delta_{\|}^2-\Delta_{\perp}^2\right)\right],\quad k_s{}^2 = k_x{}^2 + k_y{}^2,$$

$$\psi_2(E) \equiv \frac{\hbar^2 E_{\mathrm{g}}\left(E_{\mathrm{g}}+\Delta_{\perp}\right)}{\left[2m_{\perp}^*\left(E_{\mathrm{g}}+\frac{2}{3}\Delta_{\perp}\right)\right]}\left[\delta\left(E+E_{\mathrm{g}}+\frac{1}{3}\Delta_{||}\right)+(E+E_{\mathrm{g}})\right.$$
$$\left.\times\left(E+E_{\mathrm{g}}+\frac{2}{3}\Delta_{||}\right)+\frac{1}{9}\left(\Delta_{||}^2-\Delta_{||}^2\right)\right],$$

$\psi_3(E) \equiv \frac{\hbar^2 E_{\mathrm{g}}\left(E_{\mathrm{g}}+\Delta_{||}\right)}{\left[2m_{||}^*\left(E_{\mathrm{g}}+\frac{2}{3}\Delta_{||}\right)\right]}\left[(E+E_{\mathrm{g}})\left(E+E_{\mathrm{g}}+\frac{2}{3}\Delta_{||}\right)\right]$, $m_{||}^*$ and $m_{\perp}^*$ are the longitudinal and transverse effective electron masses at the edge of the conduction band respectively.

The general expression of the density-of-states (DOS) function in bulk semiconductors is given by

$$D_0(E) = \frac{2g_{\mathrm{v}}}{(2\pi)^3}\left(\frac{\partial}{\partial E}\left[V(E)\right]\right), \tag{2.3a}$$

where g_{v} is the valley degeneracy and $V(E)$ is the volume of $\mathbf{k}$ space. Using (2.2) and (2.3a), we get

$$D_0(E) = g_{\mathrm{v}}\left(3\pi^2\right)^{-1}\psi_4(E), \tag{2.3b}$$

$$\psi_4(E) \equiv \left[\frac{3}{2}\frac{\sqrt{\psi_1(E)}\,[\psi_1(E)]'}{\psi_2(E)\sqrt{\psi_3(E)}} - \frac{[\psi_2(E)]'\,[\psi_1(E)]^{3/2}}{[\psi_2(E)]^2\sqrt{\psi_3(E)}}\right.$$
$$\left. - \frac{1}{2}\frac{[\psi_3(E)]'\,[\psi_1(E)]^{3/2}}{\psi_2(E)\,[\psi_3(E)]^{3/2}}\right],$$

$$[\psi_1(E)]' \equiv [\,(2E+E_{\mathrm{g}})\,\psi_1(E)\,[E(E+E_{\mathrm{g}})]^{-1} + E(E+E_{\mathrm{g}})$$
$$\times\left(2E+2E_{\mathrm{g}}+\delta+\Delta_{||}\right)\,],$$

$$[\psi_2(E)]' \equiv \left[2m_{\perp}^*\left(E_{\mathrm{g}}+\frac{2}{3}\Delta_{\perp}\right)\right]^{-1}\left[\hbar^2 E_{\mathrm{g}}\left(E_{\mathrm{g}}+\Delta_{\perp}\right)\right]$$
$$\times\left[\delta+2E+2E_{\mathrm{g}}+\frac{2}{3}\Delta_{||}\right],$$

and $[\psi_3(E)]' \equiv \left[2m_{||}^*\left(E_{\mathrm{g}}+\frac{2}{3}\Delta_{||}\right)\right]^{-1}\left[\hbar^2 E_{\mathrm{g}}\left(E_{\mathrm{g}}+\Delta_{||}\right)\right]\left[2E+2E_{\mathrm{g}}+\frac{2}{3}\Delta_{||}\right]$, in which, the primes denote the differentiation of the differentiable functions with respect to E.

Combining (2.3b) with the Fermi–Dirac occupation probability factor and using the generalized Sommerfeld's lemma [10], the electron concentration can be written as

$$n_0 = g_{\mathrm{v}}\left(3\pi^2\right)^{-1}\left[M(E_{\mathrm{F}}) + N(E_{\mathrm{F}})\right], \tag{2.4}$$

where $M(E_{\mathrm{F}}) \equiv \left[\frac{[\psi_1(E_{\mathrm{F}})]^{\frac{3}{2}}}{\psi_2(E_{\mathrm{F}})\sqrt{\psi_3(E_{\mathrm{F}})}}\right]$, E_{F} is the Fermi energy as measured from the edge of the conduction band in the vertically upward direction in the absence of any quantization, $N(E_{\mathrm{F}}) \equiv \sum_{r=1}^{s} L(r)M(E_{\mathrm{F}})$, r is the set of real positive integers whose upper limit is s, $L(r) \equiv \left[2\left(k_{\mathrm{B}}T\right)^{2r}\left(1-2^{1-2r}\right)\xi\left(2r\right)\right] \times \left[\frac{\partial^{2r}}{\partial E_{\mathrm{F}}^{2r}}\right]$ and $\zeta(2r)$ is the Zeta function of order $2r$ [11].

Thus the use of the (2.4) and (1.11) leads to the expression of DMR as

$$\frac{D}{\mu} = \frac{1}{|e|}\frac{\left[M\left(E_{\mathrm{F}}\right)+N\left(E_{\mathrm{F}}\right)\right]}{\left[\left\{M\left(E_{\mathrm{F}}\right)\right\}'+\left\{N\left(E_{\mathrm{F}}\right)\right\}'\right]}. \tag{2.5}$$

2.1.3 Special Cases for III–V Semiconductors

(a) Under the substitutions $\delta = 0, \Delta_{||} = \Delta_{\perp} = \Delta$(the isotropic spin–orbit splitting constant) and $m_{||}^{*} = m_{\perp}^{*} = m^{*}$(the isotropic effective electron mass at the edge of the conduction band), (2.2) assumes the form [1]

$$\frac{\hbar^2 k^2}{2m^*} = \gamma(E), \quad \gamma(E) \equiv \frac{E(E+E_{\mathrm{g}})(E+E_{\mathrm{g}}+\Delta)\left(E_{\mathrm{g}}+\frac{2}{3}\Delta\right)}{E_{\mathrm{g}}(E_{\mathrm{g}}+\Delta)\left(E+E_{\mathrm{g}}+\frac{2}{3}\Delta\right)}, \tag{2.6}$$

which is the well-known three band model of Kane [1]. Equation (2.6) is the dispersion relation of the conduction electrons of III–V, ternary and quaternary materials and should be used as such for studying the electron transport in n-InAs where the spin orbit splitting constant is of the order of band gap. The III–V compounds are used in integrated optoelectronics [12, 13], passive filter devices [14], distributed feedback lasers and Bragg reflectors [15]. Besides, we shall also use n-$Hg_{1-x}Cd_xTe$ and n-$In_{1-x}Ga_xAs_yP_{1-y}$ lattice matched to InP as examples of ternary and quaternary materials respectively. The ternary alloy n-$Hg_{1-x}Cd_xTe$ is a classic narrow-gap compound and is technologically an important optoelectronic semiconductor because its band gap can be varied to cover a spectral range from 0.8 to over 30 μm by adjusting the alloy composition [16]. The n-$Hg_{1-x}Cd_xTe$ finds applications in infrared detector materials [17] and photovoltaic detector [18] arrays in the 8-12 μm wave bands. The above applications have spurred an $Hg_{1-x}Cd_xTe$ technology for the production of high mobility single crystals, with specially prepared surface layers and the same material is suitable for narrow subband physics because the relevant material constants are within experimental reach [19]. The quaternary compounds are being extensively used in optoelectronics, infrared light emitting diodes, high electron mobility transistors, visible heterostructure lasers for fiber optic systems, semiconductor lasers, [20], tandem solar cells [21], avalanche photodetectors [22], long wavelength light sources, detectors in optical fiber communications, [23] and new types of optical devices, which are being prepared from the quaternary systems [24].

Under the aforementioned limiting conditions, the density-of-states function, the electron concentration, and the DMR in accordance with the three band model of Kane assume the following forms

$$D_0(E) = 4\pi g_v \left(\frac{2m^*}{h^2}\right)^{3/2} \sqrt{\gamma(E)}\,[\gamma_1(E)], \tag{2.7}$$

$$n_0 = \frac{g_v}{3\pi^2}\left(\frac{2m^*}{\hbar^2}\right)^{3/2} [M_1(E_F) + N_1(E_F)], \tag{2.8}$$

and

$$\frac{D}{\mu} = \frac{1}{|e|}[M_1(E_F) + N_1(E_F)]\left[\{M_1(E_F)\}' + \{N_1(E_F)\}'\right]^{-1}, \tag{2.9}$$

where $\gamma_1(E) \equiv \gamma(E)\left[\frac{1}{E} + \frac{1}{E+E_g} + \frac{1}{E+E_g+\Delta} - \frac{1}{E+E_g+\frac{2}{3}\Delta}\right]$, $M_1(E_F) \equiv [\gamma(E_F)]^{3/2}$ and $N_1(E_F) \equiv \sum_{r=1}^{s} L(r)\,M_1(E_F)$.

(b) Under the inequalities $\Delta \gg E_g$ or $\Delta \ll E_g$, (2.6) gets simplified as [1]

$$\frac{\hbar^2 k^2}{2m^*} = E(1+\alpha E), \quad \alpha \equiv 1/E_g, \tag{2.10}$$

which is known as the two-band model of Kane [1]. Under the above constraints, the forms of the DOS, the electron statistics and the DMR can, respectively, be written as,

$$D_0(E) = 4\pi g_v \left(\frac{2m^*}{h^2}\right)^{3/2} \sqrt{I(E)}\,[I_1(E)], \tag{2.11}$$

$$n_0 = \frac{g_v}{3\pi^2}\left(\frac{2m^*}{\hbar^2}\right)^{3/2} [M_2(E_F) + N_2(E_F)], \tag{2.12}$$

and

$$\frac{D}{\mu} = \frac{1}{|e|}[M_2(E_F) + N_2(E_F)]\left[\{M_2(E_F)\}' + \{N_2(E_F)\}'\right]^{-1}, \tag{2.13}$$

where $I(E) \equiv E(1+\alpha E)$, $I_1(E) \equiv (1+2\alpha E)$, $M_2(E_F) \equiv [I(E_F)]^{3/2}$ and $N_2(E_F) \equiv \sum_{r=1}^{s} L(r)\,M_2(E_F)$.

(c) Under the constraints $\Delta \gg E_g$ or $\Delta \ll E_g$ together with the inequality $\alpha E_F \ll 1$, we can write [1]

$$n_0 = g_v N_c \left[F_{1/2}(\eta) + \left(\frac{15\alpha k_B T}{4}\right) F_{3/2}(\eta)\right], \tag{2.14}$$

and
$$\frac{D}{\mu} = \left[\frac{k_B T}{|e|}\right]\left[\frac{\left(F_{1/2}(\eta) + \left(\frac{15\alpha k_B T}{4}\right) F_{3/2}(\eta)\right)}{\left(F_{-1/2}(\eta) + \left(\frac{15\alpha k_B T}{4}\right) F_{1/2}(\eta)\right)}\right], \tag{2.15}$$

where $N_C \equiv 2\left(\frac{2\pi m^* k_B T}{h^2}\right)^{3/2}$ $\eta \equiv \frac{E_F}{k_B T}$ and $F_j(\eta)$ is the one parameter Fermi–Dirac integral of order j which can be written as [25],

$$F_j(\eta) = \left(\frac{1}{\Gamma(j+1)}\right) \int_0^{\infty} y^j \left(1 + \exp(y - \eta)\right)^{-1} \mathrm{d}y, \quad j > -1, \tag{2.16}$$

where $\Gamma(j+1)$ is the complete Gamma function or for all j, analytically continued as a complex contour integral around the negative axis

$$F_j(\eta) = A_j \int_{-\infty}^{(0+)} y^j \left(1 + \exp(-y - \eta)\right)^{-1} \mathrm{d}y, \tag{2.17}$$

in which $A_j \equiv \frac{\Gamma(-j)}{2\pi\sqrt{-1}}$.

(d) For relatively wide gap materials $E_g \to \infty$ and (2.14) and (2.15) assume the forms

$$n_0 = g_v N_c F_{1/2}(\eta) \tag{2.18}$$

and

$$\frac{D}{\mu} = \left(\frac{k_B T}{|e|}\right) \left[\frac{F_{1/2}(\eta)}{F_{-1/2}(\eta)}\right]. \tag{2.19}$$

Equation (2.19) was derived for the first time by Landsberg [1].

(e) Combining (2.18) and (2.19) and using the formula $\frac{\mathrm{d}}{\mathrm{d}\eta}[F_j(\eta)] = F_{j-1}(\eta)$ [25] as easily derived from (2.16) and (2.17) together with the fact that under the condition of extreme carrier degeneracy

$$F_{1/2}(\eta) = \left[\frac{4}{3\sqrt{\pi}}\right] (\eta)^{3/2}, \tag{2.20}$$

we can write

$$n_0 = \frac{g_v}{3\pi^2} \left[\frac{2m^* E_F (1 + \alpha E_F)}{\hbar^2}\right]^{3/2}, \tag{2.21}$$

and

$$\frac{D}{\mu} = \frac{1}{|e|} \left(\frac{2}{3}\right) E_F \frac{(1 + \alpha E_F)}{(1 + 2\alpha E_F)}, \tag{2.22}$$

For $\alpha \to 0$, (2.21) and (2.22) assume the forms

$$n_0 = \frac{g_v}{3\pi^2} \left[\frac{2m^* E_F}{\hbar^2}\right]^{3/2}, \tag{2.23}$$

and

$$\frac{D}{\mu} = \frac{2E_F}{3|e|}. \tag{2.24}$$

(f) Under the condition of non-degenerate electron concentration $\eta \ll 0$ and $F_j(\eta) \cong \exp(\eta)$ for all j [25]. Therefore (2.18) and (2.19) assume the well-known forms as [1]

$$n_0 = g_v N_c \exp(\eta), \tag{2.25}$$

and

$$\frac{D}{\mu} = \frac{k_B T}{|e|}. \tag{2.26}$$

2.1.4 Result and Discussions

Using n-Cd_3As_2 as an example of $A^2_{III}B^5_{II}$ compounds for the purpose of numerical computations and using (2.4) and (2.5) together with the energy band constants at $T = 4.2\,K$, as given in Table 2.1, the variation of the DMR as a function of electron concentration has been shown in curve (a) of Fig. 2.1. The circular points exhibit the same dependence and have been obtained by using (1.15) and taking the experimental values of the thermoelectric power in $n{-}Cd_3As_2$ in the presence of a classically large magnetic field [26]. The curve (b) corresponds to $\delta = 0$. The curve (c) shows the dependence of the DMR on n_0 in accordance with the three-band model of Kane using the energy band constants as $E_g = 0.095\,eV$, $m^* = \left(m^*_{||} + m^*_{\perp}\right)/2$ and $\Delta = \left(\Delta_{||} + \Delta_{\perp}\right)/2$. The curves (d) and (e) correspond to the two-band model of Kane and that of the parabolic energy bands. By comparing the curves (a) and (b) of Fig. 2.1, one can easily assess the influence of crystal field splitting on the DMR in tetragonal compounds. Figure 2.2 represents all cases of Fig. 2.1 for n-$CdGeAs_2$ which has been used as an example of ternary chalcopyrite materials where the values of the energy band constants of the said compound are given in Table 2.1.

It appears from Fig. 2.1 that, the DMR in tetragonal compounds increases with increasing carrier degeneracy as expected for degenerate semiconductors and agrees well with the suggested experimental method of determining the same ratio for materials having arbitrary carrier energy spectra. It has been observed that the tetragonal crystal field affects the DMR of the electrons quite significantly in this case. The dependence of the DMR is directly determined by the band structure because of its immediate connection with the Fermi energy. The DMR increases non-linearly with the electron concentration in other limiting cases and the rates of increase are different from that in the generalized band model.

From Fig. 2.2, one can assess that the DMR in bulk specimens of n-$CdGeAs_2$ exhibits monotonic increasing dependence with increasing electron concentration. The cases (b), (c) and (d) of Fig. 2.2 for n-$CdGeAs_2$ exhibit the similar trends with change in the respective numerical values of the DMR. The influence of spectrum constants on the DMR for n-Cd_3As_2 and n-$CdGeAs_2$ can also be assessed by comparing the respective variations as drawn in Figs. 2.1 and 2.2 respectively.

Table 2.1. The numerical values of the energy band constants of few materials

Materials n – Cd_3As_2	$\lvert E_g\rvert = 0.095\,\mathrm{eV}$, $\Delta_{\|\|} = 0.27\,\mathrm{eV}$, $\Delta_\perp = 0.25\,\mathrm{eV}$, $m^*_{\|\|} = 0.00697m_0$ (m_0 is the free electron mass), $m^*_\perp = 0.013933m_0$, $\delta = 0.085\mathrm{eV}$, $g_v = 1$ [25, 73] and $\varepsilon_{sc} = 16\varepsilon_0$ (ε_{sc} and ε_0 are the permittivity of the semiconductor and free space respectively) [74]
n – $CdGeAs_2$	$E_g = 0.57\,\mathrm{eV}$, $\Delta_{\|\|} = 0.30\,\mathrm{eV}$, $\Delta_\perp = 0.36\,\mathrm{eV}$, $m^*_{\|\|} = 0.034m_o$, $m^*_\perp = 0.039m_o$, $T = 4\,\mathrm{K}$, $\delta = -0.21\,\mathrm{eV}$, $g_v = 1$ [1, 26] and $\varepsilon_{sc} = 18.4\varepsilon_0$ [75]
n-InAs	$E_g = 0.36\,\mathrm{eV}$, $\Delta = 0.43\,\mathrm{eV}$ and $m^* = 0.026m_0$, $g_v = 1$, $\varepsilon_{sc} = 12.25\varepsilon_0$ [76]
n-InSb	$E_g = 0.2352\,\mathrm{eV}$, $\Delta = 0.81\,\mathrm{eV}$ and $m^* = 0.01359m_0$, $g_v = 1$, $\varepsilon_{sc} = 15.56\varepsilon_0$ [76]
n – $Ga_{1-x}Al_xAs$	$E_g(x) = (1.424 + 1.266x + 0.26x^2)\mathrm{eV}$, $\Delta(x) = (0.34 - 0.5x)\mathrm{eV}$, $m^*(x) = [0.066 + 0.088x]\,m_0$ $g_v = 1$, $\varepsilon_{sc}(x) = [13.18 - 3.12x]\,\varepsilon_0$ [77]
$Hg_{1-x}Cd_xTe$	$E_g(x) = \left(-0.302 + 1.93x + 5.35 \times 10^{-4}(1-2x)T - 0.810x^2 + 0.832x^3\right)\,\mathrm{eV}$, $\Delta(x) = \left(0.63 + 0.24x - 0.27x^2\right)\,\mathrm{eV}$, $m^* = 0.1m_0E_g(\mathrm{eV})^{-1}$, $g_v = 1$ and $\varepsilon_{sc} = \left[20.262 - 14.812x + 5.22795x^2\right]\varepsilon_0$ [78]
$In_{1-x}Ga_xAs_yP_{1-y}$ lattice matched to InP	$E_g = \left(1.337 - 0.73y + 0.13y^2\right)\,\mathrm{eV}$, $\Delta = \left(0.114 + 0.26y - 0.22y^2\right)\,\mathrm{eV}$, $m^* = (0.08 - 0.039y)\,m_o$, $y = (0.1896 - 0.4052x)(0.1896 - 0.0123x)^{-1}$, $g_v = 1$ [79] and $\varepsilon_{sc} = [10.65 + 0.1320y]\,\varepsilon_0$ [80]
CdS	$m^*_{\|\|} = 0.7m_o$, $m^*_\perp = 1.5m_o$ and $\bar{\lambda}_0 = 1.4 \times 10^{-10}\,\mathrm{eVm}$, $g_v = 1$ [76] and $\varepsilon_{sc} = 15.5\varepsilon_0$ [81]
n-PbTe	$m_t^- = 0.070m_0$, $m_l^- = 0.54m_0$, $m_t^+ = 0.010m_0$, $m_l^+ = 1.4m_0$, $P_{\|\|} = 141\,\mathrm{meV\,nm}$, $P_\perp = 486\,\mathrm{meV\,nm}$, $E_g = 190\,\mathrm{meV}$, $g_v = 4$ [12] and $\varepsilon_{sc} = 33\varepsilon_0$ [76, 82]
n-PbSnTe	$m_t^- = 0.063m_0$, $m_l^- = 0.41m_0$, $m_t^+ = 0.089m_0$, $m_l^+ = 1.6m_0$, $P_{\|\|} = 137\,\mathrm{meV\,nm}$, $P_\perp = 464\,\mathrm{meV\,nm}$, $E_g = 90\,\mathrm{meV}$, $g_v = 4$ [12] and $\varepsilon_{sc} = 60\varepsilon_0$ [76, 82]
n-$Pb_{1-x}Sn_xSe$	$x = 0.31$, $g_v = 4$, $m_t^- = 0.143m_0$, $m_l^- = 2.0m_0$, $m_t^+ = 0.167m_0$, $m_l^+ = 0.286m_0$, $P_{\|\|} = 3.2 \times 10^{-10}\,\mathrm{eVm}$, $P_\perp = 4.1 \times 10^{-10}\,\mathrm{eVm}$, $E_g = 0.137\mathrm{eV}$, $g_v = 4$ [12] and $\varepsilon_{sc} = 31\varepsilon_0$ [76, 83]
Stressed n-InSb	$m^* = 0.048m_o$, $E_g = 0.081\,\mathrm{eV}$, $B_2 = 9 \times 10^{-10}\,\mathrm{eVm}$, $C_1^c = 3\,\mathrm{eV}$, $C_2^c = 2\,\mathrm{eV}$, $a_0 = -10\,\mathrm{eV}$, $b_0 = -1.7\,\mathrm{eV}$, $d_0 = -4.4\,\mathrm{eV}$, $S_{xx} = 0.6 \times 10^{-3}\,(\mathrm{kbar})^{-1}$, $S_{yy} = 0.42 \times 10^{-3}\,(\mathrm{kbar})^{-1}$, $S_{zz} = 0.39 \times 10^{-3}\,(\mathrm{kbar})^{-1}$ and $S_{xy} = 0.5 \times 10^{-3}\,(\mathrm{kbar})^{-1}$, $\varepsilon_{xx} = \sigma S_{xx}$, $\varepsilon_{yy} = \sigma S_{yy}$, $\varepsilon_{zz} = \sigma S_{zz}$, $\varepsilon_{xy} = \sigma S_{xy}$ and σ is the stress in kilobar, $g_v = 1$ [44]

(*Continued*)

Table 2.1. *Continued*

$PtSb_2$	For conduction bands, along <111> direction, $\lambda_1 = 0.33\,\text{eV}$, $l_1 = 1.09\,\text{eV}$, $\nu_1 = 0.17\,\text{eV}$, $n_1 = 0.22\,\text{eV}$, $a = 0.643\,\text{nm}$, $I_0 = 0.30\ (\text{eV})^2$, $\delta_0' = 0.33\,\text{eV}$, $g_v = 8$ [56] and $\varepsilon_{sc} = 30\varepsilon_0$ [56, 84]
n-GaSb	$E_g = 0.81\,\text{eV}$, $\Delta = 0.80\,\text{eV}$, $P = 9.48 \times 10^{-10}\,\text{eVm}$, $\bar{\varsigma}_0 = -2.1$, $\bar{v}_0 = -1.49$, $\bar{\omega}_0 = 0.42$, $g_v = 1$ [53] and $\varepsilon_{sc} = 15.85\varepsilon_0$ [53, 85]
HgTe	$m_v^* = 0.028m_0$, $g_v = 1$ and $\varepsilon_\infty = 15.2\varepsilon_0$ [52]
Bismuth	$E_g = 0.0153\,\text{eV}$, $m_1 = 0.00194m_0$, $m_2 = 0.313m_0$, $m_3 = 0.00246m_0$, $m_2' = 0.36m_0$, $g_v = 3$ and $g_s = 2$ [42]

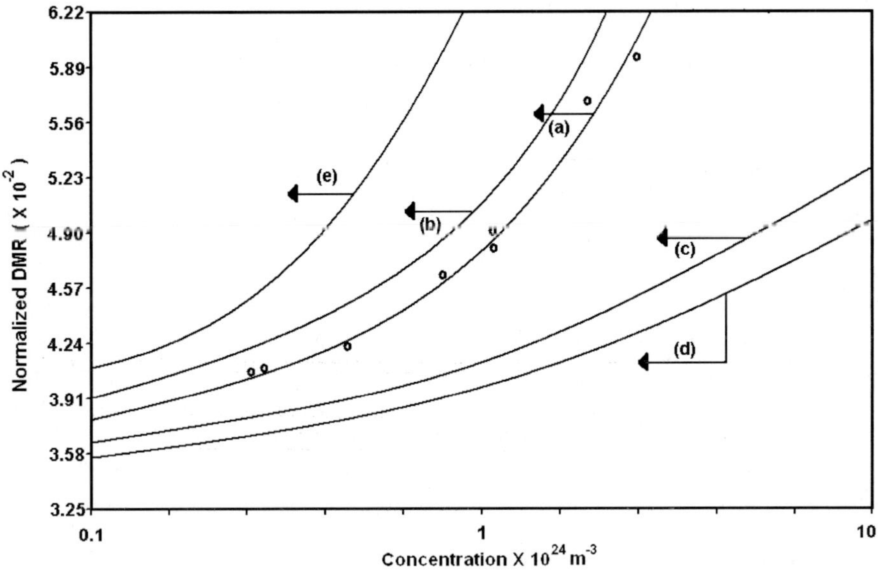

Fig. 2.1. The plot of the DMR in the bulk specimens of n-Cd_3As_2 as a function of electron concentration in accordance with (**a**) the generalized band model; (**b**) $\delta = 0$; (**c**) the three band model of Kane; (**d**) the two band model of Kane and (**e**) the parabolic energy bands. The *dotted circular points* show the same dependence which have been obtained by using (1.15) and taking the experimental values of the thermoelectric power of the electrons in bulk n-Cd_3As_2 in the presence of a classically large magnetic field [26]

Using the appropriate equations, one can numerically evaluate the DMR as a function of electron concentration for n-InAs, whose energy band constants are presented in Table 2.1. This is shown in Fig. 2.3 by curves (a), (b) and (c) respectively, in accordance with the three and two band models of Kane together with the model of parabolic energy bands. Figure 2.4 exhibits all the cases of Fig. 2.3 for n-InSb whose energy band constants are given in Table 2.1.

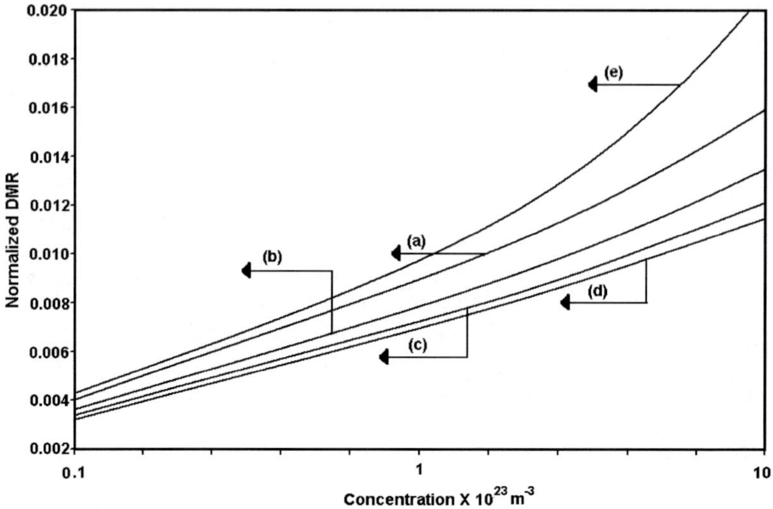

Fig. 2.2. The plot of the DMR in the bulk specimens of n-$CdGeAs_2$ as a function of electron concentration in accordance with (**a**) the generalized band model; (**b**) $\delta = 0$; (**c**) the three band model of Kane; (**d**) the two band model of Kane and (**e**) the parabolic energy bands

From Figs. 2.3 and 2.4 one can observe that the numerical values of the DMR for n-InAs and n-InSb in accordance with the three band model of Kane are less than that of the corresponding parabolic energy bands for relatively large values of the electron concentration. Also, the numerical values of the DMR in accordance with two-band model of Kane also changes as compared with the corresponding three-band model of Kane for n-InAs. The influence of energy band constants on the DMR for n-InAs and n-InSb becomes apparent by comparing Figs. 2.3 and 2.4 respectively.

For n-$Hg_{1-x}Cd_xTe$ together with the numerical values of the spectrum constants as given in Table 2.1, the DMR has been plotted at $T = 4.2\,K$ as a function of electron concentration as shown in Fig. 2.5 for all cases of Fig. 2.3. Figure 2.6 represents the variation of the DMR with respect to alloy composition x in this case.

It appears from Fig. 2.5 that the DMR of ternary materials increases with increasing carrier degeneracy. From Fig. 2.6, it appears that the DMR in ternary materials decreases with increasing alloy composition. The plots of Figs. 2.5 and 2.6 are valid for $x > 0.17$, due to the fact that for $x < 0.17$, the band gap becomes negative in n-$Hg_{1-x}Cd_xTe$ leading to a semimetallic state.

For n-$In_{1-x}Ga_xAs_yP_{1-y}$ lattice matched to InP together with the values of the energy band constants as given in Table 2.1, the DMR has been plotted as a function of electron concentration as shown in Fig. 2.7 in accordance with the three and two band models of Kane together with the isotropic parabolic energy band. It appears that the DMR monotonically increases with increasing n_0. Figure 2.8 shows the dependence of the DMR in n-$In_{1-x}Ga_xAs_yP_{1-y}$

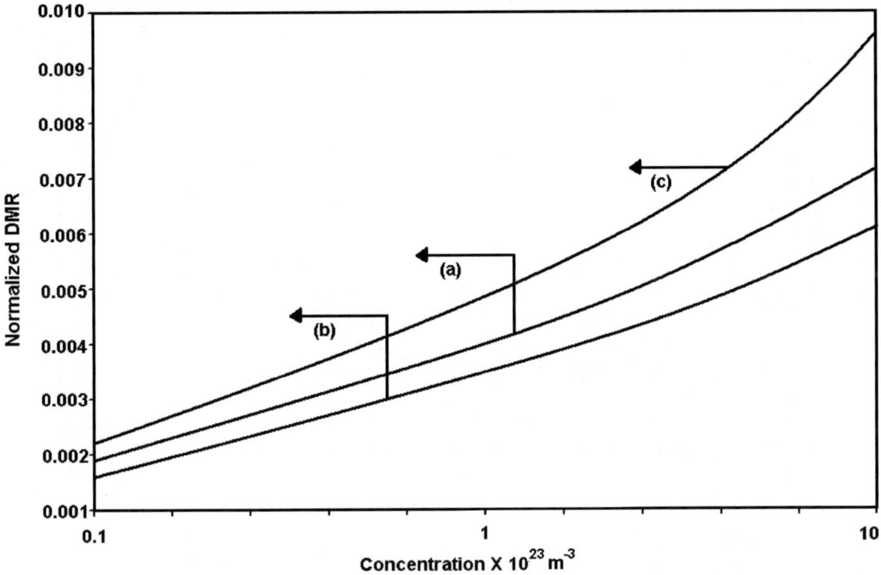

Fig. 2.3. The plot of the DMR in bulk specimens of n-InAs as a function of electron concentration in accordance with (**a**) the three band model of Kane, (**b**) the two band model of Kane, and (**c**) the parabolic energy bands

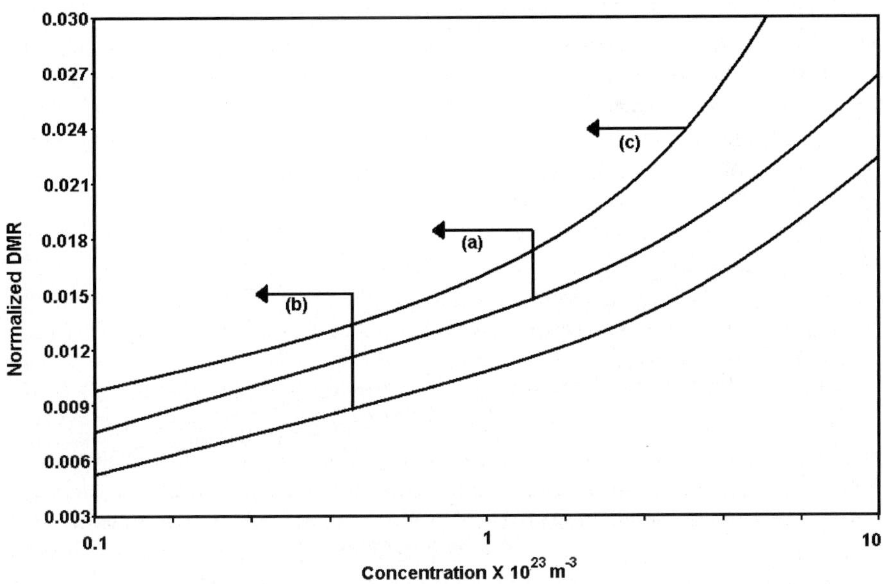

Fig. 2.4. The plot of the DMR in bulk specimens of n-InSb as a function of electron concentration in accordance with (**a**) the three-band model of Kane, (**b**) the two-band model of Kane, and (**c**) the parabolic energy bands

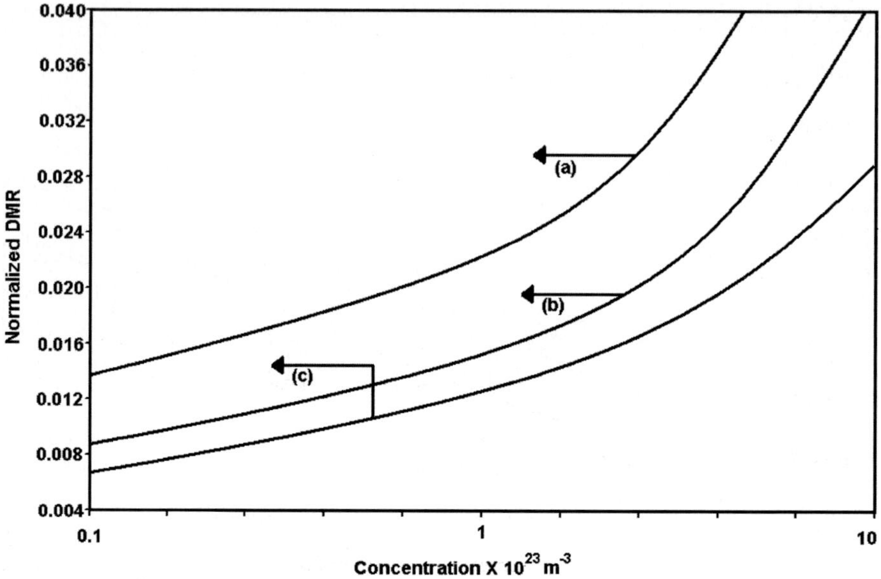

Fig. 2.5. The plot of the DMR in bulk specimens of n-$Hg_{1-x}Cd_xTe$ as a function of electron concentration in accordance with (**a**) the three band model of Kane; (**b**) the two band model of Kane and (**c**) the parabolic energy bands ($x = 0.3$)

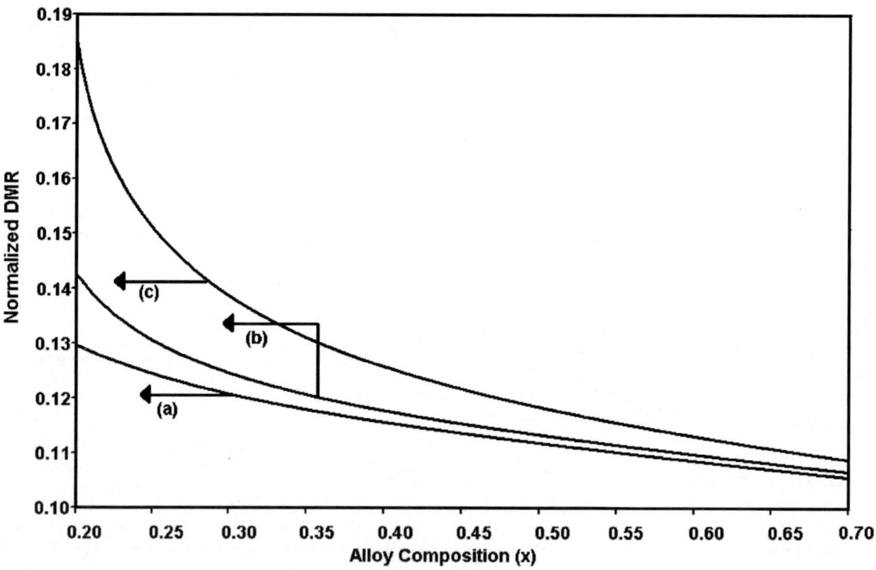

Fig. 2.6. The plot of the DMR in bulk specimens of n-$Hg_{1-x}Cd_xTe$ as a function of alloy composition (x) in accordance with (**a**) the three band model of Kane; (**b**) the two band model of Kane and (**c**) the parabolic energy bands ($n_0 = 10^{22}$ per cubic meter)

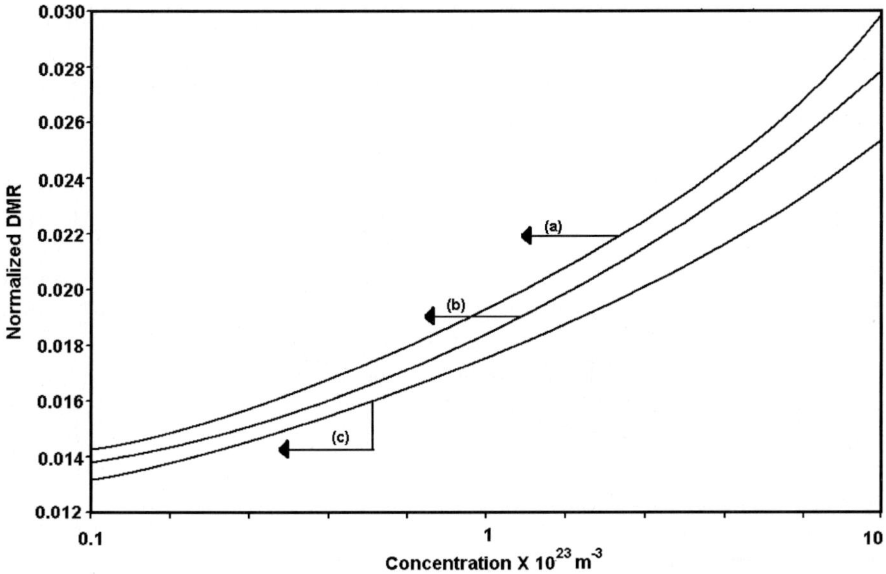

Fig. 2.7. The plot of the DMR in bulk specimens of n-$In_{1-x}Ga_xAs_yP_{1-y}$ lattice matched to InP as a function of electron concentration in accordance with (**a**) the three band model of Kane; (**b**) the two band model of Kane and (**c**) the parabolic energy bands ($y = 0.037$)

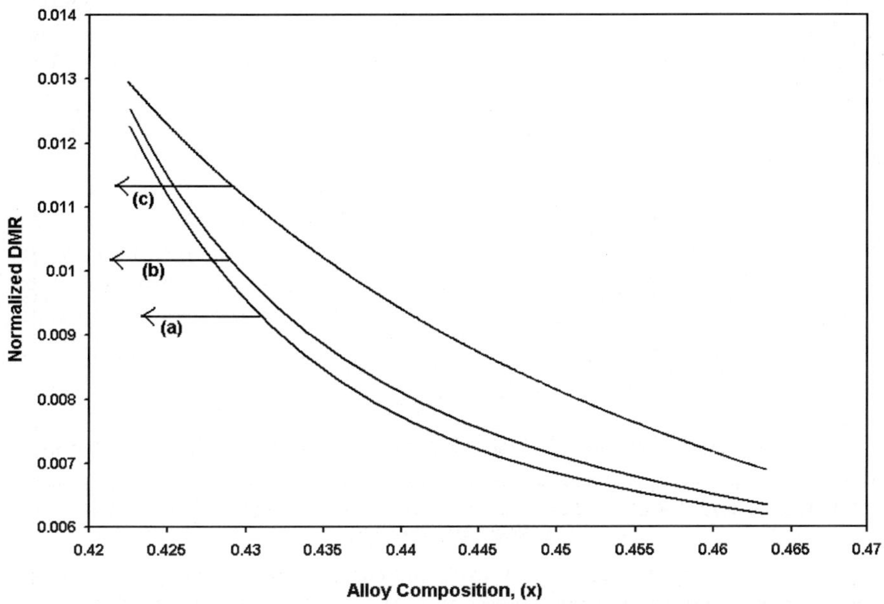

Fig. 2.8. The plot of the DMR in bulk specimens of n-$In_{1-x}Ga_xAs_yP_{1-y}$ lattice matched to InP as a function of alloy composition (x) in accordance with (**a**) the three band model of Kane; (**b**) the two band model of Kane and (**c**) the parabolic energy bands ($y = 0.037$)

lattice matched to InP on the alloy composition of x for all cases of Fig. 2.6. The DMR decreases with increasing x for all types of band models in this case. From Figs. 2.5 up to 2.8, one can infer the influence of energy band constants on the DMR for ternary and quaternary compounds respectively.

It may be noted that in recent years, the electron mobility in compound semiconductors has been extensively investigated, but the diffusion constant (a very important device quantity which cannot be easily determined experimentally) of such materials has been relatively less investigated. Therefore, the theoretical results presented in this chapter will be useful in determining the diffusion constants for even relatively wide gap materials whose energy band structures can be approximated by the parabolic energy bands.

We wish to point out that in formulating the basic dispersion relation we have taken into account the combined influences of the crystal field-splitting constant, the anisotropies in the effective electron masses, and the spin–orbit splitting constants, respectively, since these are the significant physical features of the tetragonal compounds.

In the absence of crystal-field splitting together with the assumptions of isotropic effective electron mass and isotropic spin–orbit splitting constant respectively, our basic equation (2.2) converts to the well-known form of the three-band model of Kane as given by (2.6). Many technologically important compounds obey the inequalities $\Delta \gg E_g$ or $\Delta \ll E_g$. Under these constraints, (2.6) gets simplified into (2.10) and is known as the two-band model of Kane. Finally, for $E_g \to \infty$, as for relatively wide gap materials the above equation transforms into the well-known form $E = \hbar^2 k^2 / 2m^*$. In addition, the DMR in ternary and quaternary materials has also been investigated in accordance with the three and two band models of Kane together with the parabolic energy band for the purpose of relative assessment. Therefore, the influence of energy band constants on the DMR can also be studied from the present investigation and the basic equation (2.2) covers various materials having different energy band structures. Finally, one infers that, this simplified analysis exhibits the basic features of the DMR in bulk specimens of many technologically important compounds and for n-Cd_3As_2, the theoretical result is in good agreement with the suggested experimental method of determining the same ratio.

2.2 Investigation for II–VI Semiconductors

2.2.1 Introduction

The II–VI compounds are being extensively used in infrared detectors [27], ultra high speed bipolar transistors [28], optic fiber communications [29], and advanced microwave devices [30]. These materials possess the appropriate direct band gap to produce light emitting diodes and lasers from blue to red wavelengths [31]. The Hopfield model describes the energy spectra of both the carriers of II–VI semiconductors where the splitting of the two-spin states by the spin orbit coupling and the crystalline field has been taken into account [32]. The DMR in II–VI compounds on the basis of the Hopfield model has been studied by formulating the expression of carrier concentration in Sect. 2.2.2. Section 2.2.3 contains the result and discussions for the numerical computation of the DMR taking p-CdS as an example.

2.2.2 Theoretical Background

The group theoretical analysis shows that, based on the symmetry properties of the conduction and valence band wave functions, both the energy bands of II–VI semiconductors can be written as [32]

$$E = a_0' k_s^2 + b_0' k_z^2 \pm \bar{\lambda}_0 k_s, \tag{2.27}$$

where $a_0' \equiv \frac{\hbar^2}{2m_\perp^*}$, $b_0' \equiv \frac{\hbar^2}{2m_{||}^*}$ and $\bar{\lambda}_0$ represents the splitting of the two spin-states by the spin–orbit coupling and the crystalline field.

The volume in k-space enclosed by (2.27) can be expressed as

$$V(E) = \frac{\pi}{2a_0'^2} \int_{-(E/b_0')^{1/2}}^{(E/b_0')^{1/2}} \left[\bar{\lambda}_0^2 + 2a_0'E - 2a_0'b_0'k_z^2 - \bar{\lambda}_0\left(\bar{\lambda}_0^2 - 4a_0'b_0'k_z^2 + 4a_0'E\right)^{1/2}\right] \mathrm{d}k_z, \tag{2.28}$$

From (2.28), one can write

$$V(E) = \frac{4\pi}{3a_0'\sqrt{b_0'}} \left[E^{3/2} + \frac{3}{8}\frac{(\bar{\lambda}_0)^2\sqrt{E}}{a_0'} - \left(\frac{3}{4}\frac{\bar{\lambda}_0}{\sqrt{a_0'}}\right)\left(E + \frac{(\bar{\lambda}_0)^2}{4a_0'}\right) \times \sin^{-1}\left[\frac{\sqrt{E}}{\sqrt{E + \frac{(\bar{\lambda}_0)^2}{4a_0'}}}\right]\right], \tag{2.29}$$

Hence, the density of states function can be written using (2.3a) and (2.29) as

$$D_0(E) = \frac{g_v}{2\pi^2 a_0'\sqrt{b_0'}} \left[\sqrt{E} - \left(\frac{\bar{\lambda}_0}{2\sqrt{a_0'}}\right)\sin^{-1}\left[\frac{\sqrt{E}}{\sqrt{E + \frac{(\bar{\lambda}_0)^2}{4a_0'}}}\right]\right]. \tag{2.30}$$

Combining (2.30) with the Fermi–Dirac occupation probability factor, the carrier concentration can be written as

$$n_0 = \frac{4\pi g_v}{3a_0'\sqrt{b_0'}} \left[\tau_1(E_F) + \tau_2(E_F)\right], \tag{2.31}$$

where $\tau_1(E_F) \equiv \left[E_F^{3/2} + \frac{3}{8}\frac{(\bar{\lambda}_0)^2}{a'_0}\sqrt{E_F} - \left\{\frac{3}{4}\frac{\bar{\lambda}_0}{\sqrt{a'_0}}\left(E_F + \frac{(\bar{\lambda}_0)^2}{4a'_0}\right)\right.\right.$

$\left.\left.\sin^{-1}\left[\frac{\sqrt{E_F}}{\sqrt{E_F + \frac{(\bar{\lambda}_0)^2}{4a'_0}}}\right]\right\}\right]$, and $\tau_2(E_F) \equiv \sum_{r=1}^{s} L(r)\,\tau_1(E_F)$.

Combining (2.31) and (1.11), the DMR in bulk specimens of II–VI semiconductors assumes the form

$$\frac{D}{\mu} = \frac{1}{|e|}\frac{[\tau_1(E_F) + \tau_2(E_F)]}{[[\tau_1(E_F)]' + [\tau_2(E_F)]']}, \tag{2.32}$$

Under the condition $\bar{\lambda}_0 \to 0$, (2.27) gets simplified as

$$E = a'_0 k_s^2 + b'_0 k_z^2, \tag{2.33}$$

Thus, under the condition $\bar{\lambda}_0 \to 0$, (2.32) reduces to the well-known form as given by (2.19).

2.2.3 Result and Discussions

Using (2.31) and (2.32) together with the spectrum constants as given in Table 2.1, for p-CdS, the DMR has been plotted as a function of hole concentration p_0 as shown by curve (a) of Fig. 2.9 in which the plot for $\bar{\lambda}_0 = 0$ (curve

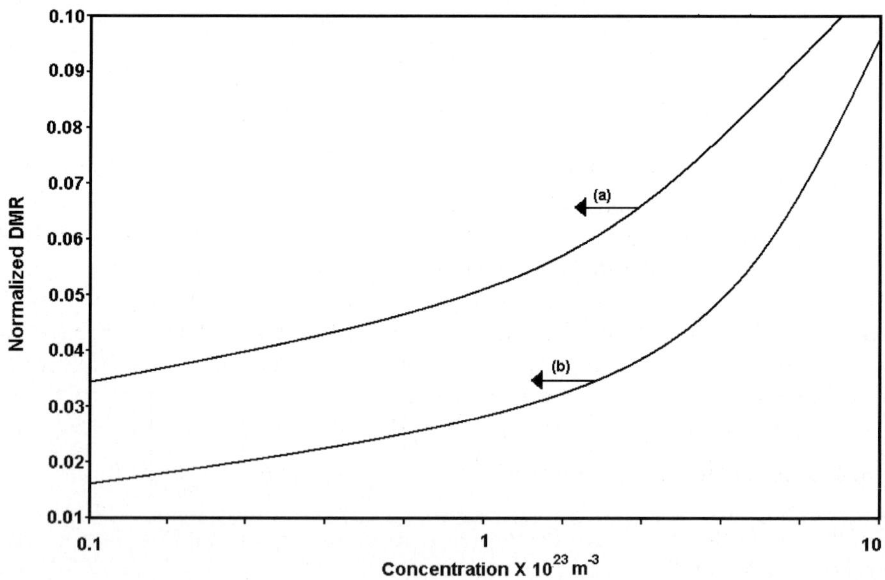

Fig. 2.9. The plot of the dependence of the DMR on hole concentration p_0 for bulk specimens of p-CdS for **(a)** $\bar{\lambda}_0 \neq 0$ and **(b)** $\bar{\lambda}_0 = 0$

(b)) has also been drawn for the purpose of assessing the influence of the splitting of the two spin states by the spin–orbit coupling and the crystalline field on DMR. From Fig. 2.8 it appears that the DMR increases with increasing hole concentration at a rate greater than that corresponding to the zero value of $\bar{\lambda}_0$. For relatively low values of p_0, the effect of $\bar{\lambda}_0$ increases whereas the same constant affects the DMR less significantly for relatively higher values of carrier degeneracy. The presence of $\bar{\lambda}_0$ enhances the numerical values of DMR in II–VI compounds for the whole range of concentration considered as compared with that corresponding to $\bar{\lambda}_0 = 0$.

2.3 Investigation for Bi in Accordance with the McClure–Choi, the Cohen, the Lax, and the Parabolic Ellipsoidal Band Models

2.3.1 Introduction

It is well-known that the carrier energy spectra in Bi differ considerably from the simple spherical energy wave vector dispersion relation of the degenerate electron gas and several models have been developed to describe the energy spectra of Bi. Earlier works [33, 34] demonstrated that the physical properties of Bi could be described by the ellipsoidal parabolic energy band model. Shoenberg [33] showed that the de Haas-Van Alphen and cyclotron resonance experiments supported the ellipsoidal parabolic model, though the latter work showed that Bi could be described by the two-band model due to the fact that the magnetic field dependence of many physical properties of Bi supports the above model [35]. The experimental results of the magneto-optical [35] and the ultrasonic quantum oscillations [36] favor the Lax ellipsoidal non-parabolic model [35]. Kao [37], Dinger and Lawson [38] and Koch and Jensen [39] observed that the Cohen model [40], is in better agreement with the experimental results. McClure and Choi [41] presented a new model of Bi, which was more accurate and general than those that were currently available. They showed that it can explain the data for a large number of magneto-oscillatory and resonance experiments. We shall study the influence of different energy band models on the DMR in bulk specimens of Bi which have been investigated by formulating the carrier concentration in Sect. 2.3.2. Section 2.3.3 contains the result and discussions in this context.

2.3.2 Theoretical Background

(a) *The McClure and Choi model*

The carrier energy spectra in Bi can be written, following McClure and Choi, [42] as

$$E(1+\alpha E)=\frac{p_x^2}{2m_1}+\frac{p_y^2}{2m_2}+\frac{p_z^2}{2m_3}+\frac{p_y^2}{2m_2}\alpha E\left\{1-\left(\frac{m_2}{m_2'}\right)\right\}+\frac{p_y^4\alpha}{4m_2m_2'}-\frac{\alpha p_x^2p_y^2}{4m_1m_2}-\frac{\alpha p_y^2p_z^2}{4m_2m_3}, \tag{2.34}$$

where m_1, m_2 and m_3 are the effective carrier masses at the band-edge along x, y and z directions respectively and m_2' is the effective- mass tensor component at the top of the valence band (for electrons) or at the bottom of the conduction band (for holes).

The area of the ellipse in the k_x–k_z plane can be expressed as

$$A(E,k_y)=\bar{T}_0\left[\frac{E(1+\alpha E)-\theta_2(E)k_y^2-\theta_3k_y^4}{1-\theta_4k_y^2}\right], \tag{2.35}$$

where $\bar{T}_0\equiv\left(\frac{2\pi\sqrt{m_1m_3}}{\hbar^2}\right)$, $\theta_2(E)\equiv\left[\left(\frac{\alpha E\hbar^2}{2m_2}\right)\left(1-\frac{m_2}{m_2'}\right)+\frac{\hbar^2}{2m_2}\right]$, $\theta_3\equiv\left(\frac{\alpha\hbar^4}{4m_2m_2'}\right)$ and $\theta_4\equiv\left(\frac{\alpha\hbar^2}{2m_2}\right)$.

The volume of k-space enclosed by (2.34) can be written as

$$V(E)=\left(\frac{4\pi\sqrt{m_1m_3}}{\hbar^2\theta_4}\right)\int_0^{p_0(E)}\left[\frac{h_9(E)}{{\theta_5}^2-{k_y}^2}+\theta_2(E)+\theta_3\left({\theta_5}^2+{k_y}^2\right)\right]dk_y, \tag{2.36}$$

where $p_0(E)\equiv\left(\frac{\sqrt{2m_2m_2'}}{\hbar^2\sqrt{\alpha}}\right)\left[-\theta_2(E)+\sqrt{\theta_2^2(E)+\theta_3E(1+\alpha E)}\right]^{1/2}$, $\theta_5\equiv(\theta_4)^{-1/2}$ and $h_9(E)\equiv\left[E(1+\alpha E)-\theta_2(E){\theta_5}^2-\theta_3{\theta_5}^4\right]$.

From (2.36) one obtains

$$V(E)=\frac{4\pi\sqrt{m_1m_3}}{\hbar^2\theta_4}\left[\frac{h_9(E)}{2\theta_5}\ln\left|\frac{\theta_5+p_0(E)}{\theta_5-p_0(E)}\right|+\left[\theta_2(E)+\theta_3{\theta_5}^2\right]p_0(E)+\frac{\theta_3}{3}[p_0(E)]^3\right]. \tag{2.37}$$

The density-of-states function in this case can be expressed using (2.3a) as

$$D_0(E)=\frac{(g_v\sqrt{m_1m_3})}{(\pi^2\hbar^2\theta_4)}\left[\frac{\{h_9(E)\}'}{2\theta_5}\ln\left|\frac{\theta_5+p_0(E)}{\theta_5-p_0(E)}\right|+\frac{h_9(E)\{p_0(E)\}'}{[\theta_5^2-p_0^2(E)]}+\{\theta_2(E)\}'p_0(E)+\left[\theta_3\{p_0(E)\}'p_0^2(E)+\left[\theta_2(E)+\theta_3\theta_5^2\right]\{p_0(E)\}'\right],\right. \tag{2.38}$$

where $\{h_9(E)\}'\equiv\left[1+2\alpha E-\left(\frac{\theta_5^2\alpha\hbar^2}{2m_2}\right)\left(1-\frac{m_2}{m_2'}\right)\right]$, $\{\theta_2(E)\}'\equiv\left(\frac{\alpha\hbar^2}{2m_2}\right)\times\left(1-\frac{m_2}{m_2'}\right)$ and

$$\{p_0(E)\}' \equiv \left[\frac{1}{2}p_0(E)\left[-\{\theta_2(E)\}' + \frac{2\theta_2(E)\{\theta_2(E)\}' + \theta_3(1+2\alpha E)}{(2)\sqrt{\theta_2^2(E) + \theta_3 E(1+\alpha E)}}\right] \times \left[-\theta_2(E) + \sqrt{\theta_2^2(E) + \theta_3 E(1+\alpha E)}\right]^{-1}\right].$$

Therefore the electron concentration is given by

$$n_0 = \theta_6\left[\overline{M_2}(E_F) + \overline{N_2}(E_F)\right]. \tag{2.39}$$

where $\theta_6 \equiv \left(\frac{g_\nu\sqrt{m_1 m_3}}{\pi^2\hbar^2\theta_4}\right)$,

$$\overline{M_2}(E_F) \equiv \left[\frac{h_9(E_F)}{2\theta_5}\ln\left|\frac{\theta_5 + p_0(E_F)}{\theta_5 - p_0(E_F)}\right| + [\,\theta_2(E_F) + \theta_3{\theta_5}^2\,]\,p_0(E_F) + \frac{\theta_3}{3}[p_0(E_F)]^3\right],$$

and $\overline{N_2}(E_F) \equiv \sum_{r=1}^{s} L(r)\left[\overline{M_2}(E_F)\right]$.

Thus, combining (2.39) and (1.11), we can write the expression of DMR in Bi in accordance with the McClure and Choi model as

$$\frac{D}{\mu} = \frac{1}{|e|}\left[\frac{\overline{M_2}(E_F) + \overline{N_2}(E_F)}{\{\overline{M_2}(E_F)\}' + \{\overline{N_2}(E_F)\}'}\right]. \tag{2.40}$$

(b) *The Cohen model*

In accordance with Cohen [40], the dispersion law of the carriers in Bi is given by

$$E(1+\alpha E) = \frac{p_x^2}{2m_1} + \frac{p_z^2}{2m_3} - \frac{\alpha E p_y^2}{2m_2'} + \left(\frac{\alpha p_y^4}{4m_2 m_2'}\right) + \frac{p_y^2}{2m_2}(1+\alpha E), \tag{2.41}$$

In this case the area of the ellipse in the k_x–k_z plane can be written as

$$A(E,k_y) = \frac{2\pi\sqrt{m_1 m_3}}{\hbar^2}\left[E(1+\alpha E) - \frac{\alpha p_y^4}{4m_2 m_2'} + \frac{\alpha E p_y^2}{2m_2'} - \frac{p_y^2}{2m_2}(1+\alpha E)\right].$$

Therefore the volume enclosed by (2.41) is given by

$$V(E) = \frac{4\pi\sqrt{m_1 m_3}}{\hbar^2}\int_0^{\overline{p_0}(E)}\left[E(1+\alpha E) - \frac{\alpha\hbar^4 k_y^4}{4m_2 m_2'} + \frac{\hbar^2 k_y^2}{2}\left[\frac{\alpha E}{m_2'} - \frac{1}{m_2}(1+\alpha E)\right]\right]\mathrm{d}k_y, \tag{2.42}$$

where $\overline{p_0}(E) \equiv \frac{1}{\hbar}\left[\frac{\alpha}{2m_2m_2'}\right]^{-1/2}\left[-\left[\frac{1+\alpha E}{2m_2}-\frac{\alpha E}{2m_2'}\right]+\left[\left[\frac{1+\alpha E}{2m_2}-\frac{\alpha E}{2m_2'}\right]^2\right.\right.$
$\left.\left.+\frac{\alpha E}{m_2m_2'}(1+\alpha E)\right]^{1/2}\right]^{1/2}$.

From (2.42) we can write

$$V(E)=\frac{4\pi\sqrt{m_1m_3}}{\hbar^2}\left[E\,(1+\alpha E)\,\overline{p_0}\,(E)-\frac{\alpha\hbar^4\,[\bar{p}_0\,(E)]^5}{20m_2m_2'}\right.$$
$$\left.+\frac{\hbar^2\,[\bar{p}_0\,(E)]^3}{6}\left[\frac{\alpha E}{m_2'}-\frac{1}{m_2}\,(1+\alpha E)\right]\right]. \quad (2.43a)$$

The density-of-states function can be expressed using (2.3a) and (2.43a) as

$$D_0\,(E)=\frac{g_{\mathrm{v}}\sqrt{m_1m_3}}{\pi^2\hbar^2}\left[\,(1+2\alpha E)\,\bar{p}_0\,(E)+E\,(1+\alpha E)\,[\bar{p}_0\,(E)]'\right.$$
$$-\frac{\alpha\hbar^4\,[\bar{p}_0\,(E)]^4\,[\bar{p}_0\,(E)]'}{5m_2m_2'}+\frac{\hbar^2}{2}\,[\bar{p}_0\,(E)]^2\,[\bar{p}_0\,(E)]'$$
$$\left.\times\left[\frac{\alpha E}{m_2'}-\frac{1}{m_2}\,(1+\alpha E)\right]+\frac{\hbar^2\,[\bar{p}_0\,(E)]^3}{6}\alpha\left(\frac{1}{m_2'}-\frac{1}{m_2}\right)\right], \quad (2.43b)$$

where $[\bar{p}_0\,(E)]' \equiv \left[\frac{m_2m_2'}{\alpha\hbar^2\bar{p}_0(E)}\right]\left[\frac{\alpha}{2}\left(\frac{1}{m_2'}-\frac{1}{m_2}\right)+\frac{1}{2}\left[\left[\frac{1+\alpha E}{2m_2}-\frac{\alpha E}{2m_2'}\right]^2+\frac{\alpha E}{m_2m_2'}\right.\right.$
$\left.\times\,(1+\alpha E)\right]^{-1/2}\cdot\left.\left[\frac{\alpha}{m_2m_2'}\,(1+2\alpha E)+\frac{\alpha}{2}\left(\frac{1}{m_2}-\frac{1}{m_2'}\right)\left(\frac{1+\alpha E}{m_2}-\frac{\alpha E}{m_2'}\right)\right]\right]$. Thus, using (2.43b) with the Fermi–Dirac occupation probability factor and using the generalized Sommerfelds lemma [10], the electron concentration in this case can be expressed as

$$n_0=\left(\frac{g_{\mathrm{v}}\sqrt{m_1m_3}}{\pi^2\hbar^2}\right)[M_3\,(E_{\mathrm{F}})+N_3\,(E_{\mathrm{F}})]\,, \quad (2.44)$$

where $M_3\,(E_{\mathrm{F}}) \equiv \left[\,E_{\mathrm{F}}\,(1+\alpha E_{\mathrm{F}})\,\bar{p}_0\,(E_{\mathrm{F}})-\frac{\alpha\hbar^4[\bar{p}_0(E_{\mathrm{F}})]^5}{20m_2m_2'}+\frac{\hbar^2[\bar{p}_0(E_{\mathrm{F}})]^3}{6}\left[\frac{\alpha E_{\mathrm{F}}}{m_2'}-\right.\right.$
$\left.\left.\frac{1}{m_2}\,(1+\alpha E_{\mathrm{F}})\right]\right]$, and $N_3\,(E_{\mathrm{F}}) \equiv \sum\limits_{r=1}^{s} L\,(r)\,[M_3\,(E_{\mathrm{F}})]$.

Thus, combining (2.44) and (1.11), we can write the expression of the DMR in bismuth in accordance with the Cohen model as

$$\frac{D}{\mu}=\frac{1}{|e|}\frac{[M_3\,(E_{\mathrm{F}})+N_3\,(E_{\mathrm{F}})]}{[\{M_3\,(E_{\mathrm{F}})\}'+\{N_3\,(E_{\mathrm{F}})\}']}. \quad (2.45)$$

(c) *The Lax model*

The carrier spectrum of Bi in accordance with the Lax model is given by [35]

$$E(1+\alpha E) = \frac{p_x{}^2}{2m_1} + \frac{p_y{}^2}{2m_2} + \frac{p_z{}^2}{2m_3}. \tag{2.46}$$

For this model, under the condition $\alpha E_F \ll 1$, the expressions for n_0 and DMR in Bi, in accordance with the Lax non-parabolic and ellipsoidal model, get simplified to (2.14) and (2.15) where

$$N_c \equiv 2\left[2\pi (m_1 m_2 m_3)^{\frac{1}{3}} \frac{k_B T}{h^2}\right]^{\frac{3}{2}}.$$

(d) *Ellipsoidal parabolic model*

The carrier energy spectra for this type of band model of Bi can be written as [33]

$$E = \frac{p_x{}^2}{2m_1} + \frac{p_y{}^2}{2m_2} + \frac{p_z{}^2}{2m_3}. \tag{2.47}$$

For this model, the expressions of carrier concentration and DMR are respectively given by (2.18) and (2.19) where N_c has been defined as above.

2.3.3 Result and Discussions

Using (2.39) and (2.40) and taking the spectrum constants from Table 2.1, in curve (a) of Fig. 2.10, the DMR in Bi has been plotted as a function of electron concentration by using the McClure and Choi model. The curves (b),

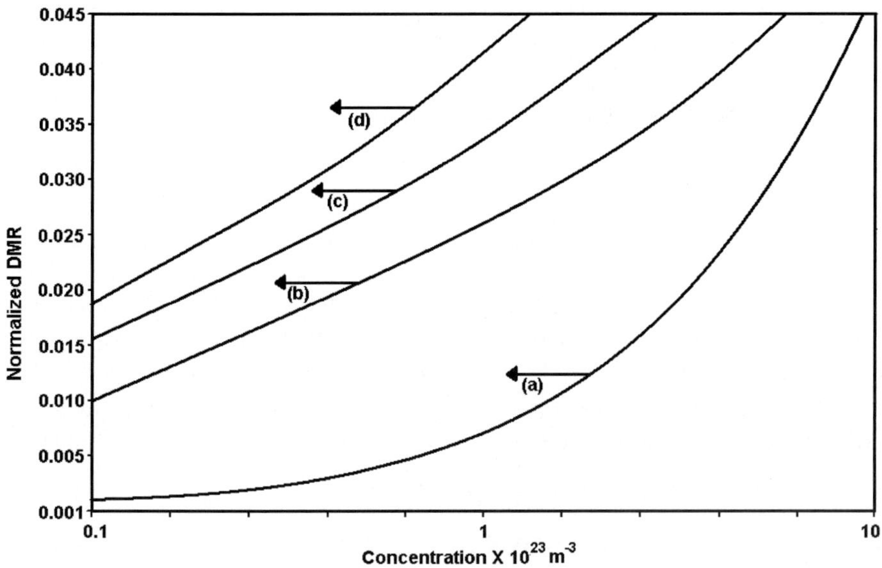

Fig. 2.10. The plot of the DMR in bulk specimens of bismuth as a function of electron concentration in accordance with the model of (**a**) the McClure Choi; (**b**) the Cohen; (**c**) the Lax and (**d**) the ellipsoidal parabolic energy bands

(c) and (d) exhibit the same dependence in accordance with the Cohen, the Lax, and the parabolic ellipsoidal models respectively.

From Fig. 2.10, it appears that the DMR increases with increasing n_0 for all the models of Bi. In accordance with the model of McClure and Choi, the DMR exhibits the least numerical values as compared to the other models of Bi. For various energy band models, the values of the DMR with respect to the electron concentration are different. The rates of variations of the DMR with respect to n_0 are also different for different types of energy band models. It should be noted that under the condition $\alpha \to 0$, the models of McClure and Choi, the Cohen and the Lax reduce to (2.47). Thus, under certain constraints, all the three energy models are reduced to the ellipsoidal parabolic energy bands and the expression for the DMR under the same condition gets simplified to the well-known equation (2.19) as given for the first time by Landsberg [1].

The Cohen model is often used to describe the dispersion relation of the carriers of IV–VI semiconductors. The model of Bi, by Lax, under the condition of the isotropic effective mass of the carriers of the band edge (i.e. $m_1 = m_2 = m_3 = m^*$.) reduces to the two-band model of Kane, which is used to investigate the physical features of III–V compounds, in general, excluding n-InAs. Thus, the analysis is valid not only for bismuth, but also for all lead chalcogenides, III–V compounds excluding n-InAs, and wide-gap materials respectively. The influence of the energy band models on the DMR of Bi can also be assessed from the Fig. 2.10. It can be noted that the present analysis is valid for the holes of Bi with the appropriate values of the energy band constants.

2.4 Investigation for IV–VI Semiconductors

2.4.1 Introduction

The IV–VI compounds are being extensively used in thermoelectric devices, superlattices, and other quantum effect devices [43]. The dispersion relation of the carriers of the IV–VI compounds could be described by the Cohen model [40], which includes the band non-parabolicity and the anisotropies of the effective masses of the carriers. The DMR in bulk specimens of IV–VI materials has been studied, taking n-PbTe, n-PbSnTe, and n-$Pb_{1-x}Sn_xSe$ as examples. Sections 2.4.2 and 2.4.3 contain the theoretical background and the result and discussions in this context.

2.4.2 Theoretical Background

The expressions of n_0 and the DMR in this case are given by (2.44) and (2.45) in which the energy band constants correspond to the IV–VI compounds.

2.4.3 Result and Discussions

Using (2.44) and (2.45) together with the spectrum constants as given in Table 2.1 for n-PbTe, n-PbSnTe, and n-$Pb_{1-x}Sn_xSe$, the DMR has been plotted as a function of electron concentration n_0 as shown in Fig. 2.11 where the plots (a), (b) and (c) correspond to the said materials respectively in accordance with the Cohen model. From Fig. 2.11, it appears that the DMR for a fixed n_0 is maximum for n-PbTe and minimum for n-$Pb_{1-x}Sn_xSe$. For relatively low values of n_0, the values of the DMR for the three materials exhibit convergence behavior whereas for relatively large values of n_0, the numerical values differ from each other. Besides, our present analysis is also valid for p-type IV–VI materials with the proper change in the energy band constants.

2.5 Investigation for Stressed Kane Type Semiconductors

2.5.1 Introduction

In recent years there has been a considerable interest in studying the various electronic properties of stressed materials because of their important physical characteristics [44]. In this section we shall study the DMR in stressed

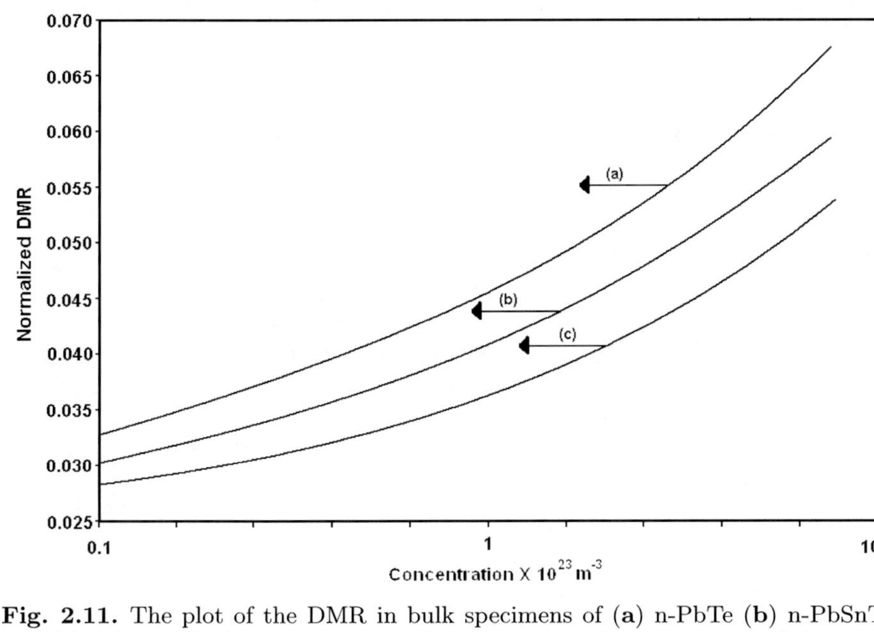

Fig. 2.11. The plot of the DMR in bulk specimens of (**a**) n-PbTe (**b**) n-PbSnTe and (**c**) n-$Pb_{1-x}Sn_xSe$ as a function of electron concentration in accordance with Cohen model

semiconductors, taking stressed n-InSb as an example for numerical computations. Sections 2.5.2 and 2.5.3 contains the theoretical background and the result and discussions in this context.

2.5.2 Theoretical Background

The electron energy spectrum in stressed Kane type semiconductors can be written [44] as

$$\left(\frac{k_x}{\bar{a}_0(E)}\right)^2+\left(\frac{k_y}{\bar{b}_0(E)}\right)^2+\left(\frac{k_z}{\bar{c}_0(E)}\right)^2=1, \tag{2.48}$$

where

$$[\bar{a}_0(E)]^2\equiv\frac{\bar{K}_0(E)}{\bar{A}_0(E)+\frac{1}{2}\bar{D}_0(E)},\quad \bar{K}_0(E)\equiv\left[E-C_1\varepsilon-\frac{2C_2^2\varepsilon_{xy}^2}{3E_g'}\right]\left(\frac{3E_g'}{2B_2^2}\right),$$

C_1 is the conduction band deformation potential, ε is the trace of the strain tensor $\hat{\varepsilon}$ which can be written as $\hat{\varepsilon}=\begin{bmatrix}\varepsilon_{xx} & \varepsilon_{xy} & 0\\ \varepsilon_{xy} & \varepsilon_{yy} & 0\\ 0 & 0 & \varepsilon_{zz}\end{bmatrix}$, C_2 is a constant which describes the strain interaction between the conduction and valance bands, $E_g'\equiv E_g+E-C_1\varepsilon$, B_2 is the momentum matrix element,

$$\bar{A}_0(E)\equiv\left[1-\frac{(\bar{a}_0+C_1)}{E_g'}+\frac{3\bar{b}_0\varepsilon_{xx}}{2E_g'}-\frac{\bar{b}_0\varepsilon}{2E_g'}\right],\ \bar{a}_0\equiv-\frac{1}{3}\left(\bar{b}_0+2\bar{m}\right),$$

$$\bar{b}_0\equiv\frac{1}{3}\left(\bar{l}-\bar{m}\right),\ \bar{d}_0\equiv\frac{2\bar{n}}{\sqrt{3}},$$

$\bar{l},\bar{m},\bar{n}$ are the matrix elements of the strain perturbation operator, $\bar{D}_0(E)\equiv\left(\bar{d}_0\sqrt{3}\right)\frac{\varepsilon_{xy}}{E_g'}$,

$$\left[\bar{b}_0(E)\right]^2\equiv\frac{\bar{K}_0(E)}{\bar{A}_0(E)-\frac{1}{2}\bar{D}_0(E)},\ [\bar{c}_0(E)]^2\equiv\frac{\bar{K}_0(E)}{\bar{L}_0(E)}$$

$$\text{and }\bar{L}_0(E)\equiv\left[1-\frac{(\bar{a}_0+C_1)}{E_g'}+\frac{3\bar{b}_0\varepsilon_{zz}}{E_g'}-\frac{\bar{b}_0\varepsilon}{2E_g'}\right],$$

The density-of-states function in this case can be written using (2.3a) and (2.48) as

$$D_0(E)=g_v\left(3\pi^2\right)^{-1}\left[\bar{a}_0(E)\bar{b}_0(E)[\bar{c}_0(E)]'+\bar{a}_0(E)\left[\bar{b}_0(E)\right]'\bar{c}_0(E)\right.$$
$$\left.+[\bar{a}_0(E)]'\bar{b}_0(E)\bar{c}_0(E)\right], \tag{2.49}$$

Combining (2.49) with the Fermi–Dirac occupation probability factor and using the generalized Sommerfelds lemma [10], the electron concentration in this case can be expressed as

$$n_0 = g_v \left(3\pi^2\right)^{-1} \left[M_4\left(E_F\right) + N_4\left(E_F\right)\right], \tag{2.50}$$

where $M_4(E_F) \equiv \left[\bar{a}_0(E_F)\,\bar{b}_0(E_F)\,\bar{c}_0(E_F)\right]$ and $N_4(E_F) \equiv \sum_{r=1}^{s} L(r)\, M_4(E_F)$.

Thus using (2.50) and (1.11) we get

$$\frac{D}{\mu} = \frac{1}{|e|}\frac{\left[M_4\left(E_F\right) + N_4\left(E_F\right)\right]}{\left[\left\{M_4\left(E_F\right)\right\}' + \left\{N_4\left(E_F\right)\right\}'\right]}. \tag{2.51}$$

In the absence of stress together with the substitution $B_2^2 \equiv \frac{3\hbar^2 E_g}{4m^*}$, (2.51) gets simplified to (2.13).

2.5.3 Result and Discussions

Using (2.50) and (2.51) together with the energy band constants as given in Table 2.1 at $T = 4.2\,\mathrm{K}$, the DMR has been plotted as a function of electron concentration n_0 in stressed n-InSb as shown in Fig. 2.12 where the curves (a), and (b) correspond to the presence and absence of stress respectively.

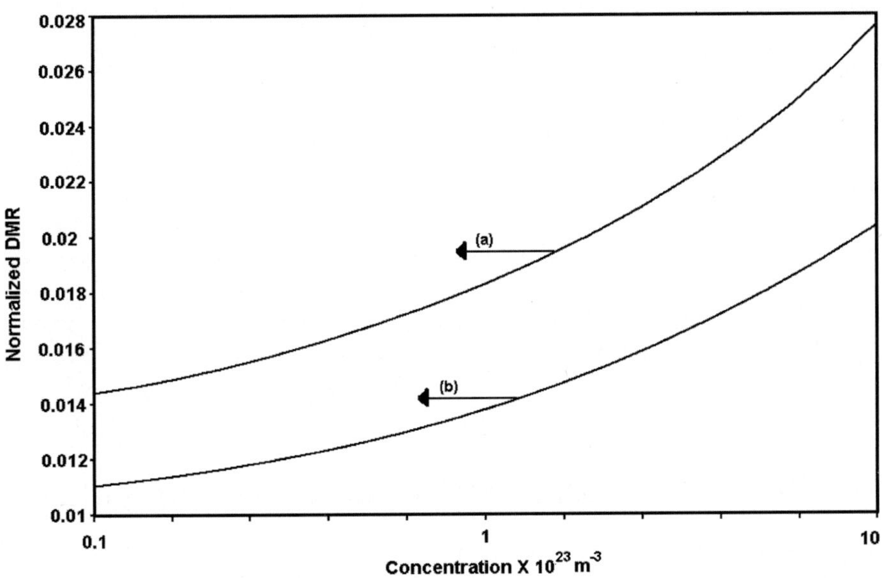

Fig. 2.12. The plot of the DMR in bulk specimens of stressed n-InSb as a function of the electron concentration both in the presence and absence of stress as shown by the curves (**a**) and (**b**) respectively

It appears that stress enhances the numerical values of the DMR to a large extent as compared to that of the stress free condition. The rates of increase of DMR for both the cases are different as concentration increases.

2.6 Summary

In Chap. 2, an attempt is made to present the DMR in tetragonal compounds on the basis of a generalized electron dispersion law by considering the anisotropies of the effective electron masses and the spin–orbit splitting constants together with the inclusion of crystal field splitting constant within the frame work of **k.p** formalism. The theoretical result is in agreement with the suggested experimental method of determining the DMR for materials having arbitrary dispersion laws. Under certain limiting conditions, the results for III–V materials as defined by the three and two band models of Kane have been obtained as special cases of the generalized analysis. The concentration dependence of the DMR has also been numerically computed for n-Cd_3As_2, n-$CdGeAs_2$, n-InAs, n-InSb, n-$Hg_{1-x}Cd_xTe$, and n-$In_{1-x}Ga_xAs_yP_{1-y}$ lattice matched to InP respectively. The II–VI compounds obey the Hopfield model and p-type CdS has been used for numerical computation. The DMR has also been investigated for Bi in accordance with the models of the McClure and Choi, the Cohen, the Lax, and the parabolic ellipsoidal energy bands respectively. The IV–VI materials obey the Cohen model and n-PbTe, n-PbSnTe, and n-$Pb_{1-x}Sn_xSe$ have been used for investigations. The chapter ends with the study of the DMR in stressed Kane type semiconductors taking stressed n-InSb as an example, which obey the dispersion relation as suggested by Seiler et al. [44].

Thus, a wide class of technologically important materials has been covered in this chapter whose energy band structures are defined by the appropriate carrier energy spectra. Under certain limiting conditions, all the results of the DMRs for different materials having various band structures lead to the well-known expression of the DMR for degenerate semiconductors having parabolic energy bands as obtained for the first time by Landsberg. For the purpose of condensed presentation, the specific electron statistics related to a particular energy dispersion law for a specific material and the Einstein relation have been presented in Table 2.2.

2.7 Open Research Problems

The problems under these sections of this book are by far the most important part for the readers. Few open and allied research problems are presented from this chapter onward to the end. The numerical values of the energy bandconstants for various materials are given in Table 2.1 *for the related computer simulations.*

Table 2.2. The carrier statistics and the Einstein relation in bulk specimens of tetragonal, III–V, ternary, quaternary, II–VI, all the models of Bismuth, IV–VI and stressed materials

Type of materials	The carrier statistics	The Einstein relation for the diffusivity mobility ratio
1. Tetragonal compounds	In accordance with the generalized dispersion relation as formulated in this chapter $n_0 = g_v \left(3\pi^2\right)^{-1} \left[M\left(E_F\right) + N\left(E_F\right)\right],$ (2.4)	$\frac{D}{\mu} = \frac{1}{\|e\|} \frac{\left[M(E_F)+N(E_F)\right]}{\left[\{M(E_F)\}'+\{N(E_F)\}'\right]}$ (2.5)
2. III–V, ternary and quaternary compounds	In accordance with the three band model of Kane which is a special case of our generalized analysis $n_0 = \frac{g_v}{3\pi^2}\left(\frac{2m^*}{\hbar^2}\right)^{3/2}\left[M_1\left(E_F\right) + N_1\left(E_F\right)\right]$ (2.8)	$\frac{D}{\mu} = \frac{1}{\|e\|}\left[M_1\left(E_F\right) + N_1\left(E_F\right)\right]\left[\{M_1\left(E_F\right)\}' + \{N_1\left(E_F\right)\}'\right]^{-1}$ (2.9)
	Equation (2.8) is a special case of (2.4) Under the conditions $\Delta \gg E_g$ or $\Delta \ll E_g$, $n_0 = \frac{g_v}{3\pi^2}\left(\frac{2m^*}{\hbar^2}\right)^{3/2}\left[M_2\left(E_F\right) + N_2\left(E_F\right)\right]$ (2.12)	Equation (2.9) is a special case of (2.5) Under the conditions $\Delta \gg E_g$ or $\Delta \ll E_g$, $\frac{D}{\mu} = \frac{1}{\|e\|}\left[M_2\left(E_F\right) + N_2\left(E_F\right)\right]\left[\{M_2\left(E_F\right)\}' + \{N_2\left(E_F\right)\}'\right]^{-1}$ (2.13)
	Equation (2.12) is a special case of (2.8) and is valid for the two band model of Kane Under the constraints $\Delta \gg E_g$ or $\Delta \ll E_g$ together with the condition $\alpha E_F \ll 1$ $n_0 = g_v N_c\left[F_{1/2}(\eta) + \left(\frac{15\alpha k_B T}{4}\right) F_{3/2}(\eta)\right]$ (2.14)	Equation (2.13) is a special case of (2.9) and is valid for the two band model of Kane Under the constraints $\Delta \gg E_g$ or $\Delta \ll E_g$ together with the condition $\alpha E_F \ll 1$ $\frac{D}{\mu} = \left[\frac{k_B T}{\|e\|}\right]\left[\frac{\left(F_{1/2}(\eta)+\left(\frac{15\alpha k_B T}{4}\right)F_{3/2}(\eta)\right)}{\left(F_{-1/2}(\eta)+\left(\frac{15\alpha k_B T}{4}\right)F_{1/2}(\eta)\right)}\right]$ (2.15)
	For $E_g \to \infty$, $n_0 = g_v N_c F_{1/2}(\eta)$ (2.18) Equation (2.18) is a special case of (2.14) and is valid for parabolic energy bands	For $E_g \to \infty$, $\frac{D}{\mu} = \left(\frac{k_B T}{\|e\|}\right)\left[\frac{F_{1/2}(\eta)}{F_{-1/2}(\eta)}\right]$ (2.19) Equation (2.19) is a special case of (2.15)
	Under the condition of extreme carrier degeneracy $n_0 = \frac{g_v}{3\pi^2}\left[\frac{2m^* E_F(1+\alpha E_F)}{\hbar^2}\right]^{3/2}$ (2.21)	Under the condition of extreme carrier degeneracy $\frac{D}{\mu} = \frac{1}{\|e\|}\left(\frac{2}{3}\right) E_F\left(1 + \alpha E_F\right)\left(1 + 2\alpha E_F\right)^{-1}$ (2.22)

(Continued)

Table 2.2. *Continued*

Type of materials	The carrier statistics	The Einstein relation for the diffusivity mobility ratio
	For $\alpha \to 0$, $n_0 = \frac{g_v}{3\pi^2}\left[\frac{2m^* E_F}{\hbar^2}\right]^{3/2}$ (2.23) Under the condition of non-degenerate electron concentration $n_0 = g_v N_c \exp(\eta)$ (2.25)	For $\alpha \to 0$, $\frac{D}{\mu} = \frac{2E_F}{3\lvert e\rvert}$ (2.24) Under the condition of non-degenerate electron concentration $\frac{D}{\mu} = \frac{k_B T}{\lvert e\rvert}$ (2.26)
3. II–VI compounds	$n_0 = \frac{4\pi g_v}{3a_0'\sqrt{b_0'}}\left[\tau_1(E_F) + \tau_2(E_F)\right]$ (2.31)	$\frac{D}{\mu} = \frac{1}{\lvert e\rvert}\frac{[\tau_1(E_F)+\tau_2(E_F)]}{[[\tau_1(E_F)]'+[\tau_2(E_F)]']}$ (2.32)
4. Bismuth	(a) The McClure and Choi model: $n_0 = \theta_6\left[\overline{M_2}(E_F) + \overline{N_2}(E_F)\right]$ (2.39)	$\frac{D}{\mu} = \frac{1}{\lvert e\rvert}\frac{[\overline{M_2}(E_F)+\overline{N_2}(E_F)]}{[\{\overline{M_2}(E_F)\}'+\{\overline{N_2}(E_F)\}']}$ (2.40)
	(b) The Cohen model: $n_0 = \frac{g_v\sqrt{m_1 m_3}}{\pi^2\hbar^2}\left[M_3(E_F) + N_3(E_F)\right]$ (2.44)	$\frac{D}{\mu} = \frac{1}{\lvert e\rvert}\frac{[M_3(E_F)+N_3(E_F)]}{[\{M_3(E_F)\}'+\{N_3(E_F)\}']}$ (2.45)
	(c) The Lax model: $n_0 = g_v N_c\left[F_{1/2}(\eta) + \left(\frac{15\alpha k_B T}{4}\right)F_{3/2}(\eta)\right]$, (2.14) where $N_c \equiv 2\left[2\pi(m_1 m_2 m_3)^{\frac{1}{3}}\frac{k_B T}{h^2}\right]^{\frac{3}{2}}$	The DMR in this case are given by (2.15)
	(d) The parabolic ellipsoidal model: The electron statistics in this case are given by (2.18), with N_c as defined above	The DMR in this case are given by (2.19)
5. IV–VI compounds	The expressions of n_0 in this case are given by (2.44) in which the constants of the energy band spectrum correspond to the carriers of the IV–VI semiconductors	The expressions of DMR in this case are given by (2.45) in which the constants of the energy band spectrum correspond to the carriers of the IV–VI semiconductors
6. Stressed compounds	$n_0 = g_v\left(3\pi^2\right)^{-1}\left[M_4(E_F) + N_4(E_F)\right]$ (2.50)	$\frac{D}{\mu} = \frac{1}{\lvert e\rvert}\frac{[M_4(E_F)+N_4(E_F)]}{[\{M_4(E_F)\}'+\{N_4(E_F)\}']}$ (2.51)

(R2.1) Investigate the Einstein relation for the materials having the respective dispersion relations as given below:

(A) The conduction electrons of n-GaP obey two different dispersion laws as given in the literature [45, 46]. In accordance with Rees [45], the electron energy spectrum is given by

$$E = \frac{\hbar^2 k_s^2}{2m_\perp^*} + \frac{\hbar^2}{2m_{||}^*}\left[\wp k_s^2 + k_z^2\right] - \left[\frac{\hbar^4 k_0^2}{m_{||}^{*2}}\left(k_s^2 + k_z^2\right) + |V_G|^2\right]^{1/2} - |V_G|, \quad \text{(R2.1a)}$$

where k_0 and $|V_G|$ are constants of the energy spectrum with $m_{||}^* = 0.92m_0, m_\perp^* = 0.25m_0, k_0 = 1.7\times10^{19}\,\mathrm{m}^{-1}, |V_G| = 0.21\,\mathrm{eV}, g_v = 6, g_s = 2$ and $\wp = 1$.

(1) In accordance with Ivchenko and Pikus [46], the electron dispersion law can be written as

$$E = \frac{\hbar^2 k_z^2}{2m_{||}^*} + \frac{\hbar^2 k_s^2}{2m_\perp^*} \mp \frac{\bar{\Delta}}{2} \pm \left[\left(\frac{\bar{\Delta}}{2}\right)^2 + P_1 k_z^2 + D_1 k_x^2 k_y^2\right]^{1/2}, \quad \text{(R2.1b)}$$

where $\bar{\Delta} = 335\,\mathrm{meV}, P_1 = 2\times10^{-10}\,\mathrm{eVm}, D_1 = P_1 a_1$ and $a_1 = 5.4\times10^{-10}\,\mathrm{m}$.

(B) In addition to the Cohen model, the dispersion relation for the conduction electrons for IV–VI compounds can also be described by the models of Dimmock [47], Bangert et al. [48], and Foley et al. [49] respectively.

(1) In accordance with the Dimmock model [47], the carrier energy spectrum of IV–VI materials assumes the form

$$\left[\in - \frac{E_g}{2} - \frac{\hbar^2 k_s^2}{2m_t^-} - \frac{\hbar^2 k_z^2}{2m_l^-}\right]\left[\in + \frac{E_g}{2} + \frac{\hbar^2 k_s^2}{2m_t^+} + \frac{\hbar^2 k_z^2}{2m_l^+}\right] = P_\perp^2 k_s^2 + P_{||}^2 k_z^2, \quad \text{(R2.2)}$$

where $\in$ is the energy as measured from the center of the band gap E_g, $m_t^\pm$ and $m_l^\pm$ represent the contribution of the transverse and longitudinal effective masses of the external L_6^+ and L_6^- bands arising from the **k.p** perturbations with the other bands taken to the second order and $g_v = 4$.

(2) In accordance with Bangert et al. [48] the dispersion relation is given by

$$\Gamma(E) = F_1(E)\,k_s^2 + F_2(E)\,k_z^2, \quad \text{(R2.3)}$$

where $\Gamma(E) \equiv 2E, F_1(E) \equiv \frac{R_1^2}{E+E_g} + \frac{S_1^2}{E+\Delta_c'} + \frac{Q_1^2}{E+E_g}, F_2(E) \equiv \frac{2C_5^2}{E+E_g} + \frac{(S_1+Q_1)^2}{E+\Delta_c''}$,

$$R_1^2 = 2.3\times10^{-19}\,(\mathrm{eVm})^2,\ C_5^2 = 0.83\times10^{-19}\,(\mathrm{eVm})^2,\ Q_1^2 = 1.3R_1^2,$$

$S_1^2 = 4.6R_1^2, \Delta_c' = 3.07\,\mathrm{eV}, \Delta_c'' = 3.28\,\mathrm{eV}$ and $g_v = 4$. It may be noted that under the substitutions $S_1 = 0, Q_1 = 0, R_1^2 \equiv \frac{\hbar^2 E_g}{m_\perp^*}, C_5^2 \equiv \frac{\hbar^2 E_g}{2m_{||}^*}$, (R2.3) assumes the form $E(1+\alpha E) = \frac{\hbar^2 k_s^2}{2m_\perp^*} + \frac{\hbar^2 k_z^2}{2m_{||}^*}$ which is the simplified Lax model.

(3) The carrier energy spectrum of IV–VI semiconductors in accordance with Foley et al. [49] can be written as

$$E + \frac{E_g}{2} = E_-(k) + \left[\left[E_+(k) + \frac{E_g}{2} \right]^2 + P_\perp^2 k_s^2 + P_{||}^2 k_z^2 \right]^{1/2}, \qquad \text{(R2.4)}$$

where $E_+(k) = \frac{\hbar^2 k_s^2}{2m_\perp^+} + \frac{\hbar^2 k_z^2}{2m_{||}^+}$, $E_-(k) = \frac{\hbar^2 k_s^2}{2m_\perp^-} + \frac{\hbar^2 k_z^2}{2m_{||}^-}$ represents the contribution from the interaction of the conduction and the valance band edge states with the more distant bands and the free electron term,

$$\frac{1}{m_\perp^\pm} = \frac{1}{2}\left[\frac{1}{m_{tc}} \pm \frac{1}{m_{tv}}\right],$$

$$\frac{1}{m_{||}^\pm} = \frac{1}{2}\left[\frac{1}{m_{lc}} \pm \frac{1}{m_{lv}}\right],$$

For n-PbTe, $P_\perp = 4.61 \times 10^{-10}$ eVm, $P_{||} = 1.48 \times 10^{-10}$ eVm, $\frac{m_0}{m_{tv}} = 10.36$, $\frac{m_0}{m_{lv}} = 0.75$, $\frac{m_0}{m_{tc}} = 11.36$, $\frac{m_0}{m_{lc}} = 1.20$ and $g_v = 4$.

(C) The importance of Germanium is well known since the inception of semiconductor physics. The conduction electrons of n-Ge obey two different dispersion laws since band non-parabolicity has been included in two different ways as given literature [50, 51]. In accordance with Cardona et al. [50] and Wang et al [51] the electron dispersion laws in Ge can respectively, be expressed as

$$E = -\frac{E_g}{2} + \frac{\hbar^2 k_z^2}{2m_{||}^*} + \left[\frac{E_g^2}{4} + E_g k_s^2 \left(\frac{\hbar^2}{2m_\perp^*}\right)\right]^{1/2}, \qquad \text{(R2.5)}$$

and

$$E = a_9 k_z^2 + l_9 k_s^2 - c_9 k_s^4 - d_9 k_s^2 k_z^2 - e_9 k_z^4, \qquad \text{(R2.6)}$$

where $a_9 = \left(\hbar^2 / 2m_{||}^*\right)$, $m_{||}^* = 1.588 m_0$, $l_9 = \left(\hbar^2 / 2m_\perp^*\right)$, $m_\perp^* = 0.0815 m_0$, $c_9 = 1.4 A_9$,

$$A_9 \equiv \frac{1}{4}\left(\hbar^4 / E_g m_\perp^{*2}\right)\left[1 - \frac{m_\perp^*}{m_0}\right]^2, E_g = 2.2\,\text{eV},$$

$$d_9 = 0.8 A_9, \quad e_9 = 0.005 A_9, \quad g_v = 4 \text{ and } g_s = 2.$$

(D) The dispersion relation of the conduction electrons of zero-gap materials (e.g. HgTe) is given by [52]

$$E = \frac{\hbar^2 k^2}{2m^*} + \frac{3e^2}{128\varepsilon_\infty} k - \left(\frac{2E_B}{\pi}\right) \ln\left|\frac{k}{k_0}\right|, \qquad \text{(R2.7)}$$

where ε_∞ is the semiconductor permittivity in the high frequency limit, $E_B \equiv \frac{m_0 e^2}{2\hbar^2 \varepsilon_\infty^2}$ and $k_0 \equiv \frac{m_0 e^2}{\hbar^2 \varepsilon_\infty}$.

(E) The conduction electrons of n-GaSb obey the following three dispersion relations:

(1) In accordance with the model of Seiler et al. [53]

$$E = \left[-\frac{E_g}{2} + \frac{E_g}{2}\left[1+\alpha_4 k^2\right]^{1/2} + \frac{\varsigma_0 \hbar^2 k^2}{2m_0} + \frac{\bar{v}_0 f_1(k)\hbar^2}{2m_0} \pm \frac{\bar{\omega}_0 f_2(k)\hbar^2}{2m_0} \right], \tag{R2.8}$$

where $\alpha_4 \equiv 4P^2\left(E_g + \frac{2}{3}\Delta\right)\left[E_g^2\left(E_g+\Delta\right)\right]^{-1}$, P is the isotropic momentum matrix element, $f_1(k) \equiv k^{-2}\left[k_x^2 k_y^2 + k_y^2 k_z^2 + k_z^2 k_x^2\right]$ represents the warping of the Fermi surface, $f_2(k) \equiv \left[\left\{k^2\left(k_x^2 k_y^2 + k_y^2 k_z^2 + k_z^2 k_x^2\right) - 9k_x^2 k_y^2 k_z^2\right\}^{1/2} k^{-1}\right]$ represents the inversion asymmetry splitting of the conduction band and $\varsigma_0, \bar{v}_0$ and $\bar{\omega}_0$ represent the constants of the electron spectrum in this case.

It should be noted that under the substitutions, $\varsigma_0 = 0, \bar{v}_0 = 0, \bar{\omega}_0 = 0$ and $P^2 \equiv \frac{\hbar^2}{2m^*}\frac{E_g(E_g+\Delta)}{\left(E_g+\frac{2}{3}\Delta\right)}$, (R2.8) assumes the form of (2.10), which represents the well known two band model of Kane.

(2) In accordance with the model of Mathur et al. [54],

$$E = \frac{\hbar^2 k^2}{2m_0} + \frac{E_{g1}}{2}\left[\left(1 + \frac{2\hbar^2 k^2}{E_{g1}}\left\{\frac{1}{m^*} - \frac{1}{m_0}\right\}\right)^{1/2} - 1\right], \tag{R2.9}$$

where $E_{g1} = \left\{E_g + \left[\left(5\times 10^{-5} T^2\right)\left(2\left(112+T\right)^{-1}\right)\right]\right\}$ eV.

(3) In accordance with the model of Zhang et al. [55]

$$E = \left[E_2^{(1)} + E_2^{(2)} K_{4,1}\right] k^2 + \left[E_4^{(1)} + E_4^{(2)} K_{4,1}\right] k^4 + k^6\left[E_6^{(1)} + E_6^{(2)} K_{4,1} + E_6^{(3)} K_{6,1}\right]. \tag{R2.10}$$

where $K_{4,1} \equiv \frac{5}{4}\sqrt{21}\left[\frac{k_x^4+k_y^4+k_z^4}{k^4} - \frac{3}{5}\right]$, $K_{6,1} \equiv \sqrt{\frac{639639}{32}}\left[\frac{k_x^2 k_y^2 k_z^2}{k^6} + \frac{1}{22}\left(\frac{k_x^4+k_y^4+k_z^4}{k^4} - \frac{3}{5}\right) - \frac{1}{105}\right]$, the coefficients are in eV, the values of k are $10\left(\frac{a}{2\pi}\right)$ times those of k in atomic units (a is the lattice constant), $E_2^{(1)} = 1.0239620, E_2^{(2)} = 0, E_4^{(1)} = -1.1320772,$

$$E_4^{(2)} = 0.05658, \quad E_6^{(1)} = 1.1072073, \quad E_6^{(2)} = -0.1134024$$
and $E_6^{(3)} = -0.0072275$.

(F) The dispersion relation of the carriers in p-type Platinum antimonide ($PtSb_2$) following Emtage [56] can be written as

$$\left(E + \lambda_1\frac{a^2}{4}k^2 - l_1 k_s^2\frac{a^2}{4}\right)\left(E + \delta_0' - \upsilon_1\frac{a^2}{4}k^2 - n_1\frac{a^2}{4}k_s^2\right) = I_0\frac{a^4}{16}k^4, \tag{R2.11}$$

where $\lambda_1, l_1, \delta_0', \nu_1, n_1$ and I_0 are the energy band constants and a is the lattice constant.

(G) In addition to the well known three band model of Kane, the conduction electrons of n-GaAs obey the following three dispersion relations:
(1) In accordance with the model of Stillman et al. [57]

$$E=\frac{\hbar^2k^2}{2m^*}-\left(1-\frac{m^*}{m_0}\right)^2\left(\frac{\hbar^2k^2}{2m^*}\right)^2\left[\frac{3E_{\mathrm{g}}+4\Delta+\left(\frac{2\Delta^2}{E_{\mathrm{g}}}\right)}{(E_{\mathrm{g}}+\Delta)(2\Delta+3E_{\mathrm{g}})}\right], \tag{R2.12}$$

(2) In accordance with the model of Newson et al. [58]

$$E=\bar{\alpha}_0k_z^4+\left[\frac{\hbar^2}{2m^*}+\left(2\bar{\alpha}_0+\bar{\beta}_0\right)k_s^2\right]k_z^2+\frac{\hbar^2}{2m^*}k_s^2+\left(2\bar{\alpha}_0+\bar{\beta}_0\right)k_x^2k_y^2$$
$$+\bar{\alpha}_0\left(k_x^4+k_y^4\right), \tag{R2.13}$$

where $\bar{\alpha}_0\left(=-1.97\times10^{-37}\,\mathrm{eVm}^4\right)$ is the non-parabolicity constant and $\bar{\beta}_0\left(=-2.3\times10^{-37}\,\mathrm{eVm}^4\right)$ is the wrapping constant.
(3) In accordance with the model of Rossler [59]

$$E=\frac{\hbar^2k^2}{2m^*}+\bar{\alpha}_{10}k^4+\bar{\beta}_{10}\left[k_x^2k_y^2+k_y^2k_z^2+k_z^2k_x^2\right]$$
$$\pm\bar{\gamma}_{10}\left[k^2\left(k_x^2k_y^2+k_y^2k_z^2+k_z^2k_x^2\right)-9k_x^2k_y^2k_z^2\right]^{1/2}, \tag{R2.14}$$

where $\bar{\alpha}_{10}=\bar{\alpha}_{11}+\bar{\alpha}_{12}k, \bar{\beta}_{10}=\bar{\beta}_{11}+\bar{\beta}_{12}k$ and $\bar{\gamma}_{10}=\bar{\gamma}_{11}+\bar{\gamma}_{12}k$, in which,

$$\bar{\alpha}_{11}=-2132\times10^{-40}\,\mathrm{eVm}^4, \bar{\alpha}_{12}=9030\times10^{-50}\,\mathrm{eVm}^5,$$
$$\bar{\beta}_{11}=-2493\times10^{-40}\,\mathrm{eVm}^4,\quad \bar{\beta}_{12}=12594\times10^{-50}\,\mathrm{eVm}^5,$$
$$\bar{\gamma}_{11}=30\times10^{-30}\,\mathrm{eVm}^3 \text{ and } \bar{\gamma}_{12}=-154\times10^{-42}\,\mathrm{eVm}^4.$$

(H) In addition to the well known three band model of Kane, the conduction electrons of n-InSb obey the following three dispersion relations:

(1) To the fourth order effective mass theory, and taking into account the interactions of the conduction, the heavy hole, the light hole, and the split-off bands, the electron energy spectrum in n-InSb is given by [60]

$$E=\frac{\hbar^2k^2}{2m^*}+\bar{b}_1k^4, \tag{R2.15}$$

where $\bar{b}_1\equiv\frac{K_2\hbar^4}{4E_{\mathrm{g}}(m^*)^2}$, $K_2\equiv-\left[\frac{\left(1+\frac{1}{2}x_1^2\right)}{1+\frac{1}{2}x_1}\right](1-y_1)^2$, $x_1\equiv\left[1+\left(\frac{\Delta}{E_{\mathrm{g}}}\right)\right]^{-1}$ and $y_1\equiv\frac{m^*}{m_0}$.
(2) In accordance with Johnson and Dickey [61], the electron energy spectrum assumes the form

$$E=-\frac{E_{\mathrm{g}}}{2}+\frac{\hbar^2k^2}{2}\left[\frac{1}{m_0}+\frac{1}{m_{\gamma b}}\right]+\frac{E_{\mathrm{g}}}{2}\left[1+4\frac{\hbar^2k^2}{2m_{\mathrm{c}}'}\frac{\bar{f}_1(E)}{E_{\mathrm{g}}}\right]^{1/2}, \tag{R2.16}$$

where $\frac{m_0}{m'_c} \equiv P_0^2 \left[\frac{\left(E_g+\frac{2\Delta}{3}\right)}{E_g(E_g+\Delta)}\right]$, P_0^2 is the energy band constant, $\bar{f}_1(E) \equiv \frac{(E_g+\Delta)\left(E+E_g+\frac{2\Delta}{3}\right)}{\left(E_g+\frac{2\Delta}{3}\right)(E+E_g+\Delta)}$, $m'_c = 0.139m_0$ and $m_{\gamma b} = \left[\frac{1}{m'_c} - \frac{2}{m_0}\right]^{-1}$.
(3) In accordance with Agafonov et al. [62], the electron energy spectrum can be written as

$$E = \frac{\bar{\eta} - E_g}{2}\left[1 - \frac{\hbar^2 k^2}{2\bar{\eta}m^*}\left\{\frac{D\sqrt{3} - 3\bar{B}}{2\left(\frac{\hbar^2}{2m^*}\right)}\right\}\left[\frac{k_x^4 + k_y^4 + k_z^4}{k^4}\right]\right], \tag{R2.17}$$

where $\bar{\eta} \equiv \left(E_g^2 + \frac{8}{3}P^2k^2\right)^{1/2}$, $\bar{B} \equiv -21\frac{\hbar^2}{2m_0}$ and $D \equiv -40\left(\frac{\hbar^2}{2m_0}\right)$.
(I) The dispersion relation of the carriers in n-type $Pb_{1-x}Ga_xTe$ with $x = 0.01$ following Vassilev [63] can be written as

$$\begin{aligned}&\left[E - 0.606k_s^2 - 0.0722k_z^2\right]\left[E + \bar{E}_g + 0.411k_s^2 + 0.0377k_z^2\right]\\ &= 0.23k_s^2 + 0.02k_z^2 \pm \left[0.06\bar{E}_g + 0.061k_s^2 + 0.0066k_z^2\right]k_s,\end{aligned} \tag{R2.18}$$

where $\bar{E}_g$ ($= 0.21$ eV) is the energy gap for the transition point, the zero of the energy E is at the edge of the conduction band of the Γ point of the Brillouin zone and is measured positively upwards, k_x, k_y and k_z are in the units of $10^9\,\mathrm{m}^{-1}$.
(J) The charge carriers of Tellurium obey two different dispersion laws as given in the literature [64, 65].
(1) The dispersion relation of the conduction electrons in Tellurium, following Bouat [64] can be written as

$$E = A_6k_z^2 + B_6k_\perp^2 \pm \left[\bar{\vartheta}k_z^2 + \bar{\ell}k_\perp^2\right]^{1/2}, \tag{R2.19a}$$

where $A_6\left(= 6.7 \times 10^{-16}\,\mathrm{meVm}^2\right)$, $B_6\left(= 4.2 \times 10^{-16}\,\mathrm{meVm}^2\right)$, $\bar{\vartheta}(= (6 \times 10^{-8}\,\mathrm{meVm})^2)$ and $\bar{\ell}\left(= \left(3.8 \times 10^{-8}\,\mathrm{meVm}\right)^2\right)$ are the band constants.
(2) The energy spectrum of the carriers in the two higher valance bands and the single lower valance band of Te can respectively be expressed as [65]

$$\begin{aligned}\bar{E} &= A_{10}k_z^2 + B_{10}k_s^2 \pm \left[\Delta_{10}^2 + (\beta_{10}k_z)^2\right]^{1/2}\\ \text{and } \bar{E} &= \Delta_{||} + A_{10}k_z^2 + B_{10}k_s^2 \pm \beta_{10}k_z\end{aligned} \tag{R2.19b}$$

where $\bar{E}$ is the energy measured within the valance bands, $A_{10} = 3.77 \times 10^{-19}\,\mathrm{eVm}^2$, $B_{10} = 3.57 \times 10^{-19}\,\mathrm{eVm}^2$, $\Delta_{10} = 0.628\,\mathrm{eV}$, $(\beta_{10})^2 = 6 \times 10^{-20}\,(\mathrm{eVm})^2$ and $\Delta_{||} = 1004 \times 10^{-5}$ eV are the spectrum constants.

(K) The dispersion relation for the electrons in graphite can be written following Brandt [66] as

$$E = \frac{1}{2}\left[E_2 + E_3\right] \pm \left[\frac{1}{4}\left(E_2 - E_3\right)^2 + \eta_2^2k^2\right]^{1/2}, \tag{R2.20}$$

where $E_2 \equiv \bar{\Delta} - 2\bar{\gamma}_1 \cos\phi_0 + 2\bar{\gamma}_5 \cos^2\phi_0, \phi_0 \equiv \frac{c_6 k_z}{2}, E_3 \equiv 2\bar{\gamma}_2 \cos^2\phi_0$, $\eta_2 \equiv \left(\frac{\sqrt{3}}{2}\right) a_6 \left(\bar{\gamma}_0 + 2\bar{\gamma}_4 \cos\phi_0\right)$, in which the band constants are $\bar{\Delta} = -0.0002\,\text{eV}, \bar{\gamma}_0 = 3\,\text{eV}, \bar{\gamma}_1 = 0.392\,\text{eV}, \bar{\gamma}_2 = -0.019\,\text{eV}, \bar{\gamma}_4 = 0.193\,\text{eV}, \bar{\gamma}_5 = 0.194\,\text{eV}, a_6 = 2.46\,\text{Å}$ and $c_6 = 6.74\,\text{Å}$.

(L) The dispersion relation of the conduction electrons in Antimony (Sb) in accordance with Ketterson [67] can be written as

$$2m_0 E = \alpha_{11} p_x^2 + \alpha_{22} p_y^2 + \alpha_{33} p_z^2 + 2\alpha_{23} p_y p_z, \qquad \text{(R2.21)}$$

and

$$2m_0 E = a_1 p_x^2 + a_2 p_y^2 + a_3 p_z^2 + a_4 p_y p_z \pm a_5 p_x p_z \pm a_6 p_x p_y, \qquad \text{(R2.22)}$$

where $a_1 = \frac{1}{4}\left(\alpha_{11} + 3\alpha_{22}\right), a_2 = \frac{1}{4}\left(\alpha_{22} + 3\alpha_{11}\right), a_3 = \alpha_{33}, a_4 = \alpha_{33}, a_5 = \sqrt{3}$ and $a_6 = \sqrt{3}\left(\alpha_{22} - \alpha_{11}\right)$ in which $\alpha_{11} = 16.7$, $\alpha_{22} = 5.98$, $\alpha_{33} = 11.61$ and $\alpha_{23} = 7.54$ are the system constants.

(M) The dispersion relation of the holes in $\text{p} - \text{Bi}_2\text{Te}_3$ can be written [68] as

$$E\left(1 + \frac{E}{E_{\text{g}}}\right) = \frac{\hbar^2}{2m_0}\left[\alpha_{11} k_x^2 + \alpha_{22} k_y^2 + \alpha_{33} k_z^2 + 2\alpha_{23} k_y k_z\right], \qquad \text{(R2.23)}$$

where x, y and z are parallel to binary, bisectrix and trigonal axes respectively, $E_g = 0.145\,\text{eV}, \alpha_{11} = 32.5, \alpha_{22} = 4.81, \alpha_{33} = 9.02, \alpha_{23} = 4.15, g_{\text{s}} = 2$ and $g_{\text{v}} = 6$.

(N) The dispersion relation of the holes in p-InSb in accordance with Cunningham [69] can be written as

$$\bar{E} = c_4\left(1 + \gamma_4 f_4\right) k^2 \pm \frac{1}{3}\left[2\sqrt{2}\sqrt{c_4}\sqrt{16 + 5\gamma_4}\sqrt{E_4 g_4} k\right], \qquad \text{(R2.24)}$$

where $\bar{E}$ is the energy of the hole as measured from the top of the valance and within it, $c_4 \equiv \frac{\hbar^2}{2m_0} + \theta_4, \theta_4 \equiv 4.7\frac{\hbar^2}{2m_0}, \gamma_4 \equiv \frac{b_4}{c_4}, b_4 \equiv \frac{3}{2} b_5 + 2\theta_4,\ \ b_5 \equiv 2.4\frac{\hbar^2}{2m_0},\ \ f_4 \equiv \frac{1}{4}\left[\sin^2 2\theta + \sin^4\theta \sin^2 2\phi\right]$, θ is measured from the positive z-axis, ϕ is measured from positive x-axis, $g_4 \equiv \sin\theta\left[\cos^2\theta + \frac{1}{4}\sin^4\theta\sin^2 2\phi\right]$ and $E_4 = 5 \times 10^{-4}\,\text{eV}$.

(O) The dispersion relation of the valance bands of II–V compounds in accordance with Yamada [70] can be written as

$$\begin{aligned} E = {} & \frac{1}{2}\left(\bar{t}_1 + \bar{t}_2\right) k_x^2 + \frac{1}{2}\left(\bar{t}_3 + \bar{t}_4\right) k_y^2 + \frac{1}{2}\left(\bar{t}_5 + \bar{t}_6\right) k_z^2 + \frac{1}{2}\left(\bar{t}_7 + \bar{t}_8\right) k_x \\ & \pm [\left\{\frac{1}{2}\left(\bar{t}_1 - \bar{t}_2\right) k_x^2 + \frac{1}{2}\left(\bar{t}_3 - \bar{t}_4\right) k_y^2 + \frac{1}{2}\left(\bar{t}_5 - \bar{t}_6\right) k_z^2 + \frac{1}{2}\left(\bar{t}_7 - \bar{t}_8\right) k_x\right\}^2 \\ & + t_9^2 k_y^2 + t_{10}^2\,]^{1/2}, \end{aligned} \qquad \text{(R2.25)}$$

where $\bar{t}_i\ (i = 1\text{to}8), t_9$ and t_{10} are the constants of the energy spectra.

For $p - \mathrm{CdSb}$, $\bar{t}_1 = -32.3 \times 10^{-20}\,\mathrm{eVm}^2, \bar{t}_2 = -60.7 \times 10^{-20}\,\mathrm{eVm}^2, \bar{t}_3 = -1.63 \times 10^{-19}\,\mathrm{eVm}^2, \bar{t}_4 = -2.44 \times 10^{-19}\,\mathrm{eVm}^2$, $\bar{t}_5 = -9.19 \times 10^{-19}\,\mathrm{eVm}^2$, $\bar{t}_6 = -10.5 \times 10^{-19}\,\mathrm{eVm}^2, \bar{t}_7 = 2.97 \times 10^{-10}\,\mathrm{eVm}, \bar{t}_8 = -3.47 \times 10^{-10}\,\mathrm{eVm}$,, $t_9 = 1.3 \times 10^{-10}\,\mathrm{eVm}$ and $t_{10} = 0.070\,\mathrm{eV}$.

(P) The energy spectrum of the valance bands of CuCl in accordance with Yekimov et al. [71] can be written as

$$E_{\mathrm{h}} = (\gamma_6 - 2\gamma_7)\frac{\hbar^2 k^2}{2m_0}, \tag{R2.26}$$

and

$$E_{l,s} = (\gamma_6 + \gamma_7)\frac{\hbar^2 k^2}{2m_0} - \frac{\Delta_1}{2} \pm \left[\frac{\Delta_1^2}{4} + \gamma_7 \Delta_1 \frac{\hbar^2 k^2}{2m_0} + 9\left(\frac{\gamma_7 \hbar^2 k^2}{2m_0}\right)^2\right]^{1/2}, \tag{R2.27}$$

where $\gamma_6 = 0.53, \gamma_7 = 0.07, \Delta_1 = 70\,\mathrm{meV}$.

(Q) In the presence of stress, χ_6 along <001> and <111> directions, the energy spectra of the holes in semiconductors having diamond structure valance bands can be respectively expressed following Roman [72] et al. as

$$E = A_6 k^2 \pm \left[\bar{B}_7^2 k^4 + \delta_6^2 + B_7 \delta_6 \left(2k_z^2 - k_s^2\right)\right]^{1/2}, \tag{R2.28}$$

and

$$E = A_6 k^2 \pm \left[\bar{B}_7^2 k^4 + \delta_7^2 + \frac{D_6}{\sqrt{3}}\delta_7 \left(2k_z^2 - k_s^2\right)\right]^{1/2}, \tag{R2.29}$$

where A_6, B_7, D_6 and C_6 are inverse mass band parameters in which $\delta_6 \equiv l_7\left(\bar{S}_{11} - \bar{S}_{12}\right)\chi_6, \bar{S}_{ij}$ are the usual elastic compliance constants, $\bar{B}_7^2 \equiv \left(B_7^2 + \frac{c_6^2}{5}\right)$ and $\delta_7 \equiv \left(\frac{d_8 S_{44}}{2\sqrt{3}}\right)\chi_6$. For gray tin,$d_8 = -4.1\,\mathrm{eV}$, $l_7 = -2.3\,\mathrm{eV}$,

$$A_6 = 19.2\frac{\hbar^2}{2m_0}, B_7 = 26.3\frac{\hbar^2}{2m_0}, D_6 = 31\frac{\hbar^2}{2m_0} \text{ and } c_6^2 = -1112\frac{\hbar^2}{2m_0}.$$

R2.2 Investigate the Einstein relation for all materials of problem (R2.1), in the presence of an arbitrarily oriented non-quantizing and (a) non-uniform electric field (b) alternating electric field respectively.

Allied Research Problems

R2.3 Investigate the Debye screening length, the carrier contribution to the elastic constants, the heat capacity, the activity coefficient, and the plasma frequency for all the materials of problem (R2.1).

R2.4 Investigate in detail, the mobility for elastic and inelastic scattering mechanisms for all the materials of problem (R2.1).

R2.5 Investigate the various transport coefficients in detail for all the materials of problem (R2.1).

References

1. K.P. Ghatak, *Influence of band structure on some quantum processes in tetragonal semiconductors*, D. Eng. Thesis, Jadavpur University, Kolkata, India, 1991; K.P. Ghatak, S. Bhattacharya, S. Bhowmik, R. Benedictus, S. Choudhury, J. Appl. Phys. **103**, 034303 (2008); J.L. Shay, J.W. Wernik, *Ternary Chalcoprite Semiconductors: Growth, Electronic Properties and Applications* (Pergamon Press, London, 1975)
2. J.L. Shay, K.G. Beckmann, E. Buchler, J. Wernik, Appl. Phys. Lett. **23**, 226 (1973); G.P. Chuiko, Sov. Phys. Semiconduct. **15**, 739 (1981); R.L. Byer, H. Kildal, R.S. Fiegelson, Appl. Phys. Lett. **19**, 237 (1971)
3. S.I. Radautsan, V.I. Morozova, A.F. Knyazev, L.S. Koval, E.K. Arushanov, A.N. Nateprov, Sov. Phys. Semiconduct. **19**, 691 (1985)
4. S.I. Radautsan, E.K. Arushanov, A.N. Nateprov, G.P. Chuiko, *Cadmium Arsenide and Phosphide (in Russian) Shtiintsa*, Kishinev (1976)
5. I.W. Rowe, I.L. Shay, Phys. Rev. **83**, 451 (1971)
6. J.J. Hopfield, I. Phys. Chem. Solids **15**, 97 (1960)
7. A. Shileika, Surf. Sci. **37**, 730 (1973); J.L. Shay, B. Tell, Surf. Sci. **37**, 748 (1973); B. Kh. Mamedov, E.O. Osmanov, Sov. Phys. Semiconduct. **5**, 1120 (1972); O.V. Emelyanenko, F.P. Kesamanly, I.K. Polushina, V.A. Skripkin, Sov. Phys. Semiconduct. **5** 304 (1971); B.L. Zlatkin, F. Yu Markov, I.K. Polushina, Sov. Phys. Semiconduct. **3**, 1336 (1970); S.D. Chemla, F.R. Beglay, L.R. Byer, IEEE J. Quantum Electron. QE-**10**, 71 (1974)
8. H. Kildal, Phys. Rev. **10B**, 5082 (1974)
9. J. Bodnar, in *Physics of Narrow-Gap Semiconductors*, ed. by J. Rautuszkiewicz, M. Gorska, E. Kaczmarek. Proceedings of International Conference, Warozawa, p. 311 (PWN-Polish Scientific Publisher, Warszwa, Poland, 1978)
10. R.K. Pathria, *Statistical Mechanics*, 2nd edn. (Butterworth-Heinmann, Oxford, 1996)
11. M. Abramowitz, I.A. Stegun, *Handbook of Mathematical Functions* (Dover Publications, New York, USA, 1965)
12. M. Meltz, M.S. Dresselhaus, Phys. Rev. **2B**, 2877 (1970)
13. T.L. Koch, E.G. Burkhardt, F.G. Storz, T.J. Bridges, T. Sizer, IEEE J.Q.E. **23**, 889 (1987); R. Bisaro, G. Laurencin, A. Friederich, M. Razeghi, Appl. Phys. Lett. **40**, 978 (1982)
14. R.C. Alferness, C.H. Joyner, M.D. Divino, M.J.R. Martyak, L.L. Buhl, Appl. Phys. Lett. **49**, 125 (1986)
15. T.E. Bell, IEEE Spectr **20**, 38 (1983)
16. P.Y. Lu, C.H. Wang, C.M. Williams, S.N.G. Chu, C.M. Stiles, Appl. Phys. Lett. **49**, 1372 (1986)
17. N.R. Taskar, I.B. Bhat, K.K. Parat, D. Terry, H. Ehsani, S.K. Gandhi, J. Vac. Sci. Tech. **7A**, 281 (1989)
18. E. Weiss, N. Mainzer, J. Vac. Sci. Tech. **7A**, 391 (1989)
19. F. Koch, in *Springer Series in Solid State Sciences*, vol. 53 (Springer-Verlag, Germany, 1984)
20. D.H. Daw, D.S. Caoand, G.B. Stringfellow, J. Appl. Phys. **69**, 2552 (1991)
21. M.L. Timmons, S.M. Bedair, R.J. Markunas, J.A. Hutchby, *Proceedings of 16th IEEE Photovoltaic Specialists Conferences*, vol 666 (IEEE San Diego, California, 1982)

22. L.R. Tomasetta, H.D. Law, R.C. Eden, I. Reyhimy, K. Nakano, IEEE Quantum Electron **14**, 800 (1978)
23. M.K. Lee, D.S. Wu, H.H. Tung, J. Appl. Phys. **62**, 3209 (1987)
24. M. Levimson, J.L. Benton, H. Temkin, L.C. Kimerling, Appl. Phys. Lett. **40**, 990 (1982)
25. J.S. Blakemore, *Semiconductor Statistics* (Dover, New York, 1987); K.P. Ghatak, S. Bhattacharya, S.K. Biswas, A. Dey, A.K. Dasgupta, Phys. Scr. **75**, 820 (2007)
26. E.A. Arushanov, A.A. Kaynzev, A.N. Natepov, S.I. Radautsan, Sov. Phys. Semicond. **15**, 828 (1981)
27. M.A. Herman, M. Pessa, J. Appl. Phys. **57**, 2671(1985)
28. K. Sakamoto, Y. Okabe, Jpn J. Appl. Phys **23**, 444 (1986)
29. T. Nguyen Duy, J. Mesloge, G. Pichard J. Crystal Growth **72**, 490 (1985)
30. D.H. Chow, T.C. McGill, Appl. Phys. Lett. **48**, 1485 (1986)
31. D.A. Cammack, R.J. Dalby, H.J. Cornelissen, J. Khurgin, J. Appl. Phys. **62**, 3071 (1987)
32. J.J. Hopfield, J. Appl. Phys. **32**, 2277 (1961)
33. D. Shoenberg, Proc. R. Soc. **A170**, 341 (1939)
34. B. Abeles, S. Mieboom, Phys. Rev. **101**, 544 (1956)
35. B. Lax, J.G. Mavroides, H.J. Ziegeer, R.J. Keyes, Phys. Rev. Lett. **5**, 241 (1960)
36. M. Cankurtarran, M. Celik, T. Alper, J. Phys. F: Metal Phys. **16**, 853 (1986)
37. Y.H. Kao, Phys. Rev. **129**, 1122 (1963)
38. R.J. Dinger, A.W. Lawson, Phys. Rev **3B**, 253 (1971)
39. J.A. Koch, J.D. Jensen, Phys. Rev. **184**, 643 (1969)
40. M.H. Cohen, Phys. Rev. **121**, 387 (1961)
41. J.W. McClure, K.H. Choi, Solid State Commun. **21**, 1015 (1977)
42. C.C. Wu, C.J. Lin, J. Low. Temp. Phys. **57**, 469 (1984)
43. D.R. Lovett, in *Semimetals and Narrow Band Gap Semiconductors* (Pion Limited, London, 1977); M. Kriecghbaum, P. Kocevar, H. Pascher, G. Baur, IEEE, QE, **24**, 1727 (1988); I.V. Kucherenko, A.E. Svistov, A.P. Shotov, Sov. Phys. Semi. **15**, 510 (1981)
44. D.G. Seiler, B.D. Bajaj, A.E. Stephens, Phys. Rev. **B 16**, 2822 (1977); A.V. Germaneko, G.M. Minkov, Phys. Stat. Sol. (b) **184**, 9 (1994); G.L. Bir, G.E. Pikus, *Symmetry and Strain –Induced effects in Semiconductors Nauka, Russia* (1972). (in Russian); M. Mondal, K.P. Ghatak, Phys. Stat. Sol. (b) **135**, K21 (1986)
45. G.J. Rees, in *Physics of Compounds*, ed. by F.G. Fumi. Proceedings of the 13th International Conference, pp. 1166 (North Holland Company, 1976)
46. E.L. Ivchenko, G.E. Pikus, Sov. Phys. Semicond. **13**, 579 (1979)
47. J.O. Dimmock, in *Physics of Semimetals and Narrow Gap Compounds*, ed. by D.L. Carter, R.T. Bates (Pergamon Press, Oxford, 1971, pp. 319)
48. E. Bangert, P. Kastner, Phys. Stat. Sol. (b) **61**, 503 (1974)
49. G.M.T. Foley, P.N. Langenberg, Phys. Rev. B, **15B**, 4850 (1977)
50. M. Cardona, W. Paul, H. Brooks Helv, Acta Phys. **33**, 329 (1960); A.F. Gibson, in *Proceeding of International School of Physics "ENRICO FERMI"* course XIII, ed. By R.A Smith (Academic Press, London, 1963), pp. 171
51. C.C. Wang, N.W. Ressler, Phys. Rev. **2**, 1827 (1970)
52. V.I. Ivanov-Omskii, A.Sh. Mekhtisev, S.A. Rustambekova, E.N. Ukraintsev, Phys. Stat. Sol. (b) **119**, 159 (1983)

53. D.G. Seiler, W.M. Beeker, L.M. Roth, Phys. Rev. **1**, 764 (1970)
54. P.C. Mathur, S. Jain, Phys. Rev. **19**, 3159 (1979)
55. H.I. Zhang, Phys. Rev. **1B**, 3450 (1970)
56. P.R. Emtage, Phys. Rev. A **246**, 138 (1965)
57. G.E. Stillman, C.M. Wolfe, J.C. Dimmock, in *Semiconductors and Semimetals*, ed. by R.K. Willardson, A.C. Beer, vol. 12 (1977), p. 169
58. D.J. Newson, A. Kurobe, Semicond. Sci. Technol. **3**, 786 (1988)
59. U. Rossler, Solid State Commun. **49**, 943 (1984)
60. E.D. Palik, G.S. Picus, S. Teither, R.F. Wallis, Phys. Rev. **122**, 475 (1961)
61. E.J. Johnson, D.H. Dicley, Phys. Rev. **1**, 2676 (1970)
62. V.G. Agafonov, P.M. Valov, B.S. Ryvkin, I.D. Yaroshetskii, Sov. Phys. Semicond. **12**, 1182 (1978)
63. L.A. Vassilev, Phys. Stat. Sol. (b) **121**, 203 (1984)
64. J. Bouat, J.C. Thuillier, Surf. Sci.**73**, 528 (1978)
65. N.S. Averkiev, V.M. Asnin, A.A. Bakun, A.M. Danishevskii, E.L. Ivchenko, G.E. Pikus, A.A. Rogachev, Sov. Phys. Semiconduct. **18**, 379 and 402 (1984)
66. N.B. Brandt, V.N. Davydov, V.A. Kulbachinskii, O.M. Nikitina, Sov. Phys. Sol. Stat. **29**, 1014 (1987)
67. J.B. Ketterson, Phys. Rev. **129**, 18 (1963)
68. M. Stordeur, W. Kuhnberger, Phys. Stat. Sol. (b) **69**, 377 (1975); H. Kohler, Phys. Stat. Sol. (b) **73**, 95 (1976)
69. R.W. Cunningham, Phys. Rev. **167**, 761 (1968)
70. Y. Yamada, J. Phys. Soc. Jpn **37**, 606 (1974)
71. A.I. Yekimov, A.A. Onushchenko, A.G. Plyukhin, Al.L. Efros, J. Expt. Theor. Phys. **88**, 1490 (1985)
72. B.J. Roman, A.W. Ewald, Phys. Rev. **B5**, 3914 (1972)
73. O. Madelung, *Semiconductors: Data Handbook*, 3rd edn. (Springer, Berlin, 2004); M. Krieehbaum, P. Kocevar, H. Pascher, G. Bauer, *IEEE QE*, **24**, 1727 (1988); G.P. Chuiko, Sov. Phys. Semiconduct. **19**, 1279 (1985)
74. M.J. Gelten, C.V.M. VanEs, F.A.P. Blom, J.W.F. Jongencelen, Solid State Commun. **33**, 833 (1980); A.A. El-Shazly, H.S. Soliman, H.E.A. El-Sayed, D.A.A. El-Hady, J. Vac. **47**, 53 (1996)
75. K.S. Hong, R.F. Speyer, R.A. Condrate, J. Phys. Chem. Solids, **51**, 969 (1990)
76. B.R. Nag, *Electron Transport in Compound Semiconductors* (Springer-Verlag, Germany, 1980); O. Madelung, *Semiconductors: Data Handbook*, 3rd edn. (Springer, Berlin, 2004); M. Krieehbaum, P. Kocevar, H. Pascher, G. Bauer, *IEEE QE*, **24** 1727 (1988)
77. S. Adachi, J. Appl. Phys. **58**, R1 (1985)
78. G.L. Hansen, J.L. Schmit, T.N. Casselman, J. Appl. Phys. **63**, 7079 (1982); R.S. Kim, S. Narita, Phys. Stat. Sol. (b) **73**, 741 (1976); V.G. Sredin, V.G. Savitskii, Ya. V. Danilyuk, M.V. Miliyanchuk, I.V. Petrovich, Sov. Phys. Semicond. **15**, 249 (1981)
79. S.K. Sutradhar, D. Chattopadhyay, B.R. Nag, Phys. Rev. (b) **25**, 4069 (1982)
80. S. Adachi, J. Appl. Phys. **53**, 8775 (1982)
81. S. Tiwari, S. Tiwari, Cryst. Res. Technol. **41**, 78 (2006)
82. J.R. Lowney, S.D. Senturia, J. Appl. Phys. **47**, 1771 (1976)
83. I.V. Kucherenko, Yu.A. Mityagin, L.K. Vodop'yanov, A.P. Shotov, Sov. Phys. Semiconduct. **11**, 282 (1977)
84. R.A. Reynolds, M.J. Brau, R.A. Chapman, J. Phys. Chem. Solids **29**, 755 (1968)
85. I.V. Skryabinskii, Yu.I. Ukhanov, Sov. Phys. Solid State, **14**, 2838 (1973)

3

The Einstein Relation in Compound Semiconductors Under Magnetic Quantization

3.1 Introduction

It is well known that the band structure of electronic materials can be dramatically changed by applying external fields [1]. The effects of the quantizing magnetic field on the band structure of compound semiconductors are more striking and can be observed easily in experiments. Under magnetic quantization, the motion of the electron parallel to the magnetic field remains unaltered while the area of the wave vector space perpendicular to the direction of the magnetic field gets quantized in accordance with Landau's rule of area quantization in the wave-vector space [2]. The energy levels of the carriers in a magnetic field (with the component of the wave-vector parallel to the direction of magnetic field equated with zero) are termed as the Landau levels and the quantized energies are known as the Landau sub-bands. It is important to note that the same conclusion may be arrived either by solving the single-particle time independent Schrödinger differential equation in the presence of a quantizing magnetic field or by using the operator method. The quantizing magnetic field tends to remove the degeneracy and increases the band gap. A semiconductor, placed in a magnetic field B, can absorb radiated energy with the frequency ω_0 $(= (|e|\, B/m^*))$. This phenomenon is known as cyclotron or diamagnetic resonance. The effect of energy quantization is experimentally noticeable when the separation between any two consecutive Landau levels is greater than $k_B T$. A number of interesting transport phenomena originate from the change in the basic band structure of the semiconductor in the presence of a quantizing magnetic field. These have been widely been investigated and have also served as diagnostic tools for characterizing the different materials having various band structures. The discreteness in the Landau levels leads to a whole crop of magneto-oscillatory phenomena, important among which are (a) Shubnikov-de Haas oscillations in magneto-resistance; (b) de Haas-Van Alphen oscillations in magnetic susceptibility; (c) magneto-phonon oscillations in thermoelectric power, etc.

In Sect. 3.2.1, of the theoretical background, the Einstein relation has been investigated in tetragonal materials in the presence of an arbitrarily oriented quantizing magnetic field by formulating the density-of-states function. Section 3.2.2 contains the results of III–V, ternary and quaternary compounds in accordance with the three and the two band models of Kane and forms the special case of Sect. 3.2.1. In the same section the well known result of DMR in relatively wide gap materials has been presented. Section 3.2.3 contains the study of the Einstein relation for the II–VI semiconductors under magnetic quantization. In Sect. 3.2.4, the magneto-DMR for Bismuth has been investigated in accordance with the models of McClure and Choi, Cohen, Lax non-parabolic ellipsoidal and the parabolic ellipsoidal respectively. In Sect. 3.2.5, the Einstein relation in IV–VI materials has been discussed. In Sect. 3.2.6, the magneto-DMR for the stressed Kane type semiconductors has been investigated. Section 3.3 contains the result and discussions in this context.

3.2 Theoretical Background

3.2.1 Tetragonal Materials

In the presence of an arbitrarily oriented quantizing magnetic field B along k_{z1} direction which makes an angle θ with k_z axis and lies in the $k_x - k_z$ plane, the magneto-dispersion law of the conduction electrons in tetragonal compounds can be expressed extending the method given by Wallace [3] as

$$\psi_1(E) = \bar{A}_{\pm}(n, E, \theta) + a_0(E, \theta)(k_{z1})^2, \tag{3.1}$$

where

$$\begin{aligned}\bar{A}_{\pm}(n, E, \theta) \equiv{}& \frac{2|e|B}{\hbar}\left(n + \frac{1}{2}\right)\left[\psi_2(E)\left\{\psi_2(E)\cos^2\theta + \psi_3(E)\sin^2\theta\right\}\right]^{\frac{1}{2}} \\ &\pm \left[\frac{|e|B\hbar E_g}{6}\left\{\frac{(E_g + \Delta_\perp)}{m_\perp^*\left(E_g + \frac{2}{3}\Delta_\perp\right)}\right\}^{\frac{1}{2}}\right] \\ &\times\left[\left(E + E_g + \delta + \left[\frac{\Delta_{\|}^2 - \Delta_\perp^2}{3\Delta_{\|}}\right]\right)^2\left\{\frac{\Delta_{\|}^2(E_g + \Delta_\perp)\cos^2\theta}{m_\perp^*\left(E_g + \frac{2}{3}\Delta_\perp\right)}\right\}\right. \\ &\left. + \left\{\frac{(E + E_g)^2(E_g + \Delta_{\|})\Delta_\perp^2\sin^2\theta}{m_{\|}^*\left(E_g + \frac{2}{3}\Delta_{\|}\right)}\right\}\right]^{\frac{1}{2}},\end{aligned}$$

$n\,(= 0, 1, 2, 3, \ldots)$ is the Landau quantum number and $a_0(E, \theta) \equiv \frac{(\psi_2(E)\psi_3(E))}{(\psi_2(E)\cos^2\theta + \psi_3(E)\sin^2\theta)}$.

It is interesting to note that three important concepts are in disguise in the apparently simple (3.1), which can, briefly be described as follows:

(1) *Effective electron mass.* The effective mass of the carriers in semiconductors, being connected with the mobility, is known to be one of the most important physical quantities used for the analysis of the semiconductor devices under different operating conditions [4]. It must be noted that among the various definitions of the effective electron mass [5], it is the effective momentum mass that should be regarded as the basic quantity [6]. This is due to the fact that it is this mass which appears in the description of transport phenomena and all other properties of the conduction electrons of the semiconductors having arbitrary dispersion laws [7]. It is the effective momentum mass in various transport coefficients which plays the most dominant role in explaining the experimental results under different scattering mechanisms [8,9]. The carrier degeneracy in semiconductors influences the effective mass when it is energy dependent. Under degenerate conditions, only the electrons at the Fermi surface of n-type semiconductors participate in the conduction process and hence, the effective momentum mass of the electrons (EMM) corresponding to the Fermi level would be of interest in electron transport under such conditions. The Fermi energy is again determined by the carrier energy spectrum and the carrier concentration and therefore these two features would determine the dependence of the EMM in degenerate materials on the degree of carrier degeneracy. In recent years, the EMM in such materials under different external conditions has been studied extensively [10–20]. It has, therefore, different values in different materials and varies with electron concentration, with the magnitude of the reciprocal quantizing magnetic field under magnetic quantization, with the quantizing electric field as in inversion layers, with the nanothickness as in quantum wells and quantum well wires and with the superlattice period as in the quantum confined superlattices having various carrier energy spectra. From (3.1), the EMM at the Fermi level along the direction of the quantizing magnetic field, can be expressed as $m^*_{k_{z1}}(E_{\mathrm{FB}}n,\theta) = \hbar^2 k_{z1} \left.\frac{\partial k_{z1}}{\partial E}\right|_{E=E_{\mathrm{FB}}}$

$$= \left(\frac{\hbar^2}{2}\right)\left\{\left[\frac{[\psi_1(E_{\mathrm{FB}})]' - [\bar{A}_\pm(n,E_{\mathrm{FB}},\theta)]'}{a_0(E_{\mathrm{FB}},\theta)}\right] - \left[\frac{[a_0(E_{\mathrm{FB}},\theta)]'}{a_0^2(E_{\mathrm{FB}},\theta)}\right]\right.$$
$$\left.\times\left[\psi_1(E_{\mathrm{FB}}) - \bar{A}_\pm(n,E_{\mathrm{FB}},\theta)\right]\right\}, \tag{3.2}$$

where E_{FB} is the Fermi energy in the presence of magnetic quantization as measured from the edge of the conduction band in the vertically upward direction in the absence of any field. From (3.2), it appears that EMM is a function of the Fermi energy, the angle of orientation of the quantizing magnetic field, the magnetic quantum number, and the electron spin for tetragonal materials due to the combined influence of the crystal field splitting and the anisotropic spin orbit splitting constant. The dependence of the oscillatory mobility on the spin dependent EMM in addition to Fermi energy is an important physical feature of tetragonal compounds.

(2) *The Landau sub-band.* The idea of Landau sub-bands is a key-concept in the study of the magneto transport of compound semiconductors [1]. The Landau singularity is the signature of the concept of branch-cut. The Landau levels can be obtained by substituting $k_{z1} = 0$ and $E = E_n$ in (3.1) for tetragonal materials. Thus, the Landau energy E_n, can be expressed through the equation

$$\psi_1 (E_n) - \bar{A}_{\pm} (n, E_n, \theta) = 0. \tag{3.3}$$

From (3.3), we infer that the difference between any two consecutive Landau levels for tetragonal compounds is not a constant quantity.

(3) *The Period of Shubnikov-de Haas Oscillation.* The SdH oscillation determines the period of all the oscillatory plots of any electronic quantity of any electronic material under magnetic quantization. In tetragonal materials, the period of SdH oscillation, $\Delta\left(\frac{1}{B}\right)$, can be expressed from (3.1) as

$$\Delta\left(\frac{1}{B}\right) = \frac{2|e|}{\hbar}\left[\psi_2 (E_{\text{FB}}) \left\{\psi_2 (E_{\text{FB}}) \cos^2\theta + \psi_3 (E_{\text{FB}}) \sin^2\theta\right\}\right]^{\frac{1}{2}} \times \left[\psi_1 (E_{\text{FB}})\right]^{-1}. \tag{3.4}$$

Thus, we observe that the SdH period is a function of the Fermi energy, the angle of orientation of the quantizing magnetic field B, and the energy band constants of tetragonal compounds, although the result is spin independent.

The formulation of the DMR needs the expression of DOS and the generalized expression of the density-of-states function under magnetic quantization ($D_{\text{B}}(E)$) can be written, including spin and extending the method as given in Nag [1], as

$$D_{\text{B}} (E) = \frac{g_{\text{v}} |e| B}{2\pi^2\hbar}\left(\sum_{n=0}^{n_{\max}} \frac{\partial k_{z1}}{\partial E} H (E - E_n)\right), \tag{3.5}$$

where $H(E - E_n)$ is the Heaviside step function and E_n is the Landau energy.

Using (3.1) and (3.5), one obtains,

$$D_{\text{B}} (E) = \left(\frac{g_{\text{v}} |e| B}{4\pi^2\hbar}\right) \sum_{n=0}^{n_{\max}} \left[\frac{\psi_1 (E) - \bar{A}_{\pm} (n, E, \theta)}{a_0 (E, \theta)}\right]^{\frac{-1}{2}} \left[a_0 (E, \theta)\right]^{-2} \Big[a_0 (E, \theta) \left\{\{\psi_1 (E)\}' - \left[\bar{A}_{\pm} (n, E, \theta)\right]'\right\} - \left\{\psi_1 (E) - \bar{A}_{\pm} (n, E, \theta)\right\} \left\{a_0 (E, \theta)\right\}' \Big] H (E - E_n). \tag{3.6}$$

Thus, combining (3.6) with the Fermi–Dirac occupation probability factor and using the generalized Sommerfeld's lemma [21], the electron concentration assumes the form

$$n_0 = \frac{g_v |e| B}{2\pi^2 \hbar} \sum_{n=0}^{n_{\max}} [T_{31}(n, E_{FB}) + T_{32}(n, E_{FB})], \tag{3.7}$$

where

$$T_{31}(n, E_{FB}) \equiv \left[\frac{\psi_1(E_{FB}) - \bar{A}_{\pm}(n, E_{FB}, \theta)}{a_0(E_{FB}, \theta)}\right]^{\frac{1}{2}},$$

and

$$T_{32}(n, E_{FB}) \equiv \sum_{r=1}^{s} L(r) [T_{31}(n, E_{FB})].$$

Thus, using (3.7) and (1.11), the Einstein relation in tetragonal compounds under magnetic quantization, can be written as

$$\frac{D}{\mu} = \frac{1}{|e|} \left[\frac{\sum_{n=0}^{n_{\max}} [T_{31}(n, E_{FB}) + T_{32}(n, E_{FB})]}{\sum_{n=0}^{n_{\max}} [\{T_{31}(n, E_{FB})\}' + \{T_{32}(n, E_{FB})\}']}\right]. \tag{3.8}$$

It is interesting to note that the electron dispersion relation for the tetragonal materials excluding the spin term in the presence of an arbitrarily oriented quantizing magnetic field B can be formulated by using the area quantization rule of Landau in the following simple way:

The area of cross section of the ellipsoid $\frac{x^2}{a^2} + \frac{y^2}{b^2} + \frac{z^2}{c^2} = 1$ by the plane $lx + my + nz = p$ is given by [22]

$$A = \frac{\pi abc}{(a^2l^2 + b^2m^2 + c^2n^2)^{1/2}} \left[1 - \frac{p^2}{(a^2l^2 + b^2m^2 + c^2n^2)}\right]. \tag{3.9}$$

In our case, the ellipsoid of the revolution can be written from (2.2) as $\frac{k_x^2}{\left(\frac{\psi_1(E)}{\psi_2(E)}\right)} + \frac{k_y^2}{\left(\frac{\psi_1(E)}{\psi_2(E)}\right)} + \frac{k_z^2}{\left(\frac{\psi_1(E)}{\psi_3(E)}\right)} = 1$ and the equation of the plane is $k_x \sin\theta + k_z \cos\theta = k_{z1}$

Therefore the use of (3.9) leads to the expression for the area of cross section as

$$A(E, k_{z1}) = \frac{\pi [\psi_1(E)]^{3/2}}{\psi_2(E)\sqrt{\psi_3(E)}} \left[\frac{\psi_1(E)\sin^2\theta}{\psi_2(E)} + \frac{\psi_1(E)\cos^2\theta}{\psi_3(E)}\right]^{-1/2} \times \left[1 - \frac{k_{z1}^2}{\frac{\psi_1(E)\sin^2\theta}{\psi_2(E)} + \frac{\psi_1(E)\cos^2\theta}{\psi_3(E)}}\right]. \tag{3.10}$$

The Landau area quantization rule is given by [2]

$$A(E, k_{z1}) = \frac{2\pi |e| B}{\hbar}\left(n + \frac{1}{2}\right). \tag{3.11}$$

Using (3.10) and (3.11) we get,

$$\psi_1(E) = \frac{2|e|B}{\hbar}\left(n+\frac{1}{2}\right)\left[\psi_2(E)\left\{\psi_2(E)\cos^2\theta+\psi_3(E)\sin^2\theta\right\}\right]^{\frac{1}{2}} + a_0(E,\theta)(k_{z1})^2 . \quad (3.12)$$

Thus the electron concentration and the DMR can, respectively, be written as

$$n_0 = \frac{g_v|e|B}{\pi^2\hbar}\sum_{n=0}^{n_{\max}}[T_{33}(n,E_{FB})+T_{34}(n,E_{FB})], \quad (3.13)$$

$$\frac{D}{\mu} = \frac{1}{|e|}\left[\frac{\sum_{n=0}^{n_{\max}}[T_{33}(n,E_{FB})+T_{34}(n,E_{FB})]}{\sum_{n=0}^{n_{\max}}[\{T_{33}(n,E_{FB})\}'+\{T_{34}(n,E_{FB})\}']}\right], \quad (3.14)$$

where

$$T_{33}(n,E_{FB}) \equiv \left\{\frac{\psi_1(E_{FB}) - \left\{\frac{2|e|B}{\hbar}\left(n+\frac{1}{2}\right)\right\}\left[\psi_2(E_{FB})\left\{\psi_2(E_{FB})\cos^2\theta+\psi_3(E_{FB})\sin^2\theta\right\}\right]^{1/2}}{a_0(E_{FB},\theta)}\right\}^{1/2},$$

and

$$T_{34}(n,E_{FB}) \equiv \sum_{r=1}^{s} L(r)\left[T_{33}(n,E_{FB})\right].$$

It is interesting to note that although the Landau area quantization rule is valid for large values of $\boldsymbol{n}$, the operator method, the Schrödinger differential equation technique, and the method of the area quantization rule of the wave vector space lead to the same result in the absence of electron spin.

3.2.2 Special Cases for III–V, Ternary and Quaternary Materials

(1) Under the conditions $\delta = 0$, $\Delta_{\|} = \Delta_{\perp} = \Delta$ and $m_{\|}^* = m_{\perp}^* = m^*$, (3.1) assumes the form

$$\gamma(E) = \left(n+\frac{1}{2}\right)\hbar\omega_0 + \frac{\hbar^2k_z^2}{2m^*} \pm |e|B\hbar\Delta\left[6m^*\left(E+E_g+\frac{2}{3}\Delta\right)\right]^{-1}, \quad (3.15)$$

where $\gamma(E)$ has already been defined in connection with (2.6) of Chap. 2. Equation (3.15) is the dispersion relation of the conduction electrons of III–V, ternary and quaternary materials in the presence of a quantizing magnetic field B along z-direction [1].

From (3.15), the EMM along the direction of magnetic quantization can be written as

$$m_{k_z}^*(E_{\mathrm{FB}}) = m^*\left[\{\gamma(E_{\mathrm{FB}})\}' \pm \frac{|e|\,B\hbar\Delta}{6m^*}\left[\left(E_{\mathrm{FB}} + E_{\mathrm{g}} + \frac{2}{3}\Delta\right)\right]^{-2}\right]. \quad (3.16)$$

Thus, the EMM is a function of the Fermi energy and the electron spin under magnetic quantization. The dependence of the EMM on the electron spin due to the presence of the spin orbit splitting constant, excluding the dependence on $\{\gamma(E_{\mathrm{FB}})\}'$, is a special property of the three band model of Kane.

The Landau energy levels (E_{n_1}) can be written from (3.15) as

$$\gamma(E_{n_1}) = \left(n + \frac{1}{2}\right)\hbar\omega_0 \pm |e|\,B\hbar\Delta\left[6m^*\left(E_{n_1} + E_{\mathrm{g}} + \frac{2}{3}\Delta\right)\right]^{-1}. \quad (3.17)$$

Thus, the solution of the Landau levels is the lowest positive root of the cubic equation where the unknown variable is E_{n_1}.

The SdH period of oscillation can be written from (3.15) as

$$\Delta\left(\frac{1}{B}\right) = (\hbar\,|e|)\,[m^*\gamma(E_{\mathrm{FB}})]^{-1}. \quad (3.18)$$

Thus, the SdH period is a function of E_{FB} and other physical constants.

Using (3.15) and (3.5), the density-of-states function in this case can be expressed as

$$D_{\mathrm{B}}(E) = \frac{g_{\mathrm{v}}|e|B\sqrt{2m^*}}{4\pi^2\hbar^2}\sum_{n=0}^{n_{\max}}\left[\gamma(E) - \left(n+\tfrac{1}{2}\right)\hbar\omega_0 \mp \frac{|e|B\hbar\Delta}{6m^*\left(E+E_{\mathrm{g}}+\frac{2}{3}\Delta\right)}\right]^{-1/2}$$
$$\left[\{\gamma(E_{\mathrm{FB}})\}' \pm \frac{|e|B\hbar\Delta}{6m^*\left(E+E_{\mathrm{g}}+\frac{2}{3}\Delta\right)^2}\right] H(E - E_{n1}). \quad (3.19)$$

Thus, the electron concentration assumes the form

$$n_0 = \frac{g_{\mathrm{v}}\,|e|\,B\sqrt{2m^*}}{2\pi^2\hbar^2}\sum_{n=0}^{n_{\max}}[T_{35}(n, E_{\mathrm{FB}}) + T_{36}(n, E_{\mathrm{FB}})], \quad (3.20)$$

where

$$T_{35}(n, E_{\mathrm{FB}}) \equiv \left[\gamma(E_{\mathrm{FB}}) - \left(n + \frac{1}{2}\right)\hbar\omega_0 \mp \frac{|e|\,B\hbar\Delta}{6m^*\left(E_{\mathrm{FB}} + E_{\mathrm{g}} + \frac{2}{3}\Delta\right)}\right]^{\frac{1}{2}},$$

and

$$T_{36}(n, E_{\mathrm{FB}}) \equiv \sum_{r=1}^{s} L(r)T_{35}(n, E_{\mathrm{FB}}).$$

Using (3.20) and (1.11), the DMR in this case can be written as

$$\frac{D}{\mu}=\frac{1}{|e|}\left[\frac{\sum\limits_{n=0}^{n_{\max}}\left[T_{35}\left(n,E_{\mathrm{FB}}\right)+T_{36}\left(n,E_{\mathrm{FB}}\right)\right]}{\sum\limits_{n=0}^{n_{\max}}\left[\left\{T_{35}\left(n,E_{\mathrm{FB}}\right)\right\}'+\left\{T_{36}\left(n,E_{\mathrm{FB}}\right)\right\}'\right]}\right]. \tag{3.21}$$

In the absence of spin, the electron concentration and the DMR, assume the forms

$$n_0=\frac{g_{\mathrm{v}}\left|e\right|B\sqrt{2m^*}}{\pi^2\hbar^2}\sum_{n=0}^{n_{\max}}\left[T_{37}\left(n,E_{\mathrm{FB}}\right)+T_{38}\left(n,E_{\mathrm{FB}}\right)\right], \tag{3.22}$$

and

$$\frac{D}{\mu}=\frac{1}{|e|}\left[\frac{\sum\limits_{n=0}^{n_{\max}}\left[T_{37}\left(n,E_{FB}\right)+T_{38}\left(n,E_{FB}\right)\right]}{\sum\limits_{n=0}^{n_{\max}}\left[\left\{T_{37}\left(n,E_{FB}\right)\right\}'+\left\{T_{38}\left(n,E_{FB}\right)\right\}'\right]}\right] \tag{3.23}$$

where

$$T_{37}\left(n,E_{\mathrm{FB}}\right)\equiv\left[\gamma\left(E_{\mathrm{FB}}\right)-\left(n+\frac{1}{2}\right)\hbar\omega_0\right]^{\frac{1}{2}},$$

and

$$T_{38}\left(n,E_{\mathrm{FB}}\right)\equiv\sum_{r=1}^{s}L\left(r\right)T_{37}\left(n,E_{\mathrm{FB}}\right).$$

(2) Under the condition $\Delta\gg E_{\mathrm{g}}$, (3.15) can be expressed as

$$E\left(1+\alpha E\right)=\left(n+\frac{1}{2}\right)\hbar\omega_0+\left(\hbar^2k_z^2/2m^*\right)\pm\frac{1}{2}\mu_0 g^*B, \tag{3.24}$$

where $\mu_0=\left(|e|\,\hbar/2m_0\right)$ is known as the Bohr magnetron, g^* is the magnitude of the band edge g-factor and is equal to $\left(m_0/m^*\right)$ in accordance with the two band model of Kane.

From (3.24), the EMM along the direction of magnetic quantization can be expressed as

$$m_{k_z}^*\left(E_{\mathrm{FB}}\right)=m^*\left[1+2\alpha E_{\mathrm{FB}}\right]. \tag{3.25}$$

Thus, the EMM is a function of Fermi energy only due to the presence of the band non-parabolicity factor α and is independent of the electron spin under magnetic quantization.

The Landau energy levels E_{n_2} can be written from (3.24) as

$$E_{n_2}=\left(2\alpha\right)^{-1}\left[-1+\sqrt{1+4\alpha\left\{\left(n+\frac{1}{2}\right)\hbar\omega_0\pm\frac{1}{2}g^*\mu_0B\right\}}\right]. \tag{3.26}$$

Thus, the difference between any two consecutive Landau levels is a function of the Landau quantum number and the electron spin in accordance with the two band model of Kane.

The SdH period can be written from (3.24) as

$$\Delta\left(\frac{1}{B}\right) = (\hbar\,|e|)\,[m^* E_{\mathrm{FB}}\,(1+\alpha E_{\mathrm{FB}})]^{-1}\,. \tag{3.27}$$

Thus, the SdH period decreases due to the presence of band non-parabolicity.

In accordance with the two-band model of Kane, the density-of-states function assumes the form

$$D_{\mathrm{B}}\,(E) = \frac{g_{\mathrm{v}}\,|e|\,B\sqrt{2m^*}}{4\pi^2\hbar^2}\sum_{n=0}^{n_{\max}}[1+2\alpha E] \times\left[E\,(1+\alpha E) - \left(n+\frac{1}{2}\right)\hbar\omega_0 \mp \frac{1}{2}g^*\mu_0 B\right]^{-\frac{1}{2}} H\,(E-E_{n_2})\,. \tag{3.28}$$

Therefore the electron concentration and the DMR can be written as

$$n_0 = \frac{g_{\mathrm{v}}\,|e|\,B\sqrt{2m^*}}{2\pi^2\hbar^2}\sum_{n=0}^{n_{\max}}[T_{39}\,(n,E_{\mathrm{FB}}) + T_{310}\,(n,E_{\mathrm{FB}})], \tag{3.29}$$

and

$$\frac{D}{\mu} = \frac{1}{|e|}\left[\frac{\sum\limits_{n=0}^{n_{\max}}[T_{39}\,(n,E_{\mathrm{FB}}) + T_{310}\,(n,E_{\mathrm{FB}})]}{\sum\limits_{n=0}^{n_{\max}}[\{T_{39}\,(n,E_{\mathrm{FB}})\}' + \{T_{310}\,(n,E_{\mathrm{FB}})\}']}\right], \tag{3.30}$$

where

$$T_{39}\,(n,E_{\mathrm{FB}}) \equiv \left[E_{\mathrm{FB}}\,(1+\alpha E_{\mathrm{FB}}) - \left(n+\frac{1}{2}\right)\hbar\omega_0 \pm \frac{1}{2}g^*\mu_0 B\right]^{\frac{1}{2}},$$

and

$$T_{310}\,(n,E_{\mathrm{FB}}) \equiv \sum_{r=1}^{s} L\,(r) T_{39}\,(n,E_{\mathrm{FB}})\,.$$

In the absence of spin, the electron concentration and the DMR assume the forms

$$n_0 = \frac{g_{\mathrm{v}}\,|e|\,B\sqrt{2m^*}}{\pi^2\hbar^2}\sum_{n=0}^{n_{\max}}[T_{311}\,(n,E_{\mathrm{FB}}) + T_{312}\,(n,E_{\mathrm{FB}})], \tag{3.31}$$

and

$$\frac{D}{\mu}=\frac{1}{|e|}\left[\frac{\sum\limits_{n=0}^{n_{\max}}\left[T_{311}\left(n,E_{\mathrm{FB}}\right)+T_{312}\left(n,E_{\mathrm{FB}}\right)\right]}{\sum\limits_{n=0}^{n_{\max}}\left[\left\{T_{311}\left(n,E_{\mathrm{FB}}\right)\right\}'+\left\{T_{312}\left(n,E_{\mathrm{FB}}\right)\right\}'\right]}\right], \tag{3.32}$$

where

$$T_{311}\left(n,E_{\mathrm{FB}}\right)\equiv\left[E_{\mathrm{FB}}\left(1+\alpha E_{\mathrm{FB}}\right)-\left(n+\frac{1}{2}\right)\hbar\omega_0\right]^{\frac{1}{2}},$$

and

$$T_{312}\left(n,E_{\mathrm{FB}}\right)\equiv\sum_{r=1}^{s}L\left(r\right)T_{311}\left(n,E_{\mathrm{FB}}\right).$$

From (3.28), under the condition $\alpha E\ll 1$, the density-of-states function can be written as

$$D_{\mathrm{B}}\left(E\right)=\frac{g_{\mathrm{v}}|e|B\sqrt{2m^*}}{4\pi^2\hbar^2}\sum_{n=0}^{n_{\max}}\left[\left(1+\tfrac{3}{2}\alpha E\right)\left[E-\left\{\frac{\left(n+\frac{1}{2}\right)\hbar\omega_0\mp\frac{1}{2}g^*\mu_0B}{1+\alpha E}\right\}\right]^{-\frac{1}{2}}\right] \times H\left(E-E_{n_2}\right). \tag{3.33}$$

Therefore the electron concentration is given by

$$n_0=\frac{g_{\mathrm{v}}\left|e\right|B\sqrt{2m^*}}{4\pi^2\hbar^2}\sum_{n=0}^{n_{\max}}\int_{E_{n2}}^{\infty}\left[E-\left\{\frac{\left(n+\frac{1}{2}\right)\hbar\omega_0\mp\frac{1}{2}g^*\mu_0B}{1+\alpha E}\right\}\right]^{-\frac{1}{2}} \times\left(1+\frac{3}{2}\alpha E\right)f_0\,\mathrm{d}E. \tag{3.34}$$

Let us substitute,

$$y=E-\left\{\left[\left(n+\frac{1}{2}\right)\hbar\omega_0\mp\frac{1}{2}g^*\mu_0B\right]\left(1+\alpha E\right)^{-1}\right\}, \tag{3.35}$$

where y is a new variable.

Since, E_{n2} is the root of (3.35), we can write

$y\left(1+\alpha E_{n_2}\right)=0$ and because, $\left(1+\alpha E_{n_2}\right)\neq 0$, $y=0$. Again when, $E\to\infty$, $y\to\infty$. Therefore, from (3.35), after binomial expansion, neglecting the terms of the order of $\left(\alpha E\right)^2$ and more, we can write

$$E=\frac{y}{a_{01}}+b_{01}, \tag{3.36}$$

where

$$a_{01}\equiv\left[1+\alpha\left(n+\frac{1}{2}\right)\hbar\omega_0\pm\frac{1}{2}g^*\mu_0B\right],$$

and

$$b_{01} \equiv (a_{01})^{-1}\left[\left(n+\frac{1}{2}\right)\hbar\omega_0 \pm \frac{1}{2}g^*\mu_0 B\right].$$

Therefore combining (3.34) and (3.36) we get,

$$n_0 = \frac{g_\mathrm{v}|e|B\sqrt{2m^*}}{4\pi^2\hbar^2}\sum_{n=0}^{n_{\max}}\frac{1}{a_{01}}\int_0^\infty (y)^{-1/2}\left[1+\frac{3}{2}\alpha\left(\frac{y}{a_{01}}+b_{01}\right)\right]$$
$$\times\left[1+\mathrm{e}^{\frac{\frac{y}{a_{01}}+b_{01}-E_{\mathrm{FB}}}{k_\mathrm{B}T}}\right]^{-1}\mathrm{d}y. \tag{3.37}$$

Let us substitute,

$$\beta_{01} = \frac{y}{a_{01}k_\mathrm{B}T},$$

and

$$\eta_\mathrm{B} = \frac{E_{\mathrm{FB}} - b_{01}}{k_\mathrm{B}T}.$$

Using (3.37) and the Fermi–Dirac integrals, the electron concentration in this case assumes the form

$$n_0 = \frac{g_\mathrm{v}N_\mathrm{C}\theta_{\mathrm{B}1}}{2}\left[\sum_{n=0}^{n_{\max}}\frac{1}{\sqrt{a_{01}}}\left[\left(1+\frac{3}{2}\alpha b_{01}\right)F_{\frac{-1}{2}}(\eta_\mathrm{B})+\frac{3}{4}\alpha k_\mathrm{B}TF_{\frac{1}{2}}(\eta_\mathrm{B})\right]\right], \tag{3.38a}$$

where

$$\theta_{\mathrm{B}1} \equiv \frac{\hbar\omega_0}{k_\mathrm{B}T}.$$

Using (3.38a) and (1.11) the DMR in this case can be expressed as

$$\frac{D}{\mu} = \frac{k_\mathrm{B}T}{|e|}\left[\frac{\sum_{n=0}^{n_{\max}}\frac{1}{\sqrt{a_{01}}}\left[\left(1+\frac{3}{2}\alpha b_{01}\right)F_{\frac{-1}{2}}(\eta_\mathrm{B})+\frac{3}{4}\alpha k_\mathrm{B}TF_{\frac{1}{2}}(\eta_\mathrm{B})\right]}{\sum_{n=0}^{n_{\max}}\frac{1}{\sqrt{a_{01}}}\left[\left(1+\frac{3}{2}\alpha b_{01}\right)F_{\frac{-3}{2}}(\eta_\mathrm{B})+\frac{3}{4}\alpha k_\mathrm{B}TF_{\frac{-1}{2}}(\eta_\mathrm{B})\right]}\right]. \tag{3.38b}$$

In the absence of spin (3.38a) and (3.38b) assume the forms [23],

$$n_0 = g_\mathrm{v}N_\mathrm{C}\theta_{\mathrm{B}1}\left[\sum_{n=0}^{n_{\max}}\frac{1}{\sqrt{a_{01}^*}}\left[\left(1+\frac{3}{2}\alpha b_{01}^*\right)F_{\frac{-1}{2}}(\bar{\eta}_{\mathrm{B}_1})+\frac{3}{4}\alpha k_\mathrm{B}TF_{\frac{1}{2}}(\bar{\eta}_{\mathrm{B}_1})\right]\right], \tag{3.39a}$$

$$\frac{D}{\mu} = \frac{k_\mathrm{B}T}{|e|}\left[\frac{\sum_{n=0}^{n_{\max}}\frac{1}{\sqrt{a_{01}^*}}\left[\left(1+\frac{3}{2}\alpha b_{01}^*\right)F_{\frac{-1}{2}}(\bar{\eta}_{\mathrm{B}_1})+\frac{3}{4}\alpha k_\mathrm{B}TF_{\frac{1}{2}}(\bar{\eta}_{\mathrm{B}_1})\right]}{\sum_{n=0}^{n_{\max}}\frac{1}{\sqrt{a_{01}^*}}\left[\left(1+\frac{3}{2}\alpha b_{01}^*\right)F_{\frac{-3}{2}}(\bar{\eta}_{\mathrm{B}_1})+\frac{3}{4}\alpha k_\mathrm{B}TF_{\frac{-1}{2}}(\bar{\eta}_{\mathrm{B}_1})\right]}\right], \tag{3.39b}$$

where

$$a_{01}^* \equiv 1 + \alpha \left(n + \frac{1}{2}\right) \hbar\omega_0, \; b_{01}^* \equiv \left[\left(n + \frac{1}{2}\right) \hbar\omega_0\right] \left(a_{01}^*\right)^{-1}$$

and

$$\bar{\eta}_{B_1} \equiv \frac{E_{FB} - b_{01}^*}{k_B T}.$$

(3) Under the condition $\alpha \to 0$ (3.24) becomes

$$E = \left(n + \frac{1}{2}\right) \hbar\omega_0 + \left(\hbar^2 k_z^2/2m^*\right) \pm \frac{1}{2} g^* \mu_0 B. \tag{3.40}$$

From (3.40), the EMM along the direction of the quantizing magnetic field can be expressed as

$$m_{k_z}^* \left(E_{\mathrm{FB}}\right) = m^*. \tag{3.41}$$

Thus, the quantizing magnetic field cannot influence the EMM in relatively wide gap semiconductors having parabolic energy bands.

The Landau energy levels E_{n_3} in this case can be written from (3.40) as

$$E_{n_3} = \left(n + \frac{1}{2}\right) \hbar\omega_0. \tag{3.42}$$

Equation (3.42) is well known in the literature [1].

The use of (3.40) leads to the well known expression of the SdH period for semiconductors having isotropic parabolic energy bands as

$$\Delta \left(\frac{1}{B}\right) = (\hbar\,|e|)\, [m^* E_{\mathrm{FB}}]^{-1}. \tag{3.43}$$

In this case, the SdH period increases in the absence of non-parabolicity.

The expression for the electron concentration and the DMR under the condition $\alpha \to 0$ can be written as

$$n_0 = \frac{g_{\mathrm{v}} N_{\mathrm{C}} \theta_{\mathrm{B1}}}{2} \sum_{n=0}^{n_{\max}} F_{\frac{-1}{2}} \left(\bar{\eta}_{\mathrm{B}}\right), \tag{3.44}$$

and

$$\frac{D}{\mu} = \frac{k_{\mathrm{B}} T}{|e|} \left[\sum_{n=0}^{n_{\max}} F_{-\frac{1}{2}} \left(\bar{\eta}_{\mathrm{B}}\right)\right] \left[\sum_{n=0}^{n_{\max}} F_{\frac{-3}{2}} \left(\bar{\eta}_{\mathrm{B}}\right)\right]^{-1}, \tag{3.45}$$

where

$$\bar{\eta}_{\mathrm{B}} \equiv (k_{\mathrm{B}} T)^{-1} \left[E_{\mathrm{FB}} - \left(n + \frac{1}{2}\right) \hbar\omega_0 \mp \frac{1}{2} g^* \mu_0 B\right].$$

In the absence of spin, (3.44) and (3.45) assume the forms [24]

$$n_0 = g_{\rm v} N_{\rm C} \theta_{\rm B1} \sum_{n=0}^{n_{\rm max}} F_{\frac{-1}{2}} (\eta_{\rm B1}), \tag{3.46}$$

and

$$\frac{D}{\mu} = \frac{k_{\rm B}T}{|e|} \left[\sum_{n=0}^{n_{\rm max}} F_{\frac{-1}{2}} (\eta_{\rm B1}) \right] \left[\sum_{n=0}^{n_{\rm max}} F_{\frac{-3}{2}} (\eta_{\rm B1}) \right]^{-1}, \tag{3.47}$$

where

$$\eta_{\rm B1} \equiv (k_{\rm B}T)^{-1} \left[E_{\rm FB} - \left(n + \frac{1}{2} \right) \hbar \omega_0 \right].$$

Under the condition of non-degeneracy, (3.45) and (3.47) get simplified to the well-known form given by (2.26).

3.2.3 II–VI Semiconductors

The Hamiltonian of the conduction electron of II–VI semiconductors in the presence of a quantizing magnetic field B along the z direction assumes the form

$$\hat{H}_{\rm B} = \frac{(\hat{p}_x)^2}{2m_\perp^*} + \frac{(\hat{p}_y - |e| B\hat{x})^2}{2m_\perp^*} \pm \frac{\bar{\lambda}_0}{\hbar} \left[(\hat{p}_x)^2 + (\hat{p}_y - |e| B\hat{x})^2 \right]^{1/2} + \frac{(\hat{p}_z)^2}{2m_\parallel^*}, \tag{3.48}$$

where the "hats" denote the respective operators.

The application of the operator method leads to the magneto-dispersion relation of the carriers of II–VI semiconductors, including spin, as

$$E = \frac{\hbar |e| B}{m_\perp^*} \left(n + \frac{1}{2} \right) + \frac{\hbar^2 k_z^2}{2m_\parallel^*} \pm \bar{\lambda}_0 \left[\frac{2 |e| B}{\hbar} \left(n + \frac{1}{2} \right) \right]^{1/2} \pm \frac{1}{2} g^* \mu_0 B. \tag{3.49}$$

From (3.49), the EMM along the direction of the magnetic quantization can be expressed as

$$m_{k_z}^* (E_{\rm FB}) = m_\parallel^*. \tag{3.50}$$

The EMM in this case is a constant quantity and is not affected by the magnetic field. The Landau energy levels E_{n_4} can be written from (3.49) as

$$E_{n_4} = \frac{\hbar |e| B}{m_\perp^*} \left(n + \frac{1}{2} \right) \pm \bar{\lambda}_0 \left[\frac{2 |e| B}{\hbar} \left(n + \frac{1}{2} \right) \right]^{1/2} \pm \frac{1}{2} g^* \mu_0 B. \tag{3.51}$$

Thus, the difference between the consecutive Landau levels is a function of the Landau quantum number and is independent of the electron spin in accordance with the magneto-Hopfield model.

The SdH period can be expressed from (3.49) as

$$\Delta \left(\frac{1}{B} \right) = \theta_5 (n + 1, m_\perp^*, g^*) - \theta_5 (n, m_\perp^*, g^*), \tag{3.52}$$

where

$$\theta_5\left(n,m_\perp^*,g^*\right)\equiv\left[2\theta_{3,\pm}\left(n,m_\perp^*,g^*\right)\right]\left[\theta_{4,\pm}\left(E_{\mathrm{FB}},n,\bar{\lambda}_0\right)\right.$$
$$\left.-\left\{\theta_{4,\pm}^2\left(E_{\mathrm{FB}},n,\bar{\lambda}_0\right)-4E_{\mathrm{FB}}^2\theta_{3,\pm}\left(E_{\mathrm{FB}},n,\bar{\lambda}_0\right)\right\}^{1/2}\right]^{-1},$$

in which

$$\theta_{3,\pm}\left(n,m_\perp^*,g^*\right)\equiv\left[\left(\frac{\hbar\left|e\right|}{m_\perp^*}\right)^2\left(n+\frac{1}{2}\right)^2+\frac{1}{4}\left(g^*\mu_0\right)^2\pm\frac{\hbar\left|e\right|g^*\mu_0}{m_\perp^*}\left(n+\frac{1}{2}\right)\right],$$

and

$$\theta_{4,\pm}\left(E_{\mathrm{FB}},n,\bar{\lambda}_0\right)\equiv\left[\left(\frac{2E_{\mathrm{FB}}\hbar\left|e\right|}{m_\perp^*}\right)\left(n+\frac{1}{2}\right)\right.$$
$$\left.\pm E_{\mathrm{FB}}g^*\mu_0+\frac{2\left(\bar{\lambda}_0\right)^2\left|e\right|}{\hbar}\left(n+\frac{1}{2}\right)\right].$$

Thus the SdH period changes with the energy band constants and the electron spin in addition to the magnetic quantum number.

Equation (3.49) can be written as

$$E=\varphi_\pm\left(n\right)+\frac{\hbar^2k_z^2}{2m_{||}^*},\tag{3.53}$$

where

$$\phi_\pm\left(n\right)\equiv\frac{\hbar\left|e\right|B}{m_\perp^*}\left(n+\frac{1}{2}\right)\pm\frac{1}{2}g^*\mu_0B\pm\bar{\lambda}_0\left[\frac{2\left|e\right|B}{\hbar}\left(n+\frac{1}{2}\right)\right]^{1/2}.$$

The use of (3.5) and (3.53) leads to the expression of the density-of-state function as

$$D_{\mathrm{B}}\left(E\right)=\frac{g_{\mathrm{v}}\left|e\right|B\sqrt{2m_{||}^*}}{4\pi^2\hbar^2}\sum_{n=0}^{n_{\max}}\frac{H\left(E-\phi_\pm\left(n\right)\right)}{\sqrt{E-\phi_\pm\left(n\right)}}.\tag{3.54}$$

Thus, combining (3.54) with the Fermi–Dirac occupation probability factor, the electron concentration in this case assumes the form

$$n_0=\frac{g_{\mathrm{v}}\left|e\right|B\sqrt{2m_{||}^*}}{4\pi^2\hbar^2}\sum_{n=0}^{n_{\max}}\int_{\phi_\pm(n)}^{\infty}\frac{f_0\,\mathrm{d}E}{\sqrt{E-\phi_\pm\left(n\right)}}.\tag{3.55}$$

Therefore

$$n_0=\frac{g_{\mathrm{v}}\left|e\right|B\sqrt{2\pi m_{||}^*k_{\mathrm{B}}T}}{h^2}\sum_{n=0}^{n_{\max}}F_{\frac{-1}{2}}\left(\theta_3\right),\ \theta_3\equiv\frac{E_{\mathrm{FB}}-\phi_\pm\left(n\right)}{k_{\mathrm{B}}T}.\tag{3.56}$$

Using (3.56) and (1.11), the DMR for the II–VI materials in the presence of a quantizing magnetic field along the z-direction can be expressed as

$$\frac{D}{\mu}=\frac{k_{\mathrm{B}}T}{|e|}\left[\sum_{n=0}^{n_{\max}}F_{\frac{-1}{2}}\left(\theta_3\right)\right]\left[\sum_{n=0}^{n_{\max}}F_{\frac{-3}{2}}\left(\theta_3\right)\right]^{-1}. \tag{3.57}$$

It should be noted that in the absence of the spin, the electron concentration and the DMR can be written as

$$n_0=\left(\frac{2g_v\,|e|\,B\sqrt{2\pi m_{||}^{*}k_{\mathrm{B}}T}}{h^2}\right)\left[\sum_{n=0}^{n_{\max}}F_{\frac{-1}{2}}\left(\eta_{\mathrm{B}2}\right)\right], \tag{3.58}$$

and

$$\frac{D}{\mu}=\frac{k_{\mathrm{B}}T}{|e|}\left[\sum_{n=0}^{n_{\max}}F_{\frac{-1}{2}}\left(\eta_{\mathrm{B}_2}\right)\right]\left[\sum_{n=0}^{n_{\max}}F_{\frac{-3}{2}}\left(\eta_{\mathrm{B}_2}\right)\right]^{-1}, \tag{3.59}$$

where

$$\eta_{\mathrm{B}2}\equiv\left[E_{\mathrm{FB}}-\left(n+\frac{1}{2}\right)\left(\hbar\frac{|e|\,B}{m_{\perp}^{*}}\right)\mp\bar{\lambda}_0\left[\frac{2\,|e|\,B}{\hbar}\left(n+\frac{1}{2}\right)\right]^{1/2}\right]\left(k_{\mathrm{B}}T\right)^{-1}. \tag{3.60}$$

3.2.4 The Formulation of DMR in Bi

(a) The McClure and Choi model

The Hamiltonian in the presence of a quantizing magnetic field B along the z-direction in accordance with this model can be written as

$$\begin{aligned}\widehat{H}_{\mathrm{B}}&=\frac{\left(\widehat{p}_x\right)^2}{2m_1}+\frac{\left(\widehat{p}_y-|e|\,B\widehat{x}\right)^2}{2m_2}\left[1+\alpha E\left(1-\frac{m_2}{m_2'}\right)\right]+\frac{\left(\widehat{p}_z\right)^2}{2m_3}\\&+\frac{\alpha\left(\widehat{p}_y-|e|\,B\widehat{x}\right)^4}{4m_2m_2'}-\alpha\left(\widehat{p}_y-|e|\,B\widehat{x}\right)^2\left[\frac{\left(\widehat{p}_x\right)^2}{4m_1m_2}\right]+\frac{\alpha\left(\widehat{p}_y-|e|\,B\widehat{x}\right)^4}{4m_2m_3}\\&-\alpha\left(\widehat{p}_y-|e|\,B\widehat{x}\right)^2\left[\frac{\left(\widehat{p}_x\right)^2}{4m_1m_2}+\frac{\left(\widehat{p}_z\right)^2}{4m_2m_3}\right].\end{aligned} \tag{3.61a}$$

Thus the modified carrier energy spectrum in accordance with McClure and Choi model up to the first order, by including spin effects, can be expressed as [25]

$$\begin{aligned}E\left(1+\alpha E\right)&=\left(n+\frac{1}{2}\right)\hbar\omega\left(E\right)+\left(n^2+1+n\right)\frac{\alpha\hbar^2\omega^2\left(E\right)}{4}+\frac{\hbar^2k_z^2}{2m_3}\\&\times\left[1-\frac{\alpha\left(n+\frac{1}{2}\right)\hbar\omega\left(E\right)}{2}\right]\pm\frac{1}{2}g^{*}\mu_0B,\end{aligned} \tag{3.61b}$$

where

$$\omega(E) \equiv \frac{|e|\, B}{\sqrt{m_1 m_2}} \left[1 + \alpha E \left(1 - \frac{m_2}{m_2'}\right)\right]^{1/2}.$$

From (3.61a), the EMM along the direction of magnetic quantization assumes the form

$$\begin{aligned} m^*_{k_z}(n, E_{\mathrm{FB}}) = m_3 & \left[\left[1 - \frac{\alpha}{2}\left(n + \frac{1}{2}\right)\hbar\omega(E_{\mathrm{FB}})\right]^{-1} \right. \\ & \times \left[1 + 2\alpha E_{\mathrm{FB}} - \left(n + \frac{1}{2}\right)\hbar\omega'(E_{\mathrm{FB}}) \right. \\ & \left. - \frac{1}{2}\left(n^2 + n + 1\right)\alpha\hbar^2\omega(E_{\mathrm{FB}})\,\omega'(E_{\mathrm{FB}})\right] \\ & + \frac{\alpha\left(n + \frac{1}{2}\right)\hbar\omega'(E_{\mathrm{FB}})}{2}\left[1 - \frac{\alpha\left(n + \frac{1}{2}\right)\hbar\omega(E_{\mathrm{FB}})}{2}\right]^{-2} \\ & \times \left[E_{\mathrm{FB}}(1 + \alpha E_{\mathrm{FB}}) - \left(n + \frac{1}{2}\right)\hbar\omega(E_{\mathrm{FB}}) \right. \\ & \left.\left. - \frac{\alpha\hbar^2\omega^2(E_{\mathrm{FB}})}{4}\left(n^2 + 1 + n\right) \pm \frac{1}{2}g^*\mu_0 B\right]\right]. \end{aligned} \tag{3.62}$$

In the absence of band non-parabolicity, from (3.62) we get

$$m^*_{k_z}(n, E_{\mathrm{FB}}) = m_3. \tag{3.63}$$

It is interesting to note that for the two band model of Kane, the band non-parabolicity alone explains the dependence of the EMM on Fermi energy, and the EMM is independent of magnetic quantum number and the electron spin. In the case of the McClure and Choi model of Bi under magnetic quantization, the same band non-parabolicity alone explains the dependence of the EMM on the magnetic quantum number, electron spin, and the Fermi energy respectively.

The Landau energy levels E_{n_5} can be written from (3.61b) as

$$\begin{aligned} E_{n_5}(1 + \alpha E_{n_5}) = & \left(n + \frac{1}{2}\right)\hbar\omega(E_{n_5}) + \left(n^2 + 1 + n\right)\frac{\alpha\hbar^2\omega^2(E_{n_5})}{4} \\ & \pm \frac{1}{2}g^*\mu_0 B. \end{aligned} \tag{3.64}$$

Thus the difference between any two consecutive Landau levels is a function of the Landau quantum number and is dependent on the electron spin.

The SdH period can be expressed from (3.61b) as

$$\Delta\left(\frac{1}{B}\right) = \alpha_{11,\pm}(n + 1, E_{\mathrm{FB}}, g^*) - \alpha_{11,\pm}(n, E_{\mathrm{FB}}, g^*), \tag{3.65}$$

where

$$\alpha_{11,\pm}(n, E_{\mathrm{FB}}, g^*) \equiv [2\alpha_9(n, E_{\mathrm{FB}})]\Bigg[-\alpha_{10,\pm}(n, E_{\mathrm{FB}}, g^*) + \sqrt{\alpha_{10,\pm}^2(n, E_{\mathrm{FB}}, g^*) + 4E_{\mathrm{FB}}(1+\alpha E_{\mathrm{FB}})\alpha_9(n, E_{\mathrm{FB}})}\Bigg]^{-1},$$

$$\alpha_9(n, E_{\mathrm{FB}}) \equiv \left[\left(n^2+1+n\right)\left(\frac{\hbar|e|}{2}\right)^2\left(\frac{\alpha}{m_1 m_2}\right)\left[1+\alpha E_{\mathrm{FB}}\left(1-\frac{m_2}{m_2'}\right)\right]\right],$$

and

$$\alpha_{10,\pm}(n, E_{\mathrm{FB}}, g^*) \equiv \left[\left(n+\frac{1}{2}\right)\left(\frac{\hbar|e|}{\sqrt{m_1 m_2}}\right)\left[1+\alpha E_{\mathrm{FB}}\left(1-\frac{m_2}{m_2'}\right)\right]^{1/2} \pm\frac{1}{2}g^*\mu_0\right].$$

Under the condition,

$$\alpha \to 0,\ \Delta\left(\frac{1}{B}\right) = \frac{\hbar|e|}{(E_{\mathrm{FB}})\sqrt{m_1 m_2}}. \tag{3.66}$$

Thus from (3.65) we infer that the SdH period for the McClure and Choi model is a function of the magnetic quantum number, the Fermi energy, the electron spin, and the other constants of the spectrum due to the presence of band non-parabolicity only. For $\alpha \to 0$, the SdH period is independent of the magnetic quantum number and the electron spin which is apparent from (3.66).

The density-of-states function for this model under magnetic quantization is given by

$$\begin{aligned} D_{\mathrm{B}}(E) = \frac{g_{\mathrm{v}}|e|B\sqrt{2m_3}}{4\pi^2\hbar^2}\sum_{n=0}^{n_{\max}}\Bigg[\Bigg[1-\frac{\alpha\left(n+\frac{1}{2}\right)\hbar\omega(E)}{2}\Bigg]^{-3/2} &\times\left(\frac{1}{2}\right)\alpha\left(n+\frac{1}{2}\right)\hbar[\omega(E)]'\Bigg[E(1+\alpha E) \\ &-\left(n+\frac{1}{2}\right)\hbar\omega(E)-\left(n^2+1+n\right)\frac{\alpha\hbar^2\omega^2(E)}{4} \\ &\mp\frac{1}{2}g^*\mu_0 B\Bigg]^{1/2}+\Bigg[E(1+\alpha E)-\left(n+\frac{1}{2}\right) \\ &\times\hbar\omega(E)-\left(n^2+1+n\right)\frac{\alpha\hbar^2\omega^2(E)}{4} \\ &\pm\frac{1}{2}g^*\mu_0 B\Bigg]^{-1/2}\Bigg[1+2\alpha E-\left(n+\frac{1}{2}\right) \end{aligned}$$

$$\times\hbar\left\{\omega\left(E\right)\right\}' - \left(n^2+1+n\right)$$
$$\times\frac{\alpha\hbar^2\omega\left(E\right)\left\{\omega\left(E\right)\right\}'}{2}\Bigg]$$
$$\times\left[1-\frac{\alpha\left(n+\frac{1}{2}\right)\hbar\omega\left(E\right)}{2}\right]^{-1/2}\Bigg] H\left(E-E_{n_5}\right). \tag{3.67}$$

Combining (3.67) with the Fermi–Dirac occupation probability and using the generalized Sommerfeld's lemma [21], the electron concentration in this case assumes the form

$$n_0 = \frac{g_{\mathrm{v}}\left|e\right|B\sqrt{2m_3}}{2\pi^2\hbar^2}\sum_{n=0}^{n_{\max}}\left[T_{313}\left(n,E_{\mathrm{FB}}\right)+T_{314}\left(n,E_{\mathrm{FB}}\right)\right], \tag{3.68}$$

where

$$T_{313}\left(n,E_{\mathrm{FB}}\right) \equiv \left[1-\frac{\alpha\left(n+\frac{1}{2}\right)\hbar\omega\left(E_{\mathrm{FB}}\right)}{2}\right]^{-1/2}$$
$$\times\left[E_{\mathrm{FB}}\left(1+\alpha E_{\mathrm{FB}}\right)-\left(n+\frac{1}{2}\right)\hbar\omega\left(E_{\mathrm{FB}}\right)\right.$$
$$\left.-\left(n^2+n+1\right)\frac{\alpha\hbar^2\omega^2\left(E_{\mathrm{FB}}\right)}{4}\mp\frac{1}{2}g^*\mu_0 B\right]^{1/2},$$

and,

$$T_{314}\left(n,E_{\mathrm{FB}}\right) \equiv \sum_{r=1}^{s}L\left(r\right)\left[T_{313}\left(n,E_{\mathrm{FB}}\right)\right].$$

Thus using (3.68) and (1.11), the magneto-DMR in accordance with the McClure and Choi model is given by

$$\frac{D}{\mu} = \frac{1}{\left|e\right|}\left[\frac{\sum_{n=0}^{n_{\max}}\left[T_{313}\left(n,E_{\mathrm{FB}}\right)+T_{314}\left(n,E_{\mathrm{FB}}\right)\right]}{\sum_{n=0}^{n_{\max}}\left[\left\{T_{313}\left(n,E_{\mathrm{FB}}\right)\right\}'+\left\{T_{314}\left(n,E_{\mathrm{FB}}\right)\right\}'\right]}\right] \tag{3.69}$$

Under the condition $\alpha \to 0$, (3.68) and (3.69) get simplified to

$$n_0 = \frac{g_{\mathrm{v}}N_{\mathrm{C}_2}\theta_{\mathrm{B3}}}{2}\sum_{n=0}^{n_{\max}}F_{\frac{-1}{2}}\left(\eta_{\mathrm{B}_3}\right). \tag{3.70}$$

and

$$\frac{D}{\mu} = \left(\frac{k_{\mathrm{B}}T}{\left|e\right|}\right)\left[\sum_{n=0}^{n_{\max}}F_{\frac{-1}{2}}\left(\eta_{\mathrm{B}_3}\right)\right]\left[\sum_{n=0}^{n_{\max}}F_{\frac{-3}{2}}\left(\eta_{\mathrm{B}_3}\right)\right]^{-1}, \tag{3.71}$$

where

$$N_{C_2} \equiv 2\left(\frac{2\pi m^*_{D3}k_B T}{h^2}\right)^{3/2}, m^*_{D3} \equiv (m_1 m_2 m_3)^{1/3}, \theta_{B3} \equiv \frac{\hbar\omega_{03}}{k_B T},$$
$$\omega_{03} \equiv (|e|\,B)/\sqrt{m_1 m_2}$$

and

$$\eta_{B_3} \equiv (k_B T)^{-1}\left[E_{FB} - \left(n + \frac{1}{2}\right)\hbar\omega_{03} \mp \frac{1}{2}g^*\mu_0 B\right].$$

In the absence of the spin, the electron concentration for McClure and Choi model can be written as

$$n_0 = \frac{g_v\,|e|\,B\sqrt{2m_3}}{\pi^2\hbar^2}\sum_{n=0}^{n_{max}}[T_{315}(n, E_{FB}) + T_{316}(n, E_{FB})], \tag{3.72}$$

where

$$T_{315}(n, E_{FB}) \equiv \left[1 - \frac{\alpha\left(n+\frac{1}{2}\right)\hbar\omega(E_{FB})}{2}\right]^{-1/2}$$
$$\times\left[E_{FB}(1+\alpha E_{FB}) - \left(n+\tfrac{1}{2}\right)\hbar\omega(E_{FB}) - \left(n^2+n+1\right)\frac{\alpha\hbar^2\omega^2(E_{FB})}{4}\right]^{1/2},$$

and,

$$T_{316}(n, E_F) \equiv \sum_{r=1}^{s} L(r)\,[T_{315}(n, E_{FB})].$$

Thus, using (3.72) and (1.11), the DMR in accordance with McClure and Choi model in this case is given by

$$\frac{D}{\mu} = \frac{1}{|e|}\left[\frac{\sum\limits_{n=0}^{n_{max}}[T_{315}(n, E_{FB}) + T_{316}(n, E_{FB})]}{\sum\limits_{n=0}^{n_{max}}[\{T_{315}(n, E_{FB})\}' + \{T_{316}(n, E_{FB})\}']}\right]. \tag{3.73}$$

It should be noted that in the presence of a quantizing magnetic field B along y direction, the dispersion relation of the conduction electrons of Bi in accordance with the McClure and Choi model can be expressed, neglecting spin and using operator method as,

$$E(1+\alpha E) = \left(n+\frac{1}{2}\right)\hbar\omega_4 + \frac{p_y^2}{2m_2}\left(1+\alpha E\left(1-\frac{m_2}{m_2'}\right)\right) + \frac{\alpha p_y^4}{4m_2 m_2'}$$
$$-\frac{\alpha p_y^2}{2m_2}\left(n+\frac{1}{2}\right)\hbar\omega_4, \tag{3.74}$$

where

$$\omega_4 \equiv \frac{|e|\,B}{\sqrt{m_1 m_3}}.$$

The electron concentration and the magneto-DMR in this case can be written as

$$n_0 = \frac{g_v |e| B}{\sqrt{2}\pi^2\hbar^2} \sum_{n=0}^{n_{max}} [T_{317}(n, E_{FB}) + T_{318}(n, E_{FB})], \tag{3.75}$$

and

$$\frac{D}{\mu} = \frac{1}{|e|}\left[\frac{\sum_{n=0}^{n_{max}} [T_{317}(n, E_{FB}) + T_{318}(n, E_{FB})]}{\sum_{n=0}^{n_{max}} [\{T_{317}(n, E_{FB})\}' + \{T_{318}(n, E_{FB})\}']}\right], \tag{3.76}$$

where

$$T_{317}(n, E_{FB}) \equiv \left[-q_1(n, E_{FB}) + \sqrt{[q_1(n, E_{FB})]^2 + 4q_2(n, E_{FB})}\right]^{1/2},$$

$$q_1(n, E_{FB}) \equiv \left(\tfrac{2m_2'}{\alpha}\right)\left[1 + \alpha E_{FB}\left(1 - \tfrac{m_2}{m_2'}\right) - \alpha\left(n + \tfrac{1}{2}\right)\hbar\omega_4\right],$$

$$q_2(n, E_{FB}) \equiv \left(\tfrac{4m_2 m_2'}{\alpha}\right)\left[E_{FB}(1 + \alpha E_{FB}) - \left(n + \tfrac{1}{2}\right)\hbar\omega_4\right],$$

and

$$T_{318}(n, E_{FB}) \equiv \sum_{r=1}^{s} L(r)\,[T_{317}(n, E_{FB})].$$

(b) The Cohen Model

The application of the above method in the Cohen model leads to the electron energy spectrum in Bi in the presence of quantizing magnetic field B along the z-direction as [25]

$$E(1 + \alpha E) = \left(n + \frac{1}{2}\right)\hbar\omega(E) \pm \frac{1}{2}g^*\mu_0 B + \frac{3}{8}\alpha\left(n^2 + n + \frac{1}{2}\right)\hbar^2\omega^2(E) + \frac{\hbar^2 k_z^2}{2m_3}. \tag{3.77}$$

From (3.77), the EMM along the direction of the quantizing magnetic field can be expressed as

$$m^*_{k_z}(n, E_{FB}) = m_3\left[2\alpha E_{FB} + 1 - \left(n + \frac{1}{2}\right)\hbar\omega'(E_{FB}) - \frac{3}{4}\alpha\hbar^2\omega(E_{FB}) \times \omega'(E_{FB})\left(n^2 + n + \frac{1}{2}\right)\right]. \tag{3.78}$$

In absence of band non-parabolicity, (3.78) gets transformed into the well known (3.63) and the mass becomes independent of Fermi energy and magnetic quantum number.

By comparing (3.78) and (3.62), it is important to note that the band non-parabolicity has been introduced between the McClure and Choi model and the Cohen model in two different ways so that in the first case, the band non-parabolicity alone explains the dependence of the EMM on the Fermi energy, magnetic quantum number and the electron spin whereas for the Cohen model, the same band non-parabolicity alone explains the independence of the EMM on the electron spin excluding the other two dependences. In the absence of band non-parabolicity for both the models of Bi, the mass along the direction of the magnetic field is not perturbed by the magnetic quantization.

The Landau energy level E_{n6} can be expressed from (3.77) as

$$E_{n6}\left(1+\alpha E_{n6}\right)=\left(n+\frac{1}{2}\right)\hbar\omega\left(E_{n6}\right)\pm\frac{1}{2}g^{*}\mu_{0}B+\frac{3}{8}\alpha\left(n^{2}+n+\frac{1}{2}\right)\times\hbar^{2}\omega^{2}\left(E_{n6}\right). \tag{3.79}$$

Thus, the difference between the consecutive Landau levels is a function of the Landau quantum number, the electron spin and the other constants of the spectrum.

The SdH period can be written as from (3.77) as

$$\Delta\left(\frac{1}{B}\right)=\alpha_{16,\pm}\left(n+1,E_{\mathrm{FB}},g^{*}\right)-\alpha_{16,\pm}\left(n,E_{\mathrm{FB}},g^{*}\right), \tag{3.80}$$

where

$$\alpha_{16}\left(n,E_{\mathrm{FB}},g^{*}\right)\equiv\left[2\alpha_{15,\pm}\left(n,E_{\mathrm{FB}}\right)\right]\Big[-\alpha_{15,\pm}\left(n,E_{\mathrm{FB}}\right)+\sqrt{\alpha_{15,\pm}^{2}\left(n,E_{\mathrm{FB}}\right)+4\alpha_{14}\left(n,E_{\mathrm{FB}}\right)E_{\mathrm{FB}}\left(1+\alpha E_{\mathrm{FB}}\right)}\,\Big]^{-1}$$

$$\alpha_{15,\pm}\left(n,E_{\mathrm{FB}},g^{*}\right)\equiv\left[\left(n+\frac{1}{2}\right)\left(\frac{\hbar\left|e\right|}{\sqrt{m_{1}m_{2}}}\right)\left[1+\alpha E_{\mathrm{FB}}\left(1-\frac{m_{2}}{m_{2}'}\right)\right]^{1/2}\pm\frac{1}{2}g^{*}\mu_{0}\right],$$

and

$$\alpha_{14}\left(n,E_{\mathrm{FB}}\right)\equiv\left[\frac{3\alpha}{8}\left(n^{2}+n+\frac{1}{2}\right)\left(\frac{\hbar\left|e\right|}{\sqrt{m_{1}m_{2}}}\right)^{2}\left[1+\alpha E_{\mathrm{FB}}\left(1-\frac{m_{2}}{m_{2}'}\right)\right]\right].$$

Thus, when $\alpha\to 0$ (3.80) gets simplified into the form given by (3.66).

Therefore we infer that the SdH period for the Cohen model is a function of the magnetic quantum number, the electron spin, the Fermi energy, and the other constants of the spectrum due to the presence of band non-parabolicity only. In the absence of band non-parabolicity, the SdH period is independent of the magnetic quantum number and the electron spin, which is obvious by comparing (3.80) and (3.66).

The density-of-states function under magnetic quantization in accordance with the Cohen model is given by

$$D_{\mathrm{B}}(E)=\frac{g_{\mathrm{v}}|e|B\sqrt{2m_3}}{4\pi^2\hbar^2}\sum_{n=0}^{n_{\max}}\left[\left[E(1+\alpha E)-\left(n+\frac{1}{2}\right)\hbar\omega(E)\right.\right.$$
$$\left.-\left(n^2+\frac{1}{2}+n\right)\frac{3\alpha\hbar^2\omega^2(E)}{8}\mp\frac{1}{2}g^*\mu_0B\right]^{-1/2},$$
$$\times\left[1+2\alpha E-\left(n+\frac{1}{2}\right)\hbar\{\omega(E)\}'\right.$$
$$\left.\left.-\left(n^2+\frac{1}{2}+n\right)\frac{3\alpha\hbar^2\omega(E)\{\omega(E)\}'}{4}\right]\right]$$
$$\times H(E-E_{n_6}). \tag{3.81}$$

Thus, the electron concentration assumes the form

$$n_0=\frac{g_{\mathrm{v}}|e|B\sqrt{2m_3}}{2\pi^2\hbar^2}\sum_{n=0}^{n_{\max}}[T_{319}(n,E_{\mathrm{FB}})+T_{320}(n,E_{\mathrm{FB}})], \tag{3.82}$$

where

$$T_{319}(n,E_{\mathrm{FB}})\equiv\left[E_{\mathrm{FB}}(1+\alpha E_{\mathrm{FB}})-\left(n+\frac{1}{2}\right)\hbar\omega(E_{\mathrm{FB}})\right.$$
$$\left.\mp\frac{1}{2}g^*\mu_0B-\frac{3}{8}\alpha\left(n^2+n+\frac{1}{2}\right)\hbar^2\omega^2(E_{\mathrm{FB}})\right]^{1/2},$$
$$\omega(E_{\mathrm{FB}})\equiv\frac{|e|B}{\sqrt{m_1m_2}}\left[1+\alpha E_{\mathrm{FB}}\left(1-\frac{m_2}{m_2'}\right)\right]^{1/2},$$

and

$$T_{320}(n,E_{\mathrm{FB}})\equiv\sum_{r=1}^{s}L(r)[T_{319}(n,E_{\mathrm{FB}})].$$

Hence, combining (3.82) with (1.11), the DMR can be expressed as

$$\frac{D}{\mu}=\frac{1}{|e|}\left[\frac{\sum_{n=0}^{n_{\max}}[T_{319}(n,E_{\mathrm{FB}})+T_{320}(n,E_{\mathrm{FB}})]}{\sum_{n=0}^{n_{\max}}[\{T_{319}(n,E_{\mathrm{FB}})\}'+\{T_{320}(n,E_{\mathrm{FB}})\}']}\right]. \tag{3.83}$$

Under the condition $\alpha\to 0$, (3.83) gets simplified into the form given by (3.71).

In the presence of a quantizing magnetic field B along the y direction, the magneto-Cohen model can be expressed, by neglecting spin, as

$$E(1+\alpha E)=\left(n+\frac{1}{2}\right)\hbar\omega_4-\frac{\alpha Ep_y^2}{2m_2'}+\frac{p_y^2}{2m_2}(1+\alpha E)+\frac{\alpha p_y^4}{4m_2m_2'}. \tag{3.84}$$

The electron concentration and the magneto-DMR in this case can be expressed as

$$n_0 = \frac{g_v |e| B}{\sqrt{2}\pi^2\hbar^2} \sum_{n=0}^{n_{\max}} [T_{319}(n, E_{FB}) + T_{320}(n, E_{FB})]. \tag{3.85}$$

and

$$\frac{D}{\mu} = \frac{1}{|e|} \left[\frac{\sum_{n=0}^{n_{\max}} [T_{321}(n, E_{FB}) + T_{322}(n, E_{FB})]}{\sum_{n=0}^{n_{\max}} [\{T_{321}(n, E_{FB})\}' + \{T_{323}(n, E_{FB})\}']} \right], \tag{3.86}$$

where

$$T_{321}(n, E_{FB}) \equiv \left[-q_3(n, E_{FB}) + \sqrt{[q_3(n, E_{FB})]^2 + 4q_4(n, E_{FB})} \right]^{1/2},$$

and

$$T_{321}(n, E_{FB}) \equiv \sum_{r=1}^{s} L(r)\,[T_{320}(n, E_{FB})],$$

in which,

$$q_3(n, E_{FB}) \equiv \left(\frac{4m_2 m_2'}{\alpha} \right) \left[\frac{-\alpha E_{FB}}{2m_2'} + \frac{1}{2m_2}(1 + \alpha E_{FB}) \right],$$

and

$$q_4(n, E_{FB}) \equiv \left(\frac{4m_2 m_2'}{\alpha} \right) \left[E_{FB}(1 + \alpha E_{FB}) - \left(n + \frac{1}{2}\right) \hbar\omega_4 \right].$$

(c) The Lax model

For this model, the magneto-dispersion relation can be written extending (3.24) to

$$E(1 + \alpha E) = \left(n + \frac{1}{2}\right) \hbar\omega_{03} + \left(\hbar^2 k_z^2 / 2m_3^*\right) \pm \frac{1}{2} g^* \mu_0 B, \tag{3.87}$$

where

$$\omega_{03} = \frac{|e| B}{\sqrt{m_1 m_2}}.$$

The expressions of the EMM and the Landau sub-bands E_{n_7} assumes the well known forms as

$$m_{k_z}^*(n, E_{FB}) = m_3 [2\alpha E_{FB} + 1], \tag{3.88}$$

$$E_{n_7} = (2\alpha)^{-1} \left\{ -1 + \sqrt{1 + 4\alpha \left[\left(n + \frac{1}{2}\right) \hbar\omega_{03} \pm \frac{1}{2} g^* \mu_0 B \right]} \right\}, \tag{3.89}$$

The density-of-states-function, the electron concentration and the DMR for the Lax model can, respectively, be written as

$$D_{\mathrm{B}}(E) = \frac{g_{\mathrm{v}}|e|B\sqrt{2m_3}}{4\pi^2\hbar^2}\sum_{n=0}^{n_{\max}}[1+2\alpha E] \times \left[E(1+\alpha E) - \left(n+\frac{1}{2}\right)\hbar\omega_{03} \mp \frac{1}{2}g^*\mu_0 B\right]^{-\frac{1}{2}} H(E - E_{n_7}), \tag{3.90}$$

$$n_0 = \frac{g_{\mathrm{v}}|e|B\sqrt{2m_3}}{2\pi^2\hbar^2}\sum_{n=0}^{n_{\max}}[T_{323}(n,E_{\mathrm{FB}}) + T_{324}(n,E_{\mathrm{FB}})], \tag{3.91}$$

and

$$\frac{D}{\mu} = \frac{1}{|e|}\left[\frac{\sum_{n=0}^{n_{\max}}[T_{323}(n,E_{\mathrm{FB}}) + T_{324}(n,E_{\mathrm{FB}})]}{\sum_{n=0}^{n_{\max}}[\{T_{323}(n,E_{\mathrm{FB}})\}' + \{T_{324}(n,E_{\mathrm{FB}})\}']}\right], \tag{3.92}$$

where

$$T_{323}(n,E_{\mathrm{FB}}) \equiv \left[E_{\mathrm{FB}}(1+\alpha E_{\mathrm{FB}}) - \left(n+\frac{1}{2}\right)\hbar\omega_{03} \pm \frac{1}{2}g^*\mu_0 B\right]^{\frac{1}{2}},$$

and

$$T_{324}(n,E_{\mathrm{FB}}) \equiv \sum_{r=1}^{s} L(r)\,[T_{323}(n,E_{\mathrm{FB}})].$$

In the absence of spin, the expressions of n_0 and the DMR assume the forms

$$n_0 = \frac{g_{\mathrm{v}}|e|B\sqrt{2m_3}}{\pi^2\hbar^2}\sum_{n=0}^{n_{\max}}[T_{325}(n,E_{\mathrm{FB}}) + T_{326}(n,E_{\mathrm{FB}})], \tag{3.93}$$

and

$$\frac{D}{\mu} = \frac{1}{|e|}\left[\frac{\sum_{n=0}^{n_{\max}}[T_{325}(n,E_{\mathrm{FB}}) + T_{326}(n,E_{\mathrm{FB}})]}{\sum_{n=0}^{n_{\max}}[\{T_{325}(n,E_{\mathrm{FB}})\}' + \{T_{326}(n,E_{\mathrm{FB}})\}']}\right], \tag{3.94}$$

where

$$T_{325}(n,E_{\mathrm{FB}}) \equiv \left[E_{\mathrm{FB}}(1+\alpha E_{\mathrm{FB}}) - \left(n+\frac{1}{2}\right)\hbar\omega_{03}\right]^{\frac{1}{2}},$$

and

$$T_{326}(n,E_{\mathrm{FB}}) \equiv \sum_{r=1}^{s} L(r)\,[T_{325}(n,E_{\mathrm{FB}})].$$

(d) The parabolic ellipsoidal model

For this model, the magneto-dispersion relation can be written, extending (3.40), as

$$E = \left(n + \frac{1}{2}\right) \hbar\omega_{03} + \left(\hbar^2 k_z^2/2m_3\right) \pm \frac{1}{2} g^* \mu_0 B. \tag{3.95}$$

The expressions of the electron concentration and the DMR for this model are the special cases of the models of McClure and Choi, the Cohen, and the Lax, respectively.

3.2.5 IV–VI Materials

It is well known that the conduction electrons of the IV–VI compounds obey the Cohen model of bismuth, where the energy band constants correspond to the said compounds. Equations (3.82) and (3.83) are applicable in this context.

3.2.6 Stressed Kane Type Semiconductors

The simplified expression of the electron energy spectrum in stressed Kane type semiconductors in the presence of an arbitrarily oriented quantizing magnetic field B, which makes angles α_1, β_1 and γ_1 with the k_x, k_y and k_z axes respectively, can be written using (2.48), (3.9) and (3.11) as

$$1 - \left[k_z'\right]^2 \left[I_2(E)\right]^{-1} = I_3(n, E), \tag{3.96}$$

where

$$I_2(E) \equiv \left[\bar{a}_0(E)\right]^2 \cos^2\alpha_1 + \left[\bar{b}_0(E)\right]^2 \cos^2\beta_1 + \left[\bar{c}_0(E)\right]^2 \cos^2\gamma_1,$$

and

$$I_3(n, E) \equiv \left(\frac{2|e|B}{\hbar}\right)\left(n + \frac{1}{2}\right)\left[\left[\bar{a}_0(E)\right]\left[\bar{b}_0(E)\right]\left[\bar{c}_0(E)\right]\right]^{-1}\left[I_2(E)\right]^{1/2}.$$

The use of (3.96) leads to the expressions of the EMM, the Landau subbands (E_{n8}), and the SdH period as

$$m^*_{k_z'}(n, E_{FB}) = \frac{\hbar^2}{2}\Big[-\left\{I_3(n, E_{FB})\right\}' I_2(n, E_{FB}) + \left(1 - I_3(n, E_{FB})\right)\left\{I_2(n, E_{FB})\right\}'\Big], \tag{3.97}$$

$$I_3(n, E_{n8}) = 1. \tag{3.98}$$

and

$$\Delta\left(\frac{1}{B}\right) = \left(\frac{2|e|}{\hbar}\right)\left[[\bar{a}_0(E_{FB})]\left[\bar{b}_0(E_{FB})\right][\bar{c}_0(E_{FB})]\right]^{-1} \times\left[[\bar{a}_0(E_{FB})]^2\cos^2\alpha_1 + \left[\bar{b}_0(E_{FB})\right]^2\cos^2\beta_1 + [\bar{c}_0(E_{FB})]^2\cos^2\gamma_1\right]^{1/2}. \tag{3.99}$$

In the absence of stress, together with the substitution $B_2^2 \equiv \frac{3\hbar^2 E_g}{4m^*}$, (3.97)–(3.99) get simplified to (3.25)–(3.27), respectively. By comparing (3.97) and (3.25), one can observe that the stress makes the EMM quantum number dependent in stressed Kane type compounds under magnetic quantization, in addition to Fermi energy.

The density of states function in this case is given by

$$D_B(E) = \frac{g_v|e|B}{2\pi^2\hbar}\sum_{n=0}^{n_{max}}\left\{\frac{\{I_2(E)\}'}{\sqrt{I_2(E)}}[1 - I_3(n,E)]^{1/2} - [1 - I_3(n,E)]^{-1/2} \times\{I_3(n,E)\}'\sqrt{I_2(E)}\right\}H(E - E_{n_8}). \tag{3.100}$$

The use of (3.100) leads to the expression of electron concentration as

$$n_0 = \frac{g_v|e|B}{\pi^2\hbar}\sum_{n=0}^{n_{max}}[T_{327}(n,E_{FB}) + T_{328}(n,E_{FB})]. \tag{3.101}$$

Using (3.100) and (1.11), the DMR can be written as

$$\frac{D}{\mu} = \frac{1}{|e|}\left[\frac{\sum_{n=0}^{n_{max}}[T_{327}(n,E_{FB}) + T_{328}(n,E_{FB})]}{\sum_{n=0}^{n_{max}}[\{T_{327}(n,E_{FB})\}' + \{T_{328}(n,E_{FB})\}']}\right], \tag{3.102}$$

where

$$T_{327}(n,E_{FB}) \equiv \sqrt{I_2(E_{FB})}\left[\sqrt{1 - [I_3(n,E_{FB})]}\right],$$

and

$$T_{328}(n,E_{FB}) \equiv \sum_{r=1}^{s} L(r)T_{327}(n,E_{FB}).$$

Finally, we infer that under stress free condition together with the substitution $B_2^2 \equiv \frac{3\hbar^2 E_g}{4m^*}$, (3.101) and (3.102) get simplified to (3.31) and (3.32), respectively.

3.3 Result and Discussions

In Figs. 3.1 and 3.2, the normalized magneto-DMR has been plotted as a function of the inverse quantizing magnetic field for n-Cd_3As_2 and n-$CdGeAs_2$ respectively. In the same figures the plots corresponding to $\delta = 0$, the three and the two band models of Kane together with the parabolic energy bands have also been drawn for the purpose of relative assessment. It appears from both the figures that the DMR is an oscillatory function of the inverse quantizing magnetic field. The oscillatory dependence is due to the crossing over of the Fermi level by the Landau sub-bands in steps resulting in successive reduction of the number of occupied Landau levels as the magnetic field is increased. For each coincidence of a Landau level, with the Fermi level, there would be a discontinuity in the density-of-states function resulting in a peak of oscillation. Thus the peaks should occur whenever the Fermi energy is a multiple of energy separation between the two consecutive Landau levels and it may be noted that the origin of oscillations in the Einstein relation is the same as that of the Subhnikov-de Hass oscillations. With increase in magnetic field, the amplitude of the oscillation will increase and, ultimately, at very large values of the magnetic field, the conditions for the quantum limit will be reached when the DMR will be found to decrease monotonically with increase in magnetic field.

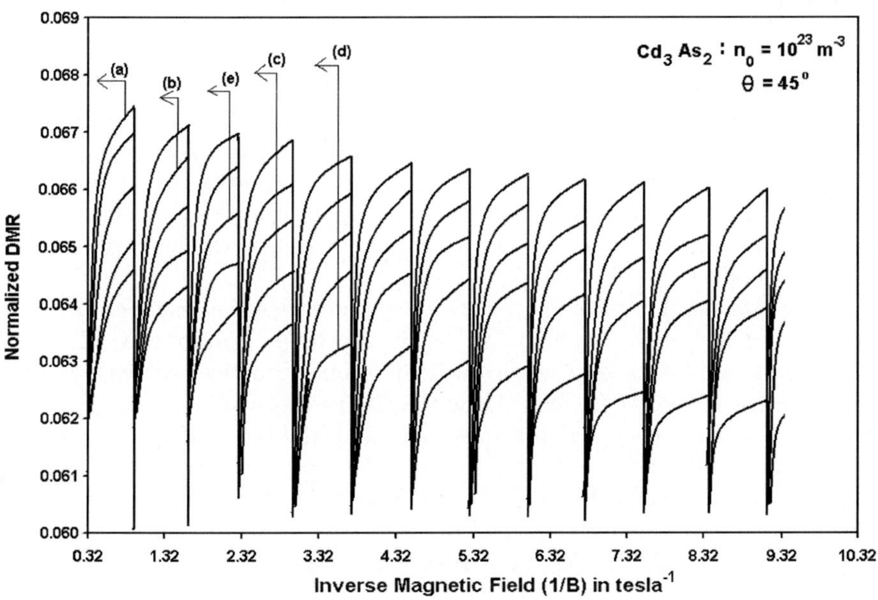

Fig. 3.1. The plot of the DMR in n-Cd_3As_2 as a function of inverse quantizing magnetic field in accordance with (**a**) the generalized band model, (**b**) $\delta = 0$, (**c**) the three band model of Kane, (**d**) the two band model of Kane and (**e**) the parabolic energy bands

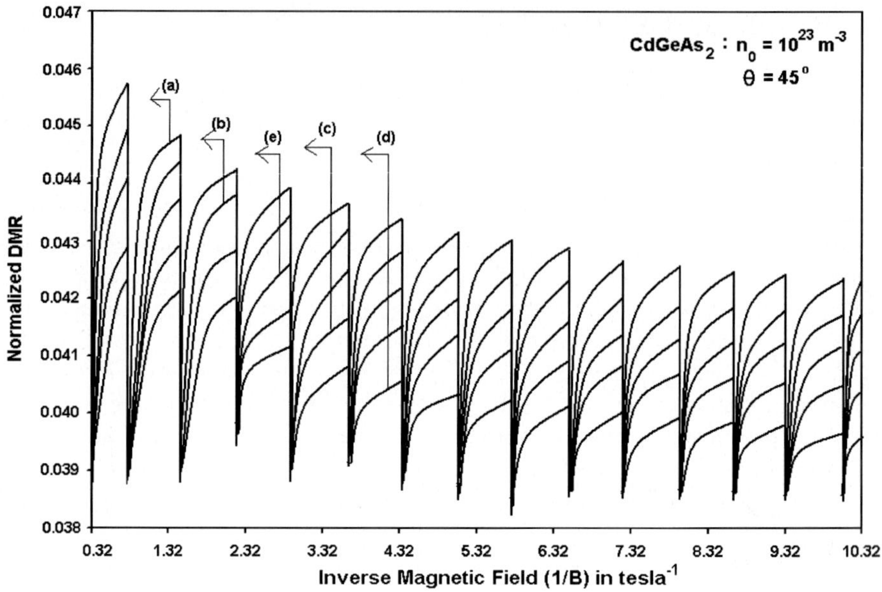

Fig. 3.2. The plot of the DMR in n-$CdGeAs_2$ as a function of inverse quantizing magnetic field in accordance with (**a**) the generalized band model, (**b**) $\delta = 0$, (**c**) the three band model of Kane, (**d**) the two band model of Kane and (**e**) the parabolic energy bands

By comparing Figs. 3.1 and 3.2, one can infer regarding the numerical value of the DMR in accordance with the various band models. The crystal field splitting enhances the peaks of the DMR in the whole range of the magnetic field considered for both the n-Cd_3As_2 and n-$CdGeAs_2$ respectively. The presence of the isotropic spin orbit splitting constant in the three band model of Kane changes the value of the DMR as compared with the corresponding two band model.

In Fig. 3.3 and 3.4, the concentration dependence of the magneto-DMR has been plotted for all the cases of Fig. 3.1 for both the compounds. The DMR again shows oscillatory dependence with different numerical values exhibiting the signature of the SdH effect. Although the rate of variations are different, the influence of the energy band constants on all the types of the band models follows the same trend as shown in Figs. 3.1 and 3.2. Figures 3.5 and 3.6 exhibit the variations of the magneto-DMR with the angle of orientation of the arbitrarily oriented quantizing field for both the n-Cd_3As_2 and n-$CdGeAs_2$. It appears that the DMR decreases with increasing θ form $0°$ to $90°$ and then increases from $90°$ to $180°$. The variation is periodically repeated and the influence of δ is to change the value of the DMR.

For the three and two band models of Kane, the DMR becomes independent of θ, since the dispersion relation of the bulk materials in accordance with the said band models is spherical in constant energy wave-vector space,

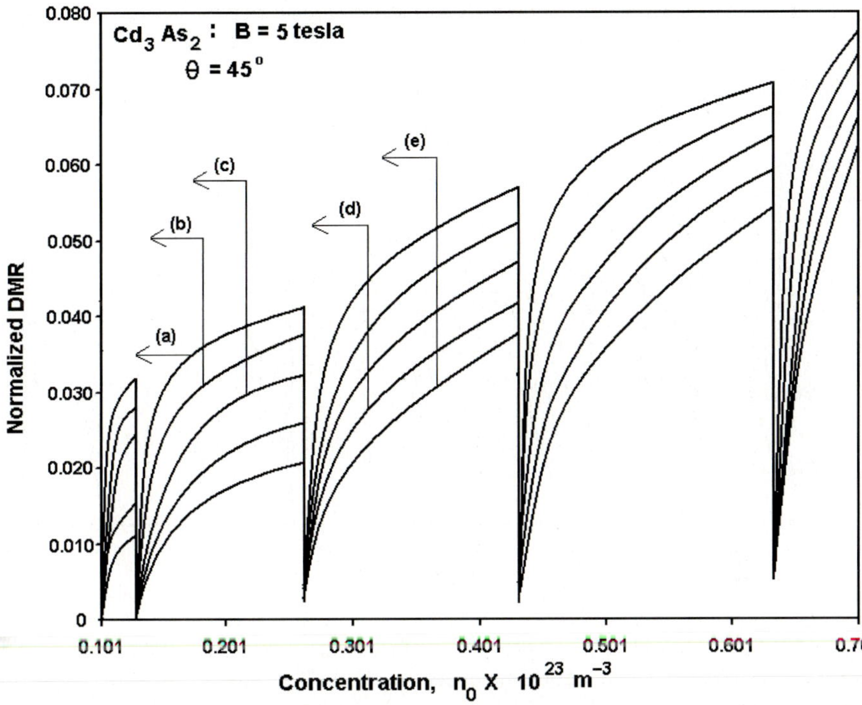

Fig. 3.3. The plot of the magneto-DMR in n-Cd_3As_2 as a function of electron concentration in accordance with (**a**) the generalized band model, (**b**) $\delta = 0$, (**c**) the three band model of Kane, (**d**) the two band model of Kane and (**e**) the parabolic energy bands

whereas, the generalized band model represents the ellipsoid of revolution in the same space. In Figs. 3.7–3.9, the normalized DMR as functions of the inverse quantizing magnetic field for GaAs, InSb and InAs has been plotted in accordance with the three and two band models of Kane together with the parabolic energy bands respectively. The variations of the DMR are periodic with the quantizing magnetic field and the influence of the energy band constants on the DMR in accordance with all the band models is apparent from the said figures.

Figures 3.10–3.12 exhibit the concentration dependence of the periodic magneto-DMR for the said materials with different numerical values. Figures 3.13 and 3.14 show the dependence of the DMR on $1/B$ for $Hg_{1-x}Cd_xTe$ and $In_{1-x}Ga_xAs_yP_{1-y}$ lattice matched to InP respectively. Figures 3.15 and 3.16 exhibit the concentration dependence of the DMR for the said materials.

It should be noted that the numerical value of the magneto-DMR is greatest for the ternary materials while it is the least for GaAs for all types of variables in accordance with all types of band models of III–V, ternary and

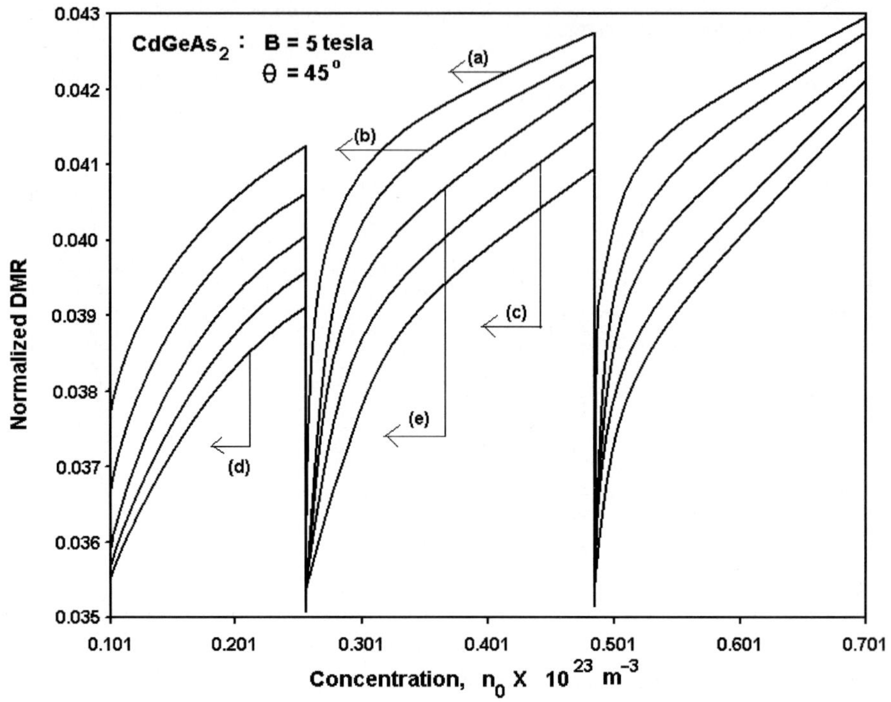

Fig. 3.4. The plot of the magneto-DMR n-$CdGeAs_2$ as a function of electron concentration in accordance with (**a**) the generalized band model (**b**) $\delta = 0$, (**c**) the three band model of Kane (**d**) the two band model of Kane and (**e**) the parabolic energy bands

quaternary materials. In Figs. 3.17 and 3.18, the magneto-DMR has been plotted as functions of alloy composition for both the said compounds and it has been observed that the DMR decreases with increasing alloy composition.

Figures 3.19 and 3.20 exhibit the dependence of the magneto-DMR on $1/B$ and n_0 respectively for p-CdS. In both the figures, the presence of the splitting of the two spin-states by the spin–orbit coupling and the crystalline field enhances the DMR in p-CdS as compared with $\bar{\lambda}_0 = 0$ for both the variables, although the nature of variations are different. The normalized magneto-DMR as functions of $1/B$ has been plotted in Figs. 3.21–3.23 for the McClure and Choi, the Cohen, and the Lax models of Bismuth. In the said plots, the $\alpha = 0$ curve indicates the magneto DMR in Bi in accordance with the parabolic ellipsoidal band model. The concentration dependence of the DMR has been plotted in Figs. 3.24–3.26 for all the band models of bismuth. The nature of oscillations and the numerical values are totally band structure dependent. The rate of oscillations and the number of spikes are the greatest for the McClure and Choi model.

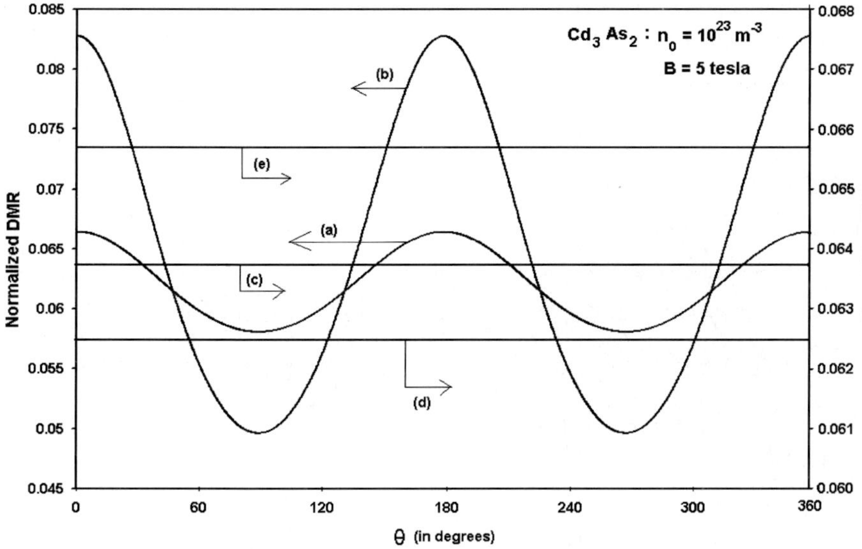

Fig. 3.5. The plot of the magneto-DMR in n-Cd_3As_2 as a function of angular orientation of the quantizing magnetic field in accordance with (**a**) the generalized band model, (**b**) $\delta = 0$, (**c**) the three band model of Kane, (**d**) the two band model of Kane and (**e**) the parabolic energy bands

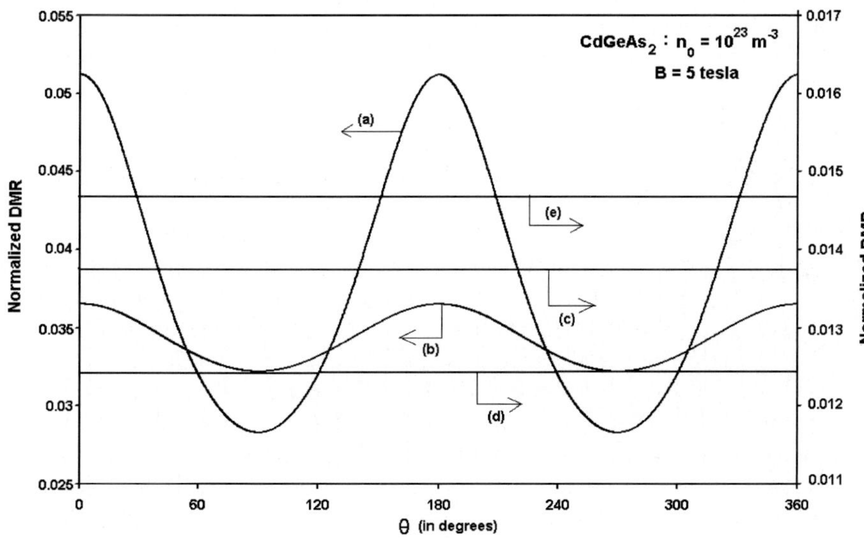

Fig. 3.6. The plot of the magneto-DMR n-$CdGeAs_2$ as a function of angular orientation of the quantizing magnetic field in accordance with (**a**) the generalized band model, (**b**) $\delta = 0$, (**c**) the three band model of Kane, (**d**) the two band model of Kane and (**e**) the parabolic energy bands

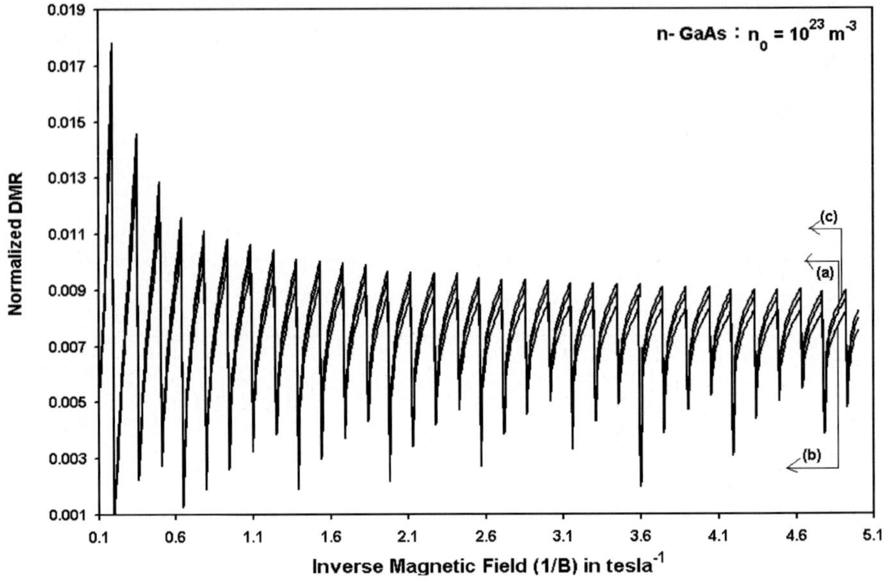

Fig. 3.7. The plot of the magneto DMR in n-GaAs as a function of inverse quantizing magnetic field in accordance with (**a**) the three band model of Kane, (**b**) the two band model of Kane and (**c**) the parabolic energy bands

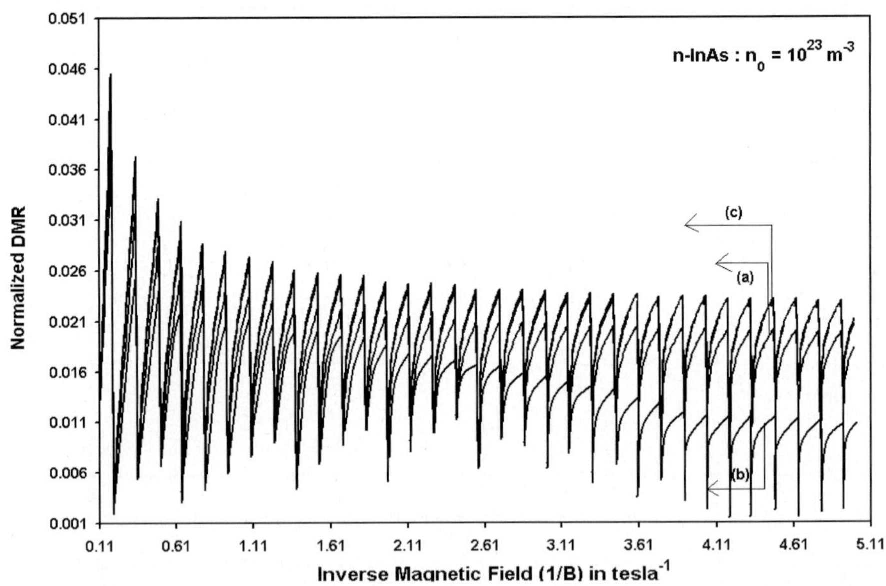

Fig. 3.8. The plot of the DMR in n-InAs as a function of inverse quantizing magnetic field in accordance with (**a**) the three band model of Kane, (**b**) the two band model of Kane and (**c**) the parabolic energy bands

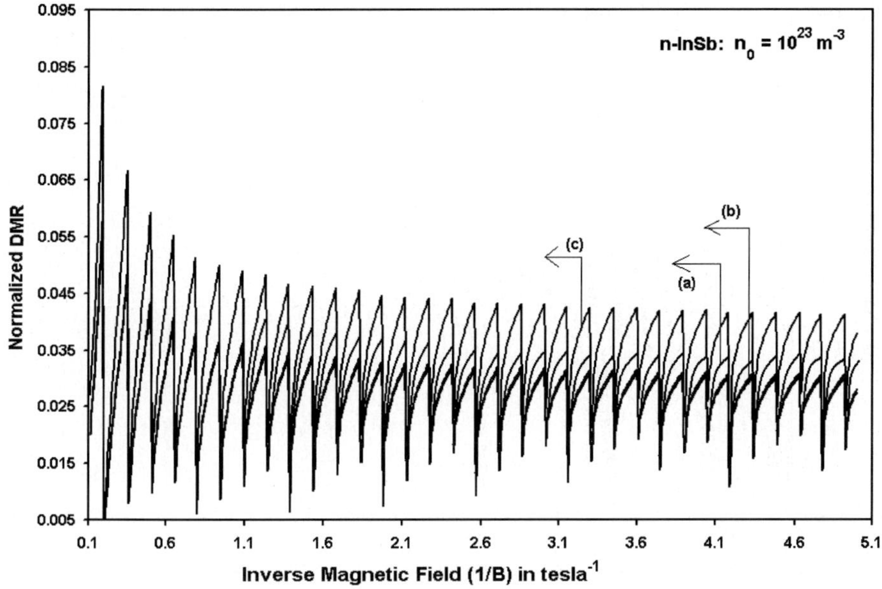

Fig. 3.9. The plot of the DMR in n-InSb as a function of inverse quantizing magnetic field in accordance with (**a**) the three band model of Kane, (**b**) the two band model of Kane and (**c**) the parabolic energy bands

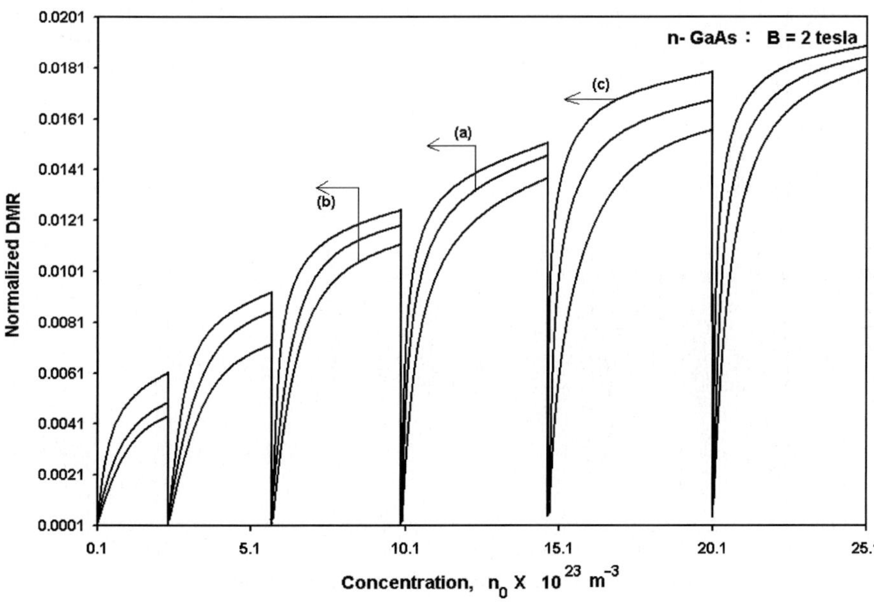

Fig. 3.10. The plot of the magneto DMR in n-GaAs as a function of electron concentration in accordance with (**a**) the three band model of Kane, (**b**) the two band model of Kane and (**c**) the parabolic energy bands

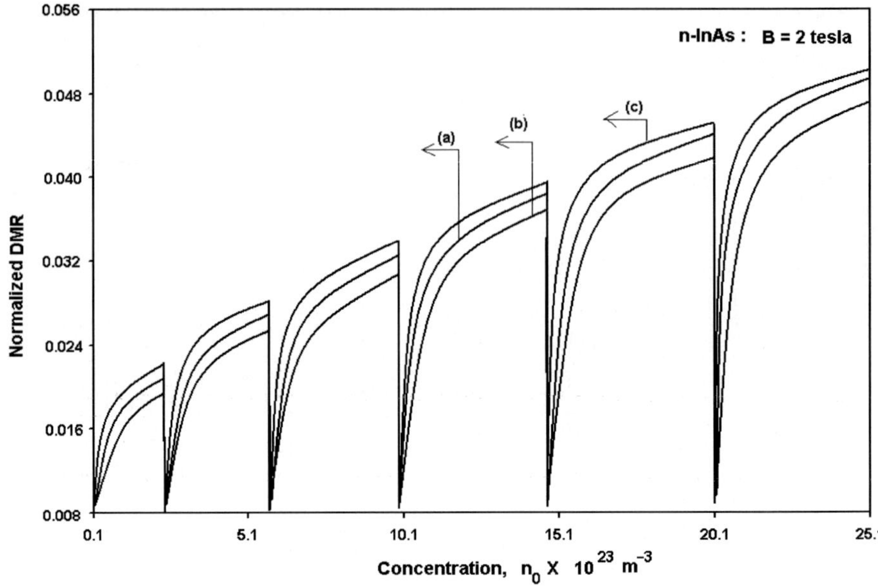

Fig. 3.11. The plot of the magneto DMR in n-InAs as a function of electron concentration in accordance with (**a**) the three band model of Kane, (**b**) the two band model of Kane and (**c**) the parabolic energy bands

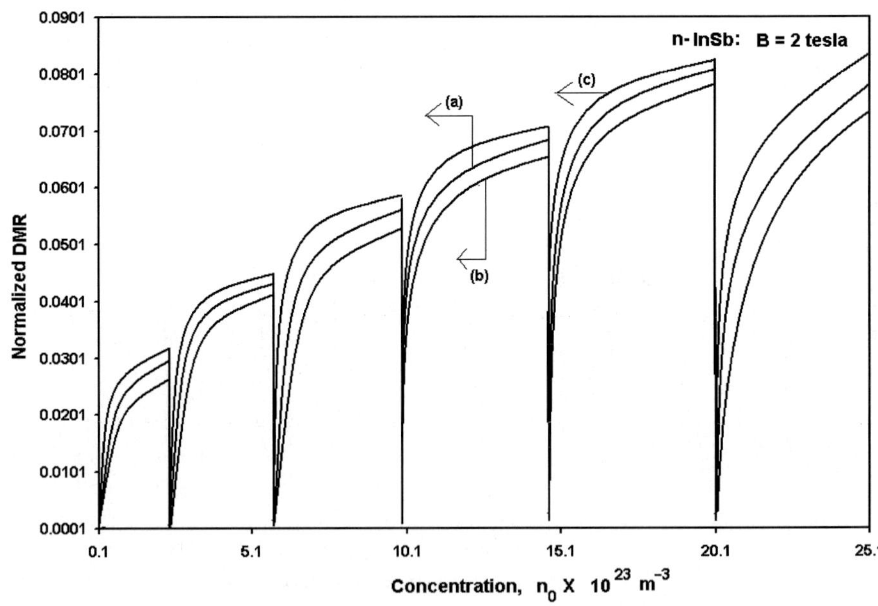

Fig. 3.12. The plot of the magneto DMR in n-InSb as a function electron concentration in accordance with (**a**) the three band model of Kane, (**b**) the two band model of Kane and (**c**) the parabolic energy bands

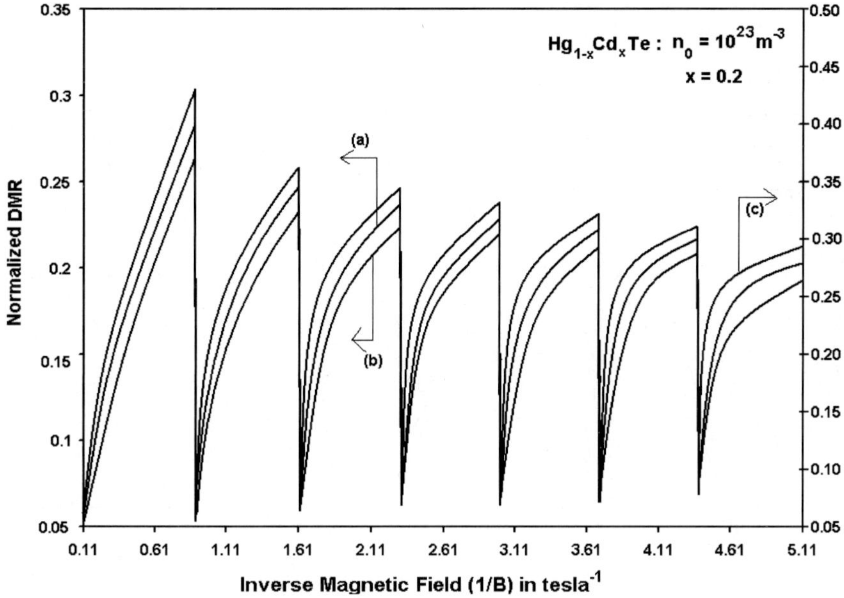

Fig. 3.13. The plot of the DMR in n-$Hg_{1-x}Cd_xTe$ as a function of inverse quantizing magnetic field in accordance with (**a**) the three band model of Kane; (**b**) the two band model of Kane and (**c**) the parabolic energy bands ($x = 0.3$)

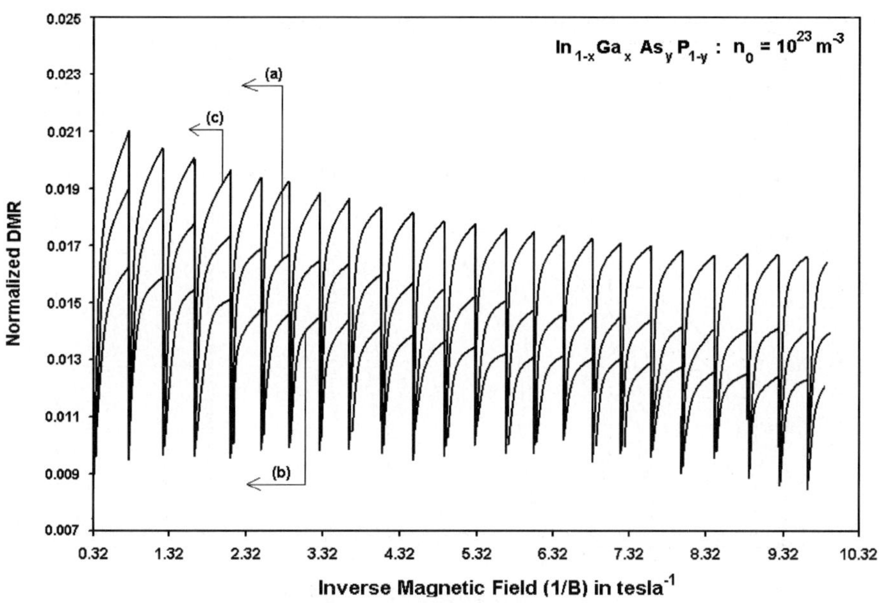

Fig. 3.14. The plot of the DMR in n-$In_{1-x}Ga_xAs_yP_{1-y}$ lattice matched to InP as a function of inverse quantizing magnetic field in accordance with (**a**) the three band model of Kane, (**b**) the two band model of Kane and (**c**) the parabolic energy bands ($y = 0.37$)

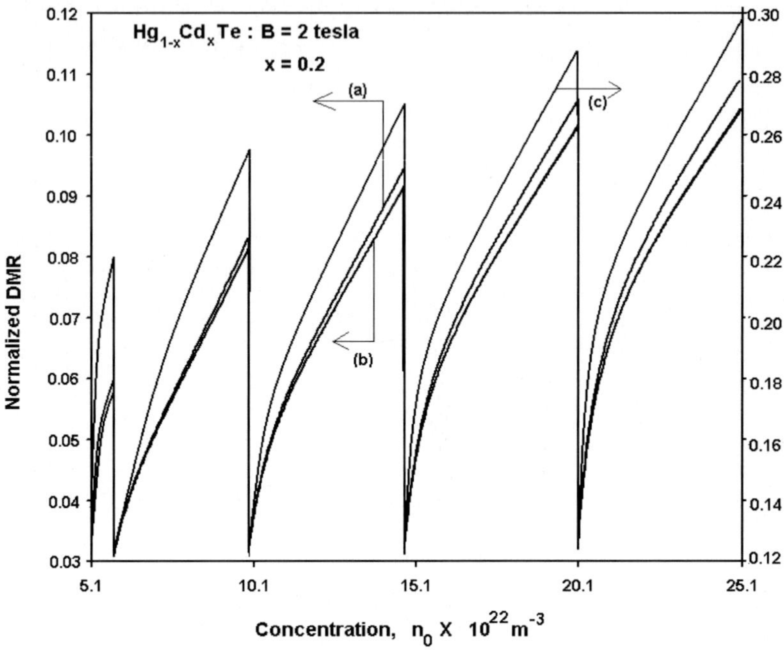

Fig. 3.15. The plot of the magneto DMR in n-$Hg_{1-x}Cd_xTe$ as a function of electron concentration in accordance with (**a**) the three band model of Kane; (**b**) the two band model of Kane and (**c**) the parabolic energy bands ($x = 0.3$)

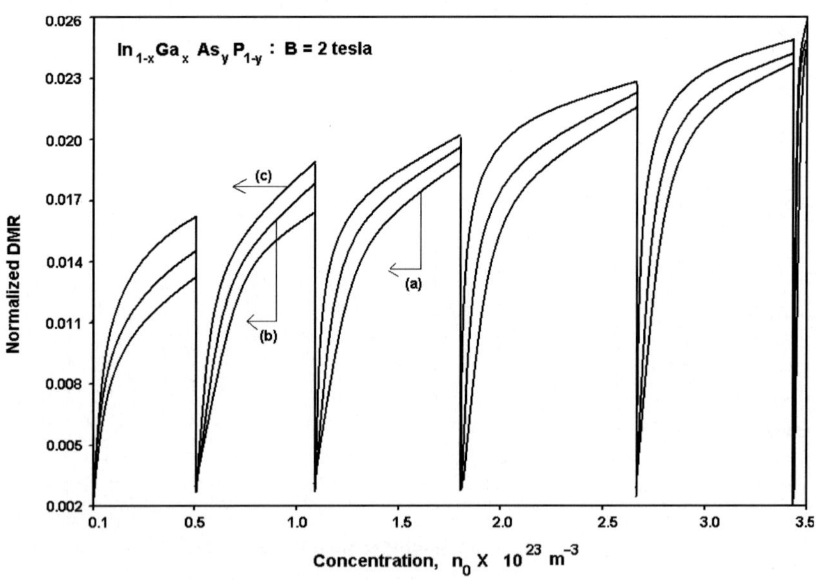

Fig. 3.16. The plot of the magneto DMR in n-$In_{1-x}Ga_xAs_yP_{1-y}$ lattice matched to InP as a function of inverse quantizing magnetic field in accordance with (**a**) the three band model of Kane, (**b**) the two band model of Kane and (**c**) the parabolic energy bands ($y = 0.37$)

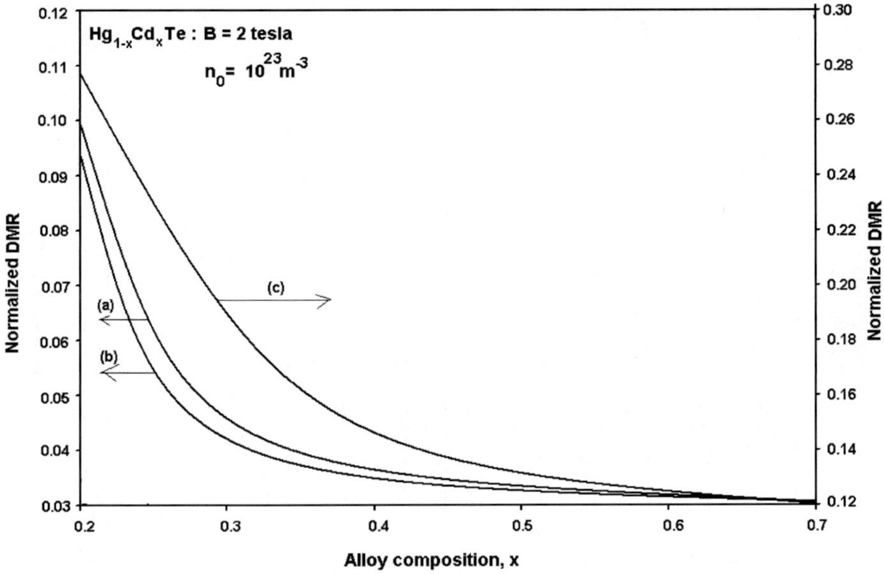

Fig. 3.17. The plot of the magneto DMR in n-$Hg_{1-x}Cd_xTe$ as a function of alloy composition (x) in accordance with (**a**) the three band model of Kane, (**b**) the two band model of Kane and (**c**) the parabolic energy bands

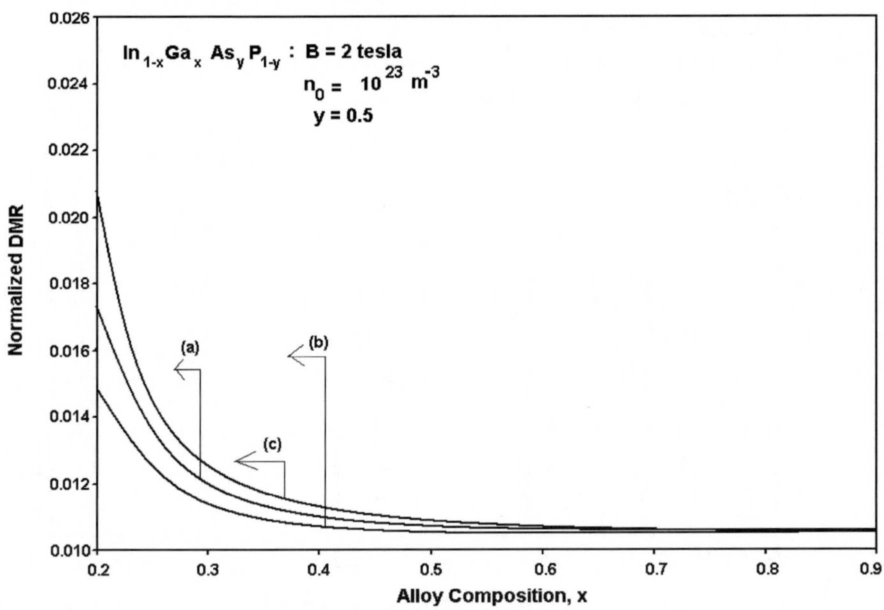

Fig. 3.18. The plot of the magneto DMR in n-$In_{1-x}Ga_xAs_yP_{1-y}$ lattice matched to InP as a function of alloy composition (x) in accordance with (**a**) the three band model of Kane, (**b**) the two band model of Kane and (**c**) the parabolic energy bands

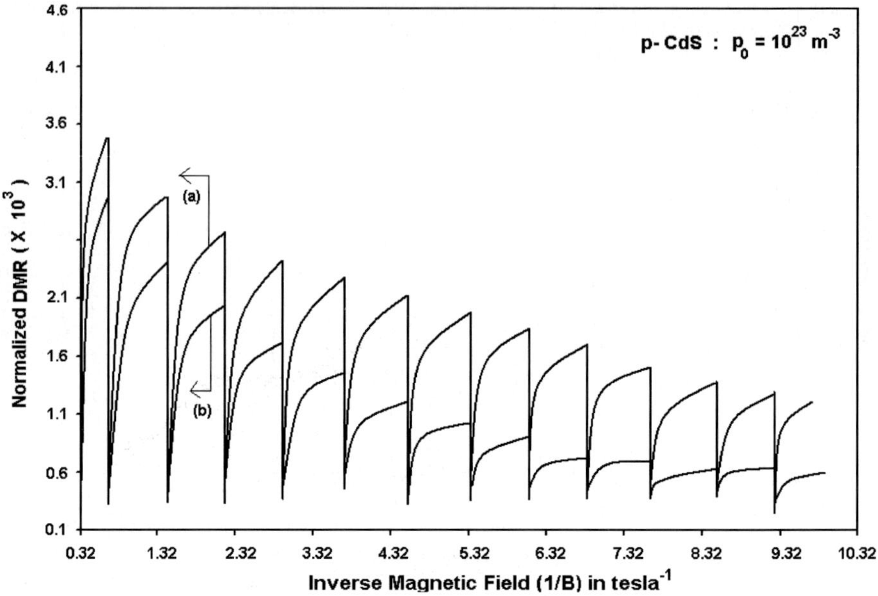

Fig. 3.19. The plot of the DMR as a function of inverse quantizing magnetic field for p-CdS for (**a**) $\bar{\lambda}_0 \neq 0$ and (b) $\bar{\lambda}_0 = 0$

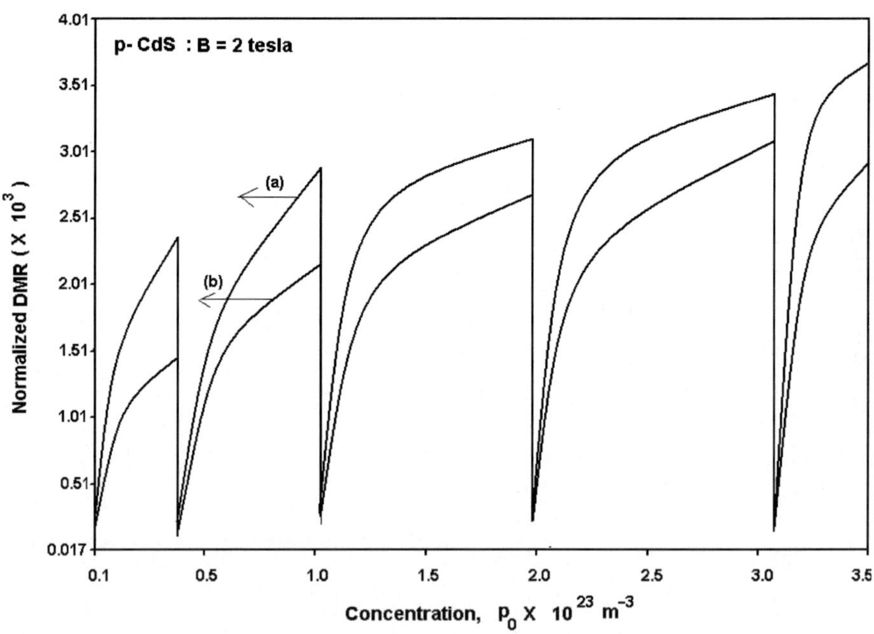

Fig. 3.20. The plot of the magneto DMR as a function of hole concentration p_0 of p-CdS for (**a**) $\bar{\lambda}_0 \neq 0$ and (**b**) $\bar{\lambda}_0 = 0$

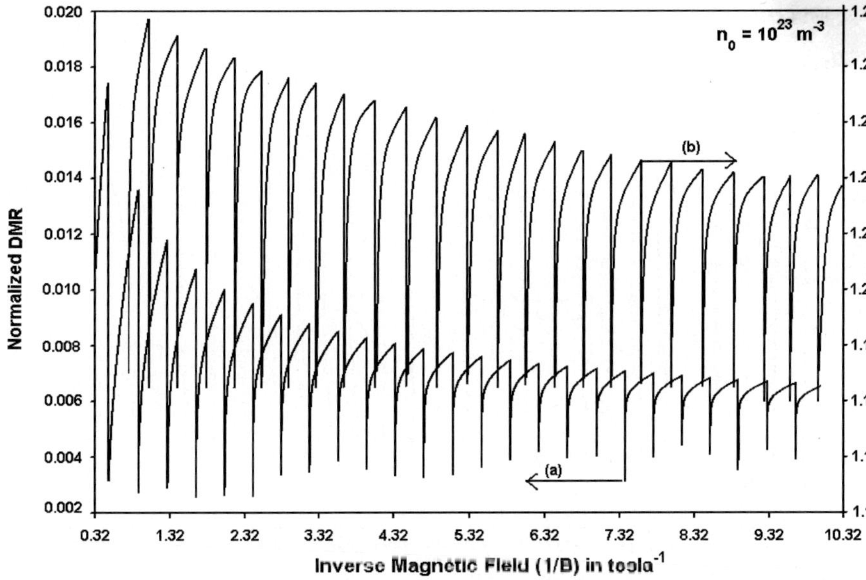

Fig. 3.21. The plot of the DMR in bismuth as a function of inverse quantizing magnetic field in accordance with the model of McClure and Choi as shown by curve (**a**) and the curve (**b**) corresponds to the ellipsoidal parabolic energy bands, where $\alpha = 0$

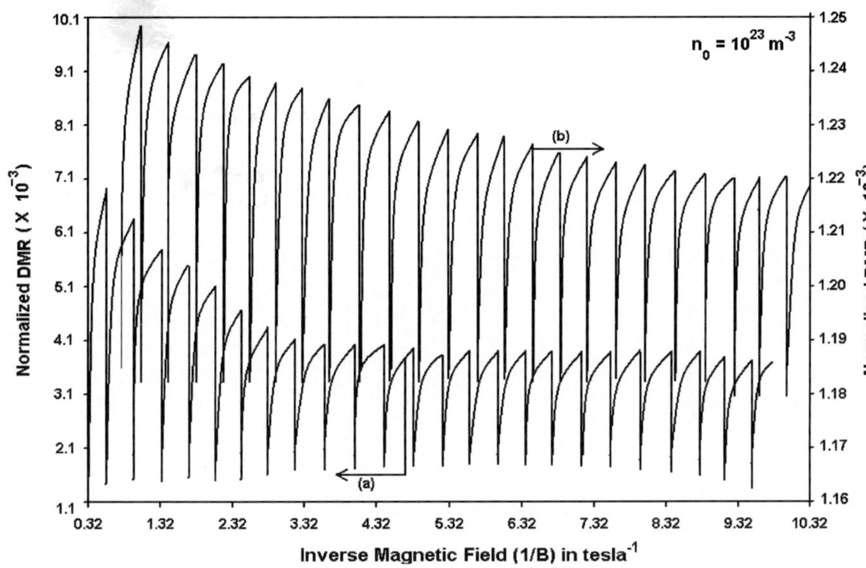

Fig. 3.22. The plot of the DMR in bismuth as a function of inverse quantizing magnetic field in accordance with the model of Cohen as shown by curve (**a**) and the curve (**b**) corresponds to the ellipsoidal parabolic energy bands, where $\alpha = 0$

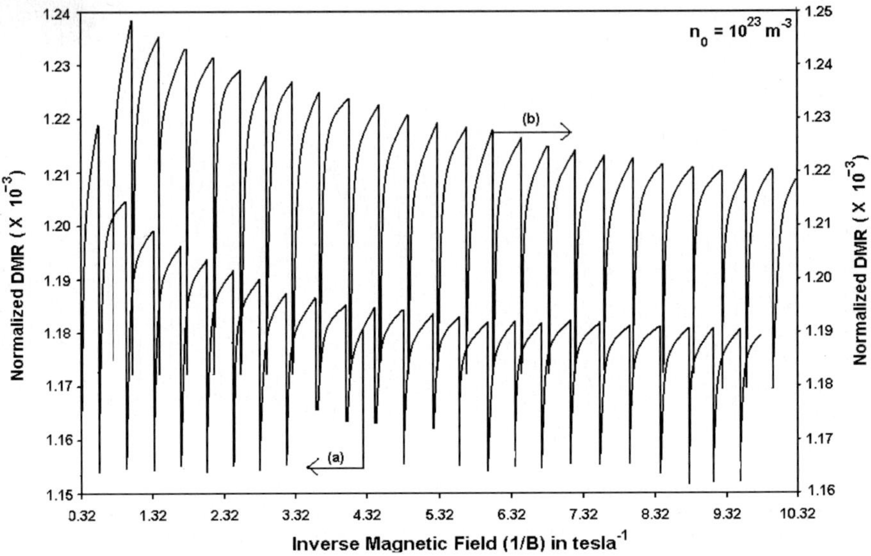

Fig. 3.23. The plot of the DMR in bismuth as a function of inverse magnetic field in accordance with the model of Lax as shown by curve (**a**) and the curve (**b**) corresponds to the ellipsoidal parabolic energy bands, where $\alpha = 0$

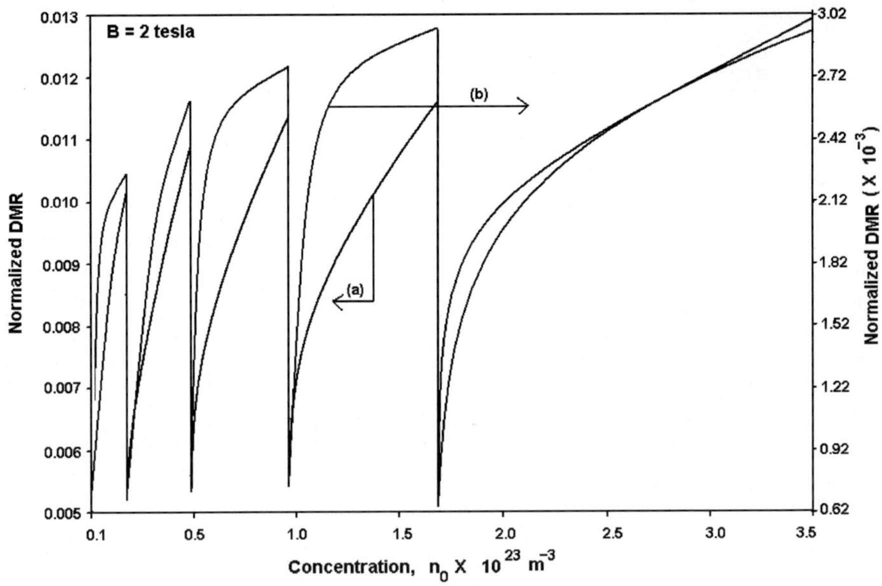

Fig. 3.24. The plot of the magneto DMR in bismuth as a function of electron concentration in accordance with the model of McClure and Choi as shown by curve (**a**) and the curve (**b**) corresponds to the ellipsoidal parabolic energy bands, where $\alpha = 0$

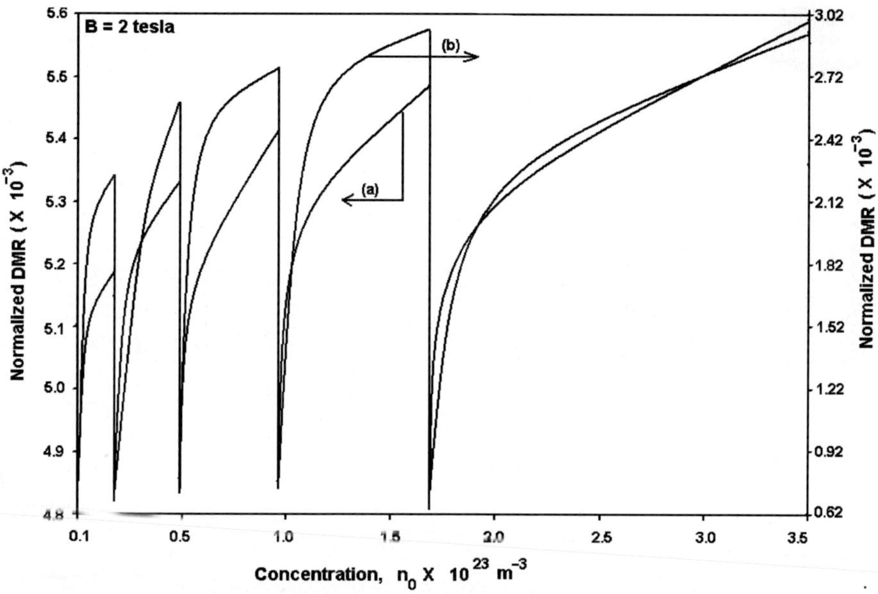

Fig. 3.25. The plot of the magneto DMR in bismuth as a function of electron concentration in accordance with the model of Cohen as shown by curve (**a**) and the curve (**b**) corresponds to the ellipsoidal parabolic energy bands, where $\alpha = 0$

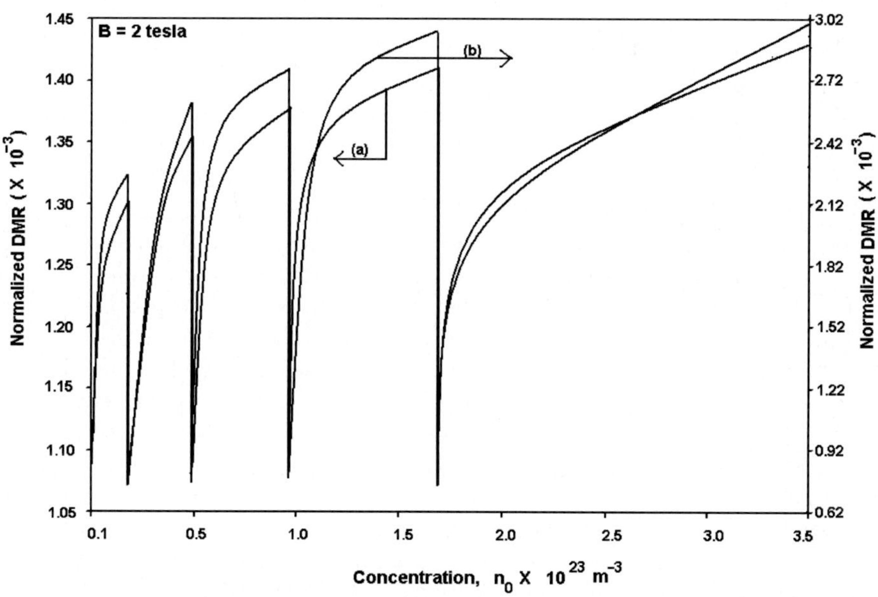

Fig. 3.26. The plot of the magneto DMR in bismuth as a function of electron concentration in accordance with the model of Lax as shown by curve (**a**) and the curve (**b**) corresponds to the ellipsoidal parabolic energy bands, where $\alpha = 0$

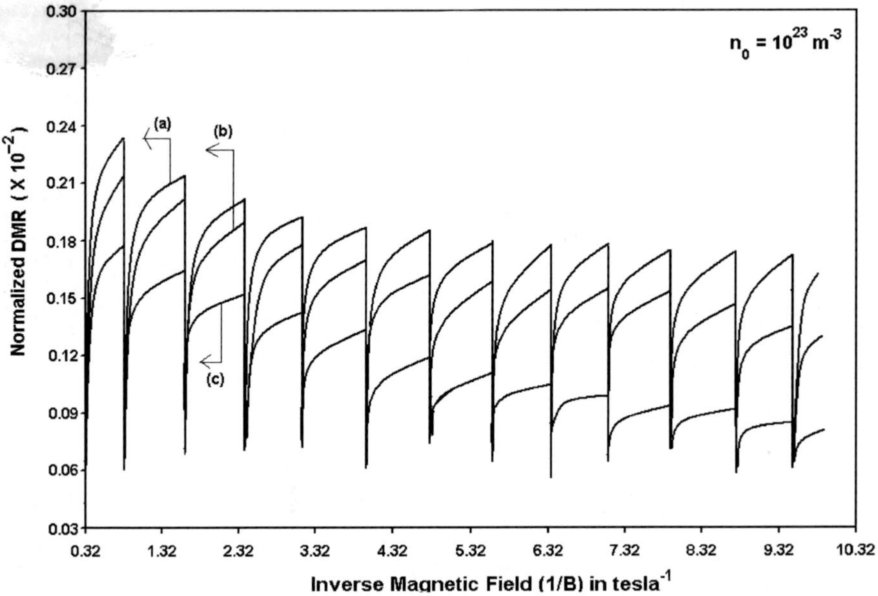

Fig. 3.27. The plot of the DMR in (**a**) PbTe (**b**) n-PbSnTe and (**c**) n-$Pb_{1-x}Sn_xSe$ as a function of inverse quantizing magnetic field in accordance with the model of Cohen

The plots of the DMR for PbTe, n-PbSnTe and n-$Pb_{1-x}Sn_xSe$ as functions of $1/B$ and n_0 have been shown in Figs. 3.27 and 3.28 respectively in accordance with the model of Cohen. Depending on the energy band constants, the values of the DMR are greatest for n-PbTe and least for n-$Pb_{1-x}Sn_xSe$. Figures 3.29–3.31 exhibit the dependence of the magneto-DMR on $1/B$, n_0 and γ_1 respectively for stressed n-InSb both in the presence and absence of stress. It appears that the value of the DMR in stressed materials is relatively large as compared with the stress-free condition for all the variables.

It may be noted that the DMR will, in general, be anisotropic under magnetic quantization. Thus for investigating the dependence of the DMR on the strength of the magnetic field, one has to determine the element $(D/\mu)_{zz}$ of the corresponding tensor under the above condition. The above conclusion will be true only under the condition of carrier degeneracy because under non-degenerate conditions, the normalized DMR will be equal to unity (neglecting magnetic freeze-out), i.e. independent of magnetic quantization. The effect of electron spin has not been considered in obtaining the oscillatory plots. The peaks in all the figures would increase in number with decrease in amplitude if spin splitting term is included in the respective numerical computations. Though, the effects of collisions are usually small at low temperatures, the sharpness of the amplitude of the oscillatory plots would be somewhat reduced by collision broadening. Nevertheless, the present analysis would remain valid

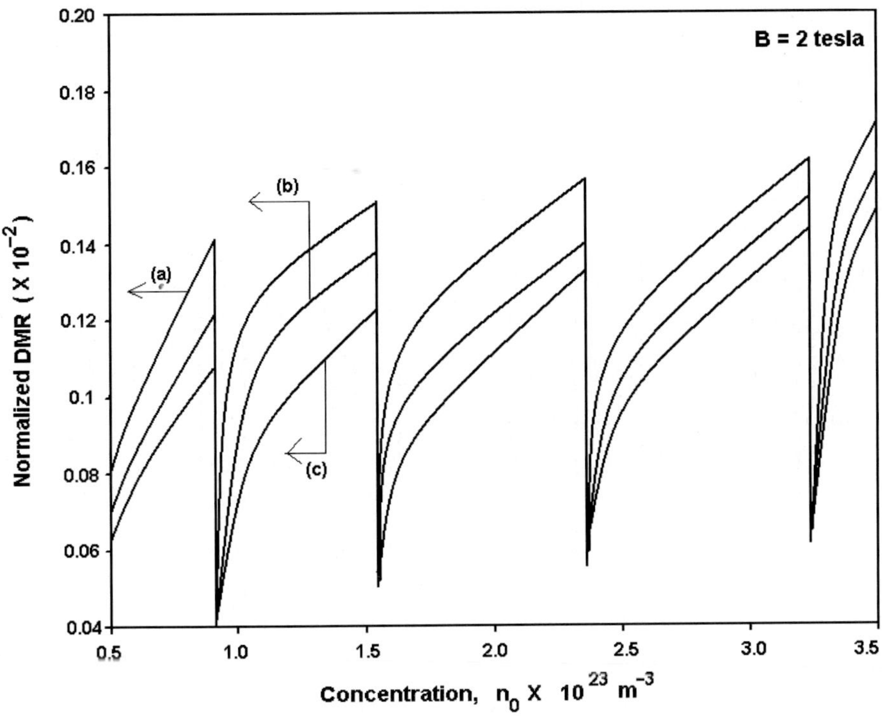

Fig. 3.28. The plot of the magneto DMR in (**a**) PbTe (**b**) n-PbSnTe and (**c**) n-$Pb_{1-x}Sn_xSe$ as a function of electron concentration in accordance with the model of Cohen

since the effects of collision broadening can usually be taken into account by an effective increase in temperature. Although in a more rigorous statement the many body effects should be considered along with the self-consistent procedure, the simplified analysis, as presented, exhibits the basic qualitative features of the DMR in degenerate materials having various band structures under the magnetic quantization with reasonable accuracy.

One important collateral understanding of this chapter is the fact that the EMM in tetragonal materials under magnetic quantization is a function of n, E_{FB}, θ and the electron spin due to the presence of crystal field splitting and the anisotropic spin orbit splitting constant in accordance with the generalized band model, whereas the EMM in the same compound in accordance with the three band model of Kane is independent of the magnetic quantum number and is a function of E_{FB} and the electron spin respectively. The spin dependence of the EMM for the three band model of Kane occurs due to the valance band spin orbit splitting constant and for the two band model of Kane the EMM is spin independent and is a function of E_{FB} only due to the presence of band non-parabolicity. For Bi, the EMM is a function of the electron spin, the magnetic quantum number, E_{FB} and other constants of the energy

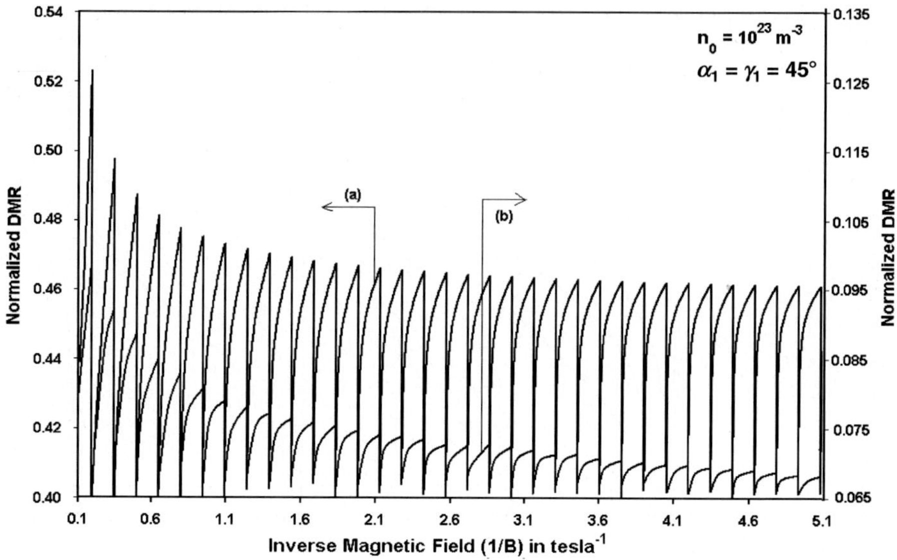

Fig. 3.29. The plot of the DMR in stressed n-InSb as a function of inverse quantizing magnetic field both in the presence and absence of stress as shown by the curves (**a**) and (**b**) respectively

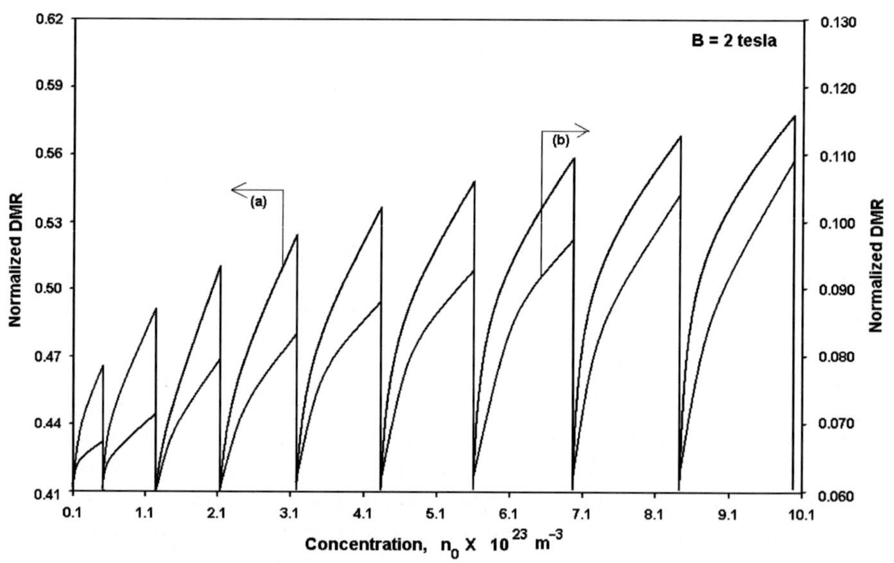

Fig. 3.30. The plot of the magneto DMR in stressed n-InSb as a function of electron concentration for both in the presence and absence of stress as shown by the curves (**a**) and (**b**) respectively (the magnetic field field lies in $k_x - k_z$ plane and $\gamma_1 = \alpha_1 = 45^0$, $\beta_1 = 0^0$)

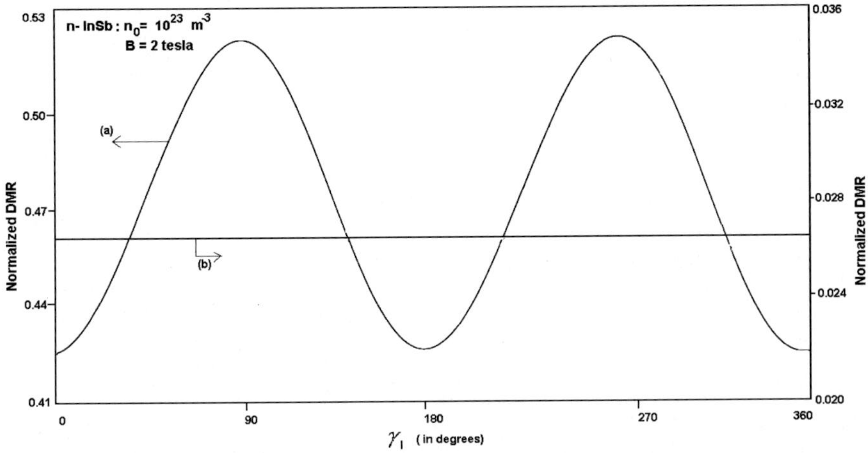

Fig. 3.31. The plot of the magneto DMR in bulk specimens of stressed n-InSb as a function of angular orientation of the quantizing magnetic field as shown in curve (**a**). The plot (**b**) refers to the stress-free case

spectrum for McClure-Choi due to the presence of band non-parabolicity only. In accordance with the Cohen model, the EMM is independent of spin and functions of n, E_{FD}, and other energy band constants of the said model again due to the presence of α, although the band non-parabolicity has been introduced in both the McClure–Choi and the Cohen models in two different ways. It is worth remarking to note that in stressed materials, the EMM is a function of the magnetic quantum number, in addition to E_{FB}, γ_1, and other system constants, due to the presence of stress only. The SdH period depends on the said variables for many dispersion relation characterizing different materials. Our suggestion for the experimental determination of the DMR of Chap. 1 is also valid under magnetic quantization. For the purpose of condensed presentation, the specific electron statistics for specific material having a particular electron energy spectrum and the corresponding Einstein relation under the magnetic quantization have been presented in Table 3.1.

3.4 Open Research Problems

R.3.1 Investigate the Einstein relation in the presence of an arbitrarily oriented alternating quantizing magnetic field in tetragonal semiconductors by including broadening and the electron spin. Study all the special cases for III–V, ternary and quaternary materials in this context.

R.3.2 Investigate the Einstein relations for all models of Bi, IV–VI, II–VI and stressed Kane type compounds in the presence of an arbitrarily oriented alternating quantizing magnetic field by including broadening and electron spin.

Table 3.1. The carrier statistics and the Einstein relation in tetragonal, III–V, ternary, quaternary, II–VI, all the models of Bismuth, IV–VI and stressed materials under the magnetic quantization

Type of materials	The carrier statistics	The Einstein relation for the diffusivity mobility ratio
1. Tetragonalcompounds	In accordance with the generalized magneto-dispersion relation (3.1) as formulated in this chapter	On the basis of (3.7),
	$$n_0 = \frac{g_{\mathrm{v}}\lvert e\rvert B}{2\pi^2\hbar}\sum_{n=0}^{n_{\max}}\left[T_{31}(n,E_{\mathrm{FB}})+T_{32}(n,E_{\mathrm{FB}})\right], \quad (3.7)$$	$$\frac{D}{\mu}=\frac{1}{\lvert e\rvert}\left[\frac{\sum_{n=0}^{n_{\max}}\left[T_{31}(n,E_{FB})+T_{32}(n,E_{FB})\right]}{\sum_{n=0}^{n_{\max}}\left[\{T_{31}(n,E_{FB})\}'+\{T_{32}(n,E_{FB})\}'\right]}\right] \quad (3.8)$$
	with electron spin In accordance with the dispersion relation (3.12) and in the absence of spin	with electron spin In accordance with (3.13) and in the absence of electron spin
	$$n_0 = \frac{g_v\lvert e\rvert B}{\pi^2\hbar}\sum_{n=0}^{n_{\max}}\left[T_{33}(n,E_{FB})+T_{34}(n,E_{FB})\right] \quad (3.13)$$	$$\frac{D}{\mu}=\frac{1}{\lvert e\rvert}\left[\frac{\sum_{n=0}^{n_{\max}}\left[T_{33}(n,E_{\mathrm{FB}})+T_{34}(n,E_{\mathrm{FB}})\right]}{\sum_{n=0}^{n_{\max}}\left[\{T_{33}(n,E_{\mathrm{FB}})\}'+\{T_{34}(n,E_{\mathrm{FB}})\}'\right]}\right] \quad (3.14)$$

2. III–V, ternary and quaternary compounds	In accordance with the magneto three band model of Kane as given by (3.15) which is a special case of (3.1) $$n_0 = \frac{g_v \lvert e \rvert B\sqrt{2m^*}}{2\pi^2\hbar^2} \sum_{n=0}^{n_{max}} [T_{35}(n, E_{FB}) + T_{36}(n, E_{FB})] \quad (3.20)$$ Equation (3.20) is a special case of (3.7) In the absence of spin, $$n_0 = \frac{g_v \lvert e \rvert B\sqrt{2m^*}}{\pi^2\hbar^2} \sum_{n=0}^{n_{max}} [T_{37}(n, E_{FB}) + T_{38}(n, E_{FB})] \quad (3.22)$$ Under the condition $\Delta \gg E_g$ for magneto two band model of Kane and in the presence of electron spin $$n_0 = \frac{g_v \lvert e \rvert B\sqrt{2m^*}}{2\pi^2\hbar^2} \sum_{n=0}^{n_{max}} [T_{39}(n, E_{FB}) + T_{310}(n, E_{FB})] \quad (3.29)$$	On the basis of (3.20) $$\frac{D}{\mu} = \frac{1}{\lvert e \rvert}\left[\frac{\sum_{n=0}^{n_{max}} [T_{35}(n, E_{FB}) + T_{36}(n, E_{FB})]}{\sum_{n=0}^{n_{max}} [\{T_{35}(n, E_{FB})\}' + \{T_{36}(n, E_{FB})\}']}\right] \quad (3.21)$$ with electron spin Equation (3.21) is a special case of (3.8) In the absence of spin, $$\frac{D}{\mu} = \frac{1}{\lvert e \rvert}\left[\frac{\sum_{n=0}^{n_{max}} [T_{37}(n, E_{FB}) + T_{38}(n, E_{FB})]}{\sum_{n=0}^{n_{max}} [\{T_{37}(n, E_{FB})\}' + \{T_{38}(n, E_{FB})\}']}\right] \quad (3.23)$$ Under the condition $\Delta \gg E_g$, from (3.29) and in the presence of electron spin $$\frac{D}{\mu} = \frac{1}{\lvert e \rvert}\left[\frac{\sum_{n=0}^{n_{max}} [T_{39}(n, E_{FB}) + T_{310}(n, E_{FB})]}{\sum_{n=0}^{n_{max}} [\{T_{39}(n, E_{FB})\}' + \{T_{310}(n, E_{FB})\}']}\right] \quad (3.30)$$

(Continued)

Table 3.1. *Continued*

Type of materials	The carrier statistics	The Einstein relation for the diffusivity mobility ratio
	In the absence of spin, $$n_0=\frac{g_v\lvert e\rvert B\sqrt{2m^*}}{\pi^2\hbar^2}\sum_{n=0}^{n_{\max}}\left[T_{311}(n,E_{FB})+T_{312}(n,E_{FB})\right] (3.31)\quad (3.31)$$	In the absence of spin and using (3.31), $$\frac{D}{\mu}=\frac{1}{\lvert e\rvert}\left[\frac{\sum_{n=0}^{n_{\max}}\left[T_{311}(n,E_{FB})+T_{312}(n,E_{FB})\right]}{\sum_{n=0}^{n_{\max}}\left[\{T_{311}(n,E_{FB})\}'+\{T_{312}(n,E_{FB})\}'\right]}\right]\quad (3.32)$$
	Equation (3.31) is a special case of (3.29) and is valid for the two band model of Kane Under the constraint $\Delta \gg E_g$ together with the condition $\alpha E_{FB} \ll 1$ (in the presence of spin) $$n_0=\frac{g_vN_C\theta_{B1}}{2}\left[\sum_{n=0}^{n_{\max}}\frac{1}{\sqrt{a_{01}}}\left[\left(1+\frac{3}{2}\alpha b_{01}\right)\times F_{-\frac{1}{2}}(\eta_B)+\frac{3}{4}\alpha k_BTF_{\frac{1}{2}}(\eta_B)\right]\right]\quad (3.38a)$$	Equation (3.32) is a special case of (3.30) and is valid for the two band model of Kane Under the constraint $\Delta \gg E_g$ together with the condition $\alpha E_{FB} \ll 1$ (in the presence of spin) $$\frac{D}{\mu}=\frac{k_BT}{\lvert e\rvert}\times\left[\frac{\sum_{n=0}^{n_{\max}}\frac{1}{\sqrt{a_{01}}}\left[\left(1+\frac{3}{2}\alpha b_{01}\right)F_{-\frac{1}{2}}(\eta_B)+\frac{3}{4}\alpha k_BTF_{\frac{1}{2}}(\eta_B)\right]}{\sum_{n=0}^{n_{\max}}\frac{1}{\sqrt{a_{01}}}\left[\left(1+\frac{3}{2}\alpha b_{01}\right)F_{-\frac{3}{2}}(\eta_B)+\frac{3}{4}\alpha k_BTF_{-\frac{1}{2}}(\eta_B)\right]}\right]\quad (3.38b)$$
	In the absence of spin, $$n_0=g_vN_C\theta_{B1}\left[\sum_{n=0}^{n_{\max}}\frac{1}{\sqrt{a_{01}^*}}\left[\left(1+\frac{3}{2}\alpha b_{01}^*\right)\times F_{\frac{-1}{2}}(\bar{\eta}_{B_1})+\frac{3}{4}\alpha k_BTF_{\frac{1}{2}}(\bar{\eta}_{B_1})\right]\right]\quad (3.39a)$$	In the absence of spin, $$\frac{D}{\mu}=\frac{k_BT}{\lvert e\rvert}\times\left[\frac{\sum_{n=0}^{n_{\max}}\frac{1}{\sqrt{a_{01}^*}}\left[\left(1+\frac{3}{2}\alpha b_{01}^*\right)F_{\frac{-1}{2}}(\bar{\eta}_{B_1})+\frac{3}{4}\alpha k_BTF_{\frac{1}{2}}(\bar{\eta}_{B_1})\right]}{\sum_{n=0}^{n_{\max}}\frac{1}{\sqrt{a_{01}^*}}\left[\left(1+\frac{3}{2}\alpha b_{01}^*\right)F_{\frac{-3}{2}}(\bar{\eta}_{B_1})+\frac{3}{4}\alpha k_BTF_{\frac{-1}{2}}(\bar{\eta}_{B_1})\right]}\right]\quad (3.39b)$$

	For $E_g \to \infty$ (in the presence of spin), $$n_0 = \frac{g_v N_C \theta_{B1}}{2} \sum_{n=0}^{n_{max}} F_{\frac{-1}{2}}(\bar{\eta}_B) \quad (3.44)$$ Equation (3.44) is a special case of (3.38a) and is valid for parabolic energy bands. In the absence of spin, $$n_0 = g_v N_C \theta_{B1} \sum_{n=0}^{n_{max}} F_{\frac{-1}{2}}(\eta_{B1}) \quad (3.46)$$	For $E_g \to \infty$ (in the presence of spin), $$\frac{D}{\mu} = \frac{k_B T}{\lvert e \rvert} \left[\sum_{n=0}^{n_{max}} F_{\frac{-1}{2}}(\bar{\eta}_B) \right] \left[\sum_{n=0}^{n_{max}} F_{\frac{-3}{2}}(\bar{\eta}_B) \right]^{-1} \quad (3.45)$$ Equation (3.45) is a special case of (3.38b). In the absence of spin, $$\frac{D}{\mu} = \frac{k_B T}{\lvert e \rvert} \left[\sum_{n=0}^{n_{max}} F_{\frac{-1}{2}}(\eta_{B1}) \right] \left[\sum_{n=0}^{n_{max}} F_{\frac{-3}{2}}(\eta_{B1}) \right]^{-1} \quad (3.47)$$
3. II–VI compounds	In the presence of spin, $$n_0 = \frac{g_v \lvert e \rvert B \sqrt{2\pi m_{\parallel}^* k_B T}}{h^2} \sum_{n=0}^{n_{max}} F_{\frac{-1}{2}}(\theta_3) \quad (3.56)$$	$$\frac{D}{\mu} = \frac{k_B T}{\lvert e \rvert} \left[\frac{\sum_{n=0}^{n_{max}} F_{\frac{-1}{2}}(\theta_3)}{\sum_{n=0}^{n_{max}} F_{\frac{-3}{2}}(\theta_3)} \right] \quad (3.57)$$

(Continued)

Table 3.1. *Continued*

Type of materials	The carrier statistics	The Einstein relation for the diffusivity mobility ratio
	In the absence of spin, $$n_0 = \frac{2g_v \lvert e\rvert B\sqrt{2\pi m_{\parallel}^{*} k_B T}}{h^2}\sum_{n=0}^{n_{max}} F_{\frac{-1}{2}}(\eta_{B2}) \quad (3.58)$$	In the absence of spin, $$\frac{D}{\mu} = \frac{k_B T}{\lvert e\rvert}\left[\frac{\sum_{n=0}^{n_{max}} F_{\frac{-1}{2}}(\eta_{B_2})}{\sum_{n=0}^{n_{max}} F_{\frac{-3}{2}}(\eta_{B_2})}\right] \quad (3.59)$$
4. Bi	(a) The McClure and Choi model: In the presence of spin and magnetic field is along z-axis, $$n_0 = \frac{g_v \lvert e\rvert B\sqrt{2m_3}}{2\pi^2\hbar^2}\sum_{n=0}^{n_{max}}\left[T_{313}(n, E_{FB}) + T_{314}(n, E_{FB})\right] \quad (3.68)$$	In the presence of the spin and magnetic field is along z-axis, $$\frac{D}{\mu} = \frac{1}{\lvert e\rvert}\left[\frac{\sum_{n=0}^{n_{max}}\left[T_{313}(n, E_{FB}) + T_{314}(n, E_{FB})\right]}{\sum_{n=0}^{n_{max}}\left[\{T_{313}(n, E_{FB})\}' + \{T_{314}(n, E_{FB})\}'\right]}\right] \quad (3.69)$$
	Under the condition $\alpha \to 0$, $$n_0 = \frac{g_v N_{C_2}\theta_{B3}}{2}\sum_{n=0}^{n_{max}} F_{\frac{-1}{2}}(\eta_{B_3}) \quad (3.70)$$	Under the condition $\alpha \to 0$, $$\frac{D}{\mu} = \frac{k_B T}{\lvert e\rvert}\left[\frac{\sum_{n=0}^{n_{max}} F_{\frac{-1}{2}}(\eta_{B_3})}{\sum_{n=0}^{n_{max}} F_{\frac{-3}{2}}(\eta_{B_3})}\right] \quad (3.71)$$

The same model without spin, $$n_0 = \frac{g_v \lvert e \rvert B\sqrt{2m_3}}{\pi^2\hbar^2} \sum_{n=0}^{n_{\max}} \left[T_{315}(n, E_{FB}) + T_{316}(n, E_{FB})\right] \quad (3.72)$$	The same model without spin, $$\frac{D}{\mu} = \frac{1}{\lvert e \rvert}\left[\frac{\sum_{n=0}^{n_{\max}} \left[T_{315}(n, E_{FB}) + T_{316}(n, E_{FB})\right]}{\sum_{n=0}^{n_{\max}} \left[\{T_{315}(n, E_{FB})\}' + \{T_{316}(n, E_{FB})\}'\right]}\right] \quad (3.73)$$
The same model without spin and the magnetic field is along y-axis $$n_0 = \frac{g_v \lvert e \rvert B}{\sqrt{2}\pi^2\hbar^2} \sum_{n=0}^{n_{\max}} \left[T_{317}(n, E_{FB}) + T_{318}(n, E_{FB})\right] \quad (3.75)$$	The same model without spin and the magnetic field is along y-axis $$\frac{D}{\mu} = \frac{1}{\lvert e \rvert}\left[\frac{\sum_{n=0}^{n_{\max}} \left[T_{317}(n, E_{FB}) + T_{318}(n, E_{FB})\right]}{\sum_{n=0}^{n_{\max}} \left[\{T_{317}(n, E_{FB})\}' + \{T_{318}(n, E_{FB})\}'\right]}\right] \quad (3.76)$$
(b) The Cohen model: With spin, and the magnetic field is along z-axis $$n_0 = \frac{g_v \lvert e \rvert B\sqrt{2m_3}}{2\pi^2\hbar^2} \sum_{n=0}^{n_{\max}} \left[T_{319}(n, E_{FB}) + T_{320}(n, E_{FB})\right] \quad (3.82)$$	With spin, and the magnetic field is along z-axis $$\frac{D}{\mu} = \frac{1}{\lvert e \rvert}\left[\frac{\sum_{n=0}^{n_{\max}} \left[T_{319}(n, E_{FB}) + T_{320}(n, E_{FB})\right]}{\sum_{n=0}^{n_{\max}} \left[\{T_{319}(n, E_{FB})\}' + \{T_{320}(n, E_{FB})\}'\right]}\right] \quad (3.83)$$

(Continued)

Table 3.1. *Continued*

Type of materials	The carrier statistics	The Einstein relation for the diffusivity mobility ratio
	The same model without spin and the quantizing magnetic field is along y-axis	The same model without spin and the quantizing magnetic field is along y-axis
	$$n_0 = \frac{g_v \lvert e \rvert B}{\sqrt{2}\pi^2\hbar^2} \sum_{n=0}^{n_{\max}} [T_{319}(n, E_{FB}) + T_{320}(n, E_{FB})] \quad (3.85)$$	$$\frac{D}{\mu} = \frac{1}{\lvert e \rvert} \left[\frac{\sum_{n=0}^{n_{\max}} [T_{321}(n, E_{FB}) + T_{322}(n, E_{FB})]}{\sum_{n=0}^{n_{\max}} [\{T_{321}(n, E_{FB})\}' + \{T_{322}(n, E_{FB})\}']} \right] \quad (3.86)$$
	(d) The Lax model: In the presence of spin	In the presence of spin
	$$n_0 = \frac{g_v \lvert e \rvert B\sqrt{2m_3}}{2\pi^2\hbar^2} \sum_{n=0}^{n_{\max}} [T_{323}(n, E_{FB}) + T_{324}(n, E_{FB})], \quad (3.91)$$	$$\frac{D}{\mu} = \frac{1}{\lvert e \rvert} \left[\frac{\sum_{n=0}^{n_{\max}} [T_{323}(n, E_{FB}) + T_{324}(n, E_{FB})]}{\sum_{n=0}^{n_{\max}} [\{T_{323}(n, E_{FB})\}' + \{T_{324}(n, E_{FB})\}']} \right] \quad (3.92)$$

	In the absence of spin $$n_0 = \frac{g_v \lvert e\rvert B\sqrt{2m_3}}{\pi^2\hbar^2} \sum_{n=0}^{n_{\max}} \left[T_{325}(n, E_{FB}) + T_{326}(n, E_{FB})\right] \quad (3.93)$$	In the absence of spin $$\frac{D}{\mu} = \frac{1}{\lvert e\rvert}\left[\frac{\sum_{n=0}^{n_{\max}} \left[T_{325}(n, E_{FB}) + T_{326}(n, E_{FB})\right]}{\sum_{n=0}^{n_{\max}} \left[\{T_{325}(n, E_{FB})\}' + \{T_{326}(n, E_{FB})\}'\right]}\right] \quad (3.94)$$
	(e) The parabolic ellipsoidal model: The expression of the electron statistics is the special case of the models of the McClure and Choi, the Cohen and the Lax respectively	The expression of the Einstein relation is the special case of the models of the McClure and Choi, the Cohen and the Lax respectively
5. IV–VI compounds	The expressions of n_0 in this case are given by (3.82) and (3.85) in which the constants of the energy band spectrum correspond to the carriers of the IV–VI semiconductors	The expressions of DMR in this case are given by (3.83) and (3.86) in which the constants of the energy band spectrum correspond to the carriers of the IV–VI semiconductors
6. Stressed compounds	$$n_0 = \frac{g_v \lvert e\rvert B}{\pi^2\hbar} \sum_{n=0}^{n_{\max}} \left[T_{327}(n, E_{FB}) + T_{328}(n, E_{FB})\right] \quad (3.101)$$	$$\frac{D}{\mu} = \frac{1}{\lvert e\rvert}\left[\frac{\sum_{n=0}^{n_{\max}} \left[T_{327}(n, E_{FB}) + T_{328}(n, E_{FB})\right]}{\sum_{n=0}^{n_{\max}} \left[\{T_{327}(n, E_{FB})\}' + \{T_{328}(n, E_{FB})\}'\right]}\right] \quad (3.102)$$

R.3.3 Investigate the Einstein relation for all the materials as stated in R.2.1 of Chap. 2 in the presence of an arbitrarily oriented alternating quantizing magnetic field by including broadening and electron spin.

Allied Research Problems

R.3.4 Investigate the EMM for all the materials as stated in R.2.1 of Chap. 2 in the presence of an arbitrarily oriented alternating quantizing magnetic field by including broadening and electron spin.

R.3.5 Investigate in details, the Debye screening length, the carrier contribution to the elastic constants, the heat capacity, the activity coefficient and the plasma frequency for all the materials covering all the cases of problems from R.3.1 to R.3.3.

R.3.6 Investigate in details, the mobility for elastic and inelastic scattering mechanisms for all the materials covering all the cases of problems from R.3.1 to R.3.3.

R.3.7 Investigate the various transport coefficients in details for all the materials covering all the cases of problems from R.3.1 to R.3.3.

R.3.8 Investigate the dia and paramagnetic susceptibilities in details for all the materials covering all the appropriate research problems of this chapter.

References

1. B.R. Nag, *Electron Transport in Compound Semiconductors* (Springer-Verlag, Germany, 1980); B.K. Ridley, *Quantum Processes in Semiconductors*, 4th edn. (Oxford Publications, Oxford, 1999); J.H. Davis, Physics of Low Dimensional Semiconductors (Cambridge University Press, UK, 1998); M. Schaden, K.F. Zhao, Z. Wu, Phys. Rev. A **76**, 062502 (2007); T. Kawarabayashi, T. Ohtsuki, Phys. Rev. B **51**, 10897 (1995); B. Laikhtman, Phys. Rev. Lett. **72**, 1060 (1994); A. Houghton, J.R. Senna, S.C. Ying, Phys. Rev. B **25**, 6468 (1982)
2. L. Landau, E.M. Liftshitz, *Statistical Physics, Part-II* (Pergamon Press, Oxford, 1980)
3. P.R. Wallace, Phys. Stat. Sol. (b) **92**, 49 (1979)
4. S.J. Adachi, J. Appl. Phys. **58**, R11 (1985)
5. R. Dornhaus, G. Nimtz, *Springer Tracts in Modern Physics*, vol. 78 (Springer, Berlin, 1976)
6. W. Zawadzki, *Handbook of Semiconductor Physics*, ed. by W. Paul, vol 1 (North Holland, Amsterdam, 1982), p. 719
7. I.M. Tsidilkovski, Cand. Thesis Leningrad University SSR (1955)
8. F.G. Bass, I.M. Tsidilkovski, Ivz. Acad. Nauk Azerb SSR **10**, 3 (1966)
9. I.M. Tsidilkovski, *Band Structures of Semiconductors* (Pergamon Press, London, 1982); K.P. Ghatak, S. Bhattacharya, S.K. Biswas, A. Dey, A.K. Dasgupta, Phys. Scr. **75**, 820 (2007)
10. P.K. Charkaborty, G.C. Dutta, K.P. Ghatak, Phys. Scr. **68**, 368 (2003); K.P. Ghatak, S.N. Biswas, Nonlin. Opt. Quant. Opts. **4**, 347 (1993)

11. A.N. Chakravarti, A.K. Choudhury, K.P. Ghatak, S. Ghosh, A. Dhar, Appl. Phys. **25**, 105 (1981); K.P. Ghatak, M. Mondal, Z. F. Physik B **B69**, 471 (1988); M. Mondal, K.P. Ghatak, Phys. Lett. **131A**, 529 (1988)
12. K.P. Ghatak, A. Ghoshal, B. Mitra, Nouvo Cimento **14D**, 903 (1992)
13. B. Mitra, A. Ghoshal, K.P. Ghatak, Nouvo Cimento D **12D**, 891 (1990); K.P. Ghatak, S.N. Biswas, Nonlin. Opt. Quant. Opts. **12**, 83 (1995)
14. B. Mitra, K.P. Ghatak, Solid State Electron. **32**, 177 (1989); K.P. Ghatak, S.N. Biswas, Proc. SPIE **1484**, 149 (1991); M. Mondal, K.P. Ghatak, *Graphite Intercalation Compounds: Science and Applications, MRS Proceedings*, ed. by M. Endo, M.S. Dresselhaus, G. Dresselhaus, MRS Fall Meeting, **EA** 16, 173 (1988)
15. M. Mondal, N. Chattapadhyay, K.P. Ghatak, J. Low Temp. Phys. **66**, 131 (1987); A.N. Chakravarti, K.P. Ghatak, K.K. Ghosh, S.Ghosh, A. Dhar, Z. Physik B. **47**, 149 (1982)
16. V.K. Arora, H. Jeafarian, Phys. Rev. B. **13** 4457 (1976)
17. M. Singh, P.R. Wallace, S.D. Jog, J.J. Erushanov, J. Phys. Chem. Solids. **45**, 409 (1984)
18. W. Zawadski, Adv. Phys. **23**, 435 (1974)
19. K.P. Ghatak, M. Mondal, Z. Fur Nature A **41A**, 881 (1986)
20. T. Ando, A.H. Fowler, F. Stern, Rev. Modern Phys. **54**, 437 (1982)
21. R.K. Pathria, *Statistical Mechanics*, 2nd edn. (Oxford, Butterworth-Heinmann, 1996)
22. R.A. Smith, *Wave Mechanics of Crystalline Solids* (Chapman & Hall, London, 1969), p. 437
23. A.N. Chakravarti, B.R. Nag, Int. J. Elect. **37**, 281 (1974)
24. P.N. Butcher, A.N. Chakravarti, S. Swaminathan, Phys. Stat. Sol. (a), **25**, K47 (1974); B.A. Aronzon, E.Z. Meilikhov, Phys. Stat. Sol. (a) **19**, 313 (1973); K.P. Ghatak, S. Bhattacharya, D. De, P.K. Bose, S.N. Mitra, S. Pahari, Phys. B, **403**, 2930 (2008)
25. C.C. Wu, C.J. Lin, J. Low Temp. Phys. **57**, 469 (1984); M.H. Chen, C.C. Wu, C.J. Lin, J. Low Temp. Phys. **55**, 127 (1984)

4

The Einstein Relation in Compound Semiconductors Under Crossed Fields Configuration

4.1 Introduction

The influence of crossed electric and quantizing magnetic fields on the transport properties of semiconductors having various band structures has relatively less investigated as compared with the corresponding magnetic quantization, although, the cross fields are fundamental with respect to the addition of new physics and the related experimental findings. It is well known that in the presence of an electric field (E_o) along the x-axis and the quantizing magnetic field (B) along the z-axis, the dispersion relations of the conduction electrons in semiconductors become modified and the electron moves in both the z and y directions. The motion along the y-direction is purely due to the presence of E_0 along the x-axis and in the absence of an electric field, the effective electron mass along the y-axis tends to infinity which indicates the fact that the electron motion along the y-axis is forbidden. The effective electron mass of the isotropic, bulk semiconductors having parabolic energy bands exhibits mass anisotropy in the presence of cross fields and this anisotropy depends on the electron energy, the magnetic quantum number, the electric and the magnetic fields respectively, although, the effective electron mass along the z-axis is a constant quantity.

In 1966, Zawadzki and Lax [1] formulated the electron dispersion law for III–V semiconductors in accordance with the two band model of Kane under cross fields configuration which has generated the interest to study this particular topic of semiconductor science in general [2–14].

In Sect. 4.2.1, theoretical background, the Einstein relation in tetragonal materials in the presence of crossed electric and quantizing magnetic fields has been investigated by formulating the electron dispersion relation. Section 4.2.2 reflects the study of the Einstein relation in III–V, ternary and quaternary compounds as a special case of Sect. 4.1. In the same section the well known result for the Einstein relation in relatively wide gap materials in the absence of the electric field as a limiting case has been discussed for the purpose of compatibility. Section 4.2.3 contains the study of the Einstein relation for the

II–VI semiconductors in the present case. In Sect. 4.2.4, the DMR under cross field configuration in Bismuth has been investigated in accordance with the McClure and Choi, the Cohen, the Lax nonparabolic ellipsoidal and the parabolic ellipsoidal models respectively. In Sect. 4.2.5, the study of the Einstein relation in IV–VI materials has been presented. In Sect. 4.2.6, the DMR for the stressed Kane type semiconductors has been investigated. Section 4.2.7 contains the result and discussions for this chapter.

4.2 Theoretical Background

4.2.1 Tetragonal Materials

Equation (2.2) of Chap. 2 can be expressed as

$$U(E) = \frac{p_s^2}{2M_\perp} + \frac{p_z^2}{2M_\parallel} V(E), \tag{4.1}$$

where

$$\begin{aligned} U(E) \equiv & \Bigg[E(1+\alpha E) \Bigg[(E+E_g)(E+E_g+\Delta_\parallel) + \delta\left(E+E_g+\frac{2}{3}\Delta_\parallel\right) \\ & + \frac{2}{9}(\Delta_\parallel^2 - \Delta_\perp^2) \Bigg] \Bigg], \\ & \times \Bigg[(E+E_g) \times \left(E+E_g+\frac{2}{3}\Delta_\parallel\right) + \delta\left(E+E_g+\frac{1}{3}\Delta_\parallel\right) \\ & + \frac{1}{9}(\Delta_\parallel^2 - \Delta_\perp^2) \Bigg]^{-1}, \\ & p_s = \hbar k_s, \quad M_\parallel = \frac{m_\parallel^*(E_g + \frac{2}{3}\Delta_\parallel)}{(E_g+\Delta_\parallel)}, \\ & M_\perp = \frac{m_\perp^*(E_g + \frac{2}{3}\Delta_\perp)}{(E_g+\Delta_\perp)}, \quad p_z = \hbar k_z \end{aligned}$$

and

$$\begin{aligned} V(E) \equiv & \left[(E+E_g)\left(E+E_g+\frac{2}{3}\Delta_\parallel\right) \right] \Bigg[(E+E_g)\left(E+E_g+\frac{2}{3}\Delta_\parallel\right) \\ & + \delta\left(E+E_g+\frac{1}{3}\Delta_\parallel\right) + \frac{1}{9}(\Delta_\parallel^2 - \Delta_\perp^2) \Bigg]^{-1}, \end{aligned}$$

We know from electromagnetic theory that

$$\vec{B} = \nabla \times \vec{A}, \tag{4.2}$$

where $\overrightarrow{A}$ is the vector potential. In the presence of quantizing magnetic field B along z direction, (4.2) assumes the form

$$0\widehat{i}+0\widehat{j}+B\widehat{k}=\begin{vmatrix} \widehat{i} & \widehat{j} & \widehat{k} \\ \frac{\partial}{\partial x} & \frac{\partial}{\partial y} & \frac{\partial}{\partial z} \\ A_x & A_y & A_z \end{vmatrix},$$

where $\widehat{i}, \widehat{j}$ and $\widehat{k}$ are orthogonal triads. Thus, we can write

$$\begin{aligned} \frac{\partial A_z}{\partial y} - \frac{\partial A_y}{\partial z} &= 0, \\ \frac{\partial A_x}{\partial z} - \frac{\partial A_z}{\partial x} &= 0, \\ \frac{\partial A_y}{\partial x} - \frac{\partial A_x}{\partial y} &= B. \end{aligned} \tag{4.3}$$

This particular set of equations is satisfied for $A_x = 0$, $A_y = Bx$ and $A_z = 0$.

Therefore in the presence of the electric field E_0 along the x axis and the quantizing magnetic field B along the z axis for the present case following [1], one can approximately write

$$U(E) + |e|\, E_o \widehat{x} \rho(E) = \frac{\widehat{p}_x^2}{2M_\perp} + \frac{\left(\widehat{p}_y - |e|\, B\widehat{x}\right)^2}{2M_\perp} + \frac{\widehat{p}_z^2}{2a(E)}, \tag{4.4}$$

where

$$\rho(E) \equiv \frac{\partial}{\partial E}[U(E)] \text{ and } a(E) \equiv M_\| [V(E)]^{-1}.$$

Let us define the operator $\widehat{\theta}$ as

$$\widehat{\theta} = -\widehat{p}_y + |e|\, B\widehat{x} - \frac{M_\perp E_0 \rho(E)}{B}. \tag{4.5}$$

Eliminating the operator $\widehat{x}$, between (4.4) and (4.5), the dispersion relation of the conduction electron in tetragonal semiconductors in the presence of cross fields configuration is given by

$$\begin{aligned} U(E) = \Bigg[& \left((n + \frac{1}{2})\hbar\omega_{01} \right) + \left(\frac{[\hbar k_z(E)]^2}{2a(E)} \right) - \left(\frac{E_0 \hbar k_y \rho(E)}{B} \right) \\ & - \left(\frac{M_\perp \rho^2(E) E_0^2}{2B^2} \right) \Bigg], \end{aligned} \tag{4.6}$$

where

$$\omega_{01} \equiv \frac{|e|\, B}{M_\perp}.$$

Therefore the EMM's along the z and y directions can, respectively, be expressed as

$$m_z^*\left(\bar{E}_{\mathrm{FB}},n,E_0,B\right)=$$
$$\times\left[\left\{a\left(\bar{E}_{\mathrm{FB}}\right)\right\}'\left\{\left\{U\left(\bar{E}_{\mathrm{FB}}\right)-\left(n+\frac{1}{2}\right)\omega_{01}+\frac{M_\perp\left[\rho\left(\bar{E}_{\mathrm{FB}}\right)\right]^2E_0^2}{2B^2}\right\}+a\left(\bar{E}_{\mathrm{FB}}\right)\right.\right.$$
$$\left.\left.\times\left[\left\{U\left(\bar{E}_{\mathrm{FB}}\right)\right\}'+\frac{M_\perp E_0^2\rho\left(\bar{E}_{\mathrm{FB}}\right)\left\{\rho\left(\bar{E}_{\mathrm{FB}}\right)\right\}'}{B^2}\right]\right\}\right], \tag{4.7}$$

$$m_y^*\left(\bar{E}_{\mathrm{FB}},n,E_0,B\right)=\left(\frac{B}{E_0}\right)^2\left(\frac{1}{\rho(\bar{E}_{\mathrm{FB}})}\right)$$
$$\times\left[U\left(\bar{E}_{\mathrm{FB}}\right)-\left(n+\frac{1}{2}\right)\hbar\omega_{01}+\frac{M_\perp\left[\rho\left(\bar{E}_{\mathrm{FB}}\right)\right]^2E_0^2}{2B^2}\right]$$
$$\times\left[\left[\rho(\bar{E}_{\mathrm{FB}})\right]^{-1}\left[\left\{U\left(\bar{E}_{\mathrm{FB}}\right)\right\}'+\frac{M_\perp\rho\left(\bar{E}_{\mathrm{FB}}\right)\left\{\rho\left(\bar{E}_{\mathrm{FB}}\right)\right\}'E_0^2}{B^2}\right]\right. \tag{4.8}$$
$$-\left[\left[\rho\left(\bar{E}_{\mathrm{FB}}\right)\right]^{-2}\left\{\rho\left(\bar{E}_{\mathrm{FB}}\right)\right\}'\left[U\left(\bar{E}_{\mathrm{FB}}\right)-\left(n+\frac{1}{2}\right)\hbar\omega_{01}\right.\right.$$
$$\left.\left.\left.+\frac{M_\perp\left[\rho\left(\bar{E}_{\mathrm{FB}}\right)\right]^2E_0^2}{2B^2}\right]\right]\right],$$

where $\bar{E}_{\mathrm{FB}}$ is the Fermi energy in the presence of cross-fields configuration as measured from the edge of the conduction band in the vertically upward direction in the absence of any quantization.

When $E_0\to0$, $m_y^*\left(\bar{E}_{\mathrm{FB}},n,E_0,B\right)\to\infty$, which is a physically justified result. The dependence of the EMM along the y direction on the Fermi energy, electric field, magnetic field, and the magnetic quantum number is an intrinsic property of cross fields. Another characteristic feature of the cross field is that various transport coefficients will be sampled dimension dependent. These conclusions are valid for even isotropic parabolic energy bands and cross fields introduce the index dependent anisotropy in the effective mass.

The Landau energy $(\bar{E}_n)$ can be written as

$$U\left(\bar{E}_n\right)=\left(n+\frac{1}{2}\right)\hbar\omega_{01}-\frac{M_\perp\left[\rho(\bar{E}_n)\right]^2E_0^2}{2B^2}. \tag{4.9}$$

The SdH period can be expressed through the equation

$$\Delta\left(\frac{1}{B}\right)=\frac{1}{B_{n+1}}-\frac{1}{B_n}. \tag{4.10}$$

Equation (4.9) at Fermi energy $\bar{E}_{\mathrm{FB}}$ generates a cubic equation in B, the real single root of which when combined with (4.10) will generate the SdH period. Thus, we observe that like the EMM's, the SdH period in the presence of cross-fields configuration depends on the Fermi energy, the magnetic quantum number, the electric field, the magnetic field, and the constants of the energy band spectrum respectively.

The formulation of DMR requires the expression of the electron concentration which can, in general, be written excluding the electron spin as

$$n_o = \frac{-g_{\mathrm{v}}}{L_x \pi^2} \sum_{n=0}^{n_{\max}} \int_{\bar{E}_0}^{\infty} I(E) \frac{\partial f_0}{\partial E} \mathrm{d}E, \tag{4.11}$$

where L_x is the sample length along the x direction, $\bar{E}_0$ is determined by the equation

$$I\left(\bar{E}_0\right) = 0,$$

where

$$I(E) = \int_{x_l(E)}^{x_h(E)} k_z(E)\, \mathrm{d}k_y, \tag{4.12}$$

in which,

$$x_l(E) \equiv \frac{-E_0 M_\perp \rho(E)}{\hbar B} \text{ and } x_h(E) \equiv \frac{|e| B L_x}{\hbar} + x_l(E).$$

Using (4.6) and (4.12) we get

$$I(E) = \frac{2}{3}\left[\frac{B\sqrt{2a(E)}}{\hbar^2 E_0 \rho(E)}\left[\left[U(E) - \left(n+\frac{1}{2}\right)\frac{\hbar|e|B}{M_\perp} + |e| E_0 L_x \rho(E) - \frac{M_\perp E_0^2 [\rho(E)]^2}{2B^2}\right]^{\frac{3}{2}}\right.\right.$$
$$\left.\left. - \left[U(E) - \left(n+\frac{1}{2}\right)\frac{\hbar|e|B}{M_\perp} - \frac{M_\perp E_0^2 [\rho(E)]^2}{2B^2}\right]^{\frac{3}{2}}\right]\right]. \tag{4.13}$$

Combining (4.11) and (4.13), the electron concentration is given by

$$n_0 = \left(\frac{2g_{\mathrm{v}} B\sqrt{2}}{3L_x \pi^2 \hbar^2 E_0}\right) \sum_{n=0}^{n_{\max}} \left[T_{41}\left(n, \bar{E}_{\mathrm{FB}}\right) + T_{42}\left(n, \bar{E}_{\mathrm{FB}}\right)\right], \tag{4.14}$$

where

$$T_{41}\left(n, \bar{E}_{\mathrm{FB}}\right) \equiv \frac{\sqrt{a\left(\bar{E}_{\mathrm{FB}}\right)}}{\rho\left(\bar{E}_{\mathrm{FB}}\right)}$$
$$\left[\left[U\left(\bar{E}_{\mathrm{FB}}\right) - \left(n+\frac{1}{2}\right)\frac{\hbar|e|B}{M_\perp} + |e| E_0 L_x \rho\left(\bar{E}_{\mathrm{FB}}\right) - \frac{M_\perp {E_0}^2 \left[\rho\left(\bar{E}_{\mathrm{FB}}\right)\right]^2}{2B^2}\right]^{3/2}\right.$$

$$-\left[U\left(\bar{E}_{\mathrm{FB}}\right)-\left(n+\frac{1}{2}\right)\frac{\hbar\left|e\right|B}{M_{\perp}}-\frac{M_{\perp}{E_0}^2\left[\rho\left(\bar{E}_{\mathrm{FB}}\right)\right]^2}{2B^2}\right]^{3/2}\Bigg] \quad \text{and}$$

$$T_{42}\left(n,\bar{E}_{\mathrm{FB}}\right)\equiv\sum_{r=1}^{s}\left[L\left(r\right)T_{41}\left(n,\bar{E}_{\mathrm{FB}}\right)\right],$$

Thus combining (4.14) and (1.11), the DMR in this case can be written as

$$\frac{D}{\mu}=\frac{1}{|e|}\left[\frac{\sum\limits_{n=0}^{n_{\max}}\left[T_{41}\left(n,\bar{E}_{\mathrm{FB}}\right)+T_{42}\left(n,\bar{E}_{\mathrm{FB}}\right)\right]}{\sum\limits_{n=0}^{n_{\max}}\left[\left\{T_{41}\left(n,\bar{E}_{\mathrm{FB}}\right)\right\}'+\left\{T_{42}\left(n,\bar{E}_{\mathrm{FB}}\right)\right\}'\right]}\right]. \tag{4.15}$$

4.2.2 Special Cases for III–V, Ternary and Quaternary Materials

(a) Under the conditions $\delta=0$, $\Delta_{\|}=\Delta_{\perp}=\Delta$ and $m_{\|}^{*}=m_{\perp}^{*}=m^{*}$, (4.6) assumes the form

$$\gamma\left(E\right)=\left(n+\frac{1}{2}\right)\hbar\omega_0+\frac{\left[\hbar k_z\left(E\right)\right]^2}{2m^*}-\frac{E_0}{B}\hbar k_y\left\{\gamma\left(E\right)\right\}'-\frac{m^*{E_0}^2\left[\left\{\gamma\left(E\right)\right\}'\right]^2}{2B^2}. \tag{4.16}$$

The use of (4.16) leads to the expressions of the EMM s' along the z and y directions as

$$m_z^*\left(\bar{E}_{\mathrm{FB}},n,E_0,B\right)=m^*\left[\left\{\gamma\left(\bar{E}_{\mathrm{FB}}\right)\right\}'+\frac{m^*E_0^2\left\{\gamma\left(\bar{E}_{\mathrm{FB}}\right)\right\}'\left\{\gamma\left(\bar{E}_{\mathrm{FB}}\right)\right\}''}{B^2}\right]. \tag{4.17}$$

$$\begin{aligned}
&m_y^*\left(\bar{E}_{\mathrm{FB}},n,E_0,B\right)\\
&=\left(\frac{B}{E_0}\right)^2\frac{1}{\left[\left\{\gamma\left(\bar{E}_{\mathrm{FB}}\right)\right\}'\right]}\left[\gamma\left(\bar{E}_{\mathrm{FB}}\right)-\left(n+\frac{1}{2}\right)\hbar\omega_0+\frac{m^*E_0^2\left[\left\{\gamma\left(\bar{E}_{\mathrm{FB}}\right)\right\}'\right]^2}{2B^2}\right]\\
&\times\left[\frac{-\left\{\gamma\left(\bar{E}_{\mathrm{FB}}\right)\right\}''}{\left[\left\{\gamma\left(\bar{E}_{\mathrm{FB}}\right)\right\}'\right]^2}\left[\gamma\left(\bar{E}_{\mathrm{FB}}\right)-\left(n+\frac{1}{2}\right)\hbar\omega_0+\frac{m^*E_0^2\left[\left\{\gamma\left(\bar{E}_{\mathrm{FB}}\right)\right\}'\right]^2}{2B^2}\right]+1\right.\\
&\left.+\frac{m^*E_0^2\left\{\gamma\left(\bar{E}_{\mathrm{FB}}\right)\right\}''}{B^2}\right].
\end{aligned} \tag{4.18}$$

The Landau energy $(\bar{E}_{n_1})$ can be written as

$$\gamma\left(\bar{E}_{n_1}\right)=\left(n+\frac{1}{2}\right)\hbar\omega_0-\left\{\frac{m^*{E_0}^2\left[\left\{\gamma\left(\bar{E}_{n_1}\right)\right\}'\right]^2}{2B^2}\right\}. \tag{4.19}$$

Equation (4.19) at the Fermi energy $\bar{E}_{\mathrm{FB}}$ generates a cubic equation in B, the real single root of which when combined with (4.10) will generate the SdH period in this case.

The electron concentration and the DMR in this case assume the forms

$$n_0=\frac{2g_{\mathrm{v}}B\sqrt{2m^*}}{3L_x\pi^2\hbar^2E_0}\sum_{n=0}^{n_{\max}}\left[T_{43}\left(n,\bar{E}_{\mathrm{FB}}\right)+T_{44}\left(n,\bar{E}_{\mathrm{FB}}\right)\right], \tag{4.20}$$

and

$$\frac{D}{\mu}=\frac{1}{|e|}\left[\frac{\sum\limits_{n=0}^{n_{\max}}\left[T_{43}\left(n,\bar{E}_{\mathrm{FB}}\right)+T_{44}\left(n,\bar{E}_{\mathrm{FB}}\right)\right]}{\sum\limits_{n=0}^{n_{\max}}\left[\left\{T_{43}\left(n,\bar{E}_{\mathrm{FB}}\right)\right\}'+\left\{T_{44}\left(n,\bar{E}_{\mathrm{FB}}\right)\right\}'\right]}\right], \tag{4.21}$$

where

$$\begin{aligned}T_{43}\left(n,\bar{E}_{\mathrm{FB}}\right)\equiv&\left[\left[\gamma\left(\bar{E}_{\mathrm{FB}}\right)-\left(n+\frac{1}{2}\right)\hbar\omega_0-\frac{m^*{E_0}^2}{2B^2}\left[\left\{\gamma\left(\bar{E}_{\mathrm{FB}}\right)\right\}'\right]^2\right.\right.\\&\left.\left.+|e|\,E_0L_x\left[\left\{\gamma\left(\bar{E}_{\mathrm{FB}}\right)\right\}'\right]\right]^{3/2}\right.\\&\left.-\left[\gamma\left(\bar{E}_{\mathrm{FB}}\right)-\left(n+\frac{1}{2}\right)\hbar\omega_0-\frac{m^*{E_0}^2}{2B^2}\left[\left\{\gamma\left(\bar{E}_{\mathrm{FB}}\right)\right\}'\right]^2\right]^{3/2}\right]\times\\&\times\frac{1}{\left[\left\{\gamma\left(\bar{E}_{\mathrm{FB}}\right)\right\}'\right]}\text{ and }T_{44}\left(n,\bar{E}_{\mathrm{FB}}\right)\equiv\sum_{r=1}^{s}\left[L\left(r\right)T_{43}\left(n,\bar{E}_{\mathrm{FB}}\right)\right].\end{aligned}$$

(b) Under the condition $\Delta\gg E_{\mathrm{g}}$, (4.16) assumes the well known form of [1]

$$\begin{aligned}E\left(1+\alpha E\right)=&\left(n+\frac{1}{2}\right)\hbar\omega_0-\frac{E_0}{B}\hbar k_y\left(1+2\alpha E\right)-\frac{m^*{E_0}^2}{2B^2}\left(1+2\alpha E\right)^2\\&+\frac{\left[\hbar k_z\left(E\right)\right]^2}{2m^*}.\end{aligned} \tag{4.22}$$

The use of (4.22) leads to the expressions of the EMM s' along z and y directions as

$$m_z^*\left(\bar{E}_{\mathrm{FB}},n,E_0,B\right)=m^*\left[\left(1+2\alpha\bar{E}_{\mathrm{FB}}\right)+\frac{2m^*E_0^2\left(1+2\alpha\bar{E}_{\mathrm{FB}}\right)\alpha}{B^2}\right], \tag{4.23}$$

$$m_y^*\left(\bar{E}_{\mathrm{FB}}, n, E_0, B\right) = \left(\frac{B}{E_0}\right)^2 \frac{1}{\left[1+2\alpha\bar{E}_{\mathrm{FB}}\right]}$$
$$\left[\bar{E}_{\mathrm{FB}}\left(1+\alpha\bar{E}_{\mathrm{FB}}\right) - \left(n+\frac{1}{2}\right)\hbar\omega_0 + \frac{m^* E_0^2\left[\left(1+2\alpha\bar{E}_{\mathrm{FB}}\right)\right]^2}{2B^2}\right],$$
$$\left[\frac{-2\alpha}{\left[\left(1+2\alpha\bar{E}_{\mathrm{FB}}\right)\right]^2}\left[\bar{E}_{\mathrm{FB}}\left(1+\alpha\bar{E}_{\mathrm{FB}}\right) - \left(n+\frac{1}{2}\right)\hbar\omega_0\right.\right.$$
$$\left.\left. + \frac{m^* E_0^2\left[\left(1+2\alpha\bar{E}_{\mathrm{FB}}\right)\right]^2}{2B^2}\right] + 1 + \frac{2\alpha m^* E_0^2}{B^2}\right]. \tag{4.24}$$

The Landau energy ($\bar{E}_{n_2}$) can be written as

$$\bar{E}_{n_2}\left(1+\alpha\bar{E}_{n_2}\right) = \left(n+\frac{1}{2}\right)\hbar\omega_0 - \frac{m^* {E_0}^2}{2B^2}\left(1+2\alpha\bar{E}_{n_2}\right)^2. \tag{4.25}$$

Equation (4.25) at the Fermi energy $\bar{E}_{\mathrm{FB}}$ generates a cubic equation in B, the real single root of which when combined with (4.10) will generate the SdH period in this case.

The expressions for n_0 and DMR in this case assume the forms

$$n_0 = \frac{2g_{\mathrm{v}} B\sqrt{2m^*}}{3L_x\pi^2\hbar^2 E_0}\sum_{n=0}^{n_{\max}}\left[T_{45}\left(n, \bar{E}_{\mathrm{FB}}\right) + T_{46}\left(n, \bar{E}_{\mathrm{FB}}\right)\right], \tag{4.26}$$

and

$$\frac{D}{\mu} = \frac{1}{|e|}\left[\frac{\sum_{n=0}^{n_{\max}}\left[T_{45}\left(n, \bar{E}_{\mathrm{FB}}\right) + T_{46}\left(n, \bar{E}_{\mathrm{FB}}\right)\right]}{\sum_{n=0}^{n_{\max}}\left[\left\{T_{45}\left(n, \bar{E}_{\mathrm{FB}}\right)\right\}' + \left\{T_{46}\left(n, \bar{E}_{\mathrm{FB}}\right)\right\}'\right]}\right], \tag{4.27}$$

where

$$T_{45}(n, \bar{E}_{\mathrm{FB}}) \equiv \left[\left[\bar{E}_{\mathrm{FB}}\left(1+\alpha\bar{E}_{\mathrm{FB}}\right) - \left(n+\frac{1}{2}\right)\hbar\omega_0\right.\right.$$
$$\left. + |e| E_0 L_x\left(1+2\alpha\bar{E}_{\mathrm{FB}}\right) - \frac{m^* {E_0}^2}{2B^2}\left(1+2\alpha\bar{E}_{\mathrm{FB}}\right)^2\right]^{3/2}$$
$$- \left[\bar{E}_{\mathrm{FB}}\left(1+\alpha\bar{E}_{\mathrm{FB}}\right) - \left(n+\frac{1}{2}\right)\hbar\omega_0\right.$$
$$\left.\left. - \frac{m^* {E_0}^2}{2B^2}\left(1+2\alpha\bar{E}_{\mathrm{FB}}\right)^2\right]^{3/2}\right]\left[1+2\alpha\bar{E}_{\mathrm{FB}}\right]^{-1},$$

and

$$T_{46}\left(n, \bar{E}_{\mathrm{FB}}\right) \equiv \sum_{r=0}^{s} L(r)\left[T_{45}\left(n, \bar{E}_{\mathrm{FB}}\right)\right].$$

(c) For parabolic energy bands, $\alpha \to 0$ and we can write

$$E = \left(n + \frac{1}{2}\right)\hbar\omega_0 + \frac{[\hbar k_z(E)]^2}{2m^*} - \frac{1}{2}m^*\left(\frac{E_0}{B}\right)^2 - \frac{E_0}{B}\hbar k_y. \tag{4.28}$$

Using (4.28), the expressions of the EMM s' along the y and z directions can be written as

$$m_z^*\left(\bar{E}_{\mathrm{FB}}, n, E_0, B\right) = m^*, \tag{4.29}$$

and

$$m_y^*\left(\bar{E}_{\mathrm{FB}}, n, E_0, B\right) = \left(\frac{B}{E_0}\right)^2\left[\bar{E}_{\mathrm{FB}} - \left(n + \frac{1}{2}\right)\hbar\omega_0 + \frac{m^* E_0^2}{2B^2}\right]. \tag{4.30}$$

The Landau energy $(\bar{E}_{n_3})$ can be written as

$$\bar{E}_{n_3} = \left(n + \frac{1}{2}\right)\hbar\omega_0 - \frac{m^* {E_0}^2}{2B^2}. \tag{4.31}$$

Equation (4.31) at the Fermi energy $\bar{E}_{\mathrm{FB}}$ generates a cubic equation in B, the real single root of which when combined with (4.10) will generate the SdH period in this case.

The electron concentration and the DMR in this case can, respectively, be expressed as

$$n_0 = N_{\mathrm{c}}\theta g_{\mathrm{v}}\left[\frac{k_{\mathrm{B}}T}{|e|\,E_0 L_x}\right]\sum_{n=0}^{n_{\max}}\left[F_{\frac{1}{2}}(\eta_1) - F_{\frac{1}{2}}(\eta_2)\right], \tag{4.32a}$$

and

$$\frac{D}{\mu} = \frac{k_{\mathrm{B}}T}{|e|}\left[\frac{\sum\limits_{n=0}^{n_{\max}}\left[F_{\frac{1}{2}}(\eta_1) - F_{\frac{1}{2}}(\eta_2)\right]}{\sum\limits_{n=0}^{n_{\max}}\left[F_{\frac{-1}{2}}(\eta_1) - F_{\frac{-1}{2}}(\eta_2)\right]}\right], \tag{4.32b}$$

where

$$\eta_1 \equiv \frac{\bar{E}_{\mathrm{FB}} - \overline{\theta_1}}{k_{\mathrm{B}}T}, \overline{\theta_1} \equiv \left[\left(n + \frac{1}{2}\right)\hbar\omega_0 + \frac{1}{2}m^*\left(\frac{E_0}{B}\right)^2 - |e|\,E_0 L_x\right],$$

$$\eta_2 \equiv \frac{\left(\bar{E}_{\mathrm{FB}} - \overline{\theta_2}\right)}{k_{\mathrm{B}}T}, \overline{\theta_2} \equiv \overline{\theta_1} + |e|\,E_0 L_x.$$

In the absence of an electric field $E_0 \to 0$ and the application of L' Hospital's rule transforms (4.32b) to the well-known form under magnetic quantization as given by (3.47) of Chap. 3.

4.2.3 II–VI Semiconductors

In the presence of an electric field along the x axis and the quantizing magnetic field B along the z axis, from (2.27) of Chap. 2 we can write

$$\hat{E} + |e|\, E_0 \hat{x} = \frac{\hat{p}_x^2}{2m_\perp^*} + \frac{\left(\hat{p}_y - eB\hat{x}^2\right)}{2m_\perp^*} + \frac{\hat{p}_z^2}{2m_{||}^*} + D\left[\frac{\hat{p}_x^2}{2m_\perp^*} + \frac{\left(\hat{p}_y - eB\hat{x}\right)^2}{2m_\perp^*}\right]^{1/2}, \tag{4.33}$$

where

$$D \equiv \pm \frac{\bar{\lambda}_0 \sqrt{2m_\perp^*}}{\hbar}.$$

Let us define the operator $\widehat{\theta}$ as

$$\widehat{\theta} = -\hat{p}_y + |e|\, B\hat{x} - \frac{E_0 m_\perp^*}{B}. \tag{4.34}$$

Eliminating $\hat{x}$ between (4.33) and (4.34), one obtains

$$\hat{E} + \frac{E_0}{B}\widehat{\theta} + \frac{E_0}{B}\hat{p}_y + \left(\frac{E_0}{B}\right)^2 m_\perp^* = \frac{\hat{p}_x^2}{2m_\perp^*} + \frac{\hat{\theta}^2}{2m_\perp^*} + \frac{{E_0}^2 m_\perp^*}{2B^2} + \frac{\hat{\theta} E_0}{B} + D\left[\frac{\hat{p}_x^2}{2m_\perp^*} + \frac{\hat{\theta}^2}{2m_\perp^*} + \frac{E_0^2 m_\perp^*}{2B^2}\right]^{1/2} + \frac{\hat{p}_z^2}{2m_{||}^*}. \tag{4.35}$$

Thus the electron energy spectrum in this case can be expressed as

$$E = \left(\beta_1\left(n, E_0\right)\right) + \left(\frac{\left[\hbar k_z\left(E\right)\right]^2}{2m_{||}^*}\right) - \left(\frac{E_0}{B}\hbar k_y\right), \tag{4.36}$$

where

$$\beta_1\left(n, E_0\right) \equiv \left[\left(n + \frac{1}{2}\right)\hbar\omega_{02} - \left(\frac{E_0^2 m_\perp^*}{2B^2}\right) + D\left\{\left(n + \frac{1}{2}\right)\hbar\omega_{02} + \left(\frac{E_0^2 m_\perp^*}{2B^2}\right)\right\}^{1/2}\right] \text{ and } \omega_{02} \equiv \frac{|e|\, B}{m_\perp^*}.$$

The use of (4.36) leads to the expressions of the EMM s' along the z and y directions as

$$m_z^*\left(\bar{E}_{\text{FB}}, n, E_0, B\right) = m_{||}^*, \tag{4.37}$$

$$m_y^*\left(\bar{E}_{\text{FB}}, n, E_0, B\right) = \left(\frac{B}{E_0}\right)^2 \left[\bar{E}_{\text{FB}} - \beta_1\left(n, E_0\right)\right]. \tag{4.38}$$

The Landau energy $(\bar{E}_{n_4})$ can be written as

$$E_{n_4} = \beta_1 (n, E_0) . \tag{4.39}$$

Equation (4.39) at the Fermi energy $\bar{E}_{\mathrm{FB}}$ when combined with (4.10) will generate the SdH period in this case. In this case

$$x_i = \frac{-E_0}{B\hbar} m_\perp^* + \frac{2D}{\hbar}\left(\frac{m_\perp^*}{2}\right)^{\frac{1}{2}}, \; x_h = \frac{|e|\, BL_x}{\hbar} + x_l \text{ and}$$

$$k_z (E) = \frac{\sqrt{2m_\parallel^*}}{\hbar}\left[E - \beta_1 (n, E_0) + \frac{E_0}{B}\hbar k_y\right]^{1/2} . \tag{4.40}$$

Equation (4.12) for II–VI semiconductors in the cross fields' configuration assumes the form

$$I (E) = \int\limits_{x_l}^{x_h} \frac{\sqrt{2m_\parallel^*}}{\hbar}\left[E - \beta_1 (n, E_0) + \frac{E_0}{B}\hbar k_y\right]^{\frac{1}{2}} \mathrm{d}k_y .$$

Therefore

$$I (E) = \frac{\sqrt{2m_\parallel^*}}{\hbar}\frac{2}{3}\frac{B}{E_0\hbar} \tag{4.41}$$
$$\times \left[\left\{E - \beta_1 (n, E_0) + \frac{E_0}{B}\hbar x_h\right\}^{3/2} - \left\{E - \beta_1 (n, E_0) + \frac{E_0}{B}\hbar x_l\right\}^{\frac{3}{2}}\right] .$$

The electron concentration, from (4.11) can be expressed as

$$n_0 = -\frac{2g_{\mathrm{v}}B\sqrt{2m_\parallel}}{3L_x\pi^2\hbar^2E_0}\sum_{n=0}^{n_{\max}}\left[\int\limits_{\theta_1}^{\infty}[E-\theta_1]^{\frac{3}{2}}\frac{\partial f_0}{\partial E}\mathrm{d}E - \int\limits_{\theta_2}^{\infty}[E-\theta_2]^{\frac{3}{2}}\frac{\partial f_0}{\partial E}\mathrm{d}E\right], \tag{4.42}$$

where

$$\theta_1 \equiv \beta_1 (n, E_0) - \frac{E_0}{B}\hbar x_h \text{ and } \theta_2 \equiv \beta_1 (n, E_0) - \frac{E_0}{B}\hbar x_l .$$

Substituting $\frac{E-\theta_1}{k_{\mathrm{B}}T} = x_1$, $\frac{E-\theta_2}{k_{\mathrm{B}}T} = x_2$, $\eta_3 = \frac{\bar{E}_{\mathrm{FB}}-\theta_1}{k_{\mathrm{B}}T}$ and $\eta_4 = \frac{\bar{E}_{\mathrm{FB}}-\theta_2}{k_{\mathrm{B}}T}$, from (4.42) we can write

$$n_0 = \frac{2g_{\mathrm{v}}B\sqrt{2m_\parallel^*}}{3E_0L_x\hbar^2\pi^2}(k_{\mathrm{B}}T)^{\frac{3}{2}}\sum_{n=0}^{n_{\max}}\left[\int\limits_0^{\infty}\frac{x_1^{3/2}\exp(x_1-\eta_3)}{[1+\exp(x_1-\eta_3)]^2}\mathrm{d}x_1\right.$$
$$\left. - \int\limits_0^{\infty}\frac{x_2^{3/2}\exp(x_2-\eta_4)}{[1+\exp(x_2-\eta_4)]^2}\mathrm{d}x_2\right] . \tag{4.43}$$

Differentiating both sides of (2.16) with respect to η, one can write

$$\Gamma(j+1)F_{j-1}(\eta)=\int_0^\infty \frac{x^j\exp(x-\eta)}{[1+\exp(x-\eta)]^2}\mathrm{d}x. \tag{4.44}$$

Using (4.43) and (4.44) the electron concentration in this case can be written as

$$n_0=\frac{g_{\mathrm{v}}B\sqrt{2m_{\parallel}^*\pi}}{2E_0L_x\hbar^2\pi^2}(k_{\mathrm{B}}T)^{\frac{3}{2}}\sum_{n=0}^{n_{\max}}\left[F_{\frac{1}{2}}(\eta_3)-F_{\frac{1}{2}}(\eta_4)\right]. \tag{4.45}$$

Using (4.45) and (1.11) the DMR in this case assumes the form

$$\frac{D}{\mu}=\frac{k_{\mathrm{B}}T}{|e|}\left[\frac{\sum\limits_{n=0}^{n_{\max}}\left[F_{\frac{1}{2}}(\eta_3)-F_{\frac{1}{2}}(\eta_4)\right]}{\sum\limits_{n=0}^{n_{\max}}\left[F_{\frac{-1}{2}}(\eta_3)-F_{\frac{-1}{2}}(\eta_4)\right]}\right]. \tag{4.46}$$

4.2.4 The Formulation of DMR in Bi

(a) The McClure and Choi model

In the presence of an electric field E_0 along the trigonal-axis (z-direction) and the quantizing magnetic field B along the bisectrix axis (y-direction) from (2.34), we can write

$$\begin{aligned}E(1+\alpha E)+|e|E_0\widehat{z}(1+2\alpha E)=&\frac{(\widehat{p}_x-|e|B\widehat{z})^2}{2m_1}+\frac{\hat{p}_y^2}{2m_2}+\frac{\hat{p}_z^2}{2m_3}+\frac{\hat{p}_y^2}{2m_2}\\&\alpha E\left(1-\frac{m_2}{m_2'}\right)-\alpha\frac{\hat{p}_y^2\hat{p}_z^2}{4m_2m_3}+\alpha\frac{\hat{p}_y^4}{4m_2m_2'}\\&-\alpha\frac{(\hat{p}_x-|e|B\widehat{z})^2\hat{p}_y^2}{4m_1m_2},\end{aligned} \tag{4.47}$$

Let us define the operator $\widehat{\theta}$ as

$$\widehat{\theta}=|e|B\widehat{z}-\widehat{p}_x-\frac{m_1E_0}{B}(1+2\alpha E). \tag{4.48}$$

Eliminating $\hat{z}$ between (4.47) and (4.48) one obtains

$$E(1+\alpha E)+\frac{E_0}{B}(1+2\alpha E)\widehat{\theta}+\frac{E_0}{B}(1+2\alpha E)\widehat{p}_x+m_1\left(\frac{E_0}{B}\right)^2(1+2\alpha E)^2$$

$$= \left(\frac{\widehat{\theta}^2}{2m_1} + \frac{\hat{p}_z^2}{2m_3}\right) + \frac{\hat{p}_y^2}{2m_2} + \frac{\hat{p}_y^2}{2m_2}\alpha E\left(1 - \frac{m_2}{m_2'}\right) + \frac{\alpha\hat{p}_y^4}{4m_2m_2'}$$

$$-\frac{\alpha\hat{p}_y^2}{2m_2}\left[\frac{\widehat{\theta}^2}{2m_1} + \frac{\hat{p}_z^2}{2m_3}\right] - \frac{\alpha\hat{p}_y^2 m_1 E_0^2}{4m_2B^2}(1+2\alpha E)^2 + \frac{E_0}{B}\widehat{\theta}(1+2\alpha E)$$

$$+\frac{1}{2}m_1\left(\frac{E_0}{B}\right)^2(1+2\alpha E)^2 . \tag{4.49}$$

Therefore the required dispersion relation is given by

$$E(1+\alpha E) = \left(n+\frac{1}{2}\right)\hbar\omega_{03} + \frac{[\hbar k_y(E)]^2}{2m_2}$$

$$-\frac{E_0}{B}(1+2\alpha E)\hbar k_x - \frac{1}{2}m_1\left(\frac{E_0}{B}\right)^2(1+2\alpha E)^2$$

$$+\frac{[\hbar k_y(E)]^2}{2m_2}\alpha E\left(1-\frac{m_2}{m_2'}\right) + \frac{\alpha[\hbar k_y(E)]^4}{4m_2m_2'} - \frac{\alpha[\hbar k_y(E)]^2}{2m_2}\times$$

$$\times\left(n+\frac{1}{2}\right)\hbar\omega_{03} - \frac{\alpha[\hbar k_y(E)]^2 m_1E_0^2}{4m_2B^2}(1+2\alpha E)^2 , \tag{4.50}$$

where

$$\omega_{03} \equiv \frac{|e|B}{\sqrt{m_1m_3}} .$$

When $\alpha \to 0$, from (4.50) we can write

$$E = \left(n+\frac{1}{2}\right)\hbar\omega_{03} + \frac{[\hbar k_y(E)]^2}{2m_2} - \frac{E_0}{B}\hbar k_x - \frac{1}{2}m_1\left(\frac{E_0}{B}\right)^2 \tag{4.51}$$

The use of (4.50) leads to the equations of the EMM s' along the x and y direction as

$$m_x^*\left(\bar{E}_{\mathrm{FB}}, n, E_0, B\right) = \left(\frac{B}{E_0}\right)^2\left(1+2\alpha\bar{E}_{\mathrm{FB}}\right)^{-3}$$

$$\times\left[\bar{E}_{\mathrm{FB}}\left(1+\alpha\bar{E}_{\mathrm{FB}}\right) - \left(n+\frac{1}{2}\right)\hbar\omega_{03} + \frac{1}{2}m_1\left(\frac{E_0}{B}\right)^2\left(1+2\alpha\bar{E}_{\mathrm{FB}}\right)^2\right]$$

$$\times\left[\left(1+2\alpha\bar{E}_{\mathrm{FB}}\right)\left\{1+2\alpha\bar{E}_{\mathrm{FB}} + 2\alpha m_1\left(1+2\alpha\bar{E}_{\mathrm{FB}}\right)\left(\frac{E_0}{B}\right)^2\right\} - 2\alpha\right.$$

$$\left.\times\left\{\bar{E}_{\mathrm{FB}}\left(1+\alpha\bar{E}_{\mathrm{FB}}\right) - \left(n+\frac{1}{2}\right)\hbar\omega_{03} + \frac{1}{2}m_1\left(\frac{E_0}{B}\right)^2\left(1+2\alpha\bar{E}_{\mathrm{FB}}\right)^2\right\}\right] \tag{4.52}$$

and

$$m_y^*\left(\bar{E}_{\mathrm{FB}},n,E_0,B\right)=\frac{1}{4}\left[\frac{[h_4(n,\bar{E}_{\mathrm{FB}})]'}{2\sqrt{h_4(n,\bar{E}_{\mathrm{FB}})}}-[h_1(n,\bar{E}_{\mathrm{FB}})]'\right], \tag{4.53}$$

where

$$h_4\left(n,\bar{E}_{\mathrm{FB}}\right)\equiv\left[h_1^2\left(n,\bar{E}_{\mathrm{FB}}\right)+4h_2\left(n,\bar{E}_{\mathrm{FB}}\right)\right],$$

$$h_1\left(n,\bar{E}_{\mathrm{FB}}\right)\equiv\frac{4m_2m_2'}{\alpha}\left[\frac{-\alpha}{2m_2}\left(n+\frac{1}{2}\right)\hbar\omega_{03}-\frac{\alpha m_1E_0^2}{4m_2B^2}\left(1+2\alpha\bar{E}_{\mathrm{FB}}\right)^2+\frac{1}{2m_2}+\frac{\alpha\bar{E}_{\mathrm{FB}}}{2m_2}\left(1-\frac{m_2}{m_2'}\right)\right],$$

and

$$h_2\left(n,\bar{E}_{\mathrm{FB}}\right)\equiv\frac{4m_2m_2'}{\alpha}\left[\bar{E}_{\mathrm{FB}}\left(1+\alpha\bar{E}_{\mathrm{FB}}\right)-\left(n+\frac{1}{2}\right)\hbar\omega_{03}+\frac{1}{2}m_1\left(\frac{E_{03}}{B}\right)^2\left(1+2\alpha\bar{E}_{\mathrm{FB}}\right)^2\right].$$

The Landau energy $(\bar{E}_{n_5})$ can be written as

$$\bar{E}_{n_5}\left(1+\alpha\bar{E}_{n_5}\right)=\left(n+\frac{1}{2}\right)\hbar\omega_{03}-\frac{1}{2}m_1\left(\frac{E_0}{B}\right)^2\left(1+2\alpha\bar{E}_{n_5}\right)^2. \tag{4.54}$$

Equation (4.54) at the Fermi energy $\bar{E}_{\mathrm{FB}}$ when combined with (4.10) will generate the S dH period. In this case

$$x_l(E)=-\frac{m_1E_0}{B\hbar}\left(1+2\alpha E\right)\text{ and }x_h\left(E\right)=\frac{|e|\,BL_z}{\hbar}+x_l\left(E\right), \tag{4.55}$$

where L_z is the sample length along z-direction.

The electron concentration in this case can be written as

$$n_0=\frac{g_{\mathrm{v}}}{L_z\pi^2}\sum_{n=0}^{n_{\max}}\int_{\bar{E}_{01}}^{\infty}J\left(E\right)\left(-\frac{\partial f_o}{\partial E}\right)\mathrm{d}E, \tag{4.56}$$

in which $\bar{E}_{01}$ is the root of the equation $J\left(\bar{E}_{01}\right)=0$ where $J\left(E\right)$ is given by

$$J(E)=\int_{x_l(E)}^{x_h(E)}k_y(E)\mathrm{d}k_x. \tag{4.57}$$

The term $k_y\left(E\right)$ in (4.57) satisfies the following equation

$$k_y\left(E\right)=\left[-h_1\left(n,E\right)+\sqrt{h_4\left(n,E\right)+h_5\left(E\right)k_x}\right]^{1/2}\left[\left(\hbar\right)\sqrt{2}\right]^{-1}, \tag{4.58}$$

where

$$h_5(E) \equiv \left(\frac{16m_2m_2'}{\alpha}\right)\left[\frac{E_0}{B}(1+2\alpha E)\hbar\right].$$

Using (4.57) and (4.58), we get,

$$\begin{aligned} J(E) = {} & \frac{2\sqrt{2}}{3\hbar}\left[\left[\frac{h_1(n,E)}{h_5(n,E)}\right]\right. \\ & \times\left[\{-h_1(n,E)+h_7(n,E)\}^{3/2}-\{-h_1(n,E)+h_6(n,E)\}^{3/2}\right] \\ & +\left[\frac{3}{5h_5(n,E)}\right]\left[\{-h_1(n,E)+h_7(n,E)\}^{5/2}\right. \\ & \left.\left.-\{-h_1(n,E)+h_6(n,E)\}^{5/2}\right]\right]. \end{aligned} \tag{4.59}$$

where

$$h_6(n,E) \equiv [h_4(n,E)+h_5(E)x_l(E)]^{1/2} \text{ and}$$
$$h_7(n,E) \equiv [h_4(n,E)+x_h(E)h_5(n,E)]^{1/2}.$$

Combining (4.56) and (4.59), the electron concentration in this case can be written as

$$n_0 = \frac{g_v}{L_z\pi^2}\frac{2\sqrt{2}}{3\hbar}\sum_{n=0}^{n_{\max}}\left[T_{47}(n,\overline{E}_{\rm FB})+T_{48}(n,\overline{E}_{\rm FB})\right], \tag{4.60}$$

where

$$\begin{aligned} T_{47}(n,\overline{E}_{\rm FB}) \equiv {} & \left(\frac{h_1(n,\overline{E}_{\rm FB})}{h_5(n,\overline{E}_{\rm FB})}\right) \\ & \times\left[\left[-h_1(n,\overline{E}_{\rm FB})+h_7(n,\overline{E}_{\rm FB})\right]^{\frac{3}{2}}-\{-h_1(n,\overline{E}_{\rm FB})+h_6(n,\overline{E}_{\rm FB})\}^{\frac{3}{2}}\right] \\ & +\left(\frac{3}{5h_5(n,\overline{E}_{\rm FB})}\right)\left[\left[-h_1(n,\overline{E}_{\rm FB})+h_7(n,\overline{E}_{\rm FB})\right]^{\frac{5}{2}}\right. \\ & \left.-\left[-h_1(n,\overline{E}_{\rm FB})+h_6(n,\overline{E}_{\rm FB})\right]^{\frac{5}{2}}\right], \end{aligned}$$

and

$$T_{48}(n,\overline{E}_{\rm FB}) \equiv \sum_{r=1}^{s}L(r)\left[T_{47}(n,\overline{E}_{\rm FB})\right].$$

Therefore the DMR assumes the form

$$\frac{D}{\mu} = \frac{1}{|e|}\frac{\sum_{n=0}^{n_{\max}}\left[T_{47}(n,\overline{E}_{\rm FB})+T_{48}(n,\overline{E}_{\rm FB})\right]}{\sum_{n=0}^{n_{\max}}\left[\{T_{47}(n,\overline{E}_{\rm FB})\}'+\{T_{48}(n,\overline{E}_{\rm FB})\}'\right]}. \tag{4.61}$$

(b) The Cohen Model

In the presence of an electric field E_0 along the trigonal axis and the quantizing magnetic field B along the bisectrix axis for this case, (2.41) assumes the form

$$E(1+\alpha E)+|e|E_0\widehat{z}(1+2\alpha E)$$
$$=\frac{(\widehat{p}_x-|e|B\widehat{z})^2}{2m_1}+\frac{\widehat{p}_z^2}{2m_3}-\frac{\alpha E\widehat{p}_y^2}{2m_2'}+\frac{\widehat{p}_y^2}{2m_2}(1+\alpha E)+\frac{\alpha\widehat{p}_y^4}{4m_2m_2'}. \quad (4.62)$$

Using the same operator $\widehat{\theta}$ as defined by (4.48) and eliminating $\widehat{z}$ between (4.48) and (4.62) one can write

$$E(1+\alpha E)=\left(\frac{\widehat{\theta}^2}{2m_1}+\frac{\widehat{p}_z^2}{2m_3}\right)-\frac{E_0}{B}\widehat{p}_x(1+2\alpha E)-\frac{E_0^2m_1}{2B^2}(1+2\alpha E)^2$$
$$-\frac{\alpha E\widehat{p}_y^2}{2m_2'}+\frac{\widehat{p}_y^2}{2m_2}(1+\alpha E)+\frac{\alpha\widehat{p}_y^4}{4m_2m_2'}.$$

Thus the electron energy spectrum can be expressed as

$$E(1+\alpha E)=\left[\left(n+\frac{1}{2}\right)\hbar\omega_{03}-\frac{E_0}{B}\hbar k_x(1+2\alpha E)-\frac{1}{2}m_1\left(\frac{E_0}{B}\right)^2(1+2\alpha E)^2\right.$$
$$\left.-\left(\frac{\alpha E[\hbar k_y(E)]^2}{2m_2'}\right)+\frac{[\hbar k_y(E)]^2}{2m_2}(1+\alpha E)+\left[\frac{\alpha[\hbar k_y(E)]^4}{4m_2m_2'}\right]\right]. \quad (4.63)$$

The use of (4.63) leads to the same expression of EMM along the x direction as given by (4.52) for the McClure and Choi model and the EMM along the y direction is given by

$$m_y^*(\bar{E}_{FB},n,E_0,B)=\frac{1}{4}\left[\frac{[\overline{H}_5(n,\bar{E}_{FB})]'}{2\sqrt{\overline{H}_5(n,\bar{E}_{FB})}}-[\overline{H}_1(n,\bar{E}_{FB})]'\right], \quad (4.64)$$

where

$$\overline{H}_5(n,\bar{E}_{FB})\equiv\left[\overline{H}_1^2(\bar{E}_{FB})+4\overline{H}_3(n,\bar{E}_{FB})\right],$$
$$\overline{H}_1(\bar{E}_{FB})\equiv\left(\frac{4m_2m_2'}{\alpha}\right)\left[\frac{1+\alpha\bar{E}_{FB}}{2m_2}-\frac{\alpha\bar{E}_{FB}}{2m_2'}\right],$$

and

$$\overline{H}_3(n,\bar{E}_{FB})\equiv\frac{4m_2m_2'}{\alpha}\left[\bar{E}_{FB}(1+\alpha\bar{E}_{FB})-\left(n+\frac{1}{2}\right)\hbar\omega_{03}\right.$$
$$\left.+\frac{1}{2}m_1\left(\frac{E_{03}}{B}\right)^2(1+2\alpha\bar{E}_{FB})^2\right].$$

The Landau energy in this case is given by the same (4.54) and the SdH period can also be determined in a similar way.

The term $k_y(E)$ of (4.57) in this case can be determined from the following equation

$$[\hbar k_y(E)]^4 + [\hbar k_y(E)]^2 \left(\frac{4m_2 m_2'}{\alpha}\right)\left[\frac{-\alpha E}{2m_2'} + \frac{1+\alpha E}{2m_2}\right] - \left[E(1+\alpha E) - \left(n+\frac{1}{2}\right)\hbar\omega_{03} + \frac{E_0}{B}\hbar k_x (1+2\alpha E) + \frac{{E_0}^2 m_1}{2B^2}(1+2\alpha E)^2\right]\left(\frac{4m_2 m_2'}{\alpha}\right) = 0. \quad (4.65)$$

Therefore,

$$k_y(E) = \left[-\overline{H}_1(E) + \sqrt{\overline{H}_5(n,E) + \overline{H}_6(E)k_x}\right]^{1/2}\left(\hbar\sqrt{2}\right)^{-1}, \quad (4.66)$$

where

$$\overline{H}_6(E) \equiv 4\overline{H}_4(E) \text{ and } \overline{H}_4(E) \equiv \left[\frac{E_0}{B}\hbar(1+2\alpha E)\right]\left(\frac{4m_2 m_2'}{\alpha}\right).$$

Using (4.66) and (4.57), the expression of $J(E)$ in this case can be written as

$$J(E) = \frac{\sqrt{2}}{\hbar}\left[\frac{\overline{H}_1(E)}{\overline{H}_6(E)}\frac{2}{3}\left\{\left(\overline{H}_8(n,E) - \overline{H}_1(E)\right)^{3/2} - \left(\overline{H}_7(n,E) - \overline{H}_1(E)\right)^{3/2}\right\} + \frac{1}{\overline{H}_6(E)}\frac{2}{5}\left[\left\{\left(\overline{H}_8(n,E) - \overline{H}_1(E)\right)\right\}^{5/2} - \left\{\left(\overline{H}_7(n,E) - \overline{H}_1(E)\right)\right\}^{5/2}\right]\right], \quad (4.67)$$

where

$$\overline{H}_8(n,E) \equiv \left[\overline{H}_5(n,E) + \overline{H}_6(E)x_h(E)\right]^{1/2},$$

$$\overline{H}_5(n,E) \equiv \left[\left[\overline{H}_1(n,E)\right]^2 + 4\overline{H}_3(n,E)\right],$$

$$x_h(E) \equiv \frac{|e|BL_z}{\hbar} + x_l(E),$$

$$x_l(E) \equiv \frac{-E_o}{B}\frac{m_1}{\hbar}(1+2\alpha E) \text{ and}$$

$$\overline{H}_7(n,E) \equiv \left[\overline{H}_5(n,E) + \overline{H}_6(E)x_l(E)\right]^{1/2}.$$

Thus using (4.67) and (4.56), the expression of the electron concentration for the Cohen model in the present case can be written as

$$n_0 = \frac{2g_v\sqrt{2}}{3L_z\pi^2\hbar}\sum_{n=0}^{n_{\max}}\left[T_{49}\left(n,\overline{E}_{\mathrm{FB}}\right) + T_{410}\left(n,\overline{E}_{\mathrm{FB}}\right)\right], \tag{4.68}$$

where

$$\begin{aligned} T_{49}\left(n,E_{\mathrm{FB}}\right) \equiv & \left[\frac{\overline{H}_1\left(\bar{E}_{\mathrm{FB}}\right)}{\overline{H}_6\left(\bar{E}_{\mathrm{FB}}\right)}\left\{\left(\overline{H}_8\left(n,\bar{E}_{\mathrm{FB}}\right)-\overline{H}_1\left(\bar{E}_{\mathrm{FB}}\right)\right)^{3/2}\right.\right.\\ & \left. - \left(\overline{H}_7\left(n,\bar{E}_{\mathrm{FB}}\right)-\overline{H}_1\left(\bar{E}_{\mathrm{FB}}\right)\right)^{3/2}\right\}\\ & + \frac{1}{\overline{H}_6\left(\bar{E}_{\mathrm{FB}}\right)}\frac{3}{5}\left[\left\{\left(\overline{H}_8\left(n,\bar{E}_{\mathrm{FB}}\right)-\overline{H}_1\left(\bar{E}_{\mathrm{FB}}\right)\right)\right\}^{5/2}\right.\\ & \left.\left. - \left\{\left(\overline{H}_7\left(n,\bar{E}_{\mathrm{FB}}\right)-\overline{H}_1\left(\bar{E}_{\mathrm{FB}}\right)\right)\right\}^{5/2}\right]\right]. \end{aligned}$$

and

$$T_{410}\left(n,\bar{E}_{\mathrm{FB}}\right) \equiv \sum_{r=1}^{s} L(r)\, T_{49}\left(n,\bar{E}_{\mathrm{FB}}\right).$$

The use of (4.68) and (1.11) leads to the expression of the DMR in accordance with the Cohen model in the presence of crossed electric and quantizing magnetic fields as

$$\frac{D}{\mu} = \frac{1}{|e|}\left[\frac{\sum_{n=0}^{n_{\max}}\left[T_{49}(n,\overline{E}_{\mathrm{FB}}) + T_{410}(n,\overline{E}_{\mathrm{FB}})\right]}{\sum_{n=0}^{n_{\max}}\left[\left\{T_{49}(n,\overline{E}_{\mathrm{FB}})\right\}' + \left\{T_{410}(n,\overline{E}_{\mathrm{FB}})\right\}'\right]}\right]. \tag{4.69}$$

(c) The Lax model

Under cross-field configuration from (2.46) of Chap. 2, one can write

$$E\left(1+\alpha E\right) + |e|\,E_0\hat{z}\left(1+2\alpha E\right) = \frac{\left(\hat{p}_x - |e|\,B\hat{z}\right)^2}{2m_1} + \frac{\hat{p}_y^2}{2m_2} + \frac{\hat{p}_z^2}{2m_3}. \tag{4.70}$$

Using the same operator $\widehat{\theta}$ as used for the McClure and Choi model we get

$$\begin{aligned} & E(1+\alpha E) + \frac{E_0}{B}(1+2\alpha E)\,\widehat{\theta} + \frac{E_0}{B}(1+2\alpha E)\,\widehat{p}_x + m_1\left(\frac{E_0}{B}\right)^2(1+2\alpha E)^2 \\ & = \frac{\widehat{\theta}^2}{2m_1} + \frac{\widehat{p}_z^2}{2m_3} + \frac{m_1}{2}\left(\frac{E_0}{B}\right)^2(1+2\alpha E)^2 + \frac{\widehat{\theta}E_0}{B}(1+2\alpha E) + \frac{\widehat{p}_y^2}{2m_2} \end{aligned} \tag{4.71}$$

Therefore the electron dispersion relation assumes the form

$$E(1+\alpha E) = \left(n+\frac{1}{2}\right)\hbar\omega_{03} - \frac{E_0}{B}(1+2\alpha E)\hbar k_x + \frac{[\hbar k_y]^2}{2m_2} - \frac{m_1}{2}\left(\frac{E_0}{B}\right)^2(1+2\alpha E)^2 \quad (4.72)$$

The EMM along the x direction in this case is given by (4.52) and the EMM along the y direction is given by

$$m_y^*\left(\bar{E}_{FB}, n, E_0, B\right) = m_2\left[1+2\alpha\bar{E}_{FB} + 2m_1\alpha\left(\frac{E_0}{B}\right)^2\left(1+2\alpha\bar{E}_{FB}\right)\right] \quad (4.73)$$

The Landau level energy (E_{n_6}) in this case can be expressed through the equation

$$E_{n_6}\left(1+\alpha E_{n_6}\right) = \left(n+\frac{1}{2}\right)\hbar\omega_{03} - \frac{m_1}{2}\left(\frac{E_0}{B}\right)^2\left(1+2\alpha E_{n_6}\right)^2. \quad (4.74)$$

Equation (4.74) at the Fermi energy $\bar{E}_{\mathrm{FB}}$ when combined with (4.10) will generate the SdH period. From (4.72) we can write

$$k_y(E) = \frac{\sqrt{2m_2}}{\hbar}\left[\bar{G}_1(n,E) + \bar{G}_2(E)k_x\right]^{1/2},$$

where

$$\bar{G}_1(n,E) \equiv \left[E(1+\alpha E) - \left(n+\frac{1}{2}\right)\hbar\omega_{03} + \frac{m_1}{2}\left(\frac{E_0}{B}\right)^2(1+2\alpha E)^2\right]$$

and $\bar{G}_2(E) \equiv \frac{E_0}{B}(1+2\alpha E)\hbar$.

Therefore the integral $J(E)$ in this case assumes the form

$$J(E) = \frac{\sqrt{2m_2}}{\hbar}\frac{2}{3}\left[\left[\bar{G}_2(E)\right]^{-1}\left[\bar{G}_1(n,E) + \bar{G}_2(E)x_h(E)\right]^{3/2} - \left[\bar{G}_2(E)\right]^{-1}\left[\bar{G}_1(n,E) + \bar{G}_2(E)x_l(E)\right]^{3/2}\right], \quad (4.75)$$

where

$$x_l(E) \equiv -\frac{E_0}{B}\frac{m_1}{\hbar}(1+2\alpha E) \text{ and } x_h(E) \equiv \frac{|e|BL_z}{\hbar} + x_l(E).$$

Using (4.75) and (4.56), the expression of the electron concentration for the Lax model in the present case can be written as

$$n_0 = \frac{2g_v\sqrt{2m_2}}{3L_z\pi^2\hbar}\sum_{n=0}^{n_{\max}}\left[T_{411}\left(n,\bar{E}_{FB}\right)+T_{412}\left(n,\bar{E}_{FB}\right)\right], \tag{4.76}$$

where

$$\begin{aligned} T_{411}\left(n,\bar{E}_{FB}\right) &\equiv \left[\bar{G}_2\left(\bar{E}_{FB}\right)\right]^{-1} \\ &\times\left[\left[\bar{G}_1\left(n,\bar{E}_{FB}\right)+\bar{G}_2\left(\bar{E}_{FB}\right)x_h\left(\bar{E}_{FB}\right)\right]^{3/2}\right. \\ &\left.-\left[\bar{G}_1\left(n,\bar{E}_{FB}\right)+\bar{G}_2\left(\bar{E}_{FB}\right)x_l(\bar{E}_{FB})\right]^{3/2}\right], \\ x_h\left(\bar{E}_{FB}\right) &\equiv \left(\frac{|e|\,BL_z}{\hbar}\right)+x_l\left(\bar{E}_{FB}\right), \\ x_l\left(\bar{E}_{FB}\right) &\equiv \frac{-E_0}{B}\frac{m_1}{\hbar}\left(1+2\alpha\bar{E}_{FB}\right), \end{aligned}$$

and

$$T_{412}\left(n,\bar{E}_{FB}\right) \equiv \sum_{r=1}^{s} L\left(r\right)T_{411}\left(n,\bar{E}_{FB}\right).$$

Thus, using (4.76) and (1.11) the DMR in this case is given by

$$\frac{D}{\mu} = \frac{1}{|e|}\left[\frac{\sum\limits_{n=0}^{n_{\max}}\left[T_{411}\left(n,\bar{E}_{FB}\right)+T_{412}\left(n,\bar{E}_{FB}\right)\right]}{\sum\limits_{n=0}^{n_{\max}}\left[\left\{T_{411}\left(n,\bar{E}_{FB}\right)\right\}'+\left\{T_{412}\left(n,\bar{E}_{FB}\right)\right\}'\right]}\right] \tag{4.77}$$

(d) The Parabolic Ellipsoidal model

For this model the electron dispersion relation for the present case assumes the form

$$E = (n+\frac{1}{2})\hbar\omega_{03} - \frac{E_0}{B}\hbar k_x + \frac{\hbar^2 k_y{}^2}{2m_2} - \frac{m_1}{2}\left(\frac{E_0}{B}\right)^2. \tag{4.78}$$

The EMM's along the y and x directions can, respectively, be expressed as

$$m_y^*\left(\bar{E}_{FB}, n, E_0, B\right) = m_2, \tag{4.79}$$

$$m_x^*\left(\bar{E}_{FB}, n, E_0, B\right) = \left(\frac{B}{E_0}\right)^2\left[\bar{E}_{FB} - \left(n+\frac{1}{2}\right)\hbar\omega_{03} + \frac{m_1}{2}\left(\frac{E_0}{B}\right)^2\right]. \tag{4.80}$$

The Landau level energy (E_{n_7}) in this case can be expressed by the equation

$$E_{n_7} = \left(n+\frac{1}{2}\right)\hbar\omega_{03} - \frac{m_1}{2}\left(\frac{E_0}{B}\right)^2. \tag{4.81}$$

Equation (4.81) at the Fermi energy $\bar{E}_{FB}$ when combined with (4.10) will generate the SdH period in this case.

For this case the electron concentration and the DMR assume the forms

$$n_0 = \frac{g_v B\sqrt{2\pi m_2}\,(k_B T)^{3/2}}{2E_0 L_z \pi^2 \hbar^2} \sum_{n=0}^{n_{\max}} \left[F_{1/2}(\bar{e}_1) - F_{1/2}(\bar{e}_2)\right], \tag{4.82}$$

$$\frac{D}{\mu} = \frac{k_B T}{|e|}\left[\frac{\sum_{n=0}^{n_{\max}}\left[F_{1/2}(\bar{e}_1) - F_{1/2}(\bar{e}_2)\right]}{\sum_{n=0}^{n_{\max}}\left[F_{-1/2}(\bar{e}_1) - F_{-1/2}(\bar{e}_2)\right]}\right], \tag{4.83}$$

where

$$\bar{e}_1 \equiv (k_B T)^{-1}\left(\bar{E}_{FB} - \bar{e}_3\right),$$
$$\bar{e}_3 \equiv \left[(n+1/2)\,\hbar\omega_{03} + \frac{m_1}{2}\left(\frac{E_0}{B}\right)^2 - |e|\,E_0 L_z\right],$$
$$\bar{e}_2 \equiv (k_B T)^{-1}\left(\bar{E}_{FB} - \bar{e}_4\right),$$

and

$$\bar{e}_4 \equiv \bar{e}_3 + |e|\,E_0 L_z.$$

4.2.5 IV–VI Materials

The conduction electrons of IV–VI semiconductors obey the Cohen model of bismuth and (4.68) and (4.69) should be used for the electron concentration and the DMR in this case along with the appropriate change of energy band constants.

4.2.6 Stressed Kane Type Semiconductors

Equation (2.48) can be written as

$$(E-\alpha_1){k_x}^2 + (E-\alpha_2){k_y}^2 + (E-\alpha_3){k_z}^2 = t_1E^3 - t_2E^2 + t_3E + t_4, \tag{4.84}$$

where

$$\alpha_1 \equiv \left[E_g - C_1\varepsilon - (\bar{a}_0 + C_1)\varepsilon + \frac{3}{2}\bar{b}_0\varepsilon_{xx} - \frac{\bar{b}_0}{2}\varepsilon + \left(\sqrt{3}/2\right)\varepsilon_{xy}\bar{d}_0\right],$$
$$\alpha_2 \equiv \left[E_g - C_1\varepsilon - (\bar{a}_0 + C_1)\varepsilon + \frac{3}{2}\bar{b}_0\varepsilon_{xx} - \frac{\bar{b}_0}{2}\varepsilon - \left(\sqrt{3}/2\right)\varepsilon_{xy}\bar{d}_0\right],$$
$$\alpha_3 \equiv \left[E_g - C_1\varepsilon - (\bar{a}_0 + C_1)\varepsilon + \frac{3}{2}\bar{b}_0\varepsilon_{zz} - \frac{\bar{b}_0}{2}\varepsilon\right],\ t_1 \equiv \left({}^{3}\!/_{2B_2^2}\right),$$
$$t_2 \equiv \left({}^{1}\!/_{2B_2^2}\right)\left[6(E_g - C_1\varepsilon) + 3C_1\varepsilon\right],$$
$$t_3 \equiv \left({}^{1}\!/_{2B_2^2}\right)\left[3(E_g - C_1\varepsilon)^2 + 6C_1\varepsilon(E_g - C_1\varepsilon) - 2C_2^2\varepsilon_{xy}^2\right] \text{ and}$$
$$t_4 \equiv \left({}^{1}\!/_{2B_2^2}\right)\left[-3C_1\varepsilon(E_g - C_1\varepsilon)^2 + 2C_2^2\varepsilon_{xy}^2\right].$$

In the presence of a quantizing magnetic field B along the z direction and the electric field along the x axis, from (4.84) one obtains

$$\frac{\hat{p}_x^2}{2M_{\|}(E)} + \frac{(\hat{p}_y - |e|\, B\hat{x})^2}{2M_{\perp}(E)} + R(E)\hat{p}_z^2 = \rho_5(E) + |e|\, E_0\hat{x}\, [\rho_5\,(E)]', \tag{4.85}$$

where

$$M_{\|}(E) \equiv \frac{1}{2P(E)},\ P(E) \equiv \frac{1}{\hbar^2}(E-\alpha_1),\ M_{\perp}(E) \equiv \frac{1}{2Q(E)},$$
$$Q(E) \equiv \frac{1}{\hbar^2}(E-\alpha_2),\ R(E) \equiv \frac{1}{\hbar^2}(E-\alpha_3) \text{ and}$$
$$\rho_5(E) \equiv \left[t_1E^3 - t_2E^2 + t_3E + t_4\right].$$

Let us define the operator $\hat{\theta}$ as

$$\hat{\theta} = -\hat{p}_y + |e|\, B\hat{x} - \frac{M_{\perp}(E)E_0\, [\rho_5\,(E)]'}{B} \tag{4.86}$$

Combining (4.85) and (4.86), we can write

$$\frac{\hat{p}_x^2}{2M_{\|}(E)} + \frac{\hat{\theta}^2}{2M_{\perp}(E)} + \frac{M_{\perp}(E){E_0}^2\left\{[\rho_5\,(E)]'\right\}^2}{2B^2} + \hat{\theta}\frac{E_0\,[\rho_5\,(E)]'}{B} + R(E){\hat{p}_z}^2$$
$$= \rho_5(E) + \frac{E_0\,[\rho_5\,(E)]'\,\hat{\theta}}{B} + \frac{E_0}{B}\,[\rho_5\,(E)]'\,\hat{p}_y + \frac{{E_0}^2}{B^2}M_{\perp}(E)\left\{[\rho_5\,(E)]'\right\}^2. \tag{4.87}$$

Therefore the electron dispersion relation in stressed Kane type semiconductors in the presence of cross field configuration can be expressed as

$$\rho_5(E) = (n+1/2)\,\hbar\bar{\omega}(E) + R(E)\,[\hbar k_z(E)]^2 - \frac{E_0}{B}\,[\rho_5\,(E)]'\,\hbar k_y(E)$$
$$-\left\{\frac{M_{\perp}(E){E_0}^2\left\{[\rho_5\,(E)]'\right\}^2}{2B^2}\right\} \tag{4.88}$$

where

$$\bar{\omega}\,(E) \equiv \frac{|e|\, B}{\sqrt{M_{\|}(E)M_{\perp}(E)}}.$$

The use of (4.88) leads to the expressions of EMM s' along the z and y directions as

$$m_z^*\left(\bar{E}_{\mathrm{FB}}, n, E_0, B\right) = \frac{1}{2}\left[R\left(\bar{E}_{\mathrm{FB}}\right)\right]^{-2}$$
$$\left\{R\left(\bar{E}_{\mathrm{FB}}\right)\left[\left[\rho_5\left(\bar{E}_{\mathrm{FB}}\right)\right]' - \left(n+\frac{1}{2}\right)\hbar\left[\bar{\omega}\left(\bar{E}_{\mathrm{FB}}\right)\right]'\right.\right.$$

$$+\frac{M_{\perp}\left(\bar{E}_{\mathrm{FB}}\right)E_0{}^2\left[\rho_5\left(\bar{E}_{\mathrm{FB}}\right)\right]'\left[\rho_5\left(\bar{E}_{\mathrm{FB}}\right)\right]''}{B^2}+\frac{\left\{M_{\perp}\left(\bar{E}_{\mathrm{FB}}\right)\right\}'E_0{}^2\left\{\left[\rho_5\left(\bar{E}_{\mathrm{FB}}\right)\right]'\right\}^2}{2B^2}\Bigg]$$

$$-\left\{R\left(\bar{E}_{\mathrm{FB}}\right)\right\}'\left[\rho_5\left(\bar{E}_{\mathrm{FB}}\right)-\left(n+\frac{1}{2}\right)\hbar\bar{\omega}\left(\bar{E}_{\mathrm{FB}}\right)+\frac{M_{\perp}\left(\bar{E}_{\mathrm{FB}}\right)E_0{}^2\left\{\left[\rho_5\left(\bar{E}_{\mathrm{FB}}\right)\right]'\right\}^2}{2B^2}\right]\Bigg]\Bigg\}, \tag{4.89}$$

and

$$m_y^*\left(\bar{E}_{\mathrm{FB}},n,E_0,B\right)=\left[\frac{B}{E_0}\right]^2\left\{\left[\rho_5\left(\bar{E}_{\mathrm{FB}}\right)\right]'\right\}^{-3}$$

$$\times\left[\rho_5\left(\bar{E}_{\mathrm{FB}}\right)-\left(n+\frac{1}{2}\right)\hbar\bar{\omega}\left(\bar{E}_{\mathrm{FB}}\right)+\frac{M_{\perp}\left(\bar{E}_{\mathrm{FB}}\right)E_0{}^2\left\{\left[\rho_5\left(\bar{E}_{\mathrm{FB}}\right)\right]'\right\}^2}{2B^2}\right]$$

$$\times\Bigg[\left[\rho_5\left(\bar{E}_{\mathrm{FB}}\right)\right]'\Bigg[\left[\rho_5\left(\bar{E}_{\mathrm{FB}}\right)\right]'-\left(n+\frac{1}{2}\right)\hbar\left[\bar{\omega}\left(\bar{E}_{\mathrm{FB}}\right)\right]'$$

$$+\frac{M_{\perp}\left(\bar{E}_{\mathrm{FB}}\right)E_0{}^2\left[\rho_5\left(\bar{E}_{\mathrm{FB}}\right)\right]'\left[\rho_5\left(\bar{E}_{\mathrm{FB}}\right)\right]''}{B^2}+\frac{\left[M_{\perp}\left(\bar{E}_{\mathrm{FB}}\right)\right]'E_0{}^2\left\{\left[\rho_5\left(\bar{E}_{\mathrm{FB}}\right)\right]'\right\}^2}{2B^2}\Bigg]$$

$$-\left[\rho_5\left(\bar{E}_{\mathrm{FB}}\right)\right]''\left[\rho_5\left(\bar{E}_{\mathrm{FB}}\right)-\left(n+\frac{1}{2}\right)\hbar\bar{\omega}\left(\bar{E}_{\mathrm{FB}}\right)+\frac{M_{\perp}\left(\bar{E}_{\mathrm{FB}}\right)E_0{}^2\left\{\left[\rho_5\left(\bar{E}_{\mathrm{FB}}\right)\right]'\right\}^2}{2B^2}\right]\Bigg]. \tag{4.90}$$

The Landau level energy (E_{n_8}) in this case can be expressed through the equation

$$\rho_5\left(E_{n_8}\right)-\left(n+\frac{1}{2}\right)\hbar\bar{\omega}\left(E_{n_8}\right)+\frac{M_{\perp}\left(E_{n_8}\right)E_0{}^2\left\{\left[\rho_5\left(\bar{E}_{\mathrm{FB}}\right)\right]'\right\}^2}{2B^2}=0. \tag{4.91}$$

Equation (4.91) at the Fermi energy $\bar{E}_{\mathrm{FB}}$ when combined with (4.10) will generate the SdH period in this case. For this case

$$x_l(E)=\frac{-M_{\perp}(E)E_0\left[\rho_5\left(\bar{E}_{\mathrm{FB}}\right)\right]'}{B},\quad x_h(E)=\frac{|e|\,BL_x}{\hbar}+x_l(E). \tag{4.92}$$

The integral $I(E)$ for stressed Kane type semiconductors in the presence of crossed electric and quantizing magnetic fields assumes the form

$$I(E)=\frac{1}{\hbar\sqrt{R(E)}}\left(\frac{B}{E_0\hbar\left[\rho_5\left(\bar{E}_{\mathrm{FB}}\right)\right]'}\right)\int_{x_l(E)}^{x_h(E)}\left[T_5(n,E)+\frac{E_0\left[\rho_5\left(\bar{E}_{\mathrm{FB}}\right)\right]'}{B}\hbar k_y\right]^{1/2}\mathrm{d}k_y, \tag{4.93}$$

where

$$T_5(n,E) \equiv \left[\rho_5(E) - \left(n+{}^1\!/_2\right)\hbar\bar{\omega}(E) + \frac{M_\perp(E){E_0}^2\left\{\left[\rho_5\left(\bar{E}_{\mathrm{FB}}\right)\right]'\right\}^2}{2B^2}\right].$$

From (4.93), we get,

$$I(E) = \frac{1}{\hbar\sqrt{R(E)}}\frac{B}{E_0\hbar\left[\rho_5\left(\bar{E}_{\mathrm{FB}}\right)\right]'}\frac{2}{3}\left[\left[T_5(n,E) + \frac{E_0}{B}\left[\rho_5\left(\bar{E}_{\mathrm{FB}}\right)\right]'\hbar\, x_h(E)\right]^{3/2} - \left[T_5(n,E) + \frac{E_0}{B}\left[\rho_5\left(\bar{E}_{\mathrm{FB}}\right)\right]'\hbar\, x_l(E)\right]^{3/2}\right]. \quad (4.94)$$

Therefore, the electron concentration can be written as

$$n_0 = \frac{2B}{3L_x\pi^2\hbar^2E_0}\sum_{n=0}^{n_{\max}}\left[T_{413}\left(n,\bar{E}_{\mathrm{FB}}\right) + T_{414}\left(n,\bar{E}_{\mathrm{FB}}\right)\right], \quad (4.95)$$

where

$$T_{413}(n,\bar{E}_{\mathrm{FB}}) \equiv \left[\frac{1}{\left(\sqrt{R\left(\bar{E}_{\mathrm{FB}}\right)}\right)\left(\rho_5\left(\bar{E}_{\mathrm{FB}}\right)\right)'}\right] \times\left[\left[T_5(n,\bar{E}_{\mathrm{FB}}) + \frac{E_0}{B}\left[\rho_5\left(\bar{E}_{\mathrm{FB}}\right)\right]'\hbar x_h(\bar{E}_{\mathrm{FB}})\right]^{3/2} - \left[T_5(n,\bar{E}_{\mathrm{FB}}) + \frac{E_0}{B}\left[\rho_5\left(\bar{E}_{\mathrm{FB}}\right)\right]'\hbar x_l(\bar{E}_{\mathrm{FB}})\right]^{3/2}\right] \text{ and}$$

$$T_{414}\left(n,\bar{E}_{\mathrm{FB}}\right) \equiv \sum_{r=1}^{s} L(r)\,T_{413}\left(n,\bar{E}_{\mathrm{FB}}\right).$$

The use of (4.95) and (1.11) leads to the expression of the DMR as

$$\frac{D}{\mu} = \frac{1}{|e|}\left[\frac{\sum_{n=0}^{n_{\max}}\left[T_{413}\left(n,\bar{E}_{\mathrm{FB}}\right) + T_{414}\left(n,\bar{E}_{\mathrm{FB}}\right)\right]}{\sum_{n=0}^{n_{\max}}\left[\left\{T_{413}\left(n,\bar{E}_{\mathrm{FB}}\right)\right\}' + \left\{T_{414}\left(n,\bar{E}_{\mathrm{FB}}\right)\right\}'\right]}\right]. \quad (4.96)$$

4.3 Result and Discussions

In Figs. 4.1 and 4.2, the normalized DMR in the presence of cross field configuration has been plotted as a function of inverse quantizing magnetic field for n-Cd_3As_2 and n-$CdGeAs_2$ respectively.

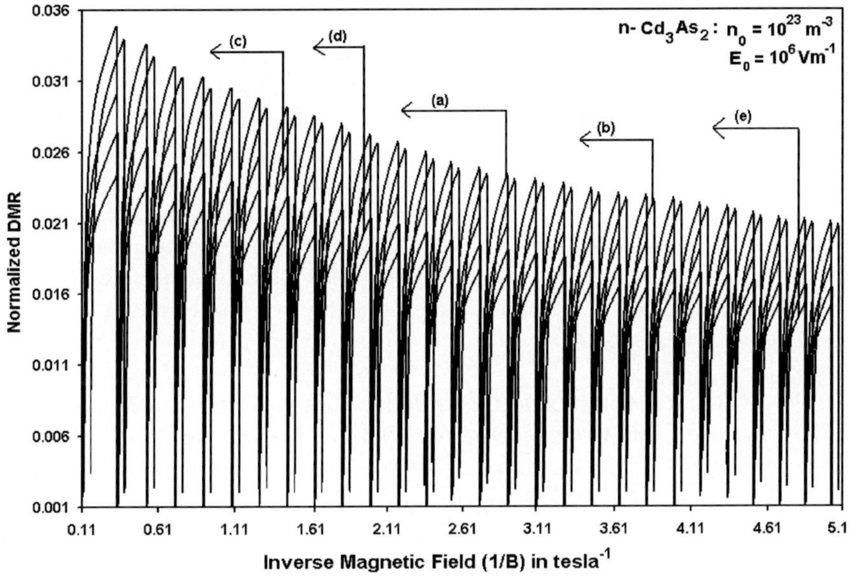

Fig. 4.1. The plot of the DMR in n-Cd_3As_2 as a function of inverse quantizing magnetic field under cross field configuration in accordance with (**a**) the generalized band model, (**b**) $\delta = 0$, (**c**) the three band model of Kane, (**d**) the two band model of Kane and (**e**) the parabolic energy bands

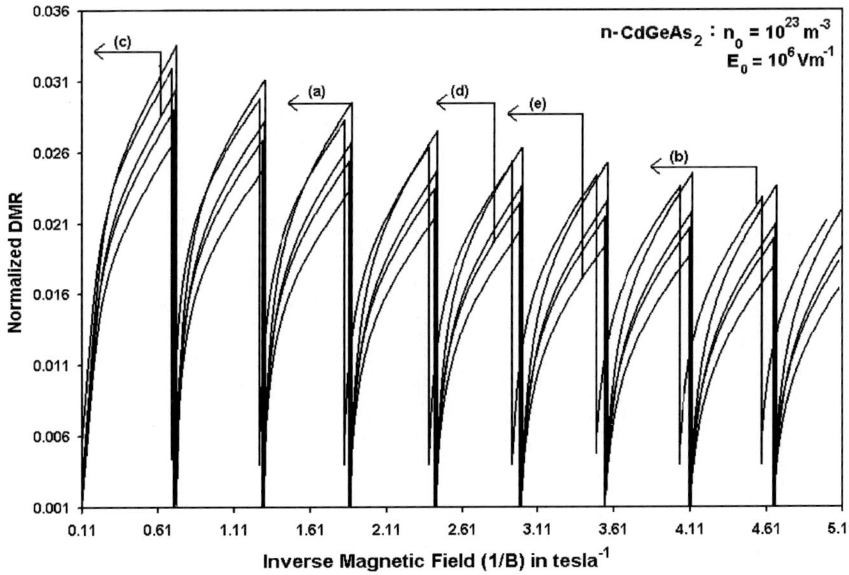

Fig. 4.2. The plot of the DMR in n-$CdGeAs_2$ as a function of inverse quantizing magnetic field under cross field configuration in accordance with (**a**) the generalized band model, (**b**) $\delta = 0$, (**c**) the three band model of Kane, (**d**) the two band model of Kane and (**e**) the parabolic energy bands

In the same figures the plots corresponding to $\delta = 0$, the three and the two band models of Kane together with the parabolic energy bands have also been included for the purpose of relative comparison. It appears from both the figures that the DMR is an oscillatory function of inverse quantizing magnetic field with various numerical values for the three and two band models of Kane together with parabolic energy band. The crystal field splitting modifies the DMR in the whole range of the magnetic field considered for both the n-Cd_3As_2 and n-$CdGeAs_2$ respectively. The presence of the isotropic spin orbit splitting constant in the three band model of Kane changes the value of the DMR as compared with the corresponding two band Kane model. The physics behind the oscillatory plots has already been described in Sect. 3.3 of Chap. 3.

In Figs. 4.3 and 4.4, the concentration dependence of the DMR under cross field configuration has been plotted for all the cases of Fig. 4.1 for both the compounds. The DMR again shows oscillatory dependence with different magnitudes. Although the rate of variations are different, the influence of the energy band constants in accordance with all the type of the band models follows the same trend as shown in Figs. 4.1 and 4.2.

Figures 4.5 and 4.6 exhibit the variations of the DMR with the electric field under cross field configuration for both the n-Cd_3As_2 and n-$CdGeAs_2$.

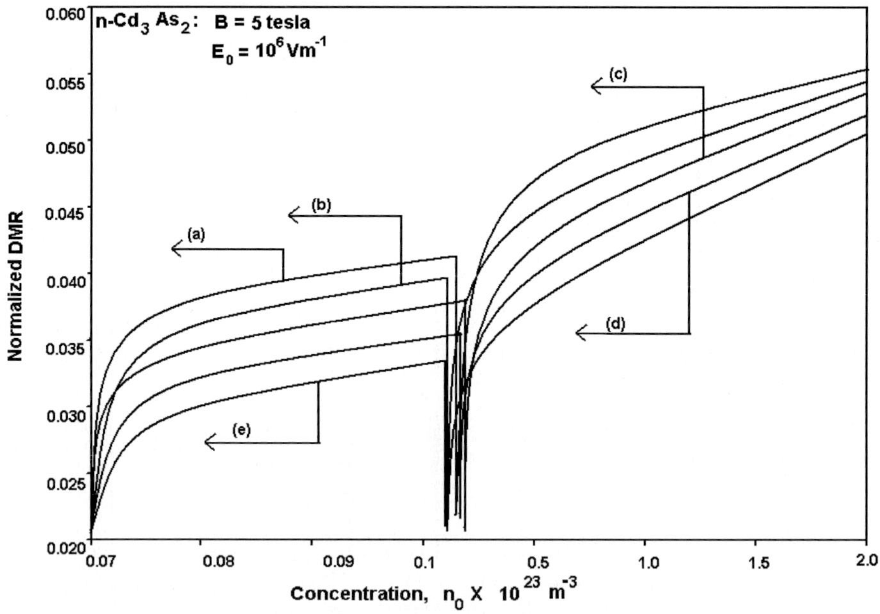

Fig. 4.3. The plot of the DMR in n-Cd_3As_2 as a function of electron concentration under cross field configuration in accordance with (**a**) the generalized band model, (**b**) $\delta = 0$ (**c**) the three band model of Kane, (**d**) the two band model of Kane and (**e**) the parabolic energy bands

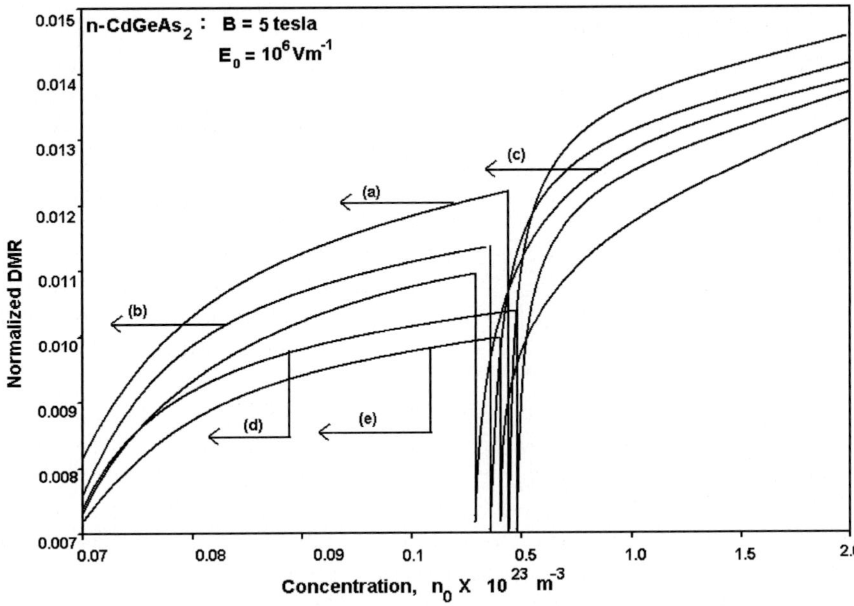

Fig. 4.4. The plot of the DMR n-$CdGeAs_2$ as a function of electron concentration under cross field configuration in accordance with (**a**) the generalized band model, (**b**) $\delta = 0$, (**c**) the three band model of Kane, (**d**) the two band model of Kane and (**e**) the parabolic energy bands

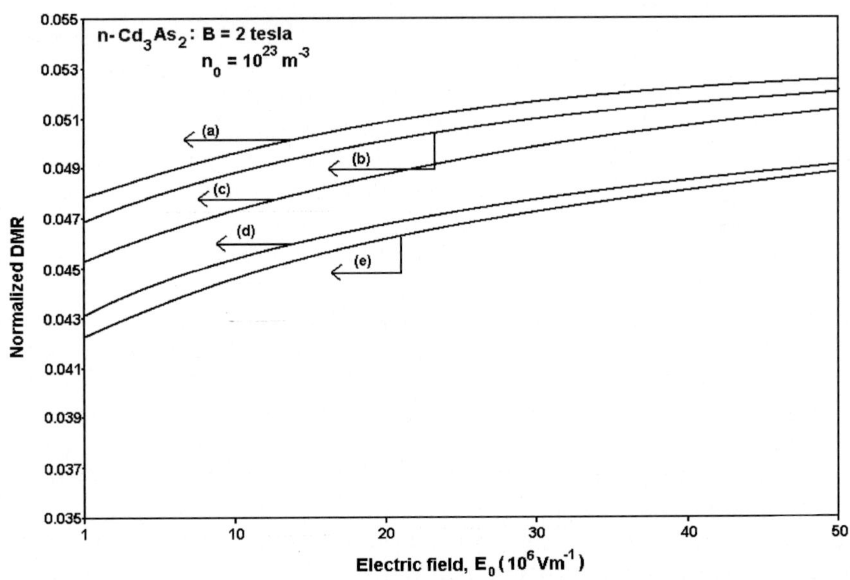

Fig. 4.5. The plot of the DMR n-Cd_3As_2 as a function of electric field under cross field configuration in accordance with (**a**) the generalized band model; (**b**) $\delta = 0$; (**c**) the three band model of Kane; (**d**) the two band model of Kane and (**e**) the parabolic energy bands

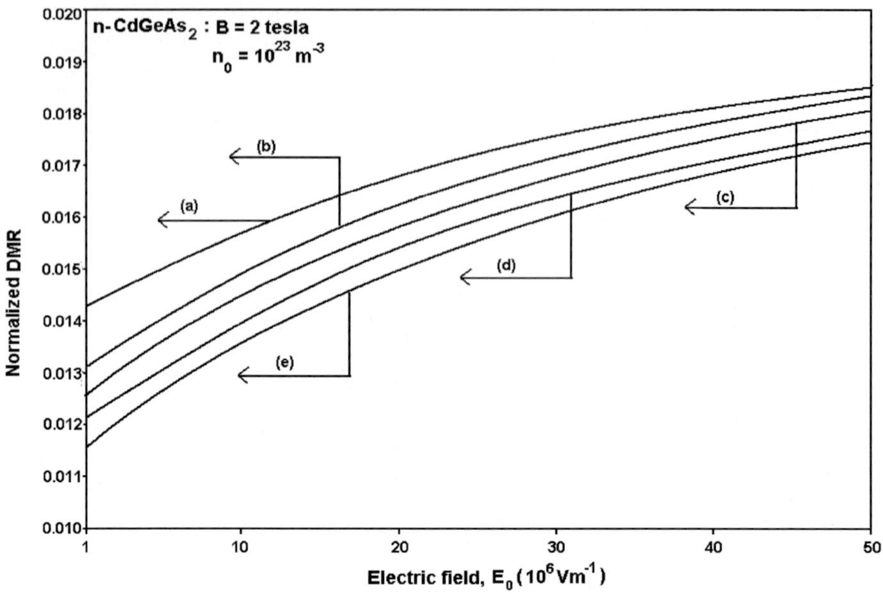

Fig. 4.6. The plot of the DMR n-$CdGeAs_2$ as a function of electric field under cross field configuration in accordance with (**a**) the generalized band model; (**b**) $\delta = 0$; (**c**) the three band model of Kane; (**d**) the two band model of Kane, and (**e**) the parabolic energy bands

It appears that the DMR increases with increasing E_0 and the presence of the crystal field splitting constant enhances the numerical values of the DMR for the whole range of E_0 for both the figures. In Figs. 4.7–4.9, the normalized DMR as functions of the inverse quantizing magnetic field under cross field configuration for GaAs, InSb and InAs has been plotted in accordance with the three and two band models of Kane together with the parabolic energy bands respectively. The variations of the DMR under cross field configuration and the influence of the energy band constants on the DMR in accordance with all the band models are apparent from the said figures.

Figures 4.10–4.12 exhibit the concentration dependence of the oscillatory DMR under cross field configuration for the said materials with different magnitudes. Figures 4.13–4.15 show the dependence of the DMR on the electric field in the presence of crossed electric and quantizing magnetic fields. The DMR increases with increasing electric field and the influence of the three and two band models of Kane together with the parabolic energy bands can also be assessed from the said figures. Figures 4.16 and 4.17 show the dependence of the DMR on $1/B$ under cross field configuration for n-$Hg_{1-x}Cd_xTe$ and n-$In_{1-x}Ga_xAs_yP_{1-y}$ lattice matched to InP respectively. Figures 4.18 and 4.19 exhibit the concentration dependence of the DMR for the said materials.

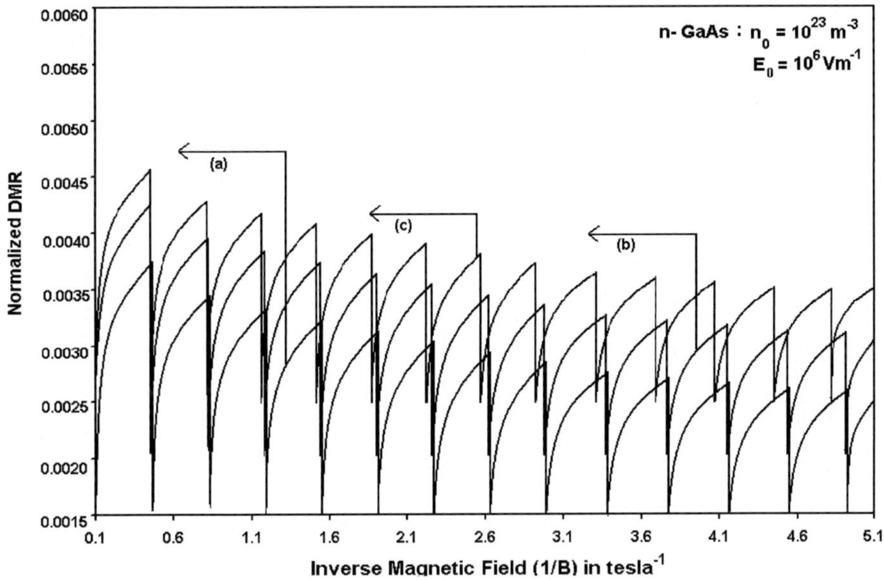

Fig. 4.7. The plot of the DMR in n-GaAs as a function of inverse quantizing magnetic field under cross field configuration in accordance with (**a**) the three band model of Kane, (**b**) the two band model of Kane, and (**c**) the parabolic energy bands

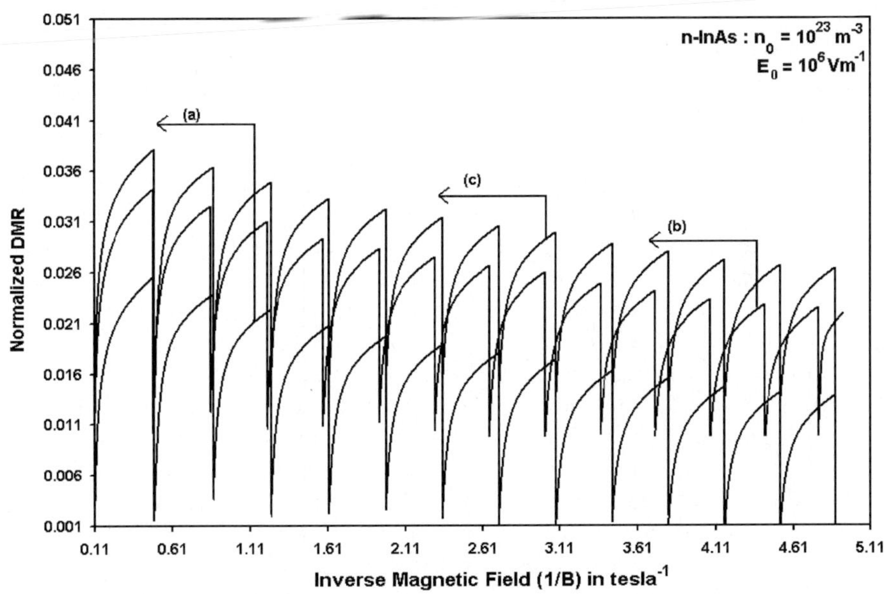

Fig. 4.8. The plot of the DMR in n-InAs as a function of inverse quantizing magnetic field under cross field configuration in accordance with (**a**) the three band model of Kane, (**b**) the two band model of Kane, and (**c**) the parabolic energy bands

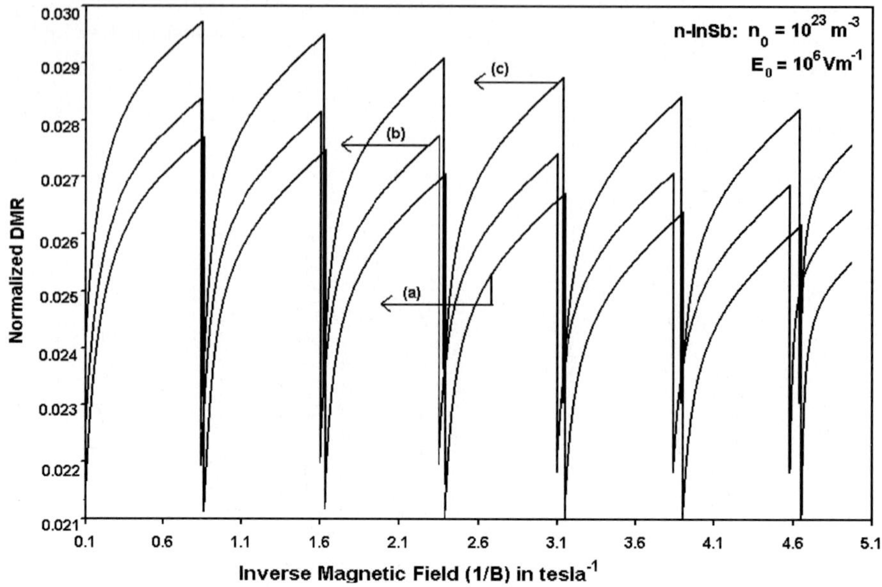

Fig. 4.9. The plot of the DMR in n-InSb as a function of inverse quantizing magnetic field under cross field configuration in accordance with (**a**) the three band model of Kane, (**b**) the two band model of Kane, and (**c**) the parabolic energy bands

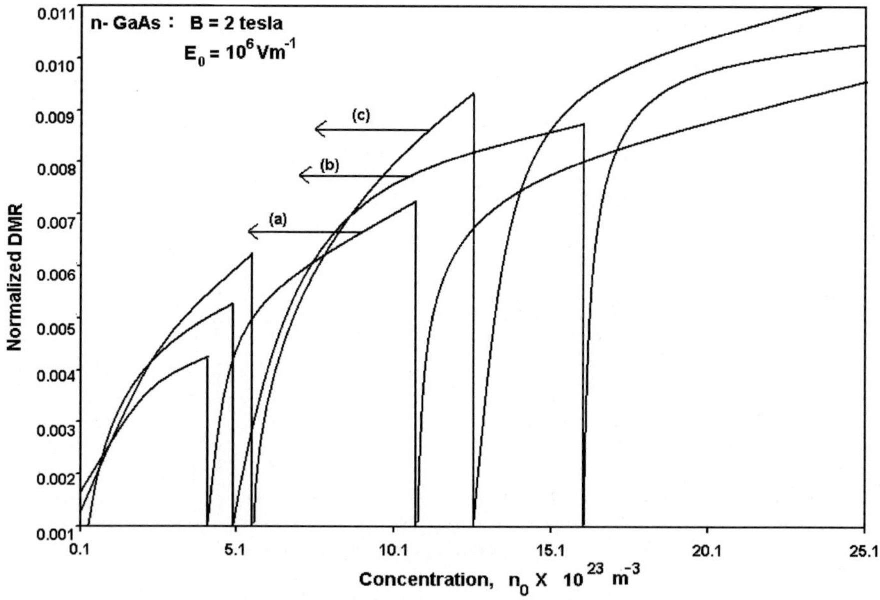

Fig. 4.10. The plot of the DMR in n-GaAs as a function of electron concentration under cross field configuration in accordance with (**a**) the three band model of Kane, (**b**) the two band model of Kane, and (**c**) the parabolic energy bands

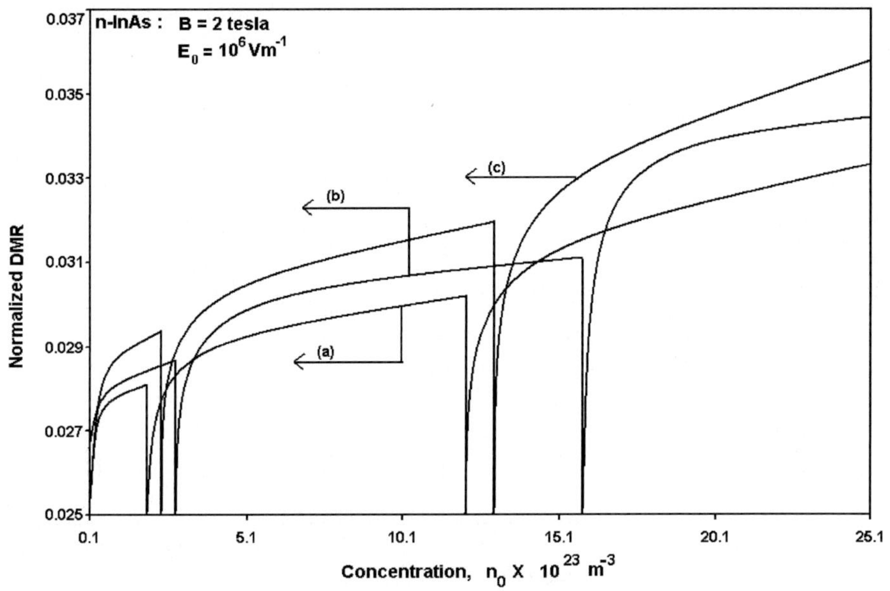

Fig. 4.11. The plot of the DMR in n-InAs as a function of electron concentration under cross field configuration in accordance with (**a**) the three band model of Kane, (**b**) the two band model of Kane, and (**c**) the parabolic energy bands

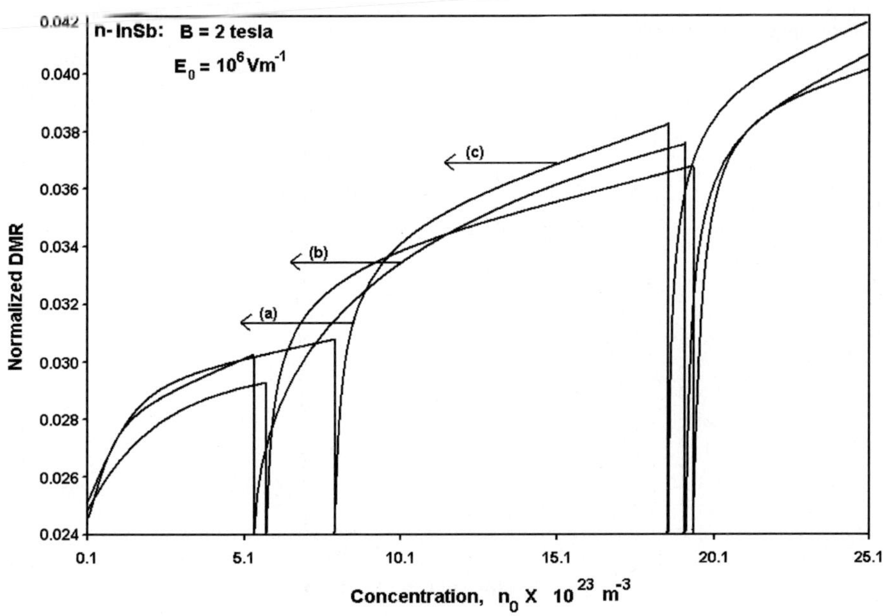

Fig. 4.12. The plot of the DMR in n-InSb as a function electron concentration under cross field configuration in accordance with (**a**) the three band model of Kane, (**b**) the two band model of Kane, and (**c**) the parabolic energy bands

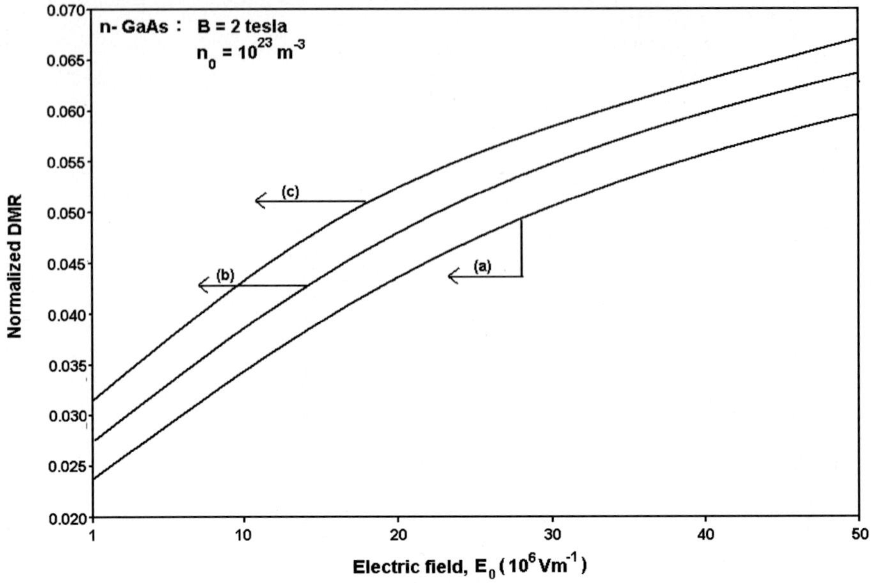

Fig. 4.13. The plot of the DMR in n-GaAs as a function electric field under cross field configuration electron concentration in accordance with (**a**) the three band model of Kane, (**b**) the two band model of Kane, and (**c**) the parabolic energy bands

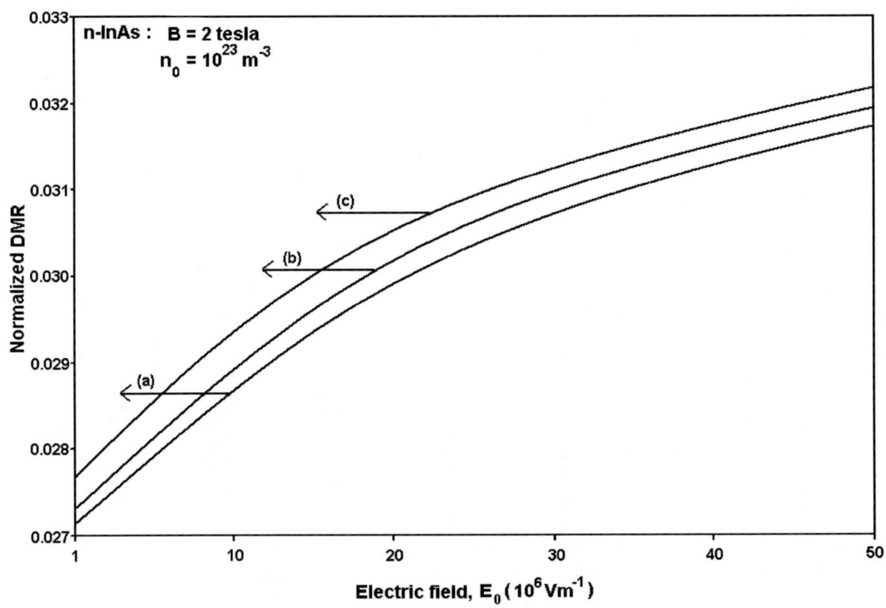

Fig. 4.14. The plot of the DMR in n-InAs as a function electric field under cross field configuration electron concentration in accordance with (**a**) the three band model of Kane, (**b**) the two band model of Kane, and (**c**) the parabolic energy bands

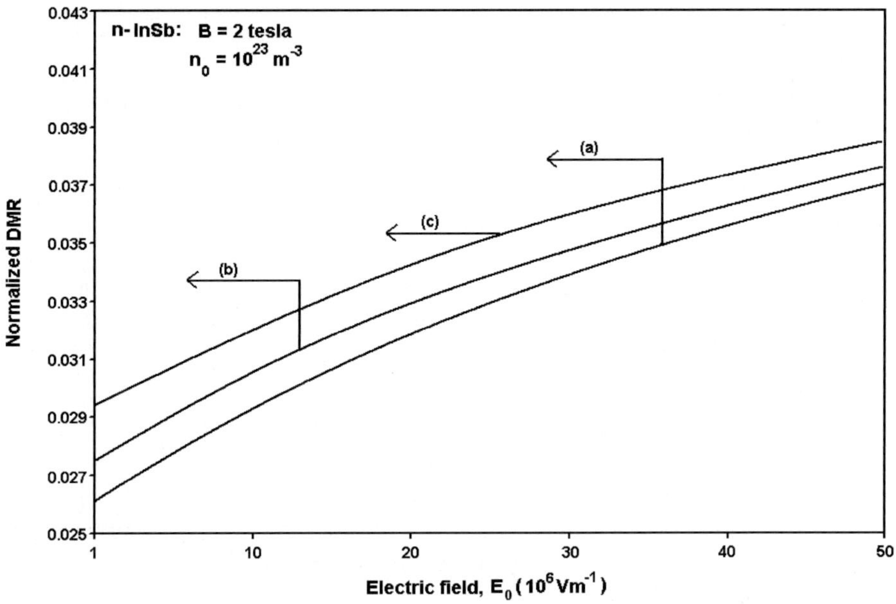

Fig. 4.15. The plot of the DMR in n-InSb as a function electric field under cross field configuration electron concentration in accordance with (**a**) the three band model of Kane, (**b**) the two band model of Kane, and (**c**) the parabolic energy bands

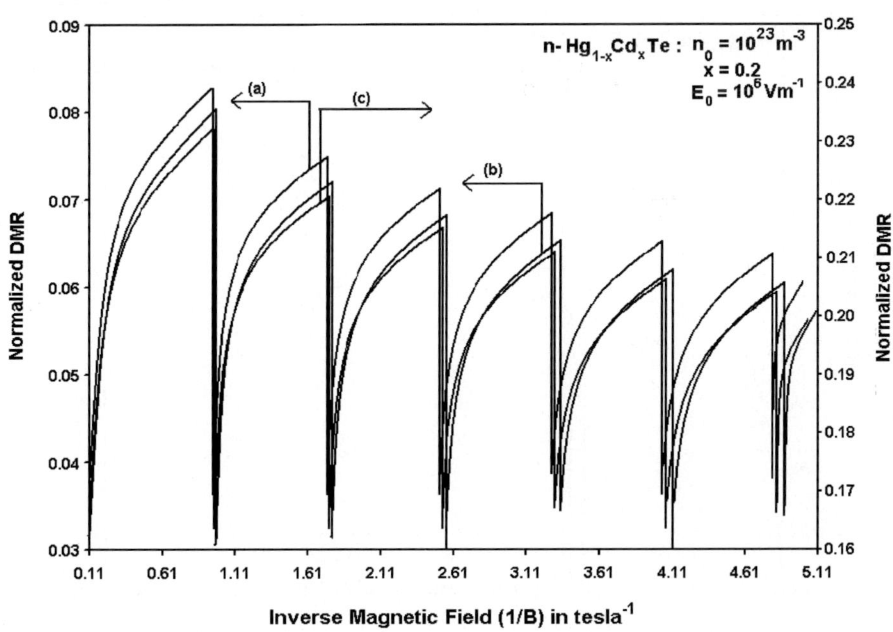

Fig. 4.16. The plot of the DMR in n-$Hg_{1-x}Cd_xTe$ as a function of inverse quantizing magnetic field under cross field configuration in accordance with (**a**) the three band model of Kane; (**b**) the two band model of Kane, and (**c**) the parabolic energy bands

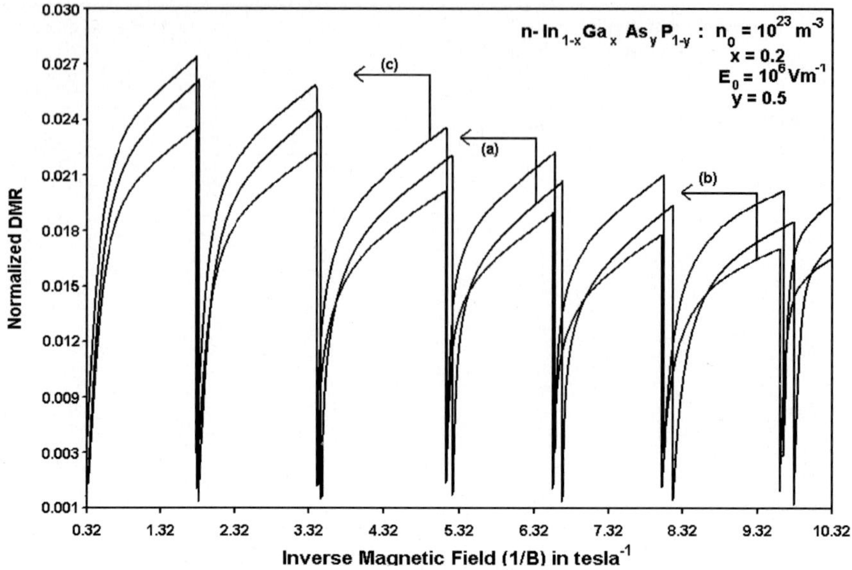

Fig. 4.17. The plot of the DMR in n-$In_{1-x}Ga_xAs_yP_{1-y}$ lattice matched to InP as a function of inverse quantizing magnetic field under cross field configuration in accordance with (**a**) the three band model of Kane; (**b**) the two band model of Kane, and (**c**) the parabolic energy bands

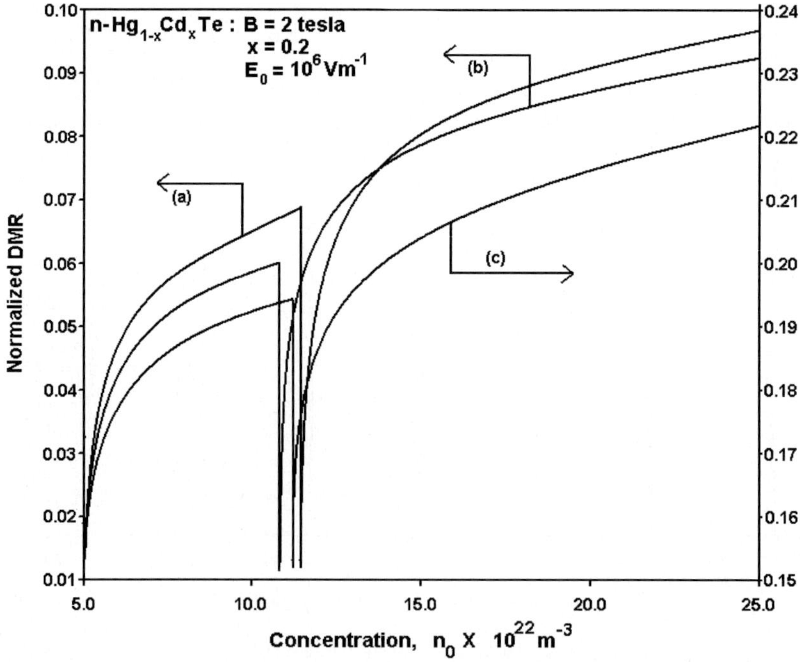

Fig. 4.18. The plot of the DMR in n-$Hg_{1-x}Cd_xTe$ as a function of electron concentration under cross field configuration in accordance with (**a**) the three band model of Kane; (**b**) the two band model of Kane, and (**c**) the parabolic energy bands

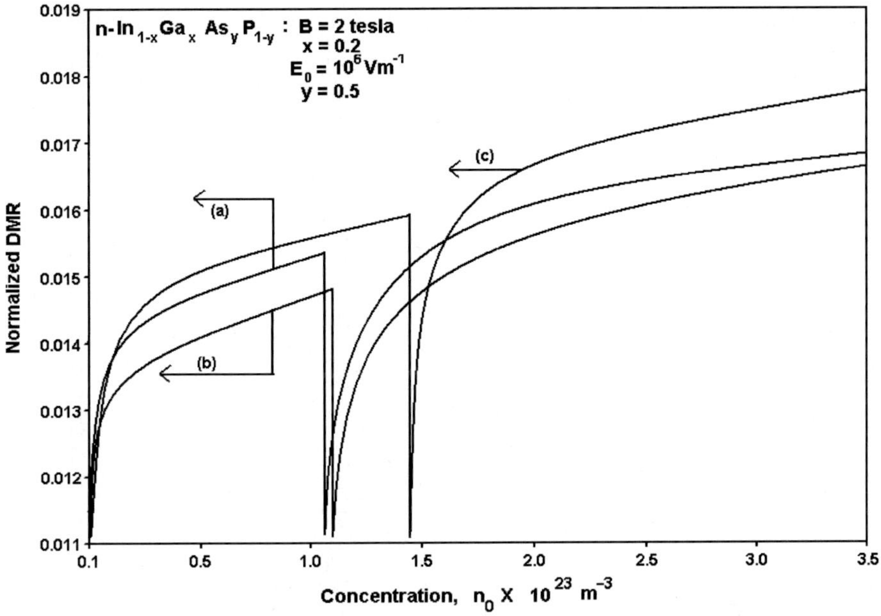

Fig. 4.19. The plot of the DMR in n-$In_{1-x}Ga_xAs_yP_{1-y}$ lattice matched to InP as a function of inverse quantizing magnetic field under cross field configuration in accordance with (**a**) the three band model of Kane, (**b**) the two band model of Kane, and (**c**) the parabolic energy bands

From Figs. 4.16–4.19, it appears that the numerical value of the DMR is greatest for the ternary materials while it is the least for quaternary materials for all types of variables in accordance with all types of band models. In Figs. 4.20 and 4.21 the DMR under cross fields has been plotted as a function of the electric field for both $Hg_{1-x}Cd_xTe$ and $In_{1-x}Ga_xAs_yP_{1-y}$ lattice matched to InP respectively. The DMR increases with increasing electric field. In Figs. 4.22 and 4.23, the DMR has been plotted as a function of alloy composition in the presence of cross fields for both the said compounds and it appears that the DMR decreases with increasing alloy composition. Figures 4.24–4.26 exhibit the dependences of the DMR under cross field configuration on $1/B$, n_0 and E_0 respectively for p-CdS. The influence of $\bar{\lambda}_0$ for all the variables is apparent from the figures. The normalized DMR as functions of $1/B$ under the cross field configuration has been plotted in Fig. 4.27 for the McClure and Choi, the Cohen, the Lax, and the ellipsoidal parabolic band models of Bismuth. The concentration dependence of the DMR has been plotted in Fig. 4.28 for all the band models of bismuth. Figure 4.29 exhibits the variation of the DMR in the present case with the electric field E_0 for all the cases of Fig. 4.27. The nature of oscillations and the numerical values are totally band structure dependent.

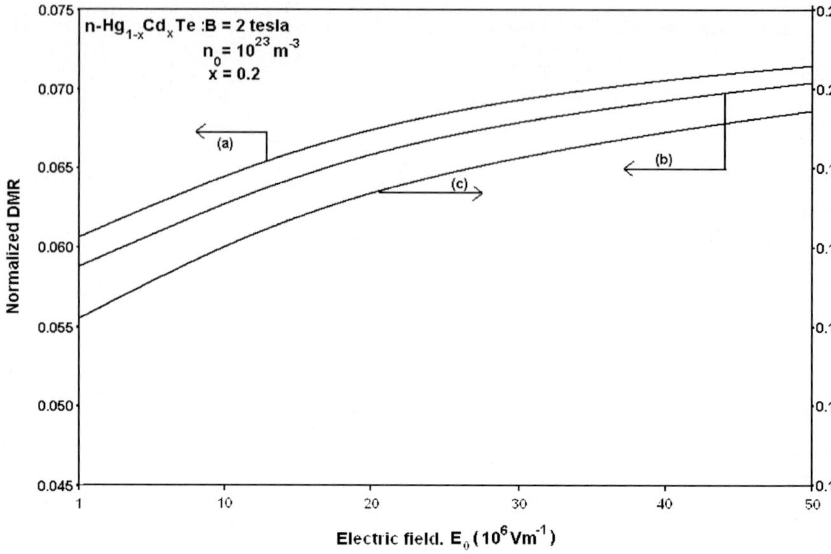

Fig. 4.20. The plot of the DMR in n-$Hg_{1-x}Cd_xTe$ as a function of electric field under cross field configuration in accordance with (**a**) the three band model of Kane; (**b**) the two band model of Kane, and (**c**) the parabolic energy bands

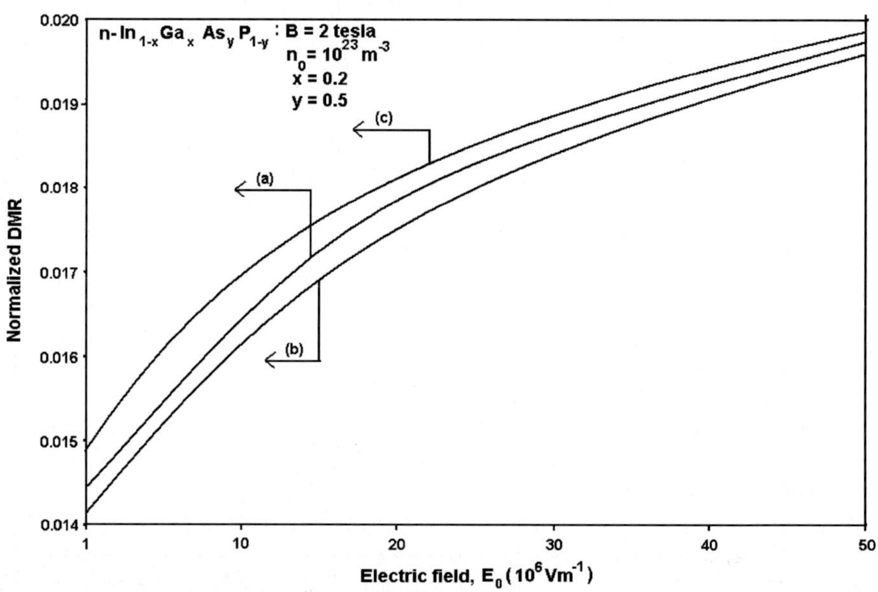

Fig. 4.21. The plot of the DMR in n-$In_{1-x}Ga_xAs_yP_{1-y}$ lattice matched to InP as a function of electron concentration under cross field configuration in accordance with (**a**) the three band model of Kane; (**b**) the two band model of Kane, and (**c**) the parabolic energy bands

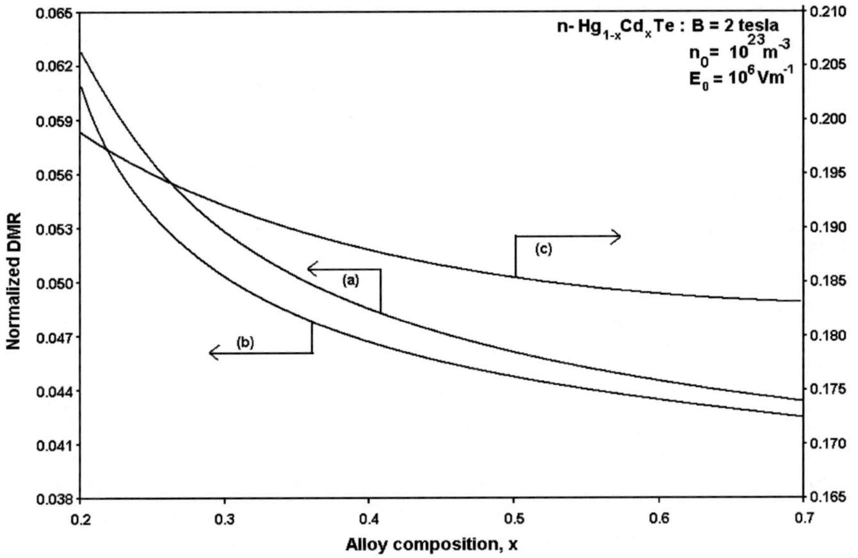

Fig. 4.22. The plot of the DMR in n-$Hg_{1-x}Cd_xTe$ as a function of alloy composition (x) under cross field in accordance with (**a**) the three band model of Kane; (**b**) the two band model of Kane and (**c**) the parabolic energy bands

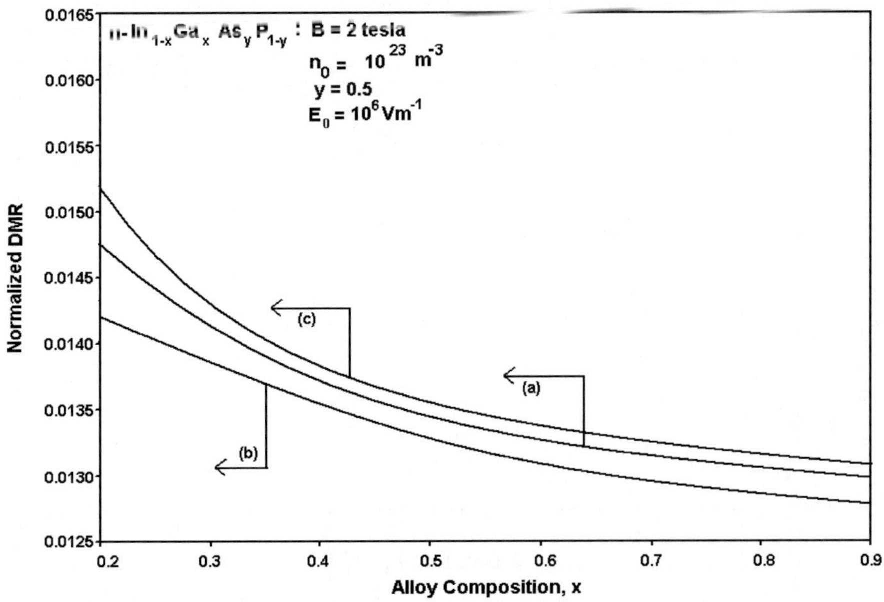

Fig. 4.23. The plot of the DMR in n-$In_{1-x}Ga_xAs_yP_{1-y}$ lattice matched to InP as a function of alloy composition (x) in accordance with (**a**) the three band model of Kane; (**b**) the two band model of Kane and (**c**) the parabolic energy bands

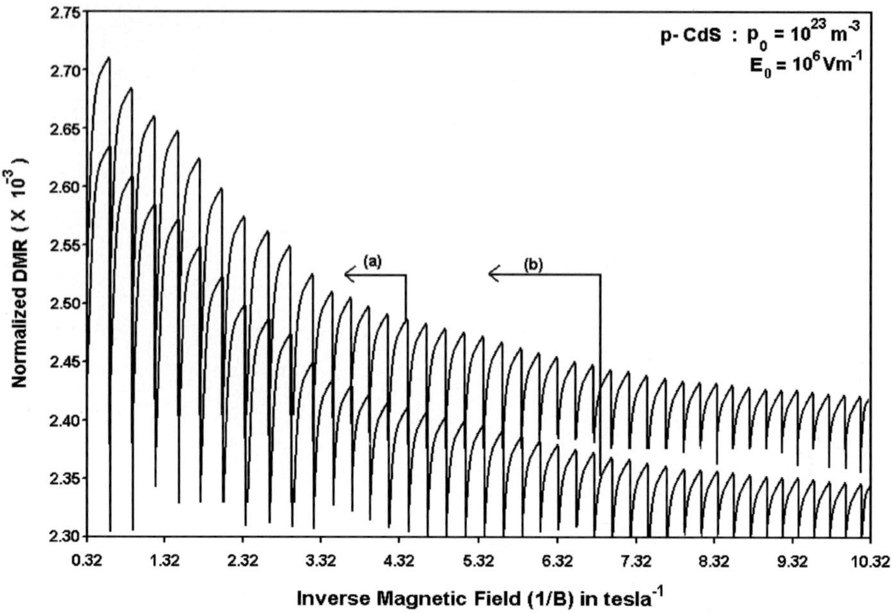

Fig. 4.24. The plot of the DMR as a function of inverse quantizing magnetic field under cross field configuration of p-CdS for (**a**) $\bar{\lambda}_0 \neq 0$ and (**b**) $\bar{\lambda}_0 = 0$

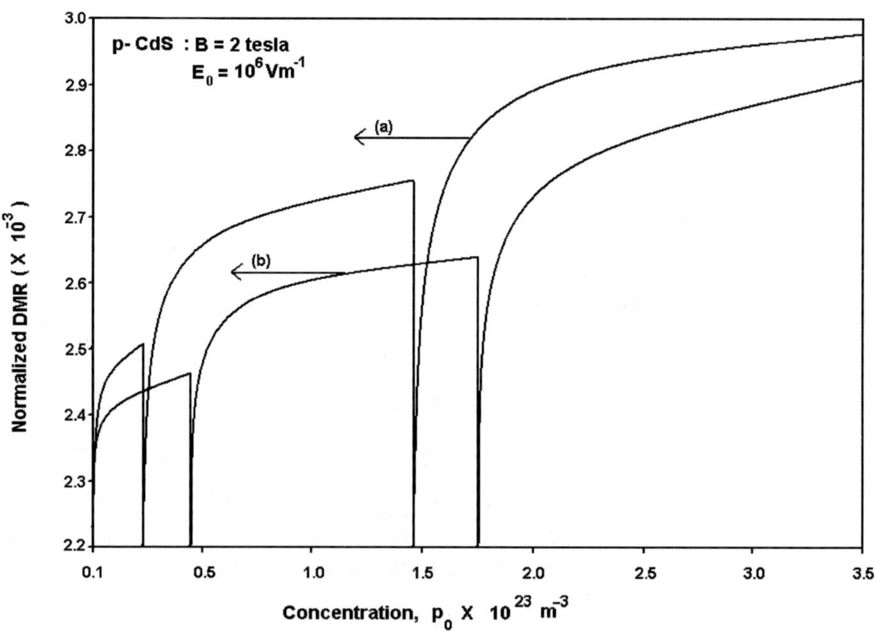

Fig. 4.25. The plot of the DMR as a function of hole concentration p_0 under cross field configuration of p-CdS for (**a**) $\bar{\lambda}_0 \neq 0$ and (**b**) $\bar{\lambda}_0 = 0$

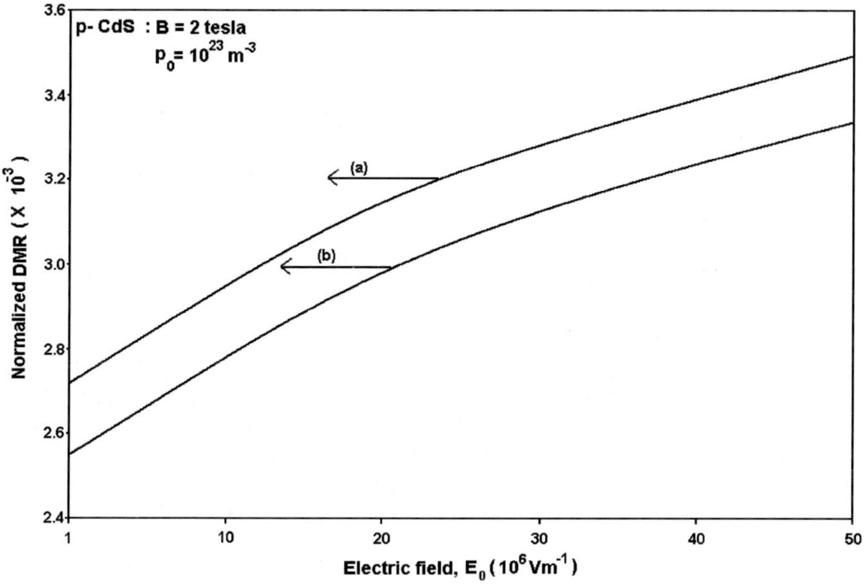

Fig. 4.26. The plot of the DMR as a function of electric field under cross field configuration of p-CdS for (**a**) $\bar{\lambda}_0 \neq 0$ and (**b**) $\bar{\lambda}_0 = 0$

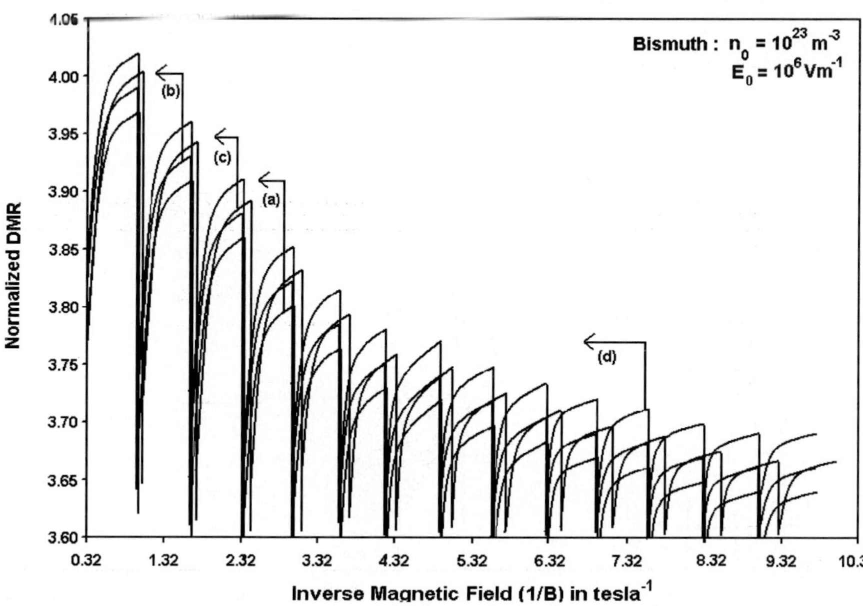

Fig. 4.27. The plot of the DMR in bismuth as a function of inverse quantizing magnetic field under the cross field configuration in accordance with (**a**) the McClure and Choi model (**b**) the Cohen model, (**c**) the Lax model and (**d**) the ellipsoidal parabolic energy bands

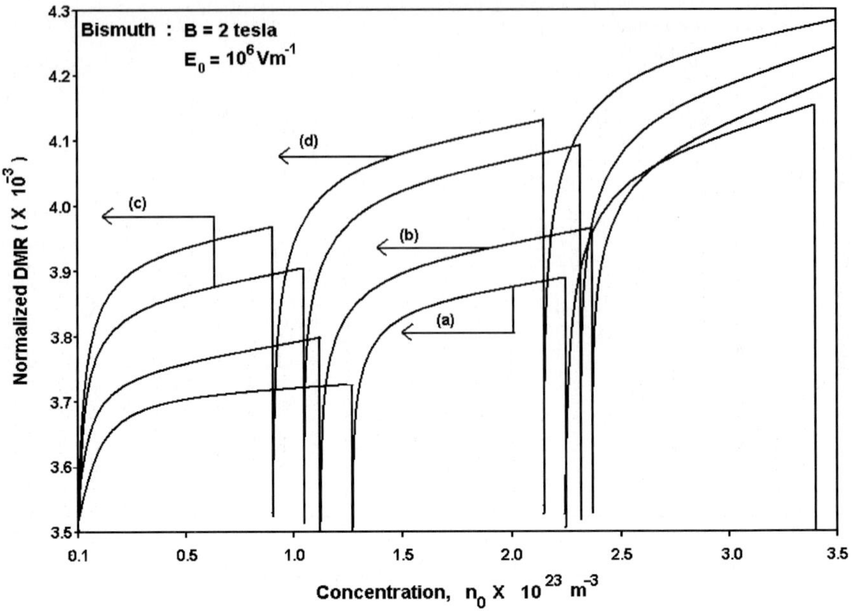

Fig. 4.28. The plot of the DMR in bismuth as a function of electron concentration under the cross field configuration in accordance with (**a**) the McClure and Choi model (**b**) the Cohen model, (**c**) the Lax model and (**d**) the ellipsoidal parabolic energy bands

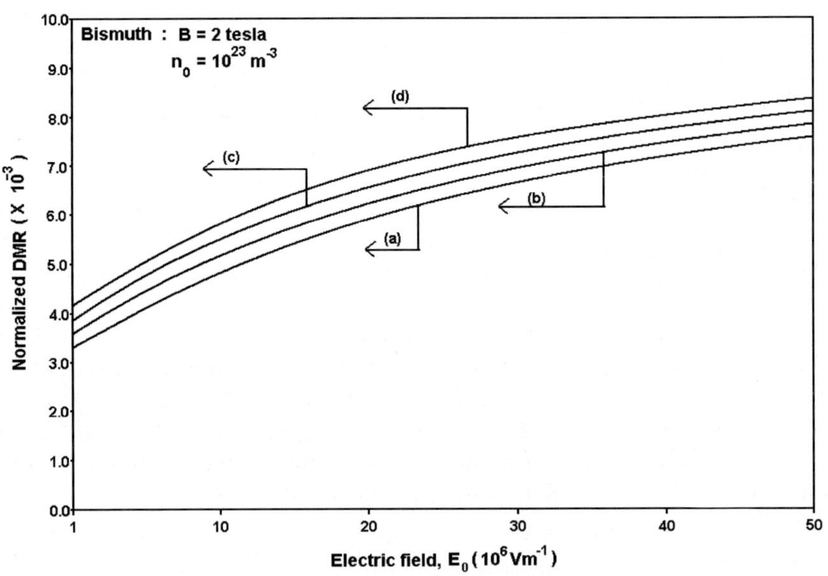

Fig. 4.29. The plot of the DMR in bismuth as a function of the electric field under the cross field configuration in accordance with (**a**) the McClure and Choi model (**b**) the Cohen model, (**c**) the Lax model and (**d**) the ellipsoidal parabolic energy bands

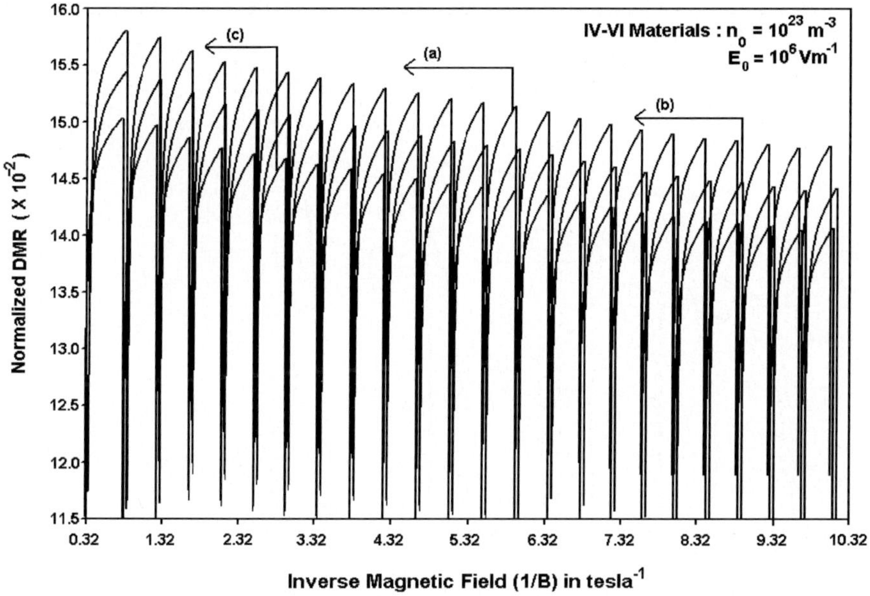

Fig. 4.30. The plot of the DMR in (**a**) n-PbTe (**b**) n-PbSnTe and (**c**) n-$Pb_{1-x}Sn_xSe$ as a function of inverse quantizing magnetic field under cross field configuration in accordance with the model of Cohen

The plots of the DMR under cross fields configuration for PbTe, n-PbSnTe and n-$Pb_{1-x}Sn_xSe$ as functions of $1/B$, n_0 and E_0 have been shown in Figs. 4.30–4.32 respectively in accordance with the model of Cohen.

Depending on the energy band constants, the values of the spikes of the DMR are greatest for n-PbTe and least for n-$Pb_{1-x}Sn_xSe$. Figures 4.33–4.35 exhibit the dependence of the magneto-DMR on $1/B$, n_0 and E_0 respectively for stressed n-InSb under cross field configuration both in the presence and absence of stress. The numerical value of the DMR in stressed materials is relatively large as compared with the stress-free condition for all the variables.

Although in a more rigorous statement the many body effects, the hot electron effects, spin and broadening should be considered along with the self-consistent procedure, the simplified analysis as presented in this chapter exhibits the basic qualitative features of the DMR in degenerate materials having various band structures in the presence of crossed electric and quantizing magnetic fields with reasonable accuracy. Our suggestion for the experimental determination of the DMR of Chap. 1 is also valid under crossed field configuration. As a collateral understanding, we have studied the EMMs along the directions of the magnetic and the electric fields. The characteristic feature of the cross fields is to introduce index-dependent oscillatory mass anisotropy. The Landau energy and the period have also been discussed.

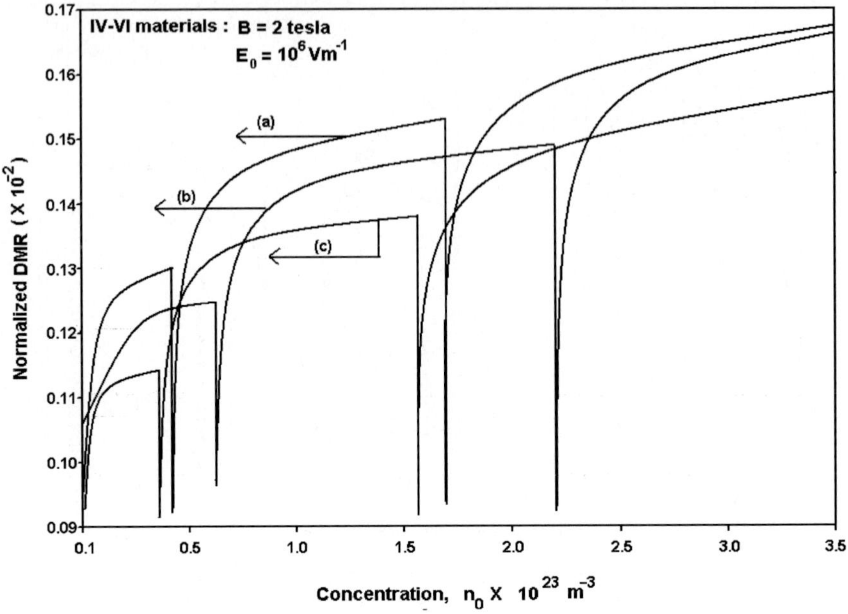

Fig. 4.31. The plot of the DMR in (**a**) PbTe (**b**) n-PbSnTe and (**c**) n-$Pb_{1-x}Sn_xSe$ as a function of electron concentration under cross field configuration in accordance with the model of Cohen

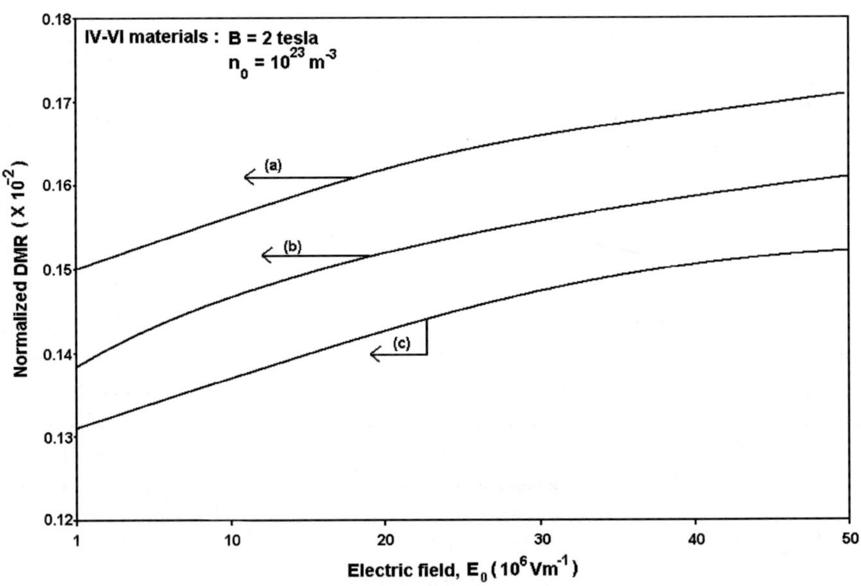

Fig. 4.32. The plot of the DMR in (**a**) PbTe (**b**) n-PbSnTe and (**c**) n-$Pb_{1-x}Sn_xSe$ as a function of electric field in accordance with the model of Cohen

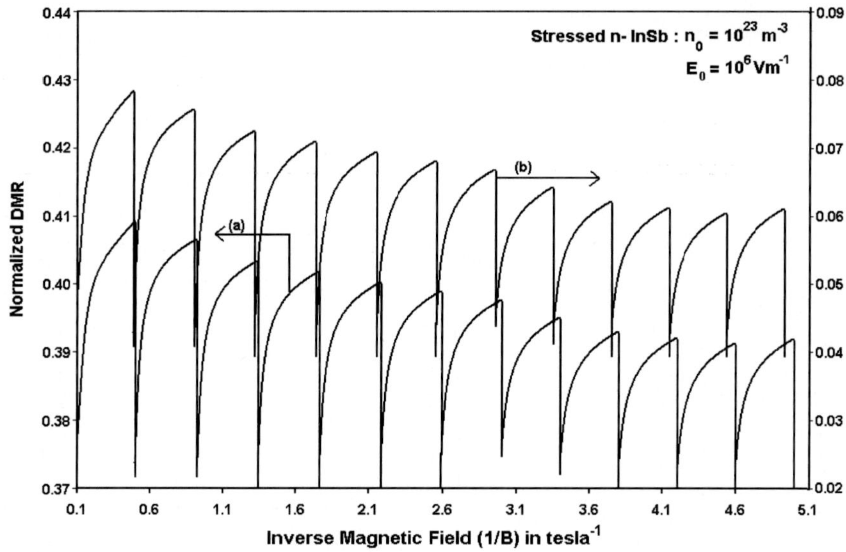

Fig. 4.33. The plot of the DMR in stressed n-InSb as a function of inverse quantizing magnetic field under cross field configuration both in the presence and absence of stress as shown by the curves (**b**) and (**a**) respectively

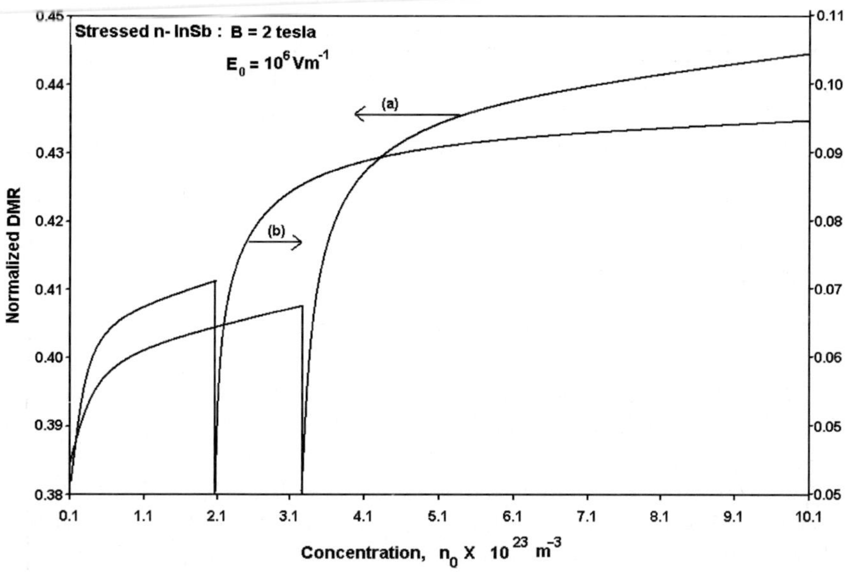

Fig. 4.34. The plot of the DMR in stressed n-InSb as a function of electron concentration under cross field configuration for both in the presence and absence of stress as shown by the curves (**b**) and (**a**) respectively

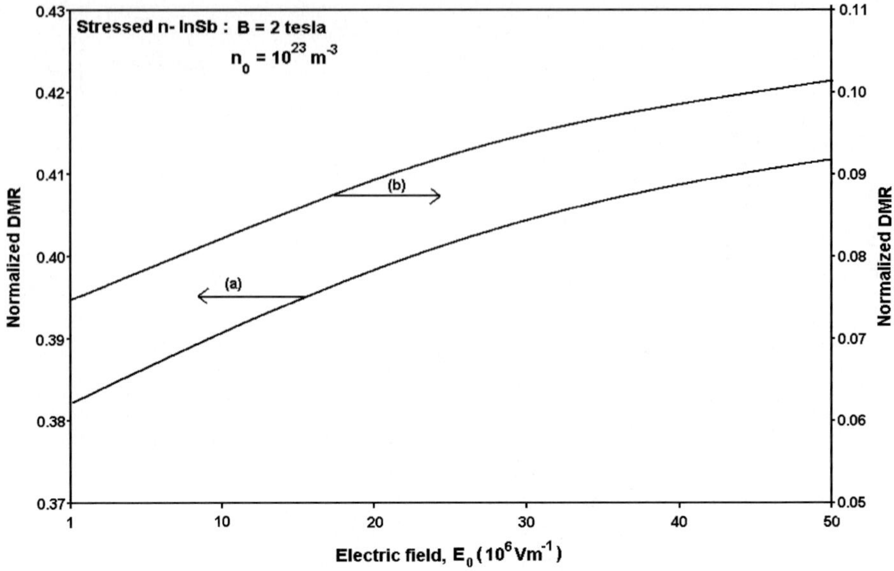

Fig. 4.35. The plot of the DMR in stressed n-InSb as a function of electric field under cross field configuration under cross field configuration for both in the presence and absence of stress as shown by the curves (**b**) and (**a**) respectively

For the purpose of condensed presentation, the related electron statistics for the specific material having a particular energy dispersion relation and the Einstein relation under the cross field configuration have been presented in Table 4.1.

4.4 Open Research Problems

R.4.1 Investigate the Einstein relation in the presence of an arbitrarily oriented quantizing alternating magnetic and crossed alternating electric fields in tetragonal semiconductors by including broadening and the electron spin. Study all the special cases for III–V, ternary and quaternary materials in this context.

R.4.2 Investigate the Einstein relations for all models of Bi, IV–VI, II–VI and stressed Kane type compounds in the presence of an arbitrarily oriented quantizing alternating magnetic and crossed alternating electric fields by including broadening and electron spin.

R.4.3 Investigate the Einstein relation for all the materials as stated in R.2.1 of Chap. 2 in the presence of arbitrarily oriented quantizing alternating magnetic and crossed alternating electric fields by including broadening and electron spin.

Table 4.1. The carrier statistics and the Einstein relation in bulk specimens of tetragonal, III–V, ternary, quaternary, II–VI, all the models of Bismuth, IV–VI and stressed materials under cross field configuration

Type of materials	The carrier statistics	The Einstein relation for the diffusivity mobility ratio
1. Tetragonal compounds	In accordance with the generalized dispersion relation (4.6) under cross field configuration as formulated in this chapter $$n_0 = \left(\frac{2g_{\mathrm{v}}B\sqrt{2}}{3L_x\pi^2\hbar^2E_o}\right)\sum_{n=0}^{n_{\max}}\left[T_{41}\left(n,\bar{E}_{\mathrm{FB}}\right)+T_{42}\left(n,\bar{E}_{\mathrm{FB}}\right)\right], \quad (4.14)$$	On the basis of (4.14), $$\frac{D}{\mu}=\frac{1}{\lvert e\rvert}\left[\frac{\sum_{n=0}^{n_{\max}}\left[T_{41}\left(n,\bar{E}_{\mathrm{FB}}\right)+T_{42}\left(n,\bar{E}_{\mathrm{FB}}\right)\right]}{\sum_{n=0}^{n_{\max}}\left[\left\{T_{41}\left(n,\bar{E}_{\mathrm{FB}}\right)\right\}'+\left\{T_{42}\left(n,\bar{E}_{\mathrm{FB}}\right)\right\}'\right]}\right] \quad (4.15)$$
2. III–V, ternary and quaternary compounds	In accordance with the three band model of Kane under cross field configuration as given by (4.16) which is a special case of (4.6) $$n_0=\frac{2g_{\mathrm{v}}B\sqrt{2m^*}}{3L_x\pi^2\hbar^2E_0}\sum_{n=0}^{n_{\max}}\left[T_{43}\left(n,\bar{E}_{\mathrm{FB}}\right)+T_{44}\left(n,\bar{E}_{\mathrm{FB}}\right)\right] \quad (4.20)$$	On the basis of (4.20) $$\frac{D}{\mu}=\frac{1}{\lvert e\rvert}\left[\frac{\sum_{n=0}^{n_{\max}}\left[T_{43}\left(n,\bar{E}_{\mathrm{FB}}\right)+T_{44}\left(n,\bar{E}_{\mathrm{FB}}\right)\right]}{\sum_{n=0}^{n_{\max}}\left[\left\{T_{43}\left(n,\bar{E}_{\mathrm{FB}}\right)\right\}'+\left\{T_{44}\left(n,\bar{E}_{\mathrm{FB}}\right)\right\}'\right]}\right] \quad (4.21)$$

(Continued)

Table 4.1. (*Continued*)

Type of materials	The carrier statistics	The Einstein relation for the diffusivity mobility ratio
	In accordance with the two band model of Kane under cross field configuration as given by (4.22) $$n_0 = \frac{2g_{\mathrm{v}}B\sqrt{2m^*}}{3L_x\pi^2\hbar^2E_0}\sum_{n=0}^{n_{\max}}\left[T_{45}\left(n,\bar{E}_{\mathrm{FB}}\right)+T_{46}\left(n,\bar{E}_{\mathrm{FB}}\right)\right] \quad (4.26)$$	On the basis of (4.26), $$\frac{D}{\mu}=\frac{1}{\lvert e\rvert}\left[\frac{\sum\limits_{n=0}^{n_{\max}}\left[T_{45}\left(n,\bar{E}_{\mathrm{FB}}\right)+T_{46}\left(n,\bar{E}_{\mathrm{FB}}\right)\right]}{\sum\limits_{n=0}^{n_{\max}}\left[\left\{T_{45}\left(n,\bar{E}_{\mathrm{FB}}\right)\right\}'+\left\{T_{46}\left(n,\bar{E}_{\mathrm{FB}}\right)\right\}'\right]}\right] \quad (4.27)$$
	In accordance with the parabolic energy bands under cross field configuration as given by (4.28) $$n_0 = N_{\mathrm{c}}\theta g_{\mathrm{v}}\left[\frac{k_{\mathrm{B}}T}{\lvert e\rvert E_0L_x}\right]\sum_{n=0}^{n_{\max}}\left[F_{\frac{1}{2}}\left(\eta_1\right)-F_{\frac{1}{2}}\left(\eta_2\right)\right] \quad (4.32\mathrm{a})$$	On the basis of (4.32a) $$\frac{D}{\mu}=\frac{k_BT}{\lvert e\rvert}\left[\frac{\sum\limits_{n=0}^{n_{\max}}\left[F_{\frac{1}{2}}(\eta_1)-F_{\frac{1}{2}}(\eta_2)\right]}{\sum\limits_{n=0}^{n_{\max}}\left[F_{-\frac{1}{2}}(\eta_1)-F_{-\frac{1}{2}}(\eta_2)\right]}\right] \quad (4.32\mathrm{b})$$
3. II–VI compounds	In accordance with (4.36) under cross field configuration $$n_0=\frac{g_{\mathrm{v}}B\sqrt{2m_{\Vert}^*\pi}}{2E_0L_x\hbar^2\pi^2}\left(k_{\mathrm{B}}T\right)^{\frac{3}{2}}\sum_{n=0}^{n_{\max}}\left[F_{\frac{1}{2}}\left(\eta_3\right)-F_{\frac{1}{2}}\left(\eta_4\right)\right] \quad (4.45)$$	On the basis of (4.45) $$\frac{D}{\mu}=\frac{k_{\mathrm{B}}T}{\lvert e\rvert}\left[\frac{\sum\limits_{n=0}^{n_{\max}}\left[F_{\frac{1}{2}}\left(\eta_3\right)-F_{\frac{1}{2}}\left(\eta_4\right)\right]}{\sum\limits_{n=0}^{n_{\max}}\left[F_{\frac{-1}{2}}\left(\eta_3\right)-F_{\frac{-1}{2}}\left(\eta_4\right)\right]}\right] \quad (4.46)$$

4. Bi	(a) The McClure and Choi model: On the basis of (4.50) under cross field configuration $$n_0 = \frac{g_v}{L_z\pi^2}\frac{2\sqrt{2}}{3\hbar}\sum_{n=0}^{n_{\max}}\left[T_{47}\left(n,\overline{E}_{FB}\right)+T_{48}\left(n,\overline{E}_{FB}\right)\right] \quad (4.60)$$	On the basis of (4.60) $$\frac{D}{\mu} = \frac{1}{\lvert e\rvert}\frac{\sum_{n=0}^{n_{\max}}\left[T_{47}\left(n,\overline{E}_{FB}\right)+T_{48}\left(n,\overline{E}_{FB}\right)\right]}{\sum_{n=0}^{n_{\max}}\left[\left\{T_{47}\left(n,\overline{E}_{FB}\right)\right\}'+\left\{T_{48}\left(n,\overline{E}_{FB}\right)\right\}'\right]} \quad (4.61)$$
	(b) The Cohen model: On the basis of (4.63) under cross field configuration $$n_0 = \frac{2g_v\sqrt{2}}{3L_z\pi^2\hbar}\sum_{n=0}^{n_{\max}}\left[T_{49}\left(n,\overline{E}_{FB}\right)+T_{410}\left(n,\overline{E}_{FB}\right)\right] \quad (4.68)$$	On the basis of (4.68) $$\frac{D}{\mu} = \frac{1}{\lvert e\rvert}\left[\frac{\sum_{n=0}^{n_{\max}}\left[T_{49}(n,\overline{E}_{FB})+T_{410}(n,\overline{E}_{FB})\right]}{\sum_{n=0}^{n_{\max}}\left[\left\{T_{49}(n,\overline{E}_{FB})\right\}'+\left\{T_{410}(n,\overline{E}_{FB})\right\}'\right]}\right] \quad (4.69)$$
	(d) The Lax model: On the basis of (4.72) under cross field configuration $$n_0 = \frac{2g_v\sqrt{2m_2}}{3L_z\pi^2\hbar}\sum_{n=0}^{n_{\max}}\left[T_{411}\left(n,\overline{E}_{FB}\right)+T_{412}\left(n,\overline{E}_{FB}\right)\right] \quad (4.76)$$	On the basis of (4.76) $$\frac{D}{\mu} = \frac{1}{\lvert e\rvert}\left[\frac{\sum_{n=0}^{n_{\max}}\left[T_{411}\left(n,\overline{E}_{FB}\right)+T_{412}\left(n,\overline{E}_{FB}\right)\right]}{\sum_{n=0}^{n_{\max}}\left[\left\{T_{411}\left(n,\overline{E}_{FB}\right)\right\}'+\left\{T_{412}\left(n,\overline{E}_{FB}\right)\right\}'\right]}\right] \quad (4.77)$$

(*Continued*)

Table 4.1. (*Continued*)

Type of materials	The carrier statistics	The Einstein relation for the diffusivity mobility ratio
	(e) The parabolic ellipsoidal model: On the basis of (4.80) under cross field configuration $$n_0 = \frac{g_v B\sqrt{2\pi m_2}\,(k_B T)^{3/2}}{2E_0 L_z \pi^2 \hbar^2} \sum_{n=0}^{n_{max}} \left[F_{1/2}(\bar{e}_1) - F_{1/2}(\bar{e}_2)\right] \quad (4.82)$$	On the basis of (4.82) $$\frac{D}{\mu} = \frac{k_B T}{\lvert e\rvert}\left[\frac{\sum_{n=0}^{n_{max}}\left[F_{1/2}(\bar{e}_1) - F_{1/2}(\bar{e}_2)\right]}{\sum_{n=0}^{n_{max}}\left[F_{-1/2}(\bar{e}_1) - F_{-1/2}(\bar{e}_2)\right]}\right] \quad (4.83)$$
5. IV–VI compounds	The expression of n_0 in this case is given by (4.68) under cross field configuration in which the constants of the energy band spectrum refers to IV–VI semiconductors	The expression of the DMR in this case is given by (4.69) under cross field configuration in which the constants of the energy band spectrum refers to IV–VI semiconductors
6. Stressed compounds	In accordance with (4.88) under cross field configuration $$n_0 = \frac{2B}{3L_x \pi^2 \hbar^2 E_0} \sum_{n=0}^{n_{max}} \left[T_{413}\left(n, \bar{E}_{FB}\right) + T_{414}\left(n, \bar{E}_{FB}\right)\right] \quad (4.95)$$	On the basis of (4.95) $$\frac{D}{\mu} = \frac{1}{\lvert e\rvert}\left[\frac{\sum_{n=0}^{n_{max}}\left[T_{413}\left(n, \bar{E}_{FB}\right) + T_{414}\left(n, \bar{E}_{FB}\right)\right]}{\sum_{n=0}^{n_{max}}\left[\left\{T_{413}\left(n, \bar{E}_{FB}\right)\right\}' + \left\{T_{414}\left(n, \bar{E}_{FB}\right)\right\}'\right]}\right] \quad (4.96)$$

Allied Research Problems

R.4.4 Investigate the EMM for all the materials as stated in R.2.1 of Chap. 2 in the presence of arbitrarily oriented quantizing alternating magnetic and crossed alternating electric fields by including broadening and electron spin.

R.4.5 Investigate the Debye screening length, the carrier contribution to the elastic constants, the heat capacity, the activity coefficient, and the plasma frequency for all the materials covering all the cases of problems from R.4.1 to R.4.3.

R.4.6 Investigate in details, the mobility for elastic and inelastic scattering mechanisms for all the materials covering all the cases of problems from R.4.1 to R.4.3.

R.4.7 Investigate the various transport coefficients for the present chapter in details for all the materials of problem R.2.1.

References

1. W. Zawadzki, B. Lax, Phys. Rev. Lett. **16**, 1001 (1966)
2. M.J. Harrison, Phys. Rev. A **29**, 2272 (1984); J. Zak, W. Zawadzki, Phys. Rev. **145**, 536 (1966)
3. W. Zawadzki, Q.H. Vrehen, B. Lax, Phys. Rev. **148**, 849 (1966); Q.H. Vrehen, W. Zawadzki, and M. Reine, Phys. Rev. **158**, 702 (1967); M.H. Weiler, W. Zawadzki and B. Lax, Phys. Rev. **163**, 733 (1967)
4. W. Zawadzki, J. Kowalski, Phys. Rev. Lett. **27**, 1713 (1971); C. Chu, M.-S. Chu, and T. Ohkawa, Phys. Rev. Lett. **41**, 653 (1978); P. Hu and C.S. Ting, Phys. Rev. B **36**, 9671 (1987)
5. E.I. Butikov, A.S. Kondratev, A.E. Kuchma, Sov. Phys. Sol. State **13**, 2594 (1972)
6. K.P. Ghatak, J.P. Banerjee, B. Goswami, B. Nag, Nonlin Opt Quantum Opt **16**, 241 (1996); M. Mondal, K.P. Ghatak, Physica Status Solidi (b) **133**, K67 (1986)
7. M. Mondal, N. Chattopadhyay, K.P. Ghatak, J. Low Temp. Phys. **66**, 131 (1987); K.P. Ghatak, M. Mondal, Zeitschrift fur Physik B **69**, 471 (1988)
8. M. Mondal, K.P. Ghatak, Phys. Lett. A **131A**, 529 (1988); M. Mondal, K.P. Ghatak, Phys. Status Solidi (b) Germany **147**, K179 (1988); B. Mitra, K.P. Ghatak, Phys. Lett. **137A**, 413 (1989)
9. B. Mitra, A. Ghoshal, K.P. Ghatak, Physica Status Solidi (b) **154**, K147 (1989)
10. B. Mitra, K.P. Ghatak, Physica Status Solidi (b), **164**, K13 (1991); K.P. Ghatak, B. Mitra, Int. J. Electron. **70**, 345 (1991); K.P. Ghatak, B. Goswami, M. Mitra, B. Nag, Nonlinear Optics **16**, 9 (1996)
11. K.P. Ghatak, M. Mitra, B. Goswami, B. Nag, Nonlinear Optics **16**, 167 (1996); K.P. Ghatak, J.P. Banerjee, B. Goswami, B. Nag, Nonlinear Optics Quant. Optics **16**, 241 (1996); K.P. Ghatak, D.K. Basu, B. Nag, J. Phys. Chem. Sol. **58**, 133 (1997)
12. K.P. Ghatak, N. Chattopadhyay, S.N. Biswas, *Proc. Society of Photo-optical and Instrumentation Engineers (SPIE)* **836**, Optoelectronic Materials, Devices, Packaging and Interconnects 203 (1988); K.P. Ghatak, M. Mondal,

S. Bhattacharyya, *SPIE* **1284**, 113 (1990); K.P. Ghatak, *SPIE* **1280**, Photonic Mater. Optical Bistability 53 (1990)
13. K.P. Ghatak, S.N. Biswas, *SPIE*, Growth Characterization Mater. Infrared Detectors Nonlinear Optical Switches, **1484**, 149 (1991)
14. K.P. Ghatak, SPIE, Fiber Optic Laser Sensors IX, **1584**, 435 (1992)

5

The Einstein Relation in Compound Semiconductors Under Size Quantization

5.1 Introduction

In recent years, with the advent of fine lithographical methods [1], molecular beam epitaxy [2], organometallic vapor-phase epitaxy [3], and other experimental techniques, the restriction of the motion of the carriers of bulk materials in one (ultrathin films, quantum wells, nipi structures, inversion layers, accumulation layers), two (quantum wires), and three (quantum dots, magnetosize quantized systems, magneto inversion layers, magneto accumulation layers, quantum dot superlattices, magneto quantum well superlattices and magneto NIPI structures) dimensions has in the last few years, attracted much attention not only for its potential in uncovering new phenomena in nanoscience but also for its interesting quantum device applications [4–6]. In ultrathin films, the restriction of the motion of the carriers in the direction normal to the film (say, the z direction) may be viewed as carrier confinement in an infinitely deep 1D rectangular potential well, leading to quantization [known as quantum size effect (QSE)] of the wave vector of the carrier along the direction of the potential well, allowing 2D carrier transport parallel to the surface of the film representing new physical features not exhibited in bulk semiconductors [7]. The low-dimensional heterostructures based on various materials are widely investigated because of the enhancement of carrier mobility [8]. These properties make such structures suitable for applications in quantum well lasers [9], heterojunction FETs [10], high-speed digital networks [11], high-frequency microwave circuits [12], optical modulators [13], optical switching systems [14], and other devices. The constant energy 3D wave-vector space of bulk semiconductors becomes 2D wave-vector surface in ultrathin films or quantum wells due to dimensional quantization. Thus, the concept of reduction of symmetry of the wave-vector space and its consequence can unlock the physics of low dimensional structures.

In Sect. 5.2.1 of this chapter, the expressions for the surface electron concentration per unit area and the 2D DMR for ultrathin films of tetragonal materials have been formulated on the basis of the generalized dispersion relation, as given by (2.2). In Sect. 5.2.2, it has been shown that the corresponding results of the 2D DMR in ultrathin films of III–V, ternary and quaternary compounds form special cases of our generalized analysis as given in Sect. 5.2.1. In Sect. 5.2.3, we have studied the same for ultrathin films of II–VI semiconductors. In Sect. 5.2.4, the 2D DMR has been derived for ultrathin films of bismuth in accordance with the McClure and Choi, the Cohen, the Lax nonparabolic ellipsoidal, and the parabolic ellipsoidal models respectively. In Sects. 5.2.5 and 5.2.6, the formulations of the 2D DMR in ultrathin films of IV–VI and stressed Kane type materials has been presented. The last Sect. 5.2.7 contains the result and discussions for this chapter.

5.2 Theoretical Background

5.2.1 Tetragonal Materials

For dimensional quantization along z-direction, the dispersion relation of the 2D electrons in tetragonal semiconductors can be written following (2.2) as

$$\psi_1(E) = \psi_2(E)\, k_s{}^2 + \psi_3(E)\,(n_z\pi/d_z)^2\,, \tag{5.1}$$

where $n_z(=1,2,3,\ldots)$ and d_z are the size quantum number and the nanothickness along the z-direction respectively.

From (5.1), the EMM in the xy-plane can be written as

$$\begin{aligned} m^*(E_{\mathrm{Fs}}, n_z) = {} & \left(\frac{\hbar^2}{2}\right)[\psi_2(E_{\mathrm{Fs}})]^{-2} \\ & \times \left[\psi_2(E_{\mathrm{Fs}})\left\{\{\psi_1(E_{\mathrm{Fs}})\}' - \{\psi_3(E_{\mathrm{Fs}})\}'\left(\frac{n_z\pi}{d_z}\right)^2\right\}\right. \\ & \left. - \left\{\psi_1(E_{\mathrm{Fs}}) - \psi_3(E_{\mathrm{Fs}})\left(\frac{n_z\pi}{d_z}\right)^2\right\}\{\psi_2(E_{\mathrm{Fs}})\}'\right], \end{aligned} \tag{5.2}$$

where E_{Fs} is the Fermi energy in the presence of size quantization as measured from the edge of the conduction band in the vertically upward direction in the absence of any quantization. Thus, we observe that the EMM is a function of size quantum number and the Fermi energy due to the combined influence of the crystal field splitting constant and the anisotropic spin–orbit splitting constants respectively.

The general expression of the total 2D density-of-states function $(N_{\mathrm{2DT}}(E))$ in this case is given by

$$N_{\mathrm{2DT}}(E) = \frac{2g_{\mathrm{v}}}{(2\pi)^2}\sum_{n_z=1}^{n_{z\max}} \frac{\partial A(E, n_z)}{\partial E} H(E - E_{n_z}), \tag{5.3a}$$

where $A(E, n_z)$ is the area of the constant energy 2D wave vector space for ultrathin films, $H(E - E_{n_z})$ is the Heaviside step function and E_{n_z} is the corresponding subband energy. Using (5.1) and (5.3a), the expression of the $N_{2DT}(E)$ for ultrathin films of tetragonal compounds can be written as

$$N_{2DT}(E) = \left(\frac{g_v}{2\pi}\right) \sum_{n_z=1}^{n_{z\max}} [\psi_2(E)]^{-2} \times \left[\psi_2(E)\left\{\{\psi_1(E)\}' - \{\psi_3(E)\}'\left(\frac{n_z\pi}{d_z}\right)^2\right\} - \left\{\psi_1(E) - \psi_3(E)\left(\frac{n_z\pi}{d_z}\right)^2\right\}\{\psi_2(E)\}'\right] H(E - E_{n_{z_1}}), \tag{5.3b}$$

where the subband energies $(E_{n_{z_1}})$ in this case is given by

$$\psi_1(E_{n_{z_1}}) = \psi_2(E_{n_{z_1}})(n_z\pi/d_z)^2. \tag{5.4}$$

The 2D carrier statistics in this case can then be expressed as

$$n_{2D} = \frac{g_v}{2\pi} \sum_{n_x=1}^{n_{x\max}} [T_{51}(E_{Fs}, n_z) + T_{52}(E_{Fs}, n_z)], \tag{5.5}$$

where

$$T_{51}(E_{Fs}, n_z) \equiv \left[\frac{\psi_1(E_{Fs}) - \psi_3(E_{Fs})(\pi n_z/d_z)^2}{\psi_2(E_{Fs})}\right] \text{ and}$$

$$T_{52}(E_{Fs}, n_z) \equiv \sum_{r=1}^{s} L(r)[T_{51}(E_{Fs}, n_z)].$$

Thus using (5.5) and (1.11), the 2D DMR for ultrathin films of tetragonal materials can be written as

$$\frac{D}{\mu} = \frac{1}{|e|} \frac{\sum_{n_z=1}^{n_{z\max}} [T_{51}(E_{Fs}, n_z) + T_{52}(E_{Fs}, n_z)]}{\sum_{n_z=1}^{n_{z\max}} [\{T_{51}(E_{Fs}, n_z)\}' + \{T_{52}(E_{Fs}, n_z)\}']}. \tag{5.6}$$

5.2.2 Special Cases for III–V, Ternary and Quaternary Materials

(a) Under the conditions $\delta = 0$, $\Delta_{\|} = \Delta_{\perp} = \Delta$ and $m_{\|}^* = m_{\perp}^* = m^*$, (5.1) assumes the form

$$\frac{\hbar^2 k_s^2}{2m^*} + \frac{\hbar^2}{2m^*}(n_z\pi/d_z)^2 = \gamma(E). \tag{5.7}$$

Using (5.7), the EMM in the $x-y$ plane for this case can be written as

$$m^*(E_{\mathrm{Fs}}) = m^*\{\gamma(E_{\mathrm{Fs}})\}'. \tag{5.8}$$

It is worth noting that the EMM in this case is a function of Fermi energy alone and is independent of the size quantum number.

The total 2D density-of-states function can be written as

$$N_{\mathrm{2DT}}(E) = \left(\frac{m^* g_{\mathrm{v}}}{\pi\hbar^2}\right)\sum_{n_z=1}^{n_{z\max}}\{[\gamma(E)]' H(E - E_{n_{z_2}})\}, \tag{5.8a}$$

where the subband energies $E_{n_{z_2}}$ can be expressed as

$$\gamma(E_{n_{z_2}}) = \frac{\hbar^2}{2m^*}(n_z\pi/d_z)^2. \tag{5.9}$$

The 2D carrier concentration assumes the form

$$n_{\mathrm{2D}} = \frac{m^* g_{\mathrm{v}}}{\pi\hbar^2}\sum_{n_z=1}^{n_{z\max}}[T_{53}(E_{\mathrm{Fs}}, n_z) + T_{54}(E_{\mathrm{Fs}}, n_z)], \tag{5.10}$$

where

$$T_{53}(E_{\mathrm{Fs}}, n_z) \equiv \left[\gamma(E_{\mathrm{Fs}}) - \frac{\hbar^2}{2m^*}\left(\frac{n_z\pi}{d_z}\right)^2\right] \quad \text{and}$$

$$T_{54}(E_{\mathrm{Fs}}, n_z) \equiv \sum_{r=1}^{s} L(r)\, T_{53}(E_{\mathrm{Fs}}, n_z).$$

The use of (5.10) and (1.11) leads to the expression of the 2D DMR in this case as

$$\frac{D}{\mu} = \frac{1}{|e|}\frac{\sum_{n_z=1}^{n_{z\max}}[T_{53}(E_{\mathrm{Fs}}, n_z) + T_{54}(E_{\mathrm{Fs}}, n_z)]}{\sum_{n_z=1}^{n_{z\max}}[\{T_{53}(E_{\mathrm{Fs}}, n_z)\}' + \{T_{54}(E_{\mathrm{Fs}}, n_z)\}']}. \tag{5.11}$$

(b) Under the inequalities $\Delta >> E_{\mathrm{g}}$ or $\Delta << E_{\mathrm{g}}$, (5.7) can be expressed as

$$E(1 + \alpha E) = \frac{\hbar^2 k_s^2}{2m^*} + \frac{\hbar^2}{2m^*}\left(\frac{n_z\pi}{d_z}\right)^2. \tag{5.12}$$

The EMM in this case can be written as

$$m^*(E_{\mathrm{Fs}}) = m^*(1 + 2\alpha E_{\mathrm{Fs}}). \tag{5.13a}$$

Thus we see that the EMM in the present case is a function of Fermi energy only due to the presence of band non-parabolicity.

The total 2D density-of-states function assumes the form

$$N_{\mathrm{2DT}}(E) = \frac{m^* g_{\mathrm{v}}}{\pi\hbar^2} \sum_{n_z=1}^{n_{z\max}} (1+2\alpha E)\, H\left(E - E_{n_{z3}}\right), \tag{5.13b}$$

where the subband energy $(E_{n_{z3}})$ can be expressed as

$$\frac{\hbar^2}{2m^*}(n_z\pi/d_z)^2 = E_{n_{z3}}\left(1+\alpha E_{n_{z3}}\right). \tag{5.14}$$

The 2D electron statistics can be written as

$$\begin{aligned} n_{\mathrm{2D}} &= \frac{m^* g_{\mathrm{v}}}{\pi\hbar^2} \sum_{n_z=1}^{n_{z\max}} \int_{E_{n_{z3}}}^{\infty} \frac{(1+2\alpha E)\,\mathrm{d}E}{1+\exp\left(\frac{E-E_{\mathrm{Fs}}}{k_{\mathrm{B}}T}\right)}, \\ &= \frac{m^* k_{\mathrm{B}} T g_{\mathrm{v}}}{\pi\hbar^2} \sum_{n_z=1}^{n_{z\max}} \left[(1+2\alpha E_{n_{z3}})F_0(\eta_{n_1}) + 2\alpha k_{\mathrm{B}} T F_1(\eta_{n_1})\right], \end{aligned} \tag{5.15}$$

where

$$\eta_{n_1} \equiv (E_{\mathrm{Fs}} - E_{n_{z3}})/k_{\mathrm{B}}T.$$

Using (5.15) and (1.11) we can write the 2D DMR in this case as

$$\frac{D}{\mu} = \frac{k_{\mathrm{B}}T}{|e|} \frac{\sum_{n_z=1}^{n_{z\max}} \left[\left(1+2\alpha E_{n_{z3}}\right) F_0(\eta_{n_1}) + 2\alpha k_{\mathrm{B}} T F_1(\eta_{n_1})\right]}{\sum_{n_z=1}^{n_{z\max}} \left[\left(1+2\alpha E_{n_{z3}}\right) F_{-1}(\eta_{n_1}) + 2\alpha k_{\mathrm{B}} T F_0(\eta_{n_1})\right]}. \tag{5.16}$$

(c) Under the condition $\alpha \to 0$, the expressions of total 2D density-of-states, for semiconducting films whose bulk electrons are defined by the isotropic parabolic energy bands, can be written from (5.13b) as

$$N_{\mathrm{2DT}}(E) = \frac{m^* g_{\mathrm{v}}}{\pi\hbar^2} \sum_{n_z=1}^{n_{z\max}} H\left(E - E_{n_{zp}}\right). \tag{5.16a}$$

The subband energy $\left(E_{n_{zp}}\right)$, n_{2D} and the DMR can respectively be expressed as

$$E_{n_{zp}} = \frac{\hbar^2}{2m^*}\left(\frac{n_z\pi}{d_z}\right)^2, \tag{5.16b}$$

$$n_{\mathrm{2D}} = \frac{m^* k_{\mathrm{B}} T g_{\mathrm{v}}}{\pi\hbar^2} \sum_{n_z=1}^{n_{z\max}} F_0(\eta_{n_2}), \tag{5.17}$$

and

$$\frac{D}{\mu}=\frac{k_{\mathrm{B}}T}{|e|}\frac{\sum\limits_{n_z=1}^{n_{z\max}}F_0(\eta_{n_2})}{\sum\limits_{n_z=1}^{n_{z\max}}F_{-1}(\eta_{n_2})}, \tag{5.18}$$

where

$$\eta_{n_2}\equiv\frac{1}{k_{\mathrm{B}}T}\left[E_{\mathrm{Fs}}-\frac{\hbar^2}{2m^*}\left(\frac{n_z\pi}{d_z}\right)^2\right].$$

It may be noted that the results of this section are already well known in the literature [15].

For bulk materials, converting the summation over n_z to the corresponding integration, (5.18) assumes the form as given by (2.19).

5.2.3 II–VI Semiconductors

The dispersion relation of the conduction electrons of ultrathin films of II–VI materials for dimensional quantization along the z-direction can be written following (2.27) as

$$E=a_0'k_s^2+b_0'\left(\frac{n_z\pi}{d_z}\right)^2\pm\bar{\lambda}_0k_s. \tag{5.19}$$

Using (5.19), the EMM in this case can be written as

$$m^*\left(E_{\mathrm{Fs}},n_z\right)=m_\perp^*\left[1\mp\frac{(\bar{\lambda}_0)}{\left[(\bar{\lambda}_0)^2-4a_0'b_0'\left(\frac{n_z\pi}{d_z}\right)^2+4a_0'E_{\mathrm{Fs}}\right]^{1/2}}\right]. \tag{5.20}$$

Thus we can infer that the EMM in the ultrathin films of II–VI compounds is a function of both the size quantum number and the Fermi energy due to the presence of the term $\bar{\lambda}_0$.

The subband energy $E_{n_{z4}}$ assumes the form

$$E_{n_{z4}}=b_o'\left(n_z\pi/d_z\right)^2. \tag{5.21}$$

From (5.19), one can write

$$N_{\mathrm{2DT}}\left(E\right)=\frac{g_{\mathrm{v}}m_\perp^*}{\pi\hbar^2}\sum_{n_z=1}^{n_{z\max}}\left[1-\frac{\left(\bar{\lambda}_0/2\sqrt{a_0'}\right)}{\sqrt{E+\delta_{51}\left(n_z\right)}}\right]H\left(E-E_{n_{z4}}\right), \tag{5.22}$$

where

$$\delta_{51}\left(n_z\right)\equiv\frac{1}{4a_0'}\left[(\bar{\lambda}_0)^2-4a_0'b_0'\left(\frac{n_z\pi}{d_z}\right)^2\right]. \tag{5.23}$$

Using (5.22), n_{2D} assumes the form

$$n_{2D} = \frac{g_v m_\perp^* k_B T}{\pi \hbar^2} \sum_{n_z=1}^{n_{z\max}} \left[F_0 \left(\eta_{n_{z3}}\right) - \frac{\bar{\lambda}_0 f_s \left(E_{Fs}, n_z\right)}{2\sqrt{a_0' k_B T}} \right], \tag{5.24}$$

where

$$\eta_{n_{z3}} \equiv \left(k_B T\right)^{-1} \left[E_{Fs} - b_0' \left(n_z \pi / d_z\right)^2 \right],$$

$$f_s \left(E_{Fs}, n_z\right) \equiv \left[2 \left(\sqrt{\eta_{n_{z3}} + \delta_{52}\left(n_z\right)} - \sqrt{\delta_{52}\left(n_z\right)} \right) + \sum_{r=1}^{s} \left[2\left(1 - 2^{1-2r}\right) \zeta(2r) \frac{(-1)^{2r-1} (2r-1)!}{\left(\eta_{n_{z3}} + \delta_{52}\left(n_z\right)\right)^{2r}} \right] \right]$$

$$\text{and } \delta_{52}\left(n_z\right) \equiv \left[\frac{b_0' \left(n_z \pi / d_z\right)^2 + \delta_{51}\left(n_z\right)}{k_B T} \right].$$

Using (5.24) and (1.11), the 2D DMR in ultrathin films of II–VI semiconductor can be expressed as

$$\frac{D}{\mu} = \frac{k_B T}{|e|} \frac{\sum_{n_z=1}^{n_{z\max}} \left[F_0 \left(\eta_{n_{z3}}\right) - \frac{\bar{\lambda}_0 f_s \left(E_{Fs}, n_z\right)}{2\sqrt{a_0' k_B T}} \right]}{\sum_{n_z=1}^{n_{z\max}} \left[F_{-1} \left(\eta_{n_{z3}}\right) - \frac{\bar{\lambda}_0 \left[f_s \left(E_{Fs}, n_z\right)\right]'}{2\sqrt{a_0' k_B T}} \right]}. \tag{5.25}$$

Under the conditions $\bar{\lambda}_0 \to 0$ and $m_\parallel^* = m_\perp^* = m^*$, (5.25) gets simplified to the well-known form as given by (5.18).

5.2.4 The Formulation of 2D DMR in Bismuth

(a) The McClure and Choi model

The dispersion relation of the conduction electrons in ultrathin films of Bi for dimensional quantization along k_z direction can be written following (2.34) for this model as

$$E(1 + \alpha E) = \frac{{p_x}^2}{2m_1} + \frac{{p_y}^2}{2m_2} + \frac{\hbar^2}{2m_3} \left(\frac{n_z \pi}{d_z} \right)^2 + \frac{{p_y}^2}{2m_2} \times \alpha E \left\{ 1 - \left(\frac{m_2}{m_2'} \right) \right\} + \frac{{p_y}^4 \alpha}{4 m_2 m_2'} - \frac{\alpha {p_x}^2 {p_y}^2}{4 m_1 m_2} - \frac{\alpha {p_y}^2 \hbar^2}{4 m_2 m_3} \left(\frac{n_z \pi}{d_z} \right)^2. \tag{5.26}$$

Equation (5.26) can be approximated as

$$\gamma_1(E,n_z) = p_1 k_x^2 + q_1(E) k_y^2 + R_1(E,n_z) k_y^4, \tag{5.27}$$

where

$$\gamma_1(E,n_z) \equiv \left[E(1+\alpha E) - \frac{\hbar^2}{2m_3}\left(\frac{n_z\pi}{d_z}\right)^2\right], \quad p_1 \equiv \frac{\hbar^2}{2m_1},$$

$$q_1(E) \equiv \frac{\hbar^2}{2m_2}\left[1+\alpha E\left(1-\frac{m_2}{m_2'}\right) - \alpha E(1+\alpha E)\right] \text{ and}$$

$$R_1(E,n_z) \equiv \left[\frac{\alpha\hbar^4}{4m_2m_2'} + \alpha\left(\frac{\hbar^2}{2m_2}\right)^2\left\{1+\alpha E\left(1-\frac{m_2}{m_2'}\right) - \frac{\alpha\hbar^2}{2m_3}\left(\frac{n_z\pi}{d_z}\right)^2\right\}\right].$$

The area enclosed by (5.27) is defined by the following integral

$$A(E,n_z) = 4\left[\frac{R_1(E,n_z)}{p_1}\right]^{1/2} J_1(E,n_z), \tag{5.28}$$

where

$$J_1(E,n_z) \equiv \int_0^{u_0(E,n_z)} \left[\frac{\gamma_1(E,n_z)}{R_1(E,n_z)} - \frac{q_1(E)k_y^2}{R_1(E,n_z)} - k_y^4\right]^{1/2} dk_y,$$

and

$$u_0(E,n_z) \equiv \left[\sqrt{\frac{q_1^2(E)}{4R_1^2(E,n_z)} + \frac{\gamma_1(E,n_z)}{R_1(E,n_z)}} - q_1(E)\right]^{1/2}.$$

Thus, the area enclosed can be written as

$$A(E,n_z) = \frac{4}{3}\left[\frac{R_1(E,n_z)}{p_1}\right]^{1/2}\left[a^2(E,n_z)+b^2(E,n_z)\right]^{1/2},$$
$$\left[a^2(E,n_z)F\left[\frac{\pi}{2},\ell(E,n_z)\right] - \left[a^2(E,n_z)-b^2(E,n_z)\right]E\left[\frac{\pi}{2},\ell(E,n_z)\right]\right], \tag{5.29}$$

where

$$a^2(E,n_z) \equiv \frac{q_1(E)}{2R_1(E,n_z)} + \frac{1}{2}\left[\frac{q_1^2(E)}{R_1^2(E,n_z)} + \frac{4\gamma_1(E,n_z)}{R_1(E,n_z)}\right]^{1/2},$$

$$b^2(E,n_z) \equiv \frac{1}{2}\left[\frac{q_1^2(E)}{R_1^2(E,n_z)} + \frac{4\gamma_1(E,n_z)}{R_1(E,n_z)}\right]^{1/2} - \left(\frac{q_1(E)}{2R_1(E,n_z)}\right),$$

$$\ell(E,n_z) \equiv \frac{b(E,n_z)}{\sqrt{a^2(E,n_z)+b^2(E,n_z)}}, \quad F\left[\frac{\pi}{2},\ell(E,n_z)\right]$$
$$\text{and } E\left[\frac{\pi}{2},\ell(E,n_z)\right]$$

are the complete elliptic integral of the first and second kinds respectively.

Using (5.29), the EMM can be written as

$$m^*\left(E_{\mathrm{Fs}}, n_z\right) = \left(\frac{2\hbar^2}{3\pi\sqrt{p_1}}\right)\left[R_3\left(E, n_z\right)\right]\Big|_{E=E_{\mathrm{Fs}}}, \tag{5.30a}$$

where

$$\begin{aligned}
R_3\left(E_{\mathrm{Fs}}, n_z\right) \equiv & \left[\frac{1}{2}\left[R_1\left(E_{\mathrm{Fs}}, n_z\right)\right]^{-1/2}\left[R_1\left(E_{\mathrm{Fs}}, n_z\right)\right]'\right. \\
& \times\left[a^2\left(E_{\mathrm{Fs}}, n_z\right)+b^2\left(E_{\mathrm{Fs}}, n_z\right)\right]^{1/2} \\
& \times\left[a^2\left(E_{\mathrm{Fs}}, n_z\right) F\left(\frac{\pi}{2}, \ell\left(E_{\mathrm{Fs}}, n_z\right)\right)\right. \\
& \left.-\left[a^2\left(E_{\mathrm{Fs}}, n_z\right)-b^2\left(E_{\mathrm{Fs}}, n_z\right)\right] E\left(\frac{\pi}{2}, \ell\left(E_{\mathrm{Fs}}, n_z\right)\right)\right] \\
& +\sqrt{R_1\left(E_{\mathrm{Fs}}, n_z\right)}\left[a^2\left(E_{\mathrm{Fs}}, n_z\right)+b^2\left(E_{\mathrm{Fs}}, n_z\right)\right]^{-1/2} \\
& \times\left[a\left(E_{\mathrm{Fs}}, n_z\right)\left(a\left(E_{\mathrm{Fs}}, n_z\right)\right)'+b\left(E_{\mathrm{Fs}}, n_z\right)\left(b\left(E_{\mathrm{Fs}}, n_z\right)\right)'\right] \\
& \times\left[a^2\left(E_{\mathrm{Fs}}, n_z\right) F\left(\frac{\pi}{2}, \ell\left(E_{\mathrm{Fs}}, n_z\right)\right)\right. \\
& \left.-\left[a^2\left(E_{\mathrm{Fs}}, n_z\right)-b^2\left(E_{\mathrm{Fs}}, n_z\right)\right] E\left(\frac{\pi}{2}, \ell\left(E_{\mathrm{Fs}}, n_z\right)\right)\right] \\
& +\sqrt{R_1\left(E_{\mathrm{Fs}}, n_z\right)}\left[a^2\left(E_{\mathrm{Fs}}, n_z\right)+b^2\left(E_{\mathrm{Fs}}, n_z\right)\right]^{1/2} \\
& \times\left[2a\left(E_{\mathrm{Fs}}, n_z\right)\left(a\left(E_{\mathrm{Fs}}, n_z\right)\right)' F\left(\frac{\pi}{2}, \ell\left(E_{\mathrm{Fs}}, n_z\right)\right)\right. \\
& +a^2\left(E_{\mathrm{Fs}}, n_z\right)\left\{F\left(\frac{\pi}{2}, \ell\left(E_{\mathrm{Fs}}, n_z\right)\right)\right\}' \\
& -\left[2a\left(E_{\mathrm{Fs}}, n_z\right)\left[a\left(E_{\mathrm{Fs}}, n_z\right)\right]'-2b\left(E_{\mathrm{Fs}}, n_z\right)\left(b\left(E_{\mathrm{Fs}}, n_z\right)\right)'\right] \\
& \times E\left(\frac{\pi}{2}, \ell\left(E_{\mathrm{Fs}}, n_z\right)\right)-\left[a^2\left(E_{\mathrm{Fs}}, n_z\right)-b^2\left(E_{\mathrm{Fs}}, n_z\right)\right] \\
& \left.\left.\times\left(E\left(\frac{\pi}{2}, \ell\left(E_{\mathrm{Fs}}, n_z\right)\right)\right)'\right]\right].
\end{aligned}$$

Thus the EMM in this case is a function of both the Fermi energy and the size quantum number due to the presence of band nonparabolicity only.

The total 2D density-of-states function can be written following (5.29), as

$$N_{2\mathrm{DT}}(E) = \left(\frac{2g_{\mathrm{v}}}{3\pi^2\sqrt{p_1}}\right)\sum_{n_z=1}^{n_{z\max}} R_3\left(E, n_z\right) H\left(E-E_{n_{z5}}\right), \tag{5.30b}$$

where the subband energies $E_{n_{z5}}$ assumes the form

$$E_{n_{z5}}\left(1+\alpha E_{n_{z5}}\right) = \frac{\hbar^2}{2m_3}\left(\frac{n_z\pi}{d_z}\right)^2. \tag{5.30c}$$

Combining (5.30b) with the Fermi–Dirac occupation probability factor, the 2D electron statistics in ultrathin films of Bi in accordance with the McClure and Choi model can be expressed as

$$n_{2D} = \left(\frac{2g_v}{3\pi^2\sqrt{p_1}}\right)\sum_{n_z=1}^{n_{z\max}}\left[\theta_1\left(E_{Fs},n_z\right)+\theta_2\left(E_{Fs},n_z\right)\right], \tag{5.31}$$

where

$$\theta_1\left(E_{Fs},n_z\right) \equiv \Big\{\sqrt{R_1\left(E_{Fs},n_z\right)}\left[a^2\left(E_{Fs},n_z\right)+b^2\left(E_{Fs},n_z\right)\right]^{1/2},$$
$$\Big[\,a^2\left(E_{Fs},n_z\right)F\left(\frac{\pi}{2},\ell\left(E_{Fs},n_z\right)\right)-\left[a^2\left(E_{Fs},n_z\right)-b^2\left(E_{Fs},n_z\right)\right]$$
$$F\left(\frac{\pi}{2},\ell\left(E_{Fs},n_z\right)\right)\Big]\Big\},$$

and

$$\theta_2\left(E_{Fs},n_z\right) \equiv \sum_{r=1}^{s}L(r)\left[\theta_1\left(E_{Fs},n_z\right)\right].$$

The use of (5.33) and (1.11) leads to the expression of the 2D DMR in this case as

$$\frac{D}{\mu} = \frac{1}{|e|}\frac{\sum\limits_{n_z=1}^{n_{z\max}}\left[\theta_1\left(E_{Fs},n_z\right)+\theta_2\left(E_{Fs},n_z\right)\right]}{\sum\limits_{n_z=1}^{n_{z\max}}\left[\left(\theta_1\left(E_{Fs},n_z\right)\right)'+\left(\theta_2\left(E_{Fs},n_z\right)\right)'\right]} \tag{5.32}$$

(b) The Cohen model

The 2D electron dispersion law in ultrathin films of Bi in accordance with this model can be written, following (2.41), as

$$E(1+\alpha E) = \frac{p_x^2}{2m_1}+\frac{\hbar^2}{2m_3}\left(\frac{n_z\pi}{d_z}\right)^2-\frac{\alpha E p_y^2}{2m_2'}+\left(\frac{\alpha p_y^4}{4m_2m_2'}\right)+\frac{p_y^2}{2m_2}(1+\alpha E). \tag{5.33}$$

Equation (5.33) can be written as

$$\gamma_1\left(E,n_z\right) = p_1k_x^2+q_2\left(E\right)k_y^2+R_2k_y^4, \tag{5.34}$$

where

$$q_2\left(E\right) \equiv \left[\frac{\hbar^2}{2m_2}\left(1+\alpha E\right)-\frac{\alpha E\hbar^2}{2m_2'}\right] \text{ and } R_2 \equiv \left(\frac{\alpha\hbar^4}{4m_2m_2'}\right).$$

The EMM in this case can be written as

$$m^*\left(E_{Fs},n_z\right) = \left(\frac{2\hbar^2}{3\pi\sqrt{p_1}}\right)\left[R_4\left(E,n_z\right)\right]\big|_{E=E_{Fs}}, \tag{5.35}$$

in which,

$$R_4(E_{Fs}, n_z) \equiv \sqrt{R_2}\left[a_1^2(E_{Fs}, n_z) + b_1^2(E_{Fs}, n_z)\right]^{-1/2}$$
$$\left[a_1(E_{Fs}, n_z)(a_1(E_{Fs}, n_z))' + b_1(E_{Fs}, n_z)(b_1(E_{Fs}, n_z))'\right]$$
$$\left[a_1^2(E_{Fs}, n_z) F\left(\frac{\pi}{2}, \ell_1(E_{Fs}, n_z)\right)\right.$$
$$\left.- \left[a_1^2(E_{Fs}, n_z) - b_1^2(E_{Fs}, n_z)\right] E\left(\frac{\pi}{2}, \ell_1(E_{Fs}, n_z)\right)\right]$$
$$+\sqrt{R_2}\left[a_1^2(E_{Fs}, n_z) + b_1^2(E_{Fs}, n_z)\right]^{1/2}$$
$$\left[2a_1(E_{Fs}, n_z)(a_1(E_{Fs}, n_z))' F\left(\frac{\pi}{2}, \ell_1(E_{Fs}, n_z)\right) + a_1^2(E_{Fs}, n_z)\right.$$
$$\left\{F\left(\frac{\pi}{2}, \ell_1(E_{Fs}, n_z)\right)\right\}'$$
$$\left[2a_1(E_{Fs}, n_z)(a_1(E_{Fs}, n_z))' - 2b_1(E_{Fs}, n_z)(b_1(E_{Fs}, n_z))'\right]$$
$$\left.E\left(\frac{\pi}{2}, \ell_1(E_{Fs}, n_z)\right) - \left[a_1^2(E_{Fs}, n_z) - b_1^2(E_{Fs}, n_z)\right]\left(E\left(\frac{\pi}{2}, \ell_1(E_{Fs}, n_z)\right)\right)'\right]\right],$$

$$a_1^2(E_{Fs}, n_z) \equiv \frac{q_2(E_{Fs})}{2R_2} + \frac{1}{2}\left[\frac{q_2^2(E_{Fs})}{R_2^2} + \frac{4\gamma_1(E_{Fs}, n_z)}{R_2}\right]^{1/2},$$

$$b_1^2(E_{Fs}, n_z) \equiv \frac{1}{2}\left[\frac{q_2^2(E_{Fs})}{R_2^2} + \frac{4\gamma_1(E_{Fs}, n_z)}{R_2}\right]^{1/2} - \left(\frac{q_2(E_{Fs})}{2R_2}\right),$$

and

$$\ell_1(E_{Fs}, n_z) \equiv \frac{b_1(E_{Fs}, n_z)}{\sqrt{a_1^2(E_{Fs}, n_z) + b_1^2(E_{Fs}, n_z)}},$$

which shows that the EMM in this present case is again a function of both the size quantum number and the Fermi energy due to the presence of the band nonparabolicity only.

The total density-of-states is given by

$$N_{2DT}(E) = \left(\frac{2g_v}{3\pi^2\sqrt{p_1}}\right)\sum_{n_z=1}^{n_{z\max}} R_4(E, n_z) H\left(E - E_{n_{z5}}\right). \tag{5.36}$$

Combining (5.36) with the Fermi–Dirac occupation probability factor, the 2D electron statistics in ultrathin films of Bi in accordance with the Cohen model can be written as

$$n_{2D} = \left(\frac{2g_v}{3\pi^2\sqrt{p_1}}\right)\sum_{n_z=1}^{n_{z\max}} \left[\theta_3(E_{Fs}, n_z) + \theta_4(E_{Fs}, n_z)\right], \tag{5.37}$$

where

$$\theta_3(E_{Fs}, n_z) \equiv \left\{\sqrt{R_2}\left[a_1^2(E_{Fs}, n_z) + b_1^2(E_{Fs}, n_z)\right]^{1/2},\right.$$
$$\left[a_1^2(E_{Fs}, n_z) F\left(\tfrac{\pi}{2}, \ell_1(E_{Fs}, n_z)\right) - \left[a_1^2(E_{Fs}, n_z) - b_1^2(E_{Fs}, n_z)\right]\right.$$
$$\left.\left.F\left(\tfrac{\pi}{2}, \ell_1(E_{Fs}, n_z)\right)\right]\right\},$$

and

$$\theta_4\left(E_{\mathrm{Fs}},n_z\right) \equiv \sum_{r=1}^{s} L\left(r\right)\left[\theta_3\left(E_{\mathrm{Fs}},n_z\right)\right].$$

The use of (5.37) and (1.11) leads to the expression of the 2D DMR in this case as

$$\frac{D}{\mu} = \frac{1}{|e|}\frac{\sum\limits_{n_z=1}^{n_{z\max}}\left[\theta_3\left(E_{\mathrm{Fs}},n_z\right)+\theta_4\left(E_{\mathrm{Fs}},n_z\right)\right]}{\sum\limits_{n_z=1}^{n_{z\max}}\left[\left(\theta_3\left(E_{\mathrm{Fs}},n_z\right)\right)'+\left(\theta_4\left(E_{\mathrm{Fs}},n_z\right)\right)'\right]}, \tag{5.38}$$

(c) The Lax model

The 2D electron dispersion law in this case can be written as

$$E\left(1+\alpha E\right) = \frac{\hbar^2 {k_x}^2}{2m_1} + \frac{\hbar^2 {k_y}^2}{2m_2} + \frac{\hbar^2}{2m_3}\left(\frac{n_z\pi}{d_z}\right)^2. \tag{5.39}$$

The EMM in this case assumes the form

$$m^*\left(E_{\mathrm{Fs}}\right) = \sqrt{m_1m_2}\left(1+2\alpha E_{\mathrm{Fs}}\right). \tag{5.40}$$

Thus, we see that the EMM for the Lax model is a function of the Fermi energy alone due to the band nonparabolicity.

The subband energy, the total density- of states function, the 2D electron statistics, and the corresponding DMR for this model can, respectively, be expressed as

$$E_{n_{z5}}\left(1+\alpha E_{n_{z5}}\right) = \frac{\hbar^2}{2m_3}\left(n_z\pi/d_z\right)^2, \tag{5.41}$$

$$N_{\mathrm{2DT}}\left(E\right) = \frac{g_{\mathrm{v}}\sqrt{m_1m_2}}{\pi\hbar^2}\sum_{n_z=1}^{n_{z\max}}\left(1+2\alpha E\right)H\left(E-E_{n_{z5}}\right), \tag{5.42}$$

$$n_{\mathrm{2D}} = \frac{g_{\mathrm{v}}\sqrt{m_1m_2}k_{\mathrm{B}}T}{\pi\hbar^2}\sum_{n_z=1}^{n_{z\max}}\left[\left(1+2\alpha E_{n_{z5}}\right)F_0\left(\eta_{y2}\right)2\alpha k_{\mathrm{B}}TF_1\left(\eta_{y2}\right)\right], \tag{5.43}$$

$$\frac{D}{\mu} = \frac{k_{\mathrm{B}}T}{|e|}\frac{\sum\limits_{n_z=1}^{n_{z\max}}\left[\left(1+2\alpha E_{n_{z5}}\right)F_0\left(\eta_{y2}\right)+2\alpha k_{\mathrm{B}}TF_1\left(\eta_{y2}\right)\right]}{\sum\limits_{n_z=1}^{n_{z\max}}\left[\left(1+2\alpha E_{n_{z5}}\right)F_{-1}\left(\eta_{y2}\right)+2\alpha k_{\mathrm{B}}TF_0\left(\eta_{y2}\right)\right]}, \tag{5.44}$$

where

$$\eta_{y2} = \frac{E_{\mathrm{Fs}}-E_{n_{z5}}}{k_{\mathrm{B}}T}.$$

(d) The ellipsoidal parabolic model

The 2D dispersion relation, the EMM, the subband energy $(E_{n_{z6}})$, the total density-of states, the 2D electron statistics and the corresponding DMR for this model can respectively be written for $\alpha \to 0$ in (5.41)–(5.44) as

$$E = \left(\frac{\hbar^2 k_x^2}{2m_1}\right) + \left(\frac{\hbar^2 k_y^2}{2m_2}\right) + \left(\frac{\hbar^2}{2m_3}\right)\left(\frac{n_z \pi}{d_z}\right)^2, \tag{5.45}$$

$$m^*(E_{\mathrm{Fs}}) = \left(\sqrt{m_1 m_2}\right), \tag{5.46a}$$

$$N_{\mathrm{2DT}}(E) = \frac{g_{\mathrm{v}}\sqrt{m_1 m_2}}{\pi\hbar^2} \sum_{n_z=1}^{n_{z\max}} H\left(E - E_{n_{z6}}\right), \tag{5.46b}$$

$$E_{n_{z6}} = \left(\frac{\hbar^2}{2m_3}\right)\left(\frac{n_z \pi}{d_z}\right)^2, \tag{5.47}$$

$$n_{\mathrm{2D}} = \left[\frac{k_{\mathrm{B}} T g_{\mathrm{v}} \sqrt{m_1 m_2}}{\pi\hbar^2}\right] \sum_{n_z=1}^{n_{z\max}} F_0\left(\eta_{y3}\right), \tag{5.48}$$

and

$$\frac{D}{\mu} = \left[\frac{k_{\mathrm{B}} T}{|e|}\right] \frac{\sum_{n_z=1}^{n_{z\max}} F_0\left(\eta_{y3}\right)}{\sum_{n_z=1}^{n_{z\max}} F_{-1}\left(\eta_{y3}\right)}, \tag{5.49}$$

where

$$\eta_{y3} \equiv (k_{\mathrm{B}} T)^{-1}\left[E_{\mathrm{Fs}} - E_{n_{z6}}\right].$$

5.2.5 IV–VI Materials

The dispersion relation of the conduction electrons in bulk specimens of IV–VI semiconductors in accordance with Dimmock model [16] can be written from (R.2.2) together with the substitutions $P_\perp^2 \equiv \left(\hbar^2 E_{\mathrm{g}}/2m_{\mathrm{t}}^*\right)$, $P_{||}^2 \equiv \left(\frac{\hbar^2 E_{\mathrm{g}}}{2m_{\mathrm{l}}^*}\right)$ and $\in \equiv \left[E + \left(\frac{E_{\mathrm{g}}}{2}\right)\right]$ (where m_{t}^* and m_{l}^* are the transverse and the longitudinal effective masses at $k = 0$) as

$$\left[E - \frac{\hbar^2 k_s^2}{2m_{\mathrm{t}}^-} - \frac{\hbar^2 k_z^2}{2m_{\mathrm{l}}^-}\right]\left[1 + \alpha E + \alpha\frac{\hbar^2 k_s^2}{2m_{\mathrm{t}}^+} + \alpha\frac{\hbar^2 k_z^2}{2m_{\mathrm{l}}^+}\right] = \frac{\hbar^2 k_s^2}{2m_{\mathrm{t}}^*} + \frac{\hbar^2 k_z^2}{2m_{\mathrm{l}}^*}. \tag{5.50}$$

The 2D dispersion relation of the conduction electrons in IV–VI materials in ultrathin films for the dimensional quantization along the z direction can be expressed as

$$E(1+\alpha E)+\alpha E\left(\frac{\hbar^2k_x^2}{2x_4}+\frac{\hbar^2k_y^2}{2x_5}\right)+\alpha E\frac{\hbar^2}{2x_6}\left(\frac{n_z\pi}{d_z}\right)^2$$
$$-(1+\alpha E)\left(\frac{\hbar^2k_x^2}{2x_1}+\frac{\hbar^2k_y^2}{2x_2}\right)$$
$$-\alpha\left(\frac{\hbar^2k_x^2}{2x_1}+\frac{\hbar^2k_y^2}{2x_2}\right)\left(\frac{\hbar^2k_x^2}{2x_4}+\frac{\hbar^2k_y^2}{2x_5}\right)-\alpha\left(\frac{\hbar^2k_x^2}{2x_1}+\frac{\hbar^2k_y^2}{2x_2}\right)\frac{\hbar^2}{2x_6}\left(\frac{n_z\pi}{d_z}\right)^2$$
$$-(1+\alpha E)\frac{\hbar^2}{2x_3}\left(\frac{n_z\pi}{d_z}\right)^2$$
$$-\alpha\frac{\hbar^2}{2x_3}\left(\frac{n_z\pi}{d_z}\right)^2\left(\frac{\hbar^2k_x^2}{2x_4}+\frac{\hbar^2k_y^2}{2x_5}\right)-\alpha\frac{\hbar^2}{2x_3}\left(\frac{n_z\pi}{d_z}\right)^2\frac{\hbar^2}{2x_6}\left(\frac{n_z\pi}{d_z}\right)^2$$
$$=\frac{\hbar^2k_x^2}{2m_1}+\frac{\hbar^2k_y^2}{2m_2}+\frac{\hbar^2}{2m_3}\left(\frac{n_z\pi}{d_z}\right)^2, \quad (5.51)$$

where

$$x_4=m_t^+,\ x_5=\frac{m_t^++2m_l^+}{3},\ x_6=\frac{3m_t^+m_l^+}{2m_l^++m_t^+},\ x_1=m_t^-,$$
$$x_2=\frac{m_t^-+2m_l^-}{3},\ x_3=\frac{3m_t^-m_l^-}{2m_l^-+m_t^-},$$
$$m_1=m_t^*, m_2=\frac{m_t^*+2m_l^*}{3} \text{ and } m_3=\frac{3m_l^*m_t^*}{m_t^*+2m_l^*}.$$

Substituting $k_x = r\mathrm{Cos}\theta$ and $k_y = r\mathrm{Sin}\theta$ (where r and θ are 2D polar coordinates in 2D wave vector space) in (5.51), we can write

$$r^4\left[\alpha\frac{1}{4}\left(\frac{\hbar^2\mathrm{Cos}^2\theta}{x_1}+\frac{\hbar^2\mathrm{Sin}^2\theta}{x_2}\right)\left(\frac{\hbar^2\mathrm{Cos}^2\theta}{x_4}+\frac{\hbar^2\mathrm{Sin}^2\theta}{x_5}\right)\right]+r^2\frac{1}{2}\left[\left(\frac{\hbar^2\mathrm{Cos}^2\theta}{m_1}+\frac{\hbar^2\mathrm{Sin}^2\theta}{m_2}\right)\right.$$
$$+\alpha\frac{\hbar^2}{2x_3}\left(\frac{n_z\pi}{d_z}\right)^2\left(\frac{\hbar^2\mathrm{Cos}^2\theta}{x_4}+\frac{\hbar^2\mathrm{Sin}^2\theta}{x_5}\right)+\alpha\left(\frac{\hbar^2\mathrm{Cos}^2\theta}{x_1}+\frac{\hbar^2\mathrm{Sin}^2\theta}{x_2}\right)\frac{\hbar^2}{2x_6}\left(\frac{n_z\pi}{d_z}\right)^2$$
$$\left.+\hbar^2(1+\alpha E)\left(\frac{\mathrm{Cos}^2\theta}{x_1}+\frac{\mathrm{Sin}^2\theta}{x_2}\right)-\hbar^2\alpha E\left(\frac{\mathrm{Cos}^2\theta}{x_4}+\frac{\mathrm{Sin}^2\theta}{x_5}\right)\right]-\left[E(1+\alpha E)\right.$$
$$\left.+\alpha E\frac{\hbar^2}{2x_6}\left(\frac{n_z\pi}{d_z}\right)^2-(1+\alpha E)\frac{\hbar^2}{2x_3}\left(\frac{n_z\pi}{d_z}\right)^2-\alpha\left(\frac{\hbar^4}{4x_3x_6}\left(\frac{n_z\pi}{d_z}\right)^4\right)\right]=0. \quad (5.52)$$

The area $A(E,n_z)$ of the 2D wave vector space can be expressed as

$$A(E,n_z)=\bar{J}_1-\bar{J}_2, \quad (5.53)$$

where

$$\bar{J}_1\equiv 2\int_0^{\pi/2}\frac{c}{b}\mathrm{d}\theta, \quad (5.54)$$

and

$$\bar{J}_2\equiv 2\int_0^{\pi/2}\frac{ac^2}{b^3}\mathrm{d}\theta, \quad (5.55)$$

in which

$$a \equiv \left[\alpha\left(\frac{\hbar^4}{4}\right)\left(\frac{\mathrm{Cos}^2\theta}{x_1}+\frac{\mathrm{Sin}^2\theta}{x_2}\right)\left(\frac{\mathrm{Cos}^2\theta}{x_4}+\frac{\mathrm{Sin}^2\theta}{x_5}\right)\right],$$

$$b \equiv \left(\frac{\hbar^2}{2}\right)\left[\left(\frac{\mathrm{Cos}^2\theta}{m_1}+\frac{\mathrm{Sin}^2\theta}{m_2}\right)+\alpha\left(\frac{\hbar^2}{2x_3}\right)\left(\frac{n_z\pi}{d_z}\right)^2\left(\frac{\mathrm{Cos}^2\theta}{x_4}+\frac{\mathrm{Sin}^2\theta}{x_5}\right)\right.$$
$$+\alpha\left(\frac{\hbar^2}{2x_6}\right)\left(\frac{n_z\pi}{d_z}\right)^2\left(\frac{\mathrm{Cos}^2\theta}{m_1}+\frac{\mathrm{Sin}^2\theta}{m_2}\right)$$
$$\left.+(1+\alpha E)\left(\frac{\mathrm{Cos}^2\theta}{x_1}+\frac{\mathrm{Sin}^2\theta}{x_2}\right)-\alpha E\left(\frac{\mathrm{Cos}^2\theta}{x_4}+\frac{\mathrm{Sin}^2\theta}{x_5}\right)\right],$$

and

$$c \equiv \left[E(1+\alpha E)+\alpha E\left(\frac{\hbar^2}{2x_6}\right)\left(\frac{n_z\pi}{d_z}\right)^2\right.$$
$$\left.-(1+\alpha E)\left(\frac{\hbar^2}{2x_3}\right)\left(\frac{n_z\pi}{d_z}\right)^2-\alpha\left(\frac{\hbar^4}{4x_3x_6}\right)\left(\frac{n_z\pi}{d_z}\right)^4\right].$$

Equation (5.54) can be expressed as $\bar{J}_1 = 2\int\limits_0^{\pi/2}\frac{t_3(E,n_z)\mathrm{d}\theta}{A_1(E,n_z)\mathrm{Cos}^2\theta+B_1(E,n_z)\mathrm{Sin}^2\theta}$ where $t_3(E,n_z)\equiv c,\ A_1(E,n_z)\equiv\frac{\hbar^2}{2m_1}t_1(E,n_z)$,

$$t_1(E,n_z)\equiv\left[1+m_1\left[\frac{1}{x_4}\frac{\alpha\hbar^2}{2x_3}\left(\frac{n_z\pi}{d_z}\right)^2+\frac{\alpha\hbar^2}{2x_1x_6}\left(\frac{n_z\pi}{d_z}\right)^2+\frac{1+\alpha E}{x_1}-\frac{\alpha E}{x_4}\right]\right],$$

$B_1(E,n_z)\equiv\frac{\hbar^2}{2m_2}t_2(E,n_z)$ and $t_2(E,n_z)$

$$\equiv\left[1+m_2\left[\frac{\alpha\hbar^2}{2x_3x_5}\left(\frac{n_z\pi}{d_z}\right)^2+\frac{\alpha\hbar^2}{2x_2x_6}\left(\frac{n_z\pi}{d_z}\right)^2+\frac{1+\alpha E}{x_2}-\frac{\alpha E}{x_5}\right]\right].$$

Performing the integration, we get

$$\bar{J}_1 = \pi t_3(E,n_z)\left[A_1(E,n_z)B_1(E,n_z)\right]^{-1/2}. \tag{5.56}$$

From (5.55) we can write

$$\bar{J}_2 = \frac{\alpha t_3^2(E,n_z)\hbar^4}{2B_1^3(E,n_z)}I, \tag{5.57}$$

where

$$I \equiv \int\limits_0^{\infty}\frac{\left(a_1+a_2z^2\right)\left(a_3+a_4z^2\right)\mathrm{d}z}{\left[(\bar{a})^2+z^2\right]^3}, \tag{5.58}$$

in which $a_1\equiv\frac{1}{x_1},\ a_2\equiv\frac{1}{x_2},\ z=\tan\theta,\theta$ is a new variable, $a_3\equiv\frac{1}{x_4},\ a_4\equiv\frac{1}{x_5}$ and $(\bar{a})^2\equiv\left(\frac{A_1(E,n_z)}{B_1(E,n_z)}\right)$.

The use of the Residue theorem leads to the evaluation of the integral in (5.58) as

$$I = \frac{\pi}{4\bar{a}}\left[a_1a_4+3a_2a_4\right]. \tag{5.59}$$

Therefore, the 2D area of the 2D wave vector space can be written as

$$A(E,n_z) = \frac{\pi t_3(E,n_z)}{\sqrt{A_1(E,n_z)B_1(E,n_z)}}\left[1-\frac{1}{x_5}\left(\frac{1}{x_1}+\frac{3}{x_2}\right)\frac{\alpha t_3(E,n_z)\hbar^4}{8B_1^2(E,n_z)}\right]. \tag{5.60}$$

The EMM for the ultrathin films of IV–VI materials can thus be written as

$$m^*(E,n_z) = \frac{\hbar^2}{2}\left[\theta_5(E,n_z)\right]\Big|_{E=E_{\mathrm{Fs}}}, \tag{5.61}$$

where

$$\begin{aligned}
\theta_5(E,n_z) \equiv {} & \left[1-\frac{1}{x_5}\left(\frac{1}{x_1}+\frac{3}{x_2}\right)\frac{\alpha t_3(E,n_z)\hbar^4}{8\left[B_1(E,n_z)\right]^2}\right]\left[A_1(E,n_z)B_1(E,n_z)\right]^{-1}\\
& \left[\sqrt{A_1(E,n_z)B_1(E,n_z)}\{t_3(E,n_z)\}' - t_3(E,n_z)\left\{\frac{1}{2}\{A_1(E,n_z)\}'\left[\frac{B_1(E,n_z)}{A_1(E,n_z)}\right]^{1/2}\right.\right.\\
& \left.\left.+\frac{1}{2}\{B_1(E,n_z)\}'\left[\frac{A_1(E,n_z)}{B_1(E,n_z)}\right]^{1/2}\right\}\right]\\
& -\frac{1}{8}\frac{t_3(E,n_z)\alpha\hbar^4}{\sqrt{A_1(E,n_z)B_1(E,n_z)}}\frac{1}{x_5}\left(\frac{1}{x_1}+\frac{3}{x_2}\right)\left[B_1(E,n_z)\right]^{-4}\left[\{B_1(E,n_z)\}^2\{t_3(E,n_z)\}'\right.\\
& \left.-2B_1(E,n_z)\{B_1(E,n_z)\}'t_3(E,n_z)\right].
\end{aligned}$$

Thus, the EMM is a function of the Fermi energy and the quantum number due to the band non-parabolicity.

The total density-of-states function can be written as

$$N_{\mathrm{2DT}}(E) = \left(\frac{g_{\mathrm{v}}}{2\pi}\right)\sum_{n_z=1}^{n_{z\max}}\theta_5(E,n_z)H\left(E-E_{n_{z7}}\right), \tag{5.62}$$

where the subband energy $\left(E_{n_{z7}}\right)$ in this case can be written as

$$\begin{aligned}
& E_{n_{z7}}\left(1+\alpha E_{n_{z7}}\right)+\alpha E_{n_{z7}}\frac{\hbar^2}{2x_6}\left(\frac{n_z\pi}{d_z}\right)^2-\left(1+\alpha E_{n_{z7}}\right)\frac{\hbar^2}{2x_3}\left(\frac{n_z\pi}{d_z}\right)^2\\
& -\alpha\frac{\hbar^2}{2x_3}\left(\frac{n_z\pi}{d_z}\right)^2\frac{\hbar^2}{2x_6}\left(\frac{n_z\pi}{d_z}\right)^2-\left[\frac{\hbar^2}{2m_3}\left(\frac{n_z\pi}{d_z}\right)^2\right]=0.
\end{aligned} \tag{5.62a}$$

The use (5.61) leads to the expression of 2D electron statistics as

$$n_{\mathrm{2D}} = \frac{g_{\mathrm{v}}}{2\pi}\sum_{n_z=1}^{n_{z\max}}\left[T_{55}(E_{\mathrm{Fs}},n_z)+T_{56}(E_{\mathrm{Fs}},n_z)\right], \tag{5.63}$$

where

$$T_{55}(E_{\mathrm{Fs}},n_z) \equiv \frac{A(E_{\mathrm{Fs}},n_z)}{\pi} \text{ and } T_{56}(E_{\mathrm{Fs}},n_z) \equiv \sum_{r=1}^{s}L(r)T_{55}(E_{\mathrm{Fs}},n_z).$$

Thus using (5.63) and (1.11), the expression for the 2D DMR for ultrathin films of IV–VI compounds can be expressed as

$$\frac{D}{\mu} = \frac{1}{|e|} \frac{\sum_{n_z=1}^{n_{z\max}} [T_{55}(E_{\mathrm{Fs}}, n_z) + T_{56}(E_{\mathrm{Fs}}, n_z)]}{\sum_{n_z=1}^{n_{z\max}} [(T_{55}(E_{\mathrm{Fs}}, n_z))' + (T_{56}(E_{\mathrm{Fs}}, n_z))']}, \tag{5.64}$$

Under the conditions $m_1^{\pm} \to \infty$, $m_t^{\pm} \to \infty$ and $m_1^* = m_t^* = m^*$, (5.64) gets simplified to the form given by (5.18).

5.2.6 Stressed Kane Type Semiconductors

The 2D electron energy spectrum in ultrathin films of stressed materials assumes the form

$$\frac{k_x^2}{[\bar{a}_0(E)]^2} + \frac{k_y^2}{[\bar{b}_0(E)]^2} + \frac{1}{[\bar{c}_0(E)]^2}(n_z\pi/d_z)^2 = 1. \tag{5.65}$$

The area of 2D wave vector space enclosed by (5.65) can be written as

$$A(E, n_z) = \pi P^2(E, n_z)\bar{a}_0(E)\bar{b}_0(E), \tag{5.66}$$

where

$$P^2(E, n_z) = \left[1 - [n_z\pi/d_z\bar{c}_0(E)]^2\right].$$

The expression of the surface EMM in this case can be written as

$$m^*(E_{\mathrm{Fs}}, n_z) = \frac{\hbar^2}{2}[\theta_6(E, n_z)]\Big|_{E=E_{\mathrm{Fs}}}, \tag{5.67}$$

in which,

$$\theta_6(E, n_z) = \Big[2P(E, n_z)\{P(E, n_z)\}'\bar{a}_0(E)\bar{b}_0(E) + \{P(E, n_z)\}^2 \{\bar{a}_0(E)\}'\bar{b}_0(E) + \{P(E, n_z)\}^2\{\bar{b}_0(E)\}'\bar{a}_0(E)\Big].$$

The EMM in this case is a function of the Fermi energy and the size quantization number due to the presence of stress only.

Thus, the total 2D density-of-states function can be expressed as

$$N_{2\mathrm{DT}}(E) = \left(\frac{g_{\mathrm{v}}}{2\pi}\right)\sum_{n_z=1}^{n_{z\max}} \theta_6(E, n_z), \tag{5.67a}$$

The subband energies $(E_{n_{z8}})$ are given by

$$\bar{c}_0(E_{n_{z8}}) = n_z\pi/d_z. \tag{5.68}$$

The 2D surface electron concentration per unit area for ultrathin films of stressed Kane type compounds can be written as

$$n_{2\mathrm{D}} = \frac{g_{\mathrm{v}}}{2\pi} \sum_{n_z=1}^{n_{z\max}} [T_{57}(E_{\mathrm{Fs}}, n_z) + T_{58}(E_{\mathrm{Fs}}, n_z)], \tag{5.69}$$

where

$$T_{57}(E_{\mathrm{Fs}}, n_z) \equiv \left[P^2(E_{\mathrm{Fs}}, n_z)\bar{a}_0(E_{\mathrm{Fs}})\bar{b}_0(E_{\mathrm{Fs}})\right] \text{ and}$$

$$T_{58}(E_{\mathrm{Fs}}, n_z) \equiv \sum_{r=1}^{s} L\,(r)\, T_{57}(E_{\mathrm{Fs}}, n_z).$$

Using (5.69) and (1.11), the 2D DMR in this case can be expressed as

$$\frac{D}{\mu} = \frac{1}{|e|} \frac{\sum\limits_{n_z=1}^{n_{z\max}} [T_{57}(E_{\mathrm{Fs}}, n_z) + T_{58}(E_{\mathrm{Fs}}, n_z)]}{\sum\limits_{n_z=1}^{n_{z\max}} \left[(T_{57}(E_{\mathrm{Fs}}, n_z))' + (T_{58}(E_{\mathrm{Fs}}, n_z))'\right]}. \tag{5.70}$$

In the absence of stress together with the substitution, $B_2^2 \equiv 3\hbar^2\left(E_{\mathrm{g}}/4m^*\right)$, (5.70) assumes the same form as given by (5.16).

5.3 Result and Discussions

Using (5.5) and (5.6), we have, in Figs. 5.1 and 5.2, plotted the normalized 2D DMR at low temperatures, where the quantum effects become prominent for the ultrathin films of n-Cd_3As_2 and n-$CdGeAs_2$ as functions of nanothickness. The curves (a) and (b) corresponds to $\delta \neq 0$ and $\delta = 0$ respectively for the purpose of assessing the influence of crystal field splitting on the 2D DMR in ultrathin films of tetragonal materials.

Using (5.10) and (5.11), we have plotted the curve (c) in accordance with the three band model of Kane. The curves (d) and (e) have been plotted using (5.15)–(5.18) in accordance with the two band model of Kane and that of the parabolic energy band models respectively.

The influence of quantum confinement is immediately apparent from all the curves of Figs. 5.1 and 5.2, since, the 2D DMR depend strongly on the nanothickness, which is in direct contrast with the corresponding bulk specimens which is also the direct signature of quantum confinement. It appears from the said figures that the 2D DMR decreases with the increasing film thickness in a step like manner as considered here although the numerical values vary widely and are determined by the constants of the energy spectra. The oscillatory dependence is due to the crossing over of the Fermi level by the size quantized levels. For each coincidence of a size quantized level with the Fermi

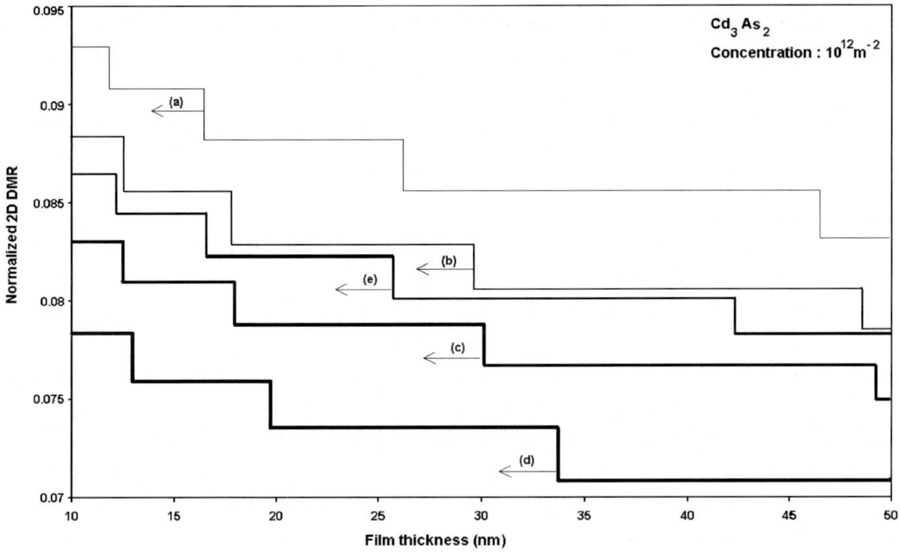

Fig. 5.1. Plot of the normalized 2D DMR as a function of film thickness for the ultrathin films of n-Cd_3As_2 in accordance with (**a**) the generalized band model with $\delta \neq 0$, (**b**) the generalized band model with $\delta = 0$, (**c**) the simplified three band model of Kane, (**d**) the two band model of Kane and (**e**) the parabolic energy bands

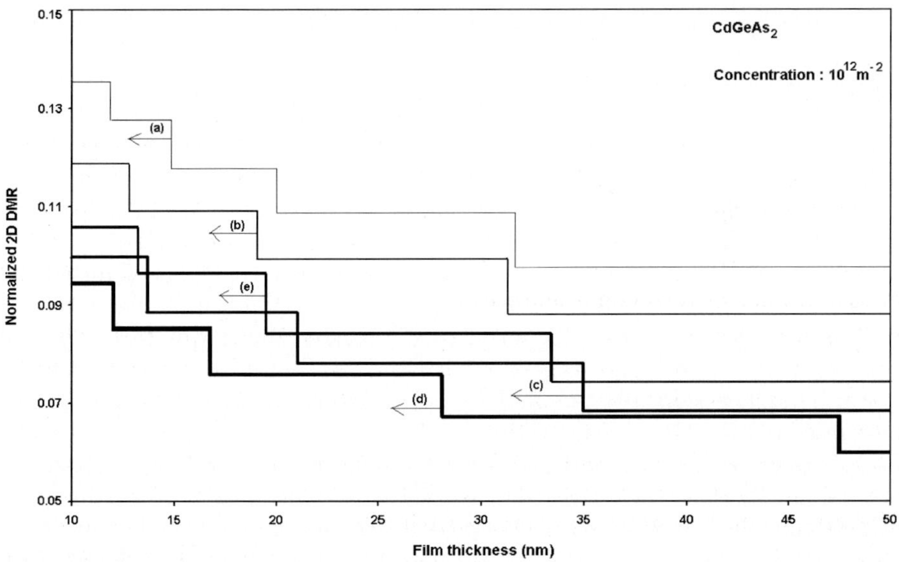

Fig. 5.2. Plot of the normalized 2D DMR as a function of film thickness for the ultrathin films of n-$CdGeAs_2$ for all cases of Fig. 5.1

level, there would be a discontinuity in the density-of-states function resulting in a peak of oscillations. With large values of film thickness, the height of the steps decreases and the DMR decreases with increasing film thickness in a non-oscillatory manner and exhibits monotonic decreasing dependence. The height of step size and the rate of decrement are totally dependent on the band structure. The influence of crystal field splitting is immediately apparent by comparing the curves (a) and (b) of Figs. 5.1 and 5.2. The crystal field splitting alters the numerical values of the 2D DMR in both the cases. The numerical values of the 2D DMR in accordance with the three band model of Kane are different compared with the corresponding two band model, which reflects that fact that the presence of the spin orbit splitting constant changes the magnitude of the 2D DMR. It may be noted that the presence of the band non-parabolicity in accordance with the two-band model of Kane further changes the peaks of the oscillatory 2D DMR for all cases of quantum confinements. The appearance of the humps of the respective curves is due to the redistribution of the electrons among the quantized energy levels when the quantum numbers corresponding to the highest occupied level changes from one fixed value to the others. With varying electron concentration, a change is reflected in the 2D DMR through the redistribution of the electrons among the quantized levels. Although the 2D DMR varies in various ways with all the variables in all the limiting cases, as evident from all the curves of Fig. 5.1 and 5.2, the rates of variations are totally band-structure dependent.

In Figs. 5.3 and 5.4, we have plotted the 2D DMR as a function of surface electron concentration per unit area for all cases of Figs. 5.1 and 5.2 respectively. It appears that the 2D DMR increases with increasing carrier degeneracy and also reflects the signature of the 1D confinement through the step like dependence with the 2D electron statistics. This oscillatory dependence will be less and less prominent with increasing carrier concentration and ultimately, for bulk specimens of the same material, the DMR will be found to increase continuously with increasing electron concentration in a non-oscillatory manner. Using (5.10), (5.11), (5.15)–(5.18), we have plotted the normalized 2D DMR as functions of nanothickness for the ultrathin films of GaAs, InAs and InSb in Figs. 5.5–5.7, where the curve (a)–(c) correspond to the three and the two band models of Kane together with the parabolic energy band respectively. The dependence of the 2D DMR on the surface electron concentration per unit area for all the said cases are shown in Figs. 5.8–5.10 respectively. Using the same set of equations as for III–V materials, we have plotted the normalized 2D DMR for $Hg_{1-x}Cd_xTe$ and $In_{1-x}Ga_xAs_yP_{1-y}$ as a function of nanothickness and 2D electron statistics as shown by Figs. 5.11–5.14 respectively in which, the curve (a)–(c) corresponds to the three and the two band models of Kane together with the parabolic energy band respectively. The numerical values of the 2D DMR depend on the energy band constants of different materials. Figures 5.15 and 5.16 exhibit the dependence of the DMR as a function of alloy composition for all the cases of the ternary and quaternary materials as considered above. The DMR decreases with increasing

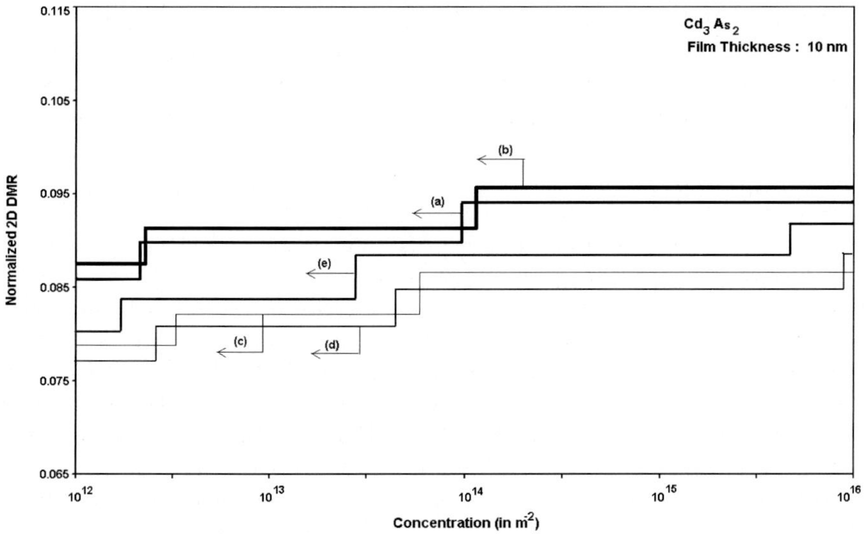

Fig. 5.3. Plot of the normalized 2D DMR as a function of surface electron concentration per unit area for the ultrathin films of n-Cd_3As_2 in accordance with (**a**) the generalized band model with $\delta \neq 0$, (**b**) the generalized band model with $\delta = 0$, (**c**) the simplified three band model of Kane, (**d**) the two band model of Kane and (**e**) the parabolic energy bands

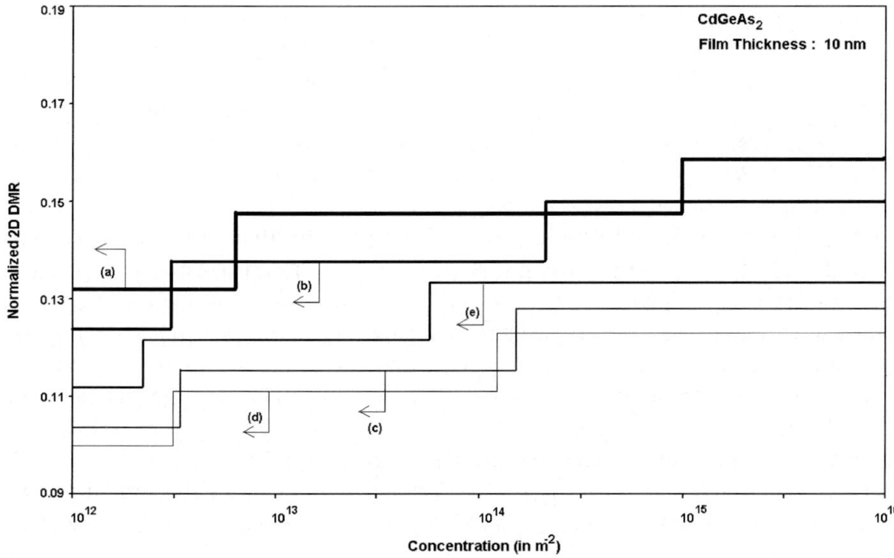

Fig. 5.4. Plot of the normalized 2D DMR as a function of surface electron concentration per unit area for the ultrathin films of n-$CdGeAs_2$ for all cases of Fig. 5.3

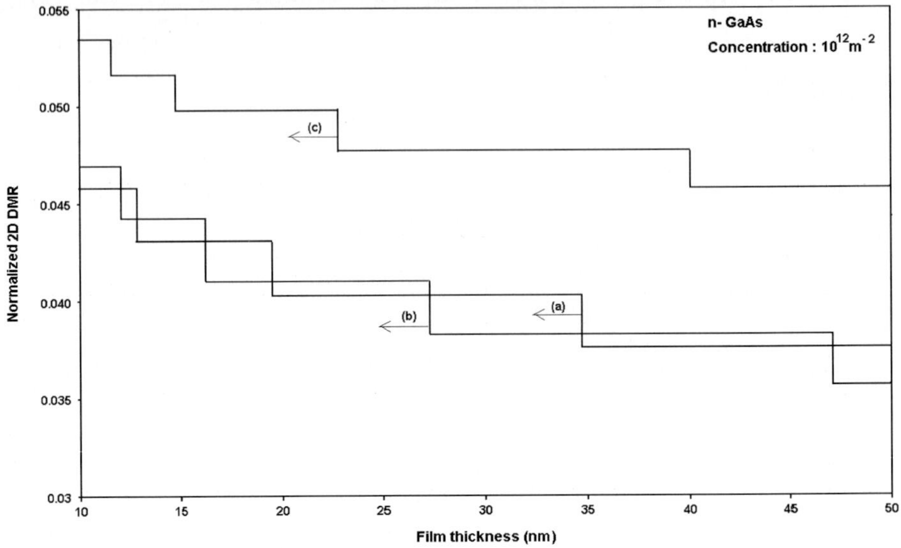

Fig. 5.5. Plot of the normalized 2D DMR as a function of film thickness for the ultrathin films of n-GaAs in accordance with (**a**) the simplified three band model of Kane, (**b**) the two band model of Kane and (**c**) the parabolic energy bands

alloy composition. Using (5.24) and (5.25), we have plotted the normalized 2D DMR, for the ultrathin films of p-CdS materials as functions of nanothickness and surface electron concentration in Figs. 5.17 and 5.18 respectively, where the curve (a) refers to $\bar{\lambda}_0 \neq 0$ and the curve (b) refers to $\bar{\lambda}_0 = 0$, which has been used for the purpose of assessing the influence of the splitting of the two-spin states by the spin orbit coupling and the crystalline field in this case. It appears from Figs. 5.17 and 5.18, that the influence of the term $\bar{\lambda}_0$ is the change of the quantum jumps of the oscillatory 2D DMR in ultrathin films of II–VI materials for both the variables.

Using (5.31), (5.32), (5.37), (5.38) and (5.43), (5.44), (5.48) and (5.49), we have plotted in Figs. 5.19 and 5.20, the normalized 2D DMR for the ultrathin films of bismuth in accordance with the McClure and Choi, the Cohen, the Lax nonparabolic ellipsoidal, and the parabolic ellipsoidal models as functions of nanothickness and surface electron concentration per unit area as shown by the curves (a)–(d) respectively. It appears from both Figs. 5.19 and 5.20, that numerically, the 2D DMR is greatest for ellipsoidal parabolic model and the least for the McClure and Choi model respectively due to the influence of the energy band constants. Using 5.63 and 5.64, Figs. 5.21 and 5.22 exhibit the normalized 2D DMR for the ultrathin films of IV–VI materials as functions of nanothickness and surface electron concentration respectively. The curves (a)–(c) correspond to PbTe, n-PbSnTe and n-$Pb_{1-x}Sn_xSe$ respectively.

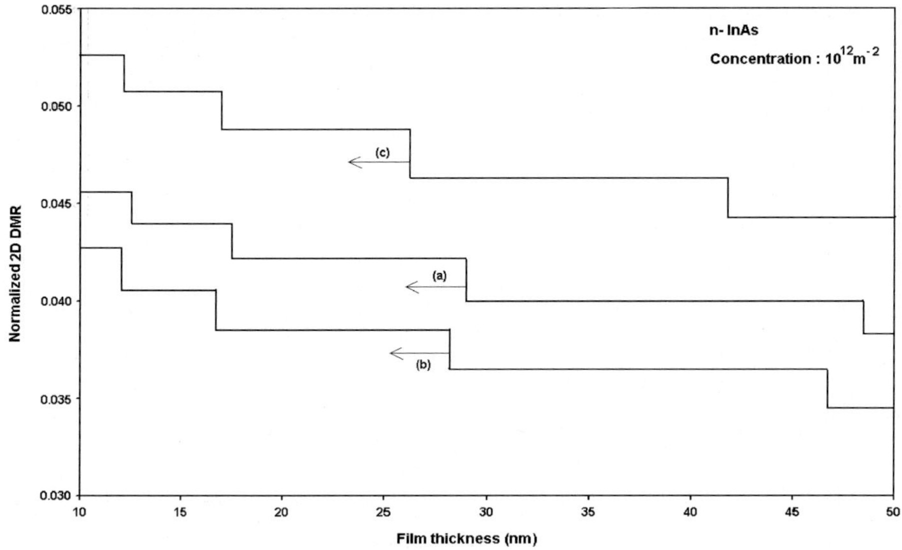

Fig. 5.6. Plot of the normalized 2D DMR as a function of film thickness for the ultrathin films of n-InAs for all cases of Fig. 5.5

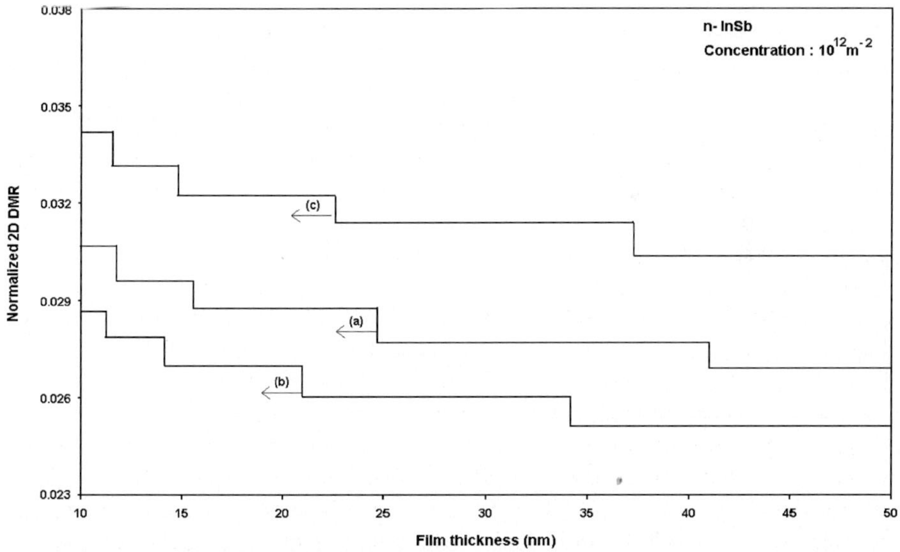

Fig. 5.7. Plot of the normalized 2D DMR as a function of film thickness for the ultrathin films of n-InAs for all cases of Fig. 5.5

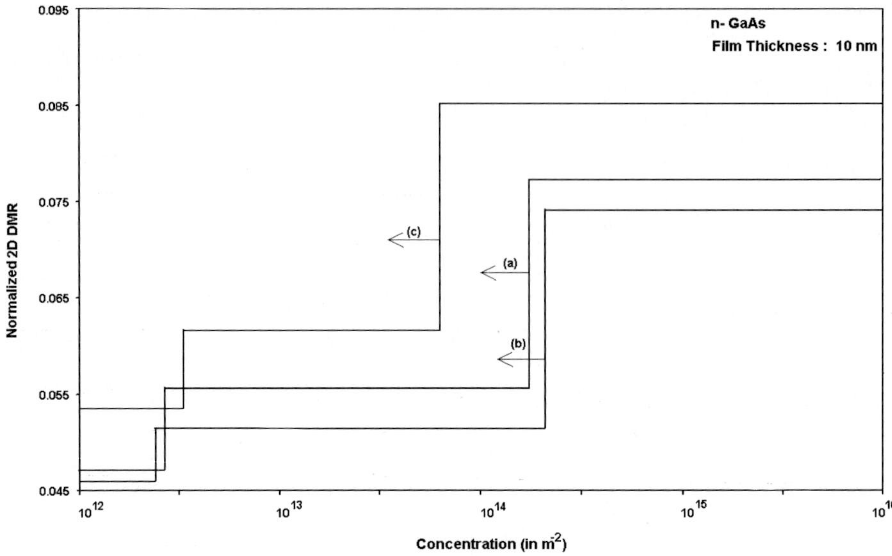

Fig. 5.8. Plot of the normalized 2D DMR as a function of surface electron concentration per unit area for the ultrathin films of GaAs in accordance with (**a**) the simplified three band model of Kane, (**b**) the two band model of Kane, and (**c**) the parabolic energy bands

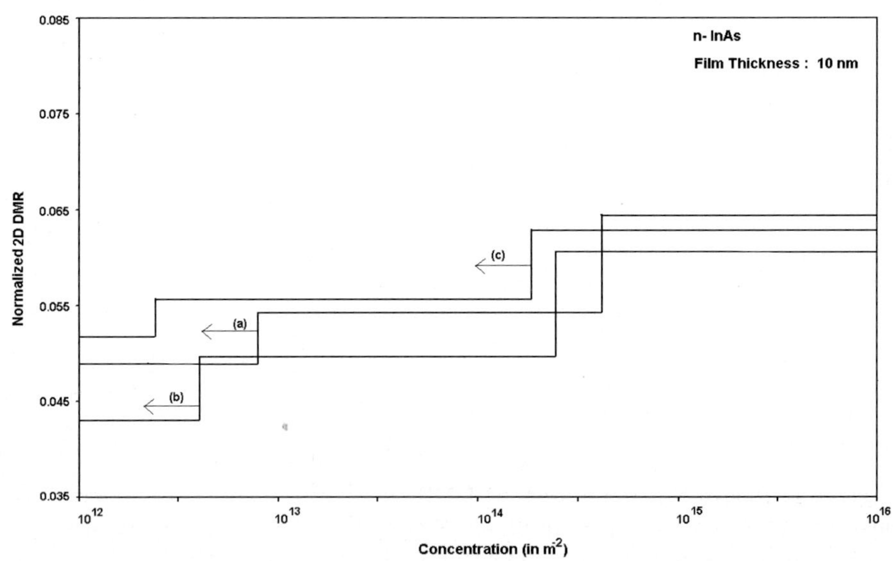

Fig. 5.9. Plot of the normalized 2D DMR as a function of surface electron concentration per unit area for the ultrathin films of n-InAs for all cases of Fig. 5.8

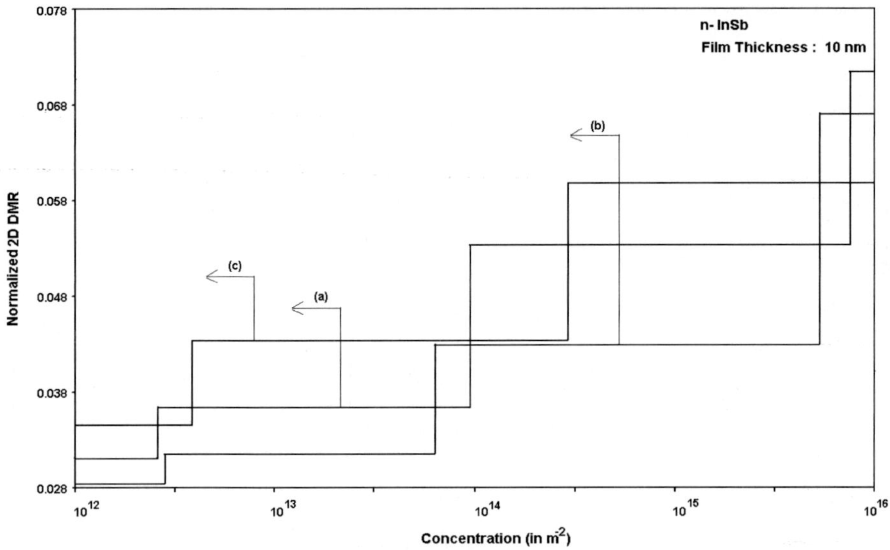

Fig. 5.10. Plot of the normalized 2D DMR as a function of surface electron concentration per unit area for the ultrathin films of n-InSb for all cases of Fig. 5.8

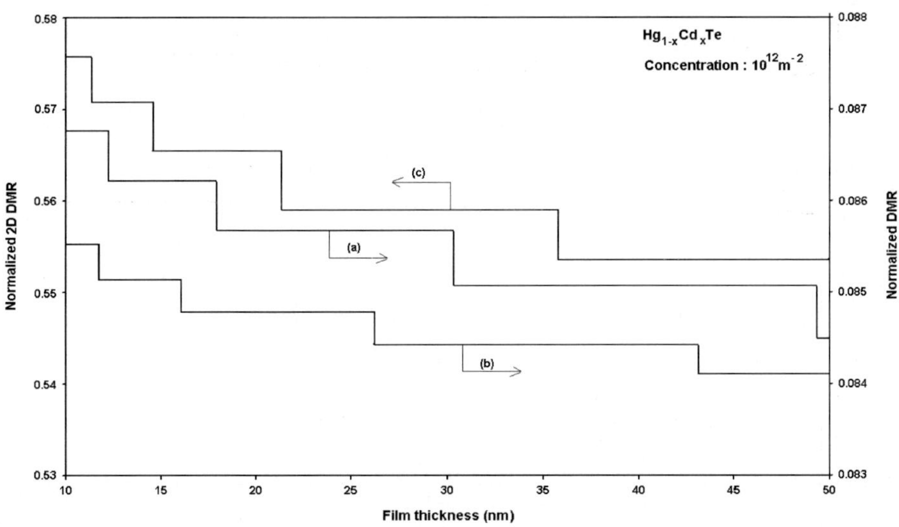

Fig. 5.11. Plot of the normalized 2D DMR as a function of film thickness for the ultrathin films of n-$Hg_{1-x}Cd_xTe$ in accordance with (**a**) the simplified three band model of Kane, (**b**) the two band model of Kane, and (**c**) the parabolic energy bands

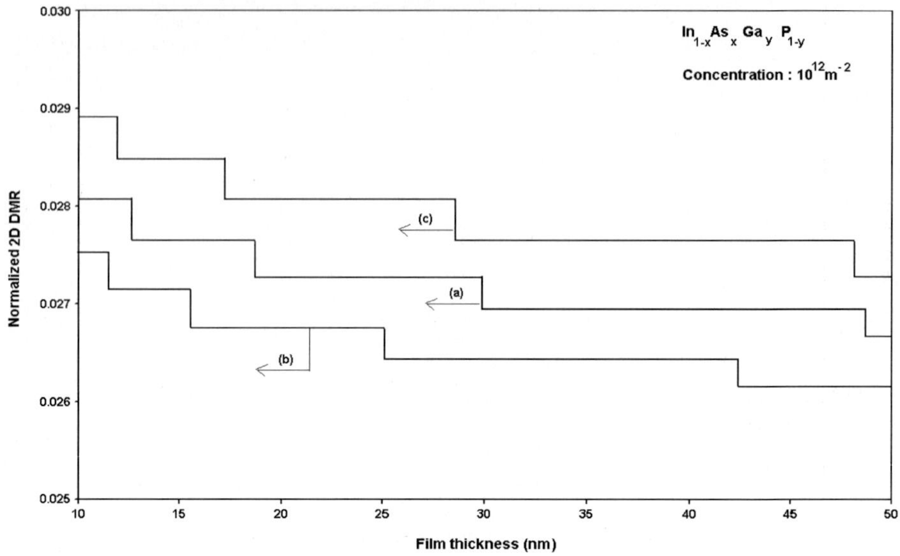

Fig. 5.12. Plot of the normalized 2D DMR as a function of film thickness for the ultrathin films of n-$In_{1-x}As_xGa_yP_{1-y}$ for all cases of Fig. 5.11

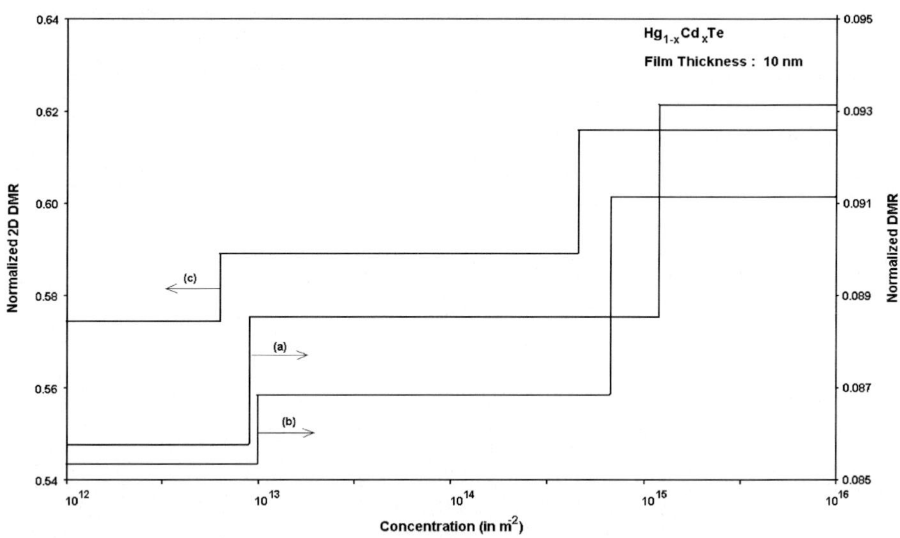

Fig. 5.13. Plot of the normalized 2D DMR as a function of surface electron concentration for the ultrathin films of n-$Hg_{1-x}Cd_xTe$ in accordance with (**a**) the simplified three band model of Kane, (**b**) the two band model of Kane, and (**c**) the parabolic energy bands

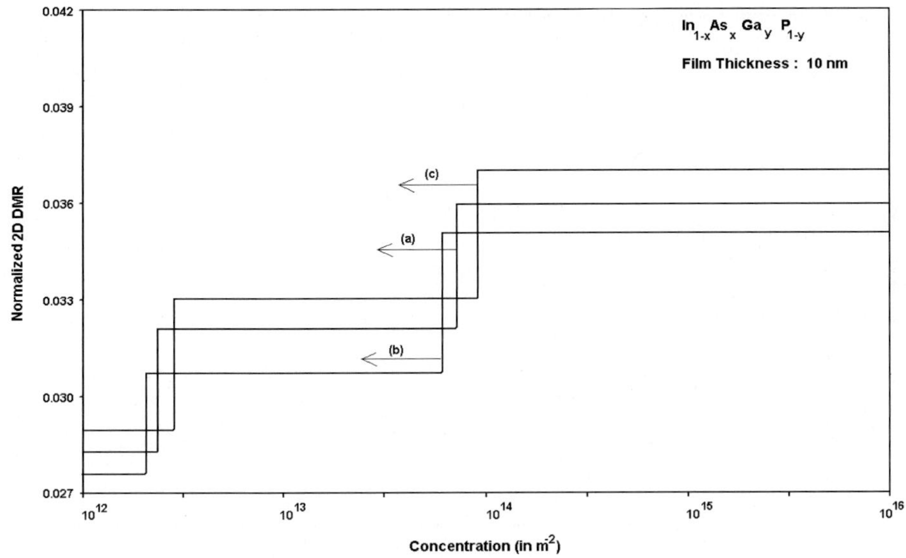

Fig. 5.14. Plot of the normalized 2D DMR as a function of surface electron concentration for the ultrathin films of n-$In_{1-x}As_xGa_yP_{1-y}$ for all cases of Fig. 5.13

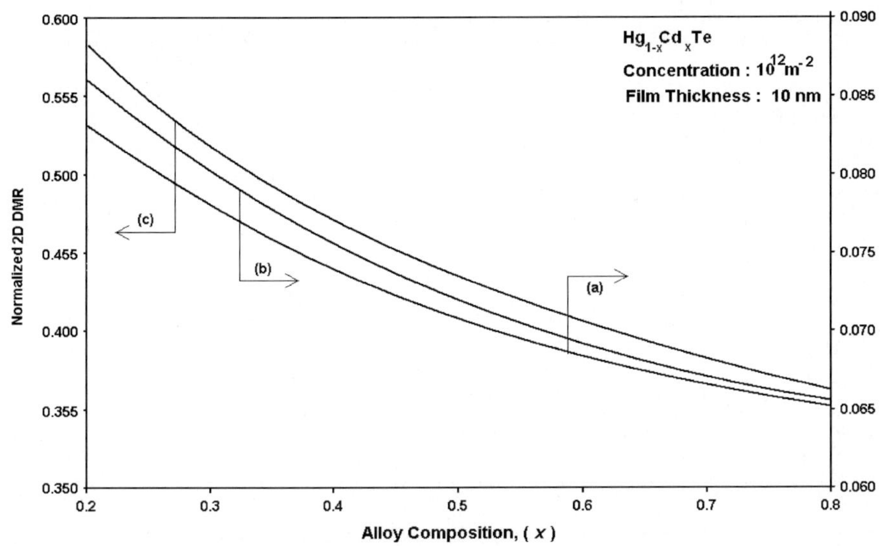

Fig. 5.15. Plot of the normalized 2D DMR as a function of alloy composition (x) for the ultrathin films of n-$Hg_{1-x}Cd_xTe$ in accordance with (**a**) the simplified three band model of Kane, (**b**) the two band model of Kane, and (**c**) the parabolic energy bands

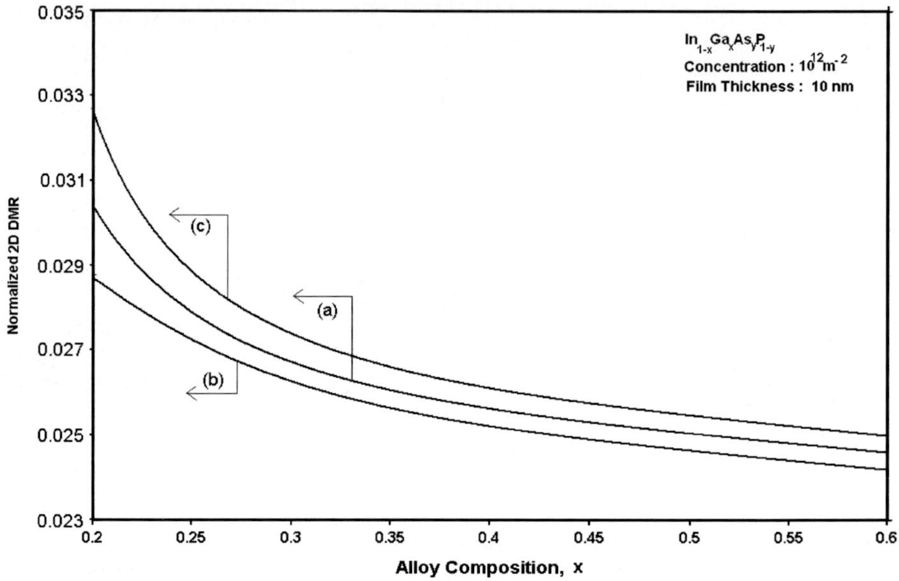

Fig. 5.16. Plot of the normalized 2D DMR as a function of alloy composition (x) for the ultrathin films of n-$In_{1-x}As_xGa_yP_{1-y}$ for all cases of Fig. 5.15

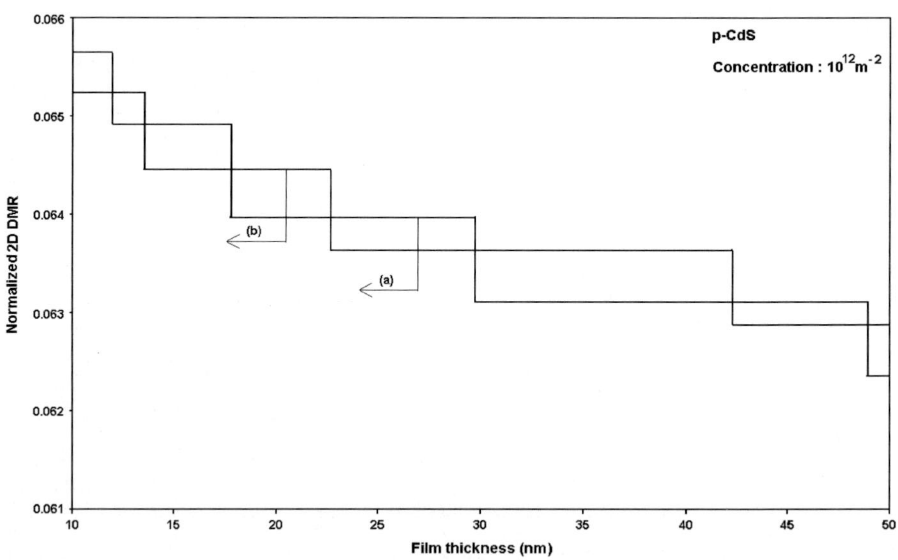

Fig. 5.17. Plot of the normalized 2D DMR as a function of film thickness for ultrathin films of p-CdS in accordance with (**a**) Hopfield model with $\bar{\lambda}_0 \neq 0$ (b) Hopfield model with $\bar{\lambda}_0 = 0$

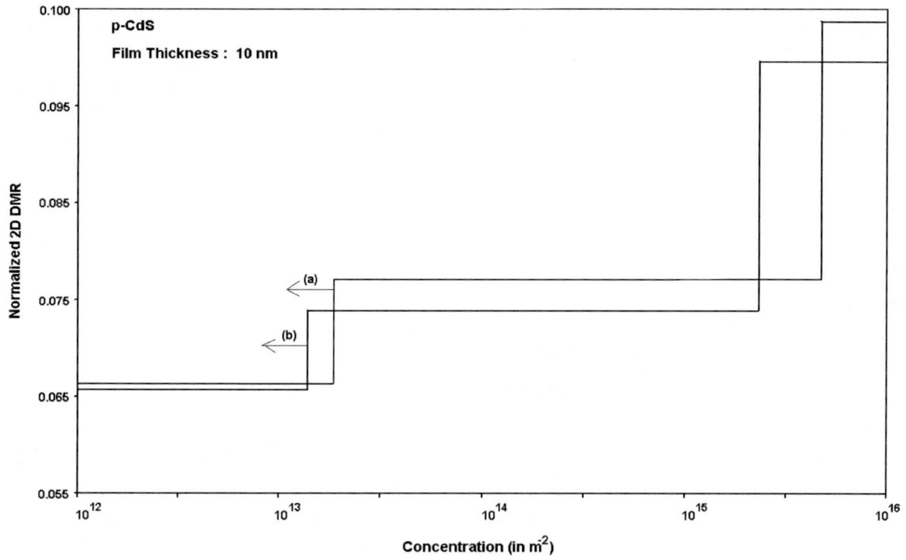

Fig. 5.18. Plot of the normalized 2D DMR for as a function of surface electron concentration for ultrathin films of p-CdS in accordance with (**a**) Hopfield model with $\bar{\lambda}_0 \neq 0$ (**b**) Hopfield model with $\bar{\lambda}_0 = 0$

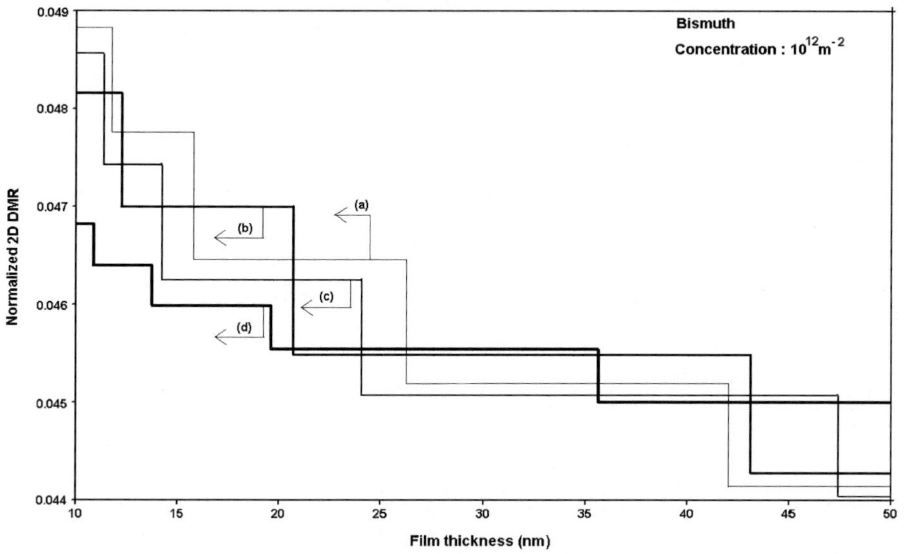

Fig. 5.19. Plot of the normalized 2D DMR for ultrathin films of bismuth as a function of film thickness in accordance with (**a**) the parabolic ellipsoidal model, (**b**) the Lax nonparabolic ellipsoidal model (**c**) the Cohen model, and (**d**) the McClure and Choi model

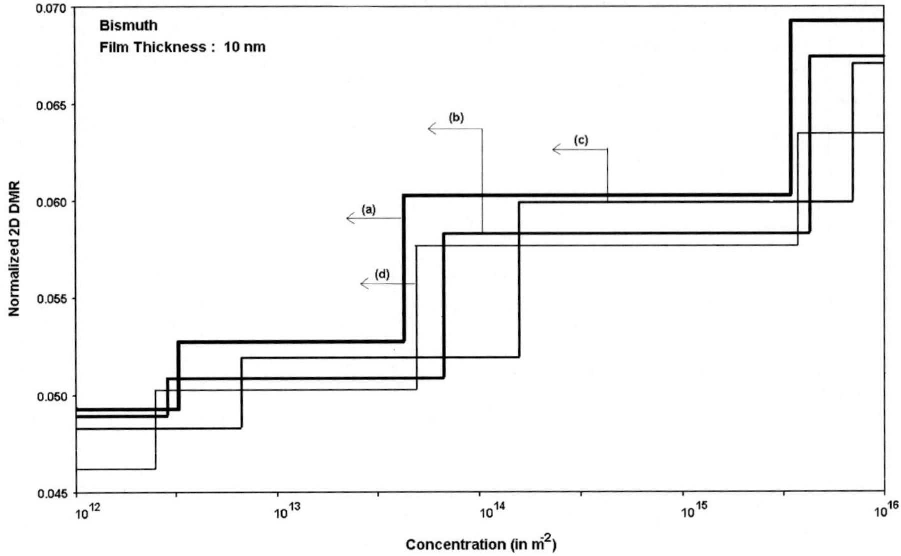

Fig. 5.20. Plot of the normalized 2D DMR for ultrathin films of bismuth as a function of surface electron concentration for all the cases of Fig. 5.19

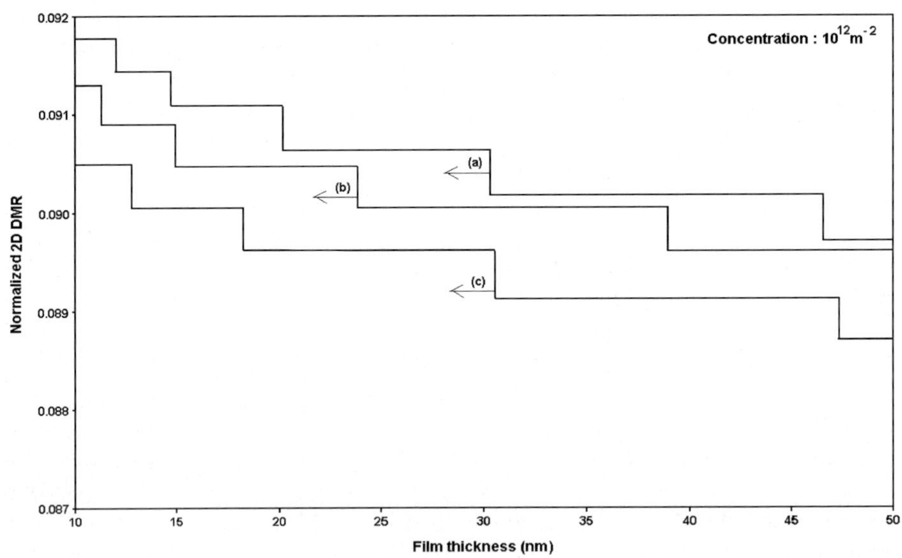

Fig. 5.21. Plot of the normalized 2D DMR as a function of film thickness for the ultrathin films of (**a**) n-PbTe, (**b**) n-PbSnTe and (**c**) n-$Pb_{1-x}Sn_xSe$

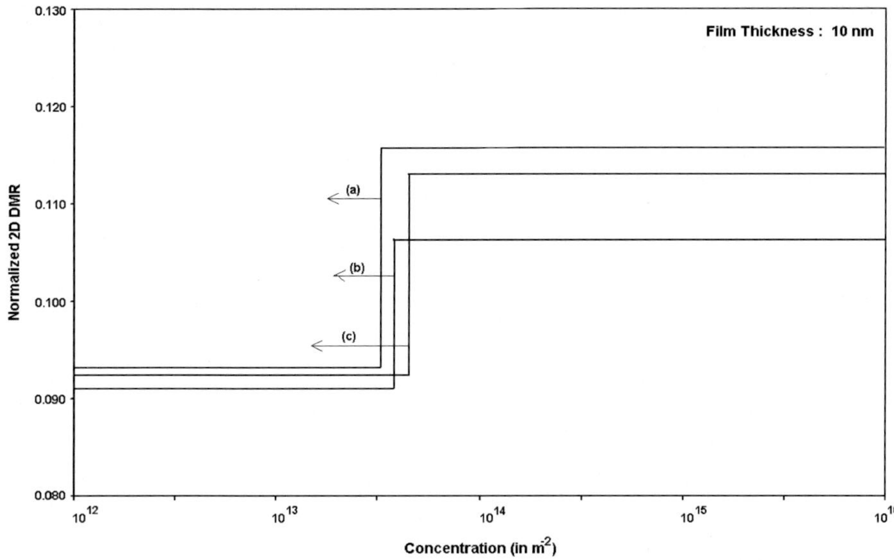

Fig. 5.22. Plot of the normalized 2D DMR as a function of surface electron concentration for the ultrathin films of (**a**) PbTe, (**b**) n-PbSnTe and (**c**) n-$Pb_{1-x}Sn_xSe$

The influence of the energy band constants on the DMR in both the cases is apparent for all the three different materials as considered here. The normalized 2D DMR for stressed Kane type n- InSb has been plotted in Figs. 5.23 and 5.24 as functions of nanothickness and surface electron concentration respectively following (5.69) and (5.70) as shown in plot (a) in the presence of stress while the plot (b) exhibits the same in the absence of stress for the purpose of assessing the influence of stress on the 2D DMR in ultrathin films of stressed n- InSb. In the presence of stress, the magnitude of the 2D DMR is being increased compared with the same under stress free condition. It may be noted that with the advent of modern experimental techniques, it is possible to fabricate quantum-confined structures with an almost defect-free surface. If the direction normal to the film was taken differently from that assumed in this work, the expressions for the 2D DMR in quasi two-dimensional structures would be different analytically, since the basic dispersion laws of many important materials are anisotropic.

It may be noted that under certain limiting conditions, all the results for all the models as derived here get simplified to transform into the well-known expression of the 2D DMR as given by (5.18). This indirect test not only exhibits the mathematical compatibility of the present formulation but also shows the fact that our simple analysis is a more generalized one, since one can obtain the corresponding results for relatively wide gap 2D materials having parabolic energy bands under certain limiting conditions from the present generalized analysis. Thus, the present investigations cover the study of 2D

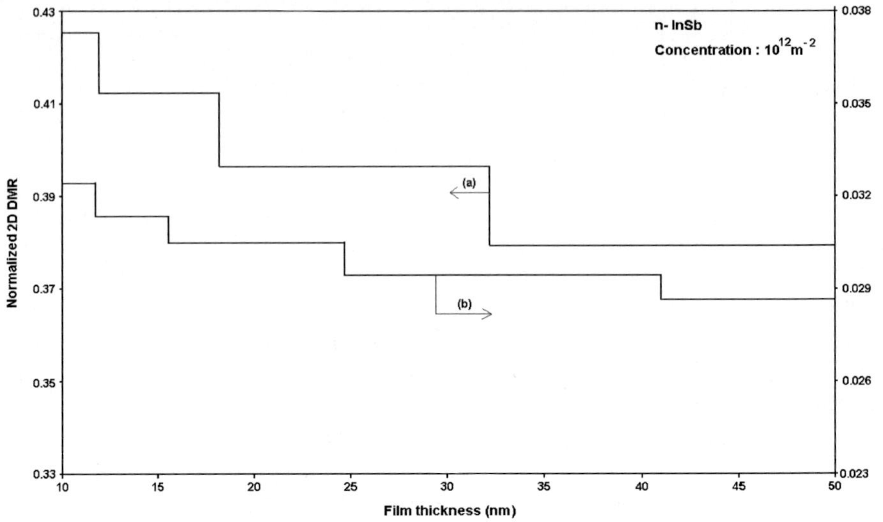

Fig. 5.23. Plot of the normalized 2D DMR as a function of film thickness for the ultrathin films of stressed n-InSb in which the curve (**a**) shows the 2D DMR in the presence of stress while the curve (**b**) is applicable in the absence of the stress

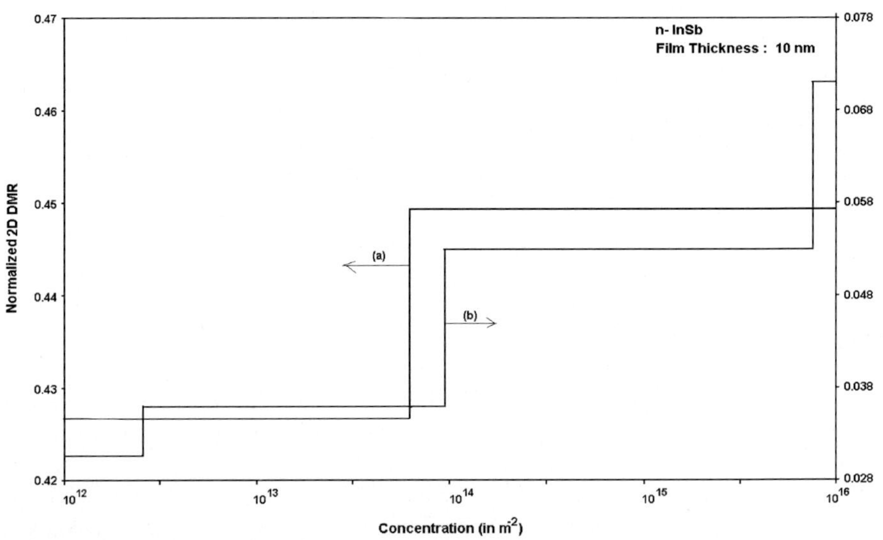

Fig. 5.24. Plot of the normalized 2D DMR as a function of surface electron concentration for the ultrathin films of stressed n-InSb in which the curve (**a**) shows the 2D DMR in the presence of stress while the curve (**b**) is applicable in the absence of the stress

DMR for ultrathin films of tetragonal, III–V, ternaries, quaternaries, II–VI, bismuth, IV–VI, and stressed compounds having different band structures.

One striking feature of the EMM in the ultrathin films of different nanomaterials, as considered here, is that, the EMM can be the function of the size quantum number, the Fermi energy and other energy band constants depending on the respective 2D dispersion laws as formulated already in the theoretical background of this chapter. It must be mentioned that a direct research application of the quantized materials is in the area of band structure. Finally, it may be noted that the theoretical results derived in this chapter can be used to determine the 2D diffusivity and the 3D diffusivity of the constituent bulk materials in the absence of quantum effects and this simplified formulation exhibits the basic qualitative features of 2D DMR for different quantum confined materials.

For the purpose of condensed presentation, the specific electron statistics related to a particular energy dispersion law for a specific material and the Einstein relation have been presented in Table 5.1.

5.4 Open Research Problems

R.5.1 Investigate the Einstein relation in the presence of an arbitrarily oriented non-quantizing alternating magnetic field for the ultrathin films of tetragonal semiconductors by including the electron spin. Study all the special cases for III–V, ternary, and quaternary materials in this context.

R.5.2 Investigate the Einstein relations in ultrathin films of all models of Bi, IV–VI, II–VI and stressed Kane type compounds in the presence of an arbitrarily oriented non-quantizing alternating magnetic field by including the electron spin.

R.5.3 Investigate the Einstein relation for ultrathin films of all the materials as stated in R.2.1 of Chap. 2.

R.5.4 Investigate the Einstein relation for all the problems from R.5.1 to R.5.3 in the presence of an arbitrarily oriented non-quantizing alternating electric field.

R.5.5 Investigate the Einstein relation for all the problems from R.5.1 to R.5.3 in the presence of an arbitrarily oriented crossed electric and magnetic fields.

R.5.6 Investigate the Einstein relation for all the problems from R.5.1 to R.5.5 in the presence of a finite potential well.

R.5.7 Investigate the Einstein relation for all the problems from R.5.1 to R.5.5 in the presence of a parabolic potential well.

R.5.8 Investigate the Einstein relation for all the problems from R.5.1 to R.5.5 in the presence of a circular potential well.

Allied Research Problems

R.5.9 Investigate the EMM for all the materials as stated in R.5.3–R.5.8.

Table 5.1. The 2D carrier statistics and the 2D Einstein relation in ultrathin films of tetragonal, III–V, ternary, quaternary, II–VI, all the models of Bismuth, IV–VI and stressed materials

Type of materials	The 2D carrier statistics	The 2D Einstein relation for the diffusivity mobility ratio
1. Tetragonal compounds	In accordance with the 2D generalized dispersion relation (5.1) for the ultrathin films as formulated in this chapter $$n_{2D} = \frac{g_v}{2\pi} \sum_{n_x=1}^{n_{x\max}} [T_{51}(E_{Fs}, n_z) + T_{52}(E_{Fs}, n_z)], \quad (5.5)$$	On the basis of (5.5), $$\frac{D}{\mu} = \frac{1}{\lvert e \rvert} \frac{\sum_{n_z=1}^{n_{z\max}} [T_{51}(E_{Fs}, n_z) + T_{52}(E_{Fs}, n_z)]}{\sum_{n_z=1}^{n_{z\max}} [\{T_{51}(E_{Fs}, n_z)\}' + \{T_{52}(E_{Fs}, n_z)\}']} \quad (5.6)$$
2. III–V, ternary and quaternary compounds	In accordance with the three band model of Kane for ultrathin films as given by (5.7) which is a special case of (5.1) $$n_{2D} = \frac{m^* g_v}{\pi \hbar^2} \sum_{n_z=1}^{n_{z\max}} [T_{53}(E_{Fs}, n_z) + T_{54}(E_{Fs}, n_z)] \quad (5.10)$$	On the basis of (5.10) $$\frac{D}{\mu} = \frac{1}{\lvert e \rvert} \frac{\sum_{n_z=1}^{n_{z\max}} [T_{53}(E_{Fs}, n_z) + T_{54}(E_{Fs}, n_z)]}{\sum_{n_z=1}^{n_{z\max}} [\{T_{53}(E_{Fs}, n_z)\}' + \{T_{54}(E_{Fs}, n_z)\}']} \quad (5.11)$$

	In accordance with the two band model of Kane for the ultrathin films as given by (5.12) $n_{2D} = \frac{m^* k_B T g_v}{\pi \hbar^2} \sum_{n_z=1}^{n_{z\max}} [(1 + 2\alpha E_{n_{z3}}) F_0(\eta_{n_1}) + 2\alpha k_B T F_1(\eta_{n_1})]$ (5.15)	On the basis of (5.15), $\frac{D}{\mu} = \frac{k_B T}{\|e\|} \frac{\sum_{n_z=1}^{n_{z\max}} [(1 + 2\alpha E_{n_{z3}}) F_0(\eta_{n_1}) + 2\alpha k_B T F_1(\eta_{n_1})]}{\sum_{n_z=1}^{n_{z\max}} [(1 + 2\alpha E_{n_{z3}}) F_{-1}(\eta_{n_1}) + 2\alpha k_B T F_0(\eta_{n_1})]}$ (5.16)
	In accordance with the parabolic energy bands for the ultrathin films under the condition $\alpha \to 0$, as given by (5.12) $n_{2D} = \frac{g_v m^* k_B T}{\pi \hbar^2} \sum_{n_z=1}^{n_{z\max}} F_0(\eta_{n_2})$ (5.17)	On the basis of (5.17) $\frac{D}{\mu} = \frac{k_B T}{\|e\|} \frac{\sum_{n_z=1}^{n_{z\max}} F_0(\eta_{n_2})}{\sum_{n_z=1}^{n_{z\max}} F_{-1}(\eta_{n_2})}$ (5.18)
3. II–VI compounds	In accordance with (5.19) for the ultrathin films, $n_{2D} = \frac{g_v m_\perp^* k_B T}{\pi \hbar^2} \sum_{n_z=1}^{n_{z\max}} \left[F_0(\eta_{n_{z3}}) - \frac{\bar{\lambda}_0 f_s(E_{Fs}, n_z)}{2\sqrt{a_0' k_B T}} \right]$ (5.24)	On the basis of (5.24) $\frac{D}{\mu} = \frac{k_B T}{\|e\|} \frac{\sum_{n_z=1}^{n_{z\max}} \left[F_0(\eta_{n_{z3}}) - \frac{\bar{\lambda}_0 f_s(E_{Fs}, n_z)}{2\sqrt{a_0' k_B T}} \right]}{\sum_{n_z=1}^{n_{z\max}} \left[F_{-1}(\eta_{n_{z3}}) - \frac{\bar{\lambda}_0 [f_s(E_{Fs}, n_z)]'}{2\sqrt{a_0' k_B T}} \right]}$ (5.25)

(Continued)

Table 5.1. *Continued*

Type of materials	The 2D carrier statistics	The 2D Einstein relation for the diffusivity mobility ratio
4. Bi	(a) The McClure and Choi model: On the basis of (5.26) for the ultrathin films, $$n_{2D} = \left(\frac{2g_v}{3\pi^2\sqrt{p_1}}\right)\sum_{n_z=1}^{n_{z\max}}\left[\theta_1(E_{Fs},n_z)+\theta_2(E_{Fs},n_z)\right] \quad (5.31)$$	On the basis of (5.31) $$\frac{D}{\mu} = \frac{1}{\lvert e\rvert}\frac{\sum_{n_z=1}^{n_{z\max}}\left[\theta_1(E_{Fs},n_z)+\theta_2(E_{Fs},n_z)\right]}{\sum_{n_z=1}^{n_{z\max}}\left[\{\theta_1(E_{Fs},n_z)\}'+\{\theta_2(E_{Fs},n_z)\}'\right]} \quad (5.32)$$
	(b) The Cohen model: On the basis of (5.33) for the ultrathin films $$n_{2D} = \left(\frac{2g_v}{3\pi^2\sqrt{p_1}}\right)\sum_{n_z=1}^{n_{z\max}}\left[\theta_3(E_{Fs},n_z)+\theta_4(E_{Fs},n_z)\right] \quad (5.37)$$	On the basis of (5.37) $$\frac{D}{\mu} = \frac{1}{\lvert e\rvert}\frac{\sum_{n_z=1}^{n_{z\max}}\left[\theta_3(E_{Fs},n_z)+\theta_4(E_{Fs},n_z)\right]}{\sum_{n_z=1}^{n_{z\max}}\left[\{\theta_3(E_{Fs},n_z)\}'+\{\theta_4(E_{Fs},n_z)\}'\right]} \quad (5.38)$$
	(d) The Lax model: On the basis of (5.39) for the ultrathin films,	On the basis of (5.43)

$$n_{2D}=\frac{g_v\sqrt{m_1m_2}k_BT}{\pi\hbar^2}\sum_{n_z=1}^{n_{z\max}}\left[\left(1+2\alpha E_{n_{z5}}\right)F_0\left(\eta_{y2}\right)\right.$$ $$\left.\times 2\alpha k_BTF_1\left(\eta_{y2}\right)\right] \quad (5.43)$$	$$\frac{D}{\mu}=\frac{k_BT}{\lvert e\rvert}\frac{\sum_{n_z=1}^{n_{z\max}}\left[\left(1+2\alpha E_{n_{z5}}\right)F_0\left(\eta_{y2}\right)+2\alpha k_BTF_1\left(\eta_{y2}\right)\right]}{\sum_{n_z=1}^{n_{z\max}}\left[\left(1+2\alpha E_{n_{z5}}\right)F_{-1}\left(\eta_{y2}\right)+2\alpha k_BTF_0\left(\eta_{y2}\right)\right]} \quad (5.44)$$
(e) The parabolic ellipsoidal model: On the basis of (5.45) for the ultrathin films $$n_{2D}=\left[\frac{g_vk_BT\sqrt{m_1m_2}}{\pi\hbar^2}\right]\sum_{n_z=1}^{n_{z\max}}F_0\left(\eta_{y3}\right) \quad (5.48)$$	On the basis of (5.48) $$\frac{D}{\mu}=\left[\frac{k_BT}{\lvert e\rvert}\right]\frac{\sum_{n_z=1}^{n_{z\max}}F_0\left(\eta_{y3}\right)}{\sum_{n_z=1}^{n_{z\max}}F_{-1}\left(\eta_{y3}\right)} \quad (5.49)$$

(Continued)

Table 5.1. *Continued*

Type of materials	The 2D carrier statistics	The 2D Einstein relation for the diffusivity mobility ratio
5. IV–VI compounds	In accordance with (5.51) for the ultrathin films, $$n_{2D} = \frac{g_v}{2\pi} \sum_{n_z=1}^{n_{z\max}} [T_{55}(E_{Fs}, n_z) + T_{56}(E_{Fs}, n_z)] \quad (5.63)$$	On the basis of (5.63) $$\frac{D}{\mu} = \frac{1}{\lvert e \rvert} \frac{\sum_{n_z=1}^{n_{z\max}} [T_{55}(E_{Fs}, n_z) + T_{56}(E_{Fs}, n_z)]}{\sum_{n_z=1}^{n_{z\max}} [\{T_{55}(E_{Fs}, n_z)\}' + \{T_{56}(E_{Fs}, n_z)\}']} \quad (5.64)$$
6. Stressed compounds	In accordance with (5.65) under cross field configuration $$n_{2D} = \frac{g_v}{2\pi} \sum_{n_z=1}^{n_{z\max}} [T_{57}(E_{Fs}, n_z) + T_{58}(E_{Fs}, n_z)] \quad (5.69)$$	On the basis of (5.69) $$\frac{D}{\mu} = \frac{1}{\lvert e \rvert} \frac{\sum_{n_z=1}^{n_{z\max}} [T_{57}(E_{Fs}, n_z) + T_{58}(E_{Fs}, n_z)]}{\sum_{n_z=1}^{n_{z\max}} [\{T_{57}(E_{Fs}, n_z)\}' + \{T_{58}(E_{Fs}, n_z)\}']} \quad (5.70)$$

R.5.10 Investigate the Debye screening length, the carrier contribution to the elastic constants, the heat capacity, the activity coefficient, and the plasma frequency for all the materials covering all the cases of problems from R.5.1 to R.5.8.

R.5.11 Investigate in detail, the mobility for elastic and inelastic scattering mechanisms for all the materials covering all the cases of problems from R.5.1 to R.5.8.

R.5.12 Investigate all transport coefficients in detail for all the materials covering all the cases of problems R.5.1–R.5.8 of this chapter.

R.5.13 Investigate the dia- and paramagnetic susceptibilities in details for all the materials covering all the appropriate research problems of this chapter.

References

1. P.M. Petroff, A.C. Gossard, W. Wiegmann, Appl. Phys. Letts. **45**, 620 (1984); J.M. Gaines, P.M. Petroff, H. Kroemar, R.J. Simes, R.S. Geels, J.H. English, J. Vac. Sci. Tech. **B6**, 1378 (1988)
2. J. Cilbert, P.M. Petroff, G.J. Dolan, S.J. Pearton, A.C. Gossard, J.H. English, Appl. Phys. Lett. **49**, 1275 (1986)
3. T. Fujui, H. Saito, Appl. Phys. Lett. **50**, 824 (1987)
4. H. Sasaki, Jpn J. Appl. Phys. **19**, 94 (1980)
5. P.M. Petroff, A.C. Gossard, R.A. Logan, W. Weigmann, Appl. Phys. Lett. **41** 635 (1982)
6. H. Temkin, G.J. Dolan, M.B. Panish, S.N.G. Chu, Appl. Phys. Lett. **50**, 413 (1988); I. Miller, A. Miller, A. Shahar, U. Koren, P.J. Corvini, Appl. Phys. Lett. **54**, 188 (1989)
7. L.L. Chang, H. Esaki, C.A. Chang, L. Esaki, *Phys. Rev. Lett.* **38**, 1489 (1977); K. Less, M.S. Shur, J.J. Drunnond, H. Morkoc, *IEEE Trans. Electron Dev.* **ED-30**, 07 (1983); G. Bastard, *Wave Mechanics Applied to Semiconductor Heterostructures*, Halsted; Les Ulis, Les Editions de Physique, New York (1988); M.J. Kelly, *Low Dimensional Semiconductors: Materials, Physics, Technology, Devices* (Oxford University Press, Oxford, 1995); C. Weisbuch, B. Vinter, *Quantum Semiconductor Structures* (Boston Academic Press, Boston, 1991)
8. N.T. Linch, *Festkorperprobleme* **23**, 27 (1985)
9. D.R. Sciferes, C. Lindstrom, R.D. Burnham, W. Streifer, T.L. Paoli, Electron. Lett. **19**, 170 (1983)
10. P.M. Solomon, Proc. IEEE, **70**, 489 (1982; T.E. Schlesinger and T. Kuech, Appl. Phys. Lett. **49**, 519 (1986)
11. D. Kasemet, C.S. Hong, N.B. Patel, P.D. Dapkus, Appl. Phys. Lett. **41**, 912 (1982); K. Woodbridge, P. Blood, E.D. Pletcher, P.J. Hulyer, Appl. Phys. Lett. **45**, 16 (1984); S. Tarucha, H.O. Okamoto, Appl. Phys. Lett. **45**, 16 (1984); H. Heiblum, D.C. Thomas, C.M. Knoedler, M.I. Nathan, Appl. Phys. Lett. **47**, 1105 (1985)
12. O. Aina, M. Mattingly, F.Y. Juan, P.K. Bhattacharyya, Appl. Phys. Lett. **50**, 43 (1987)
13. I. Suemune, L.A. Coldren, IEEE J. Quant. Electron. **24**, 1178 (1988)

14. D.A.B. Miller, D.S. Chemla, T.C. Damen, J.H. Wood, A.C. Burrus, A.C. Gossard, W. Weigmann, IEEE J. Quant. Electron. **21**, 1462 (1985)
15. A.N. Chakravarti, K.P. Ghatak, A. Dhar, K.K. Ghosh, S. Ghosh, Appl. Phys. **A26**, 169 (1981); S. Choudhury, L.J. Singh, K.P. Ghatak, Nanotechnology, **15**, 180 (2004)
16. J.O. Dimmock, in Physics of Semimetals and Narrow Gap Compounds, ed. by D.L. Carter, R.T. Bates (Pergamon Press, Oxford, 1971, pp. 319)

$$n_{1D} = \frac{2g_v}{\pi}\left(\frac{2m^*}{\hbar^2}\right)^{1/2} \sum_{n_x=1}^{n_{x\max}} \sum_{n_y=1}^{n_{y\max}} [T_{65}(E_{F_1}, n_x, n_y) + T_{66}(E_{F_1}, n_x, n_y)], \tag{6.15}$$

where

$$T_{65}(E_{F_1}, n_x, n_y) \equiv \left[E_{F1}(1 + \alpha E_{F1}) - \frac{\hbar^2}{2m^*}\phi(n_x, n_y)\right]^{1/2} \text{ and}$$

$$T_{66}(E_{F_1}, n_x, n_y) \equiv \sum_{r=1}^{s} L(r) T_{65}(E_{F_1}, n_x, n_y).$$

Using (6.15) and (1.11) we can write the 1D DMR in this case as

$$\frac{D}{\mu} = \frac{1}{|e|} \frac{\sum_{n_x=1}^{n_{x\max}} \sum_{n_y=1}^{n_{y\max}} [T_{65}(E_{F_1}, n_x, n_y) + T_{66}(E_{F_1}, n_x, n_y)]}{\sum_{n_x=1}^{n_{x\max}} \sum_{n_y=1}^{n_{y\max}} \left[\{T_{65}(E_{F_1}, n_x, n_y)\}' + \{T_{66}(E_{F_1}, n_x, n_y)\}'\right]}, \tag{6.16}$$

Under the condition, $\alpha E_F \ll 1$, the expression of the 1D electron statistics can, be written as

$$n_{1D} = \frac{2g_v\sqrt{2m^*\pi k_B T}}{h} \sum_{n_x=1}^{n_{x\max}} \sum_{n_y=1}^{n_{y\max}} \frac{1}{\sqrt{i_1}} \left[\left(1 + \frac{3}{2}\alpha i_2\right) F_{-1/2}(\eta_6) + \frac{3}{4}\alpha k_B T F_{1/2}(\eta_6)\right], \tag{6.17}$$

where

$$i_1 \equiv \left[1 + \alpha\frac{\hbar^2}{2m^*}\phi(n_x, n_y)\right], \; i_2 \equiv \left(\frac{\hbar^2}{2m^*}\right)\phi(n_x, n_y)(i_1)^{-1} \text{ and}$$

$$\eta_6 \equiv (E_{F_1} - i_2) k_B T.$$

Using (6.17) and (1.11) the 1D DMR in this case can be expressed as

$$\frac{D}{\mu} = \frac{k_B T}{|e|} \frac{\sum_{n_x=1}^{n_{x\max}} \sum_{n_y=1}^{n_{y\max}} \frac{1}{\sqrt{i_1}} \left[\left(1 + \frac{3}{2}\alpha i_2\right) F_{-1/2}(\eta_6) + \frac{3}{4}\alpha k_B T F_{1/2}(\eta_6)\right]}{\sum_{n_x=1}^{n_{x\max}} \sum_{n_y=1}^{n_{y\max}} \frac{1}{\sqrt{i_1}} \left[\left(1 + \frac{3}{2}\alpha i_2\right) F_{-3/2}(\eta_6) + \frac{3}{4}\alpha k_B T F_{-1/2}(\eta_6)\right]}. \tag{6.18}$$

(c) Under the condition $\alpha \to 0$, the expressions of n_{1D} and 1D DMR for QWs of isotropic parabolic energy bands can, respectively, be written from (6.17) and (6.18) respectively, as

$$n_{1D} = \frac{2g_v\sqrt{2\pi m^* k_B T}}{h} \sum_{n_x=1}^{n_{x\max}} \sum_{n_y=1}^{n_{y\max}} [F_{-1/2}(\eta_7)], \; \eta_7 \equiv \left(\frac{1}{k_B T}\right)\left[E_{F1} - \left\{\left(\hbar^2/2m^*\right)\phi(n_x, n_y)\right\}\right], \tag{6.19}$$

and

$$\frac{D}{\mu} = \frac{k_{\mathrm{B}}T}{|e|} \frac{\sum\limits_{n_x=1}^{n_{x\max}} \sum\limits_{n_y=1}^{n_{y\max}} \left[F_{-1/2}(\eta_7)\right]}{\sum\limits_{n_x=1}^{n_{x\max}} \sum\limits_{n_y=1}^{n_{y\max}} \left[F_{-3/2}(\eta_7)\right]}. \tag{6.20}$$

It may be noted that (6.20) is already well known in the literature [52].

For bulk materials, converting the summations over n_x and n_y to the corresponding integrations over the said variables, (6.20) assumes the well known form given for the first time by Landsberg (2.19).

6.2.3 II–VI Materials

The dispersion relation of the conduction electrons of QWs of II–VI materials for two dimensional quantization can be written, following (2.27), as

$$E = a_0'\phi(n_x, n_y) + b_0'k_z^2 \pm \bar{\lambda}_0\sqrt{\phi(n_x, n_y)}. \tag{6.21}$$

Using (6.21), the EMM along the k_z direction in this case can be written as

$$m^* = m_{\parallel}^*. \tag{6.22}$$

We observe that the EMM in the QWs of II–VI materials is a constant quantity.

The expressions of the subband energies, the 1D density-of-states function per subband and the total 1D density-of-states function ($N_{1DT}(E)$) can, respectively, be given by

$$E_{n_{xy4}} = a_0'\phi(n_x, n_y) \pm \bar{\lambda}_0\sqrt{\phi(n_x, n_y)}, \tag{6.23}$$

$$N_{1\mathrm{D}}(E, n_x, n_y) = \frac{g_{\mathrm{v}}}{\pi b_0'}\left[E - a_0'\phi(n_x, n_y) \mp \bar{\lambda}_0\sqrt{\phi(n_x, n_y)}\right]^{-1/2}, \tag{6.24}$$

$$N_{1\mathrm{DT}}(E, n_x, n_y) = \frac{g_{\mathrm{v}}}{\pi b_0'}\sum_{n_x=1}^{n_{x\max}}\sum_{n_y=1}^{n_{y\max}}\left[E - a_0'\phi(n_x, n_y) \mp \bar{\lambda}_0\sqrt{\phi(n_x, n_y)}\right]^{-1/2} H\left(E - E_{n_{xy4}}\right). \tag{6.25}$$

Using (6.25), we can write

$$n_{1\mathrm{D}} = g_{\mathrm{v}}\left(\frac{k_{\mathrm{B}}T}{\pi b_0'}\right)^{1/2}\sum_{n_x=1}^{n_{x\max}}\sum_{n_y=1}^{n_{y\max}}\left[F_{-1/2}(\eta_{2,+}) + F_{-1/2}(\eta_{2,-})\right], \tag{6.26}$$

where

$$\eta_{2,\pm} = \left(\frac{1}{k_{\mathrm{B}}T}\right)\left[E_{\mathrm{F}1} - \left[a_0'\phi\left(n_x, n_y\right) \pm \bar{\lambda}_0\sqrt{\phi\left(n_x, n_y\right)}\right]\right].$$

Using (6.26) and (1.11) the 1D DMR in QWs of II–VI materials can be expressed as

$$\frac{D}{\mu} = \frac{k_{\mathrm{B}}T}{|e|}\frac{\sum_{n_x=1}^{n_{x\max}}\sum_{n_y=1}^{n_{y\max}}\left[F_{-1/2}\left(\eta_{2,+}\right) + F_{-1/2}\left(\eta_{2,-}\right)\right]}{\sum_{n_x=1}^{n_{x\max}}\sum_{n_y=1}^{n_{y\max}}\left[F_{-3/2}\left(\eta_{2,+}\right) + F_{-3/2}\left(\eta_{2,-}\right)\right]}. \tag{6.27}$$

Under the conditions $\bar{\lambda}_0 \to 0$ and $m_{\parallel}^* = m_{\perp}^* = m^*$, (6.27) gets simplified to the well-known form given by (6.20).

6.2.4 The Formulation of 1D DMR in Bismuth

(a) The McClure and Choi model

The dispersion relation of the conduction electrons in QWs of Bi for the two dimensional quantizations along k_y and k_z directions can be written, following (2.34), as

$$k_x^2\left[\frac{\hbar^2}{2m_1} - \frac{\alpha\hbar^2}{4m_1m_2}\left(\frac{n_y\pi}{d_y}\right)^2\right] + \left(\frac{n_z\pi}{d_z}\right)^2\left[\frac{\hbar^2}{2m_3} - \frac{\alpha\hbar^2}{4m_2m_3}\left(\frac{n_y\pi}{d_y}\right)^2\right]$$
$$= E\left(1+\alpha E\right) - \frac{\alpha}{4m_2m_2'}\left(\frac{\hbar n_y\pi}{d_y}\right)^4 - \alpha E\left\{1 - \left(\frac{m_2}{m_2'}\right)\right\}\left(\frac{\hbar n_y\pi}{d_y}\right)^2, \tag{6.28}$$

where n_y $(= 1, 2, 3, \ldots)$, n_z $(= 1, 2, 3, \ldots)$, d_y and d_z are the size quantum numbers and the nanothickness along y and z-directions respectively.

The EMM along k_x direction can be written as

$$m^*\left(E_{\mathrm{F}_1}, n_y, n_z\right) = \hbar^2 T_{67}\left(E_{\mathrm{F}_1}, n_y, n_z\right)\left\{T_{67}\left(E_{\mathrm{F}_1}, n_y, n_z\right)\right\}', \tag{6.29}$$

where

$$T_{67}\left(E_{\mathrm{F}_1}, n_y, n_z\right) = \left\{\left[\left[E_{\mathrm{F}_1}\left(1+\alpha E_{\mathrm{F}_1}\right) - \frac{\alpha}{4m_2m_2'}\left(\frac{\hbar n_y\pi}{d_y}\right)^4\right.\right.\right.$$
$$\left.\left.\left.-\alpha E_{\mathrm{F}_1}\left\{1 - \left(\frac{m_2}{m_2'}\right)\right\}\left(\frac{\hbar n_y\pi}{d_y}\right)^2 - \left(\frac{n_z\pi}{d_z}\right)^2\left[\frac{\hbar^2}{2m_3} - \frac{\alpha\hbar^2}{4m_2m_3}\left(\frac{n_y\pi}{d_y}\right)^2\right]\right]\right]\right\}^{1/2}$$
$$\left[\frac{\hbar^2}{2m_1} - \frac{\alpha\hbar^2}{4m_1m_2}\left(\frac{n_y\pi}{d_y}\right)^2\right]^{-1/2}.$$

Thus, the EMM in this case is a function of both the Fermi energy and the size quantum numbers due to the presence of band non-parabolicity only.

The subband energy $(E_{n_{yz_1}})$ and the density- of states function per subband can respectively be expressed as

$$\left(\frac{n_z\pi}{d_z}\right)^2\left[\frac{\hbar^2}{2m_3}-\frac{\alpha\hbar^2}{4m_2m_3}\left(\frac{n_y\pi}{d_y}\right)^2\right]=E_{n_{yz_1}}\left(1+\alpha E_{n_{yz_1}}\right)-\frac{\alpha}{4m_2m_2'}\left(\frac{\hbar n_y\pi}{d_y}\right)^4 - \alpha E_{n_{yz_1}}\left\{1-\left(\frac{m_2}{m_2}\right)\right\}\left(\frac{\hbar n_y\pi}{d_y}\right)^2, \tag{6.30}$$

and

$$N_{1\mathrm{D}}(E,n_y,n_z)=\frac{g_{\mathrm{v}}}{\pi}\left[\frac{\hbar^2}{2m_1}-\frac{\alpha\hbar^2}{4m_1m_2}\left(\frac{n_y\pi}{d_y}\right)^2\right]^{-1/2}\left[(1+2\alpha E) - \alpha\left\{1-\left(\frac{m_2}{m_2'}\right)\right\}\left(\frac{\hbar n_y\pi}{d_y}\right)^2\right]\left\{E(1+\alpha E)-\frac{\alpha}{4m_2m_2'}\left(\frac{\hbar n_y\pi}{d_y}\right)^4 - \alpha E\left\{1-\left(\frac{m_2}{m_2'}\right)\right\}\left(\frac{\hbar n_y\pi}{d_y}\right)^2-\left(\frac{n_z\pi}{d_z}\right)^2\left[\frac{\hbar^2}{2m_3}-\frac{\alpha\hbar^2}{4m_2m_3}\left(\frac{n_y\pi}{d_y}\right)^2\right]\right\}^{-1/2}. \tag{6.31}$$

Using (6.31), the 1D electron statistics in QWs of Bi in accordance with the McClure and Choi model can be written as

$$n_{1\mathrm{D}}=\frac{2g_{\mathrm{v}}}{\pi}\sum_{n_y=1}^{n_{y\max}}\sum_{n_z=1}^{n_{z\max}}\left[T_{67}(E_{\mathrm{F}_1},n_y,n_z)+T_{68}(E_{\mathrm{F}_1},n_y,n_z)\right], \tag{6.32}$$

where

$$T_{68}(E_{\mathrm{F}_1},n_y,n_z)=\sum_{r=1}^{s}L(r)\left[T_{67}(E_{\mathrm{F}_1},n_y,n_z)\right].$$

The use of (6.32) and (1.11) leads to the expression of the 1D DMR in this case as

$$\frac{D}{\mu}=\frac{1}{|e|}\frac{\sum\limits_{n_y=1}^{n_{y\max}}\sum\limits_{n_z=1}^{n_{z\max}}\left[T_{67}(E_{\mathrm{F}_1},n_y,n_z)+T_{68}(E_{\mathrm{F}_1},n_y,n_z)\right]}{\sum\limits_{n_y=1}^{n_{y\max}}\sum\limits_{n_z=1}^{n_{z\max}}\left[\{T_{67}(E_{\mathrm{F}_1},n_y,n_z)\}'+\{T_{68}(E_{\mathrm{F}_1},n_y,n_z)\}'\right]}. \tag{6.33}$$

(b) The Cohen model

The 1D electron dispersion law in QWs of Bi in accordance with this model can be written following (2.41) as

$$E\left(1+\alpha E\right)+\frac{\alpha E\hbar^{2}}{2m_{2}'}\left(\frac{n_{y}\pi}{d_{y}}\right)^{2}-\frac{(1+\alpha E)\hbar^{2}}{2m_{2}}\left(\frac{n_{y}\pi}{d_{y}}\right)^{2}-\frac{a\hbar^{4}}{4m_{2}m_{2}'}\left(\frac{n_{y}\pi}{d_{y}}\right)^{4}$$
$$-\frac{\hbar^{2}}{2m_{3}}\left(\frac{n_{z}\pi}{d_{z}}\right)^{2}=\frac{\hbar^{2}k_{x}^{2}}{2m_{1}}. \tag{6.34}$$

The EMM along the k_x direction can be written as

$$m^{*}\left(E_{\mathrm{F}_1},n_y,n_z\right)=m_1T_{69}\left(E_{\mathrm{F}_1},n_y,n_z\right)\left\{T_{69}\left(E_{\mathrm{F}_1},n_y,n_z\right)\right\}', \tag{6.35}$$

where

$$T_{69}\left(E_{F_1},n_y,n_z\right)\equiv\left[E_{F_1}\left(1+\alpha E_{F_1}\right)+\frac{\alpha E_{F_1}\hbar^{2}}{2m_{2}'}\left(\frac{n_{y}\pi}{d_{y}}\right)^{2}\right.$$
$$\left.-\left[\frac{(1+\alpha E_{F_1})\hbar^{2}}{2m_{2}}\left(\frac{n_{y}\pi}{d_{y}}\right)^{2}\right]-\frac{a\hbar^{4}}{4m_{2}m_{2}'}\left(\frac{n_{y}\pi}{d_{y}}\right)^{4}-\frac{\hbar^{2}}{2m_{3}}\left(\frac{n_{z}\pi}{d_{z}}\right)^{2}\right]^{1/2}$$

Equation (6.35) shows that the EMM in this present case is a function of both the size quantum numbers and the Fermi energy due to the presence of the band nonparabolicity only

Following the method given above, the subband energy $(E_{n_{yz_2}})$, the density-of-states per subband, the 1D electron statistics, and the corresponding DMR for QWs of Bi in accordance with the Cohen model can, respectively, be written as

$$E_{n_{yz_2}}\left(1+\alpha E_{n_{yz_2}}\right)+\frac{\alpha E_{n_{yz_2}}\hbar^{2}}{2m_{2}'}\left(\frac{n_{y}\pi}{d_{y}}\right)^{2}-\frac{(1+\alpha E_{n_{yz_2}})\hbar^{2}}{2m_{2}}\left(\frac{n_{y}\pi}{d_{y}}\right)^{2}$$
$$-\frac{a\hbar^{4}}{4m_{2}m_{2}'}\left(\frac{n_{y}\pi}{d_{y}}\right)^{4}-\frac{\hbar^{2}}{2m_{3}}\left(\frac{n_{z}\pi}{d_{z}}\right)^{2}=0, \tag{6.36}$$

$$N_{\mathrm{1D}}\left(E\right)=\frac{g_{\mathrm{v}}\sqrt{2m_{1}}}{\pi\hbar}\left[1+2\alpha E+\left(\frac{\alpha}{2m_{2}'}-\frac{\alpha}{2m_{2}}\right)\left(\frac{\hbar n_{y}\pi}{d_{y}}\right)^{2}\right]$$
$$\left\{\left[E\left(1+\alpha E\right)+\frac{\alpha E\hbar^{2}}{2m_{2}'}\left(\frac{n_{y}\pi}{d_{y}}\right)^{2}-\frac{(1+\alpha E)\hbar^{2}}{2m_{2}}\left(\frac{n_{y}\pi}{d_{y}}\right)^{2}\right.\right.$$
$$\left.\left.-\frac{a\hbar^{4}}{4m_{2}m_{2}'}\left(\frac{n_{y}\pi}{d_{y}}\right)^{4}-\frac{\hbar^{2}}{2m_{3}}\left(\frac{n_{z}\pi}{d_{z}}\right)^{2}\right]\right\}^{-1/2}, \tag{6.37}$$

$$n_{\mathrm{1D}}=\left(\frac{2g_{\mathrm{v}}}{\pi}\frac{\sqrt{2m_{1}}}{\hbar}\right)\sum_{n_y=1}^{n_{y\max}}\sum_{n_z=1}^{n_{z\max}}\left[T_{69}\left(E_{\mathrm{F}_1},n_y,n_z\right)+T_{610}\left(E_{\mathrm{F}_1},n_y,n_z\right)\right], \tag{6.38}$$

$$\frac{D}{\mu} = \frac{1}{|e|} \frac{\sum_{n_y=1}^{n_{y\max}} \sum_{n_z=1}^{n_{z\max}} [T_{69}(E_{F_1}, n_y, n_z) + T_{610}(E_{F_1}, n_y, n_z)]}{\sum_{n_y=1}^{n_{y\max}} \sum_{n_z=1}^{n_{z\max}} [\{T_{69}(E_{F_1}, n_y, n_z)\}' + \{T_{610}(E_{F_1}, n_y, n_z)\}']}, \tag{6.39}$$

where

$$T_{610}(E_{F_1}, n_y, n_z) \equiv \sum_{r=1}^{S} L(r)[T_{69}(E_{F_1}, n_y, n_z)].$$

(c) The Lax model
The 1D electron dispersion law in this case can be written as

$$E(1+\alpha E) - \frac{\hbar^2}{2m_2}\left(\frac{n_y\pi}{d_y}\right)^2 - \frac{\hbar^2}{2m_3}\left(\frac{n_z\pi}{d_z}\right)^2 = \frac{\hbar^2 k_x^2}{2m_1}, \tag{6.40}$$

The EMM along the k_x direction can be written as

$$m^*(E_{F_1}) = m_1(1 + 2\alpha E_{F_1}). \tag{6.41}$$

Thus we see that the EMM for the Lax model is a function of the Fermi energy alone due to the band nonparabolicity.

The subband energy $(E_{n_{yz3}})$, the density- of states per subband, the 1D electron statistics, and the corresponding 1D DMR for this model can, respectively, be written as

$$E_{n_{yz3}}(1+\alpha E_{n_{yz3}}) = \frac{\hbar^2}{2m_2}\left(\frac{n_y\pi}{d_y}\right)^2 + \frac{\hbar^2}{2m_3}\left(\frac{n_z\pi}{d_z}\right)^2, \tag{6.41}$$

$$N_{1D}(E, n_y, n_z) = \frac{g_v\sqrt{2m_1}}{\pi\hbar}(1+2\alpha E)\left\{E(1+\alpha E) - \frac{\hbar^2}{2m_2}\left(\frac{n_y\pi}{d_y}\right)^2 - \frac{\hbar^2}{2m_3}\left(\frac{n_z\pi}{d_z}\right)^2\right\}^{-1/2}, \tag{6.42}$$

$$n_{1D} = \left(\frac{2g_v}{\pi}\frac{\sqrt{2m_1}}{\hbar}\right)\sum_{n_y=1}^{n_{y\max}} \sum_{n_z=1}^{n_{z\max}} [T_{611}(E_{F_1}, n_y, n_z) + T_{612}(E_{F_1}, n_y, n_z)], \tag{6.43}$$

in which

$$T_{611}(E_{F_1}, n_y, n_z) \equiv \left[E_{F_1}(1+\alpha E_{F_1}) - \frac{\hbar^2}{2m_2}\left(\frac{n_y\pi}{d_y}\right)^2 - \frac{\hbar^2}{2m_3}\left(\frac{n_z\pi}{d_z}\right)^2\right]^{1/2},$$

and

$$T_{612}\left(E_{\mathrm{F}_1}, n_y, n_z\right) \equiv \sum_{r=1}^{s} L\left(r\right)\left[T_{611}\left(E_{\mathrm{F}_1}, n_y, n_z\right)\right],$$

$$\frac{D}{\mu} = \frac{1}{|e|} \frac{\sum\limits_{n_y=1}^{n_{y\max}} \sum\limits_{n_z=1}^{n_{z\max}} \left[T_{611}\left(E_{\mathrm{F}_1}, n_y, n_z\right) + T_{612}\left(E_{\mathrm{F}_1}, n_y, n_z\right)\right]}{\sum\limits_{n_y=1}^{n_{y\max}} \sum\limits_{n_z=1}^{n_{z\max}} \left[\left\{T_{611}\left(E_{\mathrm{F}_1}, n_y, n_z\right)\right\}' + \left\{T_{612}\left(E_{\mathrm{F}_1}, n_y, n_z\right)\right\}'\right]}. \tag{6.44}$$

(d) The ellipsoidal parabolic model

The 1D dispersion relation, the EMM, the subband energy $\left(E_{n_{yz4}}\right)$, the density-of states per subband, the 1D electron statistics, and the corresponding 1D DMR for this model can, respectively, be written for $\alpha \to 0$ in (6.41)–(6.44) as

$$E = \left(\frac{\hbar^2 k_x^2}{2m_1}\right) + \left(\frac{\hbar^2}{2m_2}\right)\left(\frac{n_y \pi}{d_y}\right)^2 + \left(\frac{\hbar^2}{2m_3}\right)\left(\frac{n_z \pi}{d_z}\right)^2, \tag{6.45}$$

$$m^*\left(E_{\mathrm{F}_1}\right) = m_1, \tag{6.46}$$

$$E_{n_{yz4}} = \left(\frac{\hbar^2}{2m_2}\right)\left(\frac{n_y \pi}{d_y}\right)^2 + \left(\frac{\hbar^2}{2m_3}\right)\left(\frac{n_z \pi}{d_z}\right)^2, \tag{6.47}$$

$$N_{1\mathrm{D}}\left(E, n_y, n_z\right) = \frac{g_{\mathrm{v}}\sqrt{2m_1}}{\pi\hbar}\left\{E - \frac{\hbar^2}{2m_2}\left(\frac{n_y \pi}{d_y}\right)^2 - \frac{\hbar^2}{2m_3}\left(\frac{n_z \pi}{d_z}\right)^2\right\}^{-1/2}, \tag{6.48}$$

$$n_{1\mathrm{D}} = \frac{2g_{\mathrm{v}}\sqrt{2\pi m^* k_{\mathrm{B}} T}}{h} \sum_{n_y=1}^{n_{y\max}} \sum_{n_z=1}^{n_{z\max}} \left[F_{-1/2}\left(\eta_8\right)\right], \tag{6.49}$$

and

$$\frac{D}{\mu} = \frac{k_{\mathrm{B}} T}{|e|} \frac{\sum\limits_{n_y=1}^{n_{y\max}} \sum\limits_{n_z=1}^{n_{z\max}} \left[F_{-1/2}\left(\eta_8\right)\right]}{\sum\limits_{n_y=1}^{n_{y\max}} \sum\limits_{n_z=1}^{n_{z\max}} \left[F_{-3/2}\left(\eta_8\right)\right]}, \tag{6.50}$$

where

$$\eta_8 \equiv \left(k_{\mathrm{B}} T\right)^{-1}\left[E_{\mathrm{F}_1} - \left[\frac{\hbar^2}{2m_2}\left(\frac{n_y \pi}{d_y}\right)^2 + \frac{\hbar^2}{2m_3}\left(\frac{n_z \pi}{d_z}\right)^2\right]\right].$$

6.2.5 IV–VI Materials

The 1D dispersion relation of the conduction electrons in QWs of IV–VI materials for the two dimensional quantizations along the y and z direction can be expressed as

$$E(1+\alpha E)+\alpha E\left(\frac{x}{x_4}+\frac{\hbar^2}{2x_5}\left(\frac{n_y\pi}{d_y}\right)^2\right)+\alpha E\frac{\hbar^2}{2x_6}\left(\frac{n_z\pi}{d_z}\right)^2$$
$$-(1+\alpha E)\left(\frac{x}{x_1}+\frac{\hbar^2}{2x_2}\left(\frac{n_y\pi}{d_y}\right)^2\right)$$
$$-\alpha\left(\frac{x}{x_1}+\frac{\hbar^2}{2x_2}\left(\frac{n_y\pi}{d_y}\right)^2\right)\left(\frac{x}{x_4}+\frac{\hbar^2}{2x_5}\left(\frac{n_y\pi}{d_y}\right)^2\right)$$
$$-\alpha\left(\frac{x}{x_1}+\frac{\hbar^2}{2x_2}\left(\frac{n_y\pi}{d_y}\right)^2\right)\frac{\hbar^2}{2x_6}\left(\frac{n_z\pi}{d_z}\right)^2$$
$$-(1+\alpha E)\frac{\hbar^2}{2x_3}\left(\frac{n_z\pi}{d_z}\right)^2$$
$$-\alpha\frac{\hbar^2}{2x_3}\left(\frac{n_z\pi}{d_z}\right)^2\left(\frac{x}{x_4}+\frac{\hbar^2}{2x_5}\left(\frac{n_y\pi}{d_y}\right)^2\right)-\alpha\frac{\hbar^2}{2x_3}\left(\frac{n_z\pi}{d_z}\right)^2\frac{\hbar^2}{2x_6}\left(\frac{n_z\pi}{d_z}\right)^2$$
$$=\frac{x}{m_1}+\frac{\hbar^2}{2m_2}\left(\frac{n_y\pi}{d_y}\right)^2+\frac{\hbar^2}{2m_3}\left(\frac{n_z\pi}{d_z}\right)^2, \tag{6.51}$$

where

$$x\equiv\frac{\hbar^2k_x^2}{2}.$$

The subband energy $(E_{n_{yz5}})$ in this case can be written as

$$E_{n_{yz5}}(1+\alpha E_{n_{yz5}})+\alpha E_{n_{yz5}}\left(\frac{\hbar^2}{2x_5}\left(\frac{n_y\pi}{d_y}\right)^2\right)+\alpha E_{n_{yz5}}\frac{\hbar^2}{2x_6}\left(\frac{n_z\pi}{d_z}\right)^2$$
$$-(1+\alpha E_{n_{yz5}})\left(\frac{\hbar^2}{2x_2}\left(\frac{n_y\pi}{d_y}\right)^2\right)$$
$$-\alpha\left(\frac{\hbar^2}{2x_2}\left(\frac{n_y\pi}{d_y}\right)^2\right)\left(\frac{\hbar^2}{2x_5}\left(\frac{n_y\pi}{d_y}\right)^2\right)-\alpha\left(\frac{\hbar^2}{2x_2}\left(\frac{n_y\pi}{d_y}\right)^2\right)\frac{\hbar^2}{2x_6}\left(\frac{n_z\pi}{d_z}\right)^2$$
$$-(1+\alpha E_{n_{yz5}})\frac{\hbar^2}{2x_3}\left(\frac{n_z\pi}{d_z}\right)^2$$
$$-\alpha\frac{\hbar^2}{2x_3}\left(\frac{n_z\pi}{d_z}\right)^2\left(\frac{\hbar^2}{2x_5}\left(\frac{n_y\pi}{d_y}\right)^2\right)-\alpha\frac{\hbar^2}{2x_3}\left(\frac{n_z\pi}{d_z}\right)^2\frac{\hbar^2}{2x_6}\left(\frac{n_z\pi}{d_z}\right)^2$$
$$=\frac{\hbar^2}{2m_2}\left(\frac{n_y\pi}{d_y}\right)^2+\frac{\hbar^2}{2m_3}\left(\frac{n_z\pi}{d_z}\right)^2. \tag{6.52}$$

The EMM along the k_x direction for the QWs of IV–VI materials can thus be expressed as

$$m^*(E_{F_1},n_y,n_z)=(2\bar{g}_1)^{-1}T_{613}(E_{F_1},n_y,n_z)\{T_{613}(E_{F_1},n_y,n_z)\}', \tag{6.53}$$

where

$$T_{613}\left(E_{\mathrm{F}_1},n_y,n_z\right)\equiv\left[\sqrt{\bar{g}_2^2\left(E_{\mathrm{F}_1},n_y,n_z\right)+4\bar{g}_1c_1\left(E_{\mathrm{F}_1},n_y,n_z\right)}-\bar{g}_2\left(E_{\mathrm{F}_1},n_y,n_z\right)\right]^{1/2}$$

$$\bar{g}_1\equiv\frac{\alpha}{x_1x_4},\ \bar{g}_2\left(E_{\mathrm{F}_1},n_y,n_z\right)\equiv\left[\frac{-\alpha E_{\mathrm{F}_1}}{x_4}+\frac{1+\alpha E_{\mathrm{F}_1}}{x_1}+\alpha\left[\frac{\hbar^2}{2x_2x_4}\left(\frac{n_y\pi}{d_y}\right)^2\right.\right.$$
$$\left.\left.+\frac{\hbar^2}{2x_1x_5}\left(\frac{n_y\pi}{d_y}\right)^2\right]+\frac{\alpha\hbar^2}{2x_1x_6}\left(\frac{n_z\pi}{d_z}\right)^2+\frac{\alpha\hbar^2}{2x_3x_4}\left(\frac{n_z\pi}{d_z}\right)^2+\frac{1}{m_1}\right]$$

$$c_1\left(E_{\mathrm{F}_1},n_y,n_z\right)\equiv\left[E_{\mathrm{F}_1}\left(1+\alpha E_{\mathrm{F}_1}\right)+\alpha E_{\mathrm{F}_1}\frac{\hbar^2}{2x_5}\left(\frac{n_y\pi}{d_y}\right)^2+\alpha E_{\mathrm{F}_1}\frac{\hbar^2}{2x_6}\left(\frac{n_z\pi}{d_z}\right)^2\right.$$
$$-\left(1+\alpha E_{\mathrm{F}_1}\right)\left(\frac{\hbar^2}{2x_2}\left(\frac{n_y\pi}{d_y}\right)^2\right)-\alpha\frac{\hbar^4}{4x_2x_5}\left(\frac{n_y\pi}{d_y}\right)^4$$
$$-\alpha\frac{\hbar^4}{4x_2x_6}\left(\frac{n_y\pi}{d_y}\right)^2\left(\frac{n_z\pi}{d_z}\right)^2-\left(1+\alpha E_{\mathrm{F}_1}\right)\frac{\hbar^2}{2x_3}\left(\frac{n_z\pi}{d_z}\right)^2$$
$$-\alpha\frac{\hbar^4}{4x_3x_5}\left(\frac{n_y\pi}{d_y}\right)^2\left(\frac{n_z\pi}{d_z}\right)^2-\alpha\frac{\hbar^4}{4x_3x_6}\left(\frac{n_y\pi}{d_y}\right)^2\left(\frac{n_z\pi}{d_z}\right)^2$$
$$\left.-\frac{\hbar^2}{2m_2}\left(\frac{n_y\pi}{d_y}\right)^2-\frac{\hbar^2}{2m_3}\left(\frac{n_z\pi}{d_z}\right)^2\right].$$

Thus, the EMM in QWs of IV–VI materials depends on the size quantum numbers in addition to the Fermi energy due to band nonparabolicity. Under the conditions, $m_{\mathrm{l}}^{\pm}\to\infty$, $m_{\mathrm{t}}^{\pm}\to\infty$ and $m_{\mathrm{l}}^{*}=m_{\mathrm{t}}^{*}=m^{*}$, (6.53) gets simplified to the form given by (6.13).

The use of (6.51) leads to the expression of 1D electron statistics as

$$n_{1\mathrm{D}}=\frac{2g_{\mathrm{v}}}{\pi\hbar\sqrt{\bar{g}_1}}\sum_{n_y=1}^{n_{y\max}}\sum_{n_z=1}^{n_{z\max}}\left[T_{613}\left(E_{\mathrm{F}_1},n_y,n_z\right)+T_{614}\left(E_{\mathrm{F}_1},n_y,n_z\right)\right],\qquad(6.54)$$

where

$$T_{614}(E_{\mathrm{F}_1},n_y,n_z)\equiv\sum_{r=1}^{s}L\left(r\right)T_{613}(E_{\mathrm{F}_1},n_y,n_z).$$

Thus using (6.54) and (1.11), the expression for the 1D DMR for QWs of IV–VI compounds can be written as

$$\frac{D}{\mu}=\frac{1}{|e|}\frac{\sum\limits_{n_y=1}^{n_{y\max}}\sum\limits_{n_z=1}^{n_{z\max}}\left[T_{613}\left(E_{\mathrm{F}_1},n_y,n_z\right)+T_{614}\left(E_{\mathrm{F}_1},n_y,n_z\right)\right]}{\sum\limits_{n_y=1}^{n_{y\max}}\sum\limits_{n_z=1}^{n_{z\max}}\left[\left\{T_{613}\left(E_{\mathrm{F}_1},n_y,n_z\right)\right\}'+\left\{T_{614}\left(E_{\mathrm{F}_1},n_y,n_z\right)\right\}'\right]}.\qquad(6.55)$$

Under the conditions $m_{\mathrm{l}}^{\pm} \to \infty$, $m_{\mathrm{t}}^{\pm} \to \infty$ and $m_{\mathrm{l}}^{*} = m_{\mathrm{t}}^{*} = m^{*}$, (6.55) gets simplified to the form given by (6.18).

6.2.6 Stressed Kane Type Semiconductors

The 1D electron energy spectrum in QWs of the said materials assumes the form

$$\frac{k_x^2}{[\bar{a}_0(E)]^2} + \frac{1}{[\bar{b}_0(E)]^2}(n_y\pi/d_y)^2 + \frac{1}{[\bar{c}_0(E)]^2}(n_z\pi/d_z)^2 = 1. \quad (6.56)$$

The expression of the EMM in this case can be written as

$$m^*\left(E_{\mathrm{F}_1}, n_y, n_z\right) = \hbar^2 T_{615}(E_{\mathrm{F}_1}, n_y, n_z)\left\{T_{615}(E_{\mathrm{F}_1}, n_y, n_z)\right\}', \quad (6.57)$$

where

$$T_{615}(E_{\mathrm{F}_1}, n_y, n_z) \equiv \bar{a}_0\left(E_{\mathrm{F}_1}\right)\left[1 - \left\{\left[\bar{b}_0\left(E_{\mathrm{F}_1}\right)\right]^{-2}(\pi n_y/d_y)^2\right\} - \left\{\left[\bar{c}_0\left(E_{\mathrm{F}_1}\right)\right]^{-2}(\pi n_z/d_z)^2\right\}\right]^{1/2}.$$

The EMM in this case is a function of Fermi energy and the size quantization numbers due to the presence of stress only.

The subband energies $(E_{n_{yz6}})$ are given by

$$\frac{1}{[\bar{b}_0(E_{n_{yz6}})]^2}(n_y\pi/d_y)^2 + \frac{1}{[\bar{c}_0(E_{n_{yz6}})]^2}(n_z\pi/d_z)^2 = 1. \quad (6.58)$$

The 1D electron concentration for QWs of stressed Kane type compounds can thus be written as

$$n_{1\mathrm{D}} = \frac{2g_{\mathrm{v}}}{\pi}\sum_{n_y=1}^{n_{y\max}}\sum_{n_z=1}^{n_{z\max}}\left[T_{615}(E_{\mathrm{F}_1}, n_y, n_z) + T_{616}(E_{\mathrm{F}_1}, n_y, n_z)\right], \quad (6.59)$$

where

$$T_{616}(E_{\mathrm{F}_1}, n_y, n_z) \equiv \sum_{r=1}^{s} L(r)\, T_{615}(E_{\mathrm{F}_1}, n_y, n_z).$$

The use of (6.59) and (1.11), leads to the expression of the 1D DMR as

$$\frac{D}{\mu} = \frac{1}{|e|}\frac{\sum_{n_y=1}^{n_{y\max}}\sum_{n_z=1}^{n_{z\max}}\left[T_{615}(E_{\mathrm{F}_1}, n_y, n_z) + T_{616}(E_{\mathrm{F}_1}, n_y, n_z)\right]}{\sum_{n_y=1}^{n_{y\max}}\sum_{n_z=1}^{n_{z\max}}\left[\left\{T_{615}(E_{\mathrm{F}_1}, n_y, n_z)\right\}' + \left\{T_{616}(E_{\mathrm{F}_1}, n_y, n_z)\right\}'\right]}. \quad (6.60)$$

In the absence of stress together with the substitution, $B_2^2 \equiv 3\hbar^2\left(E_{\mathrm{g}}/4m^*\right)$, (6.60) assumes the same form as given by (6.16).

6.2.7 Carbon Nanotubes

For armchair and zigzag carbon nanotubes, the energy dispersion relations are given by [53]

$$E = t_c \left[1 + 4\cos\left(\frac{v\pi}{n}\right)\cos\left(\frac{k_y a_c\sqrt{3}}{2}\right) + 4\cos^2\left(\frac{k_y a_c\sqrt{3}}{2}\right)\right],$$
$$\left(-\pi/\sqrt{3}a_c < k_y < \pi/\sqrt{3}a_c\right) \tag{6.61}$$

$$E = t_c \left[1 + 4\cos\left(\frac{3k_y a_c}{2}\right)\cos\left(\frac{v\pi}{n}\right) + 4\cos^2\left(\frac{v\pi}{n}\right)\right]^{1/2},$$
$$\left(-\pi/3a_c < k_y < \pi/3a_c\right) \tag{6.62}$$

where t_c is the tight binding parameter, v = 1, 2... 2n, a_c is the nearest neighbor C-C bonding distance.

Using (6.61) and (6.62), the EMM for both the cases, can, respectively, be written as

$$m^*\left(E_{F_1}, i\right) = \left(\frac{4\hbar^2}{3a_c^2}\right) A_{c_1}\left(E_F, i\right)\left\{A_{c_1}\left(E_F, i\right)\right\}' \tag{6.63}$$

and

$$m^* = \left(\frac{4\hbar^2}{9a_c^2}\right) A_{c_2}\left(E_F, i\right)\left\{A_{c_2}\left(E_F, i\right)\right\}' \tag{6.64}$$

where

$A_{c_1}\left(E_{F_1}, i\right) = \cos^{-1}\left[\frac{1}{8}\left[-\left(\frac{E_i^2}{t_c^2} - 5\right) + \left[\left(\frac{E_i^2}{t_c^2} - 5\right)^2 + 16\left(\frac{E_{F_1}^2}{t_c^2} - 1\right)\right]^{1/2}\right]\right]$,

$E_i = \frac{|3i-m+n|}{2}\left|t_c\right|\frac{a_c}{r_0}$, r_0 is the radius of the nanotube, $i = 1, 2, 3 ... i_{\max}$, A_{c_2} $\left(E_{F_1}, i\right) = \cos^{-1}\left[\left[\frac{E_{F_1}^2}{t_c^2} - 1 - \left(\frac{E_i}{t_c} - 1\right)^2\right]\left[\frac{2E_i}{t_c} - 1\right]^{-1}\right]$.

It appears from (6.63) and (6.64) that the effective mass in CNTs is a function of the ith sub-band, system variables in addition to the Fermi energy which is the characteristic feature of such quantum materials.

Using (6.61) and (6.62), the electron statistics for both the cases, can, respectively be written as,

$$n_{1D} = \frac{8}{\pi a_c\sqrt{3}}\sum_{i=1}^{i_{\max}}\left[A_{c_1}\left(E_{F_1}, i\right) + B_{c_1}\left(E_{F_1}, i\right)\right] \tag{6.65}$$

$$n_{1D} = \frac{8}{3\pi a_c}\sum_{i=1}^{i_{\max}}\left[A_{c_2}\left(E_{F_2}, i\right) + B_{c_2}\left(E_{F_2}, i\right)\right] \tag{6.66}$$

where $B_{c_1}(E_{F_1},i)=\sum_{r=1}^{s}L(r)A_{c_1}(E_{F_1},i)$ and $B_{c_2}(E_{F_1},i)=\sum_{r=1}^{s}L(r)A_{c_2}(E_{F_1},i)$.

Combining (6.65), (6.66) and (1.11), the expressions of the DMRs for both the cases can be written as

$$\frac{D}{\mu}=\frac{1}{|e|}\left[\frac{\sum_{i=1}^{i_{\max}}\left[A_{c_1}(E_{F_1},i)+B_{c_1}(E_{F_1},i)\right]}{\sum_{i=1}^{i_{\max}}\left[\{A_{c_1}(E_{F_1},i)\}'+\{B_{c_1}(E_{F_1},i)\}'\right]}\right] \quad (6.67)$$

and

$$\frac{D}{\mu}=\frac{1}{|e|}\left[\frac{\sum_{i=1}^{i_{\max}}\left[A_{c_2}(E_{F_1},i)+B_{c_2}(E_{F_1},i)\right]}{\sum_{i=1}^{i_{\max}}\left[\{A_{c_2}(E_{F_1},i)\}'+\{B_{c_2}(E_{F_1},i)\}'\right]}\right] \quad (6.68)$$

6.3 Result and Discussions

Using (6.5) and (6.6), we have, in Figs. 6.1 and 6.2, plotted the 1D DMR at low temperatures, where the quantum effects become prominent for the QWs of Cd_3As_2 and $CdGeAs_2$, as a function of the nanothickness. The curves (a)

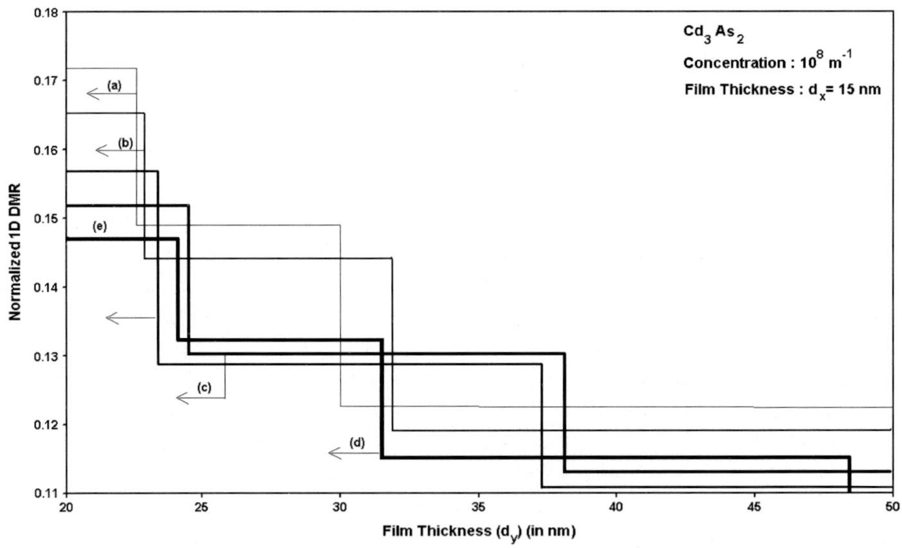

Fig. 6.1. Plot of the normalized 1D DMR as a function of film thickness for the QWs of n-Cd_3As_2 in accordance with (**a**) the generalized band model with $\delta \neq 0$, (**b**) the generalized band model with $\delta = 0$, (**c**) the simplified three band model of Kane, (**d**) the two band model of Kane, and (**e**) the parabolic energy bands

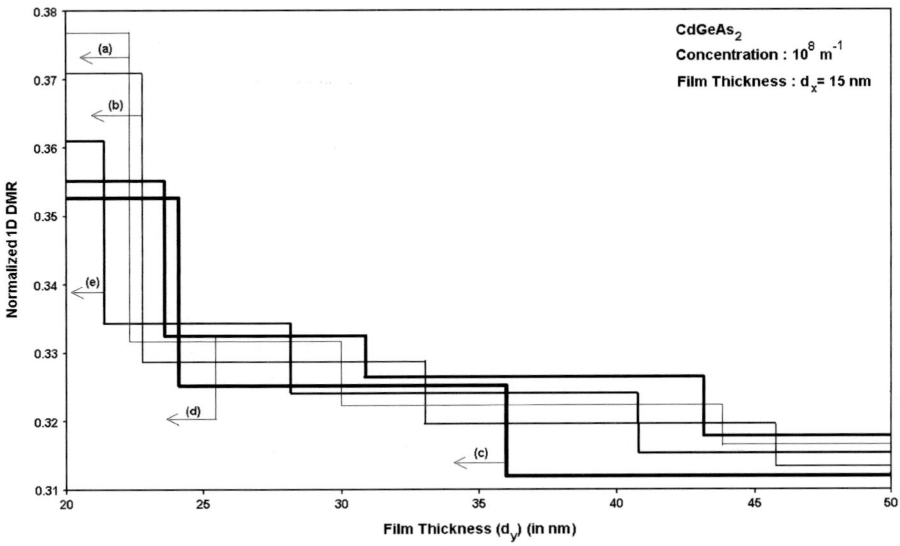

Fig. 6.2. Plot of the normalized 1D DMR as a function of film thickness for the QWs of n-$CdGeAs_2$ for all cases of Fig. 6.1

and (b) corresponds to $\delta \neq 0$ and $\delta = 0$ respectively for the purpose of assessing the influence of crystal field splitting on the 1D DMR in QWs of tetragonal materials. The crystal field splitting enhances the numerical values of the 1D DMR in both the cases. Using (6.10) and (6.11), we have plotted the curve (c) which reflects the influence of the dispersion relation of the tetragonal materials in accordance with the three band Kane model on the 1D DMR in QWs. The curve (d) and (e) have been plotted using (6.15)–6.18) in accordance with the two band model of Kane and that of the parabolic energy bands respectively.

The influence of quantum confinement is immediately apparent from all the curves of Figs. 6.1 and 6.2, since, the 1D DMR depends strongly on the nanothickness, which is in direct contrast to the corresponding bulk specimens.

It appears from the said figures that the 1D DMR decreases with the increasing film thickness in a step like manner, as considered here, although the numerical values vary widely and are determined by the constants of the energy spectra. The oscillatory dependence is due to the crossing over of the Fermi level by the quantized levels. For each coincidence of a quantized level with the Fermi level, there would be a discontinuity in the density-of-states function resulting in a peak of oscillations. With large values of film thickness, the height of the steps decreases and the DMR decreases with increasing film thickness in a non-oscillatory manner and exhibits monotonic decreasing dependence. The height of the step and the rate of decrement are totally dependent on the band structure. The numerical values of the 1D DMR in accordance with the three band model of Kane is greater than that of the corresponding two band model, which reflects the fact that the presence of the spin–orbit splitting constant enhances the magnitude of the 1D DMR. It may be noted that the presence of the band non-parabolicity in accordance with the two-band model of Kane enhances the peaks of the oscillatory 1D DMR for all cases of quantum confinements. The appearance of the humps of the respective curves is due to the redistribution of the electrons among the quantized energy levels when the quantum numbers corresponding to the highest occupied level changes from one fixed value to the others. With varying electron concentration, a change is reflected in the 1D DMR through the redistribution of the electrons among the quantized levels. Although the 1D DMR varies in various ways with all the variables in all the limiting cases as evident from the curves of Figs. 6.1 and 6.2, the rates of variations are totally band-structure dependent.

In Figs. 6.3 and 6.4, we have plotted the 1D DMR as a function of surface electron concentration per unit length for all cases of Figs. 6.1 and 6.2 respectively. It appears that the 1D DMR increases with increasing carrier degeneracy and also reflects the signature of the 1D confinement through the step-like functional dependence with the 1D electron statistics. This oscillatory dependence will be less and less prominent with increasing carrier concentration and ultimately, for bulk specimens of the same material, the DMR will be found to increase continuously with increasing electron concentration in a non-oscillatory manner.

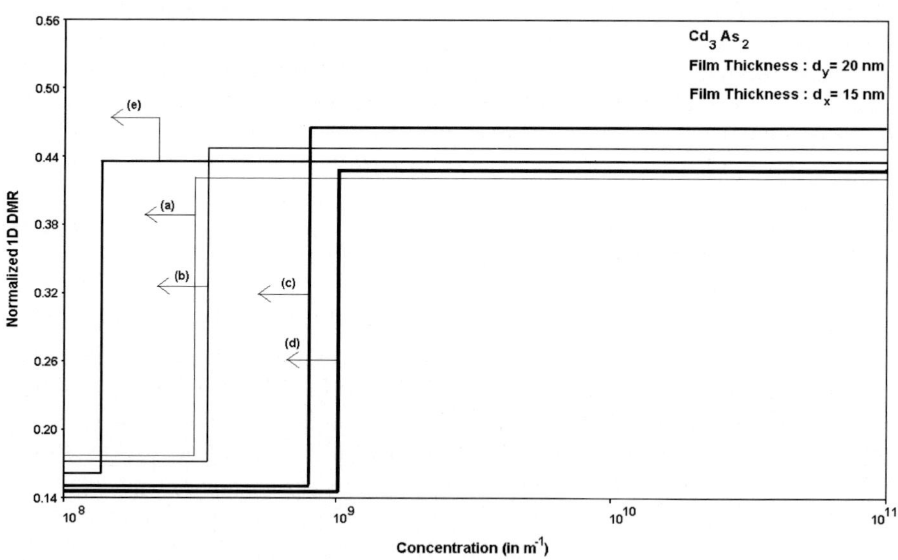

Fig. 6.3. Plot of the normalized 1D DMR as a function of electron concentration per unit length for the QWs of n-Cd_3As_2 in accordance with (**a**) the generalized band model with $\delta \neq 0$, (**b**) the generalized band model with $\delta = 0$, (**c**) the simplified three band model of Kane, (**d**) the two band model of Kane, and (e) the parabolic energy bands

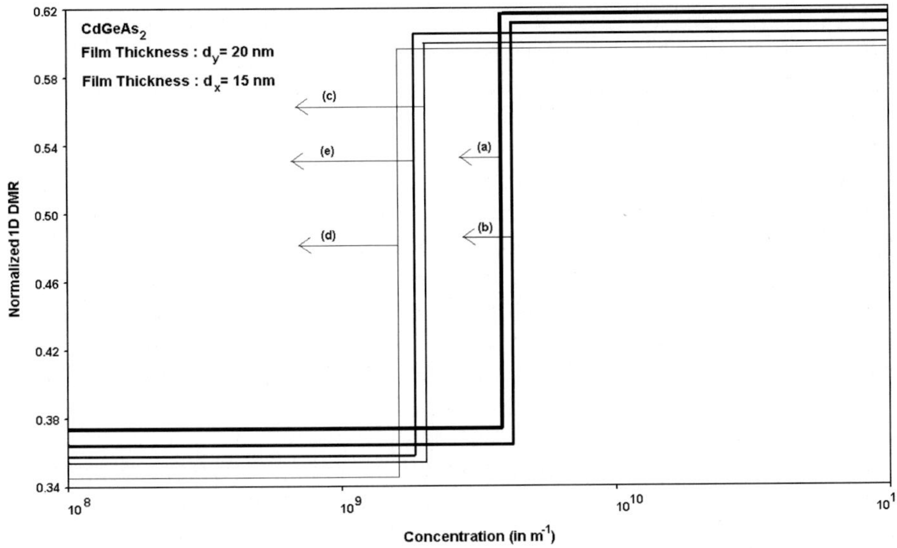

Fig. 6.4. Plot of the normalized 1D DMR as a function of electron concentration per unit length for the QWs of n-$CdGeAs_2$ for all cases of Fig. 6.3

Using (6.10), (6.11), (6.15), (6.16) and (6.19), (6.20), we have plotted the 1D DMR as a functions of the nanothickness for the QWs of GaAs, InAs and InSb in Figs. 6.5–6.7, where the curves (a)–(c) correspond to the three and the two band models of Kane together with the parabolic energy band respectively. The dependence of the 1D DMR on the electron concentration per unit length for QWs of all the said compounds is shown in Figs. 6.8–6.10 respectively. Using the same set of equations as for III–V materials, we have plotted the 1D DMR for $Hg_{1-x}Cd_xTe$ and $In_{1-x}Ga_xAs_yP_{1-y}$ as functions of nanothickness and 1D carrier degeneracy as shown by Figs. 6.11–6.14 respectively in which, the curves (a)–(c) correspond to the three and the two band models of Kane together with the parabolic energy band, respectively. The numerical values of the 1D DMR depend on the energy band constants of different materials.

Figures 6.15 and 6.16 exhibit the same dependence as a function of alloy composition for all the cases of the ternary and quaternary materials considered above. From both the figures, it is apparent that the DMR decreases with increasing alloy composition.

Using (6.26) and (6.27), we have plotted the 1D DMR, for the QWs of p-CdS materials as functions of nanothickness and electron concentration per unit length in Figs. 6.17 and 6.18 respectively, where the curve (a) refers to $\bar{\lambda}_0 \neq 0\,\mathrm{eVm}$ and the curve (b) refers to $\bar{\lambda}_0 = 0\,\mathrm{eVm}$. Both the curves have been drawn for the purpose of assessing the influence of the splitting of the two-spin states by the spin–orbit coupling and the crystalline field in this case.

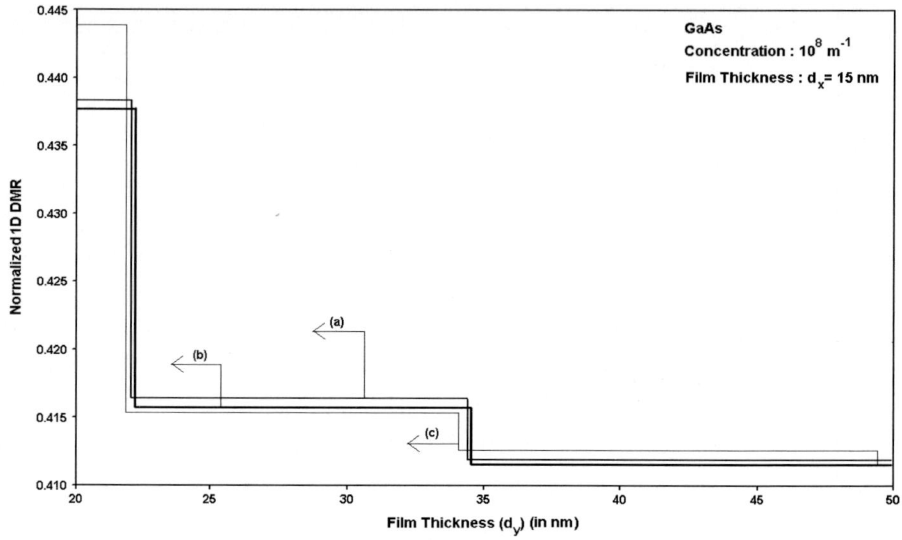

Fig. 6.5. Plot of the normalized 1D DMR as a function of film thickness for the QWs of n-GaAs in accordance with (**a**) the simplified three band model of Kane, (**b**) the two band model of Kane, and (**c**) the parabolic energy bands

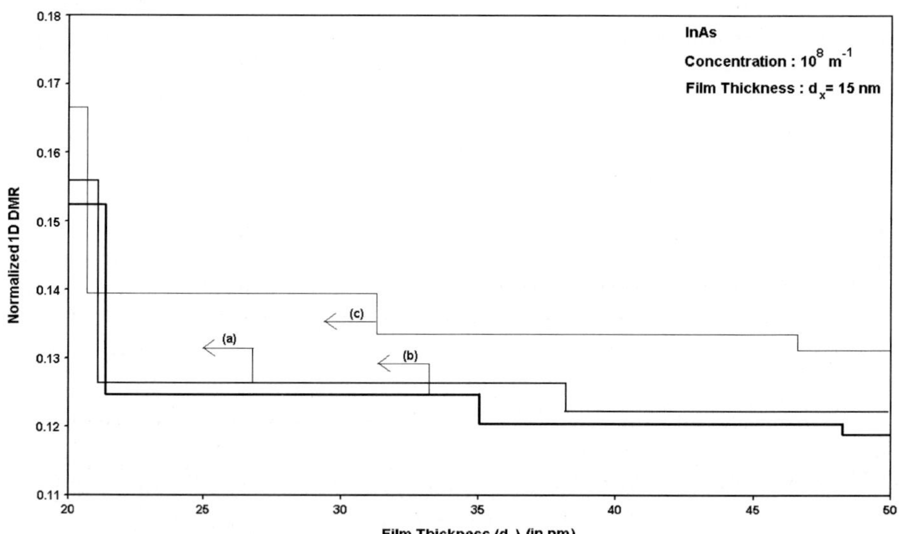

Fig. 6.6. Plot of the normalized 1D DMR as a function of film thickness for the QWs of n-InAs for all cases of Fig. 6.5

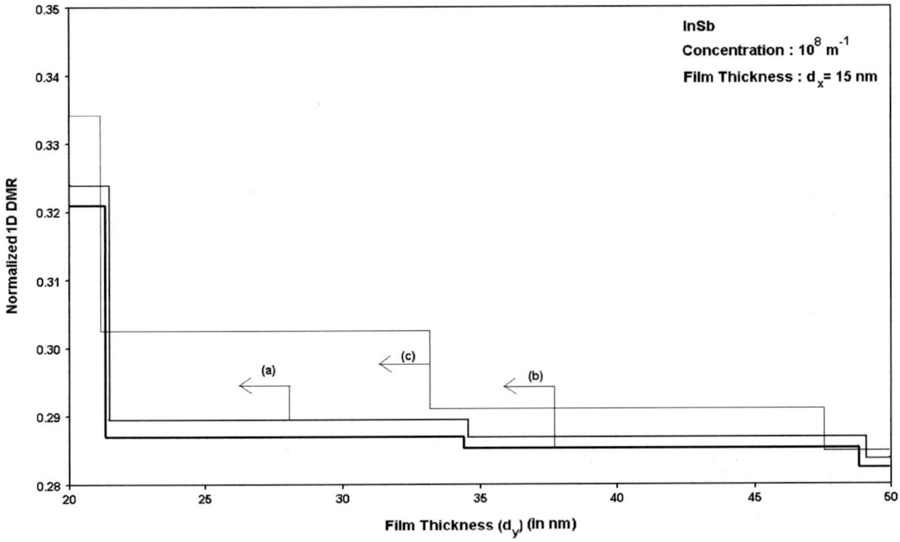

Fig. 6.7. Plot of the normalized 1D DMR as a function of film thickness for the QWs of n-InAs for all cases of Fig. 6.5

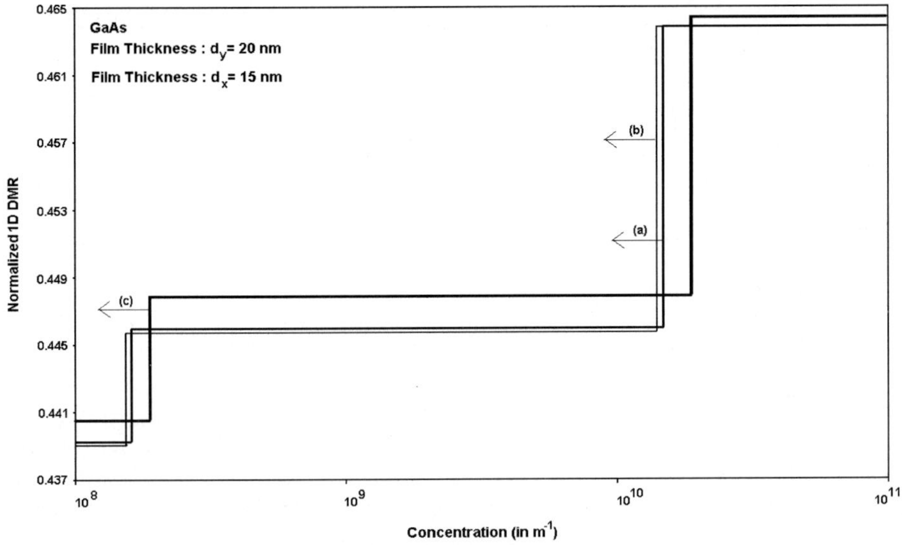

Fig. 6.8. Plot of the normalized 1D DMR as a function of electron concentration per unit length for the QWs of n-GaAs in accordance with (**a**) the simplified three band model of Kane, (**b**) the two band model of Kane, and (**c**) the parabolic energy bands

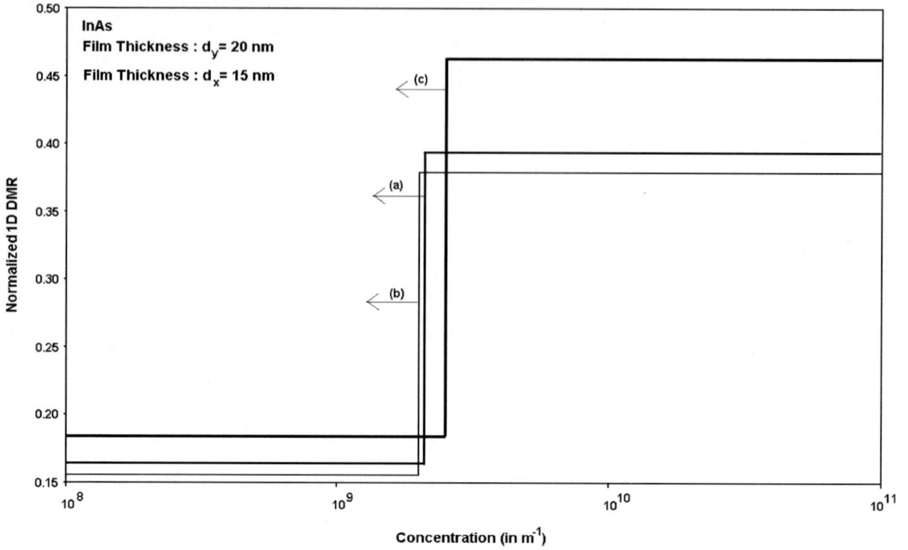

Fig. 6.9. Plot of the normalized 1D DMR as a function of electron concentration per unit length for the QWs of n-InAs for all cases of Fig. 6.8

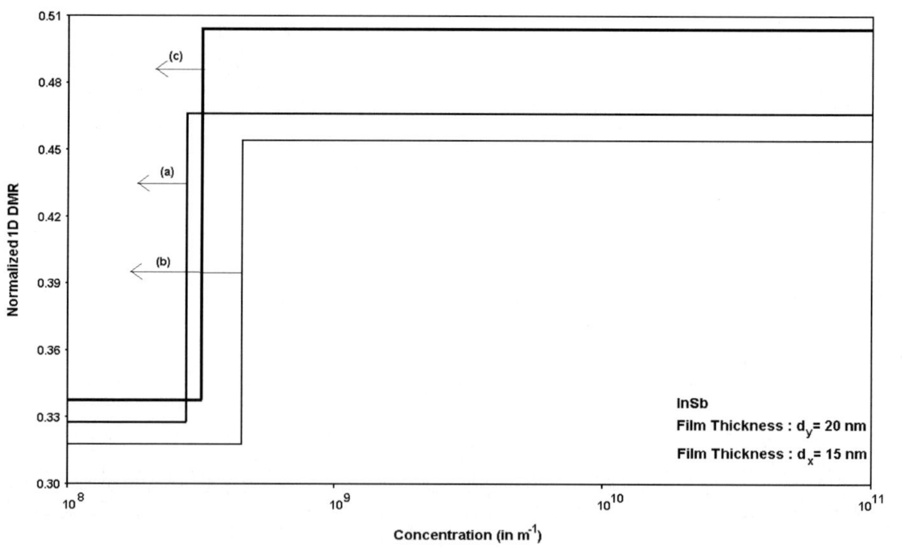

Fig. 6.10. Plot of the normalized 1D DMR as a function of electron concentration per unit length for the QWs of n-InSb for all cases of Fig. 6.8

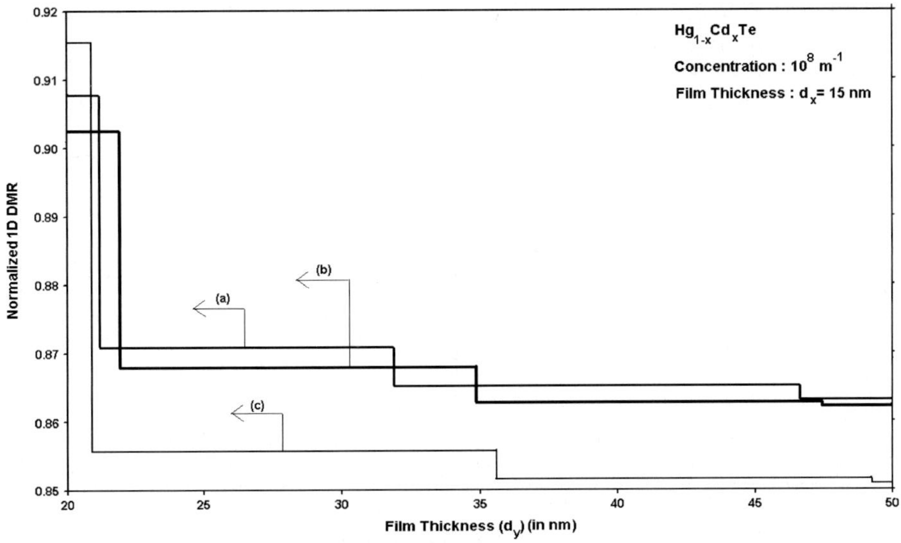

Fig. 6.11. Plot of the normalized 1D DMR as a function of film thickness for the QWs of n-$Hg_{1-x}Cd_xTe$ in accordance with (**a**) the simplified three band model of Kane, (**b**) the two band model of Kane, and (**c**) the parabolic energy bands

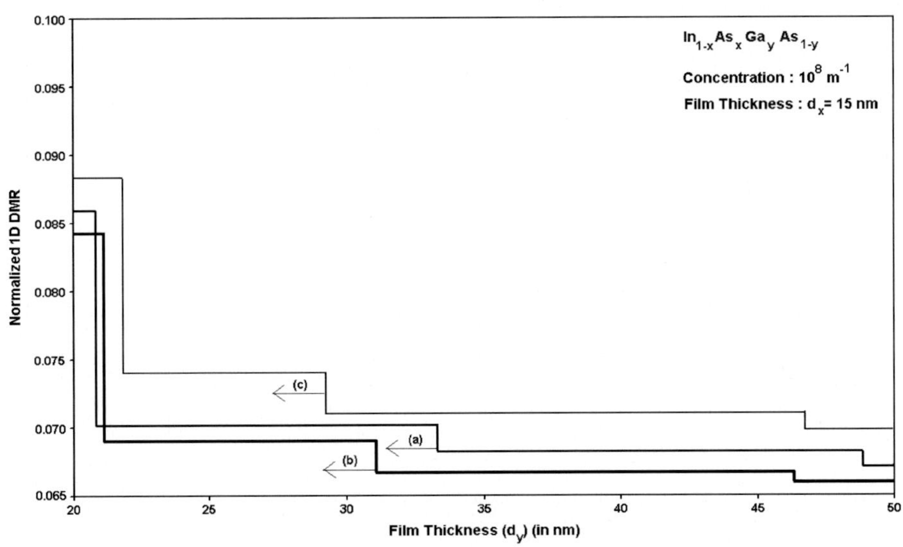

Fig. 6.12. Plot of the normalized 1D DMR as a function of film thickness for the QWs of n-$In_{1-x}As_xGa_yP_{1-y}$ for all cases of Fig. 6.11

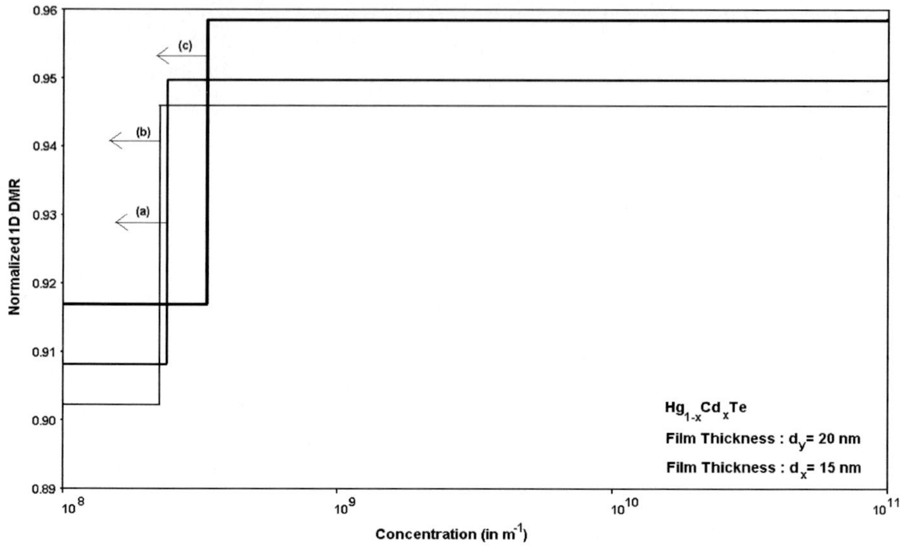

Fig. 6.13. Plot of the normalized 1D DMR as a function of electron concentration per unit length for the QWs of n-$Hg_{1-x}Cd_xTe$ in accordance with (**a**) the simplified three band model of Kane, (**b**) the two band model of Kane, and (**c**) the parabolic energy bands

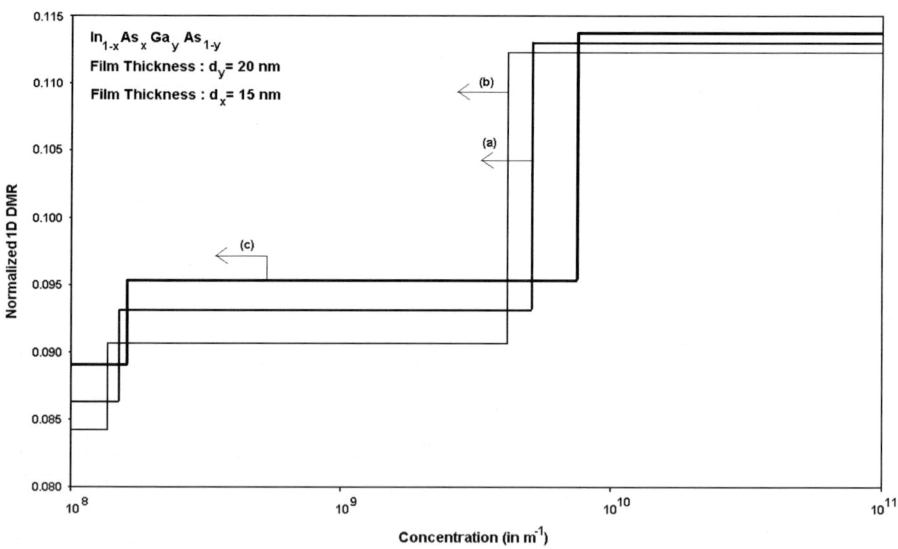

Fig. 6.14. Plot of the normalized 1D DMR as a function of electron concentration per unit length for the QWs of n-$In_{1-x}As_xGa_yP_{1-y}$ for all cases of Fig. 6.13

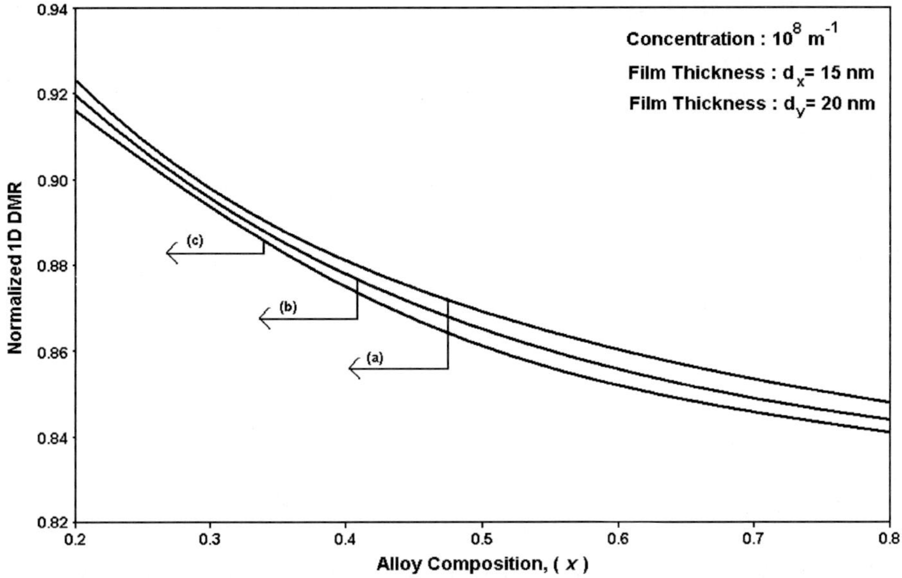

Fig. 6.15. Plot of the normalized 1D DMR as a function of alloy composition (x) for the QWs of n-$Hg_{1-x}Cd_xTe$ in accordance with (**a**) the simplified three band model of Kane, (**b**) the two band model of Kane, and (**c**) the parabolic energy bands

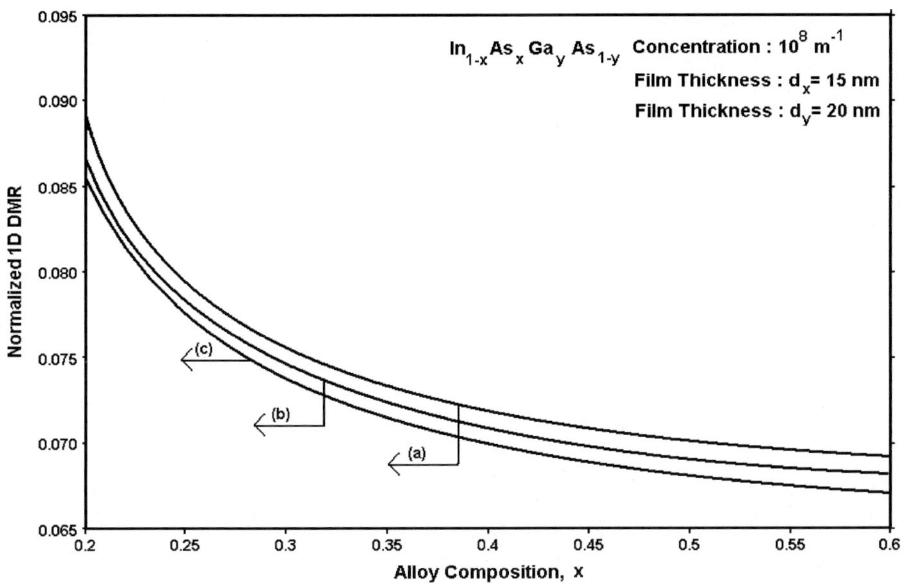

Fig. 6.16. Plot of the normalized 1D DMR as a function of alloy composition (x) for the QWs of n-$In_{1-x}As_xGa_yP_{1-y}$ for all cases of Fig. 6.15

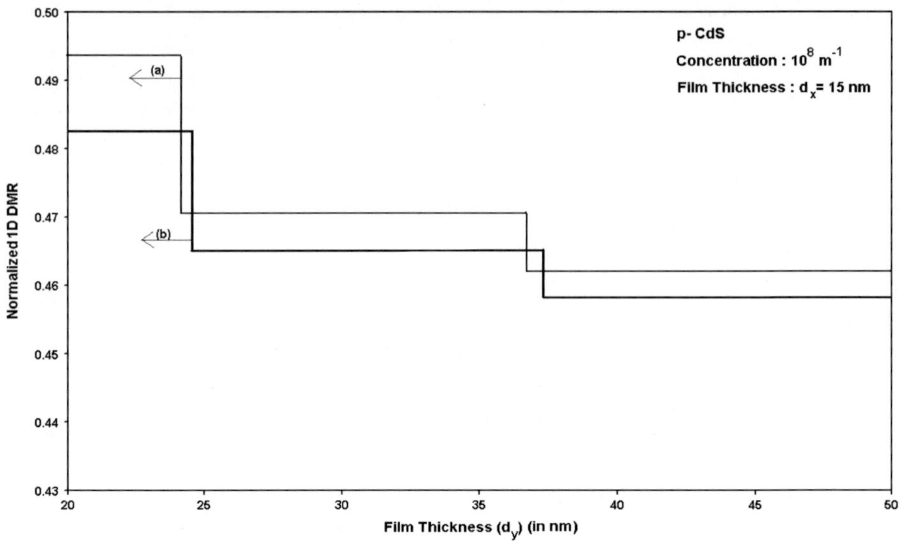

Fig. 6.17. Plot of the normalized 1D DMR for as a function of film thickness for QWs of p-CdS in accordance with (**a**) Hopfield model with $\bar{\lambda}_0 \neq 0$ (**b**) Hopfield model with $\bar{\lambda}_0 = 0$

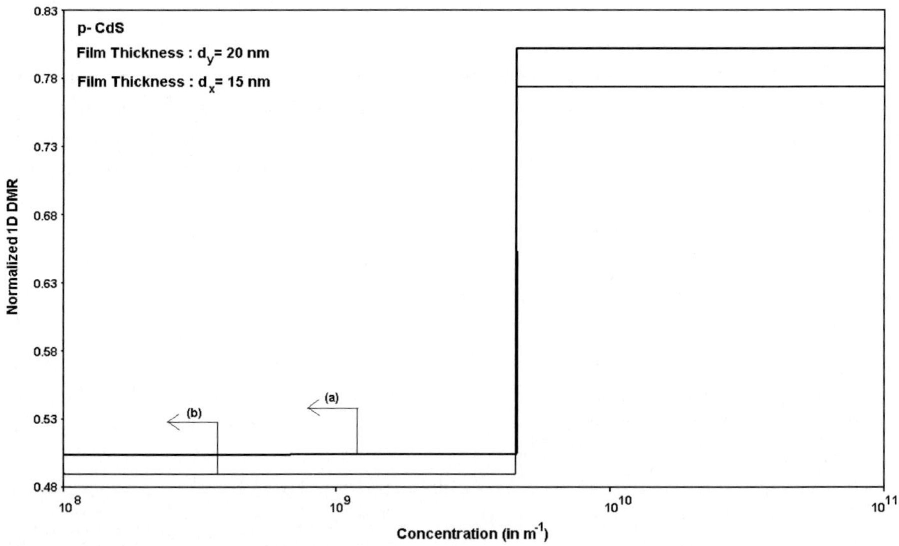

Fig. 6.18. Plot of the normalized 1D DMR for as a function of electron concentration per unit length for QWs of p-CdS in accordance with (**a**) Hopfield model with $\bar{\lambda}_0 \neq 0$ (**b**) Hopfield model with $\bar{\lambda}_0 = 0$

It appears from Figs. 6.17 and 6.18, that the influence of the term $\bar{\lambda}_0$ is the reduction of the quantum jumps of the oscillatory 1D DMR in QWs of II–VI materials for both the said variables.

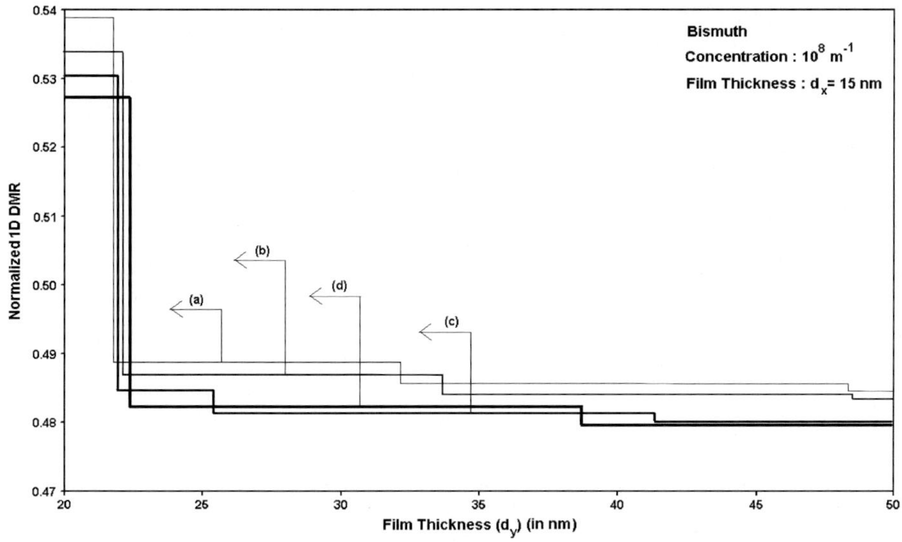

Fig. 6.19. Plot of the normalized 1D DMR for QWs of bismuth as a function of film thickness in accordance with (**a**) the parabolic ellipsoidal model, (**b**) the Lax nonparabolic ellipsoidal model (**c**) the Cohen model, and (**d**) the McClure and Choi model

Using (6.32), (6.33), (6.38), (6.39) and (6.43), (6.44), (6.49) and (6.50), we have plotted in Figs. 6.19 and 6.20, the 1D DMR for the QWs of bismuth in accordance with the McClure and Choi, the Cohen, the Lax nonparabolic ellipsoidal, and the parabolic ellipsoidal models as a function of the nanothickness and electron concentration per unit length as shown by the curves (a)–(d) respectively. It appears from Figs. 6.19 and 6.20, that numerically, the 1D DMR is greatest for ellipsoidal parabolic model and least for the McClure and Choi model, respectively, due to the influence of the energy band constants.

Using (6.54) and (6.55), Figs. 6.21 and 6.22 exhibit the 1D DMR for the QWs of IV–VI materials as a function of the nanothickness and electron concentration per unit length respectively. The curves (a)–(c) correspond to PbTe, n-PbSnTe, and n-$Pb_{1-x}Sn_xSe$ respectively. The influence of the energy band constants on the DMR in both the cases is apparent for all the three different materials as considered here.

The 1D DMR for stressed Kane type n-InSb has been plotted in Fig. 6.23 and 6.24 as functions of nanothickness and electron concentration per unit length respectively using (6.59) and (6.60) as shown in plot (a) in the presence of stress while plot (b) exhibits the same in the absence of stress for the purpose of assessing the influence of stress on the 1D DMR in QWs of stressed n-InSb. In the presence of stress, the magnitude of the 1D DMR is being increased as compared with the same under stress free condition.

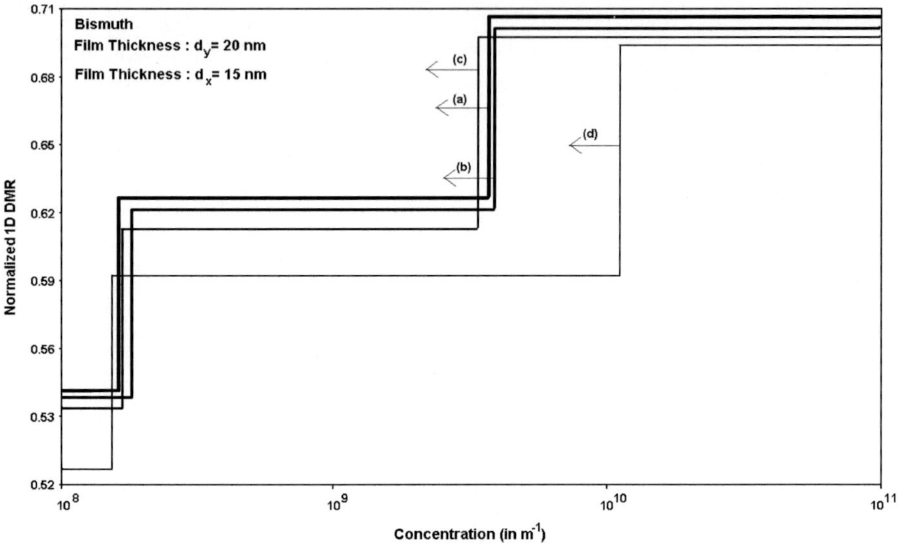

Fig. 6.20. Plot of the normalized 1D DMR for QWs of bismuth as a function of electron concentration per unit length for all the cases of Fig. 6.19

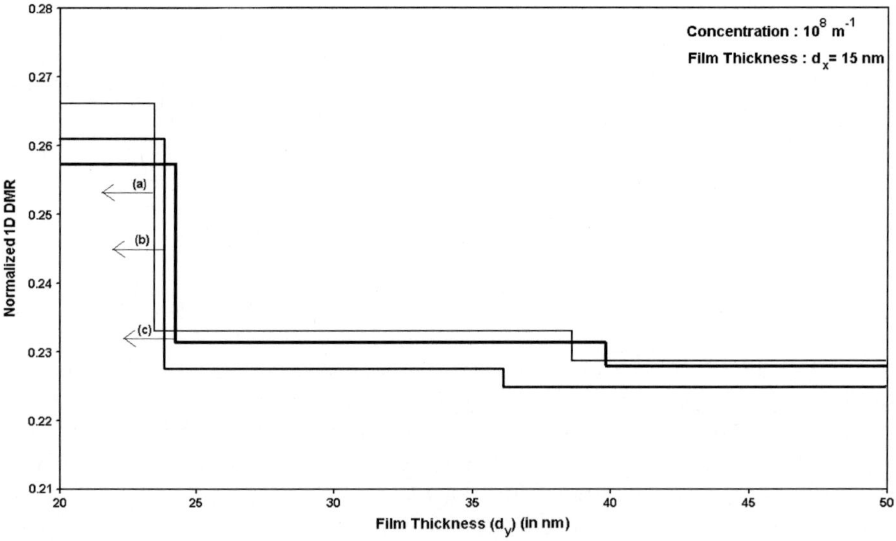

Fig. 6.21. Plot of the normalized 1D DMR as a function of film thickness for the QWs of (**a**) n-PbTe, (**b**) n-PbSnTe and (**c**) n-$Pb_{1-x}Sn_xSe$

Using (6.65) and (6.67) and (6.66) and (6.68), we have plotted the DMR for the armchair and zigzag carbon nanotubes vs. electron concentration per unit length as shown by curves (a) and (b) of Fig. 6.25 respectively. It appears

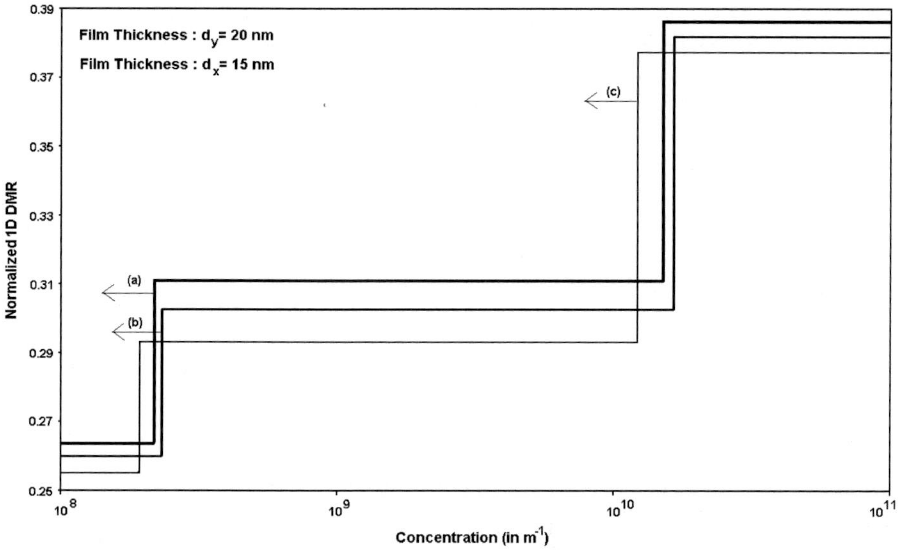

Fig. 6.22. Plot of the normalized 1D DMR as a function of electron concentration per unit length for the QWs of (**a**) PbTe, (**b**) n-PbSnTe, and (**c**) n-$Pb_{1-x}Sn_xSe$

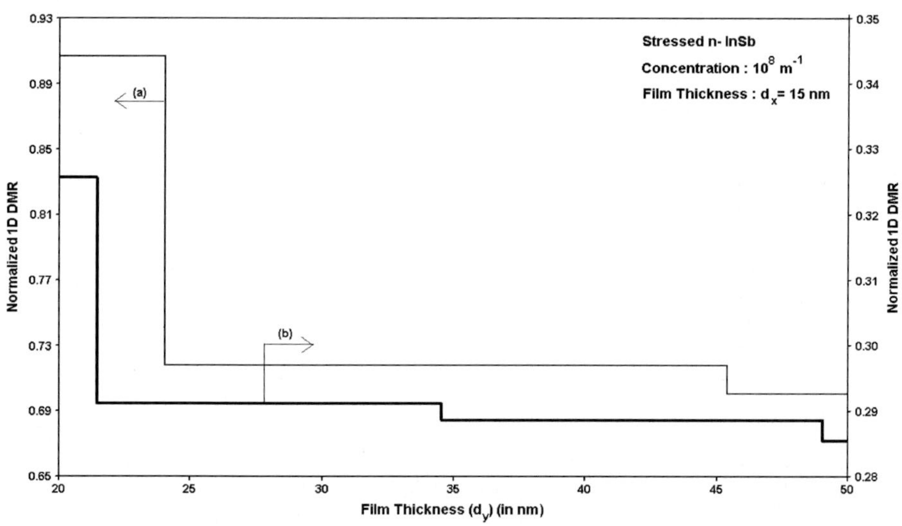

Fig. 6.23. Plot of the normalized 1D DMR as a function of film thickness for the QWs of stressed n-InSb in which curve (**a**) shows the 1D DMR in the presence of stress while curve (**b**) is applicable in the absence of the stress

that for both the CNTs, the DMR is an increasing oscillatory function of the electron concentration per unit length and the magnitude of the DMR is much higher than that of the DMR in QWs of the other materials and is a

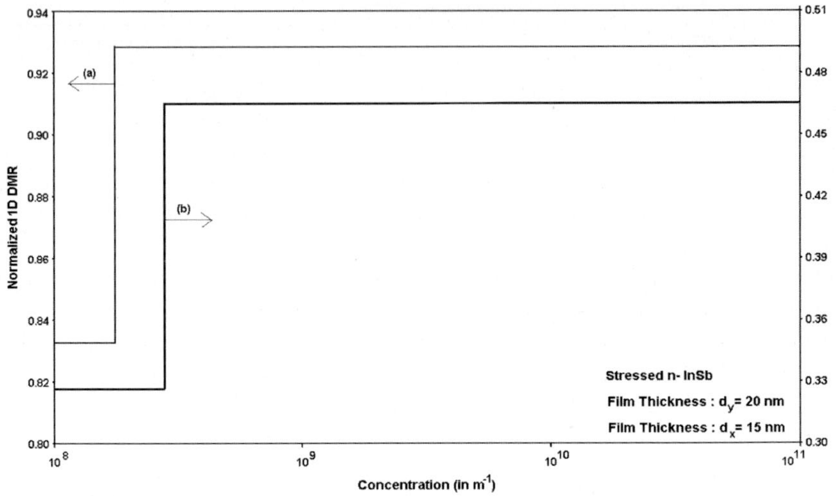

Fig. 6.24. Plot of the normalized 1D DMR as a function of electron concentration per unit length for the QWs of stressed n-InSb in which curve (**a**) shows the 1D DMR in the presence of stress while curve (**b**) is applicable in the absence of the stress

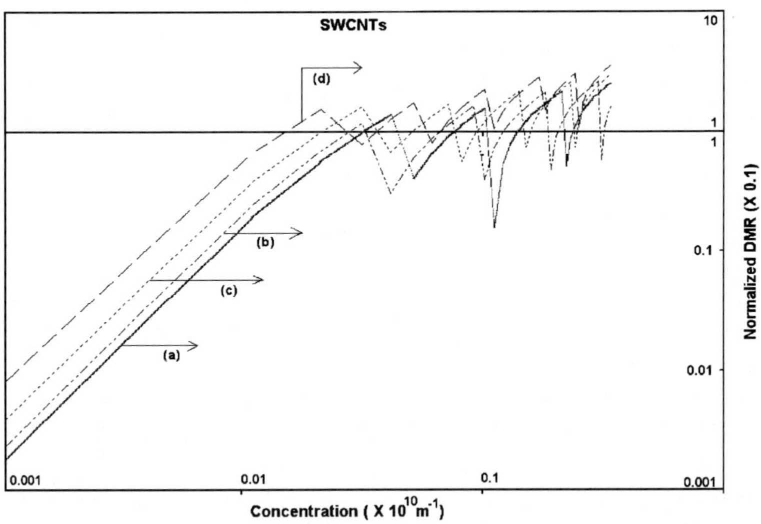

Fig. 6.25. Plot of the normalized DMR in (**a**) (13, 6) chiral semiconductor carbon nanotube, (**b**) (16, 0) zigzag semiconductor carbon nanotubes, (**c**) (10, 10) metallic armchair carbon nanotube and (**d**) (22, 19) chiral metallic carbon nanotube as function of carrier degeneracy

characteristic feature of the CNTs. The signature of the two different types of 1D electron motion is apparent by comparing the curves (a) and (b) of Fig. 6.25 with the respective curves of the other 1D systems.

It is worth remarking that the EMM in the QWs can be a function of the size quantum numbers, the Fermi energy, or other energy band constant depending on the respective 1D dispersion laws as formulated already in the theoretical background of this chapter. Since the Fermi energy changes sharply with the electron concentration per unit length and the nanothickness, the index dependent effective mass will exhibit oscillatory dependence with the said variables and this oscillatory 1D effective mass will also contribute to the oscillatory 1D mobility for CNTs.

The theoretical results as derived in this chapter can be used to determine the 1D diffusion constants in various cases and this simplified formulation exhibits the basic qualitative features of 1D DMR for different quantum confined 1D materials. For the purpose of condensed presentation, the 1D electron statistics and the 1D Einstein relation for the QWs of the respective materials have been presented in Table 6.1.

6.4 Open Research Problems

R.6.1 Investigate the Einstein relation in the presence of an arbitrarily oriented non-quantizing alternating magnetic field for the QWs of tetragonal semiconductors by including the electron spin. Study all the special cases for III–V, ternary and quaternary materials in this context.

R.6.2 Investigate the Einstein relations in QWs of all models of Bi, IV–VI, II–VI and stressed Kane type compounds in the presence of an arbitrarily oriented non-quantizing alternating magnetic field by including the electron spin.

R.6.3 Investigate the Einstein relation for QWs of all the materials as stated in R.2.1 of Chap. 2.

R.6.4 Investigate the Einstein relation for all the problems from R.6.1 to R.6.3 in the presence of an arbitrarily oriented non-quantizing non-uniform electric field.

R.6.5 Investigate the Einstein relation for all the problems from R.6.1 to R.6.3 in the presence of arbitrarily oriented crossed electric and magnetic fields.

R.6.6 Investigate the Einstein relation for all the problems from R.6.1 to R.6.5 in the presence of finite potential wells.

R.6.7 Investigate the Einstein relation for all the problems from R.6.1 to R.6.5 in the presence of two parabolic potential wells in two different orthogonal directions.

R.6.8 Investigate the Einstein relation for all the problems from R.6.1 to R.6.5 for quantum rings.

R.6.9 Investigate the Einstein relation for all the problems from R.6.1 to R.6.5 in the presence of an elliptical Hill and quantum square rings respectively.

Table 6.1. The 1D carrier statistics and the 1D Einstein relation in QWs of tetragonal, III–V, ternary, quaternary, II–VI, all the models of Bismuth, IV–VI, stressed materials and carbon nanotubes

Type of materials	The 1D carrier statistics	The 1D Einstein relation for the diffusivity mobility ratio
1. Tetragonal compounds	In accordance with the 1D generalized dispersion relation (6.1) for the QWs as formulated in this chapter $$n_{1D} = \frac{2g_v}{\pi} \sum_{n_x=1}^{n_{x\max}} \sum_{n_y=1}^{n_{y\max}} \left[T_{61}\left(E_{F_1}, n_x, n_y\right) + T_{62}\left(E_{F_1}, n_x, n_y\right)\right], \quad (6.5)$$	On the basis of (6.5), $$\frac{D}{\mu} = \frac{1}{\lvert e \rvert} \frac{\sum_{n_x=1}^{n_{x\max}} \sum_{n_y=1}^{n_{y\max}} \left[T_{61}\left(E_{F_1}, n_x, n_y\right) + T_{62}\left(E_{F_1}, n_x, n_y\right)\right]}{\sum_{n_x=1}^{n_{x\max}} \sum_{n_y=1}^{n_{y\max}} \left[\left\{T_{61}\left(E_{F_1}, n_x, n_y\right)\right\}' + \left\{T_{62}\left(E_{F_1}, n_x, n_y\right)\right\}'\right]} \quad (6.6)$$
2. III–V, ternary and quaternary compounds	In accordance with the three band model of Kane for QWs as given by (6.7) which is a special case of (6.1) $$n_{1D} = \frac{2g_v}{\pi}\left(\frac{2m^*}{\hbar^2}\right)^{1/2} \sum_{n_x=1}^{n_{x\max}} \sum_{n_y=1}^{n_{y\max}} \left[T_{63}\left(E_{F_1}, n_x, n_y\right) + T_{64}\left(E_{F_1}, n_x, n_y\right)\right] \quad (6.10)$$	On the basis of (6.10) $$\frac{D}{\mu} = \frac{1}{\lvert e \rvert} \frac{\sum_{n_z=1}^{n_{z\max}} \left[T_{53}\left(E_{F_s}, n_z\right) + T_{54}\left(E_{F_s}, n_z\right)\right]}{\sum_{n_z=1}^{n_{z\max}} \left[\left\{T_{53}\left(E_{F_s}, n_z\right)\right\}' + \left\{T_{54}\left(E_{F_s}, n_z\right)\right\}'\right]} \quad (6.11)$$
	In accordance with the two band model of Kane for the QWs as given by (6.12) $$n_{1D} = \frac{2g_v}{\pi}\left(\frac{2m^*}{\hbar^2}\right)^{1/2} \sum_{n_x=1}^{n_{x\max}} \sum_{n_y=1}^{n_{y\max}} \left[T_{65}\left(E_{F_1}, n_x, n_y\right) + T_{66}\left(E_{F_1}, n_x, n_y\right)\right] \quad (6.15)$$	On the basis of (6.15), $$\frac{D}{\mu} = \frac{1}{\lvert e \rvert} \frac{\sum_{n_x=1}^{n_{x\max}} \sum_{n_y=1}^{n_{y\max}} \left[T_{65}\left(E_{F_1}, n_x, n_y\right) + T_{66}\left(E_{F_1}, n_x, n_y\right)\right]}{\sum_{n_x=1}^{n_{x\max}} \sum_{n_y=1}^{n_{y\max}} \left[\left\{T_{65}\left(E_{F_1}, n_x, n_y\right)\right\}' + \left\{T_{66}\left(E_{F_1}, n_x, n_y\right)\right\}'\right]} \quad (6.16)$$
	In accordance with the parabolic energy bands for the QWs under the condition $\alpha \to 0$, as given by (6.12) $$n_{1D} = \frac{2g_v\sqrt{2\pi m^* k_B T}}{h} \sum_{n_x=1}^{n_{x\max}} \sum_{n_y=1}^{n_{y\max}} \left[F_{-1/2}\left(\eta_7\right)\right] \quad (6.19)$$	On the basis of (6.19) $$\frac{D}{\mu} = \frac{k_B T}{\lvert e \rvert} \frac{\sum_{n_x=1}^{n_{x\max}} \sum_{n_y=1}^{n_{y\max}} \left[F_{-1/2}\left(\eta_7\right)\right]}{\sum_{n_x=1}^{n_{x\max}} \sum_{n_y=1}^{n_{y\max}} \left[F_{-3/2}\left(\eta_7\right)\right]} \quad (6.20)$$

3. II–VI compounds

In accordance with (6.21) for the QWs,

$$n_{1D} = g_v \left(\frac{k_B T}{\pi b_0'}\right)^{1/2} \sum_{n_x=1}^{n_{x_{max}}} \sum_{n_y=1}^{n_{y_{max}}} \left[F_{-1/2}(\eta_{2,+}) + F_{-1/2}(\eta_{2,-})\right] \tag{6.26}$$

On the basis of (6.26)

$$\frac{D}{\mu} = \frac{k_B T}{|e|} \frac{\sum_{n_x=1}^{n_{x_{max}}} \sum_{n_y=1}^{n_{y_{max}}} \left[F_{-1/2}(\eta_{2,+}) + F_{-1/2}(\eta_{2,-})\right]}{\sum_{n_x=1}^{n_{x_{max}}} \sum_{n_y=1}^{n_{y_{max}}} \left[F_{-3/2}(\eta_{2,+}) + F_{-3/2}(\eta_{2,-})\right]} \tag{6.27}$$

4. Bi

(a) The McClure and Choi model:
On the basis of (6.28) for the QWs,

$$n_{1D} = \frac{2g_v}{\pi} \sum_{n_y=1}^{n_{y_{max}}} \sum_{n_z=1}^{n_{z_{max}}} \left[T_{67}(E_{F_1}, n_y, n_z) + T_{68}(E_{F_1}, n_y, n_z)\right] \tag{6.32}$$

On the basis of (6.32)

$$\frac{D}{\mu} = \frac{1}{|e|} \frac{\sum_{n_y=1}^{n_{y_{max}}} \sum_{n_z=1}^{n_{z_{max}}} \left[T_{67}(E_{F_1}, n_y, n_z) + T_{68}(E_{F_1}, n_y, n_z)\right]}{\sum_{n_y=1}^{n_{y_{max}}} \sum_{n_z=1}^{n_{z_{max}}} \left[\{T_{67}(E_{F_1}, n_y, n_z)\}' + \{T_{68}(E_{F_1}, n_y, n_z)\}'\right]} \tag{6.33}$$

(b) The Cohen model:
On the basis of (6.34) for the QWs

$$n_{1D} = \left(\frac{2g_v}{\pi}\frac{\sqrt{2m_1}}{\hbar}\right) \sum_{n_y=1}^{n_{y_{max}}} \sum_{n_z=1}^{n_{z_{max}}} \left[T_{69}(E_{F_1}, n_y, n_z) + T_{610}(E_{F_1}, n_y, n_z)\right] \tag{6.38}$$

On the basis of (6.38)

$$\frac{D}{\mu} = \frac{1}{|e|} \frac{\sum_{n_y=1}^{n_{y_{max}}} \sum_{n_z=1}^{n_{z_{max}}} \left[T_{69}(E_{F_1}, n_y, n_z) + T_{610}(E_{F_1}, n_y, n_z)\right]}{\sum_{n_y=1}^{n_{y_{max}}} \sum_{n_z=1}^{n_{z_{max}}} \left[\{T_{69}(E_{F_1}, n_y, n_z)\}' + \{T_{610}(E_{F_1}, n_y, n_z)\}'\right]} \tag{6.39}$$

(d) The Lax model:
On the basis of (6.40) for the QWs,

$$n_{1D} = \left(\frac{2g_v}{\pi}\frac{\sqrt{2m_1}}{\hbar}\right) \sum_{n_y=1}^{n_{y_{max}}} \sum_{n_z=1}^{n_{z_{max}}} \left[T_{611}(E_{F_1}, n_y, n_z) + T_{612}(E_{F_1}, n_y, n_z)\right] \tag{6.43}$$

On the basis of (6.43)

$$\frac{D}{\mu} = \frac{1}{|e|} \frac{\sum_{n_y=1}^{n_{y_{max}}} \sum_{n_z=1}^{n_{z_{max}}} \left[T_{611}(E_{F_1}, n_y, n_z) + T_{612}(E_{F_1}, n_y, n_z)\right]}{\sum_{n_y=1}^{n_{y_{max}}} \sum_{n_z=1}^{n_{z_{max}}} \left[\{T_{611}(E_{F_1}, n_y, n_z)\}' + \{T_{612}(E_{F_1}, n_y, n_z)\}'\right]} \tag{6.44}$$

(e) The parabolic ellipsoidal model:
On the basis of (6.45) for the QWs

$$n_{1D} = \frac{2g_v\sqrt{2\pi m^* k_B T}}{h} \sum_{n_y=1}^{n_{y_{max}}} \sum_{n_z=1}^{n_{z_{max}}} \left[F_{-1/2}(\eta_8)\right] \tag{6.49}$$

On the basis of (6.49)

$$\frac{D}{\mu} = \frac{k_B T}{|e|} \frac{\sum_{n_y=1}^{n_{y_{max}}} \sum_{n_z=1}^{n_{z_{max}}} \left[F_{-1/2}(\eta_8)\right]}{\sum_{n_y=1}^{n_{y_{max}}} \sum_{n_z=1}^{n_{z_{max}}} \left[F_{-3/2}(\eta_8)\right]} \tag{6.50}$$

Table 6.1. *Continued*

Type of materials	The 1D carrier statistics	The 1D Einstein relation for the diffusivity mobility ratio
5. IV–VI compounds	In accordance with (6.51) for the QWs, $$n_{1D} = \frac{2g_v}{\pi\hbar\sqrt{g_1}} \sum_{n_y=1}^{n_{y_{max}}} \sum_{n_z=1}^{n_{z_{max}}} \left[T_{613}\left(E_{F_1}, n_y, n_z\right) + T_{614}\left(E_{F_1}, n_y, n_z\right)\right] \quad (6.54)$$	On the basis of (6.54) $$\frac{D}{\mu} = \frac{1}{\lvert e\rvert} \frac{\sum_{n_y=1}^{n_{y_{max}}} \sum_{n_z=1}^{n_{z_{max}}} \left[T_{613}\left(E_{F_1}, n_y, n_z\right) + T_{614}\left(E_{F_1}, n_y, n_z\right)\right]}{\sum_{n_y=1}^{n_{y_{max}}} \sum_{n_z=1}^{n_{z_{max}}} \left[\left\{T_{613}\left(E_{F_1}, n_y, n_z\right)\right\}' + \left\{T_{614}\left(E_{F_1}, n_y, n_z\right)\right\}'\right]} \quad (6.55)$$
6. Stressed compounds	In accordance with (6.56) under QWs $$n_{1D} = \frac{2g_v}{\pi} \sum_{n_y=1}^{n_{y_{max}}} \sum_{n_z=1}^{n_{z_{max}}} \left[T_{615}(E_{F_1}, n_y, n_z) + T_{616}(E_{F_1}, n_y, n_z)\right] \quad (6.59)$$	On the basis of (6.59) $$\frac{D}{\mu} = \frac{1}{\lvert e\rvert} \frac{\sum_{n_y=1}^{n_{y_{max}}} \sum_{n_z=1}^{n_{z_{max}}} \left[T_{615}(E_{F_1}, n_y, n_z) + T_{616}(E_{F_1}, n_y, n_z)\right]}{\sum_{n_y=1}^{n_{y_{max}}} \sum_{n_z=1}^{n_{z_{max}}} \left[\left\{T_{615}(E_{F_1}, n_y, n_z)\right\}' + \left\{T_{616}(E_{F_1}, n_y, n_z)\right\}'\right]} \quad (6.60)$$
7. Carbon nanotubes	In accordance with (6.61) for armchair carbon nanotubes $$n_{1D} = \frac{8}{\pi a_c\sqrt{3}} \sum_{i=1}^{i_{max}} \left[A_{c_1}\left(E_{F_1}, i\right) + B_{c_1}\left(E_{F_1}, i\right)\right] \quad (6.65)$$ In accordance with (6.62) for zigzag carbon nanotubes $$\frac{D}{\mu} = \frac{1}{\lvert e\rvert} \left[\frac{\sum_{i=1}^{i_{max}} \left[A_{c_1}\left(E_{F_1}, i\right) + B_{c_1}\left(E_{F_1}, i\right)\right]}{\sum_{i=1}^{i_{max}} \left[\left\{A_{c_1}\left(E_{F_1}, i\right)\right\}' + \left\{B_{c_1}\left(E_{F_1}, i\right)\right\}'\right]}\right] \quad (6.67)$$	Using (6.65) $$n_{1D} = \frac{8}{3\pi a_c} \sum_{i=1}^{i_{max}} \left[A_{c_2}\left(E_{F_2}, i\right) + B_{c_2}\left(E_{F_2}, i\right)\right] \quad (6.66)$$ Using (6.67) $$\frac{D}{\mu} = \frac{1}{\lvert e\rvert} \left[\frac{\sum_{i=1}^{i_{max}} \left[A_{c_2}\left(E_{F_1}, i\right) + B_{c_2}\left(E_{F_1}, i\right)\right]}{\sum_{i=1}^{i_{max}} \left[\left\{A_{c_2}\left(E_{F_1}, i\right)\right\}' + \left\{B_{c_2}\left(E_{F_1}, i\right)\right\}'\right]}\right] \quad (6.68)$$

Allied Research Problems

R.6.10 Investigate the EMM for all the materials as stated in R.6.3–R.6.9.

R.6.11 Investigate the carrier contribution to the elastic constants, the heat capacity, the activity coefficient, and the plasma frequency for all the materials covering all the cases of problems from R.6.1 to R.6.9.

R.6.12 Investigate in detail, the mobility for elastic and inelastic scattering mechanisms for all the materials covering all the problems from R.6.1 to R.6.9.

R.6.13 Investigate all transport coefficients in detail for all the materials covering all the cases of the problems from R.6.1 to R.6.9 of this chapter.

R.6.14 Investigate the dia and paramagnetic susceptibilities in detail for all the materials covering all the appropriate research problems of this chapter.

References

1. P. Harrison, *Quantum Wells, Wires and Dots* (Wiley, New York, 2002); B.K. Ridley, *Electrons and Phonons in Semiconductors Multilayers* (Cambridge University Press, Cambridge, 1997); G. Bastard, *Wave Mechanics Applied to Semiconductor Heterostructures*, Halsted; Les Ulis, Les Editions de Physique, New York (1988); V.V. Martin, A.A. Kochelap, M.A. Stroscio, *Quantum Heterostructures* (Cambridge University Press, Cambridge, 1999)
2. C.S. Lent, D.J. Kirkner, J. Appl. Phys. **67**, 6353 (1990); F. Sols, M. Macucci, U. Ravaioli, K. Hess, Appl. Phys. Lett. **54**, 350 (1980)
3. C.S. Kim, A.M. Satanin, Y.S. Joe, R.M. Cosby, Phys. Rev. B **60**, 10962 (1999)
4. S. Midgley, J.B. Wang, Phys. Rev. B **64**, 153304 (2001)
5. T. Sugaya, J.P. Bird, M. Ogura, Y. Sugiyama, D.K. Ferry, K.Y. Jang, Appl. Phys. Lett. **80**, 434 (2002)
6. B. Kane, G. Facer, A. Dzurak, N. Lumpkin, R. Clark, L. PfeiKer, K. West, Appl. Phys. Lett. **72**, 3506 (1998)
7. C. Dekker, Phys. Today **52**, 22 (1999)
8. A. Yacoby, H.L. Stormer, N.S. Wingreen, L.N. Pfeiffer, K.W. Baldwin, K.W. West, Phys. Rev. Lett. **77**, 4612 (1996)
9. Y. Hayamizu, M. Yoshita, S. Watanabe, H.A.L. PfeiKer, K. West, Appl. Phys. Lett. **81**, 4937 (2002)
10. S. Frank, P. Poncharal, Z.L. Wang, W.A. Heer, Science **280**, 1744 (1998)
11. I. Kamiya, I.I. Tanaka, K. Tanaka, F. Yamada, Y. Shinozuka, H. Sakaki, Physica E **13**, 131 (2002)
12. A.K. Geim, P.C. Main, N. LaScala, L. Eaves, T.J. Foster, P.H. Beton, J.W. Sakai, F.W. Sheard, M. Henini, G. Hill et al., Phys. Rev. Lett. **72**, 2061 (1994)
13. A.S. Melinkov, V.M. Vinokur, Nature **415**, 60 (2002)
14. K. Schwab, E.A. Henriksen, J.M. Worlock, M.L. Roukes, Nature **404**, 974 (2000)
15. L. Kouwenhoven, Nature **403**, 374 (2000)
16. S. Komiyama, O. Astafiev, V. Antonov, H. Hirai, Nature **403**, 405 (2000)
17. E. Paspalakis, Z. Kis, E. Voutsinas, A.F. Terziz, Phys. Rev. B **69**, 155316 (2004)

18. J.H. Jefferson, M. Fearn, D.L.J. Tipton, T.P. Spiller, Phys. Rev. A **66**, 042328 (2002)
19. J. Appenzeller, C. Schroer, T. Schapers, A. Hart, A. Froster, B. Lengeler, H. Luth, Phys. Rev. B **53**, 9959 (1996)
20. J. Appenzeller, C. Schroer, J. Appl. Phys. **87**, 31659 (2002)
21. P. Debray, O.E. Raichev, M. Rahman, R. Akis, W.C. Mitchel, Appl. Phys. Lett. **74**, 768 (1999)
22. P.M. Solomon, Proc. IEEE **70**(489) (1982); T.E. Schlesinger, T. Kuech, Appl. Phys. Lett. **49**, 519 (1986)
23. D. Kasemet, C.S. Hong, N.B. Patel, P.D. Dapkus, Appl. Phys. Letts. **41**, 912 (1982); K. Woodbridge, P. Blood, E.D. Pletcher, P.J. Hulyer, Appl. Phys. Lett. **45**, 16 (1984); S. Tarucha, H.O. Okamoto, Appl. Phys. Lett. **45**, 16 (1984); H. Heiblum, D.C. Thomas, C.M. Knoedler, M.I. Nathan, Appl. Phys. Lett. **47**, 1105 (1985)
24. O. Aina, M. Mattingly, F.Y. Juan, P.K. Bhattacharyya, Appl. Phys. Lett. **50**, 43 (1987)
25. I. Suemune, L.A. Coldren, IEEE J. Quant. Electron. **24**, 1178 (1988)
26. D.A.B. Miller, D.S. Chemla, T.C. Damen, J.H. Wood, A.C. Burrus, A.C. Gossard, W. Weigmann, IEEE J. Quant. Electron. **21**, 1462 (1985)
27. S. Iijima, *Nature* **354**, 56 (1991); M. Endo, S. Iijima, M.S. Dresselhaus, Carbon Nanotubes (Pergamon Press, USA, 1996)
28. V.N. Popov, Mater. Sci. Eng. **R 43**, 61 (2004)
29. J. Sandler, M.S.P. Shaffer, T. Prasse, W. Bauhofer, K. Schulte, A.H. Windle, *Polymer* **40**, 5967 (1999)
30. D. Qian, E.C. Dickey, R. Andrews, T. Rantell, Appl. Phys. Lett. **76**, 2868 (2000)
31. J. Kong, N.R. Franklin, C.W. Zhou, M.G. Chapline, S. Peng, K.J. Cho, H.J. Dai, Science **287**, 622 (2000)
32. W.A. Deheer, A. Chatelain, D. Ugarte, Science **270**, 1179 (1995)
33. A.G. Rinzler, J.H. Hafner, P. Nikolaev, L. Lou, S.G. Kim, D. Tomanek, P. Nordlander, D.T. Olbert, R.E. Smalley, Science **269**, 1550 (1995)
34. A.C. Dillon, K.M. Jones, T.A. Bekkedahl, C.H. Kiang, D.S. Bethune, M.J. Heben, Nature **386**, 377 (1997)
35. C. Liu, Y.Y. Fan, M. Liu, H.T. Cong, H.M. Cheng, M.S. Dresselhaus, Science **286**, 1127 (1999)
36. P. Kim, C.M. Lieber, Science **286**, 2148 (1999)
37. D. Srivastava, Nanotechnology **8**, 186 (1997)
38. C. Ke, H.D. Espinosa, Appl. Phys. Lett. **85**, 681 (2004)
39. J.W. Kang, H.J. Hwang, Nanotechnology **15**, 1633 (2004); J. Cumings, A. Zettl, Science **289**, 602 (2000)
40. S.J. Tans, M.H. Devoret, H.J. Dai, A. Thess, R0020.E. Smalley, L.J. Geerligs, C. Dekker, Nature **386**, 474 (1997)
41. S.J. Tans, A.R.M. Verschueren, C. Dekker, Nature **393**, 49 (1998)
42. P. Avouris, Acc. Chem. Res. **35**, 1026 (2002)
43. P.G. Collins, A. Zettl, H. Bando, A. Thess, R.E. Smalley, Science **278**, 100 (1997)
44. S. Saito, Science **278**, 77 (1997); J.C. Charlier, Acc. Chem. Res. **35**, 1063 (2002); R.H. Baughman, A.A. Zakhidov, W.A. de Heer, Science **297**, 787 (2002)
45. Z. Yao, H.W.C. Postma, L. Balents, C. Dekker, Nature **402**, 273 (1999); C.W. Zhou, J. Kong, E. Yenilmez, H. Dai, Science **290**, 1552 (2000)

46. S.J. Tans, A.R.M. Verschueren, C. Dekker, Nature **393**, 49 (1998); H.W.Ch. Postma, T. Teepen, Z. Yao, M. Grifoni, C. Dekker, Science **293**, 76 (2001)
47. T. Rueckes, K. Kim, E. Joselevich, G.Y. Tseng, C.L. Cheung, C.M. Lieber, Science **289**, 94 (2000)
48. A. Bacthtold, P. Hadley, T. Nakanish, C. Dekker, Science **294**, 1317 (2001); A. Bacthtold, P. Hadley, T. Nakanish, C. Dekker, Phys. Rev. B **66**, 235424 (2002)
49. Q. Zheng, Q. Jiang, Phys. Rev. Lett. **88**, 045503 (2002); Q. Zheng, J.S. Liu, Q. Jiang, Phys. Rev. B. **65**, 245409 (2002); Y. Zhao, C.-C. Ma, G.H. Chen, Q. Jiang, Phys. Rev. Lett. **91**, 175504 (2003); S.B. Legoas, V.R. Coluci, S.F. Braga, P.Z. Coura, S.O. Dantus, D.S. Galvao, Phys. Rev. Lett. **90**, 055504 (2003); S.B. Legaos, V.R. Coluci, S.F. Braga, P.Z. Coura, S.O. Dantus, D.S. Galvao, Nanotechnology **15**, S184 (2004); J.W. Kang, H.W. Hwang, J. Appl. Phys. **96**, 3900 (2004)
50. W.Y. Choi, J.W. Kang, H.W. Hwang, Physica E, **23**, 125 (2004); H.J. Hwang, K.R. Byun, J.W. Kang, Physica E **23**, 208 (2004); J.W. Kang, H.J. Hwang, Physica E **23**, 36 (2004); J.W. Kang, H.J. Hwang, J. Appl. Phys. **73**, 4447 (2004); J.W. Kang, H.J. Hwang, J. Phys. Soc. Jpn **73**, 1077 (2004); J.W. Kang, Y.W. Choi, H.J. Hwang, J. Comp. Theor. Nanosci. **1**, 199 (2004)
51. M. Dequesnes, S.V. Rotkin, N.R. Aluru, Nanotechnology **13**, 120 (2002); J.M. Kinaret, T. Nord, S. Viefers, Appl. Phys. Lett. **82**, 1287 (2003); C. Ke, H.D. Espinosa, Appl. Phys. Lett. **85**, 681 (2004); L.M. Jonsson, T. Nord, J.M. Kinaret, S. Viefers, J. Appl. Phys. **96**, 629 (2004); L.M. Jonsson, S. Axelsson, T. Nord, S. Viefers, J.M. Kinaret, Nanotechnology **15**, 1497 (2004); S.W. Lee, D.S. Lee, R.E. Morjan, S.H. Jhang, M. Sveningsson, O.A. Nerushev, Y.W. Park, E.E.B. Campbell, Nano. Lett. **4**, 2027 (2004)
52. S. Chowdhury, D. De, S. Mukherjee, A. Neogi, A. Sinha, M. Pal, S.K. Biswas, S. Pahari, S. Bhattacharya, K.P. Ghatak, J. Comp. Theo. Nanoscience **5**, 375 (2008)
53. X. Yang, J. Ni, *Phys. Rev. B* **72**, 195426 (2005); M.S. Landstrom, J. Guo, Nanoscale Transistors, Device Physics, Modeling and Simulation (Springer, 2006, USA); W. Mintmire, C.T. White, Phys. Rev. Letts. **81**, 2506 (1998); R. Saito, G. Dresselhaus, M.S. Dresselhaus, Physical Properties of Carbon Nanotubes (Imperial College Press, London 1998); D. Kahn, J.P. Lu, Phys. Rev. B, **60**, 6535 (1999)

7

The Einstein Relation in Inversion Layers of Compound Semiconductors

7.1 Introduction

It is well known that the electrons in bulk semiconductors in general, have three dimensional freedom of motion. When these electrons are confined in a one dimensional potential well whose width is of the order of the carrier wavelength, the motion in that particular direction gets quantized while that along the other two directions remains as free. Thus, the energy spectrum appears in the shape of discrete levels for the one dimensional quantization, each of which has a continuum for the two dimensional free motion. The transport phenomena of such one dimensional confined carriers have recently been studied [1–20] with great interest. For the metal-oxide-semiconductor (MOS) structures, the work functions of the metal and the semiconductor substrate are different and the application of an external voltage at the metal-gate causes the change in the charge density at the oxide semiconductor interface leading to a bending of the energy bands of the semiconductor near the surface. As a result, a one dimensional potential well is formed at the semiconductor interface. The spatial variation of the potential profile is so sharp that for considerable large values of the electric field, the width of the potential well becomes of the order of the de Broglie wavelength of the carriers. The Fermi energy, which is near the edge of the conduction band in the bulk, becomes nearer to the edge of the valance band at the surface creating inversion layers. The energy levels of the carriers bound within the potential well get quantized and form electric subbands. Each of the subband corresponds to a quantized level in a plane perpendicular to the surface leading to a quasi two dimensional electron gas. Thus, the extreme band bending at low temperature allows us to observe the quantum effects at the surface.

In Sect. 7.2.1, of the theoretical background, the Einstein relation in n-channel inversion layers of tetragonal materials has been investigated for both weak and strong electric field limits. Section 7.2.2 contains the results for n-channel inversion layers of III–V, ternary and quaternary compounds for both the electric field limits whose bulk electrons obey the three and the two band models of Kane together with parabolic energy bands and they form the special cases of Sect. 7.2.1. Section 7.2.3 contains the study of the DMR for p-channel inversion layers of II–VI materials. Sections 7.2.4 and 7.2.5 contain the study of the DMR in p-channel inversion layers of IV–VI and stressed semiconductors for both the limits respectively. Section 7.3 contains the results and discussion of this chapter.

7.2 Theoretical Background

7.2.1 Formulation of the Einstein Relation in n-Channel Inversion Layers of Tetragonal Materials

In the presence of a surface electric field F_s along z direction and perpendicular to the surface, (2.2) assumes the form

$$\psi_1(E-|e|F_s z)=\psi_2(E-|e|F_s z)k_s^2+\psi_3(E-|e|F_s z)k_z^2, \tag{7.1}$$

where for this chapter, E represents the electron energy as measured from the edge of the conduction band at the surface in the vertically upward direction.

The quantization rule for the wave vector along the direction of quantization for inversion layers is given by [5]

$$\int_0^{z_t} k_z \mathrm{d}z=\frac{2}{3}(S_i)^{3/2}, \tag{7.2}$$

where z_t is the classical turning point and S_i is the zeros of the Airy function $(Ai(-S_i)=0)$.

Using (7.1) and (7.2), under the weak electric field limit, one can write

$$\int_0^{z_t}\sqrt{A_7(E)-|e|F_s z D_7(E)}\mathrm{d}z=\frac{2}{3}(S_i)^{3/2}, \tag{7.3}$$

where

$$z_t\equiv\frac{A_7(E)}{[|e|F_s D_7(E)]},\quad A_7(E)\equiv\left[\frac{\psi_1(E)-\psi_2(E)k_s^2}{\psi_3(E)}\right],$$

$$D_7(E)\equiv[B_7(E)-A_7(E)C_7(E)],\quad B_7(E)\equiv\left[\frac{(\psi_1(E))'-(\psi_2(E))'k_s^2}{\psi_3(E)}\right]$$

and $C_7(E)\equiv\left[\frac{(\psi_3(E))'}{\psi_3(E)}\right]$.

Thus, the 2D electron dispersion law in n-channel inversion layers of tetragonal materials under the weak electric field limit can approximately be written as

$$\psi_1(E)=P_7(E,i)k_s^2+Q_7(E,i), \tag{7.4}$$

where

$$P_7(E,i) \equiv \left[\psi_2(E) - \left(\frac{2t_2(E)}{3[t_1(E)]^{1/3}}\right)\psi_3(E)\,S_i\,(|e|\,F_s)^{2/3}\right],$$

$$t_2(E) \equiv \left[\frac{[\psi_2(E)]'}{\psi_3(E)} - \left(\frac{\psi_2(E)\,[\psi_3(E)]'}{[\psi_3(E)]^2}\right)\right],$$

$$t_1(E) \equiv \left[\frac{[\psi_1(E)]'}{\psi_3(E)} - \left(\frac{\psi_1(E)\,[\psi_3(E)]'}{[\psi_3(E)]^2}\right)\right] \text{ and}$$

$$Q_7(E,i) \equiv S_i\psi_3(E)\,[|e|\,F_s t_1(E)]^{2/3}.$$

The EMM in the x–y plane can be expressed as

$$m^*(E_{\mathrm{Fiw}},i) = \left(\frac{\hbar^2}{2}\right) G_7(E,i)\Big|_{E=E_{\mathrm{Fiw}}}, \tag{7.5}$$

where

$$G_7(E,i) = [P_7(E,i)]^{-2}\left[P_7(E,i)\left\{(\psi_1(E))' - (Q_7(E,i))'\right\} - \left\{\psi_1(E) - (Q_7(E,i))\right\}(P_7(E,i))'\right].$$

and E_{Fiw} is the Fermi energy under the weak electric field limit as measured from the edge of the conduction band at the surface in the vertically upward direction. Thus, we observe that the EMM is a function of subband index, the Fermi energy and other band constants due to the combined influence of the crystal field splitting constant and the anisotropic spin–orbit splitting constants respectively.

The subband energy $(E_{n_{iw1}})$ for the weak electric field limit can be obtained from (7.4) as

$$\psi_1(E_{n_{iw1}}) = Q_7(E_{n_{iw1}},i), \tag{7.6}$$

The general expression of the 2D total density-of-states function in this case can be written as

$$N_{2\mathrm{D}_i}(E) = \frac{2g_{\mathrm{v}}}{(2\pi)^2}\sum_{i=0}^{i_{\max}} \frac{\partial}{\partial E}[A(E,i)\,H(E-E_{n_i})], \tag{7.7}$$

where $A(E,i)$ is the area of the constant energy 2D wave vector space for inversion layers and E_{n_i} is the corresponding subband energy.

Using (7.4) and (7.7), the total 2D density-of-states function under the weak electric field limit can be expressed as

$$N_{2\mathrm{D}_i}(E) = \frac{g_{\mathrm{v}}}{(2\pi)}\sum_{i=0}^{i_{\max}} [G_7(E,i)\,H(E-E_{n_{iw1}})], \tag{7.8}$$

Using (7.8) and the Fermi–Dirac occupation probability factor, the 2D surface electron concentration in n-channel inversion of tetragonal materials under the weak electric field limit ($n_{2\mathrm{Dw}}$) can be written as

$$n_{2\mathrm{Dw}} = g_{\mathrm{v}} (2\pi)^{-1} \sum_{i=0}^{i_{\max}} \left[\bar{P}_{7\mathrm{w}} (E_{\mathrm{Fiw}}, i) + \bar{Q}_{7\mathrm{w}} (E_{\mathrm{Fiw}}, i)\right], \tag{7.9}$$

where

$$\bar{P}_{7\mathrm{w}} (E_{\mathrm{Fiw}}, i) \equiv [\psi_1 (E_{\mathrm{Fiw}}, i) - Q_7 (E_{\mathrm{Fiw}}, i)] \{P_7 (E_{\mathrm{Fiw}}, i)\}^{-1} \text{ and}$$

$$\bar{Q}_{7\mathrm{w}} (E_{\mathrm{Fiw}}, i) \equiv \sum_{r=1}^{s} \{L (r) [\bar{P}_7 (E_{\mathrm{Fiw}}, i)]\}.$$

$$F_{\mathrm{s}} \equiv \frac{|e|\, n_{2\mathrm{Dw}}}{\varepsilon_{\mathrm{sc}}},$$

and $\varepsilon_{\mathrm{sc}}$ is the semiconductor permittivity.

Thus the surface electron concentration under the weak electric field quantum limit at low temperatures assumes the form

$$\bar{n}_{2\mathrm{Dw}} = g_{\mathrm{v}} (2\pi)^{-1} \left[P_{7\mathrm{w}} \left(\bar{E}_{\mathrm{Fw}}, 0\right)\right], \tag{7.10}$$

where $\bar{E}_{\mathrm{Fw}}$ is the Fermi energy under the weak electric field quantum limit as measured from the edge of the conduction band at the surface.

Using (7.10) and (1.12), the DMR in this case can be expressed as

$$\frac{D}{\mu} = \frac{\bar{n}_{2\mathrm{Dw}}}{|e|} \left(x_{71} \left(\bar{E}_{\mathrm{Fw}}, 0\right) - y_{71} \left(\bar{E}_{\mathrm{0w1}}, 0\right)\right), \tag{7.11}$$

where

$$x_{71} \left(\bar{E}_{\mathrm{Fw}}, 0\right) \equiv \frac{\omega_{71} \left(\bar{E}_{\mathrm{Fw}}, 0\right)}{\omega_{72} \left(\bar{E}_{\mathrm{Fw}}, 0\right)},$$

$$\omega_{71} \left(\bar{E}_{\mathrm{Fw}}, 0\right) \equiv \left[1 - \left(\frac{(\bar{n}_{2\mathrm{Dw}})^{-1/3}}{3P_7 \left(\bar{E}_{\mathrm{Fw}}, 0\right)}\right) \left\{2\bar{n}_{2\mathrm{Dw}} V_{72} \left(\bar{E}_{\mathrm{Fw}}\right) - \frac{g_{\mathrm{v}}}{\pi} V_{71} \left(\bar{E}_{\mathrm{Fw}}\right)\right\}\right],$$

$$V_{72} \left(\bar{E}_{\mathrm{Fw}}\right) \equiv \frac{2}{3} \left[\frac{t_2 \left(\bar{E}_{\mathrm{Fw}}\right)}{\left[t_1 \left(\bar{E}_{\mathrm{Fw}}\right)\right]^{1/3}}\right] \psi_3 \left(\bar{E}_{\mathrm{Fw}}\right) S_0 \left(\frac{|e|^2}{\varepsilon_{\mathrm{sc}}}\right)^{2/3},$$

$$V_{71} \left(\bar{E}_{\mathrm{Fw}}\right) \equiv \psi_3 \left(\bar{E}_{\mathrm{Fw}}\right) S_0 \left(\frac{|e|^2}{\varepsilon_{\mathrm{sc}}} t_1 \left(\bar{E}_{\mathrm{Fw}}\right)\right)^{2/3},$$

$$y_{71} \left(\bar{E}_{\mathrm{0w1}}, 0\right) \equiv \left[\frac{2}{3} V_{71} \left(\bar{E}_{\mathrm{01w}}\right) (\bar{n}_{2\mathrm{Dw}})^{-1/3}\right] \times \left[\left\{\psi_1 \left(\bar{E}_{\mathrm{0w1}}\right)\right\}' - \left\{V_{71} \left(\bar{E}_{\mathrm{01w}}\right)\right\}' (\bar{n}_{2\mathrm{Dw}})^{2/3}\right]^{-1},$$

$\bar{E}_{0\mathrm{w}1}$ is the electron subband energy in the low electric field quantum limit and can be determined from the equation $\psi_1\left(\bar{E}_{0\mathrm{w}1}\right) = V_{71}\left(\bar{E}_{0\mathrm{w}1}\right)\left(n_{2\mathrm{Dw}}\right)^{2/3}$.

Using (7.1) and (7.2), the 2D electron dispersion law in n-channel inversion layers of tetragonal materials under the strong electric field limit can be written as

$$k_\mathrm{s}^2 = P_2\left(E, i\right), \tag{7.12}$$

$$P_2\left(E, i\right) \equiv \left[F_7\left(E\right)\right]^{-1}\left[F_6\left(E\right) - F_8\left(E, i\right)\right],$$

$$F_6\left(E\right) \equiv \left[\frac{\psi_1\left(E\right)}{\psi_3\left(E\right)}\right]\left[1 + \frac{\left[\psi_1\left(E\right)\right]'\left[\psi_3\left(E\right)\right]'}{\psi_3\left(E\right)\left[\psi_1\left(E\right)\right]''} + \frac{\psi_1\left(E\right)\left[\psi_3\left(E\right)\right]''}{2\left[\psi_1\left(E\right)\right]''\left[\psi_3\left(E\right)\right]}\right],$$

$$\begin{aligned} F_7\left(E\right) \equiv & \left[\left(\frac{\psi_2\left(E\right)}{\psi_3\left(E\right)}\right) - \left(\frac{\psi_1\left(E\right)\left[\psi_2\left(E\right)\right]''}{2\left[\psi_1\left(E\right)\right]''\left[\psi_3\left(E\right)\right]}\right)\right. \\ & + \left(\frac{\psi_2\left(E\right)}{\psi_3\left(E\right)}\right)\left[\frac{\left[\psi_1\left(E\right)\right]'\left[\psi_3\left(E\right)\right]'}{\psi_3\left(E\right)\left[\psi_1\left(E\right)\right]''} + \frac{\psi_1\left(E\right)\left[\psi_3\left(E\right)\right]''}{2\left[\psi_1\left(E\right)\right]''\left[\psi_3\left(E\right)\right]}\right] \\ & \left. + \left(\frac{\psi_1\left(E\right)}{\psi_3\left(E\right)}\right)\left[\frac{\left[\psi_2\left(E\right)\right]'\left[\psi_3\left(E\right)\right]'}{\psi_3\left(E\right)\left[\psi_1\left(E\right)\right]''} + \frac{\psi_2\left(E\right)\left[\psi_3\left(E\right)\right]''}{2\left[\psi_1\left(E\right)\right]''\left[\psi_3\left(E\right)\right]}\right]\right] \end{aligned}$$ and

$$F_8\left(E, i\right) \equiv \left[\frac{2\sqrt{2}}{3}\left(S_i\right)^{3/2}\left(|e|\,F_\mathrm{s}\right)\sqrt{\left[\psi_1\left(E\right)\right]''}\right].$$

The EMM in the x–y plane can be written in this case as

$$m^*\left(E_{\mathrm{Fis}}, i\right) = \left(\frac{\hbar^2}{2}\right)\left[P_2\left(E, i\right)\right]'\Big|_{E=E_{\mathrm{Fis}}}, \tag{7.13}$$

where E_{Fis} is the Fermi energy under the strong electric field limit as measured from the edge of the conduction band at the surface. Thus, we note that the EMM is a function of subband index and the Fermi energy due to the combined influence of the crystal filed splitting constant and the anisotropic spin–orbit splitting constants respectively.

The subband energy $\left(E_{n_{is1}}\right)$ in this case can be obtained from (7.12) as

$$P_2\left(E_{n_{is1}}, i\right) = 0. \tag{7.14}$$

The total 2D density-of-states function under the strong electric field limit can be written as

$$N_{2\mathrm{D}_i}\left(E\right) = \frac{g_\mathrm{v}}{\left(2\pi\right)}\sum_{i=0}^{i_{\max}}\left[\left(P_2\left(E, i\right)\right)' H\left(E - E_{n_{is1}}\right)\right]. \tag{7.15}$$

Using (7.15) and the Fermi–Dirac occupation probability factor, the 2D surface electron concentration ($n_{2\mathrm{Ds}}$) in this case can be expressed as

$$n_{2\mathrm{Ds}} = g_{\mathrm{v}} (2\pi)^{-1} \sum_{i=0}^{i_{\max}} [P_2 (E_{\mathrm{Fis}}, i) + Q_2 (E_{\mathrm{Fis}}, i)], \tag{7.16}$$

where

$$Q_2 (E_{\mathrm{Fis}}, i) \equiv \sum_{r=1}^{s} \{L(r) [P_2 (E_{\mathrm{Fis}}, i)]\}.$$

Thus, the surface electron concentration under the strong electric quantum limit at low temperatures assumes the form

$$\bar{n}_{2\mathrm{Ds}} = g_{\mathrm{v}} (2\pi)^{-1} \left[P_2 \left(\bar{E}_{\mathrm{Fs}}, 0\right)\right], \tag{7.17}$$

where $\bar{E}_{\mathrm{Fs}}$ is the Fermi energy in the n-channel inversion layer under the strong electric field quantum limit as measured from the edge of the conduction band at the surface.

Using (7.17) and (1.12), the DMR under the strong electric field quantum limit in this case can be expressed as

$$\frac{D}{\mu} = \frac{\bar{n}_{2\mathrm{Ds}}}{|e|} \left(x_{72} \left(\bar{E}_{\mathrm{Fs}}, 0\right) - y_{72} \left(\bar{E}_{0\mathrm{s}1}, 0\right)\right), \tag{7.18}$$

where

$$x_{72} \left(\bar{E}_{\mathrm{Fs}}, 0\right) \equiv \frac{\omega_{73} \left(\bar{E}_{\mathrm{Fs}}, 0\right)}{\omega_{74} \left(\bar{E}_{\mathrm{Fs}}, 0\right)}, \quad \omega_{73} \left(\bar{E}_{\mathrm{Fs}}, 0\right) \equiv \left[\frac{2\pi}{g_{\mathrm{v}}} + \beta_{71} \left(\bar{E}_{\mathrm{Fs}}\right)\right],$$

$$\beta_{71} \left(\bar{E}_{\mathrm{Fs}}\right) \equiv \left[\frac{2\sqrt{2}}{3} \left(\frac{|e|^2}{\varepsilon_{\mathrm{sc}}}\right) (S_0)^{3/2} \frac{\sqrt{\{\psi_1 \left(\bar{E}_{\mathrm{Fs}}\right)\}''}}{F_7 \left(\bar{E}_{\mathrm{Fs}}\right)}\right],$$

$$\omega_{74} \left(\bar{E}_{\mathrm{Fs}}\right) \equiv \left[\frac{\{F_6 \left(\bar{E}_{\mathrm{Fs}}\right)\}'}{F_7 \left(\bar{E}_{\mathrm{Fs}}\right)} - \left[\frac{\{F_6 \left(\bar{E}_{\mathrm{Fs}}\right)\}}{\{F_7 \left(\bar{E}_{\mathrm{Fs}}\right)\}^2}\right] \left[F_7 \left(\bar{E}_{\mathrm{Fs}}\right)\right]' - \left\{ \left[\beta_{71} \left(\bar{E}_{\mathrm{Fs}}\right)\right]' \bar{n}_{2\mathrm{Ds}} \right\}\right],$$

$$y_{72} \left(\bar{E}_{0\mathrm{s}1}, 0\right) \equiv \frac{\beta_{72} \left(\bar{E}_{0\mathrm{s}1}\right)}{\left[\{F_6 \left(\bar{E}_{0\mathrm{s}1}\right)\}' - \left\{\left[\beta_{72} \left(\bar{E}_{0\mathrm{s}1}\right)\right]' \bar{n}_{2\mathrm{Ds}}\right\}\right]}.$$

and

$$\beta_{72} \left(\bar{E}_{0\mathrm{s}1}\right) \equiv \frac{2\sqrt{2}}{3} \left(\frac{|e|^2}{\varepsilon_{\mathrm{sc}}}\right) (S_0)^{3/2} \sqrt{\{\psi_1 \left(\bar{E}_{0\mathrm{s}1}\right)\}''}.$$

$\bar{E}_{0\mathrm{s}1}$ is the subband energy under the strong electric field quantum limit in this case and is given by the equation

$$F_6 \left(\bar{E}_{0\mathrm{s}1}\right) = \beta_{72} \left(\bar{E}_{0\mathrm{s}1}\right) \bar{n}_{2\mathrm{Ds}},$$

7.2.2 Formulation of the Einstein Relation in n-Channel Inversion Layers of III–V, Ternary and Quaternary Materials

Using the substitutions $\delta = 0, \Delta_{\parallel} = \Delta_{\perp} = \Delta$ and $m_{\parallel}^{*} = m_{\perp}^{*} = m^{*}$, (7.4) under the condition of weak electric field limit, assumes the form

$$I(E) = \frac{\hbar^2 k_s^2}{2m^*} + S_i \left[\frac{\hbar |e| F_s [I(E)]'}{\sqrt{2m^*}} \right]^{2/3}, \tag{7.19}$$

where $I(E) \equiv \gamma(E)$ of (2.6)

Equation (7.19) represents the dispersion relation of the 2D electrons in n-channel inversion layers of III–V, ternary and quaternary materials under the weak electric field limit whose bulk electrons obey the three band model of Kane.

The EMM can be expressed as

$$m^*(E_{\mathrm{Fiw}}, i) = m^* \left[P_3(E, i) \right] \big|_{E=E_{\mathrm{Fiw}}}, \tag{7.20}$$

where

$$P_3(E, i) \equiv \left\{ [I(E)]' - \left\{ \frac{2}{3} S_i \left[\frac{\hbar |e| F_s}{\sqrt{2m^*}} \right]^{2/3} \left\{ [I(E)]' \right\}^{-1/3} [I(E)]'' \right\} \right\}.$$

Thus, one can observe that the EMM is a function of the subband index, surface electric field, the Fermi energy and the other spectrum constants due to the combined influence of E_g and Δ.

The subband energy $(E_{n_{\mathrm{iw2}}})$ in this case can be obtained from (7.19) as

$$I(E_{n_{\mathrm{iw2}}}) = S_i \left[\frac{\hbar |e| F_s [I(E_{n_{\mathrm{iw2}}})]'}{\sqrt{2m^*}} \right]^{2/3}. \tag{7.21a}$$

Using (7.19) and (7.7), the 2D total density-of-states function in weak electric field limit can be expressed as

$$N_{2\mathrm{D}_i}(E) = \frac{m^* g_v}{\pi \hbar^2} \sum_{i=0}^{i_{\max}} [P_3(E, i) H(E - E_{n_{\mathrm{iw2}}})]. \tag{7.21b}$$

Using (7.21b) and the occupation probability, the $n_{2\mathrm{Dw}}$ in the present case can be written as

$$n_{2\mathrm{Dw}} = \frac{g_v m^*}{\pi \hbar^2} \sum_{i=0}^{i_{\max}} [P_{4\mathrm{w}}(E_{\mathrm{Fiw}}, i) + Q_{4\mathrm{w}}(E_{\mathrm{Fiw}}, i)], \tag{7.21c}$$

where

$$P_{4\mathrm{w}}(E_{\mathrm{Fiw}}, i) \equiv \left\{ I(E_{\mathrm{Fiw}}) - S_i \left[\frac{\hbar e F_s [I(E_{\mathrm{Fiw}})]'}{\sqrt{2m^*}} \right]^{2/3} \right\} \text{ and}$$

$$Q_4(E_{\mathrm{Fiw}}, i) \equiv \sum_{r=1}^{s} \{ L(r) [P_4(E_{\mathrm{Fiw}}, i)] \}.$$

The electron concentration under the weak electric field quantum limit and at low temperatures assumes the form and at low temperatures

$$\bar{n}_{2\mathrm{Dw}} = \frac{g_{\mathrm{v}} m^*}{\pi \hbar^2} \left[P_{4\mathrm{w}} \left(\bar{E}_{\mathrm{Fw}}, 0 \right) \right]. \tag{7.22}$$

Using (7.22) and (1.12), the DMR under the electric field quantum limit in this case can be expressed as

$$\frac{D}{\mu} = \frac{\bar{n}_{2\mathrm{Dw}}}{|e|} \left(x_{73} \left(\bar{E}_{\mathrm{Fw}}, 0 \right) - y_{73} \left(\bar{E}_{0\mathrm{w}2}, 0 \right) \right), \tag{7.23}$$

where

$$x_{73} \left(\bar{E}_{\mathrm{Fw}}, 0 \right) \equiv \left[\frac{\left(\frac{\pi \hbar^2}{m^* g_{\mathrm{v}}} \right) + \frac{2}{3} \left(\bar{n}_{2\mathrm{Dw}} \right)^{-1/3} V_{73} \left(\bar{E}_{\mathrm{Fw}} \right)}{\left\{ I \left(\bar{E}_{\mathrm{Fw}} \right) \right\}' - \left\{ V_{73} \left(\bar{E}_{\mathrm{Fw}} \right) \right\}' \left(\bar{n}_{2\mathrm{Dw}} \right)^{1/3}} \right],$$

$$V_{73} \left(\bar{E}_{\mathrm{Fw}} \right) \equiv S_0 \left[\frac{\hbar |e|^2}{\varepsilon_{\mathrm{sc}} \sqrt{2m^*}} \left\{ I \left(\bar{E}_{\mathrm{Fw}} \right) \right\}' \right]^{2/3},$$

$$y_{73} \left(\bar{E}_{0\mathrm{w}2}, 0 \right) \equiv \frac{\frac{2}{3} V_{73} \left(\bar{E}_{0\mathrm{w}2}, 0 \right) \left(\bar{n}_{2\mathrm{Dw}} \right)^{-1/3}}{\left[\left\{ I \left(\bar{E}_{0\mathrm{w}2} \right) \right\}' - \left\{ V_{73} \left(\bar{E}_{0\mathrm{w}2}, 0 \right) \right\}' \left(\bar{n}_{2\mathrm{Dw}} \right)^{2/3} \right]}.$$

$\bar{E}_{0\mathrm{w}2}$ is the subband energy in the n-channel inversion layers of III–V, ternary and quaternary materials under the weak electric field quantum limit and is determined by the equation $I \left(\bar{E}_{0\mathrm{w}2} \right) = V_{73} \left(\bar{E}_{0\mathrm{w}2} \right) \left(\bar{n}_{2\mathrm{Dw}} \right)^{2/3}$.

Using the substitutions $\delta = 0$, $\Delta_{||} = \Delta_{\perp} = \Delta$ and $m^*_{||} = m^*_{\perp} = m^*$, (7.12) under the condition of strong electric field limit, assumes the form

$$\left[I(E) - \left\{ \frac{|e| F_{\mathrm{s}} \hbar}{\sqrt{2m^*}} \left(\frac{2\sqrt{2} \left(S_i \right)^{3/2}}{3} \right) \sqrt{[I(E)]''} \right\} \right] = \frac{\hbar^2 k_{\mathrm{s}}^2}{2m^*}. \tag{7.24}$$

Equation (7.24) represents the dispersion relation of the 2D electrons in n-channel inversion layers of III–V, ternary and quaternary materials under the strong electric field limit whose bulk conduction electrons are defined by the three band model of Kane.

The EMM can be expressed as

$$m^* \left(E_{\mathrm{Fis}}, i \right) = m^* \left[P_5 (E, i) \right] |_{E = E_{\mathrm{Fis}}}, \tag{7.25}$$

where

$$P_5 (E, i) \equiv \left\{ \{ I(E) \}' - \left\{ \frac{|e| F_{\mathrm{s}} \hbar}{\sqrt{2m^*}} \left(\frac{\sqrt{2} \left(S_i \right)^{3/2}}{3} \right) \left([I(E)]'' \right)^{-1/2} [I(E)]''' \right\} \right\}.$$

Thus, one can observe that the EMM is a function of the subband index, surface electric field, the Fermi energy and the other spectrum constants due to the combined influence of E_{g} and Δ.

The subband energy ($E_{n_{\text{is2}}}$) in this case can be obtained from (7.24) as

$$I\left(E_{n_{\text{is2}}}\right)-\left\{\frac{|e|\,F_{\text{s}}\hbar}{\sqrt{2m^*}}\left(\frac{2\sqrt{2}\left(S_i\right)^{3/2}}{3}\right)\sqrt{\left[I\left(E_{n_{is2}}\right)\right]''}\right\}=0. \tag{7.26}$$

Using (7.24) and (7.7), the total 2D density-of-states function under the strong electric field limit can be expressed as

$$N_{2\text{D}_i}\left(E\right)=\frac{m^*g_{\text{v}}}{\pi\hbar^2}\sum_{i=0}^{i_{\max}}\left[P_5\left(E,i\right)H\left(E-E_{n_{is2}}\right)\right]. \tag{7.27}$$

Using (7.27) and the Fermi–Dirac occupation probability factor, the n_{2Ds} in the present case under the strong electric field can be written as

$$n_{\text{2Ds}}=\frac{g_{\text{v}}m^*}{\pi\hbar^2}\sum_{i=0}^{i_{\max}}\left[P_{6\text{s}}\left(E_{\text{Fis}},i\right)+Q_{6\text{s}}\left(E_{\text{Fis}},i\right)\right], \tag{7.28}$$

where

$$P_{6\text{s}}\left(E_{\text{Fis}},i\right)\equiv\left\{I\left(E_{\text{Fis}}\right)-\left[\frac{2\sqrt{2}}{3}\left(S_i\right)^{3/2}\frac{\hbar\,|e|\,F_{\text{s}}\sqrt{\left[I\left(E_{\text{Fis}}\right)\right]''}}{\sqrt{2m^*}}\right]\right\}\quad\text{and}$$

$$Q_6\left(E_{\text{Fis}},i\right)\equiv\sum_{r=1}^{s}\left\{L\left(r\right)\left[P_6\left(E_{\text{Fis}},i\right)\right]\right\}.$$

The electron concentration under the strong electric field quantum limit assumes the form

$$\bar{n}_{\text{2Ds}}=\frac{g_{\text{v}}m^*}{\pi\hbar^2}\left[P_{6\text{s}}\left(\bar{E}_{\text{Fs}},0\right)\right]. \tag{7.29}$$

Using (7.29) and (1.12), the DMR under the electric field quantum limit in this case can be expressed as

$$\frac{D}{\mu}=\frac{\bar{n}_{\text{2Ds}}}{|e|}\left(x_{74}\left(\bar{E}_{\text{Fs}},0\right)-y_{74}\left(\bar{E}_{\text{0s2}},0\right)\right). \tag{7.30}$$

where

$$x_{74}\left(\bar{E}_{\text{Fs}},0\right)\equiv\left[\frac{\left(\frac{\pi\hbar^2}{m^*g_{\text{v}}}\right)+V_{74}\left(\bar{E}_{\text{Fs}}\right)}{\left\{I\left(\bar{E}_{\text{Fs}}\right)\right\}'-\left\{V_{74}\left(\bar{E}_{\text{Fs}}\right)\right\}'\left(\bar{n}_{\text{2Ds}}\right)}\right],$$

$$V_{74}\left(\bar{E}_{\text{Fs}}\right)\equiv\left(\frac{|e|^2}{\varepsilon_{\text{sc}}\sqrt{2m^*}}\right)\left(\frac{2\sqrt{2}}{3}S_0\right)\sqrt{\left\{I\left(\bar{E}_{\text{Fs}}\right)\right\}''},$$

$$y_{74}\left(\bar{E}_{\text{0s2}},0\right)\equiv\frac{V_{74}\left(\bar{E}_{\text{0s2}}\right)}{\left[\left\{I\left(\bar{E}_{\text{0s2}}\right)\right\}'-\bar{n}_{\text{2Ds}}\left\{V_{74}\left(\bar{E}_{\text{0s2}}\right)\right\}'\right]}.$$

$\bar{E}_{0s2}$ is the subband energy in the present case and is determined by the equation

$$I\left(\bar{E}_{0s2}\right) - \left[\frac{2\sqrt{2}}{3}\left(S_0\right)^{3/2}\frac{\left(\hbar\left|e\right|F_s\right)\sqrt{\left[I\left(\bar{E}_{0s2}\right)\right]''}}{\sqrt{2m^*}}\right] = 0.$$

Using the constraints $\Delta \gg E_g$ or $\Delta \ll E_g$, (7.19) under the low electric field limit assumes the form

$$E\left(1+\alpha E\right) = \frac{\hbar^2 k_s^2}{2m^*} + S_i\left[\frac{\hbar\left|e\right|F_s\left(1+2\alpha E\right)}{\sqrt{2m^*}}\right]^{2/3}. \tag{7.31}$$

For large values of i, $S_i \to \left[\frac{3\pi}{2}\left(\mathrm{i}+\frac{3}{4}\right)\right]^{2/3}$ [5], and (7.31) gets simplified to

$$E\left(1+\alpha E\right) = \frac{\hbar^2 k_s^2}{2m^*} + \left[\frac{3\pi\hbar\left|e\right|F_s}{2}\left(\mathrm{i}+\frac{3}{4}\right)\frac{\left(1+2\alpha E\right)}{\sqrt{2m^*}}\right]^{2/3}. \tag{7.32}$$

Equation (7.32) was derived for the first time by Antcliffe et al. [3].
The EMM in this case is given by

$$m^*\left(E_{\mathrm{Fiw}}, i\right) = m^*\left[P_6\left(E,i\right)\right]\big|_{E=E_{\mathrm{Fiw}}}, \tag{7.33}$$

where

$$P_6\left(E,i\right) \equiv \left\{1+2\alpha E - \frac{4\alpha}{3}S_i\left[\frac{\hbar\left|e\right|F_s}{\sqrt{2m^*}}\right]^{2/3}\left\{1+2\alpha E\right\}^{-1/3}\right\}.$$

Thus, one can observe that the EMM is a function of the subband index, surface electric field and the Fermi energy due to the presence of band nonparabolicity only.

The subband energies $(E_{n_{\mathrm{iw3}}})$ are given by

$$E_{n_{\mathrm{iw3}}}\left(1+\alpha E_{n_{\mathrm{iw3}}}\right) = S_i\left[\frac{\hbar\left|e\right|F_s\left(1+2\alpha E_{n_{\mathrm{iw3}}}\right)}{\sqrt{2m^*}}\right]^{2/3}. \tag{7.34}$$

The total 2D density-of-states function can be written as

$$N_{2D}\left(E\right) = \frac{m^* g_v}{\pi\hbar^2}\sum_{i=0}^{i_{\max}}\left\{\left[1+2\alpha E - \frac{4\alpha}{3}S_i\left[\frac{\hbar\left|e\right|F_s}{\sqrt{2m^*}}\right]^{2/3}\left(1+2\alpha E\right)^{-1/3}\right]\right.$$
$$\left.\times H\left(E-E_{n_{\mathrm{iw3}}}\right)\right\}. \tag{7.35}$$

Under the condition $\alpha E \ll 1$, the use of (7.35) and the Fermi–Dirac integral leads to the expression of $n_{2\mathrm{Dw}}$ as

$$n_{2\mathrm{Dw}} = \left(\frac{g_v m^* k_B T}{\pi\hbar^2}\right)\sum_{i=0}^{i_{\max}}\left\{\left[1+D_i+2\alpha E_{n_{\mathrm{iw3}}}\right]F_0\left(\eta_{\mathrm{iw}}\right)+2\alpha k_B T F_1\left(\eta_{\mathrm{iw}}\right)\right\}, \tag{7.36}$$

where

$$D_i \equiv \frac{4\alpha S_i}{3}\left(\frac{\hbar\,|e|\,F_{\mathrm{s}}}{\sqrt{2m^*}}\right)^{2/3} \text{ and } \eta_{\mathrm{iw}} \equiv \left[\frac{E_{\mathrm{Fiw}} - E_{n_{\mathrm{iw3}}}}{k_{\mathrm{B}}T}\right].$$

For all values of αE_{Fiw}, n_{2Dw} can be written by using (7.35) as

$$n_{\mathrm{2Dw}} = \left(\frac{g_{\mathrm{v}} m^*}{\pi\hbar^2}\right) \sum_{i=0}^{i_{\max}} \left[P_{5\mathrm{w}}\left(E_{\mathrm{Fiw}}, i\right) + Q_{5\mathrm{w}}\left(E_{\mathrm{Fiw}}, i\right)\right], \tag{7.37}$$

where

$$P_{5\mathrm{w}}\left(E_{\mathrm{Fiw}}, i\right) \equiv \left[E_{\mathrm{Fiw}}\left(1+\alpha E_{\mathrm{Fiw}}\right) - S_i\left[\frac{\hbar\,|e|\,F_{\mathrm{s}}}{\sqrt{2m^*}}\left(1+2\alpha E_{\mathrm{Fiw}}\right)\right]^{2/3}\right] \text{ and}$$

$$Q_{5\mathrm{w}}\left(E_{\mathrm{Fiw}}, i\right) \equiv \sum_{r=1}^{s} L\left(r\right) P_{5\mathrm{w}}\left(E_{\mathrm{Fiw}}, i\right).$$

The electron concentration under the weak electric field quantum limit and at low temperatures assumes the form

$$\bar{n}_{\mathrm{2Dw}} = \frac{g_{\mathrm{v}} m^*}{\pi\hbar^2}\left\{\bar{E}_{\mathrm{Fw}}\left(1+\alpha\bar{E}_{\mathrm{Fw}}\right) - S_0\left(\frac{\hbar\,|e|\,F_{\mathrm{s}}}{\sqrt{2m^*}}\right)^{2/3}\left(1+2\alpha\bar{E}_{\mathrm{Fw}}\right)^{2/3}\right\}. \tag{7.38}$$

Using (7.38) and (1.12), the DMR in the present case can be expressed as

$$\frac{D}{\mu} = \frac{\bar{n}_{\mathrm{2Dw}}}{|e|}\left(x_{75}\left(\bar{E}_{\mathrm{Fw}}, 0\right) - y_{75}\left(\bar{E}_{\mathrm{0w3}}, 0\right)\right), \tag{7.39}$$

where,

$$x_{75}\left(\bar{E}_{\mathrm{Fw}}, 0\right) \equiv \frac{\omega_{75}\left(\bar{E}_{\mathrm{Fw}}, 0\right)}{\omega_{76}\left(\bar{E}_{\mathrm{Fw}}, 0\right)},$$

$$\omega_{75}\left(\bar{E}_{\mathrm{Fw}}, 0\right) \equiv \left[\frac{\pi\hbar^2}{m^* g_{\mathrm{v}}} + \frac{2}{3} S_0\left(\bar{n}_{\mathrm{2Dw}}\right)^{-1/3}\left(\frac{\hbar\,|e|^2}{\varepsilon_{\mathrm{sc}}\sqrt{2m^*}}\left(1+2\alpha\bar{E}_{\mathrm{Fw}}\right)\right)^{2/3}\right],$$

$$\omega_{76}\left(\bar{E}_{\mathrm{Fw}}, 0\right) \equiv \left[1+2\alpha\bar{E}_{\mathrm{Fw}} - \left\{\frac{4S_0\alpha}{3}\left(1+2\alpha\bar{E}_{\mathrm{Fw}}\right)^{-1/3}\right\}\left(\frac{\bar{n}_{\mathrm{2Dw}}\hbar\,|e|^2}{\varepsilon_{\mathrm{sc}}\sqrt{2m^*}}\right)^{2/3}\right],$$

$$y_{75}\left(\bar{E}_{\mathrm{0w3}}, 0\right) \equiv \left[\frac{2}{3}\frac{\bar{E}_{\mathrm{0w3}}\left(1+\alpha\bar{E}_{\mathrm{0w3}}\right)}{\bar{n}_{\mathrm{2Dw}}}\right]\left[1+2\alpha\bar{E}_{\mathrm{0w3}} - \left\{\frac{4\alpha\bar{E}_{\mathrm{0w3}}\left(1+\alpha\bar{E}_{\mathrm{0w3}}\right)}{3\left(1+2\alpha\bar{E}_{\mathrm{0w3}}\right)}\right\}\right]^{-1},$$

$\bar{E}_{\mathrm{0w3}}$ is the subband energy in the present case and is determined by the equation

$$\bar{E}_{\mathrm{0w3}}\left(1+\alpha\bar{E}_{\mathrm{0w3}}\right) - S_0\left(\frac{\hbar\,|e|\,F_{\mathrm{s}}}{\sqrt{2m^*}}\right)^{2/3}\left(1+2\alpha\bar{E}_{\mathrm{0w3}}\right)^{2/3} = 0.$$

For $\alpha \to 0$, as for inversion layers, whose bulk electrons are defined by the parabolic energy bands, from (7.32), we can write,

$$E = \frac{\hbar^2 k_s^2}{2m^*} + S_i \left[\frac{\hbar |e| F_s}{\sqrt{2m^*}} \right]^{2/3}. \tag{7.40}$$

Equation (7.40) is valid for all values of the surface electric field [1]. The electric subband energy ($E_{n_{i4}}$) assumes the form, from (7.40) as

$$E_{n_{i4}} = S_i \left[\frac{\hbar |e| F_s}{\sqrt{2m^*}} \right]^{2/3}. \tag{7.41}$$

The total density-of-states function can be written using (7.40) as

$$N_{2D}(E) = \frac{m^* g_v}{\pi \hbar^2} \sum_{i=0}^{i_{max}} H(E - E_{n_{i4}}). \tag{7.42}$$

The use of (7.42) leads to the expression of n_{2D} [1] given by

$$n_{2D} = \frac{g_v m^* k_B T}{\pi \hbar^2} \sum_{i=0}^{i_{max}} F_0(\eta_i), \tag{7.43}$$

where,

$$\eta_i \equiv (k_B T)^{-1} \left[E_{Fi} - S_i \left[\frac{\hbar |e| F_s}{\sqrt{2m^*}} \right]^{2/3} \right]$$

and E_{Fi} is the Fermi energy measured from the edge of the conduction band at the surface.

The Einstein relation in the case $i = 0$ and $T \neq 0$ assumes the form

$$\frac{D}{\mu} = \frac{k_B T}{|e|} \left(\frac{F_0(\eta_0)}{F_{-1}(\eta_0)} \right), \tag{7.44}$$

where

$$\eta_0 \equiv \frac{(\bar{E}_{F0} - \bar{E}_0)}{k_B T},$$

$\bar{E}_{F0}$ and $\bar{E}_0$ are the Fermi energy and the electric subband energy under the electric field quantum limit respectively.

For $i = 0$ and $T \to 0$, (7.44) gets simplified to

$$\frac{D}{\mu} = \frac{1}{|e|} (\bar{E}_{F0} - \bar{E}_0). \tag{7.45}$$

Using the constraints $\Delta \gg E_g$ or $\Delta \ll E_g$, (7.24) under the strong electric field limit assumes the form

$$E(1+\alpha E) = \frac{\hbar^2 k_s^2}{2m^*} + \frac{2\sqrt{2}}{3}(S_i)^{3/2}\left[\frac{\hbar|e|F_s}{\sqrt{m^* E_g}}\right]. \tag{7.46}$$

For large values of i, $S_i \to \left[\frac{3\pi}{2}\left(\mathrm{i}+\frac{3}{4}\right)\right]^{2/3}$ [5] and (7.46) gets simplified to

$$E(1+\alpha E) = \frac{\hbar^2 k_s^2}{2m^*} + \left[\frac{\pi\hbar|e|F_s\sqrt{2}}{\sqrt{m^* E_g}}\left(\mathrm{i}+\frac{3}{4}\right)\right]. \tag{7.47}$$

Equation (7.47) was derived for the first time by Antcliffe et al. [3].

From (7.46), we observe that under the condition $E_g \to 0$, one cannot obtain the corresponding parabolic case, since under high electric field limit, the band becomes permanently nonparabolic.

The EMM is given by

$$m^*(E_{\mathrm{Fis}}, i) = m^*(1+2\alpha E_{\mathrm{Fis}}). \tag{7.48}$$

Thus, in the high electric field limit, the EMM is a function of Fermi energy due to the presence of band nonparabolicity only and is independent of the subband index.

The electric subband energy $(E_{n_{iw5}})$ in the high electric field limit is given by

$$E_{n_{iw5}}(1+\alpha E_{n_{iw5}}) = \left[\frac{\pi\hbar|e|F_s\sqrt{2}}{\sqrt{m^* E_g}}\left(\mathrm{i}+\frac{3}{4}\right)\right]. \tag{7.49}$$

The 2D total density-of-states function in this case can be written as

$$N_{2\mathrm{D}_i}(E) = \frac{m^* g_v}{\pi\hbar^2}\sum_{i=0}^{i_{\max}}\{[1+2\alpha E]H(E-E_{n_{iw5}})\}. \tag{7.50}$$

The surface electron concentration for all values of αE_{Fis} in this case assumes the form

$$n_{2\mathrm{Dw}} = \left(\frac{g_v m^* k_B T}{\pi\hbar^2}\right)\sum_{i=0}^{i_{\max}}\{[1+2\alpha k_B T]F_0(\eta_{\mathrm{is}}) + 2\alpha k_B T E_{n_{\mathrm{iw5}}}F_1(\eta_{\mathrm{is}})\}, \tag{7.51}$$

where

$$\eta_{\mathrm{is}} \equiv \left[\frac{E_{\mathrm{Fis}} - E_{n_{\mathrm{iw5}}}}{k_B T}\right].$$

The surface electron concentration under high electric field quantum limit can be written as

$$\bar{n}_{2\mathrm{Ds}} = \frac{g_v m^*}{\pi\hbar^2}\left\{\bar{E}_{\mathrm{Fs}}(1+\alpha\bar{E}_{\mathrm{Fs}}) - N_7\bar{n}_{2\mathrm{Ds}}\right\}, \tag{7.52}$$

where

$$N_7 \equiv \frac{2\sqrt{2}}{3}(S_0)^{3/2}\left(\frac{\hbar|e|^2}{\varepsilon_{sc}\sqrt{m^* E_g}}\right).$$

Using (7.52) and (1.12), the DMR in this case can be expressed as

$$\frac{D}{\mu} = \frac{\bar{n}_{2Ds}}{|e|}\left(\frac{\pi\hbar^2}{m^* g_v\left(1+2\alpha\bar{E}_{Fs}\right)}\right)\cdot \bar{t}_{77}. \tag{7.53}$$

where

$$\bar{t}_{77} = \left[\left(\frac{1}{m^* g_v} + \frac{N_7}{\pi\hbar^2}\right) - \frac{N_7}{\pi\hbar^2}\cdot\frac{\left(1+2\alpha\bar{E}_{Fs}\right)}{\left(1+2\alpha\bar{E}_{0s}\right)}\right]$$

in which

$$\bar{E}_{0s}\left(1+\alpha\bar{E}_{0s}\right) = \left(\frac{\pi\hbar|e|^2\sqrt{2}\bar{n}_{2Ds}}{\varepsilon_{sc}\sqrt{m^* E_g}}\right)\cdot\frac{3}{4}$$

7.2.3 Formulation of the Einstein Relation in p-Channel Inversion Layers of II–VI Materials

The use of (2.27) and (7.2) leads to the expression of the quantization integral as

$$\frac{\sqrt{2m_{\|}^*}}{\hbar}\int_0^{z_t}\left[E - |e| F_s z - a_0' k_s^2 \mp \left(\bar{\lambda}_0\right) k_s\right]^{1/2} dz = \frac{2}{3}(S_i)^{3/2}, \tag{7.54}$$

where

$$z_t \equiv \left(|e| F_s\right)^{-1}\left[E - a_0' k_s^2 \mp \left(\bar{\lambda}_0\right) k_s\right].$$

Therefore, the 2D electron dispersion law for n-channel inversion layers of II–VI semiconductors can be expressed for all values of F_s as

$$E = a_0' k_s^2 \pm \left(\bar{\lambda}_0\right) k_s + S_i\left(\frac{\hbar|e| F_s}{\sqrt{2m_{\|}^*}}\right)^{2/3}. \tag{7.55}$$

The area of the 2D surface enclosed by (7.55) can be expressed as

$$A(E,i) = \frac{\pi\left(m_{\perp}^*\right)^2}{\hbar^4}\left[\left\{2\left(\bar{\lambda}_0\right)^2 - \frac{2\hbar^2}{m_{\perp}^*}S_i\left(\frac{\hbar|e| F_s}{\sqrt{2m_{\|}^*}}\right)^{2/3} + \frac{2\hbar^2 E}{m_{\perp}^*}\right\}\right.$$
$$\left. -2\left(\bar{\lambda}_0\right)\left[\left(\bar{\lambda}_0\right)^2 - \frac{2\hbar^2}{m_{\perp}^*}S_i\left(\frac{\hbar|e| F_s}{\sqrt{2m_{\|}^*}}\right)^{2/3} + \frac{2\hbar^2 E}{m_{\perp}^*}\right]^{1/2}\right]. \tag{7.56}$$

The EMM is given by

$$m^*\left(E_{Fi}, i\right) = m_{\perp}^*\left[1 - \frac{\rho_{71}}{\sqrt{E_{Fi}+\rho_{72}}}\right]. \tag{7.57}$$

where E_{Fi} is the Fermi energy in this case,

$$\rho_{71} \equiv \frac{\bar{\lambda}_0}{2\sqrt{a_0'}} \text{ and } \rho_{72} \equiv \left[(\rho_{71})^2 - \left(\frac{\hbar |e| F_{\mathrm{s}}}{\sqrt{2m_{||}^*}} \right)^{2/3} \right].$$

Thus, the EMM depends on both the Fermi energy and the subband index due to the presence of the term $\bar{\lambda}_0$.

The subband energy ($E_{n_{i6}}$) can be written as

$$E_{n_{i6}} = S_i \left(\frac{\hbar |e| F_{\mathrm{s}}}{\sqrt{2m_{||}^*}} \right)^{2/3}. \tag{7.58}$$

The total 2D density-of-states function can be written as

$$N_{2\mathrm{D}_i}(E) = \frac{m_{\perp}^* g_{\mathrm{v}}}{\pi \hbar^2} \sum_{i=0}^{i_{\max}} \left\{ \left[1 - \frac{\rho_{71}}{\sqrt{E + \rho_{72}}} \right] H\left(E - E_{n_{i6}}\right) \right\}. \tag{7.59}$$

The surface electron concentration assumes the form

$$n_{2\mathrm{Di}} = \frac{g_{\mathrm{v}} m_{\perp}^* k_{\mathrm{B}} T}{\pi \hbar^2} \left\{ \sum_{i=0}^{i_{\max}} \left[F_0(\eta_i) - \left\{ \frac{\bar{\lambda}_0 f_7(E_{\mathrm{Fi}}, i)}{2\sqrt{a_0'} k_{\mathrm{B}} T} \right\} \right] \right\}, \tag{7.60}$$

where

$$\eta_i \equiv \left[\frac{E_{\mathrm{Fi}} - E_{n_{i6}}}{k_{\mathrm{B}} T} \right], \; f_7(E_{\mathrm{Fi}}, i)$$

$$\equiv \left[2\left[\sqrt{\eta_i + \delta_{72}} - \sqrt{\delta_{72}} \right] + \sum_{r=1}^{s} \left\{ 2\left(1 - 2^{1-2r}\right) \zeta(2r) \frac{(-1)^{2r-1}(2r-1)!}{(\eta_i + \delta_{72})^{2r}} \right\} \right],$$

and

$$\delta_{72} \equiv \frac{(\bar{\lambda}_0)^2}{4a_0' k_{\mathrm{B}} T}.$$

The electron concentration in the electric field quantum limit can be obtained as

$$\bar{n}_{2\mathrm{D}0} = \frac{g_{\mathrm{v}}}{2a_0'} \left[\bar{E}_{\mathrm{F}0} + \frac{(\bar{\lambda}_0)^2}{4a_0'} - \bar{E}_0 - \left\{ \frac{\bar{\lambda}_0}{2\sqrt{a_0'}} \left[\bar{E}_{\mathrm{F}0} + \frac{(\bar{\lambda}_0)^2}{4a_0'} - \bar{E}_0 \right]^{1/2} \right\} \right]. \tag{7.61}$$

The DMR in this case is given by

$$\frac{D}{\mu} = \frac{\bar{n}_{2\mathrm{D}0}}{|e|} \left(x_{76}\left(\bar{E}_{\mathrm{F}0}, 0\right) - y_{76}\left(\bar{E}_0, 0\right) \right), \tag{7.62}$$

where

$$x_{76}\left(\bar{E}_{\mathrm{F0}},0\right)\equiv\left[\frac{2a_0'}{g_{\mathrm{v}}}+\frac{2}{3}\frac{\bar{E}_0}{\bar{n}_{\mathrm{2D0}}}-\left(\frac{\bar{\lambda}_0\bar{E}_0}{3\bar{n}_{\mathrm{2D0}}\sqrt{a_0'}}\right)+\left[\bar{E}_{\mathrm{F0}}+\frac{\left(\bar{\lambda}_0\right)^2}{4a_0'}-\bar{E}_0\right]^{-1/2}\right]$$
$$\times\left[1-\left(\frac{\bar{\lambda}_0}{2\sqrt{a_0'}}\right)\left\{\bar{E}_{\mathrm{F0}}+\frac{\left(\bar{\lambda}_0\right)^2}{4a_0'}-\bar{E}_0\right\}^{-1/2}\right]^{-1},$$

$$y_{76}\left(\bar{E}_0,0\right)\equiv\frac{2\bar{E}_0}{3\bar{n}_{\mathrm{2D0}}},$$

$\bar{E}_0$ is the subband energy in this case and is given by equation

$$\bar{E}_0\equiv S_0\left(\frac{\hbar\left|e\right|F_{\mathrm{s}}}{\sqrt{2m_{||}^*}}\right)^{2/3}.$$

7.2.4 Formulation of the Einstein Relation in n-Channel Inversion Layers of IV–VI Materials

In the low electric field limit (2.41) assumes the form

$$E(1+\alpha E)-\left|e\right|F_{\mathrm{s}}z\left(1+2\alpha E\right)=\frac{p_x^2}{2M_1}+\frac{p_z^2}{2M_3}+\left(\frac{\alpha p_y^4}{4M_2M_2'}\right)+\frac{p_y^2}{2M_2}$$
$$+\frac{p_y^2}{2M_2}\alpha\left(E-\left|e\right|F_{\mathrm{s}}z\right)\left(1-\frac{M_2}{M_2'}\right),\tag{7.63}$$

where $M_1=m_{\perp\mathrm{c}}$, $m_{\perp\mathrm{c}}$ is the transverse effective electron mass at the edge of the conduction band at $k=0$, $M_2=\left(\frac{m_{\perp\mathrm{c}}+2m_{||\mathrm{c}}}{3}\right)$, $m_{||\mathrm{c}}$ is the longitudinal effective electron mass at the edge of the conduction band at $k=0$, $M_3=\left(\frac{3m_{\perp\mathrm{c}}m_{||\mathrm{c}}}{2m_{||\mathrm{c}}+m_{\perp\mathrm{c}}}\right)$, $M_2'=\left(\frac{m_{\perp\mathrm{v}}+2m_{||\mathrm{v}}}{3}\right)$, $m_{\perp\mathrm{v}}$ and $m_{||\mathrm{v}}$ are the effective transverse and longitudinal hole masses at the edge of the valance band at $k=0$.

The use of (7.63) and (7.2) leads to a simplified expression of the 2D electron dispersion law in n-channel inversion layers of IV–VI materials under the weak electric field limit given by

$$\gamma_{71}\left(E,i\right)=p_{71}k_x^2+q_{71}\left(E,i\right)k_y^2+r_{71}k_y^4,\tag{7.64}$$

where

$$\gamma_{71}\left(E,i\right)\equiv\left[E\left(1+\alpha E\right)-S_i\left(\frac{\hbar\left|e\right|F_{\mathrm{s}}}{\sqrt{2M_3}}\right)^{2/3}\left(1+\frac{4}{3}\alpha E\right)\right],\ p_{71}\equiv\frac{\hbar^2}{2M_1},$$

$$q_{71}\left(E,i\right)\equiv\left(\frac{\hbar^2}{2M_2}\right)\left[1+\alpha E\left(1-\frac{M_2}{M_2'}\right)-\frac{2\alpha S_i}{3}\left(\frac{\hbar\left|e\right|F_{\mathrm{s}}}{\sqrt{2M_3}}\right)^{2/3}\left(1-\frac{M_2}{M_2'}\right)\right]$$

and $r_{71}\equiv\left(\dfrac{\alpha\hbar^4}{4M_2M_2'}\right)$.

Following the method used for obtaining (5.29), the area enclosed by (7.64) is given by

$$A(E,i)=\frac{4}{3}\left(\frac{r_{71}}{p_{71}}\right)^{1/2}\left[\{a_{71\mathrm{w}}(E,i)\}^2+\{b_{71\mathrm{w}}(E,i)\}^2\right]^{1/2}$$

$$\times\left[\{a_{71\mathrm{w}}(E,i)\}^2F\left[\frac{\pi}{2},\ell_{71\mathrm{w}}(E,i)\right]-\left(\{a_{71\mathrm{w}}(E,i)\}^2-\{b_{71\mathrm{w}}(E,i)\}^2\right)\right.$$
$$\left.\times E\left[\frac{\pi}{2},\ell_{71\mathrm{w}}(E,i)\right]\right], \tag{7.65}$$

where

$$\{a_{71\mathrm{w}}(E,i)\}^2\equiv\left[\frac{q_{71}(E,i)}{2r_{71}}+\frac{1}{2}\left[\frac{\{q_{71}(E,i)\}^2}{(r_{71})^2}+\frac{4\gamma_{71}(E,i)}{r_{71}}\right]^{1/2}\right],$$

$$\{b_{71\mathrm{w}}(E,i)\}^2\equiv\left[\frac{1}{2}\left[\frac{\{q_{71}(E,i)\}^2}{(r_{71})^2}+\frac{4\gamma_{71}(E,i)}{r_{71}}\right]^{1/2}-\left(\frac{q_1(E,i)}{2r_{71}}\right)\right],$$

$$\ell_{71\mathrm{w}}(E,i)\equiv\frac{b_{71\mathrm{w}}(E,i)}{\sqrt{\{a_{71\mathrm{w}}(E,i)\}^2+\{b_{71\mathrm{w}}(E,i)\}^2}},\quad F\left[\frac{\pi}{2},\ell_{71\mathrm{w}}(E,i)\right]\text{ and}$$

$$E\left[\frac{\pi}{2},\ell_{71\mathrm{w}}(E,i)\right]$$

are the complete elliptic integrals of the first and second kinds respectively.

Using (7.65), the EMM in this case can be expressed as

$$m^*(E_{\mathrm{Fwi}},i)=\frac{\hbar^2}{2\pi}R_{71}(E,i)\big|_{E=E_{\mathrm{Fwi}}}. \tag{7.66}$$

where

$$R_{71}(E,i)\equiv\left[\frac{4}{3}\left(\frac{r_{71}}{p_{71}}\right)^{1/2}\left[\{a_{71\mathrm{w}}(E,i)\{a_{71\mathrm{w}}(E,i)\}'\right.\right.$$
$$+b_{71\mathrm{w}}(E,i)\{b_{71\mathrm{w}}(E,i)\}'\}\left[\{a_{71\mathrm{w}}(E,i)\}^2+\{b_{71\mathrm{w}}(E,i)\}^2\right]^{-1/2}$$
$$\times\left[\{a_{71\mathrm{w}}(E,i)\}^2F\left[\frac{\pi}{2},\ell_{71\mathrm{w}}(E,i)\right]-(\{a_{71\mathrm{w}}(E,i)\}^2-\{b_{71\mathrm{w}}(E,i)\}^2)\right.$$
$$\left.\times E\left[\frac{\pi}{2},\ell_{71\mathrm{w}}(E,i)\right]\right]+\frac{4}{3}\left(\frac{r_{71}}{p_{71}}\right)^{1/2}\left[(\{a_{71\mathrm{w}}(E,i)\}^2+\{b_{71\mathrm{w}}(E,i)\}^2)^{1/2}\right.$$
$$\times\left[2a_{71\mathrm{w}}(E,i)\{a_{71\mathrm{w}}(E,i)\}'F\left[\frac{\pi}{2},\ell_{71\mathrm{w}}(E,i)\right]+\{a_{71\mathrm{w}}(E,i)\}^2\right.$$
$$\times\left\{F\left[\frac{\pi}{2},\ell_{71\mathrm{w}}(E,i)\right]\right\}'-\left\{E\left[\frac{\pi}{2},\ell_{71\mathrm{w}}(E,i)\right]\right\}'$$
$$\times(\{a_{71\mathrm{w}}(E,i)\}^2-\{b_{71\mathrm{w}}(E,i)\}^2)-E\left[\frac{\pi}{2},\ell_{71\mathrm{w}}(E,i)\right]$$
$$\left.\left.\left.\times\left[2a_{71\mathrm{w}}(E,i)\{a_{71\mathrm{w}}(E,i)\}'-2b_{71\mathrm{w}}(E,i)\{b_{71\mathrm{w}}(E,i)\}'\right]\right]\right]\right].$$

Thus, the EMM is a function of the subband index number and the Fermi energy due to the presence of band nonparabolicity only.

The subband energies $(E_{n_{\mathrm{iw7}}})$ are given by

$$\left[E_{n_{\mathrm{iw7}}}\left(1+\alpha E_{n_{\mathrm{iw7}}}\right)-S_i\left(\frac{\hbar\left|e\right|F_{\mathrm{s}}}{\sqrt{2M_3}}\right)^{2/3}\left(1+\frac{4}{3}\alpha E_{n_{\mathrm{iw7}}}\right)\right]=0. \tag{7.67}$$

The total 2D density-of-states function can be written as

$$N_{2\mathrm{D_i}}\left(E\right)=\frac{g_{\mathrm{v}}}{2\pi^2}\sum_{i=0}^{i_{\max}}\left\{R_{71}\left(E,i\right)H\left(E-E_{n_{\mathrm{iw7}}}\right)\right\}. \tag{7.68}$$

The surface electron concentration assumes the form

$$n_{2\mathrm{Dw}}=\frac{g_{\mathrm{v}}}{3\pi^2}\left(\frac{r_{71}}{p_{71}}\right)^{1/2}\left\{\sum_{i=0}^{i_{\max}}\left[P_{7\mathrm{w}}\left(E_{\mathrm{Fwi}},i\right)+Q_{7\mathrm{w}}\left(E_{\mathrm{Fwi}},i\right)\right]\right\}, \tag{7.69}$$

where

$$\begin{aligned}P_{7\mathrm{w}}\left(E_{\mathrm{Fwi}},i\right)\equiv&\left[\left\{a_{71\mathrm{w}}\left(E_{\mathrm{Fwi}},i\right)\right\}^2+\left\{b_{71\mathrm{w}}\left(E_{\mathrm{Fwi}},i\right)\right\}^2\right]^{1/2}\\&\times\left[\left\{a_{71\mathrm{w}}\left(E_{\mathrm{Fwi}},i\right)\right\}^2F\left[\frac{\pi}{2},\ell_{71\mathrm{w}}\left(E_{\mathrm{Fwi}},i\right)\right]\right.\\&\left.-\left(\left\{a_{71\mathrm{w}}\left(E_{\mathrm{Fwi}},i\right)\right\}^2-\left\{b_{71\mathrm{w}}\left(E_{\mathrm{Fwi}},i\right)\right\}^2\right)E\left[\frac{\pi}{2},\ell_{71\mathrm{w}}\left(E_{\mathrm{Fwi}},i\right)\right]\right],\end{aligned}$$

and

$$Q_{7\mathrm{w}}\left(E_{\mathrm{Fiw}},i\right)\equiv\sum_{r=1}^{s}L\left(r\right)P_{7\mathrm{w}}\left(E_{\mathrm{Fiw}},i\right).$$

From the allied definitions of (7.65), it appears that $\ell_{71\mathrm{w}}\left(E,i\right)\ll 1$. Retaining the first two terms of the expansions of $F\left[\frac{\pi}{2},\ell_{71\mathrm{w}}\left(E,i\right)\right]$ and $E\left[\frac{\pi}{2},\ell_{71\mathrm{w}}\left(E,i\right)\right]$ together with simple algebraic manipulations, one can approximately write

$$A\left(E,i\right)=\pi\left[\frac{\gamma_{71}\left(E,i\right)}{\sqrt{p_{71}q_{71}\left(E,i\right)}}\right]\left[1+\frac{\gamma_{71}\left(E,i\right)r_{71}}{\left(q_{71}\left(E,i\right)\right)^2}\right]^{-1} \tag{7.70}$$

Thus, the expression of the surface electron concentration under the weak electric field quantum limit at low temperatures assumes the form

$$\bar{n}_{2\mathrm{Dw}}=\frac{g_{\mathrm{v}}}{2\pi}\left[\frac{\gamma_{71}\left(\bar{E}_{\mathrm{Fw}},0\right)}{\sqrt{p_{71}q_{71}\left(\bar{E}_{\mathrm{Fw}},0\right)}}\right]\left[1+\frac{\gamma_{71}\left(\bar{E}_{\mathrm{Fw}},0\right)r_{71}}{\left(q_{71}\left(\bar{E}_{\mathrm{Fw}},0\right)\right)^2}\right]^{-1}. \tag{7.71}$$

Using (7.71) and (1.12), the expression of the DMR in this case can be written as

$$\frac{D}{\mu} = \frac{\bar{n}_{2\mathrm{Dw}}}{|e|}\left(x_{77}\left(\bar{E}_{\mathrm{Fw}},0\right) - y_{77}\left(\bar{E}_{0\mathrm{w}4},0\right)\right). \tag{7.72}$$

where

$$x_{77}\left(\bar{E}_{\mathrm{Fw}},0\right) \equiv \frac{\omega_{84}\left(\bar{E}_{\mathrm{Fw}},0\right)}{\omega_{85}\left(\bar{E}_{\mathrm{Fw}},0\right)}, \ \omega_{84}\left(\bar{E}_{\mathrm{Fw}},0\right) \equiv \left[1 + \frac{2}{3}S_0\left(\frac{\hbar|e|^2\bar{n}_{2\mathrm{Dw}}}{\varepsilon_{\mathrm{sc}}\sqrt{2M_3}}\right)^{2/3}\right.$$

$$\times \frac{\left[1+\frac{4}{3}\alpha\bar{E}_{\mathrm{Fw}}\right]}{\gamma_{71}\left(\bar{E}_{\mathrm{Fw}},0\right)} - \left(\frac{2\alpha S_0}{9}\right)\left(q_{71}\left(\bar{E}_{\mathrm{Fw}},0\right)\right)^{-1}\left(1-\frac{M}{M_2'}\right)\left(\frac{\hbar|e|^2\bar{n}_{2\mathrm{Dw}}}{\varepsilon_{\mathrm{sc}}\sqrt{2M_3}}\right)^{2/3}$$

$$+ \bar{n}_{2\mathrm{Dw}}\left\{q_{71}\left(\bar{E}_{\mathrm{Fw}},0\right)\right\}^{-4}\left[1+\frac{\gamma_{71}\left(\bar{E}_{\mathrm{Fw}},0\right)r_{71}}{\left\{q_{71}\left(\bar{E}_{\mathrm{Fw}},0\right)\right\}^2}\right]^{-1}$$

$$\times\left\{\frac{-2}{3}\frac{S_0}{\bar{n}_{2\mathrm{Dw}}}\left(\frac{\hbar|e|^2\bar{n}_{2\mathrm{Dw}}}{\varepsilon_{\mathrm{sc}}\sqrt{2M_3}}\right)^{2/3}\left(1+\frac{4}{3}\alpha\bar{E}_{\mathrm{Fw}}\right)\right\} + \frac{8}{9}\left(\frac{\alpha S_0}{\bar{n}_{2\mathrm{Dw}}}\right)q_{71}\left(\bar{E}_{\mathrm{Fw}},0\right)$$

$$\left. r_{71}\gamma_{71}\left(\bar{E}_{\mathrm{Fw}},0\right)\left(1-\frac{M}{M_2'}\right)\left(\frac{\hbar|e|^2\bar{n}_{2\mathrm{Dw}}}{\varepsilon_{\mathrm{sc}}\sqrt{2M_3}}\right)^{2/3}\right],$$

$$\omega_{85}\left(\bar{E}_{\mathrm{Fw}},0\right) \equiv \left[\frac{\bar{n}_{2\mathrm{Dw}}}{\gamma_{71}\left(\bar{E}_{\mathrm{Fw}},0\right)}\left[1+2\alpha\bar{E}_{\mathrm{Fw}} - \frac{4}{3}\alpha S_0\left(\frac{\hbar|e|F_{\mathrm{s}}}{\sqrt{2M_3}}\right)^{2/3}\right]\right.$$

$$-\left\{\frac{\alpha\bar{n}_{2\mathrm{Dw}}}{4M_2 q_{71}\left(\bar{E}_{\mathrm{Fw}},0\right)}\left(1-\frac{M_2}{M_2'}\right)\right\} - \left(\frac{\bar{n}_{2\mathrm{Dw}}}{\left\{q_{71}\left(\bar{E}_{\mathrm{Fw}},0\right)\right\}^4}\right)$$

$$\times\left(1+\frac{\gamma_{71}\left(\bar{E}_{\mathrm{Fw}},0\right)r_{71}}{\left\{q_{71}\left(\bar{E}_{\mathrm{Fw}},0\right)\right\}^2}\right)^{-1}\left[r_{71}\left\{q_{71}\left(\bar{E}_{\mathrm{Fw}},0\right)\right\}^2\left[1+2\alpha\bar{E}_{\mathrm{Fw}}\right.\right.$$

$$\left.\left.-\frac{4}{3}\alpha S_0\left(\frac{\hbar|e|F_{\mathrm{s}}}{\sqrt{2M_3}}\right)^{2/3}\right] - \frac{\alpha\hbar^2}{M_2}\left(1-\frac{M_2}{M_2'}\right)q_{71}\left(\bar{E}_{\mathrm{Fw}},0\right)r_{71}\gamma_{71}\left(\bar{E}_{\mathrm{Fw}},0\right)\right],$$

$$y_{77}\left(\bar{E}_{0\mathrm{w}4},0\right) \equiv \left[\frac{2}{3\bar{n}_{2\mathrm{Dw}}}\bar{E}_{0\mathrm{w}4}\left(1+\alpha\bar{E}_{0\mathrm{w}4}\right)\right]\left[\omega_{86}\left(\bar{E}_{0\mathrm{w}4}\right)\right]^{-1}$$

$$\omega_{86}\left(\bar{E}_{0\mathrm{w}4}\right) \equiv \left[1+2\alpha\bar{E}_{0\mathrm{w}4} - \left\{\frac{4\alpha S_0}{3}\left(\frac{\hbar|e|^2\bar{n}_{2\mathrm{Dw}}}{\varepsilon_{\mathrm{sc}}\sqrt{2M_3}}\right)^{2/3}\right\}\right],$$

$\bar{E}_{0\mathrm{w}4}$ is the subband energy in this case and is determined by the equation

$$\bar{E}_{0\mathrm{w}4}\left(1+\alpha\bar{E}_{0\mathrm{w}4}\right) - S_0\left(\frac{\hbar|e|^2}{\varepsilon_{\mathrm{sc}}\sqrt{2M_3}}\right)^{2/3}\left(\bar{n}_{2\mathrm{Dw}}\right)^{2/3}\left(1+\frac{4}{3}\alpha\bar{E}_{0\mathrm{w}4}\right) = 0. \tag{7.73}$$

Under the strong electric field limit, the dispersion relation assumes the form

$$\gamma_{72}(E,i) = p_{72}p_x^2 + q_{72}(E)\,k_y^2 + r_{72}k_y^4, \tag{7.74}$$

where

$$\gamma_{72}(E,i) \equiv \left[E(1+\alpha E) - \frac{2\sqrt{2\alpha}}{3}\frac{\hbar\,|e|\,F_s}{\sqrt{M_3}}(S_i)^{3/2}\right],\ p_{72} \equiv \left(\frac{\hbar^2}{2M_1}\right),$$

$$q_{72}(E) \equiv \left(\frac{\hbar^2}{2M_2}\right)\left[1+\alpha E\left(1-\frac{M_2}{M_2'}\right)\right] \text{ and } r_{72} \equiv \left[\frac{\alpha\hbar^4}{4M_2M_2'}\right].$$

Comparing (7.74) with (7.64), we observe that the forms of the equations from (7.65) to (7.70) remain unchanged provided, $\gamma_{71}(E,i)$, p_{71}, $q_{71}(E,i)$, and r_{71} are replaced by the corresponding quantities $\gamma_{72}(E,i)$, p_{72}, $q_{72}(E)$, and r_{72} respectively.

The surface electron concentration in the strong electric field quantum limit in the regime of low temperatures in this case can be written as

$$\bar{n}_{2Ds} = \frac{g_v}{\pi}\left[\frac{\gamma_{72}(\bar{E}_{Fs},0)}{\sqrt{p_{72}q_{72}(\bar{E}_{Fs})}}\right]\left[1+\frac{\gamma_{72}(\bar{E}_{Fs},0)\,r_{72}}{\{q_{72}(\bar{E}_{Fs})\}^2}\right]^{-1}. \tag{7.75}$$

Thus, using (7.75) and (1.12), the expression of the DMR in this case is given by

$$\frac{D}{\mu} = \frac{\bar{n}_{2Ds}}{|e|}\left(x_{78}(\bar{E}_{Fs},0) - y_{78}(\bar{E}_{0s4},0)\right), \tag{7.76}$$

where

$$x_{78}(\bar{E}_{Fs},0) \equiv \frac{\omega_{90}(\bar{E}_{Fs},0)}{\omega_{91}(\bar{E}_{Fs},0)},$$

$$\omega_{90}(\bar{E}_{Fs},0) \equiv \left[1+\frac{\bar{n}_{2Ds}}{\gamma_{72}(\bar{E}_{Fs},0)}\frac{2}{3\bar{n}_{2Ds}}\left(\frac{\alpha}{M_3}\right)^{1/2}\hbar\,|e|\,F_s\,(S_0)^{3/2}\right.$$
$$\left. -\frac{2}{3\{q_{72}(\bar{E}_{Fs})\}^4}\left(\frac{\alpha}{M_3}\right)^{1/2}\hbar\,|e|\,F_s\,(S_0)^{3/2}\left[1+\frac{\gamma_{72}(\bar{E}_{Fs},0)}{\{q_{72}(\bar{E}_{Fs})\}^2}\right]^{-1}\right],$$

$$\omega_{91}(\bar{E}_{Fs},0) \equiv \left[\frac{\bar{n}_{2Ds}}{\gamma_{72}(\bar{E}_{Fs},0)}(1+2\alpha\bar{E}_{Fs}) - \left(\frac{\alpha\hbar^2\bar{n}_{2Ds}}{4M_2q_{72}(\bar{E}_{Fs})}\right)\left(1-\frac{M_2}{M_2'}\right)\right.$$
$$-\left(\frac{\bar{n}_{2Ds}r_{72}}{\{q_{72}(\bar{E}_{Fs},0)\}^2}\right)\left[1+\frac{\gamma_{72}(\bar{E}_{Fs},0)\,r_{72}}{\{q_{72}(\bar{E}_{Fs})\}^2}\right]^{-1}$$
$$\left.\times\left[1+2\alpha E_{F0} - \left\{\frac{\alpha\hbar^2}{M_2}\left(1-\frac{M_2}{M_2'}\right)\frac{\gamma_{72}(\bar{E}_{Fs},0)}{q_{72}(\bar{E}_{Fs})}\right\}\right]\right],$$

$$y_{78}(\bar{E}_{0s4},0) = \frac{\bar{E}_{0s4}(1+\alpha\bar{E}_{0s4})}{(1+2\alpha\bar{E}_{0s4})\,\bar{n}_{2Ds}},$$

$\bar{E}_{0s4}$ is the subband energy under the strong electric field quantum limit in this case and is given by

$$\bar{E}_{0s4}\left(1+\alpha\bar{E}_{0s4}\right)=\frac{2\sqrt{2\alpha}}{3}\left(\frac{\hbar\left|e\right|^{2}\bar{n}_{2\mathrm{Ds}}}{\varepsilon_{\mathrm{sc}}\sqrt{M_{3}}}\right)\left(S_{0}\right)^{3/2}. \qquad (7.77)$$

7.2.5 Formulation of the Einstein Relation in n-Channel Inversion Layers of Stressed III–V Materials

The use of (4.84) and (7.2) leads to the expression of the dispersion relation of the 2D electrons in n-channel inversion layers of stressed III–V materials under the low electric field limit given by

$$\left[T_{57}(E,i)\right]k_{x}^{2}+\left[T_{67}(E,i)\right]k_{y}^{2}=T_{77}(E,i), \qquad (7.78)$$

where

$$T_{57}(E,i)\equiv\left[E-\alpha_{1}+\frac{2}{3}S_{i}\left(\frac{\left|e\right|^{2}}{\varepsilon_{\mathrm{sc}}}\right)^{2/3}\left(n_{2\mathrm{Dw}}\right)^{2/3}L_{17}(E)\right],$$

$$L_{17}(E)=\left[\frac{(E-\alpha_{1})}{(E-\alpha_{3})^{2/3}\left[\bar{T}_{47}(E)\right]^{1/3}}-(E-\alpha_{3})^{1/3}\left[\bar{T}_{47}(E)\right]^{-1/3}\right],$$

$$\left[\bar{T}_{47}(E)\right]\equiv\left[\left\{\rho_{5}(E)\right\}'-\left(\frac{\rho_{5}(E)}{E-\alpha_{3}}\right)\right],$$

$$T_{67}(E,i)\equiv\left[E-\alpha_{2}+\frac{2}{3}S_{i}\left(\frac{\left|e\right|^{2}}{\varepsilon_{\mathrm{sc}}}\right)^{2/3}\left(n_{2\mathrm{Dw}}\right)^{2/3}L_{27}(E)\right],$$

$$L_{27}(E)\equiv\left[\frac{(E-\alpha_{2})}{(E-\alpha_{3})^{2/3}\left[\bar{T}_{47}(E)\right]^{1/3}}-\left(\frac{(E-\alpha_{3})^{1/3}}{\left[\bar{T}_{47}(E)\right]^{1/3}}\right)\right],$$

$$T_{77}(E,i)\equiv\left[\rho_{5}(E)-S_{i}\left(\frac{\left|e\right|^{2}}{\varepsilon_{\mathrm{sc}}}\right)^{2/3}\left(n_{2\mathrm{Dw}}\right)^{2/3}L_{37}(E)\right] \text{ and}$$

$$L_{37}(E)\equiv(E-\alpha_{3})^{1/3}\left[\bar{T}_{47}(E)\right]^{2/3}.$$

The area of the 2D surface under the weak electric field limit can be written as

$$A(E,i)=\frac{\pi T_{77}(E,i)}{\sqrt{T_{57}(E,i)\,T_{67}(E,i)}}. \qquad (7.79)$$

The subband energies ($E_{n_{\mathrm{iw8}}}$) in this case are defined by the expression

$$\rho_{5}\left(E_{n_{\mathrm{iw8}}}\right)=S_{i}\left(\frac{\left|e\right|^{2}}{\varepsilon_{\mathrm{sc}}}\right)^{2/3}\left(n_{2\mathrm{Dw}}\right)^{2/3}L_{37}\left(E_{n_{\mathrm{iw8}}}\right). \qquad (7.80)$$

The expression of the EMM in this case can be written as

$$m^*\left(E_{\mathrm{Fiw}}, i\right)=\frac{\hbar^2}{2} L_{47}(E, i)\Big|_{E=E_{\mathrm{Fiw}}}, \tag{7.81}$$

where

$$\begin{aligned} L_{47}(E, i) \equiv & \left[\frac{1}{T_{57}(E, i) T_{67}(E, i)}\right]\left[\{T_{77}(E, i)\}^{\prime}\left[T_{57}(E, i) T_{67}(E, i)\right]^{1/2}\right. \\ & -\left(\frac{T_{77}(E, i)}{2}\right)\left\{\{T_{57}(E, i)\}^{\prime}\left[\frac{T_{67}(E, i)}{T_{57}(E, i)}\right]^{1/2}\right. \\ & \left.\left.+\{T_{67}(E, i)\}^{\prime}\left[\frac{T_{57}(E, i)}{T_{67}(E, i)}\right]^{1/2}\right\}\right] . \end{aligned}$$

The total 2D density-of-states function can be expressed as

$$N_{\mathrm{2D}}(E)=\frac{g_{\mathrm{v}}}{2 \pi} \sum_{i=0}^{i_{\max}}\left\{L_{47}(E, i) H\left(E-E_{n_{\mathrm{iw8}}}\right)\right\}. \tag{7.82}$$

The surface electron concentration under the weak electric field quantum limit assumes the form

$$n_{\mathrm{2Dw}}=\frac{g_{\mathrm{v}}}{(2 \pi)}\left\{\sum_{i=0}^{i_{\max}}\left[P_{\mathrm{8w}}\left(E_{\mathrm{Fwi}}, i\right)+Q_{\mathrm{8w}}\left(E_{\mathrm{Fwi}}, i\right)\right]\right\}, \tag{7.83}$$

where

$$P_{\mathrm{8w}}\left(E_{\mathrm{Fwi}}, i\right) \equiv \frac{T_{77}\left(E_{\mathrm{Fwi}}, i\right)}{\sqrt{T_{57}\left(E_{\mathrm{Fwi}}, i\right) T_{67}\left(E_{\mathrm{Fwi}}, i\right)}} \text{ and}$$

$$Q_{\mathrm{8w}}\left(E_{\mathrm{Fiw}}, i\right) \equiv \sum_{r=1}^{s} L(r) P_{\mathrm{8w}}\left(E_{\mathrm{Fiw}}, i\right).$$

Using (7.83), the expression of the 2D surface electron concentration under the weak electric field quantum limit at low temperatures can be written as

$$\bar{n}_{\mathrm{2Dw}}=\frac{g_{\mathrm{v}}}{2 \pi} \frac{T_{77}\left(\bar{E}_{\mathrm{Fw}}, 0\right)}{\sqrt{T_{57}\left(\bar{E}_{\mathrm{Fw}}, 0\right) T_{67}\left(\bar{E}_{\mathrm{Fw}}, 0\right)}}. \tag{7.84}$$

Using (7.84) and (1.12), the DMR in this case is given by

$$\frac{D}{\mu}=\frac{\bar{n}_{\mathrm{2Dw}}}{|e|}\left(x_{79}\left(\bar{E}_{\mathrm{Fw}}, 0\right)-y_{79}\left(\bar{E}_{\mathrm{0w5}}, 0\right)\right), \tag{7.85}$$

where

$$x_{79}\left(\bar{E}_{\mathrm{Fw}},0\right) \equiv \frac{\omega_{93}\left(\bar{E}_{\mathrm{Fw}},0\right)}{\omega_{94}\left(\bar{E}_{\mathrm{Fw}},0\right)},$$

$$\omega_{93}\left(\bar{E}_{\mathrm{Fw}},0\right) \equiv \left[\frac{2\pi}{g_{\mathrm{v}}}\left[T_{57}\left(\bar{E}_{\mathrm{Fw}},0\right)T_{67}\left(\bar{E}_{\mathrm{Fw}},0\right)\right] + \left[T_{57}\left(\bar{E}_{\mathrm{Fw}},0\right)T_{67}\left(\bar{E}_{\mathrm{Fw}},0\right)\right]^{1/2}\right.$$

$$\times \frac{2S_0}{3\bar{n}_{2\mathrm{Dw}}}\left(\frac{|e|^2\bar{n}_{2\mathrm{Dw}}}{\varepsilon_{\mathrm{sc}}}\right)L_{37}\left(\bar{E}_{\mathrm{Fw}}\right) + T_{77}\left(\bar{E}_{\mathrm{Fw}},0\right)\left(\frac{2S_0}{9\bar{n}_{2\mathrm{Dw}}}\right)$$

$$\times \left[\frac{T_{67}\left(\bar{E}_{\mathrm{Fw}},0\right)}{T_{57}\left(\bar{E}_{\mathrm{Fw}},0\right)}\right]^{1/2}\left(\frac{|e|^2\bar{n}_{2\mathrm{Dw}}}{\varepsilon_{\mathrm{sc}}}\right)^{2/3}L_{17}\left(\bar{E}_{\mathrm{Fw}}\right) + T_{77}\left(\bar{E}_{\mathrm{Fw}},0\right)\left(\frac{2S_0}{3\bar{n}_{2\mathrm{Dw}}}\right)$$

$$\left.\times \left(\frac{T_{57}\left(\bar{E}_{\mathrm{Fw}},0\right)}{T_{67}\left(\bar{E}_{\mathrm{Fw}},0\right)}\right)^{1/2}\left(\frac{|e|^2\bar{n}_{2\mathrm{Dw}}}{\varepsilon_{\mathrm{sc}}}\right)^{2/3}L_{17}\left(\bar{E}_{\mathrm{Fw}}\right)\right],$$

$$\omega_{94}\left(\bar{E}_{\mathrm{Fw}},0\right) \equiv \left[\left[T_{57}\left(\bar{E}_{\mathrm{Fw}},0\right)T_{67}\left(\bar{E}_{\mathrm{Fw}},0\right)\right]^{1/2}\left[\left\{\rho_5\left(\bar{E}_{\mathrm{Fw}},0\right)\right\}'\right.\right.$$

$$\left. - S_0\left(\frac{|e|^2 n_{2\mathrm{Dw}}}{\varepsilon_{\mathrm{sc}}}\right)^{2/3}\left\{L_{37}\left(\bar{E}_{\mathrm{Fw}}\right)\right\}'\right]$$

$$-\left(\frac{T_{77}\left(\bar{E}_{\mathrm{Fw}},0\right)}{2}\right)\left\{\left[\frac{T_{67}\left(\bar{E}_{\mathrm{Fw}},0\right)}{T_{57}\left(\bar{E}_{\mathrm{Fw}},0\right)}\right]^{1/2}\left[1+\frac{2}{3}S_0\left(\frac{|e|^2\bar{n}_{2\mathrm{Dw}}}{\varepsilon_{\mathrm{sc}}}\right)^{2/3}\right.\right.$$

$$\left.\times\left\{L_{17}\left(\bar{E}_{\mathrm{Fw}}\right)\right\}'\right] + \left[\frac{T_{57}\left(\bar{E}_{\mathrm{Fw}},0\right)}{T_{67}\left(\bar{E}_{\mathrm{Fw}},0\right)}\right]^{1/2}\left[1+\frac{2}{3}S_0\left(\frac{|e|^2\bar{n}_{2\mathrm{Dw}}}{\varepsilon_{\mathrm{sc}}}\right)^{2/3}\right.$$

$$\left.\left.\left.\times\left\{L_{27}\left(\bar{E}_{\mathrm{Fw}}\right)\right\}'\right]\right\}\right],$$

$$y_{79}\left(\bar{E}_{0\mathrm{w}5},0\right) \equiv \frac{2}{3}\frac{S_0}{\bar{n}_{2\mathrm{Dw}}}\left(\frac{|e|^2\bar{n}_{2\mathrm{Dw}}}{\varepsilon_{\mathrm{sc}}}\right)^{2/3}L_{37}\left(\bar{E}_{0\mathrm{w}5}\right)$$

$$\times\left[\left\{\rho_5\left(\bar{E}_{0\mathrm{w}5},0\right)\right\}' - S_0\left(\frac{|e|^2\bar{n}_{2\mathrm{Dw}}}{\varepsilon_{\mathrm{sc}}}\right)^{2/3}\left\{L_{37}\left(\bar{E}_{0\mathrm{w}5}\right)\right\}'\right]^{-1},$$

$\bar{E}_{0\mathrm{w}5}$ is the subband energy in this case and is given by

$$\rho_5\left(\bar{E}_{0\mathrm{w}5},0\right) - S_0\left(\frac{|e|^2\bar{n}_{2\mathrm{Dw}}}{\varepsilon_{\mathrm{sc}}}\right)^{2/3}\left\{L_{37}\left(\bar{E}_{0\mathrm{w}5},0\right)\right\} = 0. \tag{7.86}$$

The use of (4.84) and (7.2) leads to a simplified dispersion relation of the 2D electrons in n-channel inversion layers of stressed III–V materials under the high electric field limit given by

$$\left[T_{117}\left(E,i\right)\right]k_x^2 + \left[T_{127}\left(E,i\right)\right]k_y^2 = T_{137}\left(E,i\right). \tag{7.87}$$

where

$$T_{117}(E,i) \equiv \left[E - \alpha_1 + \frac{4}{3}|e|\,F_{\mathrm{s}}\,(S_i)^{2/3}\,a_{77}(E)\right],$$

$$a_{77}(E) \equiv \frac{1}{2}\left[\frac{1}{\sqrt{(E-\alpha_3)}\,[T_{97}(E)]} - \left\{\frac{(E-\alpha_1)^2}{\sqrt{T_{97}(E)}\,(E-\alpha_3)^{3/2}}\right\}\right],$$

$$[T_{97}(E)] \equiv \left[\frac{\{\rho_5(E)\}''}{2} - \left(\frac{(\rho_5(E))'}{E-\alpha_3}\right) + \left(\frac{\rho_5(E)}{(E-\alpha_3)^2}\right)\right],$$

$$T_{127}(E,i) \equiv \left[E - \alpha_2 + \frac{4}{3}|e|\,F_{\mathrm{s}}\,(S_i)^{2/3}\,a_{87}(E)\right],$$

$$a_{87}(E) \equiv \frac{1}{2}\left[\frac{1}{\sqrt{(E-a_3)}\,T_{97}(E)} - \left(\frac{(E-\alpha_2)}{\sqrt{T_{97}(E)}\,(E-\alpha_3)^{3/2}}\right)\right],$$

$$T_{137}(E,i) \equiv \left[\rho_5(E) - (S_i)^{3/2}\,|e|F_{\mathrm{s}}a_{97}(E)\right] \text{ and}$$

$$a_{97}(E) \equiv \frac{4}{3}\sqrt{T_{97}(E)\,(E-\alpha_3)}.$$

The area of the 2D surface in this case is given by

$$A(E,i) = \frac{\pi T_{137}(E,i)}{\sqrt{T_{117}(E,i)\,T_{127}(E,i)}}. \tag{7.88}$$

The subband energies ($E_{n_{\mathrm{is8}}}$) in this case can be written as

$$T_{137}(E_{n_{\mathrm{is8}}}, i) = 0. \tag{7.89}$$

The EMM in this case assumes the form

$$m^*(E_{\mathrm{Fis}}, i) = \frac{\hbar^2}{2}\,T_{147}(E,i)|_{E=E_{\mathrm{Fis}}}. \tag{7.90}$$

where

$$T_{147}(E,i) \equiv \left[\frac{1}{T_{117}(E,i)\,T_{127}(E,i)}\right]\left[\{T_{137}(E,i)\}'\,[T_{117}(E,i)\,T_{127}(E,i)]^{1/2} - \left(\frac{T_{137}(E,i)}{2}\right)\left\{\{T_{117}(E,i)\}'\left[\frac{T_{127}(E,i)}{T_{117}(E,i)}\right]^{1/2} + \{T_{127}(E,i)\}'\left[\frac{T_{117}(E,i)}{T_{127}(E,i)}\right]^{1/2}\right\}\right].$$

The expression of the total 2D density-of-states function is given by

$$N_{\mathrm{2D}_i}(E) = \frac{g_{\mathrm{v}}}{2\pi}\sum_{i=0}^{i_{\max}}\{T_{147}(E,i)\,H(E-E_{n_{\mathrm{is8}}})\}. \tag{7.91}$$

The surface electron concentration in the strong electric field limit can be expressed as

$$n_{2\mathrm{Ds}} = \frac{g_{\mathrm{v}}}{(2\pi)} \left\{ \sum_{i=0}^{i_{\max}} \left[P_{9\mathrm{s}}(E_{\mathrm{Fis}}, i) + Q_{9\mathrm{s}}(E_{\mathrm{Fis}}, i) \right] \right\}, \tag{7.92}$$

where

$$P_{9\mathrm{s}}(E_{\mathrm{Fis}}, i) \equiv \frac{T_{137}(E_{\mathrm{Fis}}, i)}{\sqrt{T_{117}(E_{\mathrm{Fis}}, i)\, T_{127}(E_{\mathrm{Fis}}, i)}} \text{ and}$$

$$Q_{9\mathrm{s}}(E_{\mathrm{Fis}}, i) \equiv \sum_{r=1}^{s} L(r)\, P_{9\mathrm{s}}(E_{\mathrm{Fis}}, i).$$

Using (7.92), the expression of the 2D surface electron concentration under the strong electric field quantum limit can be written as

$$\bar{n}_{2\mathrm{Ds}} = \frac{g_{\mathrm{v}}}{2\pi} \left(\frac{T_{137}(\bar{E}_{\mathrm{Fs}}, 0)}{\sqrt{T_{117}(\bar{E}_{\mathrm{Fs}}, 0)\, T_{127}(\bar{E}_{\mathrm{Fs}}, 0)}} \right). \tag{7.93}$$

Using (7.93) and (1.12), the DMR in this case is given by

$$\frac{D}{\mu} = \frac{\bar{n}_{2\mathrm{Ds}}}{|e|} \left(x_{710}(\bar{E}_{\mathrm{Fs}}, 0) - y_{710}(\bar{E}_{0\mathrm{s}5}, 0) \right), \tag{7.94}$$

where

$$x_{710}(\bar{E}_{\mathrm{Fs}}, 0) \equiv \frac{\omega_{98}(\bar{E}_{\mathrm{Fs}}, 0)}{\omega_{99}(\bar{E}_{\mathrm{Fs}}, 0)},$$

$$\omega_{98}(\bar{E}_{\mathrm{Fs}}, 0) = \left[\frac{2\{T_{137}(\bar{E}_{\mathrm{Fs}}, 0)\}^2}{\bar{n}_{2\mathrm{Ds}}} + \frac{4\pi(\bar{n}_{2\mathrm{Ds}})^2}{(g_{\mathrm{v}})^2} \left[\frac{4}{3} \frac{|e|\, F_{\mathrm{s}}}{\bar{n}_{2\mathrm{Ds}}} (S_0)^{3/2} a_{77}(\bar{E}_{\mathrm{Fs}}, 0) \right.\right.$$
$$\left. \times T_{127}(\bar{E}_{\mathrm{Fs}}, 0) + \frac{4}{3} \frac{|e|^2 F_{\mathrm{s}}}{\varepsilon_{\mathrm{SC}}} (S_0)^{3/2} a_{87}(\bar{E}_{\mathrm{Fs}}, 0)\, T_{117}(\bar{E}_{\mathrm{Fs}}, 0) \right]$$
$$\left. +2T_{137}(\bar{E}_{\mathrm{Fs}}, 0) \frac{a_{97}(\bar{E}_{\mathrm{Fs}})\, |e|}{\bar{n}_{2\mathrm{Ds}}} F_{\mathrm{s}} (S_0)^{3/2} \right],$$

$$\omega_{99}(\bar{E}_{\mathrm{Fs}}, 0) \equiv \left[2T_{137}(\bar{E}_{\mathrm{Fs}}, 0) \left[\{T_{47}(\bar{E}_{\mathrm{Fs}}, 0)\}' - \{a_{97}(\bar{E}_{\mathrm{Fs}}, 0)\}' |e| F_{\mathrm{s}} (S_0)^{3/2} \right] \right.$$
$$- \left[\frac{4\pi^2 \bar{n}_{2\mathrm{Ds}}}{(g_{\mathrm{v}})^2} \left[T_{127}(\bar{E}_{\mathrm{Fs}}, 0) \left[1 + \frac{4}{3} |e|\, F_{\mathrm{s}} (S_0)^{3/2} \{a_{77}(\bar{E}_{\mathrm{Fs}}, 0)\}' \right] \right.\right.$$
$$\left.\left.\left. + T_{117}(\bar{E}_{\mathrm{Fs}}, 0) \left[1 + \frac{4}{3} |e|\, F_{\mathrm{s}} (S_0)^{3/2} \{a_{87}(\bar{E}_{\mathrm{Fs}}, 0)\}' \right] \right] \right] \right]$$

$$y_{710}\left(\bar{E}_{0s5},0\right) \equiv \left[\frac{T_{47}\left(\bar{E}_{0s5},0\right)}{\bar{n}_{2Ds}}\right]\left[\left\{\rho_5\left(\bar{E}_{0s5}\right)\right\}' - \frac{\rho_5\left(\bar{E}_{0s5}\right)\left\{a_{97}\left(\bar{E}_{0s5},0\right)\right\}'}{a_{97}\left(\bar{E}_{0s5},0\right)}\right]^{-1},$$

$\bar{E}_{0s5}$ is the subband energy in this case and is given by

$$\left\{\rho_5\left(\bar{E}_{0s5}\right) - \left[\frac{|e|\,\bar{n}_{2Ds}}{\varepsilon_{sc}}\left(S_0\right)^{2/3}\left\{a_{97}\left(\bar{E}_{0s5},0\right)\right\}\right]\right\} = 0. \quad (7.95)$$

7.3 Result and Discussions

Using (7.10) and (7.11) and the spectrum constants given in Table 2.1 of Chap. 2, the plot for the normalized DMR for n-channel inversion layers of Cd_3As_2 under weak electric field limit as a function of surface electric field has been shown in curve (a) of Fig. 7.1. Curve (b) corresponds to $\delta = 0$ and curve (c) exhibits the dependency of the DMR on the surface electric field in accordance with the three-band model of Kane. Plots (d) and (e) correspond to the two-band model of Kane and that of parabolic energy bands respectively. By comparing the curves (a) and (b) of Fig. 7.1, one can assess the influence of the crystal field splitting of the DMR in inversion layers of tetragonal compounds. Figure 7.2 represents all cases of Fig. 7.1 under high electric field limit and has been obtained by using (7.17) and (7.18). It appears from Fig. 7.2 that, the DMR increases with increasing surface electric field and the numerical values of the DMR in the high electric field are much

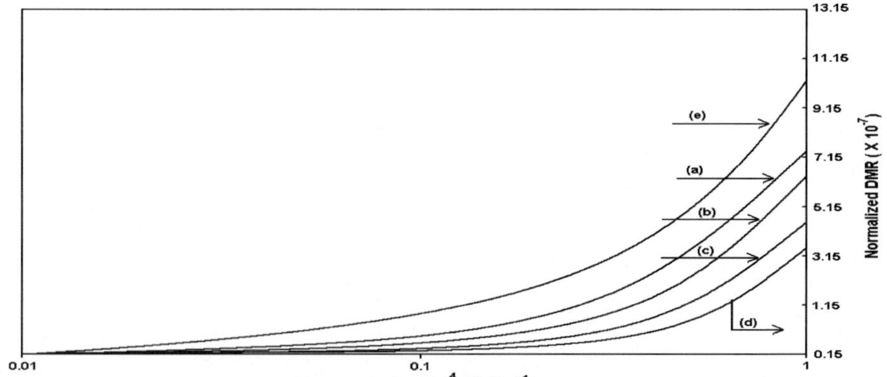

Fig. 7.1. The plot of the normalized DMR in the n-channel inversion layers of Cd_3As_2 under weak electric field quantum limit as a function of surface electric field in accordance with (**a**) the generalized band model, (**b**) $\delta = 0$, (**c**) the three band model of Kane, (**d**) the two band model of Kane and (**e**) the parabolic energy bands

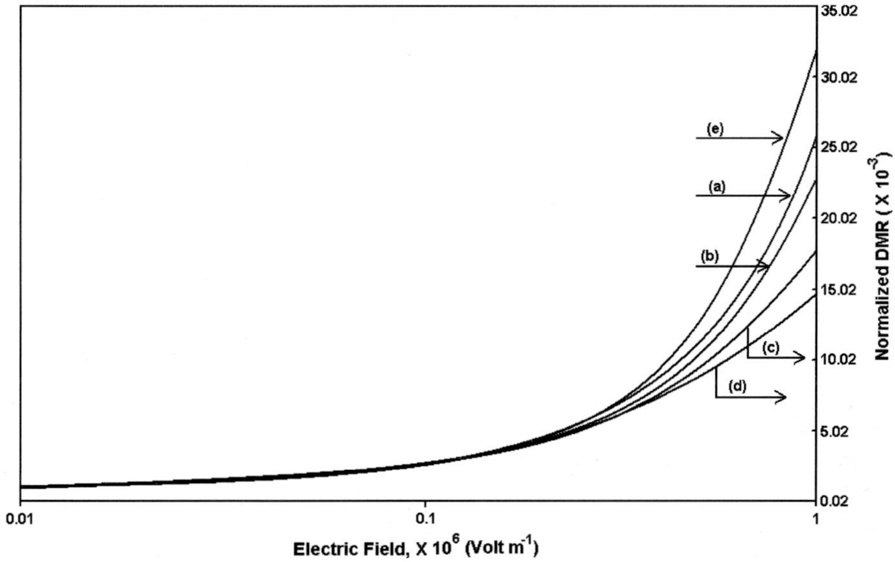

Fig. 7.2. The plot of the normalized DMR in the n-channel inversion layers of Cd_3As_2 under strong electric field quantum limit as a function of surface electric field in accordance with (**a**) the generalized band model, (**b**) $\delta = 0$, (**c**) the three band model of Kane, (**d**) the two band model of Kane and (**e**) the parabolic energy bands

greater compared to those corresponding to low electric field limit. In Figs. 7.3 and 7.4, the DMR for n-channel inversion layers of $CdGeAs_2$ has been drawn as a function of surface electric field for weak and strong electric field limits respectively for all the cases of Fig. 7.1.

The trend of variation of the DMR for n-channel inversion layers of $CdGeAs_2$ is more or less the same with different numerical magnitudes compared with n-channel inversion layers of Cd_3As_2 for both the limits.

Using ((7.22), (7.23)), ((7.29), (7.30)), ((7.38), (7.39)), ((7.52), (7.53)), and ((7.43 (for $i = 0$)), (7.44)), in Figs. 7.5 and 7.6 , the normalized DMR in n-channel inversion layers of InAs has been drawn for weak and strong electric field quantum limits respectively as function of surface electric field in accordance with (a) the three band model of Kane, (b) the two band model of Kane, and (c) the parabolic energy bands. In Figs. 7.7 and 7.8, all cases of Figs. 7.5 and 7.6 have been drawn as a function of surface electric field for n channel inversion layers of InSb for both the limits. In Figs. 7.9 and 7.10, the normalized DMR in n-channel inversion layers of $Hg_{1-x}Cd_xTe$ has been drawn for both the weak and strong electric field limits as functions of surface electric field for all the cases of Fig. 7.5. In Figs. 7.11 and 7.12, the normalized DMR in n-channel inversion layers of $In_{1-x}Ga_xAs_yP_{1-y}$ lattice matched to InP has been drawn for both the limits as a function of surface electric field for all the cases of Fig. 7.5. It appears from Figs.from 7.5 to 7.12 that the DMR for

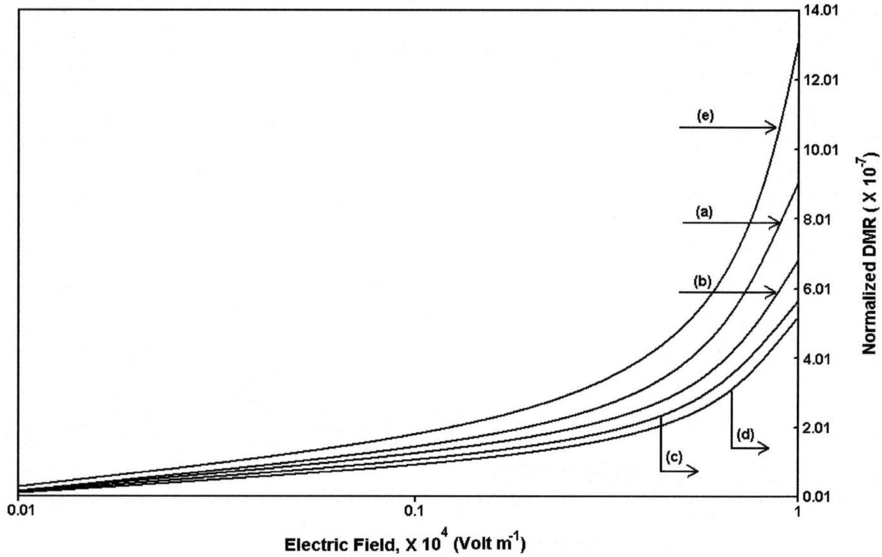

Fig. 7.3. The plot of the normalized DMR in the n-channel inversion layers of $CdGeAs_2$ under weak electric field quantum limit as a function of surface electric field in accordance with (**a**) the generalized band model, (**b**) $\delta = 0$, (**c**) the three band model of Kane, (**d**) the two band model of Kane and (**e**) the parabolic energy bands

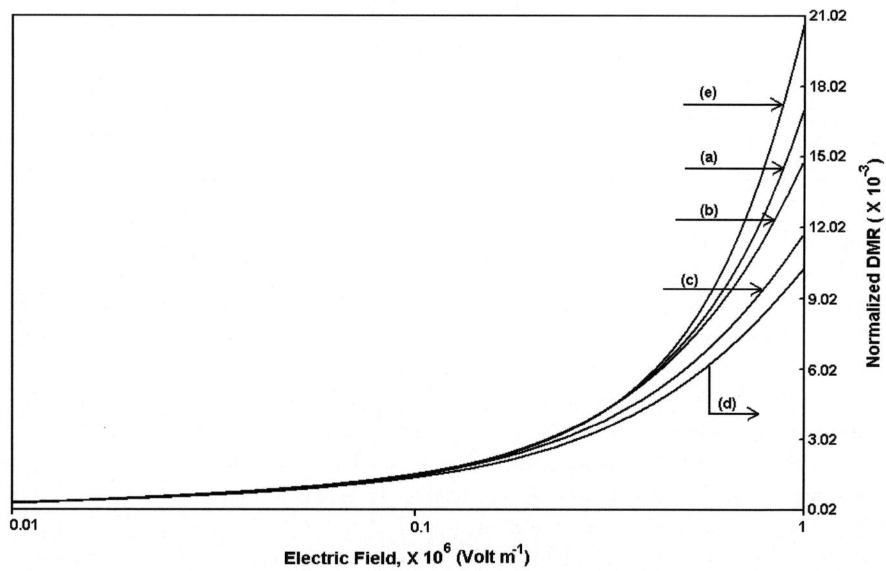

Fig. 7.4. The plot of the normalized DMR in the n-channel inversion layers of $CdGeAs_2$ under strong electric field quantum limit as a function of surface electric field in accordance with (**a**) the generalized band model, (**b**) $\delta = 0$, (**c**) the three band model of Kane, (**d**) the two band model of Kane and (**e**) the parabolic energy bands

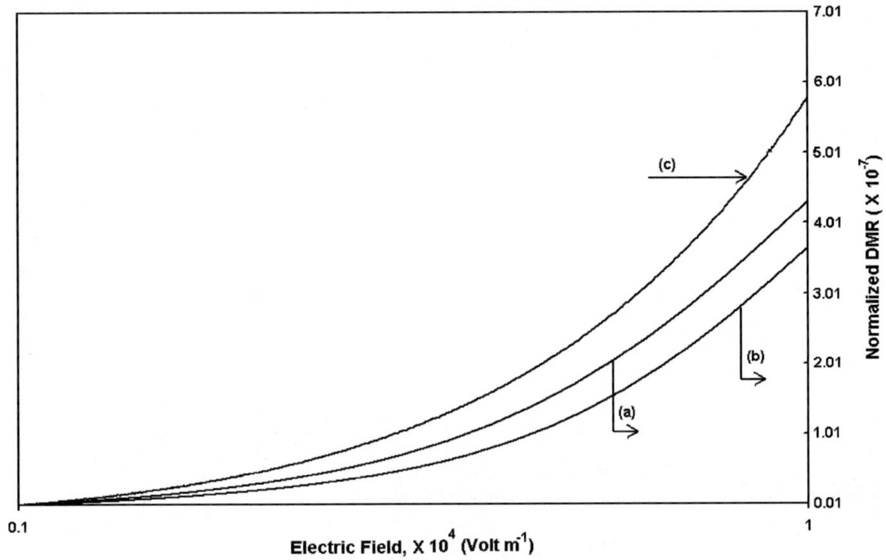

Fig. 7.5. The plot of the normalized DMR in the n-channel inversion layers of InAs under weak electric field quantum limit as a function of surface electric field in accordance with (**a**) the three band model of Kane, (**b**) the two band model of Kane and (**c**) the parabolic energy bands

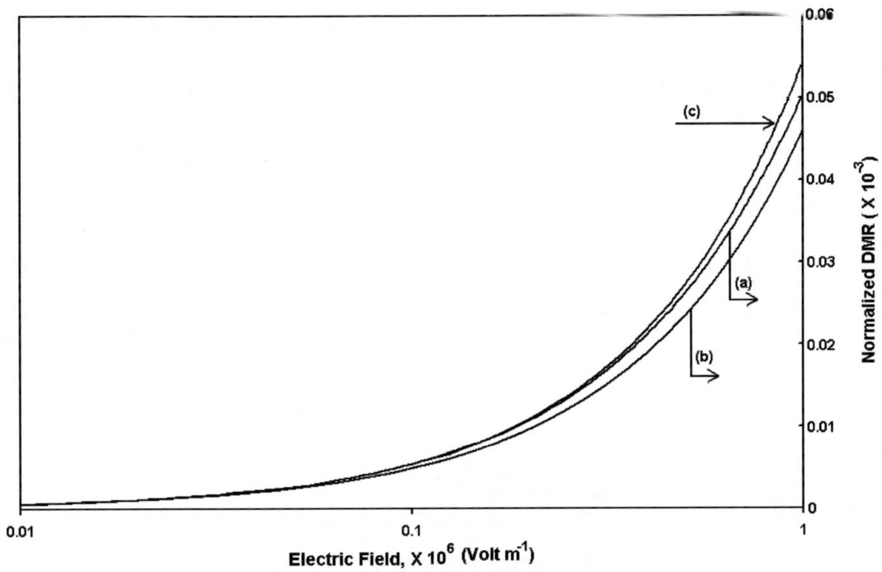

Fig. 7.6. The plot of the normalized DMR in the n-channel inversion layers of InAs under strong electric field quantum limit as a function of surface electric field in accordance with (**a**) the three band model of Kane, (**b**) the two band model of Kane and (**c**) the parabolic energy bands

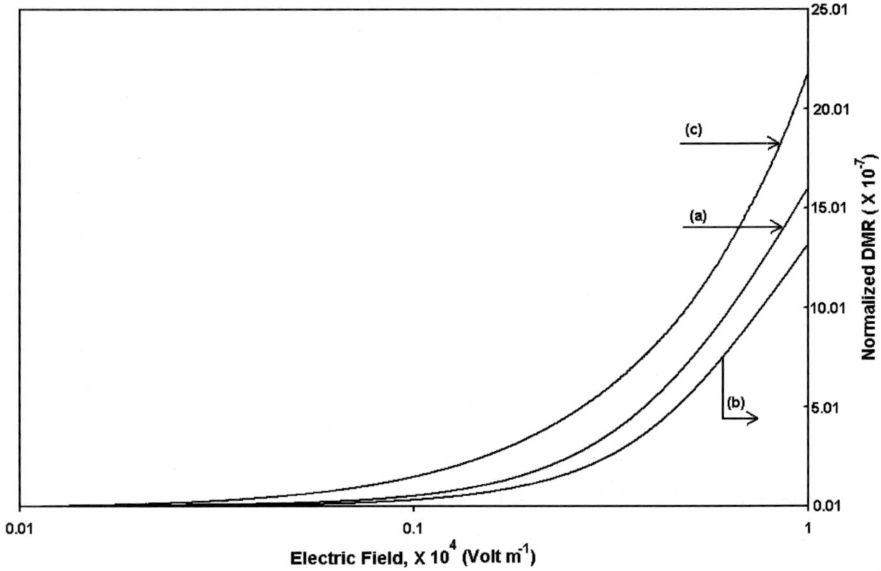

Fig. 7.7. The plot of the normalized DMR in the n-channel inversion layers of InSb under weak electric field quantum limit as a function of surface electric field in accordance with (**a**) the three band model of Kane, (**b**) the two band model of Kane and (**c**) the parabolic energy bands

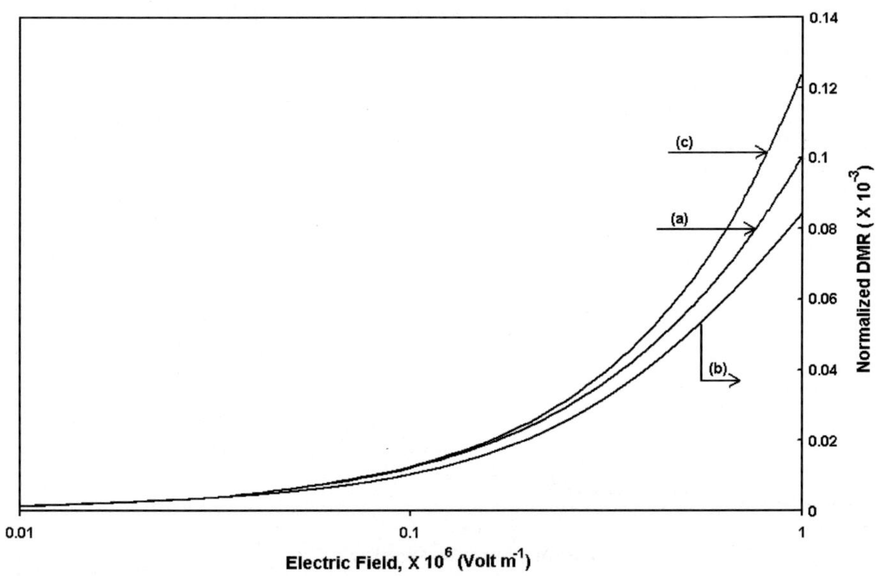

Fig. 7.8. The plot of the normalized DMR in the n-channel inversion layers of InSb under strong electric field quantum limit as a function of surface electric field in accordance with (**a**) the three band model of Kane, (**b**) the two band model of Kane and (**c**) the parabolic energy bands

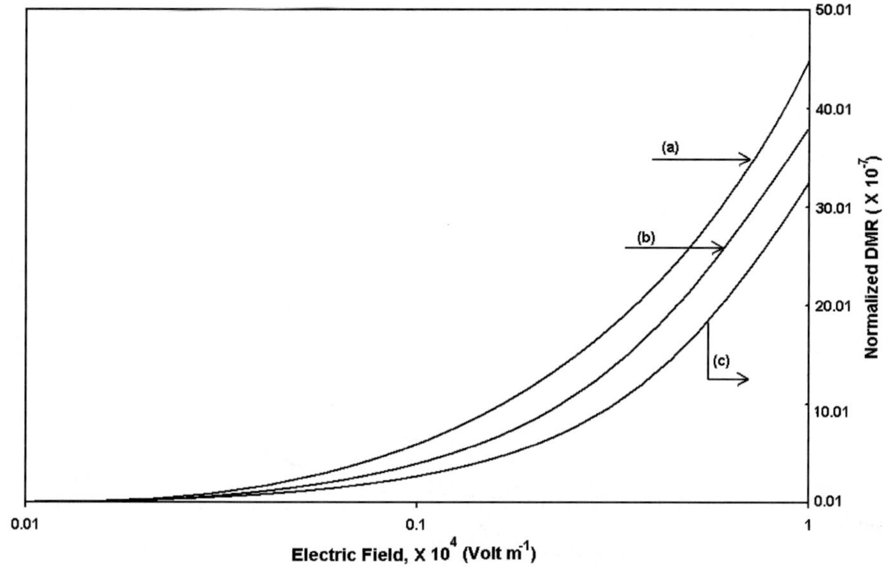

Fig. 7.9. The plot of the normalized DMR in the n-channel inversion layers of $Hg_{1-x}Cd_xTe$ under weak electric field quantum limit as a function of surface electric field in accordance with (**a**) the three band model of Kane, (**b**) the two band model of Kane and (**c**) the parabolic energy bands

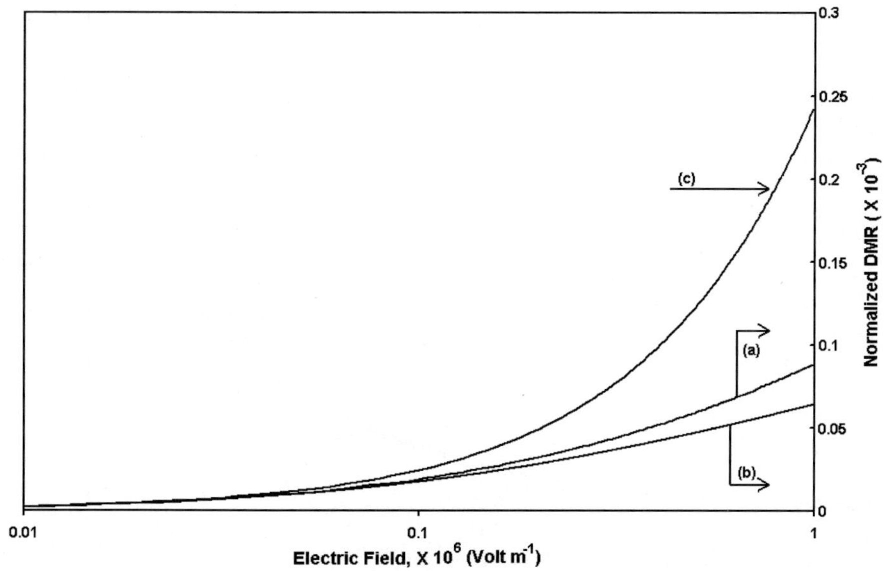

Fig. 7.10. The plot of the normalized DMR in the n-channel inversion layers of $Hg_{1-x}Cd_xTe$ under strong electric field quantum limit as a function of surface electric field in accordance with (**a**) the three band model of Kane, (**b**) the two band model of Kane and (**c**) the parabolic energy bands

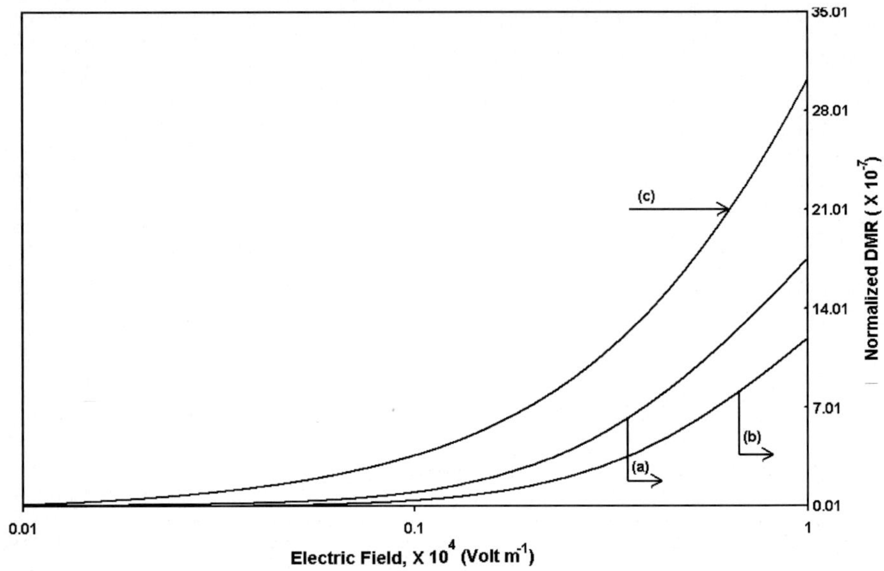

Fig. 7.11. The plot of the normalized DMR in the n-channel inversion layers of $In_{1-x}Ga_xAs_yP_{1-y}$ lattice matched to InP under weak electric field quantum limit as a function of surface electric field in accordance with (**a**) the three band model of Kane, (**b**) the two band model of Kane and (**c**) the parabolic energy bands

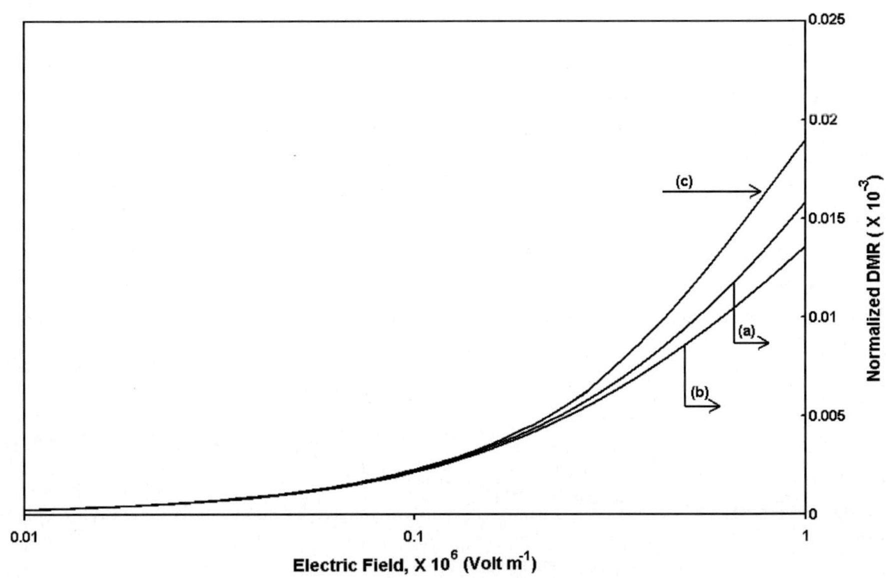

Fig. 7.12. The plot of the normalized DMR in the n-channel inversion layers of $In_{1-x}Ga_xAs_yP_{1-y}$ lattice matched to InP under strong electric field quantum limit as a function of surface electric field in accordance with (**a**) the three band model of Kane, (**b**) the two band model of Kane and (**c**) the parabolic energy bands

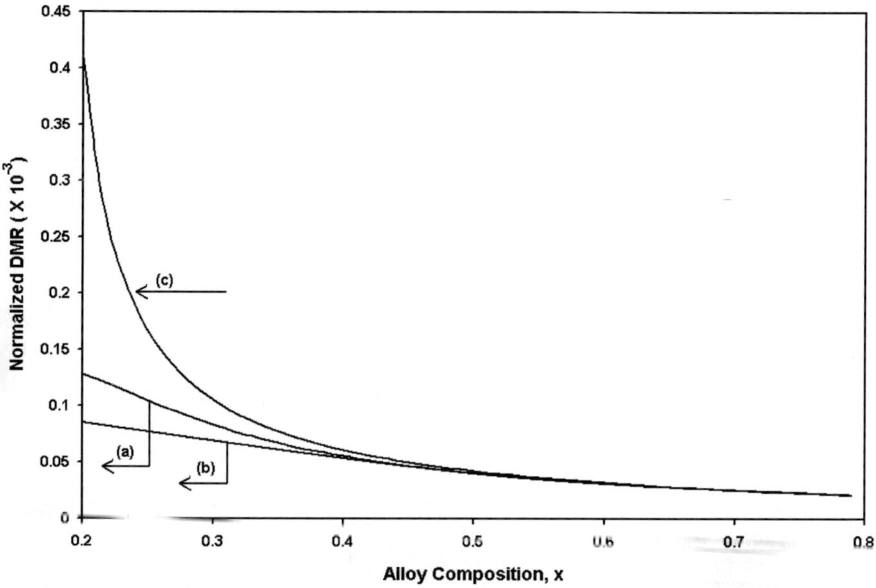

Fig. 7.13. The plot of the normalized DMR in the n-channel inversion layers of $Hg_{1-x}Cd_xTe$ under strong electric field quantum limit as a function of alloy composition in accordance with (**a**) the three band model of Kane, (**b**) the two band model of Kane and (**c**) the parabolic energy bands

n-channel inversion layers of InAs, InSb, $Hg_{1-x}Cd_x$, and $In_{1-x}Ga_xAs_yP_{1-y}$ lattice matched to InP increases with increase in surface electric field for both weak and strong electric field limits with different numerical values and the influence of the energy band constants can also be assessed from the said figures. In Figs. 7.13 and 7.14 the normalized DMR for n-channel inversion layers of $Hg_{1-x}Cd_x$ and $In_{1-x}Ga_xAs_yP_{1-y}$ lattice matched to InP has been drawn as a function of alloy composition under strong electric field limit in accordance with the three and two band models of Kane together with parabolic energy bands respectively. It appears from Figs. 7.13 and 7.14 that the DMR decreases with increasing alloy composition although the rate of decrease is determined by the respective energy band constants of the ternary and quaternary materials.

Using (7.61) and (7.62), Fig. 7.15 exhibits the plot of the DMR in p-channel inversion layers of CdS as a function of the surface electric field in accordance with (a) $\bar{\lambda}_0 \neq 0$ and (b) $\bar{\lambda}_0 = 0$. The presence of the crystal field splitting enhances the numerical values of the DMR for relatively large values of the surface electric field. Using ((7.71), (7.72)) and ((7.75), (7.76)), Figs. 7.16 and 7.17 exhibit the plots of the normalized DMR in the n-channel inversion layers of PbTe, PbSnTe, and $Pb_{1-x}Sn_xSe$ as a function of surface electric field for weak and strong electric field limits respectively. It appears

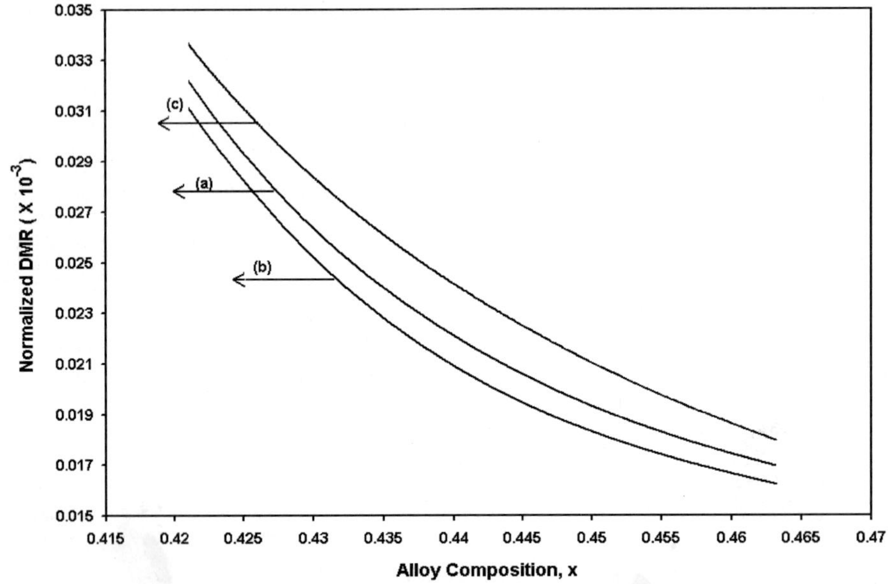

Fig. 7.14. The plot of the normalized DMR in the n-channel inversion layers of $In_{1-x}Ga_xAs_yP_{1-y}$ lattice matched to InP under strong electric field quantum limit as a function of alloy composition in accordance with (**a**) the three band model of Kane, (**b**) the two band model of Kane and (**c**) the parabolic energy bands

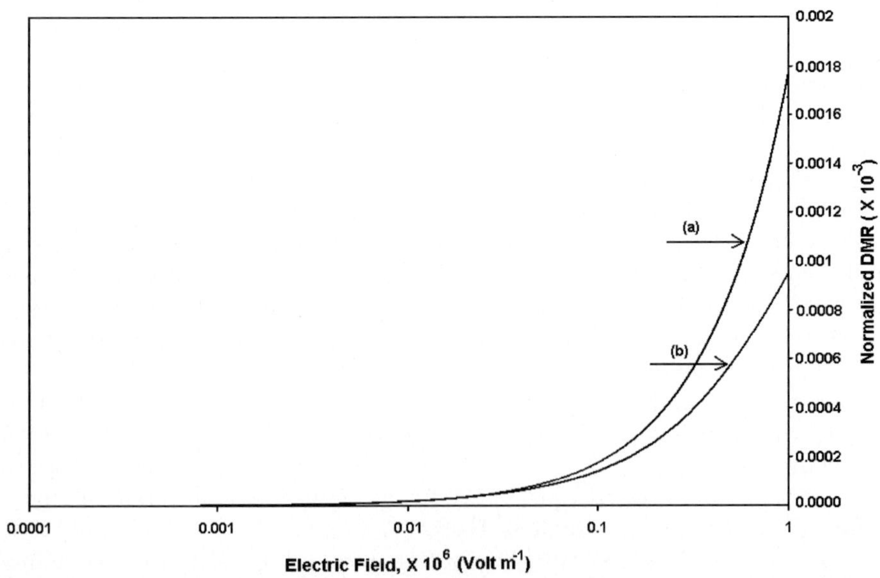

Fig. 7.15. The plot of the normalized DMR in the p-channel inversion layers of CdS as function of surface electric field in accordance with (**a**) $\bar{\lambda}_0 \neq 0$ and (**b**) $\bar{\lambda}_0 = 0$

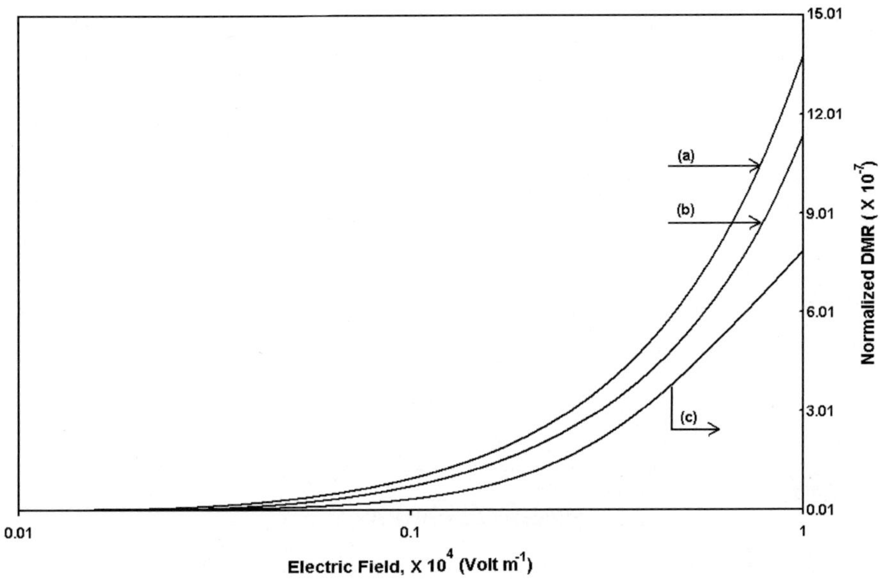

Fig. 7.16. The plot of the normalized DMR in n-channel inversion layers of (**a**) PbTe, (**b**) PbSnTe and (**c**) $Pb_{1-x}Sn_xSe$ under weak electric field quantum limit as a function of surface electric field

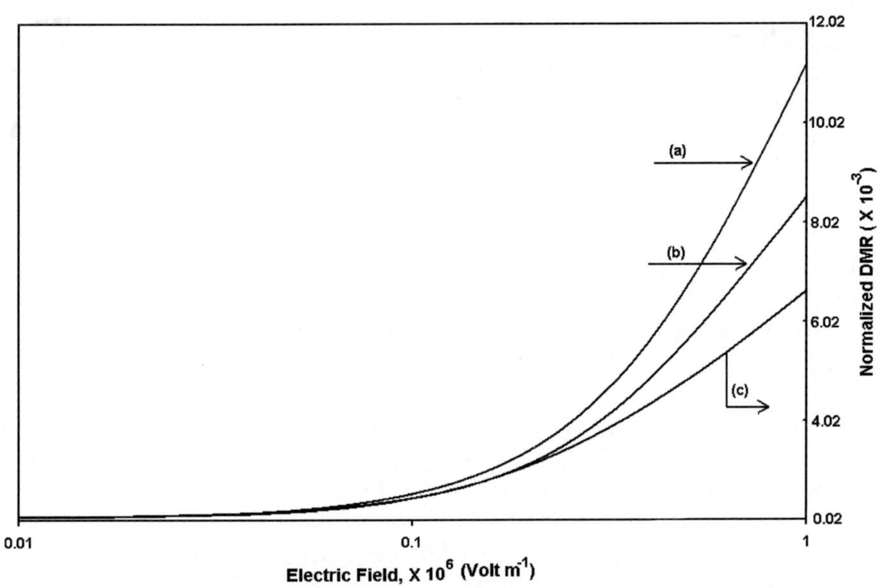

Fig. 7.17. The plot of the normalized DMR in n-channel inversion layers of (**a**) PbTe, (**b**) PbSnTe and (**c**) $Pb_{1-x}Sn_xSe$ under strong electric field quantum limit as a function of surface electric field

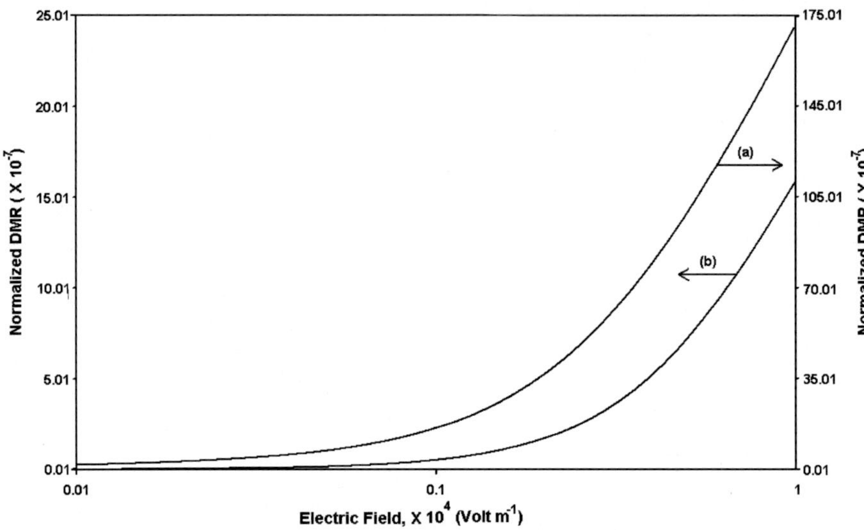

Fig. 7.18. The plot of the normalized DMR in n-channel inversion layers of stressed InSb under weak electric field quantum limit as a function of surface electric field in which the curve (**a**) is in the presence of stress and curve (**b**) is under absence of stress

that the DMR increases with increasing surface electric field with a diverging nature for relatively large values of the electric field. The numerical values of the DMR are greatest for n-channel inversion layers of PbTe and least for the corresponding $Pb_{1-x}Sn_xSe$.

Following ((7.84), (7.85)) and ((7.93), (7.94)), Figs. 7.18 and 7.19 exhibit the DMR in n-channel inversion layers of stressed n-InSb for weak and strong electric field limits respectively as a function of surface electric field, in which curve (a) is valid in the presence of stress whereas curve (b) shows the stress free condition. It appears from the said figures that the DMR exhibits the increasing dependency with increasing surface electric field and the stress enhances the value of the DMR for relatively large value of the electric field for both the limits.

In this chapter, we have investigated the Einstein relation in n-channel inversion layers of tetragonal materials for both weak and strong electric field limits on the basis of the generalized electron energy spectrum given by (2.2) of Chap. 2. The results for the n-channel inversion layers of III–V, ternary and quaternary materials whose bulk electrons obey the three and two band models of Kane together with parabolic energy bands form a special case of our generalized analysis. The Einstein relation for p-channel inversion layers of II–VI has been studied on the basis of Hopfield model for all values of surface electric field. The DMR has been investigated in n-channel inversion layers of IV–VI and stressed materials on the basis of the models of Cohen for both weak and strong electric field limits.

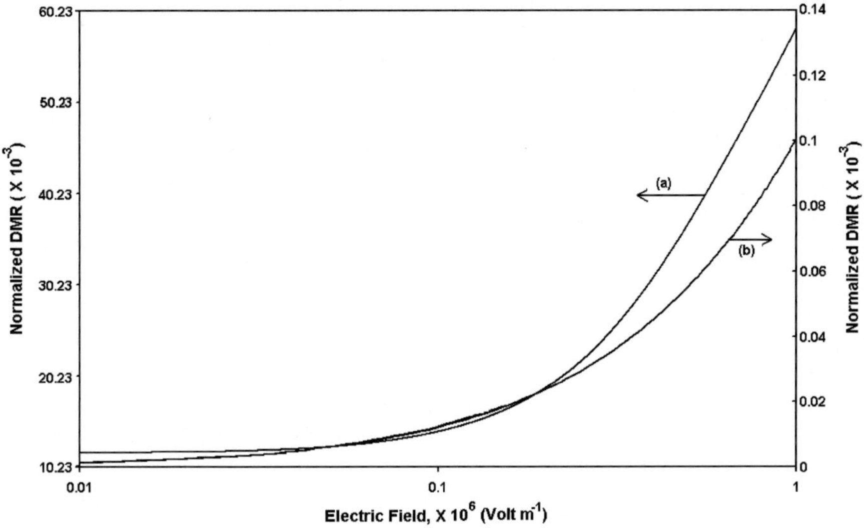

Fig. 7.19. The plot of the normalized DMR in n-channel inversion layers of stressed InSb under strong electric field quantum limit as a function of surface electric field in which the curve (**a**) is in the presence of stress and curve (**b**) is under absence of stress

It may be noted that if the direction of application of the surface electric field applied perpendicular to the surface be taken as either k_x or k_y and not as k_z as assumed in the present work, the DMR would be analytically different for both the limits. Nevertheless, the arbitrary choice of the direction normal to the surface would not result in a change of the basic qualitative feature of the DMR in n-channel inversion layers of semiconductors. The approximation of the potential well at the surface by a triangular well introduces some errors, as for instance the omission of the free charge contribution to the potential. This kind of approach is reasonable, if there are only few charge carriers in the inversion layer, but is responsible for an overestimation of the splitting when the inversion carrier density exceeds that of the depletion layer. It has been observed that the maximum error due to the triangular potential well is tolerable in the practical sense because for actual calculations, one needs a self-consistent solution which is a formidable problem, for the present generalized systems due to the non availability of the proper analytical techniques, without exhibiting a widely different qualitative behavior [1]. The second assumption of electric quantum limit in the numerical calculation is valid in the range of low temperatures, where the quantum effects become prominent. The errors which are being introduced for these assumptions are found not to be serious enough at low temperatures [4]. Thus, whenever the condition of the electric quantum limit is applied, the temperature is assumed to be low enough so that the assumption becomes well grounded because at low temperature, one can assume that nearly all electrons are at the lowest electric subband [4]. Equation (1.15) provides an experimental check of the DMR and also a technique for probing the band structures of the materials. We wish to note that many body effects, hot electron effects, the formation of band tails, arbitrary orientation of the direction of the electric quantization and the effects of surface of states have been neglected in our simplified theoretical formalism due to the lack of availability in the literature of the proper analytical techniques for including them for the generalized systems as considered in this chapter. Our simplified approach will be useful for the purpose of comparison, when, the methods of tackling the aforementioned formidable problems for the present generalized system appear. The inclusion of the said effects would certainly increase the accuracy of our results, although our suggestion for the experimental determination of the 2D DMR is independent of the inclusion of the said effects and the qualitative features of the DMR discussed in this chapter would not change in the presence of the aforementioned influences.

For the purpose of condensed presentation, the 2D electron statistics and the 2D Einstein relation for the inversion layers at the electric quantum limits of the respective materials have been presented in Table 7.1.

7.4 Open Research Problems

R.7.1 Investigate the Einstein relation in the presence of an arbitrarily oriented electric quantization for n-channel inversion layers of tetragonal semiconductors. Study all the special cases for III–V, ternary and quaternary materials in this context.

R.7.2 Investigate the Einstein relations in n-channel inversion layers of IV–VI, II–VI and stressed Kane type compounds in the presence of an arbitrarily oriented quantizing alternating electric field.

R.7.3 Investigate the Einstein relation in n-channel inversion layers of all the materials as stated in R.2.1 of Chap. 2 in the presence of an arbitrarily oriented quantizing alternating electric field.

R.7.4 Investigate the Einstein relation in the presence of an arbitrarily oriented non-quantizing alternating magnetic field in n-channel inversion layers of tetragonal semiconductors by including the electron spin. Study all the special cases for III–V, ternary and quaternary materials in this context.

R.7.5 Investigate the Einstein relations in n-channel inversion layers of IV–VI, II–VI and stressed Kane type compounds in the presence of an arbitrarily oriented non-quantizing alternating magnetic field by including the electron spin.

R.7.6 Investigate the Einstein relation in n-channel inversion layers of all the materials as stated in R.2.1 of Chap. 2 in the presence of an arbitrarily oriented non-quantizing alternating magnetic field by including electron spin.

R.7.7 Investigate the Einstein relation in n-channel inversion layers for all the problems from R.7.1 to R.7.6 in the presence of an additional arbitrarily oriented alternating electric field.

Table 7.1. The carrier statistics and the Einstein relation in the n-channel inversion layers of tetragonal, III–V, ternary, quaternary, IV–VI and stressed materials in the presence of both weak and strong electric field quantum limits. The corresponding results for II–VI materials are applicable in general without such limits

Type of materials	The carrier statistics of n-channel inversion in electric field quantum limit	The Einstein relation for the diffusivity mobility ratio in electric field quantum limit
1. Tetragonal compounds	In accordance with the generalized dispersion relation as formulated in this chapter under weak electric field	
	$\bar{n}_{2\mathrm{Dw}} = g_{\mathrm{v}} (2\pi)^{-1} \left[\bar{P}_{7\mathrm{w}} \left(\bar{E}_{\mathrm{Fw}}, 0\right)\right]$ (7.10)	$\frac{D}{\mu} = \frac{\bar{n}_{2\mathrm{Dw}}}{\lvert e \rvert} \left(x_{71} \left(\bar{E}_{\mathrm{Fw}}, 0\right) - y_{71} \left(\bar{E}_{0\mathrm{w}1}, 0\right)\right)$ (7.11)
	In accordance with the generalized dispersion relation as formulated in this chapter under strong electric field	
	$\bar{n}_{2\mathrm{Ds}} = g_{\mathrm{v}} (2\pi)^{-1} \left[P_2 \left(\bar{E}_{\mathrm{Fs}}, 0\right)\right]$ (7.17)	$\frac{D}{\mu} = \frac{\bar{n}_{2\mathrm{Ds}}}{\lvert e \rvert} \left(x_{72} \left(\bar{E}_{\mathrm{Fs}}, 0\right) - y_{72} \left(\bar{E}_{0\mathrm{s}1}, 0\right)\right)$ (7.18)
2. III–V, ternary and quaternary compounds	In accordance with the three band model of Kane which is a special case of our generalized analysis under weak electric field limit	
	$\bar{n}_{2\mathrm{Dw}} = \frac{g_{\mathrm{v}} m^*}{\pi \hbar^2} \left[P_{4\mathrm{w}} \left(\bar{E}_{\mathrm{Fw}}, 0\right)\right]$ (7.22)	$\frac{D}{\mu} = \frac{\bar{n}_{2\mathrm{Dw}}}{\lvert e \rvert} \left(x_{73} \left(\bar{E}_{\mathrm{Fw}}, 0\right) - y_{73} \left(\bar{E}_{0\mathrm{w}2}, 0\right)\right)$ (7.23)

(Continued)

Table 7.1. (*Continued*)

Type of materials	The carrier statistics of n-channel inversion in electric field quantum limit	The Einstein relation for the diffusivity mobility ratio in electric field quantum limit
	Equation (7.22) is a special case of (7.10) In accordance with the three band model of Kane which is a special case of our generalized analysis under strong electric field limit $\bar{n}_{2\mathrm{Ds}} = \frac{g_{\mathrm{v}} m^*}{\pi \hbar^2} \left[P_{6\mathrm{s}} \left(\bar{E}_{\mathrm{Fs}}, 0 \right) \right] \quad (7.29)$	Equation (7.23) is a special case of (7.11) $\frac{D}{\mu} = \frac{\bar{n}_{2\mathrm{Ds}}}{\lvert e \rvert} \left(x_{74} \left(\bar{E}_{\mathrm{Fs}}, 0 \right) - y_{74} \left(\bar{E}_{0\mathrm{s}2}, 0 \right) \right) \quad (7.30)$
	Equation (7.29) is a special case of (7.17) In accordance with the two band model of Kane under weak electric field limit $\bar{n}_{2\mathrm{Dw}} = \frac{g_{\mathrm{v}} m^*}{\pi \hbar^2} \left\{ \bar{E}_{\mathrm{Fw}} \left(1 + \alpha \bar{E}_{\mathrm{Fw}} \right) - S_0 \left(\frac{\hbar \lvert e \rvert F_{\mathrm{s}}}{\sqrt{2m^*}} \right)^{2/3} \left(1 + 2\alpha \bar{E}_{\mathrm{Fw}} \right)^{2/3} \right\} \quad (7.38)$	Equation (7.30) is a special case of (7.18) $\frac{D}{\mu} = \frac{\bar{n}_{2\mathrm{Dw}}}{\lvert e \rvert} \left(x_{75} \left(\bar{E}_{\mathrm{Fw}}, 0 \right) - y_{75} \left(\bar{E}_{0\mathrm{w}3}, 0 \right) \right) \quad (7.39)$
	Equation (7.38) is a special case of (7.22) In accordance with the two band model of Kane under strong electric field limit $\bar{n}_{2\mathrm{Ds}} = \frac{g_{\mathrm{v}} m^*}{\pi \hbar^2} \left\{ \bar{E}_{\mathrm{Fs}} \left(1 + \alpha \bar{E}_{\mathrm{Fs}} \right) - N_7 \bar{n}_{2\mathrm{Ds}} \right\} \quad (7.52)$	Equation (7.39) is a special case of (7.23) $\frac{D}{\mu} = \frac{\bar{n}_{2\mathrm{Ds}}}{\lvert e \rvert} \left(\frac{\pi \hbar^2}{m^* g_{\mathrm{v}} \left(1 + 2\alpha \bar{E}_{\mathrm{Fs}} \right)} \right) \cdot \bar{t}_{77} \quad (7.53)$
	Equation (7.52) is a special case of (7.29) In accordance with the parabolic model and is valid for all values of electric field limit	Equation (7.53) is a special case of (7.30)

	$$n_{2D} = \frac{g_v m^* k_B T}{\pi \hbar^2} \sum_{i=0}^{i_{max}} F_0(\eta_i) \Big|_{i=0} \quad (7.43)$$	$$\frac{D}{\mu} = \frac{k_B T}{\lvert e \rvert} \left(\frac{F_0(\eta_0)}{F_{-1}(\eta_0)} \right) \quad (7.44)$$
3. II–VI compounds	For all values of electric field limit $$\bar{n}_{2Di} = \frac{g_v}{2a'_0} \left[\bar{E}_{F0} + \frac{(\bar{\lambda}_0)^2}{4a'_0} - \bar{E}_0 - \left\{ \frac{\bar{\lambda}_0}{2\sqrt{a'_0}} \left[\bar{E}_{F0} + \frac{(\bar{\lambda}_0)^2}{4a'_0} - \bar{E}_0 \right]^{1/2} \right\} \right] \quad (7.61)$$	For all values of electric field limit $$\frac{D}{\mu} = \frac{\bar{n}_{2Di}}{\lvert e \rvert} \left(x_{76}(\bar{E}_{F0}, 0) - y_{76}(\bar{E}_0, 0) \right) \quad (7.62)$$
5. IV–VI compounds	For weak electric field limit $$\bar{n}_{2Dw} = \frac{g_v}{2\pi} \left[\frac{\gamma_{71}(\bar{E}_{Fw}, 0)}{\sqrt{p_{71} q_{71}(\bar{E}_{Fw}, 0)}} \right] \left[1 + \frac{\gamma_{71}(\bar{E}_{Fw}, 0)\, r_{71}}{(q_{71}(\bar{E}_{Fw}, 0))^2} \right]^{-1} \quad (7.71)$$	$$\frac{D}{\mu} = \frac{\bar{n}_{2Dw}}{\lvert e \rvert} \left(x_{77}(\bar{E}_{Fw}, 0) - y_{77}(\bar{E}_{0w4}, 0) \right) \quad (7.72)$$

(Continued)

Table 7.1. (*Continued*)

Type of materials	The carrier statistics of n-channel inversion in electric field quantum limit	The Einstein relation for the diffusivity mobility ratio in electric field quantum limit
	For strong electric field limit	
	$\bar{n}_{2\mathrm{Ds}} = \frac{g_{\mathrm{v}}}{\pi}\left[\frac{\gamma_{72}\left(\bar{E}_{\mathrm{Fs}},0\right)}{\sqrt{p_{72}q_{72}\left(\bar{E}_{\mathrm{Fs}}\right)}}\right]\left[1+\frac{\gamma_{72}\left(\bar{E}_{\mathrm{Fs}},0\right)r_{72}}{\left\{q_{72}\left(\bar{E}_{\mathrm{Fs}}\right)\right\}^{2}}\right]^{-1}$ (7.75)	$\frac{D}{\mu} = \frac{\bar{n}_{2\mathrm{Ds}}}{\lvert e\rvert}\left(x_{78}\left(\bar{E}_{\mathrm{Fs}},0\right)-y_{78}\left(\bar{E}_{0\mathrm{s}4},0\right)\right)$ (7.76)
6. Stressed compounds	For weak electric field limit	
	$\bar{n}_{2\mathrm{Dw}} = \frac{g_{\mathrm{v}}}{2\pi}\frac{T_{77}\left(\bar{E}_{\mathrm{Fw}},0\right)}{\sqrt{T_{57}\left(\bar{E}_{\mathrm{Fw}},0\right)T_{67}\left(\bar{E}_{\mathrm{Fw}},0\right)}}$ (7.84)	$\frac{D}{\mu} = \frac{\bar{n}_{2\mathrm{Dw}}}{\lvert e\rvert}\left(x_{79}\left(\bar{E}_{\mathrm{Fw}},0\right)-y_{79}\left(\bar{E}_{0\mathrm{w}5},0\right)\right)$ (7.85)
	For strong electric field limit	
	$\bar{n}_{2\mathrm{Ds}} = \frac{g_{\mathrm{v}}}{2\pi}\left(\frac{T_{137}\left(\bar{E}_{\mathrm{Fs}},0\right)}{\sqrt{T_{117}\left(\bar{E}_{\mathrm{Fs}},0\right)T_{127}\left(\bar{E}_{\mathrm{Fs}},0\right)}}\right)$ (7.93)	$\frac{D}{\mu} = \frac{\bar{n}_{2\mathrm{Ds}}}{\lvert e\rvert}\left(x_{710}\left(\bar{E}_{\mathrm{Fs}},0\right)-y_{710}\left(\bar{E}_{0\mathrm{s}5},0\right)\right)$ (7.94)

R.7.8 Investigate the Einstein relation in n-channel inversion layers for all the problems from R.7.1 to R.7.3 in the presence of arbitrarily oriented crossed electric and magnetic fields.

R.7.9 Investigate the Einstein relation in n-channel inversion layers for all the problems from R.7.1 to R.7.8 in the presence of surface states.

R.7.10 Investigate the Einstein relation in n-channel inversion layers for all the problems from R.7.1 to R.7.8 in the presence of hot electron effects.

R.7.11 Investigate the Einstein relation in n-channel inversion layers for all the problems from R.7.1 to R.7.6 by including the occupancy of the electrons in various electric subbands.

R.7.12 Investigate the problems from R.7.1 to R.7.11 for the appropriate p-channel inversion layers.

Allied Research Problems

R.7.13 Investigate the EMM for all the appropriate cases covering all the cases of the problems from R.7.1 to R.7.12.

R.7.14 Investigate the Debye screening length, the field emission, the photoemission, the gate capacitance, the carrier contribution to the elastic constants, the heat capacity, the activity coefficient, and the plasma frequency for all the materials covering all the cases of problems from R.7.1 to R.7.12.

R.7.15 Investigate in detail, the mobility for elastic and inelastic scattering mechanisms for all the materials covering all the cases of problems from R.7.1 to R.7.12.

R.7.16 Investigate the various transport coefficients in detail for all the materials covering all the cases of problems from R.7.1 to R.7.12.

R.7.17 Investigate the dia and paramagnetic susceptibilities in detail for all the materials covering all the appropriate research problems of this chapters.

R.7.18 Investigate all the problems from R.7.13 to R.7.17 for p-channel inversion layers.

References

1. T. Ando, H. Fowler, F. Stern, Rev. Mod. Phys. **54**, 437 (1982)
2. J.J. Quinn, P.J. Styles (ed.), *Electronic Properties of Quasi Two Dimensional Systems* (North Holland, Amsterdam, 1976)
3. G.A. Antcliffe, R.T. Bate, R.A. Reynolds, *Proceedings of the International Conference, Physics of Semi-Metals and Narrow-Gap Semiconductors*, ed. by D.L. Carter, R.T. Bate (Pergamon Press, Oxford, 1971), pp. 499
4. Z.A. Weinberg, Sol. Stat. Electron. **20**, 11 (1977)
5. G. Paasch, T. Fiedler, M. Kolar, I. Bartos, Phys. Stat. Sol. (b) **118**, 641 (1983)
6. S. Lamari, Phys. Rev. B, **64**, 245340 (2001)
7. T. Matsuyama, R. Kürsten, C. Meißner, U. Merkt, Phys. Rev. B, **61**, 15588 (2000)

8. P.V. Santos, M. Cardona, Phys. Rev. Lett. **72**, 432 (1994)
9. L. Bu, Y. Zhang, B.A. Mason, R.E. Doezema, J.A. Slinkman, Phys. Rev. B, **45**, 11336 (1992); S. Bhattacharya, R. Sarkar, D. De, S. Mukherjee, S. Pahari, A. Saha, S. Roy, N.C. Paul, S. Ghosh, K.P. Ghatak, J. Comp. Theo. Nano. Science (In the press) (2008)
10. P.D. Dresselhaus, C.M. Papavassiliou, R.G. Wheeler, R.N. Sacks, Phys. Rev. Lett. **68**, 106 (1992)
11. U. Kunze, Phys. Rev. B **41**, 1707 (1990)
12. E. Yamaguchi, Phys. Rev. B **32**, 5280 (1985)
13. Th. Lindner, G. Paasch, J. Appl. Phys. **102**, 054514 (2007)
14. S. Lamari, J. Appl. Phys. **91**, 1698 (2002)
15. K.P. Ghatak, M. Mondal, J. Appl. Phys. **70**, 299 (1991)
16. K.P. Ghatak, S.N. Biswas, J. Vac. Sci. Technol. **7B**, 104 (1989)
17. B. Mitra, K.P. Ghatak, Sol. State Electron. **32**, 177 (1989)
18. K.P. Ghatak, M. Mondal, J. Appl. Phys. **62**, 922 (1987)
19. M. Mondal, K.P. Ghatak, J. Magnet. Magnetic Mat. **62**, 115 (1986); M. Mondal, K.P. Ghatak, Phys. Script. **31**, 613 (1985)
20. K.P. Ghatak, M. Mondal, Z. Fur Physik B **64**, 223 (1986); K.P. Ghatak, S.N. Biswas, Sol. State Electron. **37**, 1437 (1994)

8

The Einstein Relation in Nipi Structures of Compound Semiconductors

8.1 Introduction

With the advent of modern experimental techniques of fabricating nanomaterials, it is possible to grow semiconductor superlattices (SLs) composed of alternative layers of two different degenerate layers with controlled thickness [1]. These structures have found wide applications in many new devices such as photodiodes [2], photoresistors [3], transistors [4], light emitters [5], tunneling devices [6], etc. [7–18]. The investigations of the physical properties of narrow gap SLs have increased extensively, as they are important for optoelectronic devices and also because the quality of heterostuctures involving narrow gap materials has been greatly improved. It may be noted that the nipi structures, also called the doping superlattices, are crystals with a periodic sequence of ultra-thin film layers [19, 20] of the same semiconductor with the intrinsic layer in between together with the opposite sign of doping. All the donors will be positively charged and all the acceptors negatively charged. This periodic space charge causes a periodic space charge potential which quantizes the motions of the carriers in the z-direction together with the formation of the subband energies. The electronic structures of the nipis differ radically from the corresponding bulk semiconductors as stated below:

(a) Each band is split into mini-bands,
(b) The magnitude and the spacing of these mini-bands may be designed by the choice of the superlattices parameters and,
(c) The electron energy spectrum of the nipi crystal becomes two-dimensional leading to the step functional dependence of the density-of-states function.

In Sect. 8.2.1, of the theoretical background, the Einstein relation in nipi structures of tetragonal materials has been investigated. Section 8.2.2 contains the results for nipi structures of III–V, ternary and quaternary compounds, in accordance with the three and the two band models of Kane together with parabolic energy bands and they form the special cases of Sect. 8.2.1.

Sections 8.2.3–8.2.5 contain the study of the DMR for nipis of II–VI, IV–VI, and stressed Kane type semiconductors, respectively. Section 8.3 contains the result and discussions of this chapter.

8.2 Theoretical Background

8.2.1 Formulation of the Einstein Relation in Nipi Structures of Tetragonal Materials

The dispersion relation of the conduction electrons in nipi structures of nonlinear optical materials, can be expressed by using (2.2) and following the method given in [19, 20] as

$$\psi_1(E) = \psi_2(E)\, k_s^2 + \psi_3(E)\left(n_i + \frac{1}{2}\right)\frac{2m_{||}^*}{\hbar}\omega_8(E), \tag{8.1}$$

where $\omega_8(E) \equiv \left(\frac{n_0|e|^2}{\varepsilon_{sc}[\theta_1(E)]}\right)^{1/2}$ and $\theta_1(E) \equiv \frac{\hbar^2}{2}\left\{\frac{\psi_3(E)[\psi_1(E)]' - \psi_1(E)[\psi_3(E)]'}{[\psi_3(E)]^2}\right\}$ and n_i $(= 0, 1, 2\ldots)$ is the mini-band index for nipi structures.

The EMM in this case assumes the form

$$m^*(E_{\mathrm{Fn}}, n_i) = \left(\frac{\hbar^2}{2}\right) R_{81}(E, n_i)\Big|_{E=\bar{E}_{Fn}}, \tag{8.2}$$

where $R_{81}(E, n_i) \equiv [\psi_2(E)]^{-2}\Big[\psi_2(E)\Big\{[\psi_1(E)]' - \left(\frac{2m_{||}^*}{\hbar}\right)[\psi_3(E)]'\left(n_i + \frac{1}{2}\right)$ $[\omega_8(E)] - \left(\frac{2m_{||}^*}{\hbar}\right)[\psi_3(E)]\left(n_i + \frac{1}{2}\right)[\omega_8(E)]'\Big\} - \Big\{[\psi_1(E)] - \left(\frac{2m_{||}^*}{\hbar}\right)[\psi_3(E)]$ $\left(n_i + \frac{1}{2}\right)[\omega_8(E)]\Big\}[\psi_2(E)]'\Big]$ and $\bar{E}_{\mathrm{Fn}}$ is the Fermi energy in the present case as measured from the edge of the conduction band in a vertically upward direction in the absence of any quantization.

From (8.2), we observe that the EMM is a function of the Fermi energy, nipi subband index and the other material constants, which is the characteristic feature of nipi structures of tetragonal materials.

The subband energy (E_{1ni}) can be written as

$$\psi_1(E_{1ni}) = \psi_3(E_{1ni})\left(n_i + \frac{1}{2}\right)\frac{2m_{||}^*}{\hbar}\omega_8(E_{1ni}), \tag{8.3}$$

The density-of-states function for nipi structures of tetragonal materials can be expressed as

$$N_{\mathrm{nipi}}(E) = \frac{g_{\mathrm{v}}}{2\pi d_0}\sum_{n_i=0}^{n_{i\,\max}} R_{81}(E, n_i)\, H(E - E_{1ni}), \tag{8.4}$$

in which d_0 is the superlattice period.

The electron concentration, can be written as

$$n_0 = \frac{g_{\mathrm{v}}}{2\pi d_0} \sum_{n_i=0}^{n_{i\,\max}} \left[T_{81}\left(\bar{E}_{\mathrm{Fn}}, n_i\right) + T_{82}\left(\bar{E}_{\mathrm{Fn}}, n_i\right)\right], \tag{8.5}$$

where $T_{81}\left(\bar{E}_{\mathrm{Fn}}, n_i\right) \equiv \left[\psi_1\left(\bar{E}_{\mathrm{Fn}}\right) - \psi_3\left(\bar{E}_{\mathrm{Fn}}\right)\left(n_i + \frac{1}{2}\right) \frac{2m_{\|}^{*}}{\hbar} \omega_8\left(\bar{E}_{\mathrm{Fn}}\right)\right]\left[\psi_2\left(\bar{E}_{\mathrm{Fn}}\right)\right]^{-1}$ and $T_{82}\left(\bar{E}_{\mathrm{Fn}}, n_i\right) \equiv \sum_{r=1}^{s} L(r)\, T_{81}\left(\bar{E}_{\mathrm{Fn}}, n_i\right)$.

From (8.5), the electron concentration for the nipi structures in quantum limit can be expressed as

$$\bar{n}_0 = \left[\frac{g_{\mathrm{v}}}{2\pi d_0}\right]\left[\psi_1\left(\bar{E}_{\mathrm{F0}}\right) - \psi_3\left(\bar{E}_{\mathrm{F0}}\right) \frac{m_{\|}^{*}}{\hbar} \omega_8\left(\bar{E}_{\mathrm{F0}}\right)\right]\left[\psi_2\left(\bar{E}_{\mathrm{F0}}\right)\right]^{-1}, \tag{8.6}$$

where $\bar{E}_{\mathrm{F0}}$ is the Fermi energy in the present case in the quantum limit and

$$\omega_8\left(\bar{E}_{\mathrm{F_0}}\right) \equiv \left(\frac{\bar{n}_0 |e|^2}{\varepsilon_{\mathrm{sc}}\left[\theta_1\left(\bar{E}_{\mathrm{F0}}\right)\right]}\right)^{1/2}.$$

Using (8.6) and (1.12), the DMR in this case can be written as

$$\frac{D}{\mu} = \left(\frac{\bar{n}_0}{|e|}\right)\left[x_1\left(\bar{E}_{\mathrm{F0}}\right) - x_2\left(E_{10}\right)\right], \tag{8.7}$$

where $x_1\left(\bar{E}_{\mathrm{F0}}\right) \equiv \left[\psi_2\left(\bar{E}_{\mathrm{F0}}\right) + \frac{g_{\mathrm{v}} m_{\|}^{*} \psi_3\left(\bar{E}_{\mathrm{F0}}\right)\omega_8\left(\bar{E}_{\mathrm{F0}}\right)}{4\pi d_0 \bar{n}_0 \hbar}\right]$

$$\left[-\left[\psi_2\left(\bar{E}_{\mathrm{F0}}\right)\right]' \bar{n}_0 + \frac{g_{\mathrm{v}}}{2\pi d_0}\left[\psi_1\left(\bar{E}_{\mathrm{F0}}\right)\right]' - \frac{g_{\mathrm{v}}\left[\psi_3\left(\bar{E}_{\mathrm{F0}}\right)\right]' m_{\|}^{*} \omega_8\left(\bar{E}_{\mathrm{F0}}\right)}{2\pi d_0 \hbar}\right.$$
$$\left. + \frac{g_{\mathrm{v}} m_{\|}^{*}}{4\pi d_0 \hbar} \psi_3\left(\bar{E}_{\mathrm{F0}}\right) \omega_8\left(\bar{E}_{\mathrm{F0}}\right)\left[\theta_1\left(\bar{E}_{\mathrm{F0}}\right)\right]'\left[\theta_1\left(\bar{E}_{\mathrm{F0}}\right)\right]^{-1}\right]^{-1},$$

$x_2\left(E_{10}\right) \equiv \left(\frac{\omega_8\left(E_{10}\right)}{\bar{n}_0}\right)\left[2\left\{\omega_8\left(E_{10}\right)\right\}' + \frac{\left\{\theta_1\left(E_{10}\right)\right\}' \omega_8\left(E_{10}\right)}{\left[\theta_1\left(E_{10}\right)\right]}\right]^{-1}$ and E_{10} is the sub-band energy when $n_i = 0$ and can be determined from the equation

$$\psi_1\left(E_{10}\right) = \psi_3\left(E_{10}\right) \frac{m_{\|}^{*}}{\hbar} \omega_8\left(E_{10}\right). \tag{8.8}$$

8.2.2 Einstein Relation for the Nipi Structures of III–V Compounds

(a) The electron energy spectrum in the nipi structures of III–V materials can be expressed, from (8.1) under the conditions $\Delta_{\|} = \Delta_{\perp} = \Delta$, $\delta = 0$ and $m_{\|}^{*} = m_{\perp}^{*} = m^{*}$, as

$$I(E) = \left(n_i + \frac{1}{2}\right)\hbar\omega_9(E) + \frac{\hbar^2 k_s^2}{2m^*}, \tag{8.9}$$

where $\omega_9(E) \equiv \left(\frac{n_0|e|^2}{\varepsilon_{sc} I'(E) m^*}\right)^{1/2}$.

The EMM in this case can be written as

$$m^*(E_{Fn}, n_i) = m^* R_{82}(E, n_i)|_{E=E_{Fn}}. \tag{8.10}$$

in which, $R_{82}(E, n_i) \equiv \left\{[I(E)]' - \left(n_i + \frac{1}{2}\right)\hbar[\omega_9(E)]'\right\}$.

From (8.10), we observe that the EMM in this case is a function of the Fermi energy, nipi subband index, and the other material constants, which is a characteristic feature of nipi structures of III–V compounds whose bulk dispersion relations are defined by the three band model of Kane.

The subband energies (E_{2ni}) can be written as

$$I(E_{2ni}) = \left(n_i + \frac{1}{2}\right)\hbar\omega_9(E_{2ni}). \tag{8.11}$$

The density-of-states function in this case can be expressed as

$$N_{\text{nipi}}(E) = \frac{m^* g_v}{\pi\hbar^2 d_0} \sum_{n_i=0}^{n_{i\,\max}} R_{82}(E, n_i) H(E - E_{2ni}). \tag{8.12}$$

The use of (8.12) leads to the expression of the electron concentration as

$$n_0 = \frac{m^* g_v}{\pi\hbar^2 d_0} \sum_{n_i=0}^{n_{i\,\max}} \left[T_{83}(\bar{E}_{Fn}, n_i) + T_{84}(\bar{E}_{Fn}, n_i)\right]. \tag{8.13}$$

where $T_{83}(\bar{E}_{Fn}, n_i) \equiv \left[I(\bar{E}_{Fn}) - \left(n_i + \frac{1}{2}\right)\hbar\omega_9(\bar{E}_{Fn})\right]$ and $T_{84}(\bar{E}_{Fn}, n_i) \equiv \sum_{r=1}^{s} L(r) T_{83}(\bar{E}_{Fn}, n_i)$.

Using (8.13), the electron concentration in the electric quantum limit for nipi structures of III–V ternary and quaternary materials can be written as

$$\bar{n}_0 = \left(\frac{m^* g_v}{\pi\hbar^2 d_0}\right)[I(E_{20}) - \{(1/2)\hbar\omega_9(E_{20})\}], \tag{8.14}$$

where E_{20} is determined from the equation

$$I(E_{20}) = \frac{1}{2}\hbar\omega_9(E_{20}). \tag{8.15}$$

The use of (8.14) and (1.12) leads to the expression of the DMR in this case as

$$\frac{D}{\mu} = \left(\frac{\bar{n}_0}{|e|}\right)\left[x_3(\bar{E}_{F0}) - x_4(E_{20})\right], \tag{8.16}$$

in which

$$x_3\left(\bar{E}_{\mathrm{F_0}}\right) \equiv \left[\frac{\pi\hbar^2 d_0}{m^* g_{\mathrm{v}}} + \left\{\frac{\hbar}{4}\frac{\omega_9\left(\bar{E}_{\mathrm{F_0}}\right)}{\bar{n}_0}\right\}\right] \times \left[\left\{I\left(\bar{E}_{\mathrm{F_0}}\right)\right\}' + \left\{\frac{\hbar}{4}\frac{\left\{I\left(\bar{E}_{\mathrm{F_0}}\right)\right\}'' \omega_9\left(\bar{E}_{\mathrm{F_0}}\right)}{\left\{I\left(\bar{E}_{\mathrm{F_0}}\right)\right\}'}\right\}\right]^{-1}$$

and

$$x_4\left(E_{20}\right) \equiv \left(\frac{\omega_9\left(E_{20}\right)}{\bar{n}_0}\right)\left[2\left[\omega_9\left(E_{20}\right)\right]' + \left\{\frac{\omega_9\left(E_{20}\right)\left\{I\left(\bar{E}_{20}\right)\right\}''}{\left\{I\left(\bar{E}_{20}\right)\right\}'}\right\}\right]^{-1}.$$

(b) For the two band model of Kane, the expressions of the dispersion relation, the EMM, the subband energies, the density-of-states function, n_0, $\bar{n}_0$ and the DMR as given by (8.9)–(8.14) and (8.16) remain the same where

$$I(E) = E(1+\alpha E),\ \{I(E)\}' = (1+2\alpha E) \text{ and } \{I(E)\}'' = 2\alpha.$$

The EMM in this case can be written as

$$m^*\left(E_{\mathrm{Fn}}, n_i\right) = m^*\left\{(1+2\alpha E_{\mathrm{Fn}}) + \left(n_i + \frac{1}{2}\right)\hbar\left[\omega_9\left(E_{\mathrm{Fn}}\right)\right]\frac{\alpha}{(1+2\alpha E_{\mathrm{Fn}})}\right\}. \tag{8.17}$$

From (8.17), we observe that the EMM in this case is a function of the Fermi energy, nipi subband index, and the other material constants due to the band non-parabolicity only.

(c) For parabolic energy bands, the expressions of dispersion relation, the EMM, the subband energies, the density-of-states function, n_0, $\bar{n}_0$, and the DMR as given by (8.9)–(8.14) and (8.16) remain the same, where $I(E) = E$, $\{I(E)\}' = 1$ and $\{I(E)\}'' = 0$.

The EMM can be written as

$$m^*\left(E_{\mathrm{Fn}}, n_i\right) = m^*. \tag{8.18}$$

From (8.18), we observe that the EMM in this case is a constant.

8.2.3 Einstein Relation for the Nipi Structures of II–VI Compounds

The carrier dispersion law in nipi structures of II–VI compounds can be expressed by (2.27) as

$$E = a_0' k_{\mathrm{s}}^2 + \left(n_i + \frac{1}{2}\right)\hbar\omega_{10} \pm \bar{\lambda}_0 k_{\mathrm{s}},\ \omega_{10} \equiv \left(\frac{n_0 |e|^2}{\varepsilon_{\mathrm{sc}} m_{||}^*}\right)^{1/2}. \tag{8.19}$$

Using (8.19), the EMM in this case can be written as

$$m^*(E_{\mathrm{Fn}}, n_i) = m_\perp^* \left\{1 - \bar{\lambda}_0 \left[(\bar{\lambda}_0)^2 + 4a_0' E_{\mathrm{Fn}} - 4a_0' \left(n_i + \frac{1}{2}\right)\hbar\omega_{10}\right]^{-1/2}\right\}. \tag{8.20}$$

Thus, the EMM in this case is a function of the Fermi energy, the nipi subband index number, and the energy spectrum constants due to the presence of $\bar{\lambda}_0$ only.

The subband energies (E_{3ni}) can be written as

$$E_{3ni} = \left(n_i + \frac{1}{2}\right)\hbar\omega_{10}. \tag{8.21}$$

The density-of-states function in this case can be expressed as

$$N_{\mathrm{nipi}}(E) = \frac{m_\perp^* g_{\mathrm{v}}}{\pi\hbar^2 d_0} \sum_{n_i=0}^{n_{i\,\max}} \left[1 - \frac{a_{81}}{\sqrt{E + b_{81}(n_i)}}\right] H(E - E_{3ni}), \tag{8.22}$$

in which, $a_{81} \equiv \frac{\bar{\lambda}_0}{2\sqrt{a_0'}}$ and $b_{81}(n_i) \equiv \left[\frac{1}{4a_0'}\left[(\bar{\lambda}_0)^2 - 4a_0'\left(n_i + \frac{1}{2}\right)\hbar\omega_{10}\right]\right]$.

The use of (8.22) leads to the electron concentration as

$$n_0 = \frac{m_\perp^* g_{\mathrm{v}} k_{\mathrm{B}} T}{\pi\hbar^2 d_0} \sum_{n_i=0}^{n_{i\,\max}} \left[F_0(\eta_{81}) - \left(\frac{a_{81}}{\sqrt{k_{\mathrm{B}}T}}\left[2\left(\sqrt{\eta_{81} + c_{81}(n_i)} - \sqrt{c_{81}(n_i)}\right)\right]\right)\right.$$
$$\left. + \sum_{r=1}^{s} 2\left(1 - 2^{1-2r}\right)\zeta(2r)\frac{(-1)^{2r-1}(2r-1)!}{(\eta_{81} + c_{81}(n_i))^{2r}}\right], \tag{8.23}$$

where $\eta_{81} \equiv \frac{E_{\mathrm{Fn}} - E_{3ni}}{k_{\mathrm{B}}T}$ and $c_{81}(n_i) \equiv \frac{b_{81}(n_i) + E_{3ni}}{k_{\mathrm{B}}T}$.

The electron concentration at the quantum limit is given by

$$\bar{n}_0 = \frac{m_\perp^* g_{\mathrm{v}}}{\pi\hbar^2 d_0}\left[\bar{E}_{\mathrm{F0}} - \frac{1}{2}\hbar\omega_{10} - 2a_{81}\left[\sqrt{\bar{E}_{\mathrm{F0}} + \frac{1}{4a_0'}\left[(\bar{\lambda}_0)^2 - 2a_0'\hbar\omega_{10}\right]} - a_{81}\right]\right]. \tag{8.24}$$

The DMR in this case can be written using (8.24) and (1.12) as

$$\frac{D}{\mu} = \frac{\bar{n}_0}{|e|}\left[x_5(\bar{E}_{\mathrm{F_0}}) - x_6(\bar{n}_0)\right], \tag{8.25}$$

in which, $x_5(\bar{E}_{\mathrm{F_0}}) \equiv \left[\frac{d_0\pi\hbar^2}{g_{\mathrm{v}} m_\perp^*} + \frac{\hbar\omega_{10}}{4\bar{n}_0} - \left(\frac{\hbar\omega_{10} a_{81}}{4\bar{n}_0}\right)\left\{\bar{E}_{\mathrm{F_0}} + a_{81}^2 - \frac{1}{2}\hbar\omega_{10}\right\}^{-1/2}\right]$ $\left[1 - \frac{a_{81}}{\sqrt{\bar{E}_{\mathrm{F_0}} + a_{81}^2 - \frac{1}{2}\hbar\omega_{10}}}\right]^{-1}$ and $x_6(\bar{n}_0) \equiv \frac{\hbar\omega_{10}}{4\bar{n}_0}$.

8.2.4 Einstein Relation for the Nipi Structures of IV–VI Compounds

The carrier energy spectrum in nipi structures of IV–VI compounds can be written using (5.50) as

$$k_s^2 = \left(\hbar^2 S_{19}\right)^{-1}\left[-S_{20}\left(E,n_i\right)+\sqrt{S_{20}^2\left(E,n_i\right)+4S_{19}S_{21}\left(E,n_i\right)}\right], \quad (8.26)$$

in which, $S_{19} \equiv \left(\frac{\alpha}{m_t^+ m_t^-}\right)$, $S_{20}\left(E,n_i\right) \equiv \left\{\frac{1}{m_t^*} - \left(\frac{\alpha E}{m_t^+}\right) + \frac{1+\alpha E}{m_t^-} + \frac{\alpha\hbar^2}{2m_l^+ m_t^-}\left(n_i+\frac{1}{2}\right)T\left(E\right) + \frac{\alpha\hbar^2}{2m_l^- m_t^+}\left(n_i+\frac{1}{2}\right)T\left(E\right)\right\}$,

$$T\left(E\right) \equiv \frac{2m^*\left(0\right)}{\hbar}\omega_{11}\left(E\right),\ m^*\left(0\right) \equiv \left(\frac{m_l^* m_l^-}{m_l^* + m_l^-}\right),\ \omega_{11}\left(E\right) \equiv \left(\frac{n_0\left|e\right|^2}{\varepsilon_{sc} m^*\left(E\right)}\right)^{1/2},$$

$$m^*\left(E\right) \equiv \frac{1}{4t_1}\left[-\left(t_2\left(E\right)\right)' + \frac{t_2\left(E\right)\left(t_2\left(E\right)\right)' + 2t_1\left(1+2\alpha E\right)}{\sqrt{t_2^2\left(E\right)+4Et_1\left(1+\alpha E\right)}}\right],$$

$$t_1 \equiv \left(\frac{\alpha}{4m_l^+ m_l^-}\right),\ t_2\left(E\right) \equiv \frac{1}{2}\left[\left(\frac{1}{m_l^*}\right) - \left(\frac{\alpha E}{m_l^+}\right) + \left(\frac{1+\alpha E}{m_l^-}\right)\right],$$

$$\left(t_2\left(E\right)\right)' \equiv \frac{\alpha}{2}\left(\frac{1}{m_l^-} - \left(\frac{1}{m_l^+}\right)\right).$$

and $S_{21}\left(E,n_i\right) \equiv \left[E\left(1+\alpha E\right) + \frac{\alpha E\hbar^2}{2m_l^+}\left(n_i+\frac{1}{2}\right)T\left(E\right) + \frac{\hbar^2}{2m_l^-}\left(n_i+\frac{1}{2}\right)T\left(E\right)\left(1+\alpha E\right) + \frac{\hbar^4}{4m_l^- m_l^+}\left(n_i+\frac{1}{2}\right)T\left(E\right) - \left(\frac{\hbar^2}{2m_l^*}\right)T\left(E\right)\left(n_i+\frac{1}{2}\right)\right]$.

Using (8.26), the EMM in this case can be written as

$$m^*\left(E_{Fn},n_i\right) = R_{84}\left(E,n_i\right)\Big|_{E=E_{Fn}}. \quad (8.27)$$

where

$$R_{84}\left(E,n_i\right) \equiv \left(2S_{19}\right)^{-1}\left[-\left(S_{20}\left(E,n_i\right)\right)' + \frac{S_{20}\left(E,n_i\right)\left[S_{20}\left(E,n_i\right)\right]' + 2S_{19}\left[S_{21}\left(E,n_i\right)\right]'}{\left[\left\{\left[S_{20}\left(E,n_i\right)\right]\right\}^2 + 4S_{19}S_{21}\left(E,n_i\right)\right]^{1/2}}\right].$$

One can observe that the EMM in this case is a function of both the Fermi energy and the nipi subband index number together with the spectrum constants of the system due to the presence of band non-parabolicity.

The subband energies (E_{4ni}) can be written as

$$\left[E_{4ni} - \frac{\hbar^2}{2m_l^-} T\left(E_{4ni}\right)\left(n_i + \frac{1}{2}\right)\right]\left[1 + \alpha E_{4ni} + \alpha\frac{\hbar^2}{2m_l^+} T\left(E_{4ni}\right)\left(n_i + \frac{1}{2}\right)\right] = \left[\frac{\hbar^2}{2m_l^*} T\left(E_{4ni}\right)\left(n_i + \frac{1}{2}\right)\right]. \quad (8.28)$$

The density-of-states function in this case assumes the form as

$$N_{\text{nipi}}\left(E\right) = \frac{g_{\text{v}}}{\pi\hbar^2 d_0} \sum_{n_i=0}^{n_{i\,\max}} R_{84}\left(E, n_i\right) H\left(E - E_{4ni}\right). \quad (8.29)$$

The use of (8.29) leads to the expression of the electron concentration as

$$n_0 = \frac{g_{\text{v}}}{2\pi\hbar^2 S_{19} d_0} \sum_{n_i=0}^{n_{i\,\max}} \left[T_{85}\left(\bar{E}_{\text{Fn}}, n_i\right) + T_{86}\left(\bar{E}_{\text{Fn}}, n_i\right)\right]. \quad (8.30)$$

where $T_{85}\left(\bar{E}_{\text{Fn}}, n_i\right) \equiv \left[-S_{20}\left(E_{\text{Fn}}, n_i\right) + \sqrt{\left[S_{20}\left(E_{\text{Fn}}, n_i\right)\right]^2 + 4S_{19}S_{21}\left(E_{\text{Fn}}, n_i\right)}\right]$

and $T_{86}\left(\bar{E}_{\text{Fn}}, n_i\right) \equiv \sum_{r=1}^{s} L\left(r\right) T_{85}\left(\bar{E}_{\text{Fn}}, n_i\right)$.

The electron concentration at the quantum limit can be defined through the equation

$$\bar{n}_0 = \frac{g_{\text{v}}}{2\pi\hbar^2 S_{19} d_0}\left[-S_{20}\left(\bar{E}_{\text{F0}}, 0\right) + \sqrt{\left[S_{20}\left(\bar{E}_{\text{F0}}, 0\right)\right]^2 + 4S_{19}S_{21}\left(\bar{E}_{\text{F0}}, 0\right)}\right]. \quad (8.31)$$

Using (8.31) and (1.12), the DMR in this case is given by

$$\frac{D}{\mu} = \frac{\bar{n}_0}{|e|}\left[x_7\left(\bar{E}_{\text{F}_0}\right) - x_8\left(E_{40}\right)\right], \quad (8.32)$$

in which $x_7\left(\bar{E}_{\text{F}_0}\right) \equiv \frac{\omega_{90}\left(\bar{E}_{\text{F}_0}\right)}{\omega_{91}\left(\bar{E}_{\text{F}_0}\right)}$,

$$\begin{aligned}\omega_{90}\left(\bar{E}_{\text{F}_0}\right) \equiv &\left[\frac{2\pi S_{19} d_0 \hbar^2}{g_{\text{v}}} + \left(\frac{\omega_{11}\left(\bar{E}_{\text{F}_0}\right)}{2\bar{n}_0}\right)\left\{\frac{\alpha\hbar m^*\left(0\right)}{2m_t^- m_l^+} + \frac{\alpha\hbar m^*\left(0\right)}{2m_l^- m_t^+}\right\}\right.\\ &- \left\{S_{20}^2\left(E_{\text{F0}}\right) + 4S_{19}S_{21}\left(\bar{E}_{\text{F}_0}\right)\right\}^{-1/2}\\ &\left[S_{20}\left(\bar{E}_{\text{F}_0}\right)\left\{\frac{\alpha\hbar m^*\left(0\right)}{2m_t^- m_l^+} + \frac{\alpha\hbar m^*\left(0\right)}{2m_l^- m_t^+}\right\}\frac{\omega_{11}\left(\bar{E}_{\text{F}_0}\right)}{2\bar{n}_0}\right.\\ &+4S_{19}\left\{\frac{\alpha\bar{E}_{\text{F0}} m^*\left(0\right)\hbar}{2m_l^+} + \frac{\hbar m^*\left(0\right)\left(1 + 2\alpha\bar{E}_{\text{F0}}\right)}{2m_l^-} + \frac{\alpha\hbar^3 m^*\left(0\right)}{4m_l^- m_l^+}\right.\\ &\left.\left.\left.-\frac{\hbar m^*\left(0\right)}{2m_l^*}\right\}\left(\frac{\omega_{11}\left(\bar{E}_{\text{F}_0}\right)}{2\bar{n}_0}\right)\right]\right],\end{aligned}$$

$$\omega_{91}\left(\bar{E}_{F_0}\right) \equiv \left[\frac{\alpha}{m_t^+} - \frac{\alpha}{m_t^-} + \frac{\omega_{11}\left(\bar{E}_{F_0}\right)\left[m^*\left(\bar{E}_{F_0}\right)\right]'}{2m^*\left(\bar{E}_{F_0}\right)}\left\{\frac{\alpha\hbar m^*(0)}{2m_t^- m_l^+} + \frac{\alpha\hbar m^*(0)}{2m_l^- m_t^+}\right\}\right.$$

$$+\left\{\left[S_{20}\left(\bar{E}_{F_0}\right)\right]^2 + 4S_{19}S_{21}\left(\bar{E}_{F_0}\right)\right\}^{-1/2}$$

$$\times\left[S_{20}\left(\bar{E}_{F_0}\right)\left\{\frac{-\alpha}{m_t^+} + \frac{\alpha}{m_t^-} - \left[\frac{\omega_{11}\left(\bar{E}_{F_0}\right)\left[m^*\left(\bar{E}_{F_0}\right)\right]'}{2m^*\left(\bar{E}_{F_0}\right)}\right]\right.\right.$$

$$\left.\times\left[\frac{\alpha\hbar m^*(0)}{2m_t^- m_l^+} + \frac{\alpha\hbar m^*(0)}{2m_l^- m_t^+}\right]\right\}$$

$$+4S_{19}\left\{1 + 2\alpha\bar{E}_{F_0} + \frac{\alpha m^*(0)\hbar\omega_{11}\left(\bar{E}_{F_0}\right)}{2m_l^+}\right.$$

$$+\left\{\frac{\alpha\hbar m^*(0)\,\omega_{11}\left(\bar{E}_{F_0}\right)}{2m_l^-}\right\} - \left\{\frac{\left[m^*\left(\bar{E}_{F_0}\right)\right]'\omega_{11}\left(\bar{E}_{F_0}\right)}{2m^*\left(\bar{E}_{F_0}\right)}\right\}$$

$$\times\left\{\frac{\hbar\alpha\bar{E}_{F_0}m^*(0)}{2m_l^+} + \frac{\hbar\left(1+\alpha\bar{E}_{F_0}\right)m^*(0)}{2m_l^-}\right.$$

$$\left.\left.\left.\left.+\frac{\alpha\hbar^3 m^*(0)}{4m_l^- m_l^+} - \frac{\hbar m^*(0)}{2m_l^*}\right\}\right\}\right]\right],$$

and

$$x_8\left(E_{40}\right) \equiv \left[\frac{\hbar m^*(0)\,\omega_{11}\left(E_{40}\right)}{4\bar{n}_0 m_l^-}\left[1 + \alpha E_{40} + \frac{\alpha\hbar^2}{4m_l^+}T\left(E_{40}\right)\right] - \left(\frac{\alpha\hbar m^*(0)\,\omega_{11}\left(E_{40}\right)}{4\bar{n}_0 m_l^*}\right)\right.$$

$$\times\left[E_{40} - \left(\frac{\hbar^2}{2m_l^-}\right)T\left(E_{40}\right) + \frac{\hbar^2 m^*(0)\,\omega_{11}\left(E_{40}\right)}{4m_l^+\bar{n}_0}\right]$$

$$\times\left[\left[1 + \frac{\hbar m^*(0)\,\omega_{11}\left(E_{40}\right)\left[m^*\left(E_{40}\right)\right]'}{4m_l^- m^*\left(E_{40}\right)}\right]\left[1 + \alpha E_{40} + \frac{\alpha\hbar^2}{4m_l^+}T\left(E_{40}\right)\right]\right.$$

$$+\left[E_{40} - \left(\frac{\hbar^2}{4m_l^-}T\left(E_{40}\right)\right)\right]\left[\alpha - \frac{\alpha\hbar m^*(0)\,\omega_{11}\left(E_{40}\right)}{4m_l^+ m^*\left(E_{40}\right)}\left[m^*\left(E_{40}\right)\right]'\right]$$

$$\left.+\frac{\hbar m^*(0)\,\omega_{11}\left(E_{40}\right)\left[m^*\left(E_{40}\right)\right]'}{4m_l^- m^*\left(E_{40}\right)}\right]^{-1}$$

E_{40} should be determined from the following equation

$$\left[E_{40} - \frac{\hbar^2}{4m_l^-}T\left(E_{40}\right)\right]\left[1 + \alpha E_{40} + \frac{\alpha\hbar^2}{4m_l^+}T\left(E_{40}\right)\right] = \frac{\hbar^2}{4m_l^*}T\left(E_{40}\right)$$

8.2.5 Einstein Relation for the Nipi Structures of Stressed Kane Type Compounds

The electron dispersion law in the nipi structures of stressed Kane type semiconductors can be written using (2.48) as

$$\frac{k_x^2}{[\bar{a}_0(E)]^2}+\frac{k_y^2}{[\bar{b}_0(E)]^2}+\frac{1}{[\bar{c}_0(E)]^2}\frac{2m_z^*(0)}{\hbar}\left(n_i+\frac{1}{2}\right)\omega_{12}(E)=1, \tag{8.33}$$

where $\omega_{12}(E)\equiv\left(\frac{n_0|e|^2}{\varepsilon_{sc}m_z^*(E)}\right)^{1/2}$ and $m_z^*(E)\equiv\hbar^2\bar{c}_0(E)\frac{\partial}{\partial E}[\bar{c}_0(E)]$.

The use of (8.33) leads to the expression of the EMM as

$$m^*(E_{\text{Fn}},n_i)=\left(\frac{\hbar^2}{2}\right)R_{85}(E,n_i)\Big|_{E=E_{\text{Fn}}}, \tag{8.34}$$

where

$$\begin{aligned}R_{85}(E,n_i)\equiv&\Bigg[\left[(\bar{a}_0(E))'\,\bar{b}_0(E)+(\bar{b}_0(E))'\,\bar{a}_0(E)\right]\\&\times\left[1-\frac{1}{[\bar{c}_0(E)]^2}\frac{2m_z^*(0)}{\hbar}\left(n_i+\frac{1}{2}\right)\omega_{12}(E)\right]\\&-\left[\frac{\bar{a}_0(E)\,\bar{b}_0(E)}{[\bar{c}_0(E)]^2}\frac{2m_z^*(0)}{\hbar}\left(n_i+\frac{1}{2}\right)[\omega_{12}(E)]'\right]\\&+\left[\frac{\bar{a}_0(E)\,\bar{b}_0(E)\,[\bar{c}_0(E)]'}{[\bar{c}_0(E)]^3}\frac{4m_z^*(0)}{\hbar}\left(n_i+\frac{1}{2}\right)[\omega_{12}(E)]\right]\Bigg].\end{aligned} \tag{8.35}$$

Thus, the EMM is a function of the Fermi energy and the nipi subband index due to the presence of stress and band non-parabolicity only.

The subband energies (E_{5ni}) can be written as

$$\frac{1}{[\bar{c}_0(E_{4ni})]^2}\frac{2m_z^*(0)}{\hbar}\left(n_i+\frac{1}{2}\right)\omega_{12}(E_{4ni})=1. \tag{8.36}$$

The density-of-states function can be written as

$$N_{\text{nipi}}(E)=\frac{g_v}{\pi\hbar^2 d_0}\sum_{n_i=0}^{n_{i\,\max}}R_{85}(E,n_i)\,H(E-E_{5ni}). \tag{8.37}$$

Using (8.37), the electron concentration in nipi structures of stressed compounds can be expressed as

$$n_0=\frac{g_v}{2\pi d_0}\sum_{n_i=0}^{n_{i\,\max}}\left[C_3(\bar{E}_{\text{Fn}},n_i)+C_4(\bar{E}_{\text{Fn}},n_i)\right], \tag{8.38}$$

where $C_3\left(\bar{E}_{\mathrm{Fn}}, n_i\right) \equiv \bar{a}_0\left(\bar{E}_{\mathrm{Fn}}\right) \bar{b}_0\left(E_{\mathrm{Fn}}\right)\left[1-\frac{2m_z^*(0)}{\hbar}\left(n_i+\frac{1}{2}\right) \frac{\omega_{12}\left(\bar{E}_{\mathrm{Fn}}\right)}{\left(\bar{c}_0\left(\bar{E}_{\mathrm{Fn}}\right)\right)^2}\right]$ and $C_4\left(\bar{E}_{\mathrm{Fn}}, n_i\right) \equiv \sum_{r=1}^{s} L(r)\, C_3\left(\bar{E}_{\mathrm{Fn}}, n_i\right)$.

The use of (8.38) leads to the expression of the electron statistics at the electric quantum limit and at low temperatures as

$$\bar{n}_0=\frac{g_{\mathrm{v}}}{2\pi d_0} \bar{a}_0\left(\bar{E}_{\mathrm{F}_0}\right) \bar{b}_0\left(\bar{E}_{\mathrm{F}_0}\right)\left[1-\frac{m_z^*(0)}{\left(\bar{c}_0\left(\bar{E}_{\mathrm{F}_0}\right)\right)^2 \hbar} \omega_{12}\left(\bar{E}_{\mathrm{F}_0}\right)\right]. \tag{8.39}$$

The use of (8.39) and (1.12) leads to the expression of the DMR as

$$\frac{D}{\mu}=\frac{\bar{n}_0}{|e|}\left[x_9\left(\bar{E}_{\mathrm{F}_0}\right)-x_{10}\left(E_{50}\right)\right], \tag{8.40}$$

where

$$x_9\left(\bar{E}_{\mathrm{F}_0}\right) \equiv \frac{\omega_{92}\left(\bar{E}_{\mathrm{F}_0}\right)}{\omega_{93}\left(\bar{E}_{\mathrm{F}_0}\right)},$$

$$\omega_{92}\left(\bar{E}_{\mathrm{F}_0}\right) \equiv\left[\frac{2\pi d_0}{g_{\mathrm{v}}}-\left\{\frac{\bar{a}_0\left(\bar{E}_{\mathrm{F}_0}\right) \bar{b}_0\left(\bar{E}_{\mathrm{F}_0}\right)\left[m_z^*(0)\right]^2 \omega_{12}\left(\bar{E}_{\mathrm{F}_0}\right)}{2\hbar^2 \bar{n}_0\left[\bar{c}_0\left(\bar{E}_{\mathrm{F}_0}\right)\right]^2}\right\}\right],$$

$$\begin{aligned}\omega_{93}\left(\bar{E}_{\mathrm{F}_0}\right) \equiv & \left[\frac{\bar{n}_0 2\pi d_0}{g_{\mathrm{v}}} \frac{\left[\bar{a}_0\left(\bar{E}_{\mathrm{F}_0}\right)\right]'}{\left[\bar{a}_0\left(\bar{E}_{\mathrm{F}_0}\right)\right]}+\frac{\bar{n}_0 2\pi d_0}{g_{\mathrm{v}}} \frac{\left[\bar{b}_0\left(\bar{E}_{\mathrm{F}_0}\right)\right]'}{\left[\bar{b}_0\left(\bar{E}_{\mathrm{F}_0}\right)\right]}+\frac{m_z^*(0)\, \bar{a}_0\left(\bar{E}_{\mathrm{F}_0}\right) \bar{b}_0\left(\bar{E}_{\mathrm{F}_0}\right)}{\hbar}\right. \\ & \left.\times\left[\left\{\frac{2\omega_{12}\left(\bar{E}_{F_0}\right)\left[\bar{c}_0\left(\bar{E}_{F_0}\right)\right]'}{\left[\bar{c}_0\left(\bar{E}_{F_0}\right)\right]^3}\right\}-\left\{\frac{\omega_{12}\left(\bar{E}_{F_0}\right) m_z^*(0)\left[m_z^*\left(\bar{E}_{F_0}\right)\right]'}{2\hbar\left[\bar{c}_0\left(\bar{E}_{F_0}\right)\right]^2\left[m_z^*\left(\bar{E}_{F_0}\right)\right]}\right\}\right]\right]\end{aligned}$$

and

$$x_{10}\left(E_{50}\right) \equiv\left[\frac{\omega_{12}\left(E_{50}\right)}{2\bar{n}_0}\right]\left[\frac{2\hbar \bar{c}_0\left(E_{50}\right)\left[\bar{c}_0\left(E_{50}\right)\right]'}{m_z^*(0)}+\frac{\omega_{12}\left(E_{50}\right)\left[m_z^*\left(E_{50}\right)\right]'}{2m_z^*\left(E_{50}\right)}\right]^{-1}$$

where $\frac{1}{\left[\bar{c}_0\left(E_{50}\right)\right]^2} \cdot \frac{m_z^*(0)}{\hbar} \omega_{12}\left(E_{50}\right)=1$.

8.3 Result and Discussions

Using (8.6) and (8.7) together with the energy band constants as given in Table 2.1 of Chap. 2, the DMR in the quantum limit has been plotted for the nipi structures of Cd_3As_2 as a function of electron concentration as shown in curve (a) of Fig. 8.1. The curve (b) corresponds to $\delta = 0$ and the curve (c) exhibits the dependence of the DMR on n_0, in accordance with the three-band model of Kane, respectively. The plots (d) and (e) correspond to the two-band model of Kane and that of the parabolic energy bands. By comparing the curves (a) and (b) of Fig. 8.1, one can assess the influence of crystal

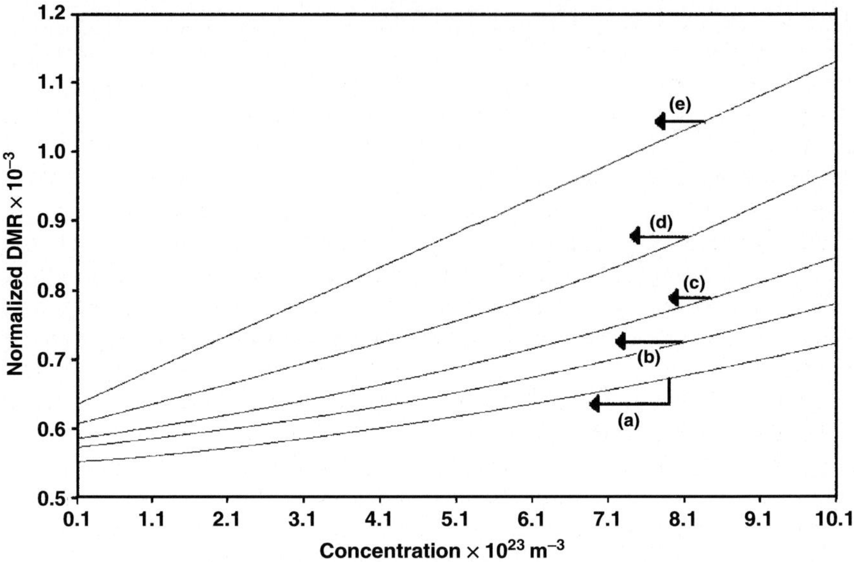

Fig. 8.1. The plot of the DMR in the quantum limit for nipi structures of Cd_3As_2 as a function of electron concentration, in accordance with (**a**) the generalized band model, (**b**) $\delta = 0$, (**c**) the three band model of Kane, (**d**) the two band model of Kane, and (**e**) the parabolic energy bands

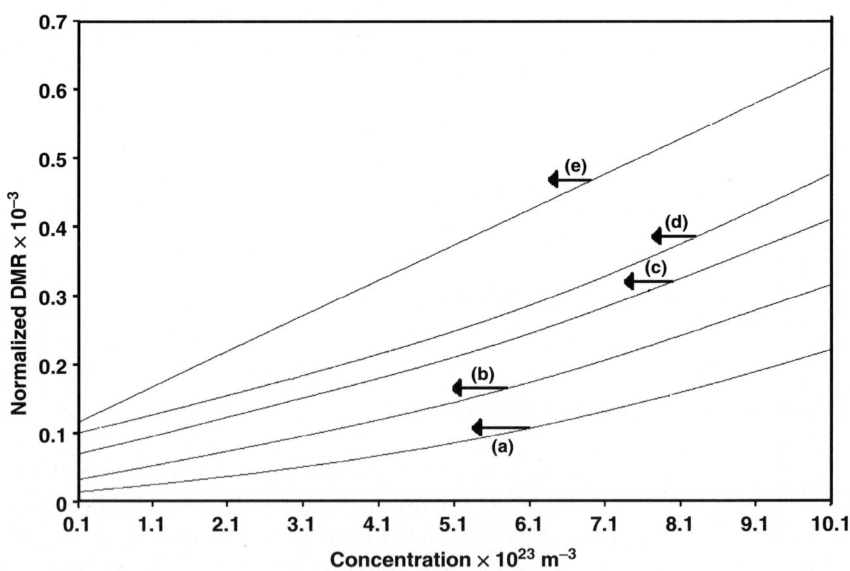

Fig. 8.2. The plot of the DMR in the quantum limit for nipi structures of $CdGeAs_2$ as a function of electron concentration, in accordance with (**a**) the generalized band model, (**b**) $\delta = 0$, (**c**) the three band model of Kane, (**d**) the two band model of Kane, and (**e**) the parabolic energy bands

field splitting of the DMR in nipi structures of Cd_3As_2. Figure 8.2 represents all cases of Fig. 8.1 for nipi structures of $CdGeAs_2$ as an example. It appears from Figs. 8.1 and 8.2 that the DMR in nipi structures of tetragonal materials increases with increasing carrier degeneracy as expected for degenerate materials.

Using appropriate equations one can numerically evaluate the DMR in the quantum limit as a function of electron concentration in nipi structures of III–V compounds by using the nipi structures of InAs, and InSb as shown in Figs. 8.3 and 8.4 by curves (a)–(c) respectively, in accordance with three and two band models of Kane together with the model of parabolic energy bands.

Taking nipi structures of $Hg_{1-x}Cd_xTe$ as an example of ternary compounds, the DMR has been plotted as a function of electron concentration as shown in Fig. 8.5, for all cases of Fig. 8.3. It appears from Fig. 8.5 that the DMR in the quantum limit in both cases of nipi structures of ternary compounds, increases with increasing electron concentration as usual for the degenerate compounds. Taking nipi structures of $In_{1-x}Ga_xAs_yP_{1-y}$ lattice matched to InP as an example of quaternary compounds, the DMR in the quantum limit has been further been plotted as a function of electron concentration as shown in Fig. 8.6, in accordance with the three and two band models of Kane together with the isotropic parabolic energy band model for both the cases.

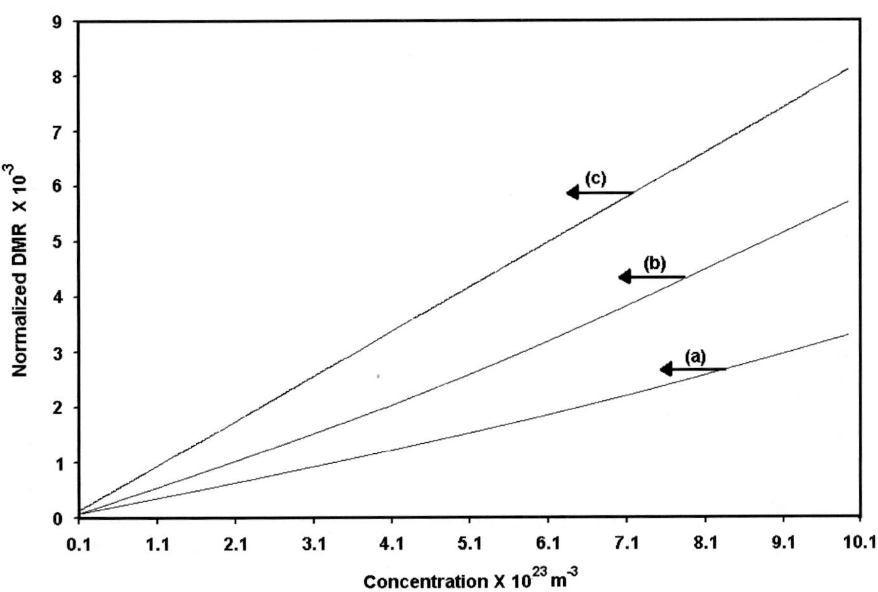

Fig. 8.3. The plot of the DMR in the quantum limit for nipi structures of InAs as a function of electron concentration, in accordance with (**a**) the three band model of Kane, (**b**) the two band model of Kane, and (**c**) the parabolic energy bands

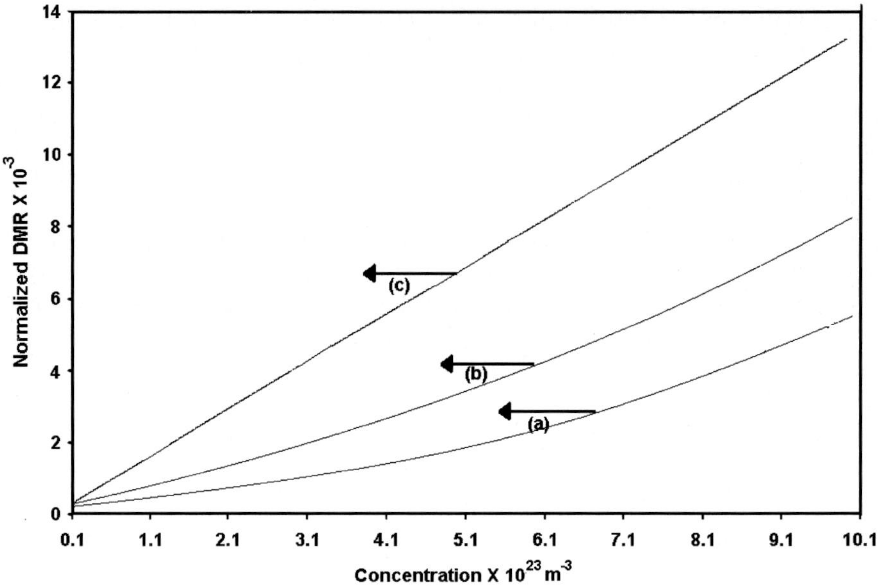

Fig. 8.4. The plot of the DMR in the quantum limit for nipi structures of InSb as a function of electron concentration, in accordance with (**a**) the three band model of Kane, (**b**) the two band model of Kane, and (**c**) the parabolic energy bands

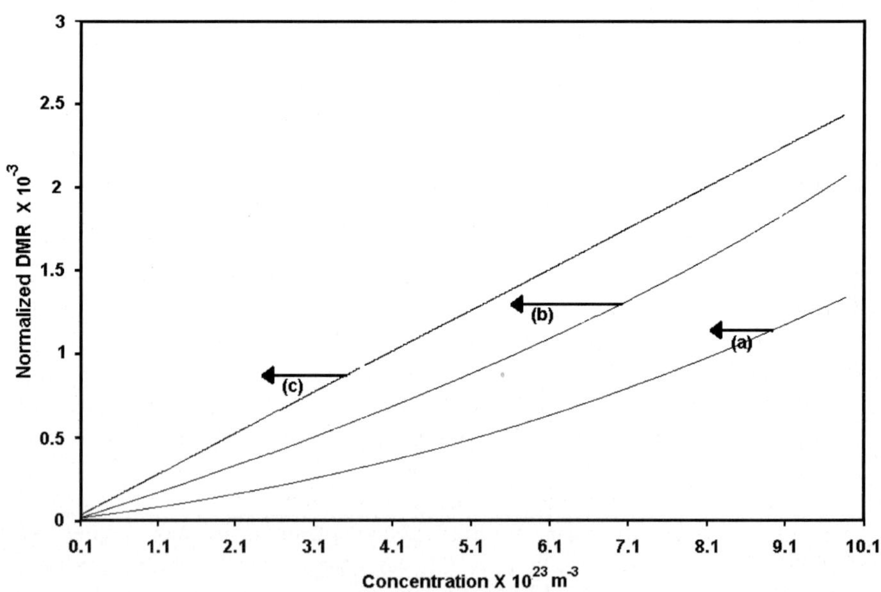

Fig. 8.5. The plot of the DMR in the quantum limit for nipi structures of $Hg_{1-x}Cd_xTe$ as a function of electron concentration in accordance with (**a**) the three band model of Kane, (**b**) the two band model of Kane, and (**c**) the parabolic energy bands

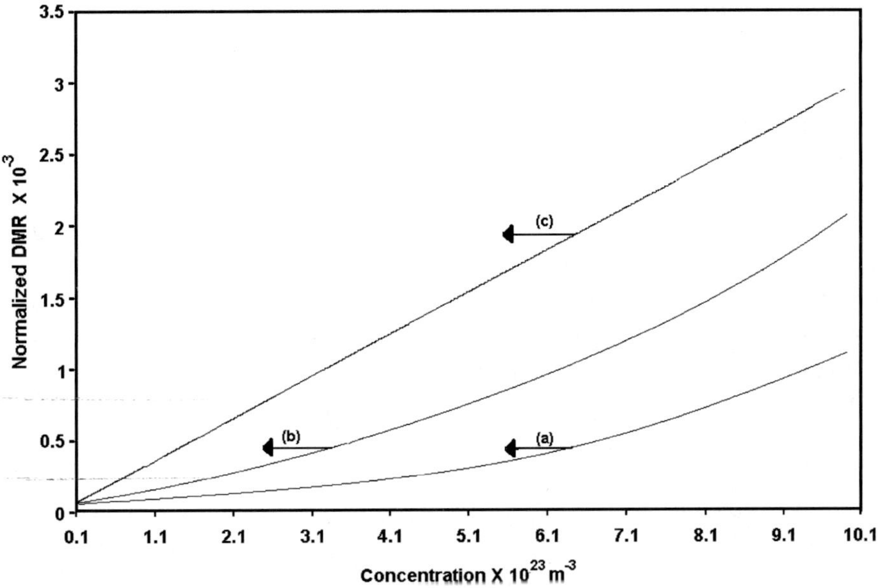

Fig. 8.6. The plot of the DMR in the quantum limit for nipi structures of $In_{1-x}Ga_xAs_yP_{1-y}$ lattice matched to InP as a function of electron concentration, in accordance with (**a**) the three band model of Kane, (**b**) the two band model of Kane, and (**c**) the parabolic energy bands

It appears that the DMR increases with increasing n_0 as usual. From Figs. 8.5 and 8.6, one can assess the influence of the energy band constants on the DMR for nipi structures of ternary and quaternary materials, respectively.

Using (8.24) and (8.25), the DMR in the quantum limit has been plotted for the nipi structures of CdS, as a function of carrier concentration as shown by curves (a) and (b) in Fig. 8.7, for both $\bar{\lambda}_0 \neq 0$ and $\bar{\lambda}_0 = 0$, respectively. This has been presented for the purpose of assessing the influence of the splitting of the two spin states by the spin–orbit coupling and the crystalline field on the DMR for nipi structures of II–VI materials. Using (8.31) and (8.32), in Fig. 8.8, the DMR in the quantum limit has been plotted for the nipi structures of (a) PbTe, (b) PbSnTe, and (c) $Pb_{1-x}Sn_xSe$ as a function of electron concentration, in accordance with the Dimmock model. For relatively low values of electron concentration, the values of the DMR for the three materials exhibit convergence behavior whereas for relatively large values of n_0, the numerical values differ widely from each other in this case. Using (8.39) and (8.40), in Fig. 8.9, the DMR in the quantum limit has been plotted for the nipi structures of stressed InSb as a function of electron concentration.

Plot (a) of Fig. 8.9 exhibits the DMR in the presence of stress while plot (b) shows the same in the absence of stress. In the presence of stress, the magnitude of the DMR increases as compared to the same under stress-free

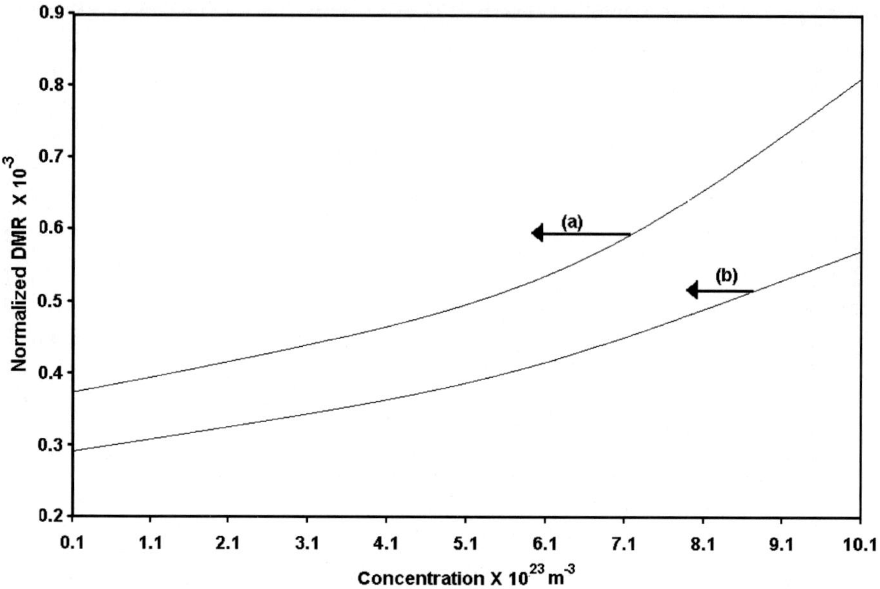

Fig. 8.7. The plot of the DMR in the quantum limit for nipi structures of CdS as a function of carrier concentration, in accordance with (**a**) $\bar{\lambda}_0 \neq 0$ and (**b**) $\bar{\lambda}_0 = 0$

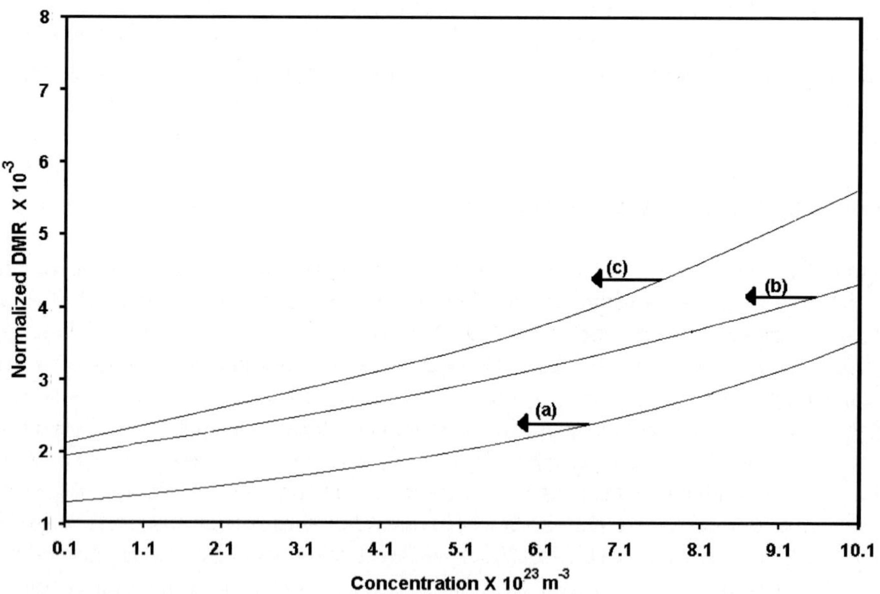

Fig. 8.8. The plot of the DMR in the quantum limit as a function of electron concentration for the nipi structures of (**a**) PbTe, (**b**) PbSnTe, and (**c**) $Pb_{1-x}Sn_xSe$

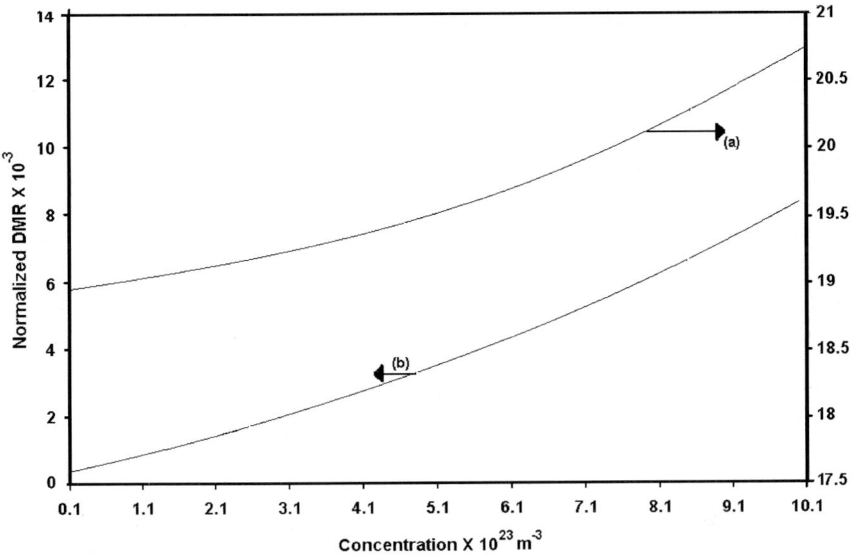

Fig. 8.9. The plot of the DMR in the quantum limit as a function of electron concentration for the nipi structures of stressed InSb in which curve (**a**) is in the presence of stress, and curve (**b**) is under absence of stress

conditions. For the purpose of condensed presentation, the specific electron statistics related to a particular energy dispersion law for specific materials and the Einstein relation in the quantum limit have been presented in Table 8.1.

8.4 Open Research Problems

R.8.1 Investigate the Einstein relation in the presence of an arbitrarily oriented non-quantizing alternating magnetic field for nipi structures of tetragonal semiconductors, by including the electron spin. Study all the special cases for III–V, ternary and quaternary materials in this context.

R.8.2 Investigate the Einstein relations in nipi structures of IV–VI, II–VI and stressed Kane type compounds, in the presence of an arbitrarily oriented non-quantizing alternating magnetic field by including the electron spin.

R.8.3 Investigate the Einstein relation for nipi structures of all the materials as stated in R.2.1 of Chap. 2 by considering the occupancy of the electrons in various minibands.

R.8.4 Investigate the Einstein relation for all the problems from R.8.1 to R.8.3, in the presence of an additional arbitrarily oriented non-uniform electric field.

Table 8.1. The carrier statistics and the Einstein relation in the nipi structures of tetragonal, III–V, ternary, quaternary, II–VI, IV–VI and stressed materials

Type of materials	The carrier statistics in quantum limit	The Einstein relation for the diffusivity mobility ratio in quantum limit
1. Tetragonal compounds	In accordance with the generalized dispersion relation as formulated in this chapter $$\bar{n}_0 = \left[\frac{g_v}{2\pi d_0}\right]\left[\psi_1\left(\bar{E}_{F0}\right) - \psi_3\left(\bar{E}_{F0}\right)\frac{m^*_{\|}}{\hbar}\omega_8\left(\bar{E}_{F0}\right)\right] \times \left[\psi_2\left(\bar{E}_{F0}\right)\right]^{-1} \quad (8.6)$$	$$\frac{D}{\mu} = \left(\frac{\bar{n}_0}{\lvert e\rvert}\right)\left[x_1\left(\bar{E}_{F0}\right) - x_2\left(E_{F0}\right)\right] \quad (8.7)$$
2. III–V, ternary and quaternary compounds	(a) In accordance with the three band model of Kane which is a special case of our generalized analysis $$\bar{n}_0 = \left(\frac{m^* g_v}{\pi\hbar^2 d_0}\right)\left[I\left(E_{20}\right) - \left\{(1/2)\,\hbar\omega_9\left(E_{20}\right)\right\}\right] \quad (8.14)$$ (8.14) is a special case of (8.6) (b) For the two band model of Kane the form of the expression of $\bar{n}_0$ is given by (8.14) where $I(E) = E(1+\alpha E)$, $I'(E) = (1+2\alpha E)$ and $I''(E) = 2\alpha$. (c) For parabolic energy bands, the form of the expression of $\bar{n}_0$ is given by (8.14) where $I(E) = E$, $I'(E) = 1$ and $I''(E) = 0$.	(a) $$\frac{D}{\mu} = \left(\frac{\bar{n}_0}{\lvert e\rvert}\right)\left[x_3\left(\bar{E}_{F0}\right) - x_4\left(E_{20}\right)\right] \quad (8.16)$$ (8.16) is a special case of (8.7). (b) For the two band model of Kane the form of the expression of DMR is given by (8.16) where $I(E) = E(1+\alpha E)$, $I'(E) = (1+2\alpha E)$ and $I''(E) = 2\alpha$. (c) For parabolic energy bands, the form of the expression of DMR is given by (8.14) where $I(E) = E$, $I'(E) = 1$ and $I''(E) = 0$

3. II–VI compounds	$\bar{n}_0 = \frac{m_{\perp}^{*} g_{\mathrm{v}}}{\pi \hbar^2 d_0} \left[\bar{E}_{F0} - \frac{1}{2} \hbar \omega_{10} - 2a_{81} \left[\sqrt{\bar{E}_{\mathrm{F0}} + \frac{1}{4a_0'} \left[\left(\bar{\lambda}_0 \right)^2 - 2a_0' \hbar \omega_{10} \right]} - a_{81} \right] \right]$ (8.24)	$\frac{D}{\mu} = \frac{\bar{n}_0}{\lvert e \rvert} \left[x_5 \left(\bar{E}_{\mathrm{F0}} \right) - x_6 \left(\bar{n}_0 \right) \right]$ (8.25)
4. IV–VI compounds	$\bar{n}_0 = \frac{g_{\mathrm{v}}}{2\pi \hbar^2 S_{19} d_0} \left[- S_{20} \left(\bar{E}_{\mathrm{F0}}, 0 \right) + \sqrt{\left[S_{20} \left(\bar{E}_{\mathrm{F0}}, 0 \right) \right]^2 + 4 S_{19} S_{21} \left(\bar{E}_{\mathrm{F0}}, 0 \right)} \right]$ (8.31)	$\frac{D}{\mu} = \frac{\bar{n}_0}{\lvert e \rvert} \left[x_7 \left(\bar{E}_{\mathrm{F}_0} \right) - x_8 \left(E_{40} \right) \right]$ (8.32)
5. Stressed compounds	$\bar{n}_0 = \frac{g_{\mathrm{v}}}{2\pi d_0} \bar{a}_0 \left(\bar{E}_{\mathrm{F}_0} \right) \bar{b}_0 \left(\bar{E}_{\mathrm{F}_0} \right) \left[1 - \frac{m_z^{*}(0)}{\left(\bar{c}_0 \left(\bar{E}_{\mathrm{F}_0} \right) \right)^2 \hbar} \omega_{12} \left(\bar{E}_{\mathrm{F}_0} \right) \right]$ (8.39)	$\frac{D}{\mu} = \frac{\bar{n}_0}{\lvert e \rvert} \left[x_9 \left(\bar{E}_{\mathrm{F}_0} \right) - x_{10} \left(E_{50} \right) \right]$ (8.40)

R.8.5 Investigate the Einstein relation for all the problems from R.8.1 to R.8.3, in the presence of arbitrarily oriented crossed electric and magnetic fields.

Allied Research Problems

R.8.6 Investigate the EMM for all the problems from R.8.1 to R.8.5.

R.8.7 Investigate the Debye screening length, the carrier contribution to the elastic constants, heat capacity, activity coefficient, plasma frequency, and the thermoelectric power for all the materials, covering all the cases of problems R.8.1–R.8.5 of this chapter.

R.8.8 Investigate in detail, the mobility for elastic and inelastic scattering mechanisms for all the materials, covering all the cases of problems R.8.1–R.8.5 of this chapter.

R.8.9 Investigate the various transport coefficients in detail for all the materials, covering all the cases of problems R.8.1–R.8.5 of this chapter.

R.8.10 Investigate the dia- and paramagnetic susceptibilities in detail for all the materials, covering all the appropriate problems of this chapter.

References

1. N.G. Anderson, W.D. Laidig, R.M. Kolbas, Y.C. Lo, J. Appl. Phys. **60**, 2361 (1986)
2. F. Capasso, Semiconduct. Semimetals **22**, 2 (1985)
3. F. Capasso, K. Mohammed, A.Y. Cho, R. Hull, A.L. Hutchinson, Appl. Phys. Lett. **47**, 420 (1985)
4. F. Capasso, R.A. Kiehl, J. Appl. Phys. **58**, 1366 (1985)
5. K. Ploog, G.H. Doheler, Adv. Phys. **32**, 285 (1983)
6. F. Capasso, K. Mohammed, A.Y. Cho, Appl. Phys. Lett. **478** (1986)
7. R. Grill, C. Metzner, G.H. Döhler, Phys. Rev. B **63**, 235316 (2001); J. Z. Wang, Z.G. Wang, Z.M. Wang, S.L. Feng, Z. Yang, Phys. Rev. B **62**, 6956 (2000); J. Z. Wang, Z.G. Wang, Z.M. Wang, S.L. Feng, Z. Yang, Phys. Rev. B **61**, 15614 (2000)
8. A.R. Kost, M.H. Jupina, T.C. Hasenberg, E.M. Garmire, J. Appl. Phys. **99**, 023501 (2006)
9. A.G. Smirnov, D.V. Ushakov, V.K. Kononenko, Proc. SPIE **4706**, 70 (2002)
10. D.V. Ushakov, V.K. Kononenko, I.S. Manak, Proc. SPIE **4358**, 171 (2001)
11. J.Z. Wang, Z.G. Wang, Z.M. Wang, S.L. Feng, Z. Yang, Phys. Rev. B **62**, 6956 (2000).
12. A.R. Kost, L. West, T.C. Hasenberg, J.O. White, M. Matloubian, G.C. Valley, Appl. Phys. Lett. **63**, 3494 (1993)
13. S. Bastola, S.J. Chua, S.J. Xu, J. Appl. Phy. **83**, 1476 (1998)
14. Z.J. Yang, E.M. Garmire, D. Doctor, J. Appl. Phys. **82**, 3874 (1997)
15. G.H. Avetisyan, V.B. Kulikov, I.D. Zalevsky, P.V. Bulaev, Proc. SPIE **2694**, 216 (1996)

16. U. Pfeiffer, M. Kneissl, B. Knüpfer, N. Müller, P. Kiesel, G.H. Döhler, J.S. Smith, Appl. Phys. Lett. **68**, 1838 (1996)
17. H.L. Vaghjiani, E.A. Johnson, M.J. Kane, R. Grey, C.C. Phillips, J. Appl. Phys. **76**, 4407 (1994)
18. P. Kiesel, K.H. Gulden, A. Hoefler, M. Kneissl, B. Knuepfer, S.U. Dankowski, P. Riel, X.X. Wu, J.S. Smith, G.H. Doehler, Proc. SPIE **1985**, 278 (1993)
19. G.H. Doheler, Phys. Script. **24** 430 (1981)
20. S. Mukherjee, S.N. Mitra, P.K. Bose, A.R. Ghatak, A. Neoigi, J.P. Banerjee, A. Sinha, M. Pal, S. Bhattacharya, K.P. Ghatak, J. Comput. Theor. Nanosci. **4**, 550 (2007)

9

The Einstein Relation in Superlattices of Compound Semiconductors in the Presence of External Fields

9.1 Introduction

It is well known that Keldysh [1] first suggested the fundamental concept of a superlattice (SL), although it was successfully experimentally realized by Esaki and Tsu [2]. The importance of SLs in the field of nanoelectronics have already been described in [3–5]. The most extensively studied III–V SL is the one consisting of alternate layers of GaAs and $Ga_{1-x}Al_xAs$ owing to the relative ease of fabrication. The GaAs layers form quantum wells and $Ga_{1-x}Al_xAs$ form potential barriers. The III–V SL's are attractive for the realization of high speed electronic and optoelectronic devices [6]. In addition to SLs with the usual structure, SLs with more complex structures such as II–VI [7], IV–VI [8] and HgTe/CdTe [9] SL's have also been proposed. The IV–VI SLs exhibit quite different properties as compared to the III–V SL due to the peculiar band structure of the constituent materials [10]. The epitaxial growth of II–VI SL is a relatively recent development and the primary motivation for studying these SLs made of materials with large band gap is their potential for optoelectronic operation in the blue [10]. HgTe/CdTe SL's have attracted a great deal of attention since 1979 as promising new materials for long wavelength infrared detectors and other electro-optical applications [11]. Interest in Hg-based SL's has increased further as new properties with potential device applications have been revealed [11, 12]. These features arise from the unique zero band gap material HgTe [13] and the direct band gap semiconductor CdTe, which can be described by the three band mode of Kane [14]. The combination of the aforementioned materials with specified dispersion relation makes HgTe/CdTe SL very attractive, especially because of the possibility of tailoring the material properties for various applications by varying the energy band constants of the SLs. In addition, for effective mass SLs, the electronic subbands appear continually in real space [15].

We note that all the SLs have been proposed on the assumption that the interfaces between the layers are sharply defined and are of zero thickness, i.e., devoid of any interface effects. The SL potential distribution may be then considered as a one dimensional array of rectangular potential wells. The advanced experimental techniques may produce SLs with physical crystallographically abrupt interfaces between the two materials; adjoining their interface will change at least on an atomic scale. As the potential form changes from a well (barrier) to a barrier (well), an intermediate potential region exists for the electrons. The influence of the finite thickness of the interfaces on the electron dispersion law is very important, as the electron energy spectrum governs the electron transport in SLs.

In this chapter, we shall study the Einstein relation under magnetic quantization in III–V, II–VI, IV–VI and HgTe/CdTe SLs with graded interfaces in Sects. 9.2.1–9.2.4. From Sects. 9.2.5 to 9.2.8, we shall investigate the DMR in III–V, II–VI, IV–VI and HgTe/CdTe effective mass SLs in the presence of a quantizing magnetic field, respectively. In Sects. 9.2.9–9.2.16, we shall investigate the DMR in the SLs in the presence of two dimensional size quantizations. In Sect. 9.3, the doping and magnetic field dependences of the DMRs have been studied by taking $GaAs/Ga_{1-x}Al_xAs$, CdS/CdTe, PbTe/PbSnTe and HgTe/CdTe SLs and the corresponding effective mass SLs as examples. In the same section, we also discuss the doping and thickness dependences of the DMR for the said quantum wire SLs.

9.2 Theoretical Background

9.2.1 Einstein Relation Under Magnetic Quantization in III–V Superlattices with Graded Interfaces

The energy spectrum of the conduction electrons in bulk specimens of the constituent materials of III–V SLs whose energy band structures are defined by the three band model of Kane can be written from (2.6) as

$$(\hbar^2 k^2) \,/\, (2m_i^*) = EG(E, E_{\mathrm{gi}}, \Delta_i), \tag{9.1}$$

$i = 1, 2, \ldots, G(E, E_{\mathrm{gi}}, \Delta_i) \equiv \frac{\left(E_{\mathrm{gi}}+\frac{2}{3}\Delta_i\right)(E+E_{\mathrm{gi}}+\Delta_i)(E+E_{\mathrm{gi}})}{\left[E_{\mathrm{gi}}(E_{\mathrm{gi}}+\Delta_i)\left(E+E_{\mathrm{gi}}+\frac{2}{3}\Delta_i\right)\right]}$. Therefore, the dispersion law of the electrons of III–V SLs with graded interfaces can be expressed, following Jiang and Lin [16], as

$$\cos\left(L_0 k\right) = \frac{1}{2}\Phi\left(E, k_{\mathrm{s}}\right), \tag{9.2}$$

where $L_0\,(\equiv a_0 + b_0)$ is the period length, a_0 and b_0 are the widths of the barrier and the well respectively,

$$\Phi\left(E, k_{\mathrm{s}}\right) \equiv \left[2\cosh\left\{\beta\left(E, k_{\mathrm{s}}\right)\right\}\cos\left\{\gamma\left(E, k_{\mathrm{s}}\right)\right\} + \varepsilon\left(E, k_{\mathrm{s}}\right)\sinh\left\{\beta\left(E, k_{\mathrm{s}}\right)\right\}\sin\left\{\gamma\left(E, k_{\mathrm{s}}\right)\right\}\right.$$

$$+\Delta_0\left[\left(\frac{K_1^2(E,k_s)}{K_2(E,k_s)}-3K_2(E,k_s)\right)\cosh\{\beta(E,k_s)\}\sin\{\gamma(E,k_s)\}+\left(3K_1(E,k_s)\right.\right.$$

$$\left.\left.-\frac{\{K_2(E,k_s)\}^2}{K_1(E,k_s)}\right)\sinh\{\beta(E,k_s)\}\cos\{\gamma(E,k_s)\}\right]$$

$$+\Delta_0\left[2\left(\{K_1(E,k_s)\}^2-\{K_2(E,k_s)\}^2\right)\cosh\{\beta(E,k_s)\}\cos\{\gamma(E,k_s)\}\right.$$

$$+\frac{1}{12}\left[\frac{5\{K_2(E,k_s)\}^3}{K_1(E,k_s)}+\frac{5\{K_1(E,k_s)\}^3}{K_2(E,k_s)}-34K_2(E,k_s)K_1(E,k_s)\right]$$

$$\left.\left.\times\sinh\{\beta(E,k_s)\}\sin\{\gamma(E,k_s)\}\right]\right],$$

$\varepsilon(E,k_s)\equiv\left[\frac{K_1(E,k_s)}{K_2(E,k_s)}-\frac{K_2(E,k_s)}{K_1(E,k_s)}\right]$, $\beta(E,k_s)\equiv K_1(E,k_s)[a_0-\Delta_0]$, Δ_0 is the interface width, $\gamma(E,k_s)=K_2(E,k_s)[b_0-\Delta_0]$, $K_1(E,k_s)\equiv\left[\frac{2m_2^*E'}{\hbar^2}G(E-V_0,\alpha_2,\Delta_2)+k_s^2\right]^{1/2}$, $E'\equiv V_0-E$, V_0 is the potential barrier encountered by the electron $(V_0\equiv|E_{g_2}-E_{g_1}|)$, $\alpha_i\equiv 1/E_{gi}$, $K_2(E,k_s)\equiv\left[\frac{2m_1^*E}{\hbar^2}G(E,\alpha_1,\Delta_1)-k_s^2\right]^{1/2}$ and $k_s^2=k_x^2+k_y^2$.

In the presence of a quantizing magnetic field B along the z-direction, the simplified magneto-dispersion relation can be written following (3.11) as

$$k_z=\frac{1}{L_0}\left[\rho(n,E)-\left\{\frac{2|e|B}{\hbar}L_0^2\left(n+\frac{1}{2}\right)\right\}\right]^{1/2},\tag{9.3}$$

where $\rho(n,E)=\left[\cos^{-1}\left\{\frac{1}{2}\psi(n,E)\right\}\right]^2$, n is the Landau quantum number,

$$\psi(n,E)=[2\cosh\{\beta(n,E)\}\cos\{\gamma(n,E)\}+\varepsilon(n,E)\sinh\{\beta(n,E)\}$$

$$\sin\{\gamma(n,E)\}+\Delta_0\left[\left(\frac{\{K_1(n,E)\}^2}{K_2(n,E)}-3K_2(n,E)\right)\cosh\{\beta(n,E)\}\right.$$

$$\left.\sin\{\gamma(n,E)\}+\left(3K_1(n,E)-\frac{\{K_2(n,E)\}^2}{K_1(n,E)}\right)\sinh\{\beta(n,E)\}\cos\{\gamma(n,E)\}\right.$$

$$+\Delta_0\left[2\left(\{K_1(n,E)\}^2-\{K_2(n,E)\}^2\right)\cosh\{\beta(n,E)\}\cos\{\gamma(n,E)\}\right.$$

$$+\frac{1}{12}\left(\frac{5\{K_1(n,E)\}^3}{K_2(n,E)}+\frac{5\{K_2(n,E)\}^3}{K_1(n,E)}-\{34K_2(n,E)K_1(n,E)\}\right)$$

$$\left.\left.\times\sinh\{\beta(n,E)\}\sin\{\gamma(n,E)\}\right]\right]$$

$$\varepsilon(n,E)\equiv\left[\frac{K_1(n,E)}{K_2(n,E)}-\frac{K_2(n,E)}{K_1(n,E)}\right],\ \beta(n,E)\equiv K_1(n,E)[a_0-\Delta_0],$$

$$K_1(n,E)\equiv\left[\frac{2m_2^*E'}{\hbar^2}G(E-V_0,\alpha_2,\Delta_2)+\frac{2|e|B}{\hbar}\left(n+\frac{1}{2}\right)\right]^{1/2}$$

The electron concentration in this case can be expressed as

$$n_0 = \left(\frac{|e|\,Bg_{\mathrm{v}}}{\pi^2\hbar L_0}\right)\sum_{n=0}^{n_{\max}}\left[T_{95}\left(n, E_{\mathrm{FSL}}\right)+T_{96}\left(n, E_{\mathrm{FSL}}\right)\right], \tag{9.21}$$

where

$$T_{95}\left(n, E_{\mathrm{FSL}}\right) \equiv \left[\rho_2\left(n, E_{\mathrm{FSL}}\right)-\left\{\frac{2\,|e|\,B}{\hbar}\left(n+\frac{1}{2}\right)L_0^2\right\}\right]^{1/2}$$

and

$$T_{96}\left(n, E_{\mathrm{FSL}}\right) \equiv \sum_{r=1}^{s} L(r)\left[T_{95}\left(n, E_{\mathrm{FSL}}\right)\right].$$

The use of (9.21) and (1.11) leads to the expression of the DMR as

$$\frac{D}{\mu} = \frac{1}{|e|}\frac{\sum\limits_{n=0}^{n_{\max}}\left[T_{95}\left(n, E_{\mathrm{FSL}}\right)+T_{96}\left(n, E_{\mathrm{FSL}}\right)\right]}{\sum\limits_{n=0}^{n_{\max}}\left[\left\{T_{95}\left(n, E_{\mathrm{FSL}}\right)\right\}'+\left\{T_{96}\left(n, E_{\mathrm{FSL}}\right)\right\}'\right]}. \tag{9.22}$$

9.2.4 Einstein Relation Under Magnetic Quantization in HgTe/CdTe Superlattices with Graded Interfaces

The dispersion relation of the conduction electrons of the constituent materials of HgTe/CdTe SLs can be expressed [13] as

$$E = \frac{\hbar^2 k^2}{2m_1^*}+\frac{3\,|e|^2\,k}{128\varepsilon_{\mathrm{sc}}}, \tag{9.23}$$

$$\frac{\hbar^2 k^2}{2m_2^*} = EG\left(E_1 E_{\mathrm{g}2}, \Delta_2\right). \tag{9.24}$$

The electron energy dispersion law in HgTe/CdTe SL is given by

$$\cos\left(L_0 k\right) = \frac{1}{2}\Phi_3\left(E, k_{\mathrm{s}}\right), \tag{9.25}$$

where

$$\Phi_3\left(E, k_{\mathrm{s}}\right)$$
$$\equiv \left[2\cosh\left\{\beta_3\left(E, k_{\mathrm{s}}\right)\right\}\cos\left\{\gamma_3\left(E, k_{\mathrm{s}}\right)\right\}+\varepsilon_3\left(E, k_{\mathrm{s}}\right)\sinh\left\{\beta_3\left(E, k_{\mathrm{s}}\right)\right\}\sin\left\{\gamma_3\left(E, k_{\mathrm{s}}\right)\right\}\right.$$

$$+\Delta_0\left[\left(\frac{\{K_7(E,k_s)\}^2}{K_8(E,k_s)}-3K_8(E,k_s)\right)\cosh\{\beta_3(E,k_s)\}\sin\{\gamma_3(E,k_s)\}\right.$$
$$\left.+\left(3K_7(E,k_s)-\frac{\{K_8(E,k_s)\}^2}{K_7(E,k_s)}\right)\sinh\{\beta_3(E,k_s)\}\cos\{\gamma_3(E,k_s)\}\right]$$
$$+\Delta_0\left[2\left(\{K_7(E,k_s)\}^2-\{K_8(E,k_s)\}^2\right)\cosh\{\beta_3(E,k_s)\}\cos\{\gamma_3(E,k_s)\}\right.$$
$$+\frac{1}{12}\left[\frac{5\{K_8(E,k_s)\}^3}{K_7(E,k_s)}+\frac{5\{K_7(E,k_s)\}^3}{K_8(E,k_s)}-34K_7(E,k_s)K_8(E,k_s)\right]$$
$$\left.\left.\times\sinh\{\beta_3(E,k_s)\}\sin\{\gamma_3(E,k_s)\}\right]\right],$$

$$\varepsilon_3(E,k_s)\equiv\left[\frac{K_7(E,k_s)}{K_8(E,k_s)}-\frac{K_8(E,k_s)}{K_7(E,k_s)}\right],\ \beta_3(E,k_s)\equiv K_7(E,k_s)\left[a_0-\Delta_0\right],$$

$$\gamma_3(E,k_s)=K_8(E,k_s)\left[b_0-\Delta_0\right],$$

$$K_8(E,k_x,k_y)\equiv\left[\frac{B_0^2+2AE-B_0\sqrt{B_0^2+4AE}}{2A^2}-k_s^2\right]^{1/2},\ B_0=\frac{3|e|^2}{128\varepsilon_{sc}},$$

$$A=\frac{\hbar^2}{2m_1^*}\ \text{and}\ K_7(E,k_s)\equiv\left[\frac{2m_2^*E'}{\hbar^2}G(E-V_0,E_{g_2},\Delta_2)+k_s^2\right]^{1/2}.$$

Following the same method given in Sect. 9.2.2, we get

$$k_z=\frac{1}{L_0}\left[\rho_3(n,E)-\left\{\frac{2|e|B}{\hbar}L_0^2\left(n+\frac{1}{2}\right)\right\}\right]^{1/2},\qquad(9.26)$$

where $\rho_3(n,E)=\left[\cos^{-1}\left\{\frac{1}{2}\psi_3(n,E)\right\}\right]^2$, $\psi_3(n,E)=[\,2\cosh\{\beta_3(n,E)\}$ $\cos\{\gamma_3(n,E)\}+\varepsilon_3(n,E)\sinh\{\beta_3(n,E)\}$,

$$\sin\{\gamma_3(n,E)\}+\Delta_0\left[\left(\frac{\{K_7(n,E)\}^2}{K_8(n,E)}-3K_8(n,E)\right)\cosh\{\beta_3(n,E)\}\right.$$
$$\left.\sin\{\gamma_3(n,E)\}+\left(3K_7(n,E)-\frac{\{K_8(n,E)\}^2}{K_7(n,E)}\right)\sinh\{\beta_3(n,E)\}\cos\{\gamma_3(n,E)\}\right.$$
$$+\Delta_0\left[2\left(\{K_7(n,E)\}^2-\{K_8(n,E)\}^2\right)\cosh\{\beta_3(n,E)\}\cos\{\gamma_3(n,E)\}\right.$$
$$+\frac{1}{12}\left(\frac{5\{K_7(n,E)\}^3}{K_8(n,E)}+\frac{5\{K_8(n,E)\}^3}{K_7(n,E)}-\{34K_8(n,E)K_7(n,E)\}\right)$$
$$\left.\left.\times\sinh\{\beta_3(n,E)\}\sin\{\gamma_3(n,E)\}\right]\right],$$

$$\varepsilon_3(n,E)\equiv\left[\frac{K_7(n,E)}{K_8(n,E)}-\frac{K_8(n,E)}{K_7(n,E)}\right],\ \beta_3(n,E)\equiv K_7(n,E)\left[a_0-\Delta_0\right],$$

$$\gamma_3(n,E) = K_8(n,E)\left[b_0 - \Delta_0\right],$$

$$K_8(n,E) \equiv \left[\frac{B_0^2 + 2AE - B_0\sqrt{B_0^2 + 4AE}}{2A^2} - \left\{\frac{2|e|B}{\hbar}\left(n + \frac{1}{2}\right)\right\}\right]^{1/2},$$

and $K_7(n,E) \equiv \left[\left(\frac{2m_2^* E'}{\hbar^2}\right) G(E - V_0, \alpha_2, \Delta_2) + \left\{\frac{2|e|B}{\hbar}\left(n + \frac{1}{2}\right)\right\}\right]^{1/2}$.

The EMM in this case assumes the form

$$m^*(n, E_{\mathrm{FSL}}) = \frac{\hbar^2}{2L_0^2}\left\{\rho_3(n, E_{\mathrm{FSL}})\right\}'. \tag{9.27}$$

The EMM in HgTe/CdTe SL under magnetic quantization depends on both the Fermi energy and magnetic quantum number.

The Landau subband energies ($E_{n_{\mathrm{SL4}}}$) can be written as

$$\rho_3(n, E_{n_{\mathrm{SL4}}}) = \left[\left\{\frac{2|e|B}{\hbar}L_0^2\left(n + \frac{1}{2}\right)\right\}\right]. \tag{9.28}$$

The electron concentration in this case can be expressed as

$$n_0 = \left(\frac{|e|Bg_{\mathrm{v}}}{\pi^2\hbar L_0}\right)\sum_{n=0}^{n_{\max}}\left[T_{97}(n, E_{\mathrm{FSL}}) + T_{98}(n, E_{\mathrm{FSL}})\right], \tag{9.29}$$

where

$$T_{97}(n, E_{\mathrm{FSL}}) \equiv \left[\rho_3(n, E_{\mathrm{FSL}}) - \left\{\frac{2|e|B}{\hbar}\left(n + \frac{1}{2}\right)L_0^2\right\}\right]^{1/2}$$

and

$$T_{98}(n, E_{\mathrm{FSL}}) \equiv \sum_{r=1}^{s} L(r)\left[T_{97}(n, E_{\mathrm{FSL}})\right].$$

The use of (9.29) and (1.11) leads to the expression of the DMR as

$$\frac{D}{\mu} = \frac{1}{|e|}\frac{\sum_{n=0}^{n_{\max}}\left[T_{97}(n, E_{\mathrm{FSL}}) + T_{98}(n, E_{\mathrm{FSL}})\right]}{\sum_{n=0}^{n_{\max}}\left[\left\{T_{97}(n, E_{\mathrm{FSL}})\right\}' + \left\{T_{98}(n, E_{\mathrm{FSL}})\right\}'\right]}. \tag{9.30}$$

9.2.5 Einstein Relation Under Magnetic Quantization in III–V Effective Mass Superlattices

Following Sasaki [15], the electron dispersion law in III–V effective mass superlattices (EMSLs) can be written as

$$k_x^2 = \left[\frac{1}{L_0^2}\left\{\cos^{-1}\left(f(E, k_y, k_z)\right)\right\}^2 - k_\perp^2\right], \tag{9.31}$$

$$+\Delta_0\left[\left(\frac{K_1^2(E,k_s)}{K_2(E,k_s)}-3K_2(E,k_s)\right)\cosh\{\beta(E,k_s)\}\sin\{\gamma(E,k_s)\}+\left(3K_1(E,k_s)\right.\right.$$
$$\left.\left.-\frac{\{K_2(E,k_s)\}^2}{K_1(E,k_s)}\right)\sinh\{\beta(E,k_s)\}\cos\{\gamma(E,k_s)\}\right]$$
$$+\Delta_0\left[2\left(\{K_1(E,k_s)\}^2-\{K_2(E,k_s)\}^2\right)\cosh\{\beta(E,k_s)\}\cos\{\gamma(E,k_s)\}\right.$$
$$+\frac{1}{12}\left[\frac{5\{K_2(E,k_s)\}^3}{K_1(E,k_s)}+\frac{5\{K_1(E,k_s)\}^3}{K_2(E,k_s)}-34K_2(E,k_s)K_1(E,k_s)\right]$$
$$\left.\left.\times\sinh\{\beta(E,k_s)\}\sin\{\gamma(E,k_s)\}\right]\right],$$

$\varepsilon(E,k_s) \equiv \left[\frac{K_1(E,k_s)}{K_2(E,k_s)}-\frac{K_2(E,k_s)}{K_1(E,k_s)}\right]$, $\beta(E,k_s) \equiv K_1(E,k_s)[a_0-\Delta_0]$, Δ_0 is the interface width, $\gamma(E,k_s) = K_2(E,k_s)[b_0-\Delta_0]$, $K_1(E,k_s) \equiv \left[\frac{2m_2^*E'}{\hbar^2}G(E-V_0,\alpha_2,\Delta_2)+k_s^2\right]^{1/2}$, $E' \equiv V_0-E$, V_0 is the potential barrier encountered by the electron $(V_0 \equiv |E_{g_2}-E_{g_1}|)$, $\alpha_i \equiv 1/E_{gi}$, $K_2(E,k_s) = \left[\frac{2m_1^*E}{\hbar^2}G(E,\alpha_1,\Delta_1)-k_s^2\right]^{1/2}$ and $k_s^2 = k_x^2+k_y^2$.

In the presence of a quantizing magnetic field B along the z-direction, the simplified magneto-dispersion relation can be written following (3.11) as

$$k_z=\frac{1}{L_0}\left[\rho(n,E)-\left\{\frac{2|e|B}{\hbar}L_0^2\left(n+\frac{1}{2}\right)\right\}\right]^{1/2}, \tag{9.3}$$

where $\rho(n,E)=\left[\cos^{-1}\left\{\frac{1}{2}\psi(n,E)\right\}\right]^2$, n is the Landau quantum number,

$$\psi(n,E)=[2\cosh\{\beta(n,E)\}\cos\{\gamma(n,E)\}+\varepsilon(n,E)\sinh\{\beta(n,E)\}$$
$$\sin\{\gamma(n,E)\}+\Delta_0\left[\left(\frac{\{K_1(n,E)\}^2}{K_2(n,E)}-3K_2(n,E)\right)\cosh\{\beta(n,E)\}\right.$$
$$\left.\sin\{\gamma(n,E)\}+\left(3K_1(n,E)-\frac{\{K_2(n,E)\}^2}{K_1(n,E)}\right)\sinh\{\beta(n,E)\}\cos\{\gamma(n,E)\}\right.$$
$$+\Delta_0\left[2\left(\{K_1(n,E)\}^2-\{K_2(n,E)\}^2\right)\cosh\{\beta(n,E)\}\cos\{\gamma(n,E)\}\right.$$
$$+\frac{1}{12}\left(\frac{5\{K_1(n,E)\}^3}{K_2(n,E)}+\frac{5\{K_2(n,E)\}^3}{K_1(n,E)}-\{34K_2(n,E)K_1(n,E)\}\right)$$
$$\left.\left.\times\sinh\{\beta(n,E)\}\sin\{\gamma(n,E)\}\right]\right]$$

$$\varepsilon(n,E)\equiv\left[\frac{K_1(n,E)}{K_2(n,E)}-\frac{K_2(n,E)}{K_1(n,E)}\right],\ \beta(n,E)\equiv K_1(n,E)[a_0-\Delta_0],$$

$$K_1(n,E)\equiv\left[\frac{2m_2^*E'}{\hbar^2}G(E-V_0,\alpha_2,\Delta_2)+\frac{2|e|B}{\hbar}\left(n+\frac{1}{2}\right)\right]^{1/2}$$

$$\gamma(n,E) = K_2(n,E_s)[b_0-\Delta_0] \text{ and } K_2(n,E) \equiv \left[\frac{2m_1^*E}{\hbar^2}G(E,\alpha_1,\Delta_1) - \left\{\frac{2|e|B}{\hbar}\left(n+\frac{1}{2}\right)\right\}\right]^{1/2}.$$

The EMM in this case assumes the form

$$m^*(n,E_{\mathrm{FSL}}) = \frac{\hbar^2}{2L_0^2}\{\rho(n,E_{\mathrm{FSL}})\}', \tag{9.4}$$

where E_{FSL} is the Fermi energy in this case. Thus the EMM is a function of both Fermi energy and magnetic quantum number which is an intrinsic property of III–V SLs.

The Landau subband energies $(E_{n_{\mathrm{SL1}}})$ can be expressed as

$$\rho(n,E_{n_{\mathrm{SL1}}}) = \left[\left\{\frac{2|e|B}{\hbar}L_0^2\left(n+\frac{1}{2}\right)\right\}\right]. \tag{9.5}$$

Considering only the lowest miniband, as in an actual SL only the lowest miniband is significantly populated at low temperatures, where the quantum effects become prominent, the electron concentration (n_0) can be written as,

$$n_0 = \left(\frac{|e|Bg_{\mathrm{v}}}{\pi^2\hbar L_0}\right)\sum_{n=0}^{n_{\max}}[T_{91}(n,E_{\mathrm{FSL}})+T_{92}(n,E_{\mathrm{FSL}})], \tag{9.6}$$

where $T_{91}(n,E_{\mathrm{FSL}}) \equiv \left[\rho(E_{\mathrm{FSL}},n) - \left\{\frac{2|e|B}{\hbar}\left(n+\frac{1}{2}\right)L_0^2\right\}\right]^{1/2}$ and $T_{92}(n,E_{\mathrm{FSL}}) \equiv \sum_{r=1}^{s} L(r)[T_{91}(n,E_{\mathrm{FSL}})]$.

The use of (9.6) and (1.11) leads to the expression of the DMR in this case as

$$\frac{D}{\mu} = \frac{1}{|e|}\frac{\sum_{n=0}^{n_{\max}}[T_{91}(n,E_{\mathrm{FSL}})+T_{92}(n,E_{\mathrm{FSL}})]}{\sum_{n=0}^{n_{\max}}[\{T_{91}(n,E_{\mathrm{FSL}})\}'+\{T_{92}(n,E_{\mathrm{FSL}})\}']}. \tag{9.7}$$

9.2.2 Einstein Relation Under Magnetic Quantization in II–VI Superlattices with Graded Interfaces

The energy spectrum of the conduction electrons of the constituent materials of II–VI SLs are given by [17]

$$E = \frac{\hbar^2 k_{\mathrm{s}}^2}{2m_{\perp,1}^*} + \frac{\hbar^2 k_z^2}{2m_{\parallel,1}^*} \pm \bar{\lambda}_0 k_{\mathrm{s}}, \tag{9.8}$$

and

$$\frac{\hbar^2 k^2}{2m_2^*} = EG\left(E, E_{g2}, \Delta_2\right), \tag{9.9}$$

where $m_{\perp,1}^*$ and $m_{\parallel,1}^*$ are the transverse and longitudinal effective electron masses respectively at the edge of the conduction band for the first material. The electron dispersion law in II–VI SLs with graded interfaces can be expressed as

$$\cos\left(L_0 k\right) = \frac{1}{2}\Phi_1\left(E, k_s\right), \tag{9.10}$$

where

$$\Phi_1\left(E,k_s\right) \equiv \Bigg[2\cosh\left\{\beta_1\left(E,k_s\right)\right\}\cos\left\{\gamma_1\left(E,k_s\right)\right\} + \varepsilon_1\left(E,k_s\right)\sinh\left\{\beta_1\left(E,k_s\right)\right\}$$
$$\times\sin\left\{\gamma_1\left(E,k_s\right)\right\} + \Delta_0\left[\left(\frac{\left\{K_3\left(E,k_s\right)\right\}^2}{K_4\left(E,k_s\right)} - 3K_4\left(E,k_s\right)\right)\right.$$
$$\times\cosh\left\{\beta_1\left(E,k_s\right)\right\}\sin\left\{\gamma_1\left(E,k_s\right)\right\} + \left(3K_3\left(E,k_s\right)\right.$$
$$\left.\left.\frac{\left\{K_4\left(E,k_s\right)\right\}^2}{K_3\left(E,k_s\right)}\right)\sinh\left\{\beta_1\left(E,k_s\right)\right\}\cos\left\{\gamma_1\left(E,k_s\right)\right\}\right]$$
$$+\Delta_0\Bigg[2\left(\left\{K_3\left(E,k_s\right)\right\}^2 - \left\{K_4\left(E,k_s\right)\right\}^2\right)\cosh\left\{\beta_1\left(E,k_s\right)\right\}$$
$$\times\cos\left\{\gamma_1\left(E,k_s\right)\right\} + \frac{1}{12}\left[\frac{5\left\{K_3\left(E,k_s\right)\right\}^3}{K_4\left(E,k_s\right)} + \frac{5\left\{K_4\left(E,k_s\right)\right\}^3}{K_3\left(E,k_s\right)}\right.$$
$$\left.-34K_4\left(E,k_s\right)K_3\left(E,k_s\right)\right]\sinh\left\{\beta_1\left(E,k_s\right)\right\}\sin\left\{\gamma_1\left(E,k_s\right)\right\}\Bigg]\Bigg],$$

$$\varepsilon_1\left(E,k_s\right) \equiv \left[\frac{K_3\left(E,k_s\right)}{K_4\left(E,k_s\right)} - \frac{K_4\left(E,k_s\right)}{K_3\left(E,k_s\right)}\right],$$

$$\beta_1\left(E,k_s\right) \equiv K_3\left(E,k_s\right)\left[a_0 - \Delta_0\right],\ \gamma_1\left(E,k_s\right) = K_4\left(E,k_s\right)\left[b_0 - \Delta_0\right],$$

$$K_3\left(E,k_s\right) \equiv \left[\frac{2m_2^* E'}{\hbar^2}G(E - V_0, \alpha_2, \Delta_2) + k_s^2\right]^{1/2} \quad \text{and}$$

$$K_4\left(E,k_s\right) \equiv \left[\frac{2m_{\parallel,1}^*}{\hbar^2}\left[E - \frac{\hbar^2 k_s^2}{2m_{\perp,1}^*} \mp \bar{\lambda}_0 k_s\right]\right]^{1/2}.$$

In the presence of a quantizing magnetic field B along the z-direction, the simplified magneto-dispersion relation can be, written as

$$k_z^2 = \frac{1}{L_0^2}\left[\rho_1\left(n,E\right) - \left\{\frac{2\left|e\right|B}{\hbar}L_0^2\left(n + \frac{1}{2}\right)\right\}\right], \tag{9.11}$$

where $\rho_1(n,E) = \left[\cos^{-1}\left\{\frac{1}{2}\psi_1(n,E)\right\}\right]^2$, $\psi_1(n,E) = [\,2\cosh\{\beta_1(n,E)\}$ $\cos\{\gamma_1(n,E)\} + \varepsilon_1(n,E)\sinh\{\beta_1(n,E)\}\times$

$$\times\sin\{\gamma_1(n,E)\} + \Delta_0\left[\left(\frac{\{K_3(n,E)\}^2}{K_4(n,E)} - 3K_4(n,E)\right)\cosh\{\beta_1(n,E)\}\right.$$

$$\sin\{\gamma_1(n,E)\} + \left(3K_3(n,E) - \frac{\{K_4(n,E)\}^2}{K_3(n,E)}\right)\sinh\{\beta_1(n,E)\}\cos\{\gamma_1(n,E)\}$$

$$+\Delta_0\left[2\left(\{K_3(n,E)\}^2 - \{K_4(n,E)\}^2\right)\cosh\{\beta_1(n,E)\}\cos\{\gamma_1(n,E)\}\right.$$

$$+\frac{1}{12}\left(\frac{5\{K_3(n,E)\}^3}{K_4(n,E)} + \frac{5\{K_4(n,E)\}^3}{K_3(n,E)} - \{34K_4(n,E)\,K_3(n,E)\}\right)$$

$$\left.\left.\sinh\{\beta_1(n,E)\}\sin\{\gamma_1(n,E)\}\right]\right],$$

$$\varepsilon_1(n,E) \equiv \left[\frac{K_3(n,E)}{K_4(n,E)} - \frac{K_4(n,E)}{K_3(n,E)}\right],$$

$$\beta_1(n,E) \equiv K_3(n,E)[a_0 - \Delta_0],\ \gamma_1(n,E) = K_4(n,E)[b_0 - \Delta_0],$$

$$K_3(n,E) \equiv \left[\frac{2m_2^*}{\hbar^2}E'G(E - V_0,\alpha_2,\Delta_2) + \frac{2|e|B}{\hbar}\left(n+\frac{1}{2}\right)\right]^{1/2} \text{ and}$$

$$K_4(n,E) \equiv \left[\frac{2m_{\|,1}^*}{\hbar^2}\left[E - \frac{\hbar|e|B}{m_{\perp,1}^*}\left(n+\frac{1}{2}\right) \mp \bar{\lambda}_0\left\{\frac{2|e|B}{\hbar}\left(n+\frac{1}{2}\right)\right\}^{1/2}\right]\right]^{1/2}.$$

The EMM in this case assumes the form

$$m^*(n,E_{\text{FSL}}) = \frac{\hbar^2}{2L_0^2}\{\rho_1(n,E_{\text{FSL}})\}'. \tag{9.12}$$

The EMM in II–VI SLs under magnetic quantization depends on both the Fermi energy and magnetic quantum number, which is the intrinsic property of such SLs.

The Landau subband energies ($E_{n_{\text{SL2}}}$) can be written as

$$\rho_1(n,E_{n_{\text{SL2}}}) = \left[\left\{\frac{2|e|B}{\hbar}L_0^2\left(n+\frac{1}{2}\right)\right\}\right]. \tag{9.13}$$

The electron concentration in this case can be expressed as

$$n_0 = \left(\frac{|e|Bg_{\text{v}}}{2\pi^2\hbar L_0}\right)\sum_{n=0}^{n_{\max}}[T_{93}(n,E_{\text{FSL}}) + T_{94}(n,E_{\text{FSL}})], \tag{9.14}$$

where $T_{93}(n, E_{\mathrm{FSL}}) \equiv \left[\rho_1(n, E_{\mathrm{FSL}}) - \left\{\frac{2|e|B}{\hbar}\left(n+\frac{1}{2}\right)L_0^2\right\}\right]^{1/2}$ and $T_{94}(n, E_{\mathrm{FSL}}) \equiv \sum\limits_{r=1}^{s} L(r)\left[T_{93}(n, E_{\mathrm{FSL}})\right]$.

The use of (9.14) and (1.11) leads to the expression of the DMR as

$$\frac{D}{\mu} = \frac{1}{|e|}\frac{\sum\limits_{n=0}^{n_{\max}}\left[T_{93}(n, E_{\mathrm{FSL}}) + T_{94}(n, E_{\mathrm{FSL}})\right]}{\sum\limits_{n=0}^{n_{\max}}\left[\{T_{93}(n, E_{\mathrm{FSL}})\}' + \{T_{94}(n, E_{\mathrm{FSL}})\}'\right]}. \tag{9.15}$$

9.2.3 Einstein Relation Under Magnetic Quantization in IV–VI Superlattices with Graded Interfaces

The **E–k** dispersion relation of the conduction electrons of the constituent materials of the IV–VI SLs can be expressed [18] as

$$E = a_i k_s^2 + b_i k_z^2 + \left[\left[c_i k_s^2 + d_i k_z^2\right] + \left(e_i k_s^2 + f_i k_z^2 + \frac{E_{g_i}}{2}\right)^2\right]^{1/2} - \frac{E_{g_i}}{2}, \tag{9.16}$$

where $a_i \equiv \left[\frac{\hbar^2}{2m_{\perp,i}^-}\right]$, $b_i \equiv \left(\frac{\hbar^2}{2m_{\|,i}^-}\right)$, $c_i \equiv P_{\|,i}^2$, $d_i \equiv P_{\|,i}^2$, $e_i \equiv \left[\frac{\hbar^2}{2m_{\perp,i}^+}\right]$ and $f_i \equiv \left(\frac{\hbar^2}{2m_{\|,i}^+}\right)$.

The electron dispersion law in IV–VI SLs with graded interfaces can be expressed as

$$\cos(L_0 k) = \frac{1}{2}\Phi_2(E, k_s), \tag{9.17}$$

where

$$\begin{aligned}
\Phi_2(E, k_s) \equiv {} & \left[\, 2\cosh\{\beta_2(E, k_s)\}\cos\{\gamma_2(E, k_s)\} + \varepsilon_2(E, k_s)\sinh\{\beta_2(E, k_s)\}\right. \\
& \times \sin\{\gamma_2(E, k_s)\} + \Delta_0\left[\left(\frac{\{K_5(E, k_s)\}^2}{K_6(E, k_s)} - 3K_6(E, k_s)\right)\right. \\
& \times \cosh\{\beta_2(E, k_s)\}\sin\{\gamma_2(E, k_s)\} + \left(3K_5(E, k_s) - \frac{\{K_6(E, k_s)\}^2}{K_5(E, k_s)}\right) \\
& \left.\times \sinh\{\beta_2(E, k_s)\}\cos\{\gamma_2(E, k_s)\}\right] \\
& + \Delta_0\left[2\left(\{K_5(E, k_s)\}^2 - \{K_6(E, k_s)\}^2\right)\cosh\{\beta_2(E, k_s)\}\cos\{\gamma_2(E, k_s)\}\right. \\
& + \frac{1}{12}\left[\frac{5\{K_5(E, k_s)\}^3}{K_6(E, k_s)} + \frac{5\{K_6(E, k_s)\}^3}{K_5(E, k_s)} - 34K_6(E, k_s)K_5(E, k_s)\right] \\
& \left.\left.\times \sinh\{\beta_2(E, k_s)\}\sin\{\gamma_2(E, k_s)\}\right]\right],
\end{aligned}$$

$$\varepsilon_2(E, k_s) \equiv \left[\frac{K_5(E, k_s)}{K_6(E, k_s)} - \frac{K_6(E, k_s)}{K_5(E, k_s)}\right],$$

$\beta_2(E, k_s) \equiv K_5(E, k_s)\left[a_0 - \Delta_0\right]$, $\gamma_2(E, k_s) = K_6(E, k_s)\left[b_0 - \Delta_0\right]$,

$$K_6(E,k_x,k_y) \equiv \left[\,[EH_{11}+H_{21}(k_x,k_y)] - \left[H_{31}E^2+EH_{41}(k_x,k_y)+H_{51}(k_x,k_y)\right]^{1/2}\right]^{1/2},$$

$$H_{1i} = \left(b_i^2 - f_i^2\right)^{-1},$$

$$H_{2i}(k_x,k_y) = [2H_{1i}]^{-1}\left[E_{g_i}b_i + d_i + f_iE_{g_i} + 2(e_if_i - a_ib_i)\left(k_x^2+k_y^2\right)\right],$$

$$H_{3i} = \frac{f_i^2}{\left(b_i^2-f_i^2\right)^2},$$

$$H_{4i}(k_x,k_y) = \left[4H_{1i}^2\right]^{-1}\left[4b_id_i+4b_if_iE_{g_i}+4f_i^2E_{g_i}+8\left(k_x^2+k_y^2\right)\left[e_if_ib_i-a_if_i^2\right]\right],$$

$$\begin{aligned}H_{5i}(k_x,k_y) \equiv \left[4H_{1i}^2\right]^{-1}&\left[\left(k_x^2+k_y^2\right)^2\left[-8a_ib_ie_if_i+4b_i^2e_i^2+4f_i^2a_i^2\right]+\left(k_x^2+k_y^2\right)\right.\\ &\times\left[4e_if_iE_{g_i}b_i - 4e_if_id_i + 4e_if_i^2E_{g_i} - 4a_ib_i^2E_{g_i} - 4a_ib_id_i\right.\\ &- 4a_ib_if_iE_{g_i} + 4b_i^2e_iE_{g_i} + 4b_i^2c_i + 4b_i^2E_{g_i}a_i\\ &\left.- 4f_i^2e_iE_{g_i} - 4f_i^2c_i - 4f_i^2E_{g_i}a_i\right]\\ &\left.+\left[E_{g_i}^2b_i^2 + d_i^2 + f_i^2g_i^2 + 2E_{g_i}b_id_i + 2E_{g_i}^2b_if_i + 2d_if_iE_{g_i}\right]\right].\end{aligned}$$

and

$$\begin{aligned}K_5(E,k_x,k_y) \equiv \Bigg[&\left[(E-V_0)^2H_{32}+(E-V_0)H_{42}(k_x,k_y)+H_{52}(k_x,k_y)\right]^{1/2}\\ &- [(E-V_0)H_{12}+H_{22}(k_x,k_y)]\Bigg]^{1/2}.\end{aligned}$$

Following the method given in Sect. 9.2.2, the simplified magneto-dispersion relation in this case can be written as

$$k_z = \frac{1}{L_0}\left[\rho_2(n,E) - \left\{\frac{2|e|B}{\hbar}L_0^2\left(n+\frac{1}{2}\right)\right\}\right]^{1/2}, \tag{9.18}$$

where

$$\rho_2(n,E) = \left[\cos^{-1}\left\{\frac{1}{2}\psi_2(n,E)\right\}\right]^2,$$

$$\begin{aligned}\psi_2(n,E) = {}&[2\cosh\{\beta_2(n,E)\}\cos\{\gamma_2(n,E)\} + \varepsilon_2(n,E)\sinh\{\beta_2(n,E)\}\\ &\times\sin\{\gamma_2(n,E)\} + \Delta_0\left[\left(\frac{\{K_5(n,E)\}^2}{K_6(n,E)} - 3K_6(n,E)\right)\right.\\ &\times\cosh\{\beta_2(n,E)\}\sin\{\gamma_2(n,E)\} + \left(3K_5(n,E) - \frac{\{K_6(n,E)\}^2}{K_5(n,E)}\right)\end{aligned}$$

$$\times \sinh\{\beta_2(n,E)\}\cos\{\gamma_2(n,E)\}+\Delta_0\left[2\left(\{K_5(n,E)\}^2-\{K_6(n,E)\}^2\right)\right.$$
$$\times \cosh\{\beta_2(n,E)\}\cos\{\gamma_2(n,E)\}$$
$$+\frac{1}{12}\left(\frac{5\{K_5(n,E)\}^3}{K_6(n,E)}+\frac{5\{K_6(n,E)\}^3}{K_5(n,E)}-\{34K_6(n,E)\,K_5(n,E)\}\right)$$
$$\left.\left.\times \sinh\{\beta_2(n,E)\}\sin\{\gamma_2(n,E)\}\right]\right],$$

$$\varepsilon_2(n,E)\equiv\left[\frac{K_5(n,E)}{K_6(n,E)}-\frac{K_6(n,E)}{K_5(n,E)}\right],\ \beta_2(n,E)\equiv K_5(n,E)\left[a_0-\Delta_0\right],$$

$$K_6(n,E)\equiv\left[\left[EH_{11}+H_{21}(n)\right]-\left[E^2H_{31}+EH_{41}(n)+H_{51}(n)\right]^{1/2}\right]^{1/2},$$

$$H_{2i}(n)=\left[2H_{1i}\right]^{-1}\left[E_{g_i}b_i+d_i+f_iE_{g_i}+2\left(e_if_i-a_ib_i\right)\left\{\frac{2|e|B}{\hbar}\left(n+\frac{1}{2}\right)\right\}\right],$$

$$H_{4i}(n)=\left[4H_{1i}^2\right]^{-1}$$
$$\times\left[4b_id_i+4b_if_iE_{g_i}+4f_i^2E_{g_i}+8\left\{\frac{2|e|B}{\hbar}\left(n+\frac{1}{2}\right)\right\}\left[e_if_ib_i-a_if_i^2\right]\right],$$

$$H_{5i}(n)\equiv\left[4H_{1i}^2\right]^{-1}\left[\left\{\frac{2|e|B}{\hbar}\left(n+\frac{1}{2}\right)\right\}^2\left[-8a_ib_ie_if_i+4b_i^2e_i^2+4f_i^2a_i^2\right]\right.$$
$$+\left\{\frac{2|e|B}{\hbar}\left(n+\frac{1}{2}\right)\right\}\left[4e_if_iE_{g_i}b_i-4e_if_id_i+4e_if_i^2E_{g_i}-4a_ib_i^2E_{g_i}\right.$$
$$-4a_ib_id_i-4a_ib_if_iE_{g_i}+4b_i^2e_iE_{g_i}+4b_i^2c_i+4b_i^2E_{g_i}a_i-4f_i^2e_iE_{g_i}$$
$$\left.-4f_i^2c_i-4f_i^2E_{g_i}a_i\right]+\left[E_{g_i}^2b_i^2+d_i^2+f_i^2g_i^2\right.$$
$$\left.\left.+2E_{g_i}b_id_i+2E_{g_i}^2b_if_i+2d_if_iE_{g_i}\right]\right],$$

and

$$K_5(n,E)\equiv\left[\left[(E-V_0)^2H_{32}+(E-V_0)H_{42}(n)+H_{52}(n)\right]^{1/2}\right.$$
$$\left.-\left[(E-V_0)H_{12}+H_{22}(n)\right]\right]^{1/2}.$$

The EMM in this case assumes the form

$$m^*(n,E_{\mathrm{FSL}})=\frac{\hbar^2}{2L_0^2}\{\rho_2(n,E_{\mathrm{FSL}})\}'. \qquad (9.19)$$

The EMM in IV–VI SLs under magnetic quantization depends on both the Fermi energy and magnetic quantum number, which is the intrinsic property of such SLs.

The Landau subband energies ($E_{n\mathrm{SL3}}$) can be written as

$$\rho_2(n,E_{n\mathrm{SL3}})=\left[\left\{\frac{2|e|B}{\hbar}L_0^2\left(n+\frac{1}{2}\right)\right\}\right]. \qquad (9.20)$$

The electron concentration in this case can be expressed as

$$n_0 = \left(\frac{|e|\, B g_{\mathrm{v}}}{\pi^2 \hbar L_0}\right) \sum_{n=0}^{n_{\max}} \left[T_{95}\left(n, E_{\mathrm{FSL}}\right) + T_{96}\left(n, E_{\mathrm{FSL}}\right)\right], \tag{9.21}$$

where

$$T_{95}\left(n, E_{\mathrm{FSL}}\right) \equiv \left[\rho_2\left(n, E_{\mathrm{FSL}}\right) - \left\{\frac{2\,|e|\, B}{\hbar}\left(n + \frac{1}{2}\right) L_0^2\right\}\right]^{1/2}$$

and

$$T_{96}\left(n, E_{\mathrm{FSL}}\right) \equiv \sum_{r=1}^{s} L\left(r\right)\left[T_{95}\left(n, E_{\mathrm{FSL}}\right)\right].$$

The use of (9.21) and (1.11) leads to the expression of the DMR as

$$\frac{D}{\mu} = \frac{1}{|e|} \frac{\sum\limits_{n=0}^{n_{\max}} \left[T_{95}\left(n, E_{\mathrm{FSL}}\right) + T_{96}\left(n, E_{\mathrm{FSL}}\right)\right]}{\sum\limits_{n=0}^{n_{\max}} \left[\left\{T_{95}\left(n, E_{\mathrm{FSL}}\right)\right\}' + \left\{T_{96}\left(n, E_{\mathrm{FSL}}\right)\right\}'\right]}. \tag{9.22}$$

9.2.4 Einstein Relation Under Magnetic Quantization in HgTe/CdTe Superlattices with Graded Interfaces

The dispersion relation of the conduction electrons of the constituent materials of HgTe/CdTe SLs can be expressed [13] as

$$E = \frac{\hbar^2 k^2}{2m_1^*} + \frac{3\,|e|^2\, k}{128\varepsilon_{\mathrm{sc}}}, \tag{9.23}$$

$$\frac{\hbar^2 k^2}{2m_2^*} = EG\left(E_1 E_{\mathrm{g2}}, \Delta_2\right). \tag{9.24}$$

The electron energy dispersion law in HgTe/CdTe SL is given by

$$\cos\left(L_0 k\right) = \frac{1}{2} \Phi_3\left(E, k_{\mathrm{s}}\right), \tag{9.25}$$

where

$$\Phi_3\left(E, k_{\mathrm{s}}\right) \equiv \Big[2\cosh\left\{\beta_3\left(E, k_{\mathrm{s}}\right)\right\}\cos\left\{\gamma_3\left(E, k_{\mathrm{s}}\right)\right\} + \varepsilon_3\left(E, k_{\mathrm{s}}\right)\sinh\left\{\beta_3\left(E, k_{\mathrm{s}}\right)\right\}\sin\left\{\gamma_3\left(E, k_{\mathrm{s}}\right)\right\}$$

in which, $f(E, k_y, k_z) = a_1 \cos[a_0 C_1(E, k_\perp) + b_0 D_1(E, k_\perp)] - a_2 \cos[a_0 C_1(E, k_\perp) - b_0 D_1(E, k_\perp)]$, $k_\perp^2 = k_y^2 + k_z^2$,

$$a_1 = \left[\sqrt{\frac{m_2^*}{m_1^*}} + 1\right]^2 \left[4\left(\frac{m_2^*}{m_1^*}\right)^{1/2}\right]^{-1}, \; a_2 = \left[-1 + \sqrt{\frac{m_2^*}{m_1^*}}\right]^2 \left[4\left(\frac{m_2^*}{m_1^*}\right)^{1/2}\right]^{-1},$$

$$C_1(E, k_\perp) \equiv \left[\left(\frac{2m_1^* E}{\hbar^2}\right) G(E, E_{g_1}, \Delta_1) - k_\perp^2\right]^{1/2},$$

and $D_1(E, k_\perp) \equiv \left[\left(\frac{2m_2^* E}{\hbar^2}\right) G(E, E_{g_2}, \Delta_2) - k_\perp^2\right]^{1/2}$.

In the presence of an external quantizing magnetic field along the x-direction, the simplified magneto dispersion law in this case can be written as

$$k_x^2 = [\rho_4(n, E)]. \tag{9.32}$$

in which, $\rho_4(n, E) = \frac{1}{L_0^2}\left[\cos^{-1}(f(n, E))\right]^2 - \left\{\frac{2|e|B}{\hbar}\left(n + \frac{1}{2}\right)\right\}$,

$$f(n, E) = a_1 \cos[a_0 C_1(n, E) + b_0 D_1(n, E)] - a_2 \cos[a_0 C_1(n, E) - b_0 D_1(n, E)],$$

$$C_1(n, E) \equiv \left[\left(\frac{2m_1^* E}{\hbar^2}\right) G(E, E_{g_1}, \Delta_1) - \left\{\frac{2|e|B}{\hbar}\left(n + \frac{1}{2}\right)\right\}\right]^{1/2} \text{ and}$$

$$D_1(n, E) \equiv \left[\left(\frac{2m_2^* E}{\hbar^2}\right) G(E, E_{g_2}, \Delta_2) - \left\{\frac{2|e|B}{\hbar}\left(n + \frac{1}{2}\right)\right\}\right]^{1/2}.$$

The EMM in this case assumes the form

$$m^*(n, E_{\mathrm{FSL}}) = \frac{\hbar^2}{2}\{\rho_4(n, E_{\mathrm{FSL}})\}'. \tag{9.33}$$

The EMM in III–V EMSLs under magnetic quantization depends on both the Fermi energy and magnetic quantum number, which is the intrinsic property of such SLs.

The Landau subband energies $(E_{n_{SL5}})$ can be written as

$$\rho_4(E_{n_{\mathrm{SL5}}}, n) = 0. \tag{9.34}$$

The electron concentration in this case can be expressed as

$$n_0 = \left(\frac{|e| B g_{\mathrm{v}}}{\pi^2 \hbar}\right) \sum_{n=0}^{n_{\max}} [T_{99}(n, E_{\mathrm{FSL}}) + T_{910}(n, E_{\mathrm{FSL}})]. \tag{9.35}$$

where $T_{99}(n, E_{\mathrm{FSL}}) \equiv [\rho_4(n, E_{\mathrm{FSL}})]^{1/2}$ and $T_{910}(n, E_{\mathrm{FSL}}) \equiv \sum_{r=1}^{s} L(r) [T_{99}(n, E_{\mathrm{FSL}})]$.

The use of (9.35) and (1.11) leads to the expression of the DMR as

$$\frac{D}{\mu} = \frac{1}{|e|} \frac{\sum_{n=0}^{n_{\max}} [T_{99}(n, E_{\mathrm{FSL}}) + T_{910}(n, E_{\mathrm{FSL}})]}{\sum_{n=0}^{n_{\max}} [\{T_{99}(n, E_{\mathrm{FSL}})\}' + \{T_{910}(n, E_{\mathrm{FSL}})\}']}. \tag{9.36}$$

9.2.6 Einstein Relation Under Magnetic Quantization in II–VI Effective Mass Superlattices

Following Sasaki [15], the electron dispersion law in II–VI EMSLs can be written as

$$k_z^2 = \left[\frac{1}{L_0^2}\left\{\cos^{-1}\left(f_1\left(E, k_x, k_y\right)\right)\right\}^2 - k_s^2\right], \tag{9.37}$$

in which, $f_1\left(E, k_x, k_y\right) = a_3 \cos\left[a_0 C_2\left(E, k_s\right) + b_0 D_2\left(E, k_s\right)\right] - a_4 \cos\left[a_0 C_2\left(E, k_s\right) - b_0 D_2\left(E, k_s\right)\right]$, $k_s^2 = k_x^2 + k_y^2$,

$$a_3 = \left[\sqrt{\frac{m_2^*}{m_{\|,1}^*}} + 1\right]^2 \left[4\left(\frac{m_2^*}{m_{\|,1}^*}\right)^{1/2}\right]^{-1},$$

$$a_4 = \left[-1 + \sqrt{\frac{m_2^*}{m_{\|,1}^*}}\right]^2 \left[4\left(\frac{m_2^*}{m_{\|,1}^*}\right)^{1/2}\right]^{-1},$$

$$C_2\left(E, k_s\right) \equiv \left(\frac{2m_{\|,1}^*}{\hbar^2}\right)^{1/2}\left[E - \frac{\hbar^2 k_s^2}{2m_{\perp,1}^*} \mp \bar{\lambda}_0 k_s\right]^{1/2},$$

and $D_2\left(E, k_s\right) \equiv \left[\left(\frac{2m_2^*}{\hbar^2}\right) EG\left(E, E_{g_2}, \Delta_2\right) - k_s^2\right]^{1/2}$.

Under magnetic quantization along the z-direction, the simplified magneto dispersion law can be expressed as

$$k_z^2 = \left[\rho_5\left(n, E\right)\right], \tag{9.38}$$

in which, $\rho_5\left(n, E\right) = \frac{1}{L_0^2}\left[\cos^{-1}\left(f_1\left(n, E\right)\right)\right]^2 - \left\{\frac{2|e|B}{\hbar}\left(n + \frac{1}{2}\right)\right\}$,

$$f_1\left(n, E\right) = a_3 \cos\left[a_0 C_2\left(n, E\right) + b_0 D_2\left(n, E\right)\right] - a_4 \cos\left[a_0 C_2\left(n, E\right) - b_0 D_2\left(n, E\right)\right],$$

$$C_2\left(E, n\right) \equiv \left(\frac{2m_{\|,1}^*}{\hbar^2}\right)^{1/2}\left[E - \frac{\hbar\,|e|\,B}{m_{\perp,1}^*}\left(n + \frac{1}{2}\right) \mp \bar{\lambda}_0\left\{\frac{2\,|e|\,B}{\hbar}\left(n + \frac{1}{2}\right)\right\}^{1/2}\right]^{1/2}$$

and $D_2\left(n, E\right) \equiv \left[\left(\frac{2m_2^*}{\hbar^2}\right) EG\left(E, E_{g_2}, \Delta_2\right) - \frac{2|e|B}{\hbar}\left(n + \frac{1}{2}\right)\right]^{1/2}$.

The EMM in this case assumes the form

$$m^*\left(n, E_{\mathrm{FSL}}\right) = \frac{\hbar^2}{2}\left\{\rho_5\left(n, E_{\mathrm{FSL}}\right)\right\}'. \tag{9.39}$$

Thus we observe that the EMM in this case depends on both the Fermi energy and magnetic quantum number, which is the intrinsic property of such SLs.

The Landau subband energies ($E_{n_{\mathrm{SL6}}}$) can be written as

$$\rho_5\left(E_{n_{\mathrm{SL6}}}, n\right) = 0. \tag{9.40}$$

The electron concentration in this case can be expressed as

$$n_0 = \left(\frac{|e|\, B g_{\mathrm{v}}}{2\pi^2\hbar}\right) \sum_{n=0}^{n_{\max}} [T_{911}(n, E_{\mathrm{FSL}}) + T_{912}(n, E_{\mathrm{FSL}})], \tag{9.41}$$

where $T_{911}(n, E_{\mathrm{FSL}}) \equiv [\rho_5(n, E_{\mathrm{FSL}})]^{1/2}$ and $T_{912}(n, E_{\mathrm{FSL}}) \equiv \sum_{r=1}^{s} L(r)$ $[T_{911}(n, E_{\mathrm{FSL}})]$.

The use of (9.41) and (1.11) leads to the expression of the DMR as

$$\frac{D}{\mu} = \frac{1}{|e|} \frac{\sum_{n=0}^{n_{\max}} [T_{911}(n, E_{\mathrm{FSL}}) + T_{912}(n, E_{\mathrm{FSL}})]}{\sum_{n=0}^{n_{\max}} [\{T_{911}(n, E_{\mathrm{FSL}})\}' + \{T_{912}(n, E_{\mathrm{FSL}})\}']}. \tag{9.42}$$

9.2.7 Einstein Relation Under Magnetic Quantization in IV–VI Effective Mass Superlattices

Following Sasaki [15], the electron dispersion law in IV–VI, EMSLs can be written as

$$k_x^2 = \left[\frac{1}{L_0^2}\left\{\cos^{-1}(f_2(E, k_y, k_z))\right\}^2 - k_\perp^2\right], \tag{9.43}$$

in which, $f_2(E, k_y, k_z) = a_5 \cos[a_0 C_3(E, k_y, k_z) + b_0 D_3(E, k_y, k_z)] - a_6 \cos[a_0 C_3(E, k_y, k_z) - b_0 D_3(E, k_y, k_z)]$, $a_5 = \left[\sqrt{\frac{m_2^*}{m_1^*}} + 1\right]^2 \left[4\left(\frac{m_2^*}{m_1^*}\right)^{1/2}\right]^{-1}$,

$$m_i^* = \left[\frac{2\hbar^2}{\{a_i^2 - e_i^2\}}\right] \Big[a_i + [a_i C_i + a_i e_i E_{\mathrm{g}_i} - e_i^2 E_{\mathrm{g}_i}] \, [E_{\mathrm{g}_i}^2 a_i^2 + C_i^2 + e_i^2 E_{\mathrm{g}_i}^2 + 2C_i e_i E_{\mathrm{g}_i} - 2E_{\mathrm{g}_i} a_i C_i - 2e_i a_i E_{\mathrm{g}_i}^2]^{-1/2}\Big]$$

$$C_3(E, k_y, k_z) \equiv \Big[[E H_{11} + H_{21}(k_y, k_z)] - [E^2 H_{31} + E H_{41}(k_y, k_z) + H_{51}(k_y, k_z)]^{1/2}\Big]^{1/2},$$

$$D_3(E, k_y, k_z) \equiv \Big[[E H_{12} + H_{22}(k_y, k_z)] - [E^2 H_{32} + E H_{42}(k_y, k_z) + H_{52}(k_y, k_z)]^{1/2}\Big]^{1/2}$$ and

$$a_6 = \left[-1 + \sqrt{\frac{m_2^*}{m_1^*}}\right]^2 \left[4\left(\frac{m_2^*}{m_1^*}\right)^{1/2}\right]^{-1}.$$

Thus, in the presence of a quantizing magnetic field along the x-direction, the simplified magneto dispersion law in this case can be written as

$$k_x^2 = [\rho_6(n, E)], \tag{9.44}$$

in which, $\rho_6(n,E) = \frac{1}{L_0^2}\left[\cos^{-1}(f_2(n,E))\right]^2 - \left\{\frac{2|e|B}{\hbar}\left(n+\frac{1}{2}\right)\right\}$,

$$f_2(n,E) = a_5\cos\left[a_0C_3(n,E)+b_0D_3(n,E)\right] \\ - a_6\cos\left[a_0C_3(n,E)-b_0D_3(n,E)\right],$$

$$C_3(n,E) \equiv \left[\left[EH_{11}+H_{21}(n)\right]-\left[E^2H_{31}+EH_{41}(n)+H_{51}(n)\right]^{1/2}\right]^{1/2},$$

$$D_3(n,E) \equiv \left[\left[EH_{12}+H_{22}(n)\right]-\left[E^2H_{32}+EH_{42}(n)+H_{52}(n)\right]^{1/2}\right]^{1/2}.$$

The EMM in this case assumes the form

$$m^*(n,E_{\mathrm{FSL}}) = \frac{\hbar^2}{2}\left\{\rho_6(n,E_{\mathrm{FSL}})\right\}'. \quad (9.45)$$

Thus, we observe that the EMM in this case depends on both the Fermi energy and magnetic quantum number, which is the intrinsic property of such SLs.

The Landau subband energies $(E_{n_{SL7}})$ can be written as

$$\rho_6(E_{n_{\mathrm{SL7}}},n) = 0. \quad (9.46)$$

The electron concentration in this case can be expressed as

$$n_0 = \left(\frac{|e|Bg_{\mathrm{v}}}{\pi^2\hbar}\right)\sum_{n=0}^{n_{\max}}\left[T_{913}(n,E_{\mathrm{FSL}})+T_{914}(n,E_{\mathrm{FSL}})\right], \quad (9.47)$$

where $T_{913}(n,E_{\mathrm{FSL}}) \equiv \left[\rho_6(n,E_{\mathrm{FSL}})\right]^{1/2}$ and $T_{914}(n,E_{\mathrm{FSL}}) \equiv \sum_{r=1}^{s}L(r)$ $[T_{913}(n,E_{\mathrm{FSL}})]$.

Using (9.47) and (1.11) we obtain the expression of the DMR as

$$\frac{D}{\mu} = \frac{1}{|e|}\frac{\sum_{n=0}^{n_{\max}}\left[T_{913}(n,E_{\mathrm{FSL}})+T_{914}(n,E_{\mathrm{FSL}})\right]}{\sum_{n=0}^{n_{\max}}\left[\{T_{913}(n,E_{\mathrm{FSL}})\}'+\{T_{914}(n,E_{\mathrm{FSL}})\}'\right]}. \quad (9.48)$$

9.2.8 Einstein Relation Under Magnetic Quantization in HgTe/CdTe Effective Mass Superlattices

Following Sasaki [15], the electron dispersion law in HgTe/CdTe EMSLs can be written as

$$k_x^2 = \left[\frac{1}{L_0^2}\left\{\cos^{-1}(f_3(E,k_y,k_z))\right\}^2 - k_\perp^2\right], \quad (9.49)$$

in which, $f_3(E,k_\perp) = a_7\cos\left[a_0C_4(E,k_\perp)+b_0D_4(E,k_\perp)\right] - a_8\cos\left[a_0 C_4(E,k_\perp)-b_0D_4(E,k_\perp)\right]$,

$$a_7 = \left[\sqrt{\frac{m_2^*}{m_1^*}} + 1\right]^2 \left[4\left(\frac{m_2^*}{m_1^*}\right)^{1/2}\right]^{-1}, \quad a_8 = \left[-1 + \sqrt{\frac{m_2^*}{m_1^*}}\right]^2 \left[4\left(\frac{m_2^*}{m_1^*}\right)^{1/2}\right]^{-1},$$

$$C_4(E, k_\perp) \equiv \left[\frac{B_0^2 + 2AE - B_0\sqrt{B_0^2 + 4AE}}{2A^2} - k_\perp^2\right]^{1/2},$$

and $D_4(E, k_\perp) \equiv \left[\left(\frac{2m_2^* E}{\hbar^2}\right) G(E, E_{g_2}, \Delta_2) - k_\perp^2\right]^{1/2}$.

In the presence of an external magnetic field along the x-direction, the simplified magneto dispersion law in this case can be written as

$$k_x^2 = [\rho_7(n, E)], \tag{9.50}$$

in which, $\rho_7(n, E) = \left\{\frac{1}{L_0^2}\left[\cos^{-1}(f_3(n, E))\right]^2\right\} - \left\{\frac{2|e|B}{\hbar}\left(n + \frac{1}{2}\right)\right\}$,

$$f_3(n, E) = a_7 \cos\left[a_0 C_4(n, E) + b_0 D_4(n, E)\right] - a_8 \cos\left[a_0 C_4(n, E) - b_0 D_4(n, E)\right],$$

$$C_4(n, E) \equiv \left[\frac{B_0^2 + 2AE - B_0\sqrt{B_0^2 + 4AE}}{2A^2} - \left\{\frac{2|e|B}{\hbar}\left(n + \frac{1}{2}\right)\right\}\right]^{1/2} \text{ and}$$

$$D_4(n, E) \equiv \left[\left(\frac{2m_2^* E}{\hbar^2}\right) G(E, E_{g_2}, \Delta_2) - \left\{\frac{2|e|B}{\hbar}\left(n + \frac{1}{2}\right)\right\}\right]^{1/2}.$$

The EMM in this case assumes the form

$$m^*(n, E_{\mathrm{FSL}}) = \frac{\hbar^2}{2}\{\rho_7(n, E_{\mathrm{FSL}})\}'. \tag{9.51}$$

The EMM in HgTe/CdTe EMSLs under magnetic quantization depends on both the Fermi energy and magnetic quantum number, which is the intrinsic property of such SLs.

The Landau subband energies ($E_{n\mathrm{SL8}}$) can be written as

$$\rho_7(E_{n\mathrm{SL8}}, n) = 0, \tag{9.52}$$

The electron concentration in this case can be expressed as

$$n_0 = \left(\frac{|e| B g_{\mathrm{v}}}{\pi^2 \hbar}\right) \sum_{n=0}^{n_{\max}} [T_{915}(n, E_{\mathrm{FSL}}) + T_{916}(n, E_{\mathrm{FSL}})], \tag{9.53}$$

where $T_{915}(n, E_{\mathrm{FSL}}) \equiv [\rho_7(n, E_{\mathrm{FSL}})]^{1/2}$ and $T_{916}(n, E_{\mathrm{FSL}}) \equiv \sum_{r=1}^{s} L(r)$ $[T_{915}(n, E_{\mathrm{FSL}})]$.

The use of (9.53) and (1.11) leads to the expression of the DMR as

$$\frac{D}{\mu} = \frac{1}{|e|} \frac{\sum_{n=0}^{n_{\max}} [T_{915}(n, E_{\mathrm{FSL}}) + T_{916}(n, E_{\mathrm{FSL}})]}{\sum_{n=0}^{n_{\max}} [\{T_{915}(n, E_{\mathrm{FSL}})\}' + \{T_{916}(n, E_{\mathrm{FSL}})\}']}. \tag{9.54}$$

9.2.9 Einstein Relation in III–V Quantum Wire Superlattices with Graded Interfaces

The electron dispersion law in III–V quantum wire superlattices (QWSLs) can be written, following (9.2), as

$$k_z^2 = \left[\frac{1}{L_0^2}\left\{\rho_8\left(n_x, n_y, E\right)\right\} - \phi\left(n_x, n_y\right)\right], \tag{9.55}$$

where $\rho_8\left(n_x, n_y, E\right) = \left[\cos^{-1}\left\{\frac{1}{2}\psi_8\left(n_x, n_y, E\right)\right\}\right]^2$, the function $\phi\left(n_x, n_y\right)$ has already been defined in Chap. 6, $\psi_8\left(n_x, n_y, E\right) = \left[\,2\cosh\left\{\beta_8\left(n_x, n_y, E\right)\right\}\right.$ $\cos\left\{\gamma_8\left(n_x, n_y, E\right)\right\} + \varepsilon_8\left(n_x, n_y, E\right)\sinh\left\{\beta_8\left(n_x, n_y, E\right)\right\}$,

$$\begin{aligned}
&\sin\left\{\gamma_8\left(n_x, n_y, E\right)\right\} + \Delta_0\left[\left(\frac{\left\{K_9\left(n_x, n_y, E\right)\right\}^2}{K_{10}\left(n_x, n_y, E\right)} - 3K_{10}\left(n_x, n_y, E\right)\right)\right.\\
&\cosh\left\{\beta_8\left(n_x, n_y, E\right)\right\}\sin\left\{\gamma_8\left(n_x, n_y, E\right)\right\}\\
&+\left(3K_9\left(n_x, n_y, E\right) - \frac{\left\{K_{10}\left(n_x, n_y, E\right)\right\}^2}{K_9\left(n_x, n_y, E\right)}\right)\sinh\left\{\beta_8\left(n_x, n_y, E\right)\right\}\\
&\cos\left\{\gamma_8\left(n_x, n_y, E\right)\right\} + \Delta_0\left[2\left(\left\{K_9\left(n_x, n_y, E\right)\right\}^2 - \left\{K_{10}\left(n_x, n_y, E\right)\right\}^2\right)\right.\\
&\cosh\left\{\beta_8\left(n_x, n_y, E\right)\right\}\cos\left\{\gamma_8\left(n_x, n_y, E\right)\right\}\\
&+\frac{1}{12}\left(\frac{5\left\{K_9\left(n_x, n_y, E\right)\right\}^3}{K_{10}\left(n_x, n_y, E\right)} + \frac{5\left\{K_{10}\left(n_x, n_y, E\right)\right\}^3}{K_9\left(n_x, n_y, E\right)}\right.\\
&\left.-\left\{34K_{10}\left(n_x, n_y, E\right)K_9\left(n_x, n_y, E\right)\right\}\right)\\
&\left.\left.\sinh\left\{\beta_8\left(n_x, n_y, E\right)\right\}\sin\left\{\gamma_8\left(n_x, n_y, E\right)\right\}\right]\right],
\end{aligned}$$

$$\varepsilon_8\left(n_x, n_y, E\right) \equiv \left[\frac{K_9\left(n_x, n_y, E\right)}{K_{10}\left(n_x, n_y, E\right)} - \frac{K_{10}\left(n_x, n_y, E\right)}{K_9\left(n_x, n_y, E\right)}\right],$$

$$\beta_8\left(n_x, n_y, E\right) \equiv K_9\left(n_x, n_y, E\right)\left[a_0 - \Delta_0\right],$$

$$K_9\left(n_x, n_y, E\right) \equiv \left[\frac{2m_2^* E'}{\hbar^2}G(E - V_0, \alpha_2, \Delta_2) + \phi\left(n_x, n_y\right)\right]^{1/2},$$

$$\gamma_8\left(n_x, n_y, E\right) = K_{10}\left(n_x, n_y, E\right)\left[b_0 - \Delta_0\right] \text{ and}$$

$$K_{10}\left(n_x, n_y, E\right) \equiv \left[\frac{2m_1^* E}{\hbar^2}G(E, \alpha_1, \Delta_1) - \phi\left(n_x, n_y\right)\right]^{1/2}.$$

The EMM in this case assumes the form

$$m^*\left(n_x, n_y, E_{\mathrm{FQWSL}}\right) = \frac{\hbar^2}{2L_0^2}\left\{\rho_8\left(n_x, n_y, E_{\mathrm{FQWSL}}\right)\right\}', \tag{9.56}$$

where E_{FQWSL} is the Fermi energy in this case. Thus, the EMM in III–V QWSLs is a function of both the Fermi energy and the size quantum numbers.

The subband energies $\left(E_{n_x,n_y\mathrm{SL9}}\right)$ can be expressed as

$$\rho_8\left(n_x,n_y,E_{n_x,n_y\mathrm{SL9}}\right)=\left[\phi\left(n_x,n_y\right)\right]. \tag{9.57}$$

Considering only the lowest miniband, as in an actual SL only the lowest miniband is significantly populated at low temperatures, where the quantum effects become prominent, the relation between the 1D electron concentration $(n_{1\mathrm{D}})$ and the Fermi energy in the present case can be written as,

$$n_{1\mathrm{D}}=\left(\frac{2g_{\mathrm{v}}}{\pi}\right)\sum_{n_x=1}^{n_{x\max}}\sum_{n_y=1}^{n_{y\max}}\left[T_{917}\left(n_x,n_y,E_{\mathrm{FQWSL}}\right)+T_{918}\left(n_x,n_y,E_{\mathrm{FQWSL}}\right)\right], \tag{9.58}$$

where $T_{917}\left(n_x,n_y,E_{\mathrm{FQWSL}}\right)=\left[\rho_8\left(n_x,n_y,E_{\mathrm{FQWSL}}\right)\cdot\left(L_0\right)^{-2}-\phi\left(n_x,n_y\right)\right]^{1/2}$ and $T_{918}\left(n_x,n_y,E_{\mathrm{FQWSL}}\right)\equiv\sum_{r=1}^{s}L(r)\left[T_{917}\left(n_x,n_y,E_{\mathrm{FQWSL}}\right)\right]$

The use of (9.58) and (1.11) leads to the expression of the DMR in this case as

$$\frac{D}{\mu}=\frac{1}{|e|}\frac{\sum_{n_x=1}^{n_{x\max}}\sum_{n_y=1}^{n_{y\max}}\left[T_{917}\left(n_x,n_y,E_{\mathrm{FQWSL}}\right)+T_{918}\left(n_x,n_y,E_{\mathrm{FQWSL}}\right)\right]}{\sum_{n_x=1}^{n_{x\max}}\sum_{n_y=1}^{n_{y\max}}\left[\left\{T_{917}\left(n_x,n_y,E_{\mathrm{FQWSL}}\right)\right\}'+\left\{T_{918}\left(n_x,n_y,E_{\mathrm{FQWSL}}\right)\right\}'\right]}. \tag{9.59}$$

9.2.10 Einstein Relation in II–VI Quantum Wire Superlattices with Graded Interfaces

The electron dispersion law in II–VI QWSLs can be written, following (9.11), as

$$k_z^2=\left[\frac{1}{L_0^2}\left\{\rho_9\left(n_x,n_y,E\right)\right\}-\phi\left(n_x,n_y\right)\right], \tag{9.60}$$

where $\rho_9\left(n_x,n_y,E\right)=\left[\cos^{-1}\left\{\frac{1}{2}\psi_9\left(n_x,n_y,E\right)\right\}\right]^2$,

$$\psi_9\left(n_x,n_y,E\right)=\Bigg[2\cosh\left\{\beta_9\left(n_x,n_y,E\right)\right\}\cos\left\{\gamma_9\left(n_x,n_y,E\right)\right\}$$
$$+\varepsilon_9\left(n_x,n_y,E\right)\sinh\left\{\beta_9\left(n_x,n_y,E\right)\right\}\sin\left\{\gamma_9\left(n_x,n_y,E\right)\right\}$$
$$+\Delta_0\Bigg[\left(\frac{\left\{K_{11}\left(n_x,n_y,E\right)\right\}^2}{K_{12}\left(n_x,n_y,E\right)}-3K_{12}\left(n_x,n_y,E\right)\right)$$

$$\cosh\{\beta_9(n_x,n_y,E)\}\sin\{\gamma_9(n_x,n_y,E)\}$$
$$+\left(3K_{11}(n_x,n_y,E)-\frac{\{K_{12}(n_x,n_y,E)\}^2}{K_{11}(n_x,n_y,E)}\right)$$
$$\sinh\{\beta_9(n_x,n_y,E)\}\cos\{\gamma_9(n_x,n_y,E)\}$$
$$+\Delta_0\left[2\left(\{K_{11}(n_x,n_y,E)\}^2-\{K_{12}(n_x,n_y,E)\}^2\right)\right.$$
$$\cosh\{\beta_9(n_x,n_y,E)\}\cos\{\gamma_9(n_x,n_y,E)\}$$
$$+\frac{1}{12}\left(\frac{5\{K_{11}(n_x,n_y,E)\}^3}{K_{12}(n_x,n_y,E)}+\frac{5\{K_{12}(n_x,n_y,E)\}^3}{K_{11}(n_x,n_y,E)}\right.$$
$$\left.-\{34K_{12}(n_x,n_y,E)\,K_{11}(n_x,n_y,E)\}\right)$$
$$\left.\left.\sinh\{\beta_9(n_x,n_y,E)\}\sin\{\gamma_9(n_x,n_y,E)\}\right]\right],$$

$$\varepsilon_9(n_x,n_y,E)\equiv\left[\frac{K_{11}(n_x,n_y,E)}{K_{12}(n_x,n_y,E)}-\frac{K_{12}(n_x,n_y,E)}{K_{11}(n_x,n_y,E)}\right],$$
$$\beta_9(n_x,n_y,E)\equiv K_{11}(n_x,n_y,E)\left[a_0-\Delta_0\right],$$
$$\gamma_9(n_x,n_y,E)=K_{12}(n_x,n_y,E)\left[b_0-\Delta_0\right],$$
$$K_{11}(n_x,n_y,E)\equiv\left[\frac{2m_2^*}{\hbar^2}E'G(E-V_0,\alpha_2,\Delta_2)+\phi(n_x,n_y)\right]^{1/2}\quad\text{and}$$
$$K_{12}(n_x,n_y,E)\equiv\left[\frac{2m_{\|,1}^*}{\hbar^2}\left[E-\frac{\hbar^2}{2m_{\perp,1}^*}\phi(n_x,n_y)\mp\bar{\lambda}_0\{\phi(n_x,n_y)\}^{1/2}\right]\right]^{1/2}.$$

The EMM in this case assumes the form

$$m^*(n_x,n_y,E_{\mathrm{FQWSL}})=\frac{\hbar^2}{2L_0^2}\{\rho_9(n_x,n_y,E_{\mathrm{FQWSL}})\}'. \tag{9.61}$$

The EMM in II–VI QWSLs depends on both the Fermi energy and size quantum numbers, which is the intrinsic property of such QWSLs.

The subband energies $\left(E_{n_x,n_y\mathrm{SL10}}\right)$ can be written as

$$\rho_9\left(n_x,n_y,E_{n_x,n_y\mathrm{SL10}}\right)=\left[\phi(n_x,n_y)\right]. \tag{9.62}$$

The electron concentration in this case can be expressed as

$$n_{1\mathrm{D}}=\left(\frac{2g_{\mathrm{v}}}{\pi}\right)\sum_{n_x=1}^{n_{x_{mas}}}\sum_{n_y=1}^{n_{y_{\max}}}\left[T_{919}(n_x,n_y,E_{\mathrm{FQWSL}})+T_{920}(n_x,n_y,E_{\mathrm{FQWSL}})\right], \tag{9.63}$$

where $T_{919}(n_x, n_y, E_{\mathrm{FQWSL}}) \equiv \left[\rho_9(n_x, n_y, E_{\mathrm{FQWSL}}) \cdot (L_0)^{-2} - \phi(n_x, n_y)\right]^{1/2}$ and $T_{920}(n_x, n_y, E_{\mathrm{FQWSL}}) \equiv \sum_{r=1}^{s} L(r)\,[T_{919}(n_x, n_y, E_{\mathrm{FQWSL}})]$.

The use of (9.63) and (1.11) leads to the expression of the DMR as

$$\frac{D}{\mu} = \frac{1}{|e|} \frac{\sum_{n_x=1}^{n_{x_{mas}}} \sum_{n_y=1}^{n_{y_{\max}}} [T_{919}(n_x, n_y, E_{\mathrm{FQWSL}}) + T_{920}(n_x, n_y, E_{\mathrm{FQWSL}})]}{\sum_{n_x=1}^{n_{x_{mas}}} \sum_{n_y=1}^{n_{y_{\max}}} [\{T_{919}(n_x, n_y, E_{\mathrm{FQWSL}})\}' + \{T_{920}(n_x, n_y, E_{\mathrm{FQWSL}})\}']}. \tag{9.64}$$

9.2.11 Einstein Relation in IV–VI Quantum Wire Superlattices with Graded Interfaces

The electron dispersion law in IV–VI QWSLs can be written from (9.17) as

$$k_z^2 = \left[\frac{1}{L_0^2}\{\rho_{10}(n_x, n_y, E)\} - \phi(n_x, n_y)\right], \tag{9.65}$$

where $\rho_{10}(n_x, n_y, E) = \left[\cos^{-1}\left\{\frac{1}{2}\psi_{10}(n_x, n_y, E)\right\}\right]^2$,

$$\begin{aligned}
\psi_{10}(n_x, n_y, E) = \Bigg[& 2\cosh\{\beta_{10}(n_x, n_y, E)\}\cos\{\gamma_{10}(n_x, n_y, E)\} \\
& + \varepsilon_{10}(n_x, n_y, E)\sinh\{\beta_{10}(n_x, n_y, E)\}\sin\{\gamma_{10}(n_x, n_y, E)\} \\
& + \Delta_0 \Bigg[\left(\frac{\{K_{11}(n_x, n_y, E)\}^2}{K_{12}(n_x, n_y, E)} - 3K_{12}(n_x, n_y, E)\right) \\
& \times \cosh\{\beta_{10}(n_x, n_y, E)\}\sin\{\gamma_{10}(n_x, n_y, E)\} \\
& + \left(3K_{11}(n_x, n_y, E) - \frac{\{K_{12}(n_x, n_y, E)\}^2}{K_{11}(n_x, n_y, E)}\right) \\
& \times \sinh\{\beta_{10}(n_x, n_y, E)\}\cos\{\gamma_{10}(n_x, n_y, E)\} \\
& + \Delta_0 \Big[2\left(\{K_{11}(n_x, n_y, E)\}^2 - \{K_{12}(n_x, n_y, E)\}^2\right) \\
& \times \cosh\{\beta_{10}(n_x, n_y, E)\}\cos\{\gamma_{10}(n_x, n_y, E)\} \\
& + \frac{1}{12}\Bigg(\frac{5\{K_{11}(n_x, n_y, E)\}^3}{K_{12}(n_x, n_y, E)} + \frac{5\{K_{12}(n_x, n_y, E)\}^3}{K_{11}(n_x, n_y, E)} \\
& - \{34K_{12}(n_x, n_y, E)\ K_{11}(n_x, n_y, E)\}\Bigg) \\
& \times \sinh\{\beta_{10}(n_x, n_y, E)\}\sin\{\gamma_{10}(n_x, n_y, E)\}\Big]\Bigg]\Bigg],
\end{aligned}$$

$$\varepsilon_{10}(n_x,n_y,E) \equiv \left[\frac{K_{11}(n_x,n_y,E)}{K_{12}(n_x,n_y,E)} - \frac{K_{12}(n_x,n_y,E)}{K_{11}(n_x,n_y,E)}\right],$$
$$\beta_{10}(n_x,n_y,E) \equiv K_{11}(n_x,n_y,E)\,[a_0-\Delta_0],$$
$$K_{12}(n_x,n_y,E) \equiv [\,[EH_{11}+H_{21}(n_x,n_y)] - \left[E^2H_{31}+EH_{41}(n_x,n_y)+H_{51}(n_x,n_y)\right]^{1/2}\,]^{1/2},$$

$$H_{2i}(n_x,n_y) = [2H_{1i}]^{-1}\,[E_{g_i}b_i+d_i+f_iE_{g_i}+2(e_if_i-a_ib_i)\,\phi(n_x,n_y)],$$

$$H_{4i}(n_x,n_y) = \left[4H_{1i}^2\right]^{-1}\left[4b_id_i+4b_if_iE_{g_i}+4f_i^2E_{g_i}+8\phi(n_x,n_y)\left[e_if_ib_i-a_if_i^2\right]\,\right],$$

$$\begin{aligned}H_{5i}(n_x,n_y) \equiv \left[4H_{1i}^2\right]^{-1}&\Big[\{\phi(n_x,n_y)\}^2\left[-8a_ib_ie_if_i+4b_i^2e_i^2+4f_i^2a_i^2\right]\\&+\{\phi(n_x,n_y)\}\,[4e_if_iE_{g_i}b_i-4e_if_id_i+4e_if_i^2E_{g_i}\\&-4a_ib_i^2E_{g_i}-4a_ib_id_i-4a_ib_if_iE_{g_i}+4b_i^2e_iE_{g_i}+4b_i^2c_i\\&+4b_i^2E_{g_i}a_i-4f_i^2e_iE_{g_i}-4f_i^2c_i-4f_i^2E_{g_i}a_i]\\&+\left[E_{g_i}^2b_i^2+d_i^2+f_i^2g_i^2+2E_{g_i}b_id_i+2E_{g_i}^2b_if_i+2d_if_iE_{g_i}\right]\Big].\end{aligned}$$

and

$$K_{11}(n_x,n_y,E) \equiv \left[\left[(E-V_0)^2H_{32}+(E-V_0)H_{42}(n_x,n_y)+H_{52}(n_x,n_y)\right]^{1/2} - [(E-V_0)H_{12}+H_{22}(n_x,n_y)]\right]^{1/2}.$$

The EMM in this case assumes the form

$$m^*(n_x,n_y,E_{\mathrm{FQWSL}}) = \frac{\hbar^2}{2L_0^2}\{\rho_{10}(n_x,n_y,E_{\mathrm{FQWSL}})\}'. \tag{9.66}$$

The EMM in IV–VI QWSLs under magnetic quantization depends on both the Fermi energy and magnetic quantum number, which is the intrinsic property of such SLs.

The subband energies $(E_{n_x,n_y\mathrm{SL11}})$ can be written as

$$\rho_{10}(n,E_{n_x,n_y\mathrm{SL11}}) = [\phi(n_x,n_y)]. \tag{9.67}$$

The electron concentration in this case can be expressed as

$$n_{1\mathrm{D}} = \left(\frac{2g_{\mathrm{v}}}{\pi}\right)\sum_{n_x=1}^{n_{x\max}}\sum_{n_y=1}^{n_{y\max}}[T_{921}(n_x,n_y,E_{\mathrm{FQWSL}})+T_{922}(n_x,n_y,E_{\mathrm{FQWSL}})], \tag{9.68}$$

where $T_{921}(n_x, n_y, E_{\mathrm{FQWSL}}) \equiv \left[\rho_{10}(n_x, n_y, E_{\mathrm{FQWSL}}) \cdot (L_0)^{-2} - \{\phi(n_x, n_y)\}\right]^{1/2}$ and $T_{922}(n_x, n_y, E_{\mathrm{FQWSL}}) \equiv \sum_{r=1}^{s} L(r)\,[T_{921}(n_x, n_y, E_{\mathrm{FQWSL}})]$.

The use of (9.68) and (1.11) leads to the expression of the DMR as

$$\frac{D}{\mu} = \frac{1}{|e|} \frac{\sum_{n_x=1}^{n_{x\max}} \sum_{n_y=1}^{n_{y\max}} [T_{921}(n_x, n_y, E_{\mathrm{FQWSL}}) + T_{922}(n_x, n_y, E_{\mathrm{FQWSL}})]}{\sum_{n_x=1}^{n_{x\max}} \sum_{n_y=1}^{n_{y\max}} [\{T_{921}(n_x, n_y, E_{\mathrm{FQWSL}})\}' + \{T_{922}(n_x, n_y, E_{\mathrm{FQWSL}})\}']}. \tag{9.69}$$

9.2.12 Einstein Relation in HgTe/CdTe Quantum Wire Superlattices with Graded Interfaces

The electron dispersion law in HgTe/CdTe QWSLs can be written from (9.25) as

$$k_z = \left[\frac{1}{L_0^2}\{\rho_{11}(n_x, n_y, E)\} - \{\phi(n_x, n_y)\}\right]^{1/2}, \tag{9.70}$$

where $\rho_{11}(n_x, n_y, E) = \left[\cos^{-1}\left\{\frac{1}{2}\psi_{11}(n_x, n_y, E)\right\}\right]^2$,

$$\begin{aligned}
\psi_{11}(n_x, n_y, E) = \Bigg[& 2\cosh\{\beta_{11}(n_x, n_y, E)\}\cos\{\gamma_{11}(n_x, n_y, E)\} \\
& + \varepsilon_{11}(n_x, n_y, E)\sinh\{\beta_{11}(n_x, n_y, E)\}\sin\{\gamma_{11}(n_x, n_y, E)\} \\
& + \Delta_0 \Bigg[\Bigg(\frac{\{K_{13}(n_x, n_y, E)\}^2}{K_{14}(n_x, n_y, E)} - 3K_{14}(n_x, n_y, E) \Bigg) \\
& \times \cosh\{\beta_{11}(n_x, n_y, E)\}\sin\{\gamma_{11}(n_x, n_y, E)\} \\
& + \Bigg(3K_{13}(n_x, n_y, E) - \frac{\{K_{14}(n_x, n_y, E)\}^2}{K_{13}(n_x, n_y, E)} \Bigg) \\
& \times \sinh\{\beta_{11}(n_x, n_y, E)\}\cos\{\gamma_{11}(n_x, n_y, E)\} \\
& + \Delta_0 \Big[2\Big(\{K_{13}(n_x, n_y, E)\}^2 - \{K_{14}(n_x, n_y, E)\}^2 \Big) \\
& \times \cosh\{\beta_{11}(n_x, n_y, E)\}\cos\{\gamma_{11}(n_x, n_y, E)\} \\
& + \frac{1}{12} \Bigg(\frac{5\{K_{13}(n_x, n_y, E)\}^3}{K_{14}(n_x, n_y, E)} + \frac{5\{K_{14}(n_x, n_y, E)\}^3}{K_{13}(n_x, n_y, E)} \\
& - \{34K_{14}(n_x, n_y, E)\,K_{13}(n_x, n_y, E)\} \Bigg) \\
& \times \sinh\{\beta_{11}(n_x, n_y, E)\}\sin\{\gamma_{11}(n_x, n_y, E)\} \Big] \Bigg] \Bigg],
\end{aligned}$$

$$\varepsilon_{11}(n_x,n_y,E) \equiv \left[\frac{K_{13}(n_x,n_y,E)}{K_{14}(n_x,n_y,E)} - \frac{K_{14}(n_x,n_y,E)}{K_{13}(n_x,n_y,E)}\right],$$

$$\beta_{11}(n_x,n_y,E) \equiv K_{13}(n_x,n_y,E)\,[a_0 - \Delta_0],$$

$$\gamma_{11}(n_x,n_y,E) = K_{14}(n_x,n_y,E)\,[b_0 - \Delta_0],$$

$$K_{14}(n_x,n_y,E) \equiv \left[\frac{B_0^2 + 2AE - B_0\sqrt{B_0^2 + 4AE}}{2A^2} - \{\phi(n_x,n_y)\}\right]^{1/2}, \text{ and}$$

$$K_{13}(n_x,n_y,E) \equiv \left[\left(\frac{2m_2^* E'}{\hbar^2}\right) G(E - V_0, \alpha_2, \Delta_2) + \{\phi(n_x,n_y)\}\right]^{1/2}.$$

The EMM in this case assumes the form

$$m^*(n_x,n_y,E_{\mathrm{FQWSL}}) = \frac{\hbar^2}{2L_0^2}\{\rho_{11}(n_x,n_y,E_{\mathrm{FQWSL}})\}'. \tag{9.71}$$

Thus the EMM in this case depends on both the Fermi energy and size quantum numbers.

The subband energies $\left(E_{n_x,n_y\mathrm{SL12}}\right)$ can be written as

$$\rho_{11}\left(n_x,n_y,E_{n_x,n_y\mathrm{SL12}}\right) = [\phi(n_x,n_y)]. \tag{9.72}$$

The electron concentration in this case can be expressed as

$$n_{1\mathrm{D}} = \left(\frac{2g_{\mathrm{v}}}{\pi}\right)\sum_{n_x=1}^{n_{\max}}\sum_{n_y=1}^{n_{y\max}}[T_{923}(n_x,n_y,E_{\mathrm{FQWSL}}) + T_{924}(n_x,n_y,E_{\mathrm{FQWSL}})] \tag{9.73}$$

where $T_{923}(n_x,n_y,E_{\mathrm{FQWSL}}) \equiv \left[\rho_{11}(n_x,n_y,E_{\mathrm{FQWSL}})\cdot(L_0)^{-2} - \{\phi(n_x,n_y)\}\right]^{1/2}$, and

$$T_{924}(n_x,n_y,E_{\mathrm{FQWSL}}) \equiv \sum_{r=1}^{s} L(r)\,[T_{923}(n_x,n_y,E_{\mathrm{FQWSL}})].$$

The use of (9.73) and (1.11) leads to the expression of the DMR as

$$\frac{D}{\mu} = \frac{1}{|e|}\frac{\sum_{n_x=1}^{n_{\max}}\sum_{n_y=1}^{n_{y\max}}[T_{923}(n_x,n_y,E_{\mathrm{FQWSL}}) + T_{924}(n_x,n_y,E_{\mathrm{FQWSL}})]}{\sum_{n_x=1}^{n_{\max}}\sum_{n_y=1}^{n_{y\max}}[\{T_{923}(n_x,n_y,E_{\mathrm{FQWSL}})\}' + \{T_{924}(n_x,n_y,E_{\mathrm{FQWSL}})\}']}. \tag{9.74}$$

9.2.13 Einstein Relation in III–V Effective Mass Quantum Wire Superlattices

The electron dispersion law in III–V effective mass quantum wire superlattices (EMQWSLs) can be written, following (9.31), as

$$k_x^2 = [\rho_{12}(n_y, n_z, E)], \tag{9.75}$$

in which, $\rho_{12}(n_y, n_z, E) = \frac{1}{L_0^2}\left[\cos^{-1}(f_{12}(n_y, n_z, E))\right]^2 - \{\phi(n_y, n_z)\}$, $\phi(n_y, n_z) = \left\{\left(\frac{n_y \pi}{d_y}\right)^2 + \left(\frac{n_z \pi}{d_z}\right)^2\right\}$,

$$f_{12}(n_y, n_z, E) = a_7 \cos[a_0 C_5(n_y, n_z, E) + b_0 D_5(n_y, n_z, E)] - a_8 \cos[a_0 C_5(n_y, n_z, E) - b_0 D_5(n_y, n_z, E)],$$

$$C_5(n_y, n_z, E) \equiv \left[\left(\frac{2m_1^* E}{\hbar^2}\right) G(E, E_{g_1}, \Delta_1) - \{\phi(n_y, n_z)\}\right]^{1/2} \text{ and}$$

$$D_5(n_y, n_z, E) \equiv \left[\left(\frac{2m_2^* E}{\hbar^2}\right) G(E, E_{g_2}, \Delta_2) - \{\phi(n_y, n_z)\}\right]^{1/2}.$$

The EMM in this case assumes the form

$$m^*(n_y, n_z, E_{FQWSL}) = \frac{\hbar^2}{2}\{\rho_{12}(n_y, n_z, E_{FQWSL})\}'. \tag{9.76}$$

The EMM in III–V EMQWSLs depends on both the Fermi energy and size quantum numbers, which is the intrinsic property of such SLs.

The subband energies $\left(E_{n_y, n_z SL13}\right)$ can be written as

$$\rho_{12}\left(n_x, n_y, E_{n_y, n_z SL13}\right) = 0. \tag{9.77}$$

The electron concentration in this case can be expressed as

$$n_{1D} = \left(\frac{2g_v}{\pi}\right) \sum_{n_y=1}^{n_{y\max}} \sum_{n_z=1}^{n_{z\max}} [T_{925}(n_y, n_z, E_{FQWSL}) + T_{926}(n_y, n_z, E_{FQWSL})], \tag{9.78}$$

where $T_{925}(n_y, n_z, E_{FQWSL}) \equiv [\rho_{12}(n_y, n_z, E_{FQWSL})]^{1/2}$ and

$$T_{926}(n_y, n_z, E_{FQWSL}) \equiv \sum_{r=1}^{s} L(r)\,[T_{925}(n_y, n_z, E_{FQWSL})].$$

The use of (9.78) and (1.11) leads to the expression of the DMR as

$$\frac{D}{\mu} = \frac{1}{|e|} \frac{\sum_{n_y=1}^{n_{y\max}} \sum_{n_z=1}^{n_{z\max}} [T_{925}(n_y, n_z, E_{FQWSL}) + T_{926}(n_y, n_z, E_{FQWSL})]}{\sum_{n_y=1}^{n_{y\max}} \sum_{n_z=1}^{n_{z\max}} [\{T_{925}(n_y, n_z, E_{FQWSL})\}' + \{T_{926}(n_y, n_z, E_{FQWSL})\}']}. \tag{9.79}$$

9.2.14 Einstein Relation in II–VI Effective Mass Quantum Wire Superlattices

The dispersion law in II–VI EMQWSLs can be written, following (9.37), as

$$k_z^2 = [\rho_{13}(n_x, n_y, E)], \tag{9.80}$$

in which, $\rho_{13}(n_x, n_y, E) = \frac{1}{L_0^2}\left[\cos^{-1}(f_{13}(n_x, n_y, E))\right]^2 - \{\phi(n_x, n_y)\}$,

$$f_{13}(n_x, n_y, E) = a_3 \cos\left[a_0 C_6(n_x, n_y, E) + b_0 D_6(n_x, n_y, E)\right] - a_4 \cos\left[a_0 C_6(n_x, n_y, E) - b_0 D_6(n_x, n_y, E)\right],$$

$$C_6(n_x, n_y, E) \equiv \left(\frac{2m_{\parallel,1}^*}{\hbar^2}\right)^{1/2}\left[E - \left\{\frac{\hbar^2}{2m_{\perp,1}^*}\phi(n_x, n_y)\right\} \mp \bar{\lambda}_0\{\phi(n_x, n_y)\}^{1/2}\right]^{1/2}$$

and

$D_6(n_x, n_y, E) \equiv \left[\left(\frac{2m_2^*}{\hbar^2}\right) EG(E, E_{g_2}, \Delta_2) - \phi(n_x, n_y)\right]^{1/2}$.

The EMM in this case assumes the form

$$m^*(n_x, n_y, E_{\mathrm{FQWSL}}) = \frac{\hbar^2}{2}\{\rho_{13}(n_x, n_y, E_{\mathrm{FQWSL}})\}'. \tag{9.81}$$

We observe that the EMM in this case depends on both the Fermi energy and size quantum numbers, which is the intrinsic property of such SLs.

The subband energies $(E_{n_y,n_z\mathrm{SL14}})$ can be written as

$$\rho_{13}\left(n_x, n_y, E_{n_x,n_y\mathrm{SL14}}\right) = 0. \tag{9.82}$$

The electron concentration in this case can be expressed as

$$n_{1\mathrm{D}} = \left(\frac{2g_{\mathrm{v}}}{\pi}\right)\sum_{n_x=1}^{n_{x\max}}\sum_{n_y=1}^{n_{y\max}}[T_{927}(n_x, n_y, E_{\mathrm{FQWSL}}) + T_{928}(n_x, n_y, E_{\mathrm{FQWSL}})], \tag{9.83}$$

where $T_{927}(n_x, n_y, E_{\mathrm{FQWSL}}) \equiv [\rho_{13}(n_x, n_y, E_{\mathrm{FQWSL}})]^{1/2}$ and $T_{928}(n_x, n_y, E_{\mathrm{FQWSL}}) \equiv \sum_{r=1}^{s} L(r)[T_{927}(n_x, n_y, E_{\mathrm{FQWSL}})]$.

Thus using (9.83) and (1.11) leads to the expression of the DMR as

$$\frac{D}{\mu} = \frac{1}{|e|}\frac{\sum_{n_x=1}^{n_{x\max}}\sum_{n_y=1}^{n_{y\max}}[T_{927}(n_x, n_y, E_{\mathrm{FQWSL}}) + T_{928}(n_x, n_y, E_{\mathrm{FQWSL}})]}{\sum_{n_x=1}^{n_{x\max}}\sum_{n_y=1}^{n_{y\max}}[\{T_{927}(n_x, n_y, E_{\mathrm{FQWSL}})\}' + \{T_{928}(n_x, n_y, E_{\mathrm{FQWSL}})\}']}. \tag{9.84}$$

9.2.15 Einstein Relation in IV–VI Effective Mass Quantum Wire Superlattices

The dispersion law in IV–VI, EMQWSLs can be written following (9.43) as

$$k_x^2 = [\rho_{14}(n_y, n_z, E)], \tag{9.85}$$

in which, $\rho_{14}(n_y, n_z, E) = \frac{1}{L_0^2}\left[\cos^{-1}(f_{14}(n_y, n_z, E))\right]^2 - \{\phi(n_y, n_z)\}$,

$$f_{14}(n_y, n_z, E) = a_5 \cos\left[a_0 C_7(n_y, n_z, E) + b_0 D_7(n_y, n_z, E)\right] - a_6 \cos\left[a_0 C_7(n_y, n_z, E) - b_0 D_7(n_y, n_z, E)\right],$$

$$C_7(n_y, n_z, E) \equiv \left[[EH_{11} + H_{21}(n_y, n_z)] - \left[E^2 H_{31} + EH_{41}(n_y, n_z) + H_{51}(n_y, n_z)\right]^{1/2}\right]^{1/2},$$

$$D_7(n_y, n_z, E) \equiv \left[[EH_{12} + H_{22}(n_y, n_z)] - \left[E^2 H_{32} + EH_{42}(n_y, n_z) + H_{52}(n_y, n_z)\right]^{1/2}\right]^{1/2}$$

The EMM in this case assumes the form

$$m^*(n_y, n_z, E_{\mathrm{FQWSL}}) = \frac{\hbar^2}{2}\{\rho_{14}(n_y, n_z, E_{\mathrm{FQWSL}})\}'. \tag{9.86}$$

Thus, we observe that the EMM in this case depends on both the Fermi energy and size quantum numbers, which is the intrinsic property of such SLs.

The subband energies $(E_{n_y,n_z\mathrm{SL}15})$ can be written as

$$\rho_{14}(E_{n_y,n_z\mathrm{SL}15}, n_y, n_z) = 0. \tag{9.87}$$

The electron concentration in this case can be expressed as

$$n_{1\mathrm{D}} = \left(\frac{2g_{\mathrm{v}}}{\pi}\right)\sum_{n_y=1}^{n_{y\max}}\sum_{n_z=1}^{n_{z\max}}[T_{929}(n_y, n_z, E_{\mathrm{FQWSL}}) + T_{930}(n_y, n_z, E_{\mathrm{FQWSL}})], \tag{9.88}$$

where $T_{929}(n_y, n_z, E_{\mathrm{FQWSL}}) \equiv [\rho_{14}(n_y, n_z, E_{\mathrm{FQWSL}})]^{1/2}$ and

$$T_{930}(n_y, n_z, E_{\mathrm{FQWSL}}) \equiv \sum_{r=1}^{s} L(r)\,[T_{929}(n_y, n_z, E_{\mathrm{FQWSL}})].$$

Thus using (9.88) and (1.11) leads to the expression of the DMR as

$$\frac{D}{\mu} = \frac{1}{|e|}\frac{\sum_{n_y=1}^{n_{y\max}}\sum_{n_z=1}^{n_{z\max}}[T_{929}(n_y, n_z, E_{\mathrm{FQWSL}}) + T_{930}(n_y, n_z, E_{\mathrm{FQWSL}})]}{\sum_{n_y=1}^{n_{y\max}}\sum_{n_z=1}^{n_{z\max}}\left[\{T_{929}(n_y, n_z, E_{\mathrm{FQWSL}})\}' + \{T_{930}(n_y, n_z, E_{\mathrm{FQWSL}})\}'\right]}. \tag{9.89}$$

9.2.16 Einstein Relation in HgTe/CdTe Effective Mass Quantum Wire Superlattices

The dispersion law in HgTe/CdTe EMQWSLs can be written, following (9.49), as

$$k_x^2 = [\rho_{15}(n_y, n_z, E)], \tag{9.90}$$

in which, $\rho_{15}(n_y, n_z, E) = \frac{1}{L_0^2}\left[\cos^{-1}(f_{15}(n_y, n_z, E))\right]^2 - \{\phi(n_y, n_z)\}$,

$$f_{15}(n_y, n_z, E) = a_7 \cos\left[a_0 C_8(n_y, n_z, E) + b_0 D_8(n_y, n_z, E)\right] \\ - a_8 \cos\left[a_0 C_8(n_y, n_z, E) - b_0 D_8(n_y, n_z, E)\right],$$

$$C_8(n_y, n_z, E) \equiv \left[\frac{B_0^2 + 2AE - B_0\sqrt{B_0^2 + 4AE}}{2A^2} - \{\phi(n_y, n_z)\}\right]^{1/2} \text{ and}$$

$$D_8(n_y, n_z, E) \equiv \left[\left(\frac{2m_2^* E}{\hbar^2}\right) G(E, E_{g_2}, \Delta_2) - \{\phi(n_y, n_z)\}\right]^{1/2}.$$

The EMM in this case assumes the form

$$m^*(n_y, n_z, E_{\mathrm{FQWSL}}) = \frac{\hbar^2}{2}\{\rho_{15}(n_y, n_z, E_{\mathrm{FQWSL}})\}'. \tag{9.91}$$

The EMM in HgTe/CdTe QWEMSLs depends on both the Fermi energy and size quantum numbers.

The subband energies $\left(E_{n_y,n_z\mathrm{SL}16}\right)$ can be written as

$$\rho_{14}\left(n_y, n_z, E_{n_y,n_z\mathrm{SL}16}\right) = 0. \tag{9.92}$$

The electron concentration in this case can be expressed as

$$n_{1\mathrm{D}} = \left(\frac{2g_{\mathrm{v}}}{\pi}\right)\sum_{n_y=1}^{n_{y\max}}\sum_{n_z=1}^{n_{z\max}}\left[T_{931}(n_y, n_z, E_{\mathrm{FQWSL}}) + T_{932}(n_y, n_z, E_{\mathrm{FQWSL}})\right], \tag{9.93}$$

where $T_{931}(n_y, n_z, E_{\mathrm{FQWSL}}) \equiv [\rho_{15}(n_y, n_z, E_{\mathrm{FQWSL}})]^{1/2}$ and

$$T_{932}(n_y, n_z, E_{\mathrm{FQWSL}}) \equiv \sum_{r=1}^{s} L(r)\left[T_{931}(n_y, n_z, E_{\mathrm{FQWSL}})\right].$$

The use of (9.93) and (1.11) leads to the expression of the DMR as

$$\frac{D}{\mu} = \frac{1}{|e|}\frac{\sum\limits_{n_y=1}^{n_{y\max}}\sum\limits_{n_z=1}^{n_{z\max}}\left[T_{931}(n_y, n_z, E_{\mathrm{FQWSL}}) + T_{932}(n_y, n_z, E_{\mathrm{FQWSL}})\right]}{\sum\limits_{n_y=1}^{n_{y\max}}\sum\limits_{n_z=1}^{n_{z\max}}\left[\{T_{931}(n_y, n_z, E_{\mathrm{FQWSL}})\}' + \{T_{932}(n_y, n_z, E_{\mathrm{FQWSL}})\}'\right]}. \tag{9.94}$$

9.3 Result and Discussions

Using Table 2.1 and $\Delta_0 = 0.4\,\text{nm}$ [16] together with (9.6), (9.7), (9.14), (9.15), (9.21), (9.22) and (9.29), (9.30), we have plotted the DMR in Fig. 9.1 as a function of inverse quantizing magnetic field for GaAs/$Ga_{1-x}Al_{-}As$, CdS/CdTe, PbTe/PbSnTe and HgTe/CdTe superlattices with graded interfaces as shown by curves (a)–(d) respectively (Fig. 9.2).

Using the same equations, we have plotted the DMR as a function of electron concentration for GaAs/$Ga_{1-x}Al_xAs$, CdS/CdTe superlattices, as shown by curves (a) and (b) in Fig. 9.2, whereas, in Fig. 9.3, the same has been drawn for PbTe/PbSnTe and HgTe/CdTe superlattices as shown by curves (c) and (d) respectively.

Using (9.35), (9.36), (9.41), (9.42), (9.47), (9.48) and (9.53), (9.54), we have plotted the DMR in Fig. 9.4 as a function of inverse quantizing magnetic field for the effective mass superlattices. Using the same equations, we have plotted the DMR as a function of electron concentration for GaAs/$Ga_{1-x}Al_xAs$, CdS/CdTe effective mass superlattices, as shown by curves (a) and (b) in Fig. 9.5, whereas the same has been drawn for PbTe/PbSnTe and HgTe/CdTe effective mass superlattices, as shown by curves (c) and (d) in Fig. 9.6. It appears from the said figures that the DMR oscillates both with $1/B$ and n_0 due to the SdH effect, although the rates of variations are totally band structure-dependent for all types of superlattices as considered here. Using (9.58), (9.59),

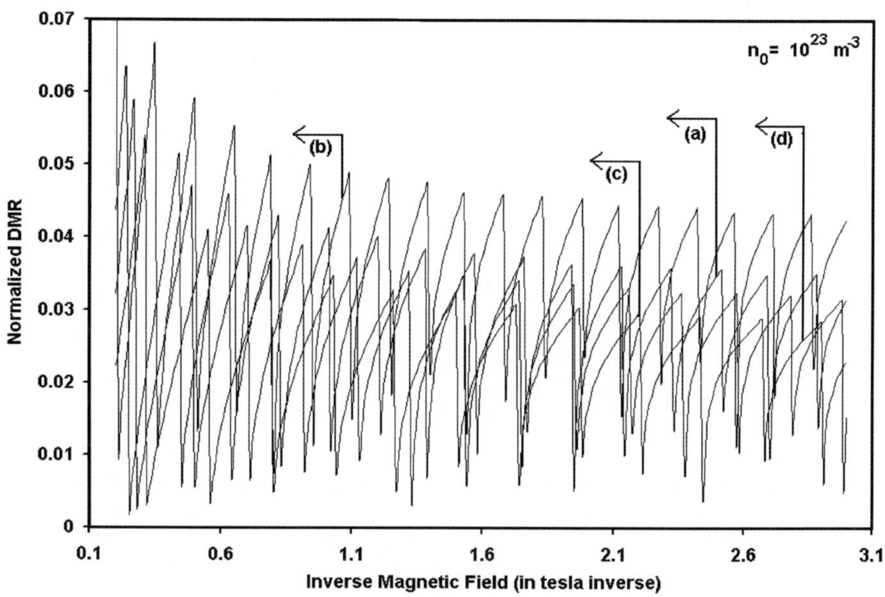

Fig. 9.1. The plot of the DMR as a function of inverse quantizing magnetic field for (**a**) GaAs/$Ga_{1-x}Al_xAs$, (**b**) CdS/CdTe, (**c**) PbTe/PbSnTe, and (**d**) HgTe/CdTe superlattices with graded interfaces

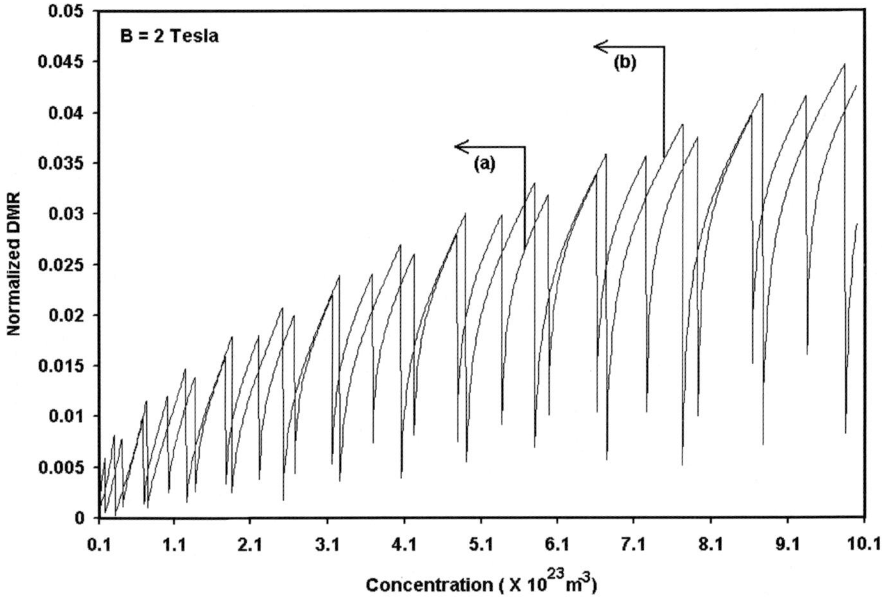

Fig. 9.2. The plot of the DMR as a function of electron concentration for (**a**) $GaAs/Ga_{1-x}Al_xAs$ and (**b**) CdS/CdTe superlattices with graded interfaces

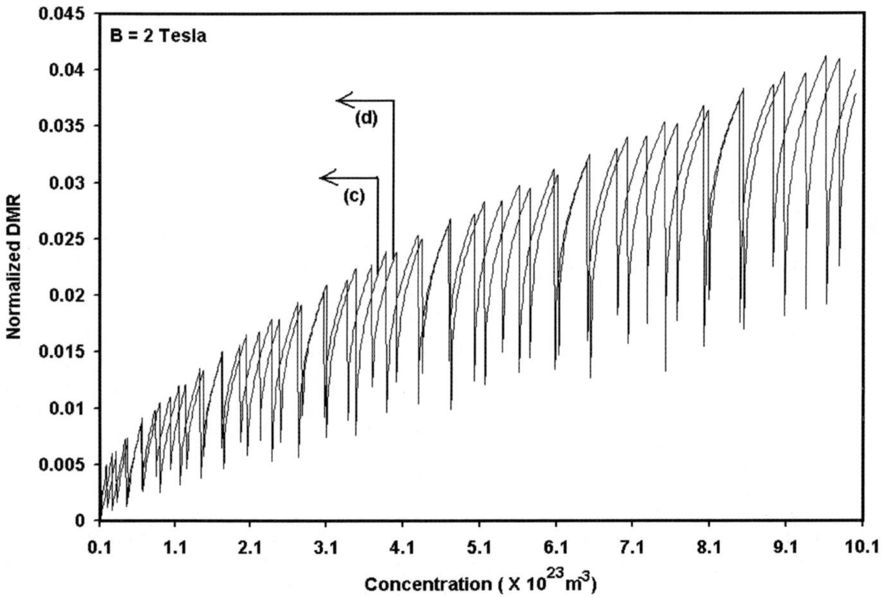

Fig. 9.3. The plot of the DMR as a function of electron concentration for (**c**) PbTe/PbSnTe and (**d**) HgTe/CdTe superlattices with graded interfaces

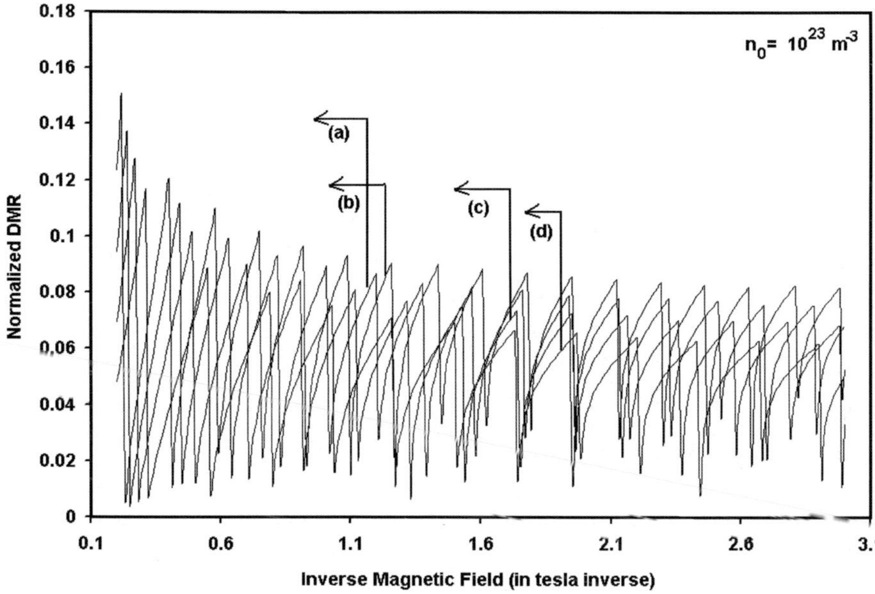

Fig. 9.4. The plot of the DMR as a function of inverse quantizing magnetic field for (**a**) $GaAs/Ga_{1-x}Al_xAs$, (**b**) CdS/CdTe, (**c**) PbTe/PbSnTe, and (**d**) HgTe/CdTe effective mass superlattices

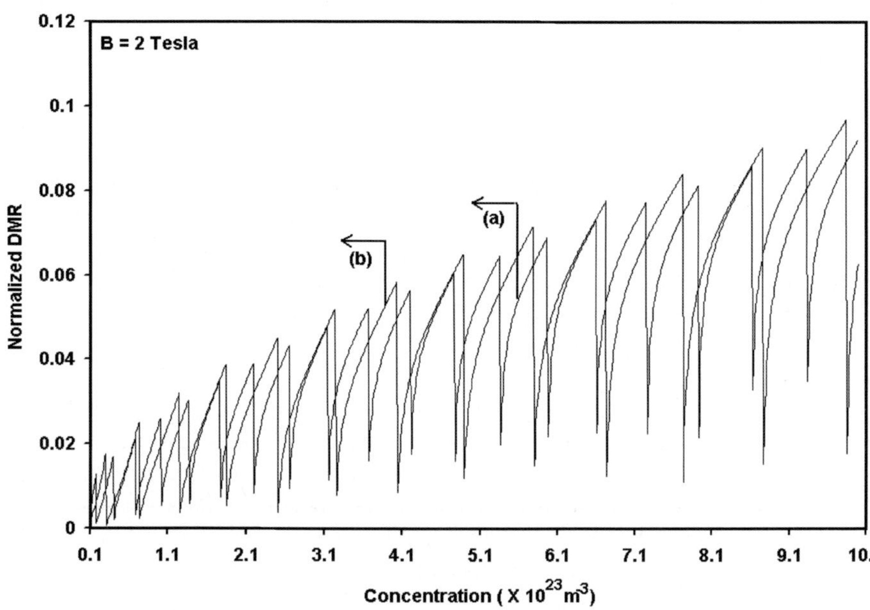

Fig. 9.5. The plot of the DMR as a function of electron concentration for (**a**) $GaAs/Ga_{1-x}Al_xAs$ and (**b**) CdS/CdTe effective mass superlattices

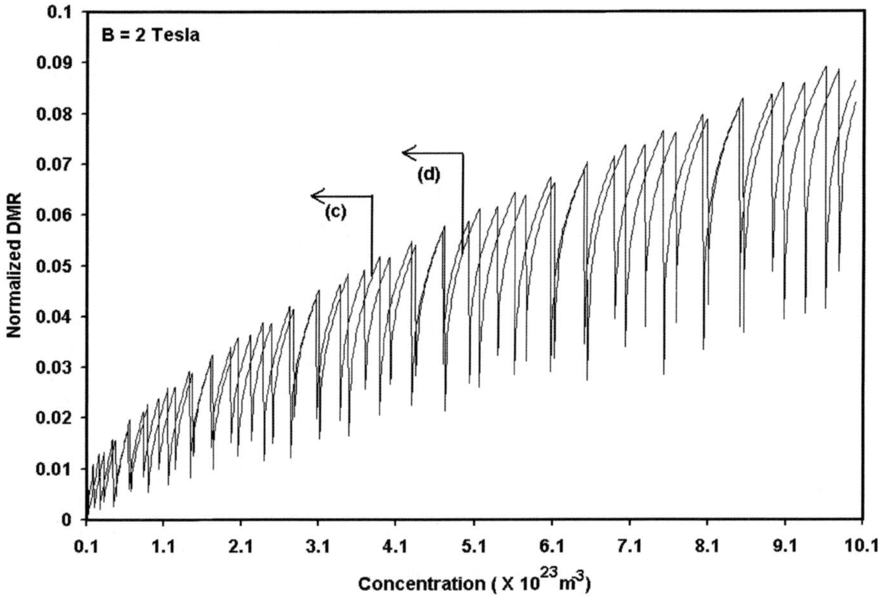

Fig. 9.6. The plot of the DMR as a function of electron concentration for (**c**) PbTe/PbSnTe and (**d**) HgTe/CdTe effective mass superlattices

(9.63), (9.64), (9.68), (9.69) and (9.73), (9.74), we have plotted the normalized 1D DMR as a function of film thickness for GaAs/$Ga_{1-x}Al_xAs$, CdS/CdTe, PbTe/PbSnTe and HgTe/CdTe quantum wire superlattices with graded interfaces as shown by curves (a)–(d) of Fig. 9.7. In Fig. 9.8, we have plotted all cases of Fig. 9.7 as a function of electron concentration per unit length. It appears from both Figs. 9.7 and 9.8 that the DMR increases with decreasing film thickness and increasing electron concentration per unit length, although the numerical values are totally band structure-dependent.

Using (9.78), (9.79), (9.83), (9.84), (9.88), (9.89) and (9.93), (9.94), we gave plotted the normalized 1D DMR in quantum wire effective mass superlattices of the aforementioned materials as functions of film thickness and electron concentration per unit length as shown by curves (a)–(d) in Figs. 9.9 and 9.10 respectively.

It appears that the DMR in this case decreases with film thickness and increases with electron concentration in all the cases. The nature of the discussions of Chaps. 3 and 6 are also applicable for the present chapter. The experimental suggestion for the determination of the DMR is also valid for all the cases of this chapter. For the purpose of condensed presentation, the specific electron statistics related to a particular energy dispersion law for specific materials and the Einstein relation have been presented in Table 9.1.

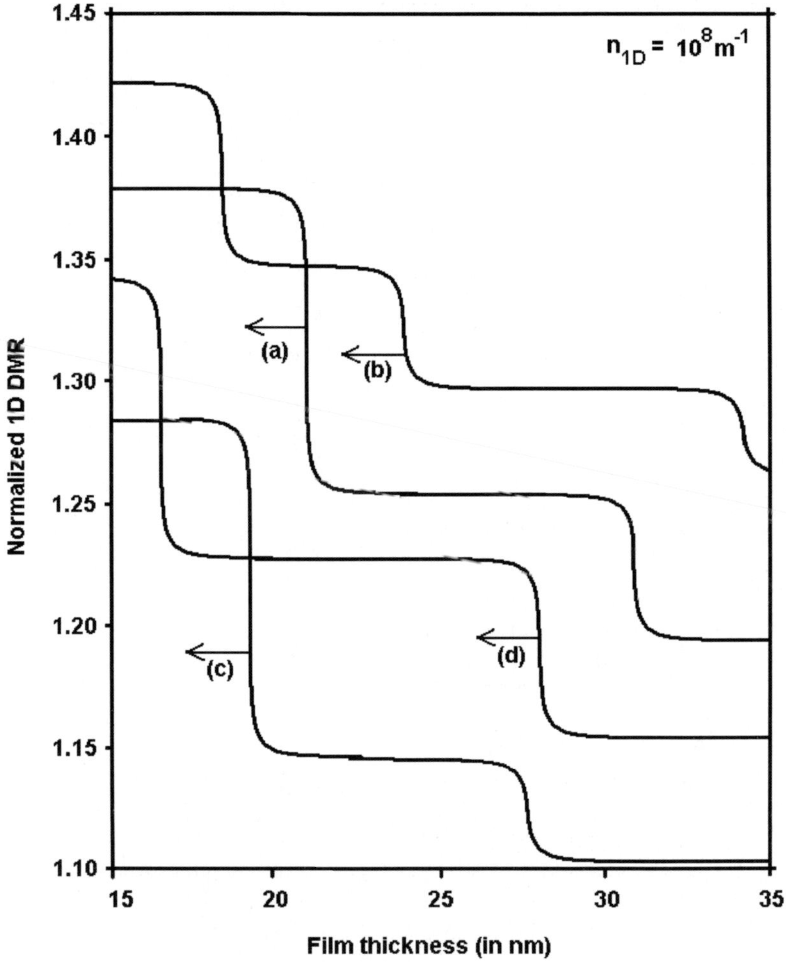

Fig. 9.7. The plot of the 1D DMR as a function of film thickness for (**a**) $GaAs/Ga_{1-x}Al_xAs$, (**b**) CdS/CdTe, (**c**) PbTe/PbSnTe, and (**d**) HgTe/CdTe superlattices with graded interfaces

9.4 Open Research Problems

R.9.1 Investigate the Einstein relation in the presence of an arbitrarily oriented non-quantizing alternating magnetic field for all types of superlattices as considered in this chapter by considering the electron spin.

R.9.2 Investigate the Einstein relation in the presence of an additional arbitrarily oriented non-uniform electric field for all types of superlattices.

R.9.3 Investigate the Einstein relation for all types of superlattices as considered in this chapter under arbitrarily oriented crossed electric and magnetic fields.

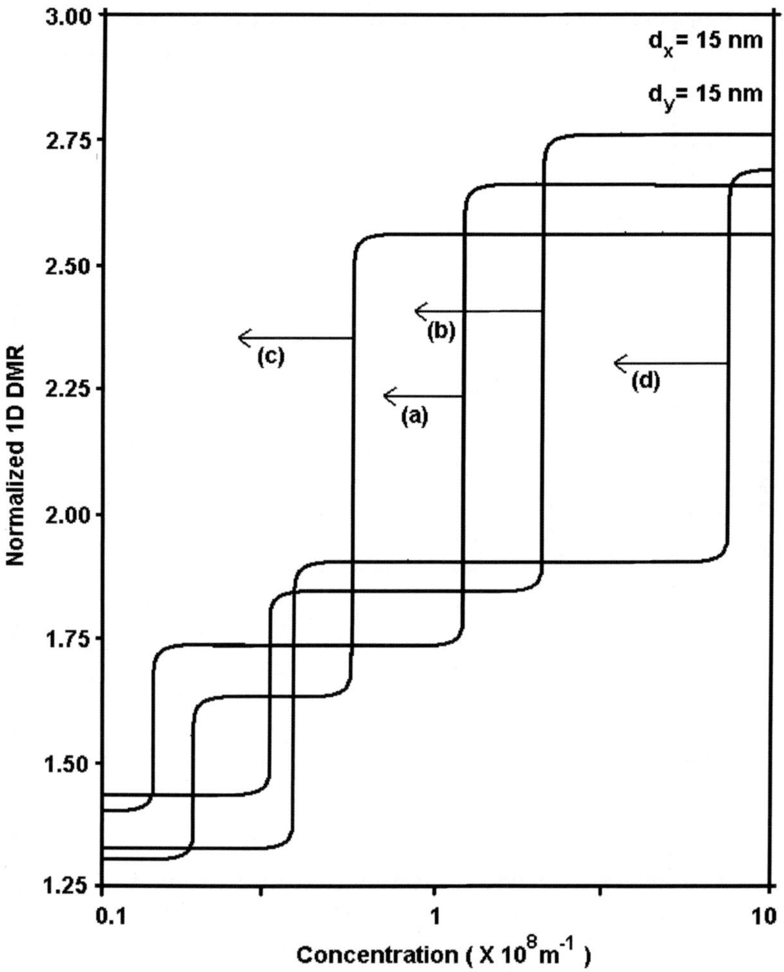

Fig. 9.8. The plot of the 1D DMR as a function of surface electron concentration for (**a**) GaAs/$Ga_{1-x}Al_xAs$, (**b**) CdS/CdTe, (**c**) PbTe/PbSnTe, and (**d**) HgTe/CdTe superlattices with graded interfaces

R.9.4 Investigate the Einstein relation in III–V, II–VI, IV–VI and HgTe/CdTe quantum well superlattices with graded interfaces.

R.9.5 Investigate the Einstein relation for all problems of R.9.1–R.9.3 for III–V, II–VI, IV–VI and HgTe/CdTe quantum well superlattices with graded interfaces.

R.9.6 Investigate the Einstein relation for III–V, II–VI, IV–VI and HgTe/CdTe quantum well effective mass superlattices.

R.9.7 Investigate the Einstein relation for all problems of R.9.1–R.9.3 for III–V, II–VI, IV–VI and HgTe/CdTe quantum well effective mass superlattices.

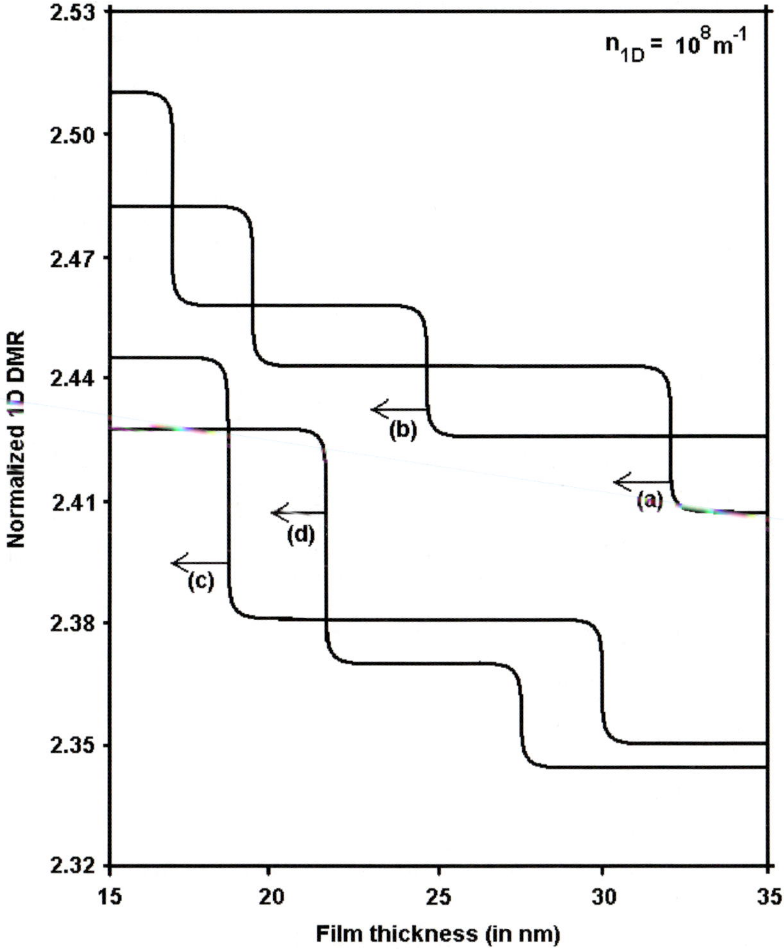

Fig. 9.9. The plot of the 1D DMR as a function of film thickness for (**a**) $GaAs/Ga_{1-x}Al_xAs$, (**b**) CdS/CdTe, (**c**) PbTe/PbSnTe, and (**d**) HgTe/CdTe effective mass superlattices

R.9.8 Investigate the Einstein relation for short period, strained layer, random and Fibonacci superlattices.

R.9.9 Investigate the Einstein relation for short period, strained layer, random and Fibonacci superlattices in the presence of an arbitrarily oriented alternating magnetic field by considering electron spin and broadening.

R.9.10 Investigate the Einstein relation for strained layer, random, Fibonacci, polytype and sawtooth superlattices in the presence of an arbitrarily oriented non-uniform electric field.

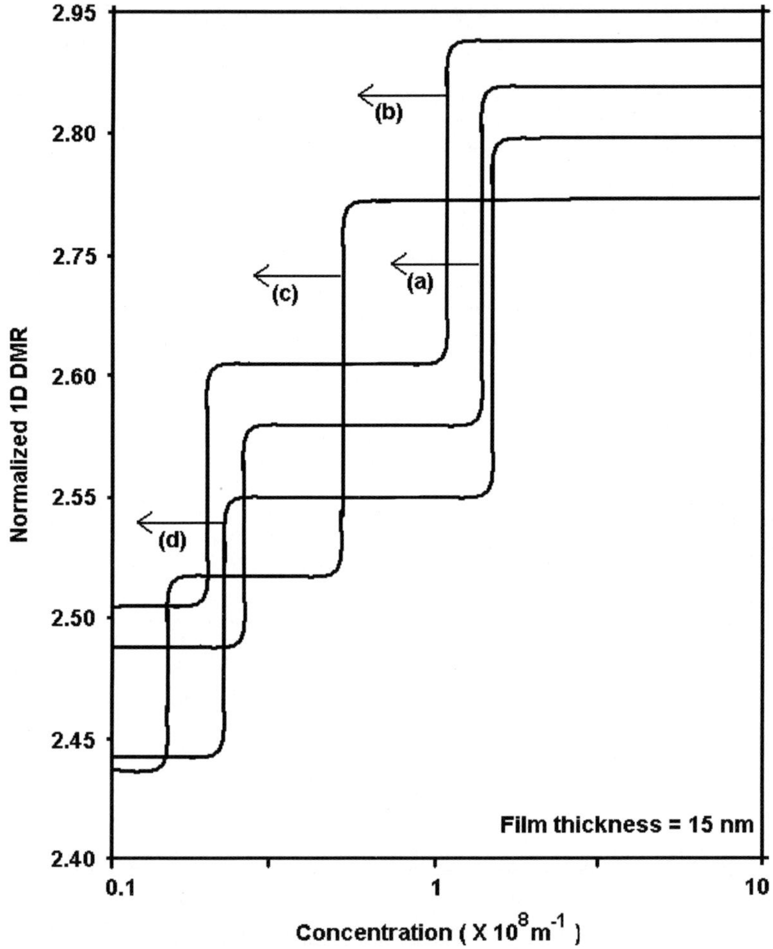

Fig. 9.10. The plot of the 1D DMR as a function of surface electron concentration for (**a**) $GaAs/Ga_{1-x}Al_xAs$, (**b**) CdS/CdTe, (**c**) PbTe/PbSnTe, and (**d**) HgTe/CdTe effective mass superlattices

R.9.11 Investigate the Einstein relation for strained layer, random, Fibonacci, polytype and sawtooth superlattices in the presence of an arbitrarily oriented crossed electric and magnetic field.

R.9.12 Investigate the Einstein relation for strained layer, random, Fibonacci, polytype and sawtooth quantum wells and quantum wires superlattices in the presence of an arbitrarily oriented alternating electric field.

R.9.13 Investigate the Einstein relation for strained layer, random, Fibonacci, polytype and sawtooth quantum wells and quantum wires superlattices in the presence of arbitrarily oriented crossed electric and quantizing magnetic fields.

Table 9.1. The carrier statistics and the Einstein relation in superlattices

Confinement of wave vector space	The carrier statistics	The Einstein relation for the diffusivity mobility ratio
Magnetic quantization	III–V superlattices with graded interfaces $n_0 = \left(\frac{\lvert e\rvert B g_v}{\pi^2\hbar L_0}\right)\sum_{n=0}^{n_{max}}[T_{91}(n,E_{FSL}) + T_{92}(n,E_{FSL})]$ (9.6)	$\frac{D}{\mu} = \frac{1}{\lvert e\rvert}\frac{\sum_{n=0}^{n_{max}}[T_{91}(n,E_{FSL})+T_{92}(n,E_{FSL})]}{\sum_{n=0}^{n_{max}}[\{T_{91}(n,E_{FSL})\}'+\{T_{92}(n,E_{FSL})\}']}$ (9.7)
	II–VI superlattices with graded interfaces $n_0 = \left(\frac{\lvert e\rvert B g_v}{2\pi^2\hbar L_0}\right)\sum_{n=0}^{n_{max}}[T_{93}(n,E_{FSL}) + T_{94}(n,E_{FSL})]$ (9.14)	$\frac{D}{\mu} = \frac{1}{\lvert e\rvert}\frac{\sum_{n=0}^{n_{max}}[T_{93}(n,E_{FSL})+T_{94}(n,E_{FSL})]}{\sum_{n=0}^{n_{max}}[\{T_{93}(n,E_{FSL})\}'+\{T_{94}(n,E_{FSL})\}']}$ (9.15)
	IV–VI superlattices with graded interfaces $n_0 = \left(\frac{\lvert e\rvert B g_v}{\pi^2\hbar L_0}\right)\sum_{n=0}^{n_{max}}[T_{95}(n,E_{FSL}) + T_{96}(n,E_{FSL})]$ (9.21)	$\frac{D}{\mu} = \frac{1}{\lvert e\rvert}\frac{\sum_{n=0}^{n_{max}}[T_{95}(n,E_{FSL})+T_{96}(n,E_{FSL})]}{\sum_{n=0}^{n_{max}}[\{T_{95}(n,E_{FSL})\}'+\{T_{96}(n,E_{FSL})\}']}$ (9.22)
	HgTe/CdTe superlattices with graded interfaces $n_0 = \left(\frac{\lvert e\rvert B g_v}{\pi^2\hbar L_0}\right)\sum_{n=0}^{n_{max}}[T_{97}(n,E_{FSL}) + T_{98}(n,E_{FSL})]$ (9.29)	$\frac{D}{\mu} = \frac{1}{\lvert e\rvert}\frac{\sum_{n=0}^{n_{max}}[T_{97}(n,E_{FSL})+T_{98}(n,E_{FSL})]}{\sum_{n=0}^{n_{max}}[\{T_{97}(n,E_{FSL})\}'+\{T_{98}(n,E_{FSL})\}']}$ (9.30)
	III–V effective mass superlattices $n_0 = \left(\frac{\lvert e\rvert B g_v}{\pi^2\hbar}\right)\sum_{n=0}^{n_{max}}[T_{99}(n,E_{FSL}) + T_{910}(n,E_{FSL})]$ (9.35)	$\frac{D}{\mu} = \frac{1}{\lvert e\rvert}\frac{\sum_{n=0}^{n_{max}}[T_{99}(n,E_{FSL})+T_{910}(n,E_{FSL})]}{\sum_{n=0}^{n_{max}}[\{T_{99}(n,E_{FSL})\}'+\{T_{910}(n,E_{FSL})\}']}$ (9.36)
	II–VI effective mass superlattices $n_0 = \left(\frac{\lvert e\rvert B g_v}{2\pi^2\hbar}\right)\sum_{n=0}^{n_{max}}[T_{911}(n,E_{FSL}) + T_{912}(n,E_{FSL})]$ (9.41)	$\frac{D}{\mu} = \frac{1}{\lvert e\rvert}\frac{\sum_{n=0}^{n_{max}}[T_{911}(n,E_{FSL})+T_{912}(n,E_{FSL})]}{\sum_{n=0}^{n_{max}}[\{T_{911}(n,E_{FSL})\}'+\{T_{912}(n,E_{FSL})\}']}$ (9.42)
	IV–VI effective mass superlattices $n_0 = \left(\frac{\lvert e\rvert B g_v}{2\pi^2\hbar}\right)\sum_{n=0}^{n_{max}}[T_{911}(n,E_{FSL}) + T_{912}(n,E_{FSL})]$ (9.41)	$\frac{D}{\mu} = \frac{1}{\lvert e\rvert}\frac{\sum_{n=0}^{n_{max}}[T_{911}(n,E_{FSL})+T_{912}(n,E_{FSL})]}{\sum_{n=0}^{n_{max}}[\{T_{911}(n,E_{FSL})\}'+\{T_{912}(n,E_{FSL})\}']}$ (9.42)
	HgTe/CdTe effective mass superlattices $n_0 = \left(\frac{\lvert e\rvert B g_v}{\pi^2\hbar}\right)\sum_{n=0}^{n_{max}}[T_{915}(n,E_{FSL}) + T_{916}(n,E_{FSL})]$ (9.53)	$\frac{D}{\mu} = \frac{1}{\lvert e\rvert}\frac{\sum_{n=0}^{n_{max}}[T_{915}(n,E_{FSL})+T_{916}(n,E_{FSL})]}{\sum_{n=0}^{n_{max}}[\{T_{915}(n,E_{FSL})\}'+\{T_{916}(n,E_{FSL})\}']}$ (9.54)

(*Continued*)

Table 9.1. (*Continued*)

Confinement of wave vector space	The carrier statistics	The Einstein relation for the diffusivity mobility ratio
Quantum wires	III–V superlattices with graded interfaces $n_{1D} = \left(\frac{2g_V}{\pi}\right) \sum_{n_x=1}^{n_{x\max}} \sum_{n_y=1}^{n_{y\max}} [T_{917}(n_x, n_y, E_{FQWSL}) + T_{918}(n_x, n_y, E_{FQWSL})]$ (9.58)	$\frac{D}{\mu} = \frac{1}{\lvert e\rvert} \frac{\sum_{n_x=1}^{n_{x\max}} \sum_{n_y=1}^{n_{y\max}} [T_{917}(n_x,n_y,E_{FQWSL})+T_{918}(n_x,n_y,E_{FQWSL})]}{\sum_{n_x=1}^{n_{x\max}} \sum_{n_y=1}^{n_{y\max}} [\{T_{917}(n_x,n_y,E_{FQWSL})\}'+\{T_{918}(n_x,n_y,E_{FQWSL})\}']}$ (9.59)
	II–VI superlattices with graded interfaces $n_{1D} = \left(\frac{2g_V}{\pi}\right) \sum_{n_x=1}^{n_{xmas}} \sum_{n_y=1}^{n_{y\max}} [T_{919}(n_x, n_y, E_{FQWSL}) + T_{920}(n_x, n_y, E_{FQWSL})]$ (9.63)	$\frac{D}{\mu} = \frac{1}{\lvert e\rvert} \frac{\sum_{n_x=1}^{n_{xmas}} \sum_{n_y=1}^{n_{y\max}} [T_{919}(n_x,n_y,E_{FQWSL})+T_{920}(n_x,n_y,E_{FQWSL})]}{\sum_{n_x=1}^{n_{xmas}} \sum_{n_y=1}^{n_{y\max}} [\{T_{919}(n_x,n_y,E_{FQWSL})\}'+\{T_{920}(n_x,n_y,E_{FQWSL})\}']}$ (9.69)
	IV–VI superlattices with graded interfaces $n_{1D} = \left(\frac{2g_V}{\pi}\right) \sum_{n_x=1}^{n_{x\max}} \sum_{n_y=1}^{n_{y\max}} [T_{921}(n_x, n_y, E_{FQWSL}) + T_{922}(n_x, n_y, E_{FQWSL})]$ (9.68)	$\frac{D}{\mu} = \frac{1}{\lvert e\rvert} \frac{\sum_{n_x=1}^{n_{x\max}} \sum_{n_y=1}^{n_{y\max}} [T_{921}(n_x,n_y,E_{FQWSL})+T_{922}(n_x,n_y,E_{FQWSL})]}{\sum_{n_x=1}^{n_{x\max}} \sum_{n_y=1}^{n_{y\max}} [\{T_{921}(n_x,n_y,E_{FQWSL})\}'+\{T_{922}(n_x,n_y,E_{FQWSL})\}']}$ (9.69)
	HgTe/CdTe superlattices with graded interfaces $n_{1D} = \left(\frac{2g_V}{\pi}\right) \sum_{n_x=1}^{n_{\max}} \sum_{n_y=1}^{n_{y\max}} [T_{923}(n_x, n_y, E_{FQWSL}) + T_{924}(n_x, n_y, E_{FQWSL})]$ (9.73)	$\frac{D}{\mu} = \frac{1}{\lvert e\rvert} \frac{\sum_{n_x=1}^{n_{\max}} \sum_{n_y=1}^{n_{y\max}} [T_{923}(n_x,n_y,E_{FQWSL})+T_{924}(n_x,n_y,E_{FQWSL})]}{\sum_{n_x=1}^{n_{\max}} \sum_{n_y=1}^{n_{y\max}} [\{T_{923}(n_x,n_y,E_{FQWSL})\}'+\{T_{924}(n_x,n_y,E_{FQWSL})\}']}$ (9.74)
	III–V effective mass superlattices $n_{1D} = \left(\frac{2g_V}{\pi}\right) \sum_{n_y=1}^{n_{y\max}} \sum_{n_z=1}^{n_{z\max}} [T_{925}(n_y, n_z, E_{FQWSL}) + T_{926}(n_y, n_z, E_{FQWSL})]$ (9.78)	$\frac{D}{\mu} = \frac{1}{\lvert e\rvert} \frac{\sum_{n_y=1}^{n_{y\max}} \sum_{n_z=1}^{n_{z\max}} [T_{925}(n_y,n_z,E_{FQWSL})+T_{926}(n_y,n_z,E_{FQWSL})]}{\sum_{n_y=1}^{n_{y\max}} \sum_{n_z=1}^{n_{z\max}} [\{T_{925}(n_y,n_z,E_{FQWSL})\}'+\{T_{926}(n_y,n_z,E_{FQWSL})\}']}$ (9.79)
	II–VI effective mass superlattices $n_{1D} = \left(\frac{2g_V}{\pi}\right) \sum_{n_x=1}^{n_{x\max}} \sum_{n_y=1}^{n_{y\max}} T_{927}(n_x, n_y, E_{FQWSL}) + T_{928}(n_x, n_y, E_{FQWSL})]$ (9.83)	$\frac{D}{\mu} = \frac{1}{\lvert e\rvert} \frac{\sum_{n_x=1}^{n_{x\max}} \sum_{n_y=1}^{n_{y\max}} [T_{927}(n_x,n_y,E_{FQWSL})+T_{928}(n_x,n_y,E_{FQWSL})]}{\sum_{n_x=1}^{n_{x\max}} \sum_{n_y=1}^{n_{y\max}} [\{T_{927}(n_x,n_y,E_{FQWSL})\}'+\{T_{928}(n_x,n_y,E_{FQWSL})\}']}$ (9.84)
	IV–VI effective mass superlattices $n_{1D} = \left(\frac{2g_V}{\pi}\right) \sum_{n_y=1}^{n_{y\max}} \sum_{n_z=1}^{n_{z\max}} [T_{929}(n_y, n_z, E_{FQWSL}) + T_{930}(n_y, n_z, E_{FQWSL})]$ (9.88)	$\frac{D}{\mu} = \frac{1}{\lvert e\rvert} \frac{\sum_{n_y=1}^{n_{y\max}} \sum_{n_z=1}^{n_{z\max}} [T_{929}(n_y,n_z,E_{FQWSL})+T_{930}(n_y,n_z,E_{FQWSL})]}{\sum_{n_y=1}^{n_{y\max}} \sum_{n_z=1}^{n_{z\max}} [\{T_{929}(n_y,n_z,E_{FQWSL})\}'+\{T_{930}(n_y,n_z,E_{FQWSL})\}']}$ (9.89)
	HgTe/CdTe Effective mass superlattices $n_{1D} = \left(\frac{2g_V}{\pi}\right) \sum_{n_y=1}^{n_{y\max}} \sum_{n_z=1}^{n_{z\max}} [T_{931}(n_y, n_z, E_{FQWSL}) + T_{932}(n_y, n_z, E_{FQWSL})]$ (9.93)	$\frac{D}{\mu} = \frac{1}{\lvert e\rvert} \frac{\sum_{n_y=1}^{n_{y\max}} \sum_{n_z=1}^{n_{z\max}} [T_{931}(n_y,n_z,E_{FQWSL})+T_{932}(n_y,n_z,E_{FQWSL})]}{\sum_{n_y=1}^{n_{y\max}} \sum_{n_z=1}^{n_{z\max}} [\{T_{931}(n_y,n_z,E_{FQWSL})\}'+\{T_{932}(n_y,n_z,E_{FQWSL})\}']}$ (9.94)

Allied Research Problems

R.9.14 Investigate the EMM for all the problems from R.9.1 to R.9.13.

R.9.15 Investigate the Debye screening length, the carrier contribution to the elastic constants, heat capacity, activity coefficient, plasma frequency, and the thermoelectric power for all the materials covering all the cases of problems R.9.1–R.9.13 of this chapter.

R.9.16 Investigate in detail, the mobility for elastic and inelastic scattering mechanisms for all the materials covering all the cases of problems R.9.1–R.9.13 of this chapter.

R.9.17 Investigate the various transport coefficients in detail for all the materials covering all the cases of problems R.9.1–R.9.13 of this chapter.

R.9.18 Investigate the dia and paramagnetic susceptibilities in detail for all the materials covering all the appropriate problems of this chapter.

References

1. L.V. Keldysh, Sov. Phys. Solid State **4**, 1658 (1962)
2. L. Esaki, R. Tsu, IBM J. Res. Dev. **14**, 61 (1970)
3. R. Tsu, *Superlattices to Nanoelectronics* (Elsevier, Amsterdam, 2005); E. L. Ivchenko, G. Pikus, *Superlattices And Other Heterostructures* (Springer, Berlin, 1995); G. Bastard, *Wave Mechanics Applied to Heterostructures, Editions de Physique, Les Ulis, France* (1990)
4. P.K. Basu, *Theory of Optical Process in Semiconductors, Bulk and Microstructures*, Oxford University Press (1997)
5. K.P. Ghatak, B. Mitra, Int. J. Electron. **72**, 541 (1992); S.N. Biswas, K.P. Ghatak, Int. J. Electron. **70**, 125 (1991); B. Mitra, K.P. Ghatak, Phys. Stat. Sol. (b) **149**, K117 (1988); K.P. Ghatak, B. Mitra, A. Ghoshal, Phys. Stat. Sol. (b) **154**, K121 (1989)
6. K.V. Vaidyanathan, R.A. Jullens, C.L. Anderson, H.L. Dunlap, Solid State Electron. **26**, 717 (1983)
7. B.A. Wilson, IEEE J. Quant. Electron. **24**, 1763 (1988)
8. M. Krichbaum, P. Kocevar, H. Pascher, G. Bauer, IEEE J. Quant. Electron. **24**, 717 (1988)
9. J.N. Schulman, T.C. McGill, Appl. Phys. Lett. **34**, 663 (1979)
10. H. Kinoshita, T. Sakashita, H. Fajiyasu, J. Appl. Phys. **52**, 2869 (1981)
11. L. Ghenin, R.G. Mani, J.R. Anderson, J.T. Cheung, Phys. Rev. B **39**, 1419 (1989)
12. C.A. Hoffman, J.R. Mayer, F.J. Bartoli, J.W. Han, J.W. Cook, J.F. Schetzina, J.M. Schubman, Phys. Rev. B. **39**, 5208 (1989)
13. V.A. Yakovlev, Sov. Phys. Semiconduct. **13**, 692 (1979)
14. E.O. Kane, J. Phys. Chem. Solids **1**, 249 (1957)
15. H. Sasaki, Phys. Rev. B **30**, 7016 (1984)
16. H.X. Jiang, J.Y. Lin, J. Appl. Phys. **61**, 624 (1987)
17. J.J. Hopfield, J. Phys. Chem. Solids **15**, 97 (1960)
18. G.M.T. Foley, P.N. Langenberg, Phys. Rev. B, **15B** 4850 (1977)

10

The Einstein Relation in Compound Semiconductors in the Presence of Light Waves

10.1 Introduction

With the advent of nano-photonics, there has been a considerable interest in studying the optical processes in semiconductors and their nanostructures [1]. It appears from the literature, that the investigations have been carried out on the assumption that the carrier energy spectra are invariant quantities in the presence of intense light waves, which is not fundamentally true. The physical properties of semiconductors in the presence of light waves which change the basic dispersion relation have relatively been less investigated in the literature [2–4]. In this chapter, we shall study the Einstein relation in III–V, ternary and quaternary materials in the presence of external photo-excitation on the basis of newly formulated electron dispersion laws under different physical conditions.

In Sect. 10.1 of the theoretical background, we have formulated the dispersion relation of the conduction electrons of III–V, ternary and quaternary materials in the presence of light waves whose unperturbed electron energy spectrum is described by the three band model of Kane. In the same section, we have studied the dispersion relations for the said materials in the presence of external photo-excitation, when the unperturbed energy spectra are defined by the two band model of Kane and that of parabolic energy bands, respectively, for the purpose of relative comparison. In Sect. 10.2, we have derived the expressions of the electron statistics and the DMR for all the aforementioned cases. The DMR has been numerically investigated by taking n-InAs and n-InSb as examples of III–V compounds, n-$Hg_{1-x}Cd_xTe$ as an example of ternary compounds, and n-$In_{1-x}Ga_xAs_yP_{1-y}$ lattice matched to InP as an example of quaternary compounds, in accordance with the said band models for the purpose of relative assessment. Section 10.3 contains the result and discussion.

10.2 Theoretical Background

10.2.1 The Formulation of the Electron Dispersion Law in the Presence of Light Waves in III–V, Ternary and Quaternary Materials

The Hamiltonian $\left(\hat{H}\right)$ of an electron in the presence of light wave characterized by the vector potential $\overrightarrow{A}$ can be written following [5] as

$$\hat{H} = \left[\left|\left(\hat{p} + |e|\,\overrightarrow{A}\right)\right|^2 \Big/ 2m\right] + V\left(\overrightarrow{r}\right), \tag{10.1}$$

in which, $\hat{p}$ is the momentum operator, $V(\overrightarrow{r})$ is the crystal potential and m is the free electron mass.

Equation (10.1) can be expressed as

$$\hat{H} = \hat{H}_0 + \hat{H}', \tag{10.2}$$

where $\hat{H}_0 = \dfrac{\hat{p}^2}{2m} + V(\overrightarrow{r})$
and

$$\hat{H}' = \frac{|e|}{2m}\overrightarrow{A}\cdot\hat{p}. \tag{10.3}$$

The perturbed Hamiltonian $\hat{H}'$ can be written as

$$\hat{H}' = \left(\frac{-\mathrm{i}\hbar\,|e|}{2m}\right)\left(\overrightarrow{A}\cdot\nabla\right) \tag{10.4}$$

where $\mathrm{i} = \sqrt{-1}$ and $\hat{p} = -\mathrm{i}\hbar\nabla$.

The vector potential $(\overrightarrow{A})$ of the monochromatic light of plane wave can be expressed as

$$\overrightarrow{A} = A_0\,\overrightarrow{\varepsilon}_{\mathrm{s}}\cos(\overrightarrow{s}_0\cdot\overrightarrow{r} - \omega t), \tag{10.5}$$

where A_0 is the amplitude of the light wave, $\overrightarrow{\varepsilon}_{\mathrm{s}}$ is the polarization vector, $\overrightarrow{s}_0$ is the momentum vector of the incident photon, $\overrightarrow{r}$ is the position vector, ω is the angular frequency of light wave and t is the time scale. The matrix element of $\hat{H}'_{\mathrm{nl}}$ between initial state, $\psi_{\mathrm{l}}(\overrightarrow{q},\overrightarrow{r})$ and final state $\psi_{\mathrm{n}}\left(\overrightarrow{k},\overrightarrow{r}\right)$ in different bands can be written as

$$\hat{H}'_{\mathrm{nl}} = \frac{|e|}{2m}\left\langle n\overrightarrow{k}\left|\overrightarrow{A}\cdot\hat{p}\right|l\overrightarrow{q}\right\rangle. \tag{10.6}$$

Using (10.4) and (10.5), we can re-write (10.6) as

$$\hat{H}'_{\mathrm{nl}} = \left(\frac{-\mathrm{i}\hbar\,|e|\,A_0}{4m}\right)\overrightarrow{\varepsilon}_{\mathrm{s}}\cdot \left[\left\{\left\langle n\overrightarrow{k}\left|\mathrm{e}^{(\mathrm{i}\overrightarrow{s}_0\cdot\overrightarrow{r})}\nabla\right|l\overrightarrow{q}\right\rangle\mathrm{e}^{-\mathrm{i}\omega t}\right\} + \left\{\left\langle n\overrightarrow{k}\left|\mathrm{e}^{(-\mathrm{i}\overrightarrow{s}_0\cdot\overrightarrow{r})}\nabla\right|l\overrightarrow{q}\right\rangle\mathrm{e}^{\mathrm{i}\omega t}\right\}\right], \tag{10.7}$$

The first matrix element of (10.7) can be written as

$$\left\langle n\vec{k}\left|\mathrm{e}^{(\mathrm{i}\vec{s}_0\cdot\vec{r})}\nabla\right|l\vec{q}\right\rangle=\int\mathrm{e}^{\left(\mathrm{i}\left[\vec{q}+\vec{s}_0-\vec{k}\right]\cdot\vec{r}\right)}i\vec{q}u_{\mathrm{n}}^*\left(\vec{k},\vec{r}\right)u_{\mathrm{l}}\left(\vec{q},\vec{r}\right)\mathrm{d}^3r$$
$$+\int\mathrm{e}^{\left(\mathrm{i}\left[\vec{q}+\vec{s}_0-\vec{k}\right]\cdot\vec{r}\right)}u_{\mathrm{n}}^*\left(\vec{k},\vec{r}\right)\nabla u_{\mathrm{l}}\left(\vec{q},\vec{r}\right)\mathrm{d}^3r. \tag{10.8}$$

The functions $u_{\mathrm{n}}^*u_{\mathrm{l}}$ and $u_{\mathrm{n}}^*\nabla u_{\mathrm{l}}$ are periodic. The integral over all space can be separated into a sum over unit cells times an integral over a single unit cell. It is assumed that the wave length of the electromagnetic wave is sufficiently large, so that if $\vec{k}$ and $\vec{q}$ are within the Brillouin zone, $\left(\vec{q}+\vec{s}_0-\vec{k}\right)$ is not a reciprocal lattice vector.

Therefore, we can write (10.8) as

$$\begin{aligned}&\left\langle n\vec{k}\left|\mathrm{e}^{(\mathrm{i}\vec{s}_0\cdot\vec{r})}\nabla\right|l\vec{q}\right\rangle\\&=\left[\frac{(2\pi)^3}{\Omega}\right]\left\{\mathrm{i}\vec{q}\delta\left(\vec{q}+\vec{s}_0-\vec{k}\right)\delta_{\mathrm{nl}}+\delta\left(\vec{q}+\vec{s}_0-\vec{k}\right)\right.\\&\qquad\left.\times\int_{\mathrm{cell}}u_{\mathrm{n}}^*\left(\vec{k},\vec{r}\right)\nabla u_{\mathrm{l}}\left(\vec{q},\vec{r}\right)\mathrm{d}^3r\right\}\\&=\left[\frac{(2\pi)^3}{\Omega}\right]\left\{\delta\left(\vec{q}+\vec{s}_0-\vec{k}\right)\int_{\mathrm{cell}}u_{\mathrm{n}}^*\left(\vec{k},\vec{r}\right)\nabla u_{\mathrm{l}}\left(\vec{q},\vec{r}\right)\mathrm{d}^3r\right\},\end{aligned} \tag{10.9}$$

where Ω is the volume of the unit cell and $\int u_{\mathrm{n}}^*(\vec{k},\vec{r})u_{\mathrm{l}}(\vec{q},\vec{r})\mathrm{d}^3r=\delta(\vec{q}-\vec{k})\delta_{\mathrm{nl}}=0.$, since $n\neq l$.

The delta function expresses the conservation of wave vector in the absorption of light wave, and $\vec{s}_0$ is small compared to the dimension of a typical Brillouin zone and we set $\vec{q}=\vec{k}$.

From (10.8) and (10.9), we can write,

$$\hat{H}'_{\mathrm{nl}}=\frac{|e|\,A_0}{2m}\vec{\varepsilon}_{\mathrm{s}}\cdot\hat{p}_{\mathrm{nl}}(\vec{k})\delta(\vec{q}-\vec{k})\cos(\omega t), \tag{10.10}$$

where $\hat{p}_{\mathrm{nl}}(\vec{k})=-\mathrm{i}\hbar\int u_{\mathrm{n}}^*\nabla u_{\mathrm{l}}\mathrm{d}^3r=\int u_{\mathrm{n}}^*(\vec{k},\vec{r})\hat{p}u_{\mathrm{l}}(\vec{k},\vec{r})\mathrm{d}^3r$

Therefore, we can write

$$\hat{H}'_{\mathrm{nl}}=\frac{|e|\,A_0}{2m}\vec{\varepsilon}\cdot\hat{p}_{\mathrm{nl}}(\vec{k}), \tag{10.11}$$

where $\vec{\varepsilon}=\vec{\varepsilon}_{\mathrm{s}}\cos\omega t$.

When a photon interacts with a semiconductor, the carriers (i.e., electrons) are generated in the bands which are followed by the interband transitions. For example, when the carriers are generated in the valence band, the carriers then make interband transition to the conduction band. The transition of the electrons within the same band, i.e., $\hat{H}'_{\mathrm{nn}}=\left\langle n\vec{k}|\hat{H}'|n\vec{k}\right\rangle$ is neglected.

Because, in such a case, i.e., when the carriers are generated within the same bands, the photons are lost by recombination within the aforementioned band resulting in zero carriers.

Therefore,

$$\langle n\vec{k}\left|\hat{H}'\right|n\vec{k}\rangle = 0. \tag{10.12}$$

With $n = c$ stands for conduction band and $l = v$ stand for valance band, the energy equation for the conduction electron can approximately be written as

$$\gamma(E) = \left(\frac{\hbar^2k^2}{2m^*}\right) + \frac{\left(\frac{|e|A_0}{2m}\right)^2 \left\langle\left|\vec{\varepsilon}\cdot\hat{p}_{cv}(\vec{k})\right|^2\right\rangle_{av}}{E_c(\vec{k}) - E_v(\vec{k})}, \tag{10.13}$$

where $\gamma(E) \equiv E(aE+1)(bE+1)/(cE+1), a \equiv 1/E_{g_0}$, E_{g_0} is the un-perturbed band-gap, $b \equiv 1/(E_{g_0} + \Delta)$, $c \equiv 1/(E_{g_0} + 2\Delta/3)$, and $\left\langle\left|\vec{\varepsilon}\cdot\hat{p}_{cv}(\vec{k})\right|^2\right\rangle_{av}$ represents the average of the square of the optical matrix element (OME).

For the three-band model of Kane, we can write,

$$\xi_{1k} = E_c(\vec{k}) - E_v(\vec{k}) = (E_{g_0}^2 + E_{g_0}\hbar^2k^2/m_r)^{1/2}, \tag{10.14}$$

where m_r is the reduced mass and is given by $m_r^{-1} = (m^*)^{-1} + m_v^{-1}$, and m_v is the effective mass of the heavy hole at the top of the valance band in the absence of any field.

The doubly degenerate wave functions $u_1(\vec{k},\vec{r})$ and $u_2(\vec{k},\vec{r})$ can be expressed as [6]

$$u_1(\vec{k},\vec{r}) = a_{k+}\left[(\mathrm{i}s)\downarrow'\right] + b_{k+}\left[\frac{X' - \mathrm{i}Y'}{\sqrt{2}}\uparrow'\right] + c_{k+}\left[Z'\downarrow'\right], \tag{10.15}$$

and

$$u_2(\vec{k},\vec{r}) = a_{k-}\left[(\mathrm{i}s)\uparrow'\right] - b_{k-}\left[\frac{X' + \mathrm{i}Y'}{\sqrt{2}}\downarrow'\right] + c_{k-}\left[Z'\uparrow'\right]. \tag{10.16}$$

s is the s-type atomic orbital in both unprimed and primed coordinates, $\downarrow'$ indicates the spin down function in the primed coordinates, $a_{k\pm} \equiv \beta\left[E_{g_0} - (\gamma_{0k\pm})^2(E_{g_0} - \delta')\right]^{1/2}(E_{g_0} + \delta')^{-1/2}\Big]$, $\beta \equiv [(6(E_{g_0} + 2\Delta/3)(E_{g_0} + \Delta))/\chi]^{1/2}$, $\chi \equiv (6E_{g_0}^2 + 9E_{g_0}\Delta + 4\Delta^2)$, $\gamma_{0k\pm} \equiv \left[\frac{(\xi_{1k} \mp E_{g_0})}{2(\xi_{1k} + \delta')}\right]^{1/2}$, $\xi_{1k} \equiv E_c(\vec{k}) - E_v(\vec{k}) = E_{g_0}\left[1 + 2\left(1 + \frac{m_c}{m_v}\right)\frac{\gamma(E)}{E_{g_0}}\right]^{1/2}$, $\delta' \equiv (E_{g_0}^2\Delta)(\chi)^{-1}$, X', Y', and Z' are the p-type atomic orbitals in the primed coordinates, $\uparrow'$ indicates the spin-up function in the primed coordinates, $b_{k\pm} \equiv \rho\gamma_{0k\pm}$, $\rho \equiv$

$(4\Delta^2/3\chi)^{1/2}$, $c_{k\pm} \equiv t\gamma_{0k\pm}$ and $t \equiv \left[6(E_{\mathrm{go}} + 2\Delta/3)^2/\chi\right]^{1/2}$.

We can, therefore, write the expression for the optical matrix element (OME) as

$$\mathrm{OME} = \hat{p}_{\mathrm{cv}}\left(\overrightarrow{k}\right) = < u_1(\overrightarrow{k}, \overrightarrow{r}) \left|\hat{p}\right| u_2(\overrightarrow{k}, \overrightarrow{r}) > \tag{10.17}$$

Since the photon vector has no interaction in the same band for the study of inter-band optical transition, we can therefore, write

$$\langle S\,|\hat{p}\,|S\,\rangle = \langle X\,|\hat{p}\,|X\,\rangle = \langle Y\,|\hat{p}\,|Y\,\rangle = \langle Z\,|\hat{p}\,|Z\,\rangle = 0,$$

and $\langle X\,|\hat{p}\,|Y\,\rangle = \langle Y\,|\hat{p}\,|Z\,\rangle = \langle Z\,|\hat{p}\,|X\,\rangle = 0$.

There are finite interactions between the conduction band (CB) and the valance band (VB) and we can obtain

$$\left\langle S|\hat{P}|X\right\rangle = \hat{i}\cdot\hat{P}_x$$
$$\left\langle S|\hat{P}|Y\right\rangle = \hat{j}\cdot\hat{P}_y$$
$$\left\langle S|\hat{P}|Z\right\rangle = \hat{k}\cdot\hat{P}_z$$

where $\hat{i}, \hat{j}$ and $\hat{k}$ are the unit vectors along x, y and z axes respectively.

It is well known that

$$\begin{bmatrix}\uparrow' \\ \downarrow'\end{bmatrix} = \begin{bmatrix} \mathrm{e}^{-\mathrm{i}\phi/2}\cos(\theta/2) & \mathrm{e}^{\mathrm{i}\phi/2}\sin(\theta/2) \\ -\mathrm{e}^{-\mathrm{i}\phi/2}\sin(\theta/2) & \mathrm{e}^{\mathrm{i}\phi/2}\cos(\theta/2)\end{bmatrix}\begin{bmatrix}\uparrow \\ \downarrow\end{bmatrix} \quad \text{and}$$

$$\begin{bmatrix}X' \\ Y' \\ Z'\end{bmatrix} = \begin{bmatrix}\cos\theta\cos\phi & \cos\theta\sin\phi & -\sin\theta \\ -\sin\phi & \cos\phi & 0 \\ \sin\theta\cos\phi & \sin\theta\sin\phi & \cos\theta\end{bmatrix}\begin{bmatrix}X \\ Y \\ Z\end{bmatrix}.$$

Besides, the spin vector can be written as

$$\overrightarrow{S} = \frac{\hbar}{2}\overrightarrow{\sigma}, \text{where, } \sigma_x = \begin{bmatrix}0 & 1 \\ 1 & 0\end{bmatrix}, \sigma_y = \begin{bmatrix}0 & -\mathrm{i} \\ \mathrm{i} & 0\end{bmatrix} \text{ and } \sigma_z = \begin{bmatrix}1 & 0 \\ 0 & -1\end{bmatrix}.$$

From above, we can write

$$\begin{aligned}
\hat{p}_{\mathrm{CV}}\left(\overrightarrow{k}\right) &= \left\langle u_1\left(\overrightarrow{k}, \overrightarrow{r}\right)|\hat{P}|u_2\left(\overrightarrow{k}, \overrightarrow{r}\right)\right\rangle \\
&= \left\langle\left\{a_{k_+}\left[(\mathrm{i}S)\downarrow'\right] + b_{k_+}\left[\left(\frac{X' - \mathrm{i}Y'}{\sqrt{2}}\right)\uparrow'\right]\right.\right. \\
&\quad \left. + c_{k_+}\left[Z'\downarrow'\right]\right\}|\hat{P}|\left\{a_{k_-}\left[(\mathrm{i}S)\uparrow'\right]\right. \\
&\quad \left.\left. - b_{k_-}\left[\left(\frac{X' + \mathrm{i}Y'}{\sqrt{2}}\right)\downarrow' + c_{k_-}\left[Z'\uparrow'\right]\right]\right\}\right\rangle.
\end{aligned}$$

Using above relations, we get

$$\begin{aligned}\hat{p}_{\mathrm{CV}}\left(\overrightarrow{k}\right) &= \left\langle u_1\left(\overrightarrow{k},\overrightarrow{r}\right)|\hat{P}|u_2\left(\overrightarrow{k},\overrightarrow{r}\right)\right\rangle \\ &= \frac{b_{k_+}a_{k_-}}{\sqrt{2}}\left\{\left\langle\left(X'-\mathrm{i}Y'\right)|\hat{P}|\mathrm{i}S\right\rangle\left\langle\uparrow'|\uparrow'\right\rangle\right\} \\ &+c_{k_+}a_{k_-}\left\{\left\langle Z'|\hat{P}|\mathrm{i}S\right\rangle\left\langle\downarrow'|\uparrow'\right\rangle\right\} \\ &-\frac{a_{k_+}b_{k_-}}{\sqrt{2}}\left\{\left\langle\mathrm{i}S|\hat{P}|\left(X'+\mathrm{i}Y'\right)\right\rangle\left\langle\downarrow'|\downarrow'\right\rangle\right\} \\ &+a_{k_+}c_{k_-}\left\{\left\langle\mathrm{i}S|\hat{P}|Z'\right\rangle\left\langle\downarrow'|\uparrow'\right\rangle\right\}.\end{aligned} \tag{10.18}$$

From (10.18), we can write

$$\begin{aligned}&\left\langle\left(X'-\mathrm{i}Y'\right)|\hat{P}|\mathrm{i}S\right\rangle = \left\langle\left(X'\right)|\hat{P}|\mathrm{i}S\right\rangle - \left\langle\left(\mathrm{i}Y'\right)|\hat{P}|\mathrm{i}S\right\rangle \\ &= \mathrm{i}\int u_{X'}^{*}\hat{P}S - \int -\mathrm{i}u_{Y'}^{*}\hat{P}\mathrm{i}u_X = \mathrm{i}\left\langle X'|\hat{P}|S\right\rangle - \left\langle Y'|\hat{P}|S\right\rangle.\end{aligned}$$

From the above relations, for X', Y' and Z', we get

$$|X'\rangle = \cos\,\theta\cos\,\phi\,|X\rangle + \cos\,\theta\sin\,\phi\,|Y\rangle - \sin\,\theta\,|Z\rangle\,.$$

Thus, $\left\langle X'|\hat{P}|S\right\rangle = \cos\theta\cos\phi\left\langle X|\hat{P}|S\right\rangle + \cos\theta\sin\phi\left\langle Y|\hat{P}|S\right\rangle - \sin\theta\left\langle Z|\hat{P}|S\right\rangle$,
$= \hat{P}\hat{r}_1$
where $\hat{r}_1 = \hat{\imath}\cos\theta\cos\phi + \hat{\jmath}\cos\theta\sin\phi - \hat{k}\sin\theta$,

$$|Y'\rangle = -\sin\,\phi\,|X\rangle + \cos\,\phi\,|Y\rangle + 0\,|Z\rangle\,.$$

Thus, $\left\langle Y'|\hat{P}|S\right\rangle = -\sin\phi\left\langle X|\hat{P}|S\right\rangle + \cos\phi\left\langle Y|\hat{P}|S\right\rangle + 0\left\langle Z|\hat{P}|S\right\rangle = \hat{P}\hat{r}_2$,
where $\hat{r}_2 = -\hat{\imath}\sin\phi + \hat{\jmath}\cos\phi$,
so that $\left\langle\left(X'-\mathrm{i}Y'\right)|\hat{P}|S\right\rangle = \hat{P}\left(\mathrm{i}\hat{r}_1 - \hat{r}_2\right)$,
Thus,

$$\frac{a_{k_-}b_{k_+}}{\sqrt{2}}\left\langle\left(X'-\mathrm{i}Y'\right)|\hat{P}|S\right\rangle\left\langle\uparrow'|\uparrow'\right\rangle = \frac{a_{k_-}b_{k_+}}{\sqrt{2}}\hat{P}\left(\mathrm{i}\hat{r}_1 - \hat{r}_2\right)\left\langle\uparrow'|\uparrow'\right\rangle. \tag{10.19}$$

Now since,

$$\left\langle\mathrm{i}S|\hat{P}|\left(X'+\mathrm{i}Y'\right)\right\rangle = \mathrm{i}\left\langle S|\hat{P}|X'\right\rangle - \left\langle S|\hat{P}|Y'\right\rangle = \hat{P}\left(\mathrm{i}\hat{r}_1 - \hat{r}_2\right).$$

We can write,

$$-\left[\frac{a_{k_+}b_{k_-}}{\sqrt{2}}\left\{\left\langle\mathrm{i}S|\hat{P}|\left(X'+\mathrm{i}Y'\right)\right\rangle\left\langle\downarrow'|\downarrow'\right\rangle\right\}\right] = -\left[\frac{a_{k_+}b_{k_-}}{\sqrt{2}}\hat{P}\left(\mathrm{i}\hat{r}_1 - \hat{r}_2\right)\left\langle\downarrow'|\downarrow'\right\rangle\right]. \tag{10.20}$$

Similarly, we get

$$|Z'\rangle = \sin\,\theta\cos\,\phi\,|X\rangle + \sin\,\theta\sin\,\phi\,|Y\rangle + \cos\,\theta\,|Z\rangle\,,$$

So that, $\left\langle Z'|\hat{P}|\mathrm{i}S\right\rangle = \mathrm{i}\left\langle Z'|\hat{P}|S\right\rangle = \mathrm{i}\hat{P}\left\{\sin\theta\cos\phi\hat{i} + \sin\theta\sin\phi\hat{j} + \cos\theta\hat{k}\right\}$
$= \mathrm{i}\hat{P}\hat{r}_3$,
where $\hat{r}_3 = \hat{i}\sin\theta\cos\phi + \hat{j}\sin\theta\sin\phi + \hat{k}\cos\theta$.

Thus,

$$c_{k_+}a_{k_-}\left\langle Z'|\hat{P}|\mathrm{i}S\right\rangle\left\langle\downarrow'|\uparrow'\right\rangle = c_{k_+}a_{k_-}\mathrm{i}\hat{P}\hat{r}_3\left\langle\downarrow'|\uparrow'\right\rangle. \tag{10.21}$$

Similarly, we can write,

$$c_{k_-}a_{k_+}\left\langle \mathrm{i}S|\hat{P}|Z'\right\rangle\left\langle\downarrow'|\uparrow'\right\rangle = c_{k_-}a_{k_+}\mathrm{i}\hat{P}\hat{r}_3\left\langle\downarrow'|\uparrow'\right\rangle. \tag{10.22}$$

Therefore, we obtain

$$\begin{aligned}&\frac{a_{k_-}b_{k_+}}{\sqrt{2}}\left\{\left\langle(X'-\mathrm{i}Y')\,|\hat{P}|S\right\rangle\langle\uparrow'|\uparrow'\rangle\right\} - \frac{a_{k_+}b_{k_-}}{\sqrt{2}}\left\{\left\langle\mathrm{i}S|\hat{P}|\,(X'+\mathrm{i}Y')\right\rangle\langle\downarrow'|\downarrow'\rangle\right\}\\ &= \frac{\hat{P}}{\sqrt{2}}\left(-a_{k_+}b_{k_-}\langle\downarrow'|\downarrow'\rangle + a_{k_-}b_{k_+}\langle\uparrow'|\uparrow'\rangle\right)(\mathrm{i}\hat{r}_1 - \hat{r}_2)\,.\end{aligned} \tag{10.23}$$

Also, we can write,

$$\begin{aligned}&c_{k_+}a_{k_-}\left\langle Z'|\hat{P}|\mathrm{i}S\right\rangle\left\langle\downarrow'|\uparrow'\right\rangle + c_{k_-}a_{k_+}\left\langle\mathrm{i}S|\hat{P}|Z'\right\rangle\left\langle\downarrow'|\uparrow'\right\rangle\\ &\quad = \mathrm{i}\hat{P}\left(c_{k_+}a_{k_-} + c_{k_-}a_{k_+}\right)\hat{r}_3\left[\langle\downarrow'|\downarrow'\rangle\right].\end{aligned} \tag{10.24}$$

Combining (10.23) and (10.24), we find

$$\begin{aligned}\hat{p}_{\mathrm{CV}}\left(\overrightarrow{k}\right) &= \frac{\hat{P}}{\sqrt{2}}(\mathrm{i}\hat{r}_1 - \hat{r}_2)\left\{\left(b_{k_+}a_{k_-}\right)\left\langle\uparrow'|\uparrow'\right\rangle - \left(b_{k_-}a_{k_+}\right)\langle\downarrow'|\downarrow'\rangle\right\}\\ &\quad + \mathrm{i}\hat{P}\hat{r}_3\left(c_{k_+}a_{k_-} - c_{k_-}a_{k_+}\right)\left\langle\downarrow'|\uparrow'\right\rangle.\end{aligned} \tag{10.25}$$

From the above relations, we obtain,

$$\left.\begin{aligned}\uparrow' &= \mathrm{e}^{-\mathrm{i}\phi/2}\cos\left(\theta/2\right)\uparrow + \mathrm{e}^{\mathrm{i}\phi/2}\sin\left(\theta/2\right)\downarrow\\ \downarrow' &= -\mathrm{e}^{-\mathrm{i}\phi/2}\sin\left(\theta/2\right)\uparrow + \mathrm{e}^{\mathrm{i}\phi/2}\cos\left(\theta/2\right)\downarrow\end{aligned}\right\}. \tag{10.26}$$

Therefore,

$$\begin{aligned}\langle\downarrow'|\uparrow'\rangle_x &= -\sin\left(\theta/2\right)\cos\left(\theta/2\right)\langle\uparrow\mid\uparrow\rangle_x + \mathrm{e}^{-\mathrm{i}\phi}\cos^2\left(\theta/2\right)\langle\downarrow\mid\uparrow\rangle_x - \mathrm{e}^{\mathrm{i}\phi}\sin^2\left(\theta/2\right)\\ &\qquad\langle\uparrow\mid\downarrow\rangle_x + \sin\left(\theta/2\right)\cos\left(\theta/2\right)\langle\downarrow\mid\downarrow\rangle_x\,.\end{aligned} \tag{10.27}$$

But we know from above that

$$\langle\uparrow \mid \uparrow\rangle_x = 0, \langle\downarrow \mid \uparrow\rangle_x = \frac{1}{2}, \text{ and } \langle\downarrow \mid \downarrow\rangle_x = 0.$$

Thus, from (10.27), we get

$$\begin{aligned}
\langle\downarrow'|\uparrow'\rangle_x &= \frac{1}{2}\left[\mathrm{e}^{-\mathrm{i}\phi}\cos^2(\theta/2) - \mathrm{e}^{\mathrm{i}\phi}\sin^2(\theta/2)\right] \\
&= \frac{1}{2}\left[(\cos\phi - \mathrm{i}\sin\phi)\cos^2(\theta/2) - (\cos\phi + \mathrm{i}\sin\phi)\sin^2(\theta/2)\right] \\
&= \frac{1}{2}\left[\cos\phi\cos\theta - \mathrm{i}\sin\phi\right].
\end{aligned} \tag{10.28}$$

Similarly, we obtain

$$\langle\downarrow'|\uparrow'\rangle_y = \frac{1}{2}\left[\mathrm{i}\cos\phi + \sin\phi\cos\theta\right] \text{ and } \left\langle\downarrow'|\uparrow'\right\rangle_z = \frac{1}{2}\left[-\sin\theta\right].$$

Therefore,

$$\begin{aligned}
\left\langle\downarrow'|\uparrow'\right\rangle &= \hat{i}\left\langle\downarrow'|\uparrow'\right\rangle_x + \hat{j}\left\langle\downarrow'|\uparrow'\right\rangle_y + \hat{k}\left\langle\downarrow'|\uparrow'\right\rangle_z, \\
&= \frac{1}{2}\left\{(\cos\theta\cos\phi - \mathrm{i}\sin\phi)\,\hat{i} + (\mathrm{i}\cos\phi + \sin\phi\cos\theta)\,\hat{j} - \sin\theta\hat{k}\right\} \\
&= \frac{1}{2}\left[\left\{(\cos\theta\cos\phi)\,\hat{i} + (\sin\phi\cos\theta)\,\hat{j} - \sin\theta\hat{k}\right\}\right. \\
&\qquad \left. + \mathrm{i}\left\{-\hat{i}\sin\phi + \hat{j}\cos\phi\right\}\right] \\
&= \frac{1}{2}\left[\hat{r}_1 + \mathrm{i}\hat{r}_2\right] = -\frac{1}{2}\mathrm{i}\left[\mathrm{i}\hat{r}_1 - \hat{r}_2\right].
\end{aligned}$$

Similarly, we can write

$$\begin{aligned}
\left\langle\uparrow'|\uparrow'\right\rangle &= \frac{1}{2}\left[\hat{i}\sin\theta\cos\phi + \hat{j}\sin\theta\sin\phi + \hat{k}\cos\theta\right] \\
&= \frac{1}{2}\hat{r}_3 \text{ and } \langle\downarrow'|\downarrow'\rangle = -\frac{1}{2}\hat{r}_3.
\end{aligned}$$

Using the above results and following (10.25) we can write

$$\begin{aligned}
\hat{p}_{\mathrm{CV}}\left(\overrightarrow{k}\right) &= \frac{\hat{P}}{\sqrt{2}}\left(\mathrm{i}\hat{r}_1 - \hat{r}_2\right)\left\{\left(a_{k_-}b_{k_+}\right)\left\langle\uparrow'|\uparrow'\right\rangle - \left(b_{k_-}a_{k_+}\right)\langle\downarrow'|\downarrow'\rangle\right\} \\
&\quad + \mathrm{i}\hat{P}\hat{r}_3\left\{\left(c_{k_+}a_{k_-} - c_{k_-}a_{k_+}\right)\langle\downarrow'|\uparrow'\rangle\right\} \\
&= \frac{\hat{P}}{2}\hat{r}_3\left(\mathrm{i}\hat{r}_1 - \hat{r}_2\right)\left\{\left(\frac{a_{k_-}b_{k_+}}{\sqrt{2}} + \frac{b_{k_-}a_{k_+}}{\sqrt{2}}\right)\right\} \\
&\quad + \frac{\hat{P}}{2}\hat{r}_3\left(\mathrm{i}\hat{r}_1 - \hat{r}_2\right)\left\{\left(c_{k_+}a_{k_-} + c_{k_-}a_{k_+}\right)\right\}.
\end{aligned}$$

Thus,

$$\hat{p}_{\mathrm{CV}}\left(\overrightarrow{k}\right) = \frac{\hat{P}}{2}\hat{r}_3\left(\mathrm{i}\hat{r}_1 - \hat{r}_2\right)\left\{a_{k_+}\left(\frac{b_{k_-}}{\sqrt{2}} + c_{k_-}\right) + a_{k_-}\left(\frac{b_{k_+}}{\sqrt{2}} + c_{k_+}\right)\right\}. \tag{10.29}$$

We can write that,

$|\hat{r}_1| = |\hat{r}_2| = |\hat{r}_3| = 1$, also, $\hat{P}\hat{r}_3 = \hat{P}_x \sin\theta\cos\phi\hat{i} + \hat{P}_y \sin\theta\sin\phi\hat{j} + \hat{P}_z\cos\theta\hat{k}$,

where $\hat{P} = \left\langle S|\hat{P}|X\right\rangle = \left\langle S|\hat{P}|Y\right\rangle = \left\langle S|\hat{P}|Z\right\rangle$,

$$\left\langle S|\hat{P}|X\right\rangle = \int u_{\mathrm{C}}^*(0,\overrightarrow{r})\hat{P}u_{\mathrm{VX}}$$

$$(0,\overrightarrow{r})\,\mathrm{d}^3r = \hat{P}_{\mathrm{CVX}}(0), \left\langle S|\hat{P}|Y\right\rangle = \hat{P}_{\mathrm{CVY}}(0) \text{ and } \left\langle S|\hat{P}|Z\right\rangle = \hat{P}_{\mathrm{CVZ}}(0).$$

Thus, $\hat{P} = \hat{P}_{\mathrm{CVX}}(0) = \hat{P}_{\mathrm{CVY}}(0) - \hat{P}_{\mathrm{CVZ}}(0) = \hat{P}_{\mathrm{CV}}(0)$, where $\hat{P}_{\mathrm{CV}}(0) \equiv \int u_{\mathrm{c}}^*(0,\overrightarrow{r})\hat{P}u_{\mathrm{V}}(0,\overrightarrow{r})\,\mathrm{d}^3r \equiv \hat{P}$.

For a plane polarized light wave, we have the polarization vector $\overrightarrow{\varepsilon}_{\mathrm{s}} = \hat{k}$, when the light wave vector is traveling along the z-axis. Therefore, for a plane polarized light-wave, we have considered $\overrightarrow{\varepsilon}_{\mathrm{s}} = \hat{k}$.

Then, from (10.29)

$$\left(\overrightarrow{\varepsilon}\cdot\hat{p}_{\mathrm{CV}}\left(\overrightarrow{k}\right)\right) = \overrightarrow{k}\cdot\frac{\hat{P}}{2}\hat{r}_3\left(\mathrm{i}\hat{r}_1 - \hat{r}_2\right)\left[A\left(\overrightarrow{k}\right) + B\left(\overrightarrow{k}\right)\right]\cos\,\omega t, \tag{10.30}$$

and

$$\left.\begin{aligned} A(\overrightarrow{k}) &= a_{k_-}\left(\frac{b_{k_+}}{\sqrt{2}} + c_{k_+}\right) \\ B(\overrightarrow{k}) &= a_{k+}\left(\frac{b_{k_-}}{\sqrt{2}} + c_{k-}\right)\end{aligned}\right\}. \tag{10.31}$$

Thus,

$$\begin{aligned}\left|\overrightarrow{\varepsilon}\cdot\hat{p}_{\mathrm{cv}}\left(\overrightarrow{k}\right)\right|^2 &= \left|\hat{k}\cdot\frac{\hat{P}}{2}\hat{r}_3\right|^2 |\mathrm{i}\hat{r}_1 - \hat{r}_2|^2\left[A(\overrightarrow{k}) + B(\overrightarrow{k})\right]^2\cos^2\omega t \\ &= \frac{1}{4}\left|\hat{P}_z\cos\,\theta\right|^2\left[A\left(\overrightarrow{k}\right) + B\left(\overrightarrow{k}\right)\right]^2\cos^2\omega t.\end{aligned} \tag{10.32}$$

So, the average value of $\left|\overrightarrow{\varepsilon}\cdot\hat{p}_{\mathrm{cv}}\left(\overrightarrow{k}\right)\right|^2$ for a plane polarized light-wave is given by

$$\begin{aligned}\left\langle\left|\overrightarrow{\varepsilon}\cdot\hat{p}_{\mathrm{cv}}\left(\overrightarrow{k}\right)\right|^2\right\rangle_{\mathrm{av}} &= \frac{1}{4}\left|\hat{P}_z\right|^2\left[A(\overrightarrow{k}) + B(\overrightarrow{k})\right]^2\left(\int_0^{2\pi}\mathrm{d}\phi\int_0^{\pi}\cos^2\theta\sin\theta\,\mathrm{d}\theta\right) \\ \left(\frac{1}{2}\right) &= \frac{\pi}{3}\left|\hat{P}_z\right|^2\left[A\left(\overrightarrow{k}\right) + B\left(\overrightarrow{k}\right)\right]^2,\end{aligned} \tag{10.33}$$

where $\left|\hat{P}_z\right|^2 = \left|\overrightarrow{k} \cdot \hat{p}_{\mathrm{cv}}(0)\right|^2$ and

$$\left|\overrightarrow{k} \cdot \hat{p}_{\mathrm{cv}}(0)\right|^2 = \frac{m^2}{4m_{\mathrm{r}}} \frac{E_{\mathrm{g_0}}\left(E_{\mathrm{g_0}} + \Delta\right)}{\left(E_{\mathrm{g_0}} + \frac{2}{3}\Delta\right)}. \tag{10.34}$$

We shall express $A(\overrightarrow{k})$ and $B\left(\overrightarrow{k}\right)$ in terms of constants of the energy spectra in the following way:

Substituting $a_{k_\pm}$, $b_{k_\pm}$, $c_{k_\pm}$ and $\gamma_{0k_\pm}$ in $A(\overrightarrow{k})$ and $B\left(\overrightarrow{k}\right)$ in (10.31) we get

$$A\left(\overrightarrow{k}\right) = \beta\left(t + \frac{\rho}{\sqrt{2}}\right)\left\{\left(\frac{E_{\mathrm{g_0}}}{E_{\mathrm{g_0}} + \delta'}\right)\gamma_{0k_+}^2 - \gamma_{0k_+}^2\gamma_{0k_-}^2\left(\frac{E_{\mathrm{g_0}} - \delta'}{E_{\mathrm{g_0}} + \delta'}\right)\right\}^{1/2}, \tag{10.35}$$

$$B\left(\overrightarrow{k}\right) = \beta\left(t + \frac{\rho}{\sqrt{2}}\right)\left\{\left(\frac{E_{\mathrm{g_0}}}{E_{\mathrm{g_0}} + \delta'}\right)\gamma_{0k_-}^2 - \gamma_{0k_+}^2\gamma_{0k_-}^2\left(\frac{E_{\mathrm{g_0}} - \delta'}{E_{\mathrm{g_0}} + \delta'}\right)\right\}^{1/2}, \tag{10.36}$$

in which, $\gamma_{0k_+}^2 \equiv \dfrac{\xi_{1k} - E_{\mathrm{g_0}}}{2\left(\xi_{1k} + \delta'\right)} \equiv \dfrac{1}{2}\left[1 - \left(\dfrac{E_{\mathrm{g_0}} + \delta'}{\xi_{1k} + \delta'}\right)\right]$ and $\gamma_{0k_-}^2 \equiv \dfrac{\xi_{1k} + E_{\mathrm{g_0}}}{2\left(\xi_{1k} + \delta'\right)} \equiv \dfrac{1}{2}\left[1 + \left(\dfrac{E_{\mathrm{g_0}} - \delta'}{\xi_{1k} + \delta'}\right)\right]$.

Substituting $x \equiv \xi_{1k} + \delta'$ in $\gamma_{0k_\pm}^2$, we can write,

$$A\left(\overrightarrow{k}\right) = \beta\left(t + \frac{\rho}{\sqrt{2}}\right)\left\{\left(\frac{E_{\mathrm{g_0}}}{E_{\mathrm{g_0}} + \delta'}\right)\frac{1}{2}\left(1 - \frac{E_{\mathrm{g_0}} + \delta'}{x}\right) - \frac{1}{4}\left(\frac{E_{\mathrm{g_0}} - \delta'}{E_{\mathrm{g_0}} + \delta'}\right)\left(1 - \frac{E_{\mathrm{g_0}} + \delta'}{x}\right)\left(1 + \frac{E_{\mathrm{g_0}} - \delta'}{x}\right)\right\}^{1/2}.$$

Thus, $A\left(\overrightarrow{k}\right) = \dfrac{\beta}{2}\left(t + \dfrac{\rho}{\sqrt{2}}\right)\left\{1 - \dfrac{2a_0}{x} + \dfrac{a_1}{x^2}\right\}^{1/2}$.

where $a_0 \equiv \left(E_{\mathrm{g_0}}^2 + \delta'^2\right)\left(E_{\mathrm{g_0}} + \delta'\right)^{-1}$ and $a_1 \equiv \left(E_{\mathrm{g_0}} - \delta'\right)^2$.

After tedious algebra, one can show that

$$A\left(\overrightarrow{k}\right) = \frac{\beta}{2}\left(t + \frac{\rho}{\sqrt{2}}\right)\left(E_{\mathrm{g_0}} - \delta'\right)\left[\frac{1}{\xi_{1k} + \delta'} - \frac{1}{E_{\mathrm{g_0}} + \delta'}\right]^{1/2} \times \left[\frac{1}{\xi_{1k} + \delta'} - \frac{\left(E_{\mathrm{g_0}} + \delta'\right)}{\left(E_{\mathrm{g_0}} - \delta'\right)^2}\right]^{1/2}. \tag{10.37}$$

Similarly, from (10.36), we can write,

$$B\left(\vec{k}\right)=\beta\left(t+\frac{\rho}{\sqrt{2}}\right)\left\{\left(\frac{E_{g_0}}{E_{g_0}+\delta'}\right)\frac{1}{2}\left(1+\frac{E_{g_0}-\delta'}{x}\right)\right.$$
$$\left.-\frac{1}{4}\left(\frac{E_{g_0}-\delta'}{E_{g_0}+\delta'}\right)\left(1-\frac{E_{g_0}+\delta'}{x}\right)\left(1+\frac{E_{g_0}-\delta'}{x}\right)\right\}^{1/2}.$$

So that, finally we get,

$$B\left(\vec{k}\right)=\frac{\beta}{2}\left(t+\frac{\rho}{\sqrt{2}}\right)\left(1+\frac{E_{g_0}-\delta'}{\xi_{1k}+\delta'}\right). \tag{10.38}$$

Using (10.33), (10.34), (10.37) and (10.38), we can write

$$\left(\frac{|e|\,A_0}{2m}\right)^2\frac{\left\langle\left|\vec{\varepsilon}\cdot\hat{p}_{\mathrm{cv}}(\vec{k})\right|^2\right\rangle_{\mathrm{av}}}{E_{\mathrm{c}}(\vec{k})-E_{\mathrm{v}}(\vec{k})}$$
$$=\left(\frac{|e|\,A_0}{2m}\right)^2\frac{2\pi}{3}\left|\vec{k}\cdot\hat{p}_{\mathrm{cv}}(0)\right|^2\frac{\beta^2}{4}\left(t+\frac{\rho}{\sqrt{2}}\right)^2\frac{1}{\xi_{1k}}\left\{\left(1+\frac{E_{g_0}-\delta'}{\xi_{1k}+\delta'}\right)\right.$$
$$\left.+(E_{g_0}-\delta')\left[\frac{1}{\xi_{1k}+\delta'}-\frac{1}{E_{g_0}+\delta'}\right]^{1/2}\left[\frac{1}{\xi_{1k}+\delta'}-\frac{E_{g_0}+\delta'}{(E_{g_0}-\delta')^2}\right]^{1/2}\right\}^2. \tag{10.39}$$

Following Nag [7], it can be shown that

$$A_0^2=\frac{I\lambda^2}{2\pi^2c^3\sqrt{\varepsilon_{\mathrm{sc}}\varepsilon_0}}, \tag{10.40}$$

where I is the light intensity of wavelength λ, ε_0 is the permittivity of free space, and c is the velocity of light. Thus, the simplified electron energy spectrum in III–V, ternary and quaternary materials up to the second order in the presence of light waves, can approximately be written as

$$\frac{\hbar^2k^2}{2m^*}=\beta_0\left(E,\lambda\right), \tag{10.41}$$

where $\beta_0\left(E,\lambda\right)\equiv\left[\gamma\left(E\right)-\theta_0\left(E,\lambda\right)\right]$,

$$\theta_0\left(E,\lambda\right)\equiv\frac{|e|^2}{96m_{\mathrm{r}}\pi c^3}\frac{I\lambda^2}{\sqrt{\varepsilon_{\mathrm{sc}}\varepsilon_0}}\frac{E_{g_0}\left(E_{g_0}+\Delta\right)}{\left(E_{g_0}+\frac{2}{3}\Delta\right)}\frac{\beta^2}{4}\left(t+\frac{\rho}{\sqrt{2}}\right)^2\frac{1}{\phi_0\left(E\right)}$$
$$\times\left\{\left(1+\frac{E_{g_0}-\delta'}{\phi_0\left(E\right)+\delta'}\right)+(E_{g_0}-\delta')\left[\frac{1}{\phi_0\left(E\right)+\delta'}-\frac{1}{E_{g_0}+\delta'}\right]^{1/2}\right.$$
$$\left.\times\left[\frac{1}{\phi_0\left(E\right)+\delta'}-\frac{E_{g_0}+\delta'}{(E_{g_0}-\delta')^2}\right]^{1/2}\right\}^2,$$

and $\phi_0(E) \equiv E_{g_0}\left(1 + 2\left(1 + \frac{m^*}{m_v}\right)\frac{\gamma(E)}{E_{g_0}}\right)^{1/2}$.

Thus, under the limiting condition $\overrightarrow{k} \to 0$, from (41), we observe that $E \neq 0$ and is positive. Therefore, in the presence of external light waves, the energy of the electron does not tend to zero when $\overrightarrow{k} \to 0$, whereas for the un-perturbed three band model of Kane, $I(E) = \left[\hbar^2 k^2/(2m^*)\right]$ in which $E \to 0$ for $\overrightarrow{k} \to 0$. As the conduction band is taken as the reference level of energy, the lowest positive value of E for $\overrightarrow{k} \to 0$ provides the increased band gap (ΔE_g) of the semiconductor due to photon excitation. The values of the increased band gap can be obtained by computer iteration processes for various values of I and λ respectively.

Special cases

(1) For the two-band model of Kane, we have $\Delta \to 0$. Under this condition, $\gamma(E) \to E(1 + aE) = \frac{\hbar^2 k^2}{2m^*}$. Since, $\beta \to 1,\ t \to 1,\ \rho \to 0,\ \delta' \to 0$ for $\Delta \to 0$, from (41), we can write the energy spectrum of III–V, ternary and quaternary materials in the presence of external photo-excitation whose unperturbed conduction electrons obey the two band model of Kane, as

$$\frac{\hbar^2 k^2}{2m^*} = \tau_0(E, \lambda), \tag{10.42}$$

where $\tau_0(E, \lambda) \equiv E(1 + aE) - B_0(E, \lambda)$,

$$B_0(E, \lambda) \equiv \frac{|e|^2 I \lambda^2 E_{g_0}}{384\pi c^3 m_r \sqrt{\varepsilon_{sc}\varepsilon_0}} \frac{1}{\phi_1(E)} \times \left\{\left(1 + \frac{E_{g_0}}{\phi_1(E)}\right) + E_{g_0}\left[\frac{1}{\phi_1(E)} - \frac{1}{E_{g_0}}\right]\right\}^2,$$

$$\phi_1(E) \equiv E_{g_0}\left\{1 + \frac{2m^*}{m_r} aE(1 + aE)\right\}^{1/2}.$$

(2) For relatively wide band gap semiconductors, one can write, $a \to 0, b \to 0$, $c \to 0$ and $\gamma(E) \to E$.
Thus, from (10.42), we get,

$$\frac{\hbar^2 k^2}{2m^*} = \rho_0(E, \lambda), \tag{10.43}$$

$$\rho_0(E, \lambda) \equiv E - \frac{|e|^2 I \lambda^2}{96\pi c^3 m_r \sqrt{\varepsilon_{sc}\varepsilon_0}}\left[1 + \left(\frac{2m^*}{m_r}\right) aE\right]^{-3/2}. \tag{10.44}$$

10.2.2 The Formulation of the DMR in the Presence of Light Waves in III–V, Ternary and Quaternary Materials

(1) Using (10.41), the density-of-states function for III–V, ternary and quaternary materials in the presence of light waves whose unperturbed conduction electrons obey the three band model of Kane, can be expressed as

$$D_0(E) = 4\pi g_v \left(\frac{2m^*}{h^2}\right)^{3/2} \sqrt{\beta_0(E,\lambda)}\{\beta_0(E,\lambda)\}'. \tag{10.45}$$

Combining (10.45) with the Fermi–Dirac occupation probability factor and using the generalized Sommerfeld's lemma [8], the electron concentration can be written as

$$n_0 = \left(3\pi^2\right)^{-1} g_v \left(\frac{2m^*}{\hbar^2}\right)^{3/2} [M_{10}(E_{F_L},\lambda) + N_{10}(E_{F_L},\lambda)], \tag{10.46}$$

where $M_{10}(E_{F_L},\lambda) \equiv [\beta_0(E_{F_L},\lambda)]^{3/2}$, E_{F_L} is the Fermi energy in the presence of light waves as measured from the edge of the conduction band in the vertically upward direction in the absence of any field and $N_{10}(E_{F_L},\lambda) \equiv \sum_{r=1}^{s} L(r) M_{10}(E_{F_L},\lambda)$,

Using (10.46) and (1.11), one can write

$$\frac{D}{\mu} = \frac{1}{|e|} \frac{[M_{10}(E_{F_L},\lambda) + N_{10}(E_{F_L},\lambda)]}{[\{M_{10}(E_{F_L},\lambda)\}' + \{N_{10}(E_{F_L},\lambda)\}']}. \tag{10.47}$$

In the absence of external photo-excitation, (10.46) and (10.47) get simplified to (2.8) and (2.9) of Chap. 2, respectively.
(2) The expressions of n_0 and DMR for III–V, ternary and quaternary materials in the presence of light waves whose unperturbed conduction electrons obey the two band model of Kane, can be expressed as

$$n_0 = \left(3\pi^2\right)^{-1} g_v \left(\frac{2m^*}{\hbar^2}\right)^{3/2} [M_{11}(E_{F_L},\lambda) + N_{11}(E_{F_L},\lambda)], \tag{10.48}$$

and

$$\frac{D}{\mu} = \frac{1}{|e|} \frac{[M_{11}(E_{F_L},\lambda) + N_{11}(E_{F_L},\lambda)]}{[\{M_{11}(E_{F_L},\lambda)\}' + \{N_{11}(E_{F_L},\lambda)\}']}, \tag{10.49}$$

where $M_2(E_{F_L},\lambda) \equiv [\tau_0(E_{F_L},\lambda)]^{3/2}$ and $N_{11}(E_{F_L},\lambda) \equiv \sum_{r=1}^{s} L(r) M_{11}(E_{F_L},\lambda)$,

In the absence of external photo-excitation, (10.48) and (10.49) get simplified to (2.12) and (2.13), respectively.
(3) The expressions of n_0 and DMR for III–V, ternary and quaternary materials in the presence of light waves whose unperturbed conduction electrons obey the parabolic energy bands, can be expressed as

$$n_0 = \left(3\pi^2\right)^{-1} g_v \left(\frac{2m^*}{\hbar^2}\right)^{3/2} [M_{12}(E_{F_L},\lambda) + N_{12}(E_{F_L},\lambda)], \tag{10.50}$$

and

$$\frac{D}{\mu} = \frac{1}{|e|} \frac{[M_{12}(E_{F_L},\lambda) + N_{12}(E_{F_L},\lambda)]}{[\{M_{12}(E_{F_L},\lambda)\}' + \{N_{12}(E_{F_L},\lambda)\}']}, \tag{10.51}$$

where $M_{12}(E_{F_L},\lambda) \equiv [\rho_0(E_{F_L},\lambda)]^{3/2}$ and $N_{12}(E_{F_L},\lambda) \equiv \sum_{r=1}^{s} L(r) M_{12}(E_{F_L},\lambda)$.

In the absence of external photo-excitation, (10.50) and (10.51) get simplified to (2.18) and (2.19), respectively.

10.3 Result and Discussions

Using Table 2.1 and (10.46)–(10.51), the plot of the DMR as a function of electron concentration at $T = 4.2\ K$ has been shown in Figs. 10.1–10.4 by taking n-InAs, n-InSb, $Hg_{1-x}Cd_xTe$ and n-$In_{1-x}Ga_xAs_yP_{1-y}$ lattice matched to InP in the presence of light waves, whose unperturbed conduction electrons obey the three and the two band models of Kane and that of the parabolic energy bands, respectively. From Figs. 10.44 to 10.47, it appears that the DMR increases with the increasing electron concentration and the numerical values of the DMR in the presence of light waves for all the materials, in accordance with all the band models, are relatively smaller, which can be compared with the same in the absence of the external photo-excitation as given in corresponding figures of Chap. 2. The combined influence of the energy band constants on the DMR for n-InAs and n-InSb can easily be assessed from Figs. 10.1 to 10.2. In Figs. 10.5–10.8, the DMR has been plotted as a function of light intensity for all the aforementioned materials. From Figs. 10.5 to 10.8, one can observe that, the DMR decreases with increasing light intensity for all the materials. It should be noted from the aforementioned equations, that in the absence of external photo-excitation, the DMR is independent of light intensity.

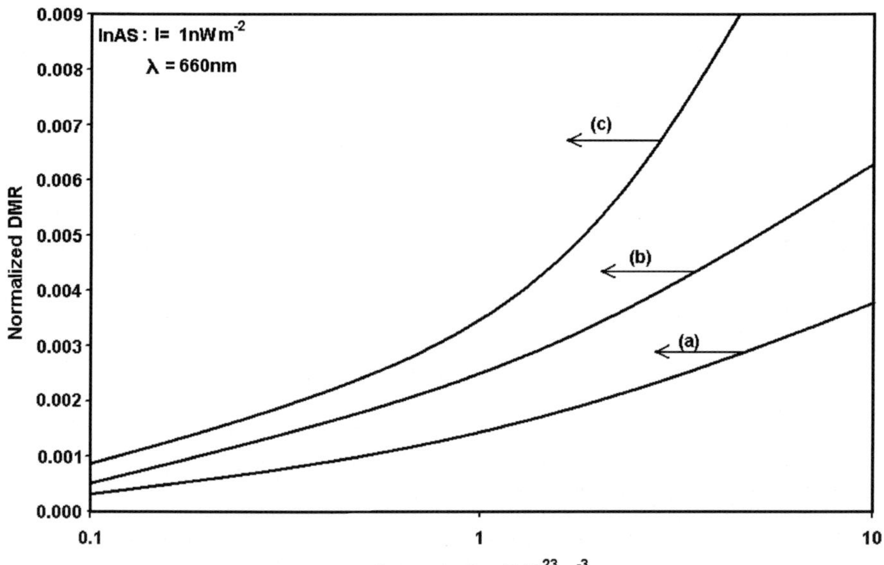

Fig. 10.1. Plot of the DMR as a function of electron concentration for bulk n-InAs in the presence of light waves in which the curves (**a**), (**b**) and (**c**) represent the perturbed three and two band models of Kane and that of the parabolic energy bands respectively

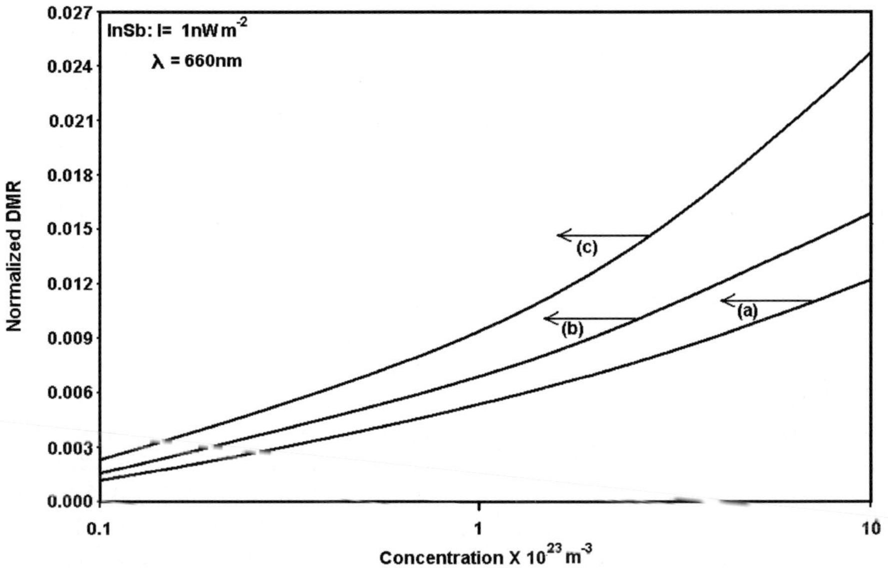

Fig. 10.2. Plot of the DMR as a function of electron concentration for bulk n-InSb in the presence of light waves in which the curves (**a**), (**b**) and (**c**) represent the perturbed three and two band models of Kane and that of the parabolic energy bands respectively

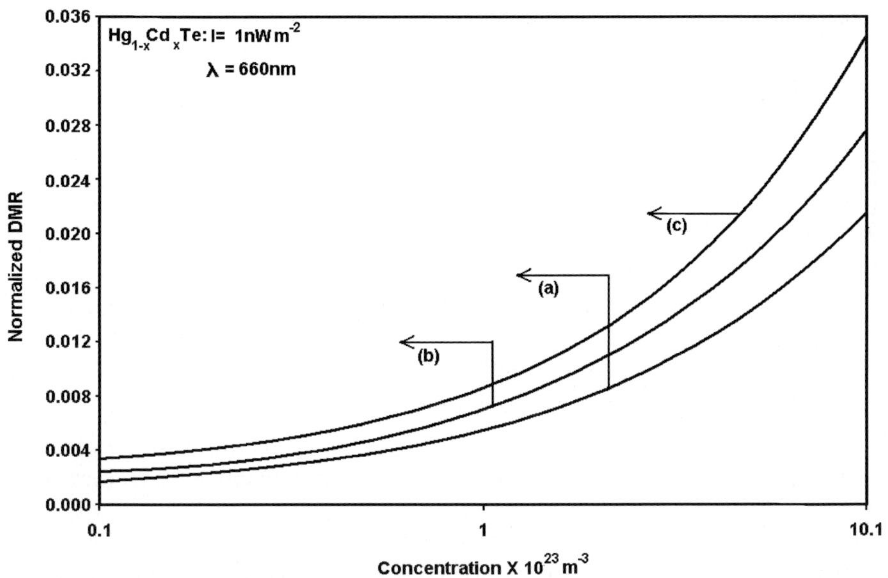

Fig. 10.3. Plot of the DMR as a function of electron concentration for bulk n-$Hg_{1-x}Cd_xTe$ in the presence of light waves in which the curves (**a**), (**b**) and (**c**) represent the perturbed three and two band models of Kane and that of the parabolic energy bands respectively

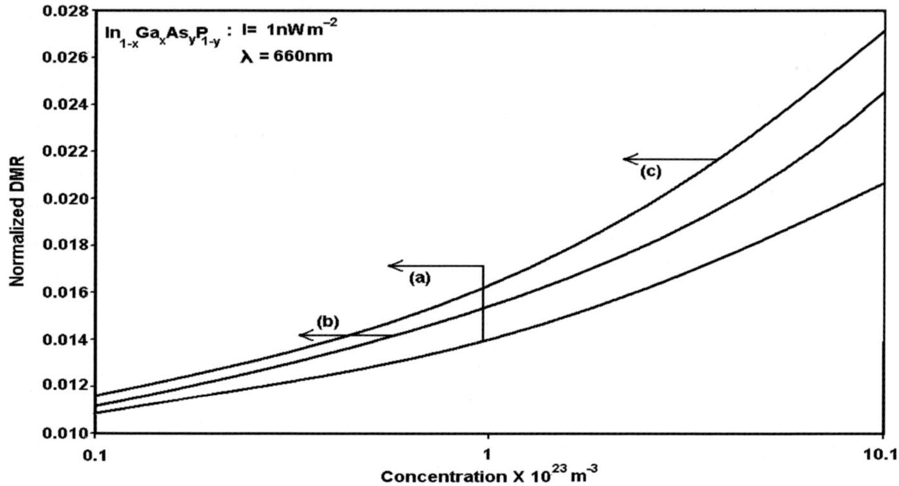

Fig. 10.4. Plot of the DMR as a function of electron concentration for bulk n-$In_{1-x}Ga_xAs_yP_{1-y}$ lattice matched to InP in the presence of light waves in which the curves (**a**), (**b**) and (**c**) represent the perturbed three and two band models of Kane and that of the parabolic energy bands respectively

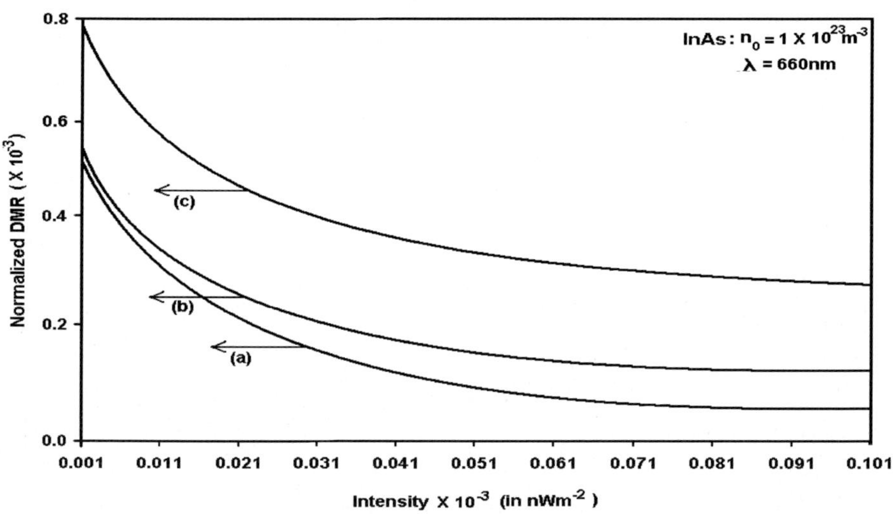

Fig. 10.5. Plot of the DMR as a function of light intensity for bulk n-InAs in which the curves (**a**), (**b**) and (**c**) represent the perturbed three and two band models of Kane and that of parabolic energy bands respectively

In Figs. 10.9–10.12, the DMR has been plotted as a function of wavelengths in the visible region for all the aforementioned materials for all the energy band models. It appears that the DMR decreases as the wavelength shifts from red color to violet.

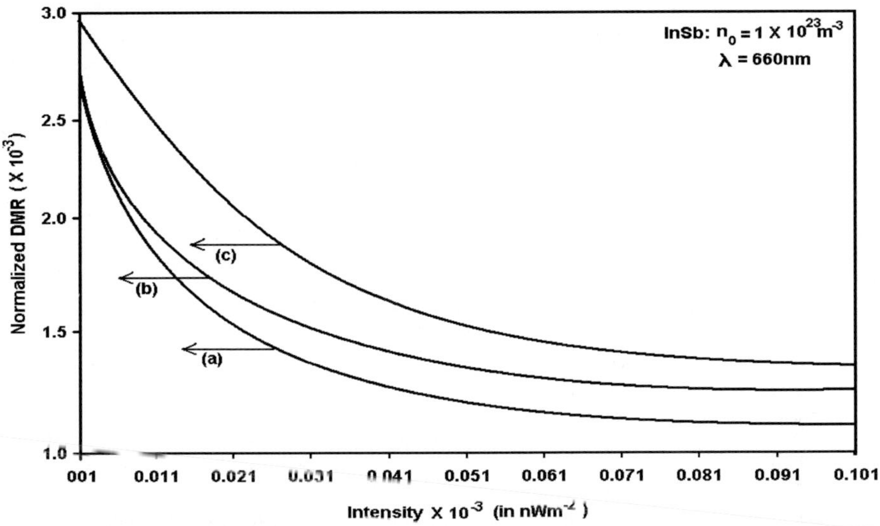

Fig. 10.6. Plot of the DMR as a function of light intensity for bulk n-InSb in which the curves (**a**), (**b**) and (**c**) represent the perturbed three and two band models of Kane and that of parabolic energy bands respectively

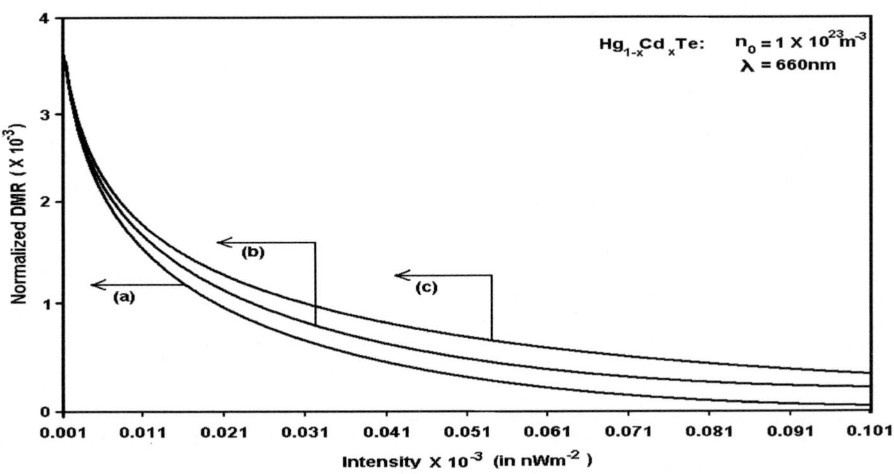

Fig. 10.7. Plot of the DMR as a function of light intensity for bulk n-$Hg_{1-x}Cd_xTe$ in which the curves (**a**), (**b**) and (**c**) represent the perturbed three and two band models of Kane and that of parabolic energy bands respectively

The influence of light is immediately apparent from the plots in Figs. 10.5–10.12, since the DMR depends strongly on I and λ in direct contrast with the corresponding bulk specimens of the said compounds. The variations of the DMRs in Figs. 10.5–10.12 reflect the direct signature of the light wave on

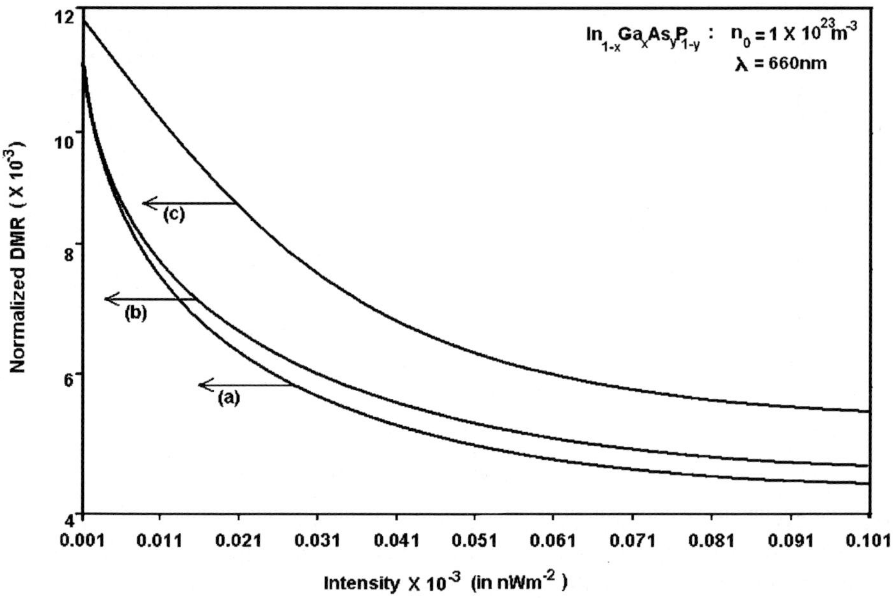

Fig. 10.8. Plot of the DMR as a function of light intensity for bulk n-$In_{1-x}Ga_xAs_yP_{1-y}$ lattice matched to InP in which the curves (**a**), (**b**) and (**c**) represent the perturbed three and two band models of Kane and that of parabolic energy bands respectively

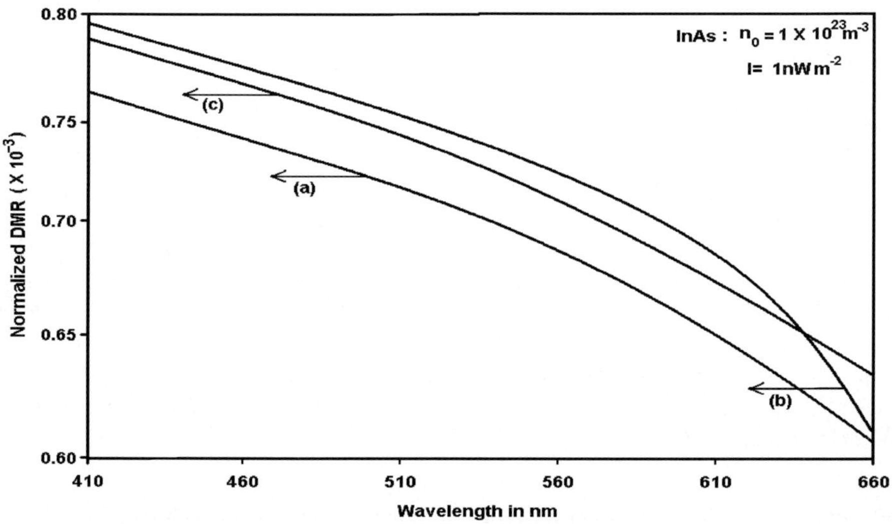

Fig. 10.9. Plot of the DMR as a function of wavelength for bulk n-InAs in which the curves (**a**), (**b**) and (**c**) represent the perturbed three and two band models of Kane and that of parabolic energy bands respectively

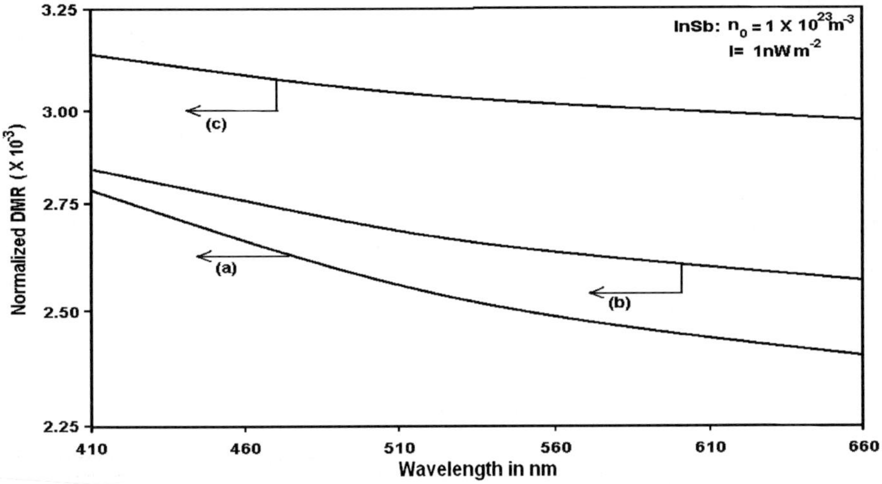

Fig. 10.10. Plot of the DMR as a function of wavelength for bulk n-InSb in which the curves (**a**), (**b**) and (**c**) represent the perturbed three and two band models of Kane and that of parabolic energy bands respectively

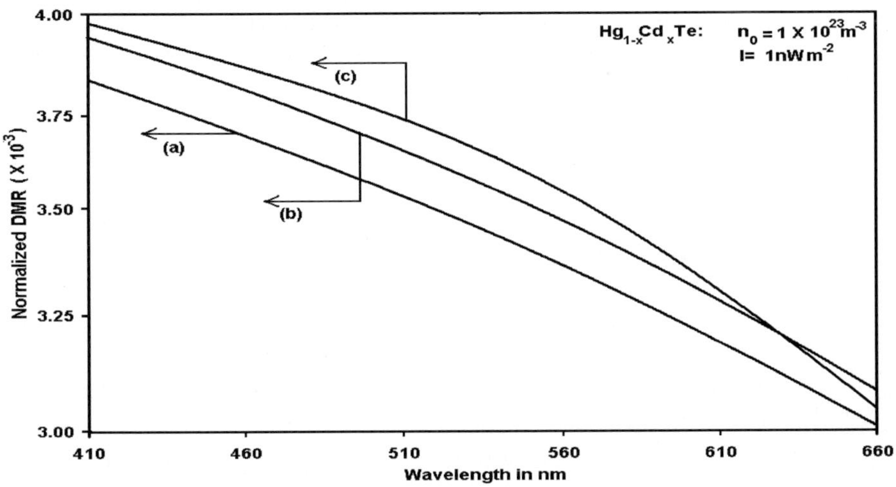

Fig. 10.11. Plot of the DMR as a function of wavelength for bulk n-$Hg_{1-x}Cd_xTe$ in which the curves (**a**), (**b**) and (**c**) represent the perturbed three and two band models of Kane and that of parabolic energy bands respectively

the band structure-dependent physical properties of semiconducting materials, in the presence of light waves and the photon-assisted transport for the corresponding photonic devices. Although the DMR tends to decrease with the intensity and the wavelength, but the rate of decrease is totally band structure-dependent.

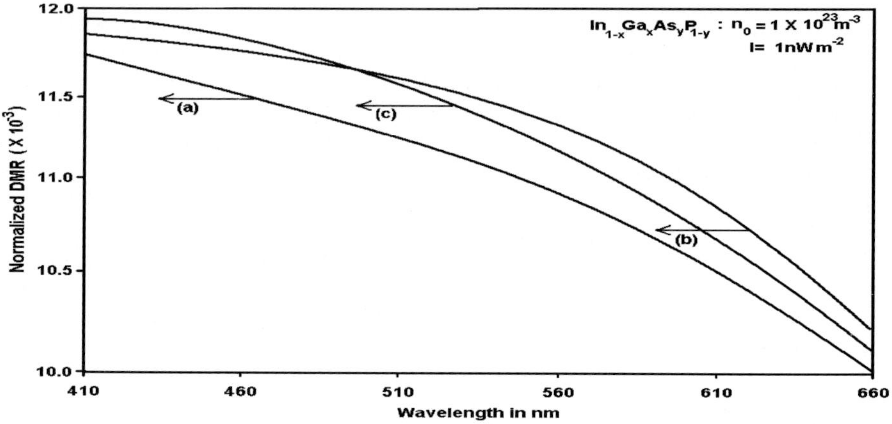

Fig. 10.12. Plot of the DMR as a function of wavelength for bulk n-$In_{1-x}Ga_xAs_yP_{1-y}$ lattice matched to InP in which the curves (**a**), (**b**) and (**c**) represent the perturbed three and two band models of Kane and that of parabolic energy bands respectively

Thus, we can conclude that the influence of an external photo-excitation is to change radically the original band structure of the material. Because of this change, the photon field causes to increase the band gap of semiconductor.

The numerical results presented in this chapter would be different for other materials, but the nature of variation would be unaltered. The theoretical results as given here would be useful in analyzing various other experimental data related to this phenomenon. Finally, we can write that this theory can be used to investigate modern semiconductor devices operated under the influence of external photon field.

10.4 The Formulation of the DMR in the Presence of Quantizing Magnetic Field Under External Photo-Excitation in III–V, Ternary and Quaternary Materials

It appears from the literature that the influence of a quantizing magnetic field on the DMR in III–V, ternary and quaternary materials under external photo-excitation has yet to be investigated. In Sect. 10.5 of the theoretical background, the same has been studied. The DMR has been investigated numerically by taking the materials as stated in Sect. 10.1. Section 10.6 contains the result and discussion.

10.5 Theoretical Background

(1) Using (3.11) and (10.41), the magneto-dispersion law, in the absence of spin, for III–V, ternary and quaternary semiconductors, in the presence of photo-excitation, whose unperturbed conduction electrons obey the three band model of Kane, is given by

$$\beta_0(E,\lambda) = \left(n+\frac{1}{2}\right)\hbar\omega_0 + \frac{\hbar^2 k_z^2}{2m^*}. \tag{10.52}$$

Using (10.52), the density-of-states function in the present case can be expressed as

$$D_B(E,\lambda) = \frac{Bg_v|e|\sqrt{2m^*}}{2\pi^2\hbar^2}\sum_{n=0}^{n_{\max}} \times\left[\{\beta_0(E,\lambda)\}'\left\{\beta_0(E,\lambda)-\left(n+\frac{1}{2}\right)\hbar\omega_0\right\}^{-1/2} H(E-E_{n_{l1}})\right], \tag{10.53}$$

where $E_{n_{l1}}$ is the Landau sub-band energies in this case and is given as

$$\beta_0(E_{n_{l1}},\lambda) = \left(n+\frac{1}{2}\right)\hbar\omega_0. \tag{10.54}$$

The EMM in this case assumes the form

$$m^*(E_{F_{BL}},\lambda) = m^*\{\beta_0(E_{F_{BL}},\lambda)\}', \tag{10.55}$$

where $E_{F_{BL}}$ is the Fermi energy under quantizing magnetic field in the presence of light waves as measured from the edge of the conduction band in the vertically upward direction in the absence of any quantization. Combining (10.53) with the Fermi–Dirac occupation probability factor and using the generalized Sommerfeld's lemma [8], the electron concentration can be written as

$$n_0 = \frac{g_v|e|B\sqrt{2m^*}}{\pi^2\hbar^2}\sum_{n=0}^{n_{\max}}[M_{13}(E_{F_{BL}},B,\lambda)+N_{13}(E_{F_{BL}},B,\lambda)], \tag{10.56}$$

where $M_{13}(E_{F_{BL}},B,\lambda) \equiv \left[\beta_0(E_{F_{BL}},\lambda)-\left(n+\frac{1}{2}\right)\hbar\omega_0\right]^{1/2}$ and $N_{13}(E_{F_{BL}}, B,\lambda) \equiv \sum_{r=1}^{s} L(r)M_{13}(E_{F_{BL}},B,\lambda)$.

Using (10.56) and (1.11) one can write

$$\frac{D}{\mu} = \frac{1}{|e|}\frac{\sum_{n=0}^{n_{\max}}[M_{13}(E_{F_{BL}},B,\lambda)+N_{13}(E_{F_{BL}},B,\lambda)]}{\sum_{n=0}^{n_{\max}}[\{M_{13}(E_{F_{BL}},B,\lambda)\}'+\{N_{13}(E_{F_{BL}},B,\lambda)\}']}. \tag{10.57}$$

In the absence of external photo-excitation, (10.56) and (10.57) get simplified to (3.22) and (3.23) of Chap. 3, respectively.
(2) Using (3.24) and (10.42), the magneto-dispersion law in the absence of spin, for III–V, ternary and quaternary semiconductors, in the presence of photo-excitation, whose unperturbed conduction electrons obey the two band model of Kane, is given by

$$\tau_0(E,\lambda) = \left(n+\frac{1}{2}\right)\hbar\omega_0 + \frac{\hbar^2 k_z^2}{2m^*}. \tag{10.58}$$

Using (10.58), the density-of-states function in this case can be written as

$$D_B(E,\lambda) = \frac{Bg_v|e|\sqrt{2m^*}}{2\pi^2\hbar^2}\sum_{n=0}^{n_{max}}\left[\{\tau_0(E,\lambda)\}'\left\{\tau_0(E,\lambda) - \left(n+\frac{1}{2}\right)\hbar\omega_0\right\}^{-1/2} H(E-E_{n_{l2}})\right], \tag{10.59}$$

where $E_{n_{l2}}$ is the Landau sub-band energies and can be expressed as

$$\tau_0(E_{n_{l2}},\lambda) = \left(n+\frac{1}{2}\right)\hbar\omega_0. \tag{10.60}$$

The EMM assumes the form

$$m^*(E_{F_{BL}},\lambda) = m^*\{\tau_0(E_{F_{BL}}\lambda)\}'. \tag{10.61}$$

Thus, the electron concentration can be written as

$$n_0 = \frac{g_v|e|B\sqrt{2m^*}}{\pi^2\hbar^2}\sum_{n=0}^{n_{max}}[M_{14}(E_{F_{BL}},B,\lambda) + N_{14}(E_{F_{BL}},B,\lambda)], \tag{10.62}$$

where $M_{14}(E_{F_{BL}},B,\lambda) \equiv \left[\tau_0(E_{F_{BL}},\lambda) - \left(n+\frac{1}{2}\right)\hbar\omega_0\right]^{1/2}$ and $N_{14}(E_{F_{BL}},B,\lambda) \equiv \sum_{r=1}^{s} L(r)M_{14}(E_{F_{BL}},B,\lambda)$.
Using (10.62) and (1.11) one can write

$$\frac{D}{\mu} = \frac{1}{|e|}\frac{\sum_{n=0}^{n_{max}}[M_{14}(E_{F_{BL}},B,\lambda) + N_{14}(E_{F_{BL}},B,\lambda)]}{\sum_{n=0}^{n_{max}}[\{M_{14}(E_{F_{BL}},B,\lambda)\}' + \{N_{14}(E_{F_{BL}},B,\lambda)\}']}. \tag{10.63}$$

In the absence of external photo-excitation, (10.62) and (10.63) get simplified to (3.31) and (3.32), respectively.

(3) Using (3.40) and (10.43), the magneto-dispersion law in the absence of spin, for III–V, ternary and quaternary semiconductors, in the presence of

photo-excitation, whose unperturbed conduction electrons obey the parabolic energy bands, is given by

$$\rho_0(E,\lambda) = \left(n+\frac{1}{2}\right)\hbar\omega_0 + \frac{\hbar^2 k_z^2}{2m^*}. \tag{10.64}$$

Using (10.64), the density-of-states function in this case can be written as

$$D_{\mathrm{B}}(E,\lambda) = \frac{Bg_{\mathrm{v}}|e|\sqrt{2m^*}}{2\pi^2\hbar^2}\sum_{n=0}^{n_{\max}} \times\left[\{\rho_0(E,\lambda)\}'\left\{\rho_0(E,\lambda)-\left(n+\frac{1}{2}\right)\hbar\omega_0\right\}^{-1/2} H(E-E_{nl3})\right], \tag{10.65}$$

where E_{nl3} is the Landau sub-band energies and is given by

$$\rho_0(E_{nl3},\lambda) - \left(n+\frac{1}{2}\right)\hbar\omega_0. \tag{10.66}$$

The EMM assumes the form

$$m^*(E_{\mathrm{F_{BL}}},\lambda) = m^*\{\tau_0(E_{\mathrm{F_{BL}}},\lambda)\}'. \tag{10.67}$$

Thus, the electron concentration in this case can be written as

$$n_0 = \frac{g_{\mathrm{v}}|e|B\sqrt{2m^*}}{\pi^2\hbar^2}\sum_{n=0}^{n_{\max}}[M_{15}(E_{\mathrm{F_{BL}}},B,\lambda)+N_{15}(E_{\mathrm{F_{BL}}},B,\lambda)], \tag{10.68}$$

where $M_{15}(E_{\mathrm{F_{BL}}},B,\lambda) \equiv \left[\rho_0(E_{\mathrm{F_{BL}}},\lambda)-\left(n+\frac{1}{2}\right)\hbar\omega_0\right]^{1/2}$ and $N_{15}(E_{\mathrm{F_{BL}}}, B,\lambda) \equiv \sum_{r=1}^{s} L(r)M_{15}(E_{\mathrm{F_{BL}}},B,\lambda)$.

Using (10.68) and (1.11) one can write

$$\frac{D}{\mu} = \frac{1}{|e|}\frac{\sum_{n=0}^{n_{\max}}[M_{15}(E_{\mathrm{F_{BL}}},B,\lambda)+N_{15}(E_{\mathrm{F_{BL}}},B,\lambda)]}{\sum_{n=0}^{n_{\max}}[\{M_{15}(E_{\mathrm{F_{BL}}},B,\lambda)\}'+\{N_{15}(E_{\mathrm{F_{BL}}},B,\lambda)\}']}. \tag{10.69}$$

In the absence of external photo-excitation, (10.68) and (10.69) get simplified to (3.46) and (3.47) respectively.

10.6 Result and Discussions

Using Table 2.2 and (10.56), (10.57), (10.62), (10.63) and (10.68), (10.69), the plot of the DMR as a function of inverse magnetic field in the presence of light waves at $T = 4.2$ K is shown in Figs. 10.13–10.16 by taking n-InAs, n-InSb, $Hg_{1-x}Cd_xTe$ and n-$In_{1-x}Ga_xAs_yP_{1-y}$ lattice matched to InP, respectively.

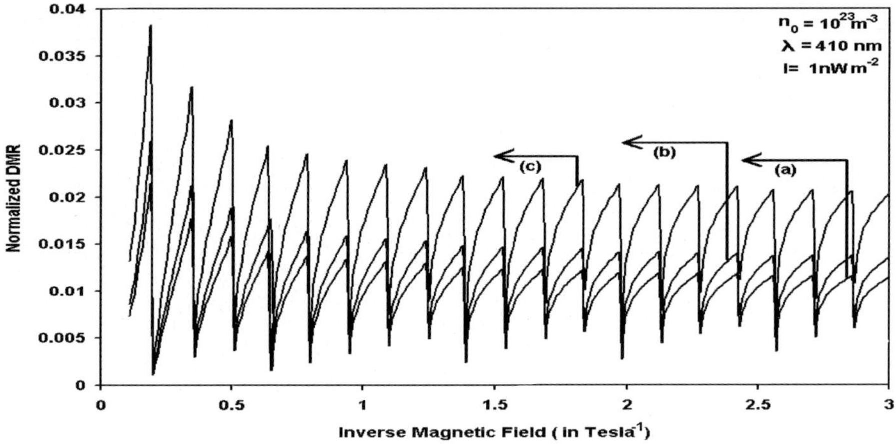

Fig. 10.13. Plot of the DMR as a function of inverse quantizing magnetic field in the presence of light waves for n-InAs, in which the curves (**a**), (**b**) and (**c**) represent the perturbed three and two band models of Kane and that of parabolic energy bands respectively

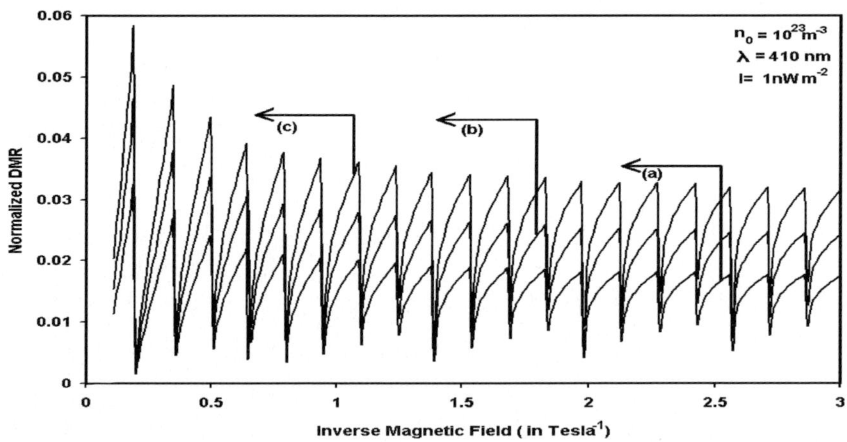

Fig. 10.14. Plot of the DMR as a function of inverse quantizing magnetic field in the presence of light waves for n-InSb, in which the curves (**a**), (**b**) and (**c**) represent the perturbed three and two band models of Kane and that of parabolic energy bands respectively

Figures 10.17–10.20 exhibit the variation of the DMR as a function of electron concentration, under quantizing magnetic field in the presence of light waves for the aforementioned materials. The DMR again shows the oscillatory dependence with different numerical magnitude emphasizing the influence of the energy band constants. The origin of the oscillation is same as that of SdH oscillations and all discussions of the relevant portions of Sect. 3.3 are applicable in this case.

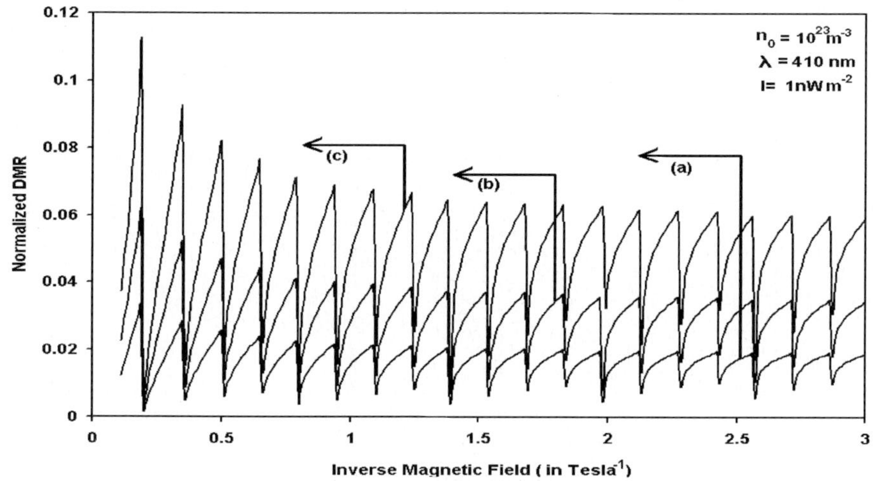

Fig. 10.15. Plot of the DMR as a function of inverse quantizing magnetic field in the presence of light waves for n-$Hg_{1-x}Cd_xTe$, in which the curves (**a**), (**b**) and (**c**) represent the perturbed three and two band models of Kane and that of parabolic energy bands respectively

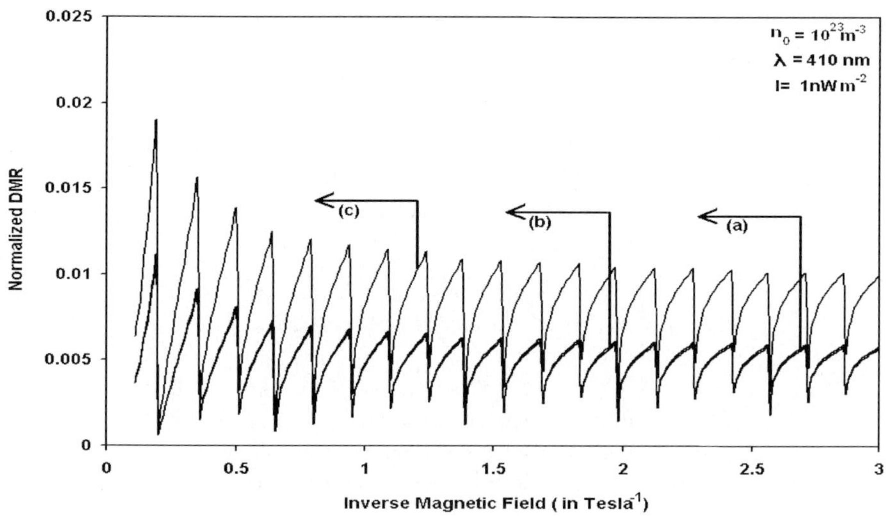

Fig. 10.16. Plot of the DMR as a function of inverse quantizing magnetic field in the presence of light waves for n-$In_{1-x}Ga_xAs_yP_{1-y}$ lattice matched to InP, in which the curves (**a**), (**b**) and (**c**) represent the perturbed three and two band models of Kane and that of parabolic energy bands respectively

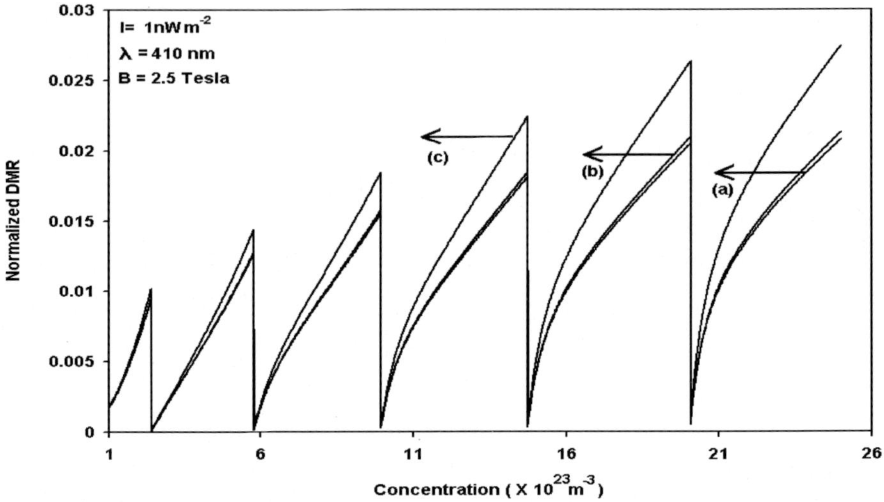

Fig. 10.17. Plot of the DMR as a function of electron concentration under quantizing magnetic field in the presence of light waves for n-InAs, in which the curves (**a**), (**b**) and (**c**) represent the perturbed three and two band models of Kane and that of parabolic energy bands respectively

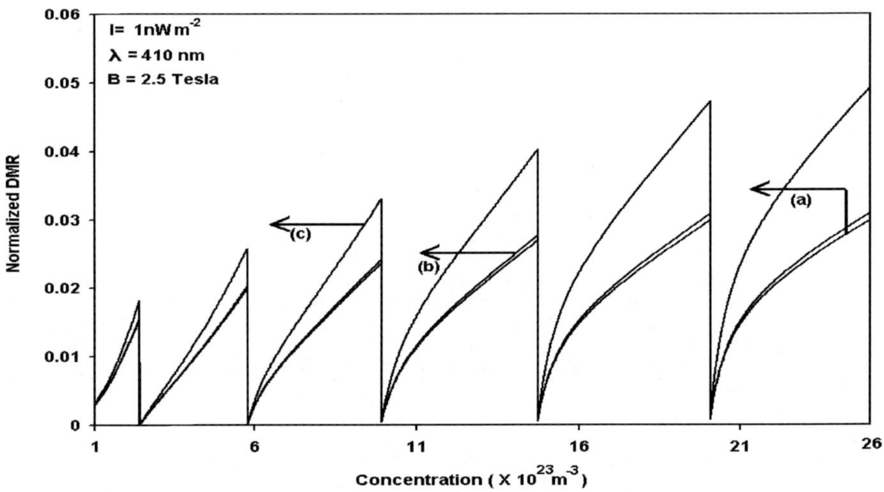

Fig. 10.18. Plot of the DMR as a function of electron concentration under quantizing magnetic field in the presence of light waves for n-InSb, in which the curves (**a**), (**b**) and (**c**) represent the perturbed three and two band models of Kane and that of parabolic energy bands respectively

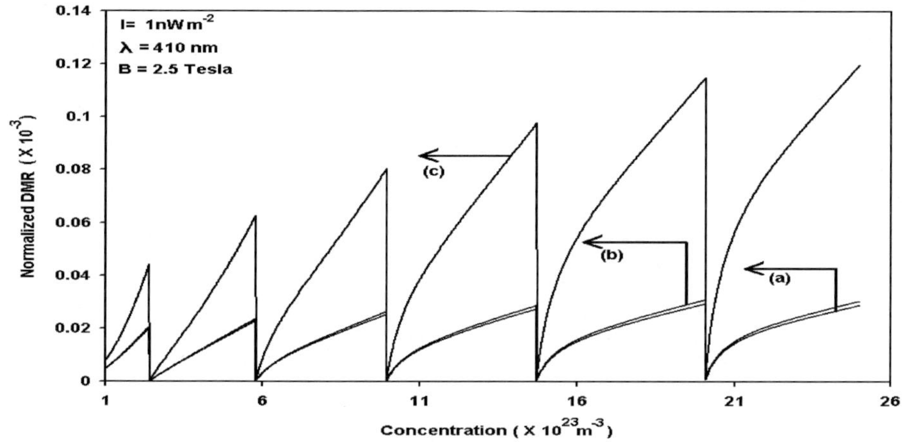

Fig. 10.19. Plot of the DMR as a function of electron concentration under quantizing magnetic field in the presence of light waves for n-$Hg_{1-x}Cd_xTe$, in which the curves (**a**), (**b**) and (**c**) represent the perturbed three and two band models of Kane and that of parabolic energy bands respectively

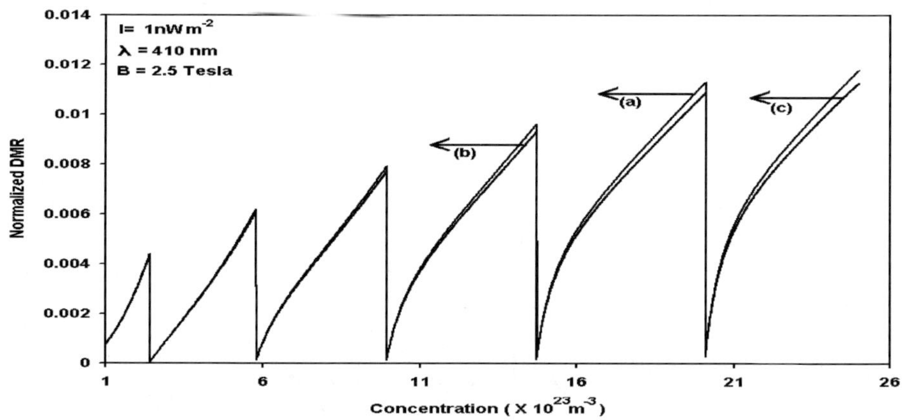

Fig. 10.20. Plot of the DMR as a function of electron concentration under quantizing magnetic field in the presence of light waves for n-$In_{1-x}Ga_xAs_yP_{1-y}$ lattice matched to InP, in which the curves (**a**), (**b**) and (**c**) represent the perturbed three and two band models of Kane and that of parabolic energy bands respectively

Figures 10.21–10.24 show the variation of the DMR as a function of light intensity in the presence of quantizing magnetic field, while Figs. 10.25–10.28 exhibit the same as a function of wavelength, in which, the variations of the wavelengths are in the zone of visible region. One can observe that the DMR decreases with increase in the light intensity and wavelengths in different ways, as appears from Figs. 10.21 to 10.28. The nature of variations in all the cases

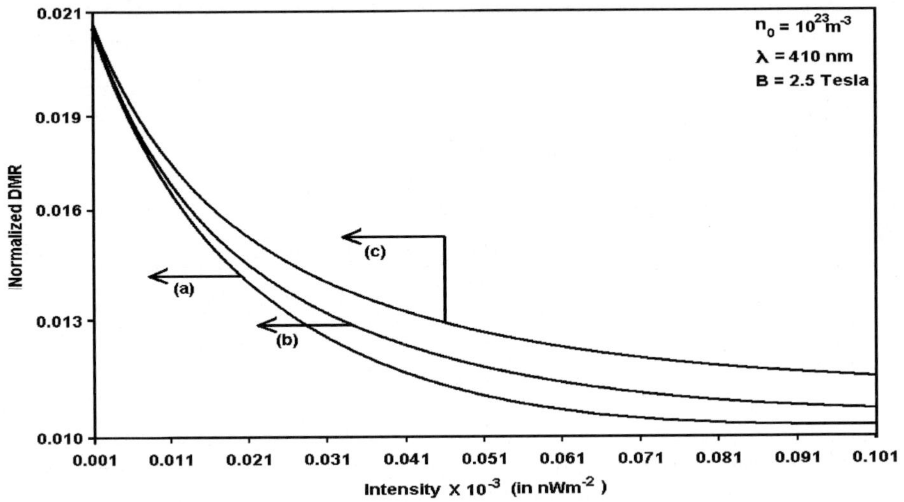

Fig. 10.21. Plot of the DMR as a function of light intensity under quantizing magnetic field for n-InAs, in which the curves (**a**), (**b**) and (**c**) represent the perturbed three and two band models of Kane and that of parabolic energy bands respectively

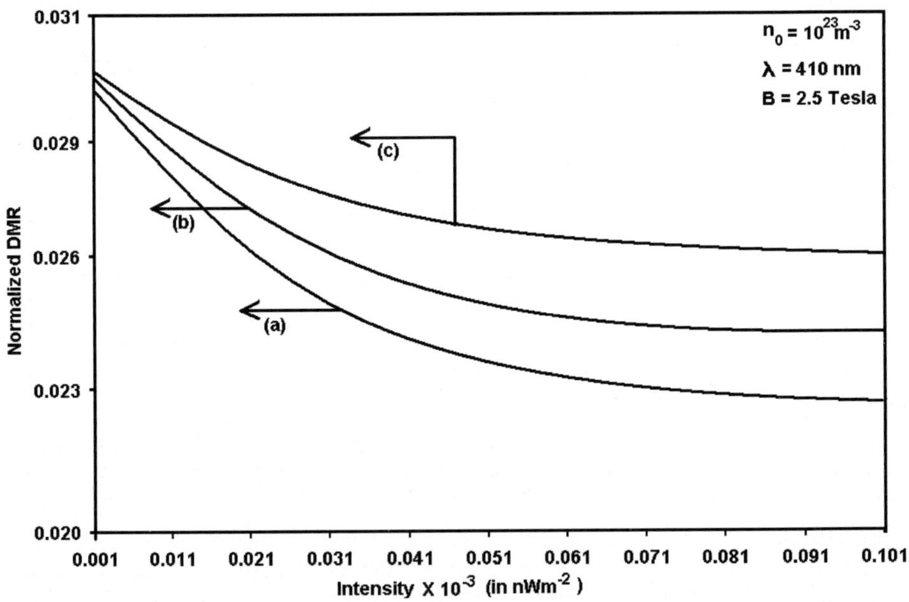

Fig. 10.22. Plot of the DMR as a function of light intensity under quantizing magnetic field for n-InSb, in which the curves (**a**), (**b**) and (**c**) represent the perturbed three and two band models of Kane and that of parabolic energy bands respectively

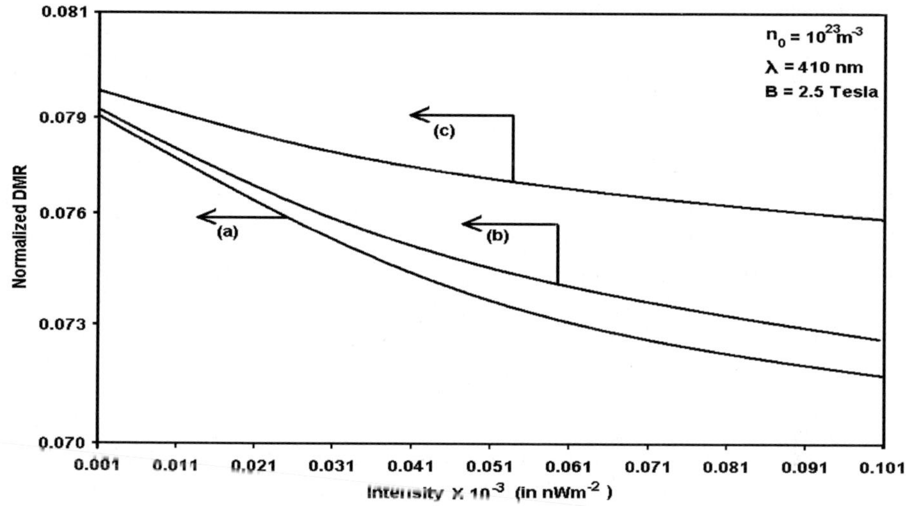

Fig. 10.23. Plot of the DMR as a function of light intensity under quantizing magnetic field for n-$Hg_{1-x}Cd_xTe$, in which the curves (**a**), (**b**) and (**c**) represent the perturbed three and two band models of Kane and that of parabolic energy bands respectively

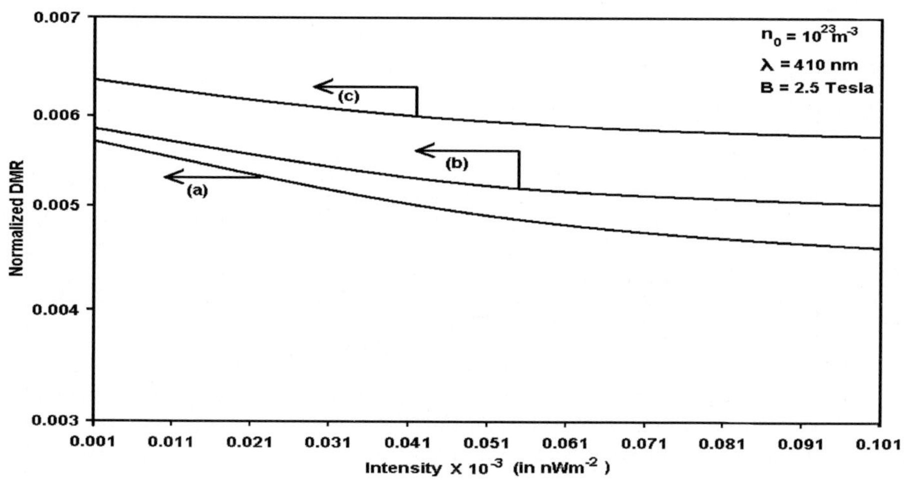

Fig. 10.24. Plot of the DMR as a function of light intensity under quantizing magnetic field for n-$In_{1-x}Ga_xAs_yP_{1-y}$ lattice matched to InP, in which the curves (**a**), (**b**) and (**c**) represent the perturbed three and two band models of Kane and that of parabolic energy bands respectively

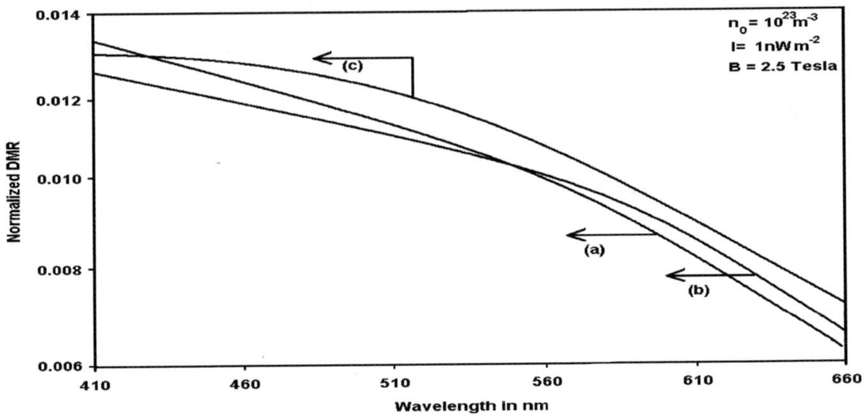

Fig. 10.25. Plot of the DMR as a function of wavelength under quantizing magnetic field for n-InAs, in which the curves (**a**), (**b**) and (**c**) represent the perturbed three and two band models of Kane and that of parabolic energy bands respectively

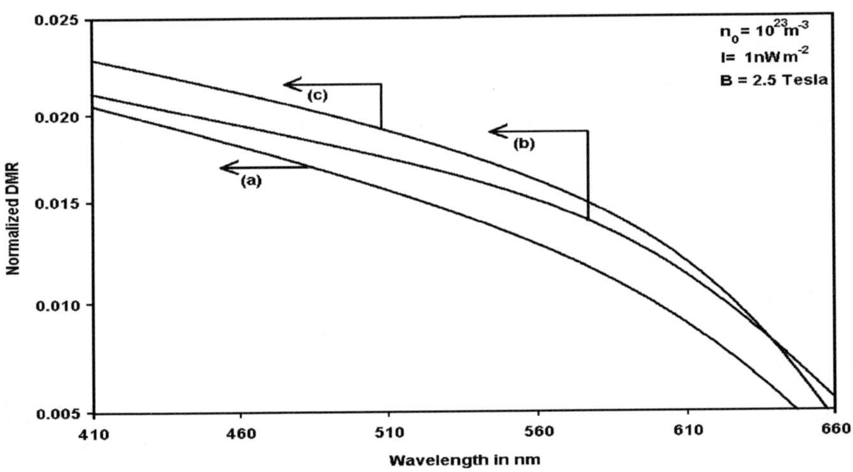

Fig. 10.26. Plot of the DMR as a function of wavelength under quantizing magnetic field for n-InSb, in which the curves (**a**), (**b**) and (**c**) represent the perturbed three and two band models of Kane and that of parabolic energy bands respectively

depends strongly on the energy spectrum constants of the respective materials and the external physical conditions. It should be noted that the numerical value of the DMR in the presence of light waves is relatively smaller even at higher value of magnetic field, than that of the corresponding figures of Chap. 3.

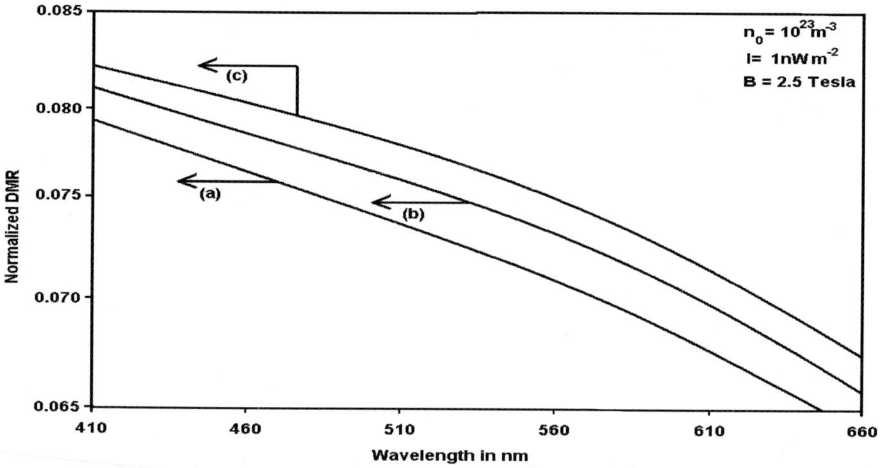

Fig. 10.27. Plot of the DMR as a function of wavelength under quantizing magnetic field for n-$Hg_{1-x}Cd_xTe$, in which the curves (**a**), (**b**) and (**c**) represent the perturbed three and two band models of Kane and that of parabolic energy bands respectively

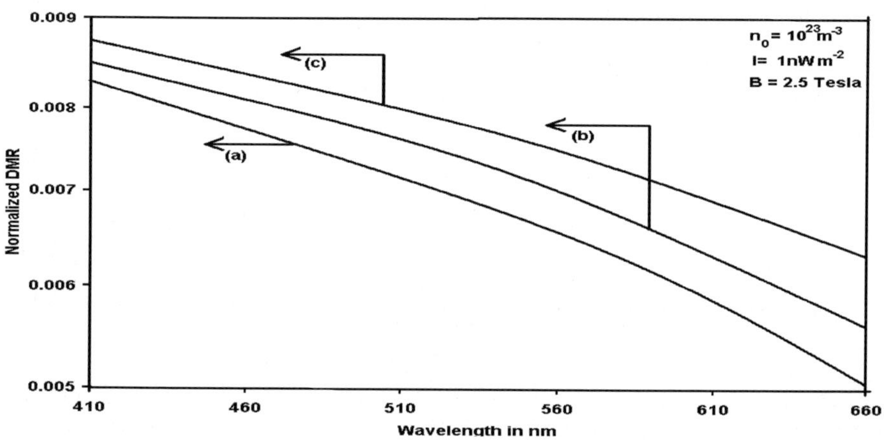

Fig. 10.28. Plot of the DMR as a function of wavelength under quantizing magnetic field for n-$In_{1-x}Ga_xAs_yP_{1-y}$ lattice matched to InP, in which the curves (**a**), (**b**) and (**c**) represent the perturbed three and two band models of Kane and that of parabolic energy bands respectively

10.7 The Formulation of the DMR in the Presence of Cross-Field Configuration Under External Photo-Excitation in III–V, Ternary and Quaternary Materials

It appears from the literature that the influence of a crossed electric and quantizing magnetic field on the DMR in III–V, ternary and quaternary materials under external photo-excitation, has yet to be investigated. In Sect. 10.8 of the theoretical background, the same has been studied. The DMR has been investigated numerically by taking the materials as stated in Sect. 10.1. Section 10.9 contains the result and discussion.

10.8 Theoretical Background

(1) Following the method as given in Sect. 4.2.2(a), the electron dispersion law in the present case is given by

$$\beta_0(E,\lambda) = \left(n+\frac{1}{2}\right)\hbar\omega_0 + \frac{[\hbar k_z(E)]^2}{2m^*} - \frac{E_0}{B}\hbar k_y\{\beta_0(E,\lambda)\}' - \left\{\frac{m^* E_0^2\left[\{\beta_0(E,\lambda)\}'\right]^2}{2B^2}\right\}. \tag{10.70}$$

The use of (4.16) leads to the expressions of the EMM s' along z and y directions as

$$m_z^*(E_{F_{BL}},n,E_0,B,\lambda) = m^*\left[\{\beta_0(E_{F_{BL}},\lambda)\}' + \frac{m^* E_0^2\{\beta_0(E_{F_{BL}},\lambda)\}'\{\beta_0(E_{F_{BL}},\lambda)\}''}{B^2}\right] \tag{10.71}$$

$$\begin{aligned} m_y^*(E_{F_{BL}},n,E_0,B,\lambda) &= \left(\frac{B}{E_0}\right)^2 \frac{1}{[\{\beta_0(E_{F_{BL}},\lambda)\}']} \\ &\times\left[\beta_0(E_{F_{BL}},\lambda) - \left(n+\frac{1}{2}\right)\hbar\omega_0 + \frac{m^* E_0^2\left[\{\beta_0(E_{F_{BL}},\lambda)\}'\right]^2}{2B^2}\right] \\ &\times\left[\frac{-\{\beta_0(E_{F_{BL}},\lambda)\}''}{\left[\{\beta_0(E_{F_{BL}},\lambda)\}'\right]^2}\times\left[\beta_0(E_{F_{BL}},\lambda) - \left(n+\frac{1}{2}\right)\hbar\omega_0\right.\right. \\ &\left.\left. + \frac{m^* E_0^2\left[\{\beta_0(E_{F_{BL}},\lambda)\}'\right]^2}{2B^2}\right] + 1 + \frac{m^* E_0^2\{\beta_0(E_{F_{BL}},\lambda)\}''}{B^2}\right] \end{aligned} \tag{10.72}$$

The Landau energy ($E_{n_{l4}}$) can be written as

$$\beta_0\left(E_{n_{l4}},\lambda\right)=\left(n+\frac{1}{2}\right)\hbar\omega_0-\left\{\frac{m^*E_0^2\left[\left\{\beta_0\left(E_{n_{l4}},\lambda\right)\right\}'\right]^2}{2B^2}\right\} \tag{10.73}$$

The electron concentration and the DMR in this case assume the forms

$$n_0=\frac{2g_{\mathrm{v}}B\sqrt{2m^*}}{3L_x\pi^2\hbar^2E_0}\sum_{n=0}^{n_{\max}}\left[M_{16}\left(n,E_{\mathrm{F_{BL}}},\lambda\right)+N_{16}\left(n,E_{\mathrm{F_{BL}}},\lambda\right)\right], \tag{10.74}$$

and

$$\frac{D}{\mu}=\frac{1}{|e|}\left[\frac{\sum\limits_{n=0}^{n_{\max}}\left[M_{16}\left(n,E_{\mathrm{F_{BL}}},\lambda\right)+N_{16}\left(n,E_{\mathrm{F_{BL}}},\lambda\right)\right]}{\sum\limits_{n=0}^{n_{\max}}\left[\left\{M_{16}\left(n,E_{\mathrm{F_{BL}}},\lambda\right)\right\}'+\left\{N_{16}\left(n,E_{\mathrm{F_{BL}}},\lambda\right)\right\}'\right]}\right], \tag{10.75}$$

where $M_{16}\left(n,E_{F_{BL}},\lambda\right)\equiv\left[\left[\beta_0\left(E_{F_{BL}},\lambda\right)-\left(n+\frac{1}{2}\right)\hbar\omega_0-\frac{m^*E_0^2}{2B^2}\right.\right.$

$\left.\left[\left\{\beta_0\left(E_{F_{BL}},\lambda\right)\right\}'\right]^2+|e|\,E_0L_x\left[\left\{\beta_0\left(E_{F_{BL}},\lambda\right)\right\}'\right]\right]^{3/2}$

$$-\left[\beta_0\left(E_{\mathrm{F_{BL}}},\lambda\right)-\left(n+\frac{1}{2}\right)\hbar\omega_0-\frac{m^*E_0^2}{2B^2}\left[\left\{\beta_0\left(E_{\mathrm{F_{BL}}},\lambda\right)\right\}'\right]^2\right]^{3/2}\Bigg]$$

$\times\dfrac{1}{\left[\left\{\beta_0\left(E_{\mathrm{F_{BL}}},\lambda\right)\right\}'\right]}$ and $N_{16}\left(n,E_{\mathrm{F_{BL}}},\lambda\right)\equiv\sum\limits_{r=1}^{s}\left[L(r)\,M_{16}\left(n,E_{\mathrm{F_{BL}}},\lambda\right)\right]$.

In the absence of external photo-excitation, (10.74) and (10.75) get simplified to (4.20) and (4.21) of Chap. 4, respectively.
(2) Similarly, following the method as given in Sect. 4.2.2(a), the electron dispersion law in this case is given by

$$\tau_0(E,\lambda)=\left(n+\frac{1}{2}\right)\hbar\omega_0-\frac{E_0}{B}\hbar k_y\left\{\tau_0(E,\lambda)\right\}'-\frac{m^*E_0^2}{2B^2}\left[\left\{\tau_0(E,\lambda)\right\}'\right]^2+\frac{\left[\hbar k_z(E)\right]^2}{2m^*}. \tag{10.76}$$

The use of (10.76) leads to the expressions of the EMM s' along z and y directions as

$$m_z^*\left(E_{\mathrm{F_{BL}}},n,E_0,B,\lambda\right)=m^*\left[\left\{\tau_0\left(E_{\mathrm{F_{BL}}},\lambda\right)\right\}'+\frac{m^*E_0^2\left\{\tau_0\left(E_{\mathrm{F_{BL}}},\lambda\right)\right\}'\left\{\tau_0\left(E_{\mathrm{F_{BL}}},\lambda\right)\right\}''}{B^2}\right] \tag{10.77}$$

$$m_y^*(E_{F_{BL}}, n, E_0, B, \lambda) = \left(\frac{B}{E_0}\right)^2 \frac{1}{\{\tau_0(E_{F_{BL}}, \lambda)\}'}$$
$$\times \left[\{\tau_0(E_{F_{BL}}, \lambda)\} - \left(n+\frac{1}{2}\right)\hbar\omega_0 + \frac{m^* E_0^2 \left[\{\tau_0(E_{F_{BL}}, \lambda)\}'\right]^2}{2B^2}\right]$$
$$\times \left[\frac{-\{\tau_0(E_{F_{BL}}, \lambda)\}''}{\left[\{\tau_0(E_{F_{BL}}, \lambda)\}'\right]^2} \times \left[\{\tau_0(E_{F_{BL}}, \lambda)\} - \left(n+\frac{1}{2}\right)\hbar\omega_0 + \frac{m^* E_0^2 \left[\{\tau_0(E_{F_{BL}}, \lambda)\}'\right]^2}{2B^2}\right] + 1 + \frac{\{\tau_0(E_{F_{BL}}, \lambda)\}'' m^* E_0^2}{B^2}\right] \quad (10.78)$$

The Landau energy $(E_{n_{l5}})$ can be written as

$$\tau_0(E_{n_{l5}}, \lambda) = \left(n+\frac{1}{2}\right)\hbar\omega_0 - \frac{m^* E_0^2}{2B^2}\left[\{\tau_0(E_{n_{l5}}, \lambda)\}'\right]^2. \quad (10.79)$$

The expressions for n_0 and DMR in this case assume the forms

$$n_0 = \frac{2g_v B\sqrt{2m^*}}{3L_x \pi^2 \hbar^2 E_0} \sum_{n=0}^{n_{\max}} \left[M_{17}(n, E_{F_{BL}}, \lambda) + N_{17}(n, E_{F_{BL}}, \lambda)\right], \quad (10.80)$$

and

$$\frac{D}{\mu} = \frac{1}{|e|}\left[\frac{\sum_{n=0}^{n_{\max}} \left[M_{17}(n, E_{F_{BL}}, \lambda) + N_{17}(n, E_{F_{BL}}, \lambda)\right]}{\sum_{n=0}^{n_{\max}} \left[\{M_{17}(n, E_{F_{BL}}, \lambda)\}' + \{N_{17}(n, E_{F_{BL}}, \lambda)\}'\right]}\right], \quad (10.81)$$

where $M_{17}(n, E_{F_{BL}}, \lambda) \equiv \left[\left[\tau_0(E_{F_{BL}}, \lambda) - \left(n+\frac{1}{2}\right)\hbar\omega_0 + |e| E_0 L_x \{\tau_0(E_{F_{BL}}, \lambda)\}' - \frac{m^* E_0^2}{2B^2}\left[\{\tau_0(E_{F_{BL}}, \lambda)\}'\right]^2\right]^{3/2} - \left[\tau_0(E_{F_{BL}}, \lambda) - \left(n+\frac{1}{2}\right)\hbar\omega_0 - \frac{m^* E_0^2}{2B^2}\left[\{\tau_0(E_{F_{BL}}, \lambda)\}'\right]^2\right]^{3/2}\right]\left[\{\tau_0(E_{F_{BL}}, \lambda)\}'\right]^{-1}$.

and $N_{17}(n, E_{F_{BL}}, \lambda) \equiv \sum_{r=0}^{s} L(r)\left[M_{17}(n, E_{F_{BL}}, \lambda)\right]$.

In the absence of external photo-excitation, (10.80) and (10.81) get simplified to (4.26) and (4.27), respectively.

(3) For perturbed parabolic energy bands, we can write,

$$\rho_0(E, \lambda) = \left(n+\frac{1}{2}\right)\hbar\omega_0 - \frac{E_0}{B}\hbar k_y \{\rho_0(E, \lambda)\}' - \frac{m^* E_0^2}{2B^2}\left[\{\rho_0(E, \lambda)\}'\right]^2 + \frac{\left[\hbar k_z(E)\right]^2}{2m^*}. \quad (10.82)$$

Using equation (10.82), the expressions of the EMM s' along y and z directions can be written as

$$m_z^*(E_{F_{BL}}, n, E_0, B, \lambda) = m^*\left[\{\rho_0(E_{F_{BL}}, \lambda)\}' + \frac{m^* E_0^2 \{\rho_0(E_{F_{BL}}, \lambda)\}' \{\rho_0(E_{F_{BL}}, \lambda)\}''}{B^2}\right] \quad (10.83)$$

and

$$m_y^*(E_{F_{BL}}, n, E_0, B, \lambda) = \left(\frac{B}{E_0}\right)^2 \frac{1}{\{\rho_0(E_{F_{BL}}, \lambda)\}'}\left[\{\rho_0(E_{F_{BL}}, \lambda)\} - \left(n+\frac{1}{2}\right)\hbar\omega_0 + \frac{m^* E_0^2 \left[\{\rho_0(E_{F_{BL}}, \lambda)\}'\right]^2}{2B^2}\right]\left[\frac{-\{\rho_0(E_{F_{BL}}, \lambda)\}''}{\left[\{\rho_0(E_{F_{BL}}, \lambda)\}'\right]^2}\left[\{\rho_0(E_{F_{BL}}, \lambda)\} - \left(n+\frac{1}{2}\right)\hbar\omega_0 + \frac{m^* E_0^2 \left[\{\rho_0(E_{F_{BL}}, \lambda)\}'\right]^2}{2B^2}\right] + 1 + \frac{\{\rho_0(E_{F_{BL}}, \lambda)\}'' m^* E_0^2}{B^2}\right] \quad (10.84)$$

The Landau energy $(E_{n_{l6}})$ can be written as

$$\rho_0(E_{n_{l6}}, \lambda) = \left(n+\frac{1}{2}\right)\hbar\omega_0 - \frac{m^* E_0^2}{2B^2} \times \left[\{\rho_0(E_{nl6}, \lambda)\}'\right]^2 \quad (10.85)$$

The electron concentration and the DMR in this case can, respectively, be expressed as

$$n_0 = \frac{2g_v B\sqrt{2m^*}}{3L_x \pi^2 \hbar^2 E_0} \sum_{n=0}^{n_{\max}} \left[\bar{M}_{17}(n, E_{F_{BL}}, \lambda) + \bar{N}_{17}(n, E_{F_{BL}}, \lambda)\right] \quad (10.86)$$

and

$$\frac{D}{\mu} = \frac{1}{|e|}\left[\frac{\sum_{n=0}^{n_{\max}} \left[\bar{M}_{17}(n, E_{F_{BL}}, \lambda) + \bar{N}_{17}(n, E_{F_{BL}}, \lambda)\right]}{\sum_{n=0}^{n_{\max}} \left[\{\bar{M}_{17}(n, E_{F_{BL}}, \lambda)\}' + \{\bar{N}_{17}(n, E_{F_{BL}}, \lambda)\}'\right]}\right] \quad (10.87)$$

where $\bar{M}_{17}(n, E_{F_{BL}}, \lambda) \equiv [[\rho_0(E_{F_{BL}}, \lambda) - (n+\frac{1}{2})\hbar\omega_0 + |e|E_0 L_x\{\rho_0(E_{F_{BL}}, \lambda)\}' - \frac{m^* E_0^2}{2B^2}[\{\rho_0(E_{F_{BL}}, \lambda)\}']^2]^{\frac{3}{2}} - [\rho_0(E_{F_{BL}}, \lambda) - (n+\frac{1}{2})\hbar\omega_0 - \frac{m^* E_0^2}{2B^2}[\{\rho_0(E_{F_{BL}}, \lambda)\}']^2]^{\frac{3}{2}}][\{\rho_0(E_{F_{BL}}, \lambda)\}']^{-1}$

and $\bar{N}_{17}(n, E_{F_{BL}}, \lambda) \equiv \sum_{r=0}^{s} L(r)\left[\bar{M}_{17}(n, E_{F_{BL}}, \lambda)\right]$.

In the absence of electric field $E_0 \to 0$ and the application of L' Hospital's rule transforms the equation (10.87) into the well-known form under magnetic quantization as given by equation (3.46) of this chapter.

In the absence of external photo-excitation, equation (10.86) and (10.87) get simplified to equations (4.32a) and (4.32b) respectively.

10.9 Result and Discussions

Using Table 2.1 and (10.74), (10.75), (10.80), (10.81) and (10.86), (10.87), the plot of the DMR as a function of inverse magnetic field under cross-field configurations in the presence of external photo-excitation at $T = 4.2$ K is shown in Figs. 10.29–10.32 by taking n-InAs, n-InSb, $Hg_{1-x}Cd_xTe$ and $n - In_{1-x}Ga_xAs_yP_{1-y}$ lattice matched to InP respectively. It appears that the DMR oscillates with the inverse quantizing magnetic field with different numerical magnitudes for all the cases.

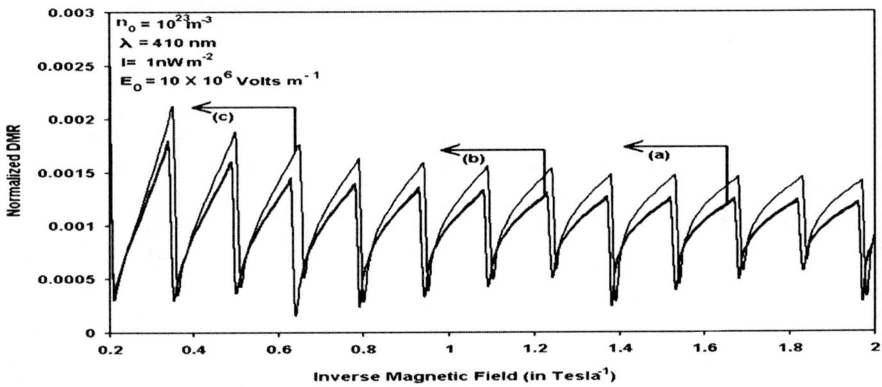

Fig. 10.29. Plot of the DMR as a function of inverse quantizing magnetic field under cross field configuration in external photo-excitation for n-InAs, in which the curves (**a**), (**b**) and (**c**) represent the perturbed three and two band models of Kane and that of parabolic energy bands respectively

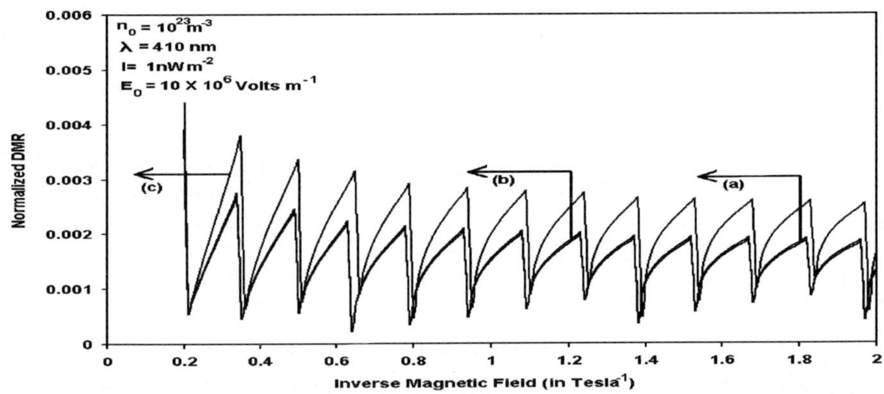

Fig. 10.30. Plot of the DMR as a function of inverse quantizing magnetic field under cross field configuration in external photo-excitation for n-InSb, in which the curves (**a**), (**b**) and (**c**) represent the perturbed three and two band models of Kane and that of parabolic energy bands respectively

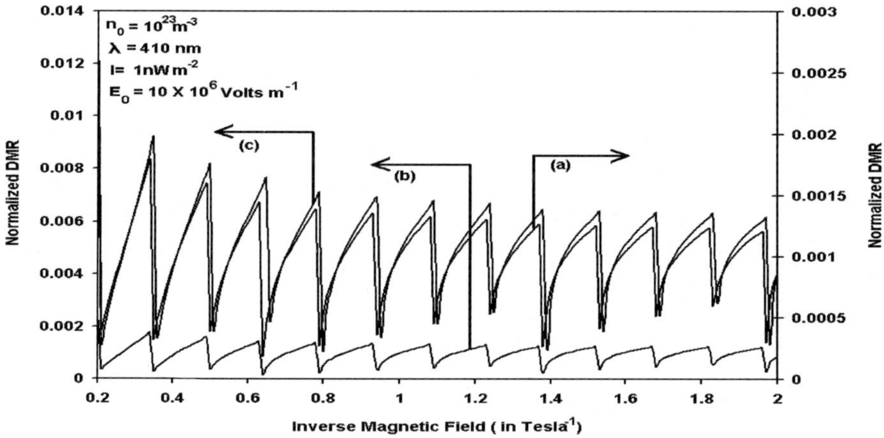

Fig. 10.31. Plot of the DMR as a function of inverse quantizing magnetic field under cross field configuration in external photo-excitation for n-$Hg_{1-x}Cd_xTe$, in which the curves (**a**), (**b**) and (**c**) represent the perturbed three and two band models of Kane and that of parabolic energy bands respectively

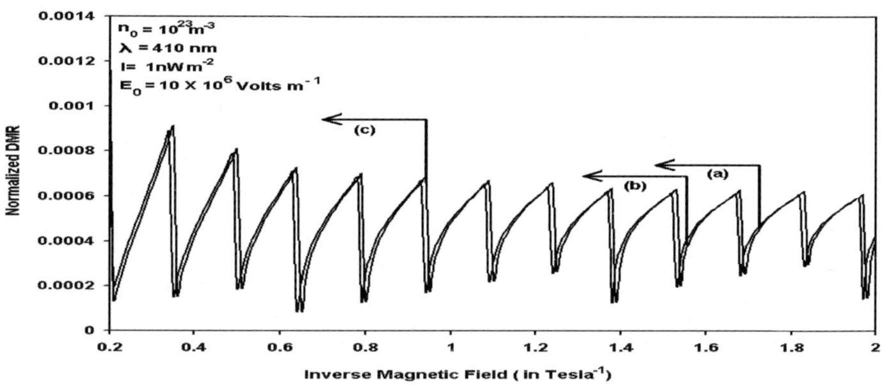

Fig. 10.32. Plot of the DMR as a function of inverse quantizing magnetic field under cross field configuration in external photo-excitation for n-$In_{1-x}Ga_xAs_yP_{1-y}$ lattice matched to InP, in which the curves (**a**), (**b**) and (**c**) represent the perturbed three and two band models of Kane and that of parabolic energy bands respectively

Figures 10.33–10.43 exhibit the variation of the DMR in this case as functions of electron concentration, light intensity and wavelength, respectively.

Figures 10.44–10.47 show the variation of the same function of electric field for all the respective cases as mentioned earlier.

It appears from Figs. 10.37–10.43, that the DMR decreases with the increase in light intensity and the wavelength which is in the visible region. From

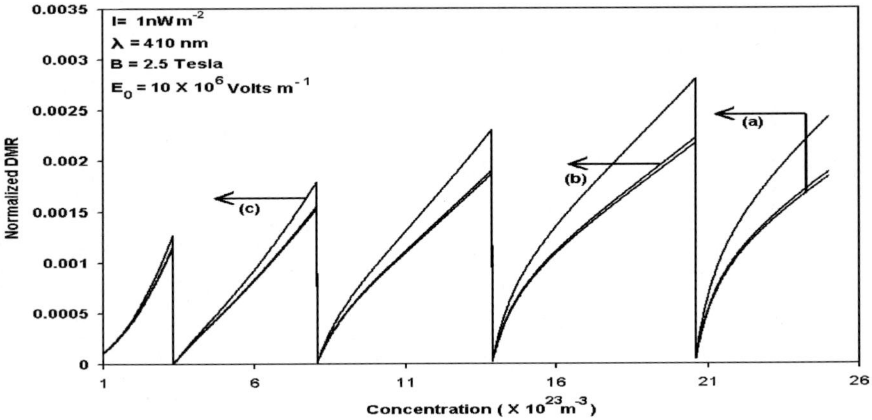

Fig. 10.33. Plot of the DMR as a function of electron concentration field under cross field configuration in external photo-excitation for n-InAs, in which the curves (**a**), (**b**) and (**c**) represent the perturbed three and two band models of Kane and that of parabolic energy bands respectively

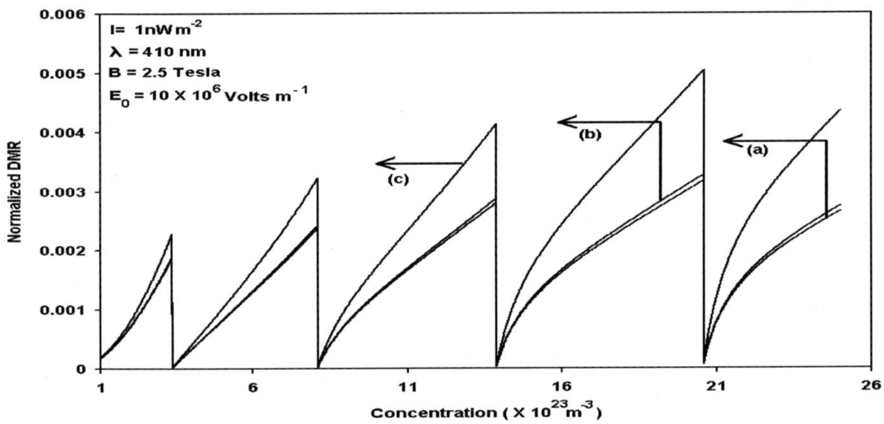

Fig. 10.34. Plot of the DMR as a function of electron concentration field under cross field configuration in external photo-excitation for n-InSb, in which the curves (**a**), (**b**) and (**c**) represent the perturbed three and two band models of Kane and that of parabolic energy bands respectively

Figs. 10.44 to 10.47, it appears that the DMR increases with the increase in the electric field. It should be noted that the rate of change of the DMR in the respective cases, are totally energy spectrum dependent. The remaining discussion of Sect. 4.4 is also applicable here in this case.

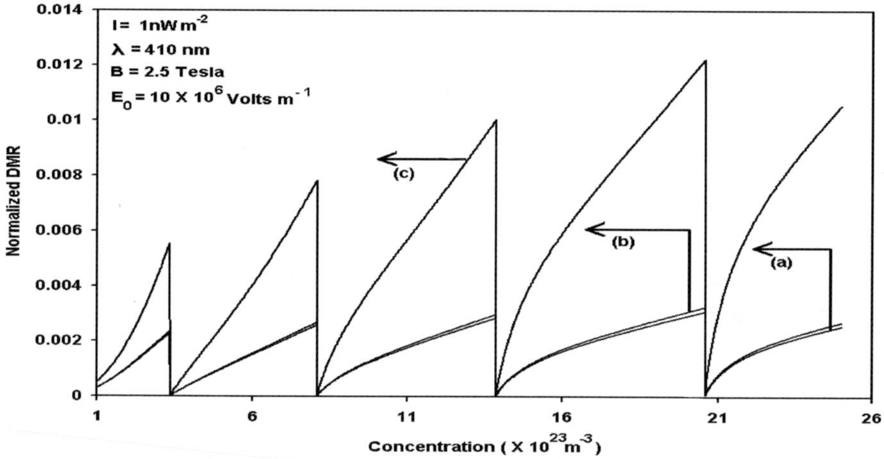

Fig. 10.35. Plot of the DMR as a function of electron concentration field under cross field configuration in external photo-excitation for n-$Hg_{1-x}Cd_xTe$ in which the curves (**a**), (**b**) and (**c**) represent the perturbed three and two band models of Kane and that of parabolic energy bands respectively

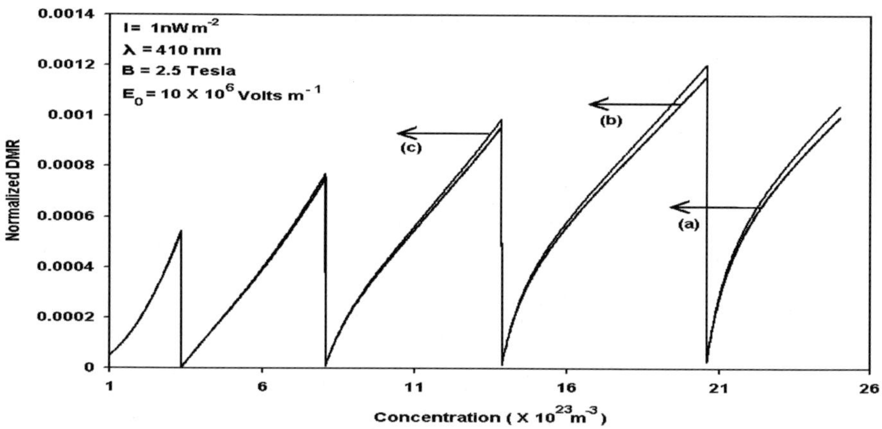

Fig. 10.36. Plot of the DMR as a function of electron concentration field under cross field configuration in external photo-excitation for n-$In_{1-x}Ga_xAs_yP_{1-y}$ lattice matched to InP in which the curves (**a**), (**b**) and (**c**) represent the perturbed three and two band models of Kane and that of parabolic energy bands respectively

10.10 The Formulation of the DMR for the Ultrathin Films of III–V, Ternary and Quaternary Materials Under External Photo-Excitation

In this section, we will study the DMR in ultra-thin films of III–V, ternary and quaternary materials under external photo-excitation. The DMR has

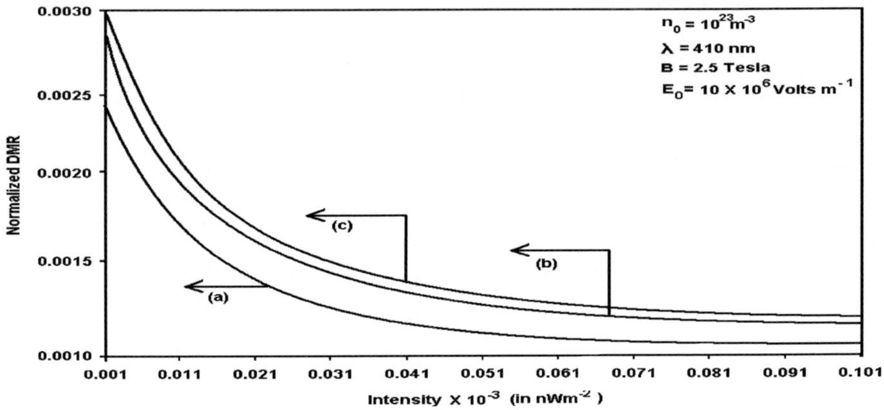

Fig. 10.37. Plot of the DMR as a function of light intensity under cross field configuration in external photo-excitation for n-InAs, in which the curves (**a**), (**b**) and (**c**) represent the perturbed three and two band models of Kane and that of parabolic energy bands respectively

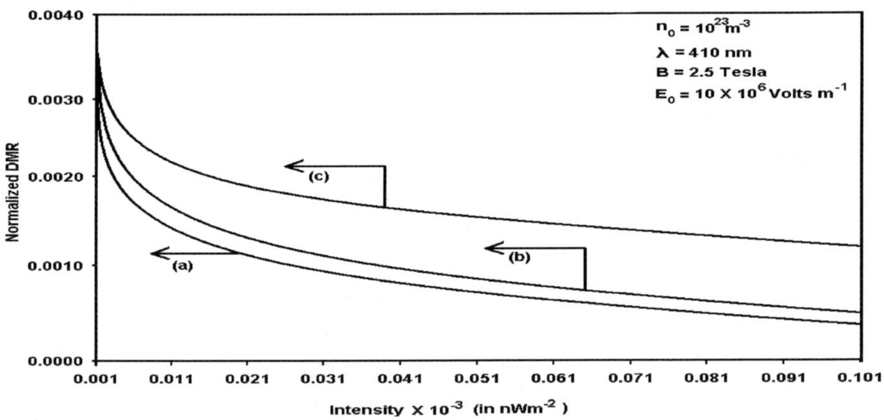

Fig. 10.38. Plot of the DMR as a function of light intensity under cross field configuration in external photo-excitation for n-InSb, in which the curves (**a**), (**b**) and (**c**) represent the perturbed three and two band models of Kane and that of parabolic energy bands respectively

been investigated numerically by taking the materials as stated in Sect. 10.1. Section 10.11 contains the result and discussion of this section.

The 2D electron energy spectrum in ultra-thin films of III–V, ternary and quaternary materials, whose unperturbed band structure is defined by the three band model of Kane, in the presence of light waves can be expressed following (10.41) as

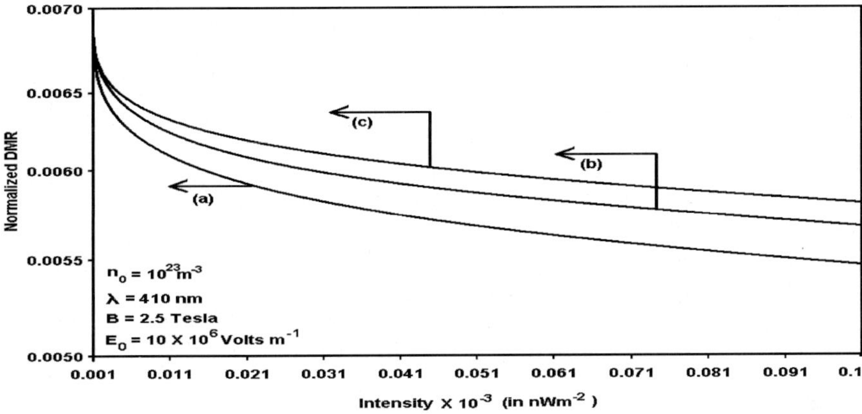

Fig. 10.39. Plot of the DMR as a function of light intensity under cross field configuration in external photo-excitation for n-$Hg_{1-x}Cd_xTe$, in which the curves (**a**), (**b**) and (**c**) represent the perturbed three and two band models of Kane and that of parabolic energy bands respectively

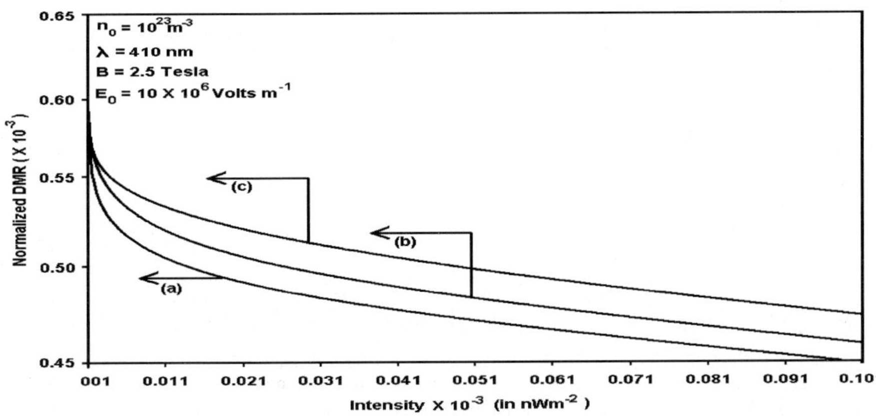

Fig. 10.40. Plot of the DMR as a function of light intensity under cross field configuration in external photo-excitation for n-$In_{1-x}Ga_xAs_yP_{1-y}$ lattice matched to InP, in which the curves (**a**), (**b**) and (**c**) represent the perturbed three and two band models of Kane and that of parabolic energy bands respectively

$$\frac{\hbar^2 k_s^2}{2m^*} + \frac{\hbar^2}{2m^*}\left(\frac{n_z \pi}{d_z}\right)^2 = \beta_0 (E, \lambda) . \tag{10.88}$$

The sub-band energies $(E_{n_{l7}})$ can be written as

$$\beta_0(E_{n_{l7}}, \lambda) = \frac{\hbar^2}{2m^*} (n_z \pi / d_z)^2 . \tag{10.89}$$

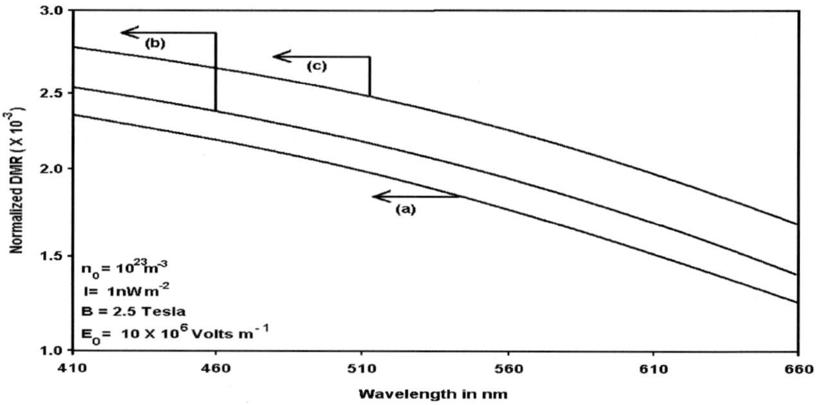

Fig. 10.41. Plot of the DMR as a function of wavelength under cross field configuration in external photo-excitation for n-InAs, in which the curves (**a**), (**b**) and (**c**) represent the perturbed three and two band models of Kane and that of parabolic energy bands respectively

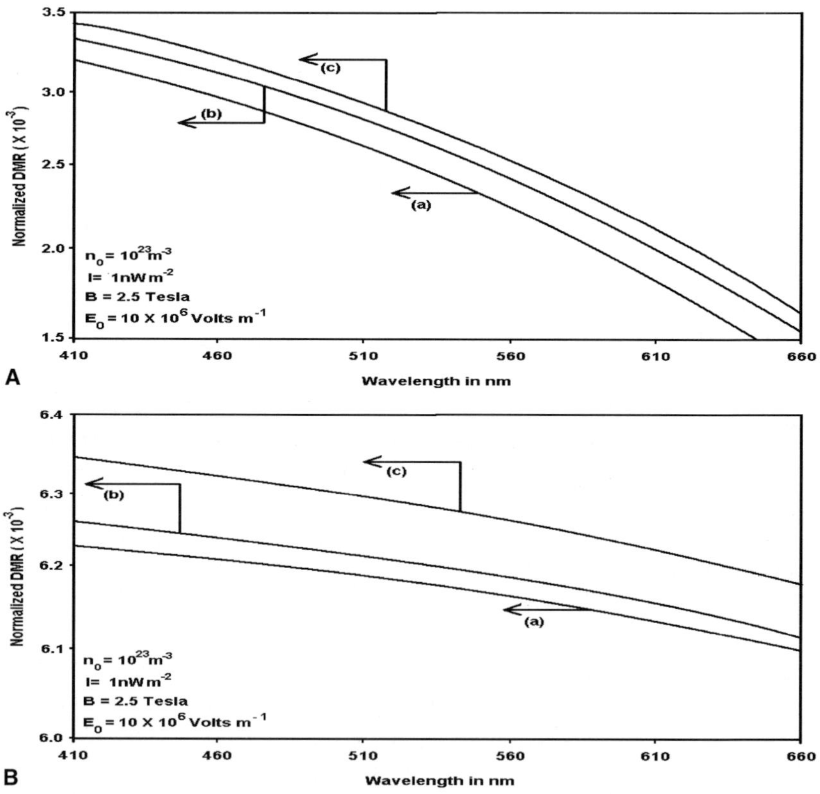

Fig. 10.42. Plot of the DMR as a function of wavelength under cross field configuration in external photo-excitation for n-InSb, and n-$Hg_{1-x}Cd_xTe$ as shown in figs. A and B respectively in which the curves (**a**), (**b**) and (**c**) represent the perturbed three and two band models of Kane and that of parabolic energy bands respectively

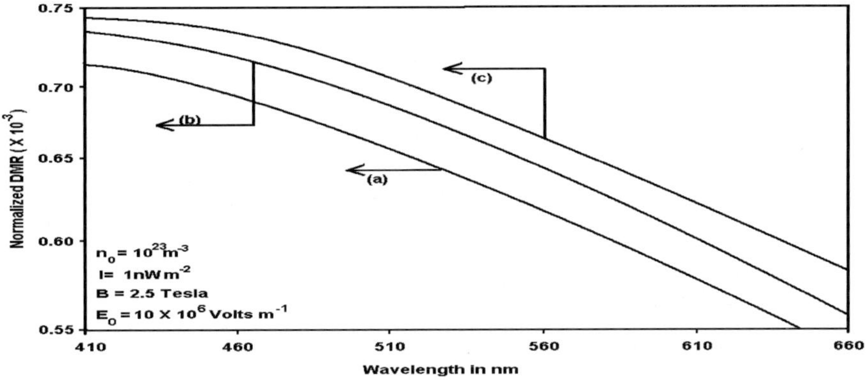

Fig. 10.43. Plot of the DMR as a function of wavelength under cross field configuration in external photo-excitation for n-$In_{1-x}Ga_xAs_yP_{1-y}$ lattice matched to InP, in which the curves (**a**), (**b**) and (**c**) represent the perturbed three and two band models of Kane and that of parabolic energy bands

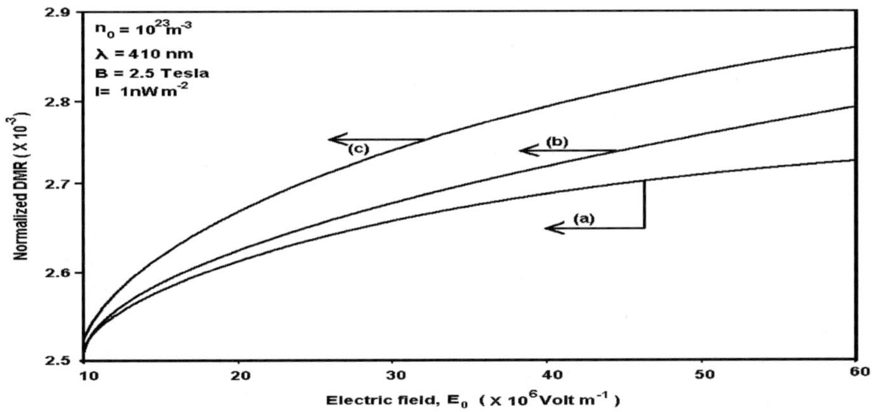

Fig. 10.44. Plot of the DMR as a function of electric field under cross field configuration in external photo-excitation for n-InAs, in which the curves (**a**), (**b**) and (**c**) represent the perturbed three and two band models of Kane and that of parabolic energy bands respectively

The expression of the EMM in this case is given by

$$m^* \left(E_{\mathrm{F2DL}}, n_z, \lambda\right) = m^* \left\{\beta_0(E_{\mathrm{F2DL}}, \lambda)\right\}', \qquad (10.90)$$

where E_{F2DL} is the Fermi energy in the present case, as measured from the edge of the conduction band in the vertically upward direction in the absence of any quantization.

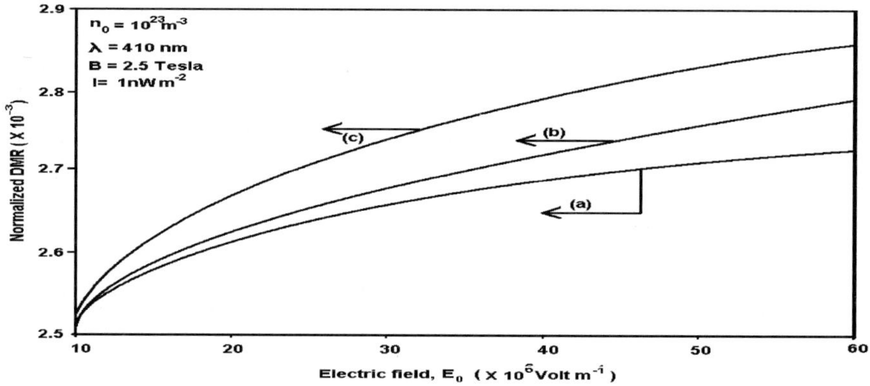

Fig. 10.45. Plot of the DMR as a function of electric field under cross field configuration in external photo-excitation for n-InSb, in which the curves (**a**), (**b**) and (**c**) represent the perturbed three and two band models of Kane and that of parabolic energy bands respectively

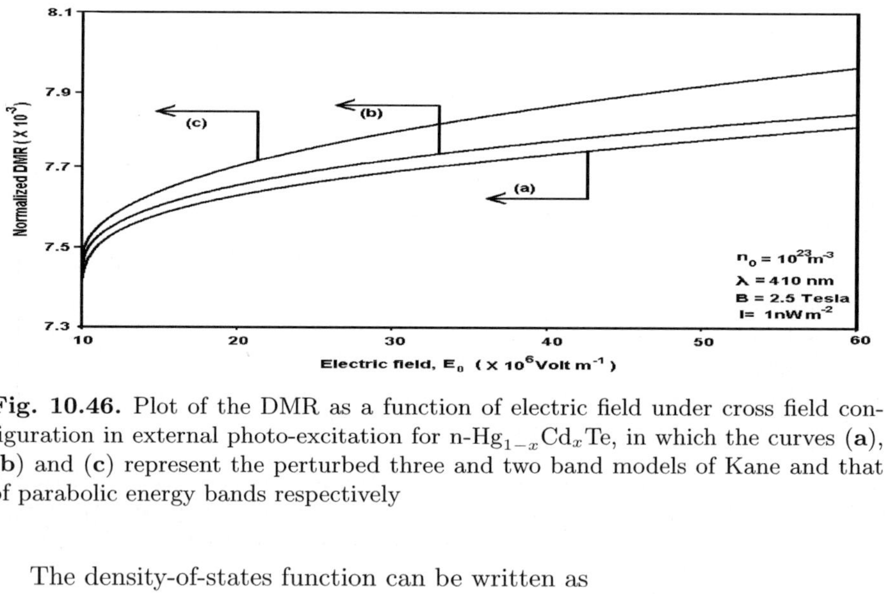

Fig. 10.46. Plot of the DMR as a function of electric field under cross field configuration in external photo-excitation for n-$Hg_{1-x}Cd_xTe$, in which the curves (**a**), (**b**) and (**c**) represent the perturbed three and two band models of Kane and that of parabolic energy bands respectively

The density-of-states function can be written as

$$N_{2D}(E,\lambda) = \left(\frac{m^* g_v}{\pi\hbar^2}\right) \sum_{n_z=1}^{n_{z\max}} [\beta_0(E,\lambda)]' H(E - E_{n_{17}}). \tag{10.91}$$

Combining (10.91) with the Fermi–Dirac, the occupation probability factor, and the two dimensional electron concentration can be expressed as

$$n_{2D} = \frac{m^* g_v}{\pi\hbar^2} \sum_{n_z=1}^{n_{z\max}} [M_{18}(n_z, E_{F2DL}, \lambda) + N_{18}(n_z, E_{F2DL}, \lambda)], \tag{10.92}$$

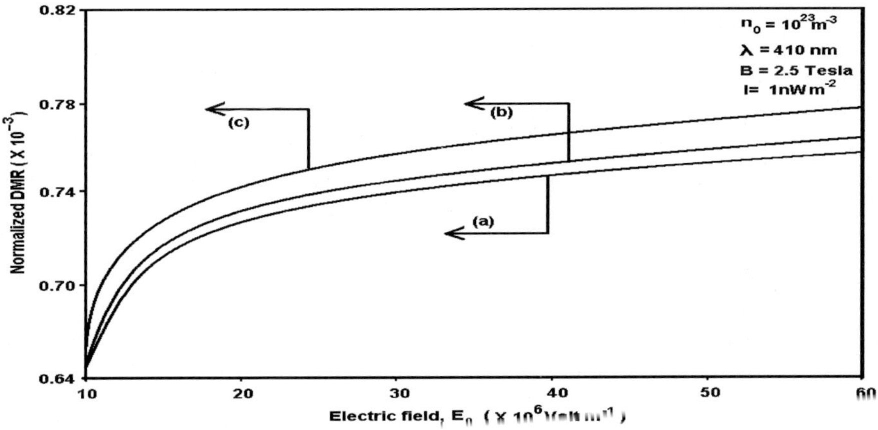

Fig. 10.47. Plot of the DMR as a function of electric field under cross field configuration in external photo-excitation for n-$In_{1-x}Ga_xAs_yP_{1-y}$ lattice matched to InP, in which the curves (**a**), (**b**) and (**c**) represent the perturbed three and two band models of Kane and that of parabolic energy bands respectively

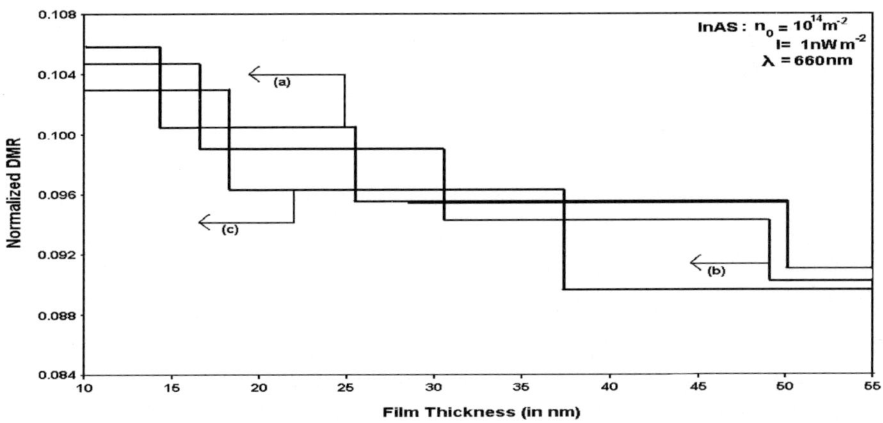

Fig. 10.48. Plot of the normalized 2D DMR as a function of film thickness for ultra-thin films of n-InAs in the presence of light waves in which the curves (**a**), (**b**) and (**c**) represent the perturbed three and two band models of Kane and that of the parabolic energy bands respectively

where $M_{18}(n_z, E_{\mathrm{F2DL}}, \lambda) \equiv \left[\beta_0(E_{\mathrm{F2DL}}, \lambda) - \frac{\hbar^2}{2m^*}\left(\frac{n_z\pi}{d_z}\right)^2\right]$ and $N_{18}(n_z, E_{\mathrm{F2DL}}, \lambda) \equiv \sum_{r=1}^{s} L(r)\, M_{18}(n_z, E_{\mathrm{F2DL}}, \lambda)$.

The use of (10.92) and (1.11) leads to the expression of the DMR in this case as

$$\frac{D}{\mu}=\frac{1}{|e|}\frac{\sum\limits_{n_z=1}^{n_{z\max}}\left[M_{18}\left(n_z,E_{\mathrm{F2DL}},\lambda\right)+N_{18}\left(n_z,E_{\mathrm{F2DL}},\lambda\right)\right]}{\sum\limits_{n_z=1}^{n_{z\max}}\left[\left\{M_{18}\left(n_z,E_{\mathrm{F2DL}},\lambda\right)\right\}'+\left\{N_{18}\left(n_z,E_{\mathrm{F2DL}},\lambda\right)\right\}'\right]}. \tag{10.93}$$

In the absence of external photo-excitation, (10.92) and (10.93) get simplified to (5.10) and (5.11) of Chap. 5, respectively.

Using (10.42), the expressions for the 2D dispersion relation, the sub-band energies, the EMM, the density of states function and the electron concentration for ultra-thin films of III–V, ternary and quaternary materials, whose unperturbed band structure is defined by the two band model of Kane, can respectively be written in the presence of photo-excitation as

$$\frac{\hbar^2 k_{\mathrm{s}}^2}{2m^*}+\frac{\hbar^2}{2m^*}\left(\frac{n_z\pi}{d_z}\right)^2=\tau_0\left(E,\lambda\right), \tag{10.94}$$

$$\tau_0\left(E_{n_{l8}},\lambda\right)=\frac{\hbar^2}{2m^*}\left(n_z\pi/d_z\right)^2, \tag{10.95}$$

$$m^*\left(E_{\mathrm{F2DL}},n_z,\lambda\right)=m^*\left\{\tau_0(E_{\mathrm{F2DL}},\lambda)\right\}', \tag{10.96}$$

$$N_{\mathrm{2D}}\left(E,\lambda\right)=\left(\frac{m^* g_{\mathrm{v}}}{\pi\hbar^2}\right)\sum_{n_z=1}^{n_{z\max}}\left[\tau_0\left(E,\lambda\right)\right]' H\left(E-E_{n_{l8}}\right), \tag{10.97}$$

$$n_{\mathrm{2D}}=\frac{m^* g_{\mathrm{v}}}{\pi\hbar^2}\sum_{n_z=1}^{n_{z\max}}\left[M_{19}\left(n_z,E_{\mathrm{F2DL}},\lambda\right)+N_{19}\left(n_z,E_{\mathrm{F2DL}},\lambda\right)\right]. \tag{10.98}$$

where $M_{19}(n_z,E_{\mathrm{F2DL}},\lambda)\equiv\left[\tau_0(E_{\mathrm{F2DL}},\lambda)\frac{\hbar^2}{2m^*}\left(\frac{n_z\pi}{d_z}\right)^2\right]$

and $N_{19}(n_z,E_{\mathrm{F2DL}},\lambda)\equiv\sum\limits_{r=1}^{s}L(r)M_{19}(n_z,E_{\mathrm{F2DL}},\lambda)$.

The use of (10.98) and (1.11) leads to the expression of the DMR in this case as

$$\frac{D}{\mu}=\frac{1}{|e|}\frac{\sum\limits_{n_z=1}^{n_{z\max}}\left[M_{19}\left(n_z,E_{\mathrm{F2DL}},\lambda\right)+N_{19}\left(n_z,E_{\mathrm{F2DL}},\lambda\right)\right]}{\sum\limits_{n_z=1}^{n_{z\max}}\left[\left\{M_{19}\left(n_z,E_{\mathrm{F2DL}},\lambda\right)\right\}'+\left\{N_{19}\left(n_z,E_{\mathrm{F2DL}},\lambda\right)\right\}'\right]}. \tag{10.99}$$

In the absence of external photo-excitation, (10.98) and (10.99) get simplified to (5.15) and (5.16), respectively.

Using (10.43), the expressions for the 2D dispersion relation, the sub-band energies, the EMM, the density of states function and the electron concentration for ultra-thin films of III–V, ternary and quaternary materials, whose unperturbed band structure is defined by the parabolic energy bands, can respectively be written in the presence of photo-excitation as

$$\frac{\hbar^2 k_{\mathrm{s}}^2}{2m^*}+\frac{\hbar^2}{2m^*}\left(\frac{n_z\pi}{d_z}\right)^2=\rho_0\left(E,\lambda\right), \tag{10.100}$$

$$\rho_0(E_{n_{l9}},\lambda) = \frac{\hbar^2}{2m^*}\left(n_z\pi/d_z\right)^2, \tag{10.101}$$

$$N_{2D}\left(E,\lambda\right) = \left(\frac{m^* g_v}{\pi\hbar^2}\right)\sum_{n_z=1}^{n_{z\max}}\left[\rho_0\left(E,\lambda\right)\right]' H\left(E - E_{n_{l9}}\right), \tag{10.102}$$

$$n_{2D} = \frac{m^* g_v}{\pi\hbar^2}\sum_{n_z=1}^{n_{z\max}}\left[M_{20}\left(n_z, E_{F2DL},\lambda\right) + N_{20}\left(n_z, E_{F2DL},\lambda\right)\right], \tag{10.103}$$

where $M_{20}\left(n_z, E_{F2DL},\lambda\right) \equiv \left[\rho_0(E_{F2DL},\lambda) - \frac{\hbar^2}{2m^*}\left(\frac{n_z\pi}{d_z}\right)^2\right]$ and $N_{20}\left(n_z, E_{F2DL},\lambda\right) \equiv \sum_{r=1}^{s} L\left(r\right) M_{20}\left(n_z, E_{F2DL},\lambda\right)$.

The use of (10.103) and (1.11) leads to the expression of the DMR in this case as

$$\frac{D}{\mu} = \frac{1}{|e|}\frac{\sum_{n_z=1}^{n_{z\max}}\left[M_{20}\left(n_z, E_{F2DL},\lambda\right) + N_{20}\left(n_z, E_{F2DL},\lambda\right)\right]}{\sum_{n_z=1}^{n_{z\max}}\left[\left\{M_{20}\left(n_z, E_{F2DL},\lambda\right)\right\}' + \left\{N_{20}\left(n_z, E_{F2DL},\lambda\right)\right\}'\right]}. \tag{10.104}$$

In the absence of external photo-excitation, (10.103) and (10.104) get simplified to (5.17) and (5.18) respectively.

10.11 Result and Discussions

Using (10.92), (10.93), (10.98), (10.99) and (10.103), (10.104), we have, in Figs. 10.48–10.51, plotted DMR as function of the film thickness in the presence of external photoemission for QWs of n-InAs, n-InSb, n-$Hg_{1-x}Cd_xTe$ and n-$In_{1-x}Ga_xAs_yP_{1-y}$, in which the curves (a)–(c) represents the respective DMR for the perturbed three and two models of Kane and the parabolic energy bands. It has been observed that the DMR, in this case, also decreases with the increase in film thickness in a step wise manner and follows the same logic, as presented in Sect. 5.3 of Chap. 5.

Figures 10.52–10.55 exhibit the plot of the DMR as a function of electron concentration in the presence of light waves for the QWs of the aforementioned materials.

The variations of the DMR against light intensity for the QWs of aforementioned materials, have been plotted in Figs. 10.56–10.59. The DMR decreases with the increase in the light intensity.

Figures 10.60–10.63 exhibit the variations of the DMR as functions of wavelength in the visible region. The DMR, in this case, decreases as the wavelength shifts from red to violet. Figures 10.64 and 10.65 show that the DMR decreases with increasing alloy composition for the ternary and quaternary materials.

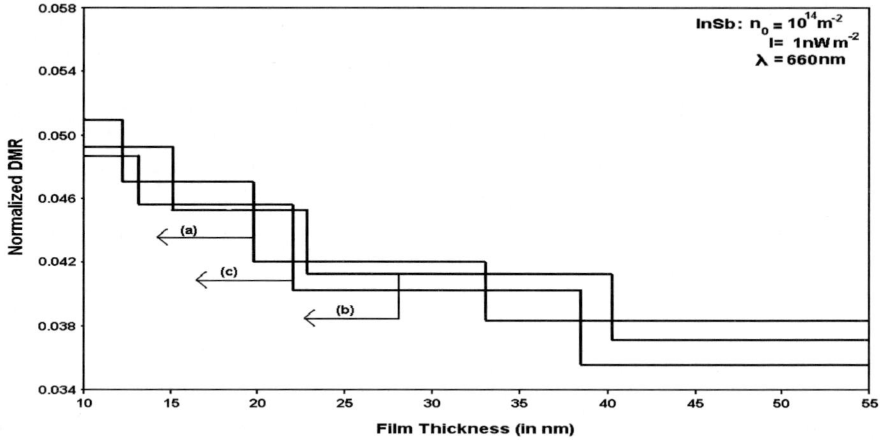

Fig. 10.49. Plot of the normalized 2D DMR as a function of film thickness for ultra-thin films of n-InSb in the presence of light waves in which the curves (**a**), (**b**) and (**c**) represent the perturbed three and two band models of Kane and that of the parabolic energy bands respectively

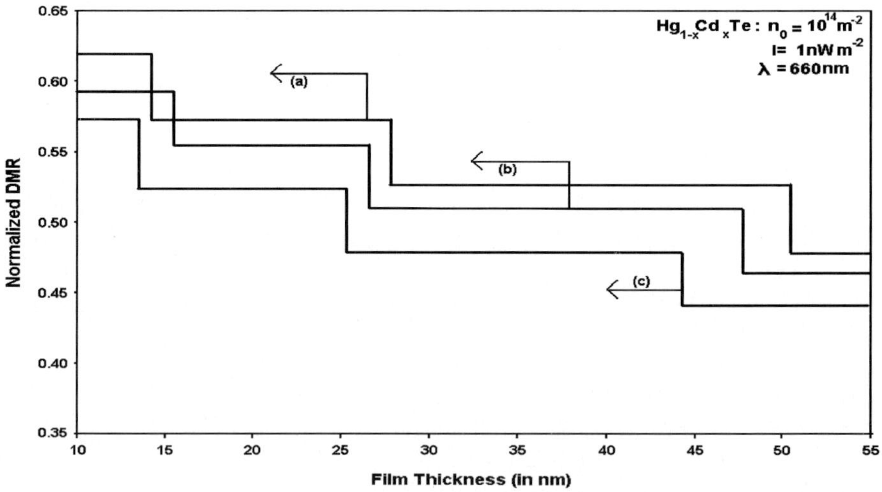

Fig. 10.50. Plot of the normalized 2D DMR as a function of film thickness for ultra-thin films of n-$Hg_{1-x}Cd_xTe$ in the presence of light waves in which the curves (**a**), (**b**) and (**c**) represent the perturbed three and two band models of Kane and that of the parabolic energy bands respectively

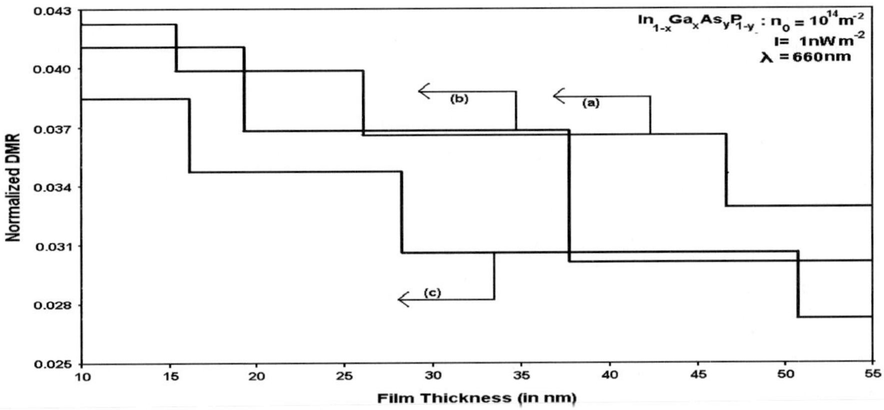

Fig. 10.51. Plot of the normalized 2D DMR as a function of film thickness for ultra-thin films of n-$In_{1-x}Ga_xAs_yP_{1-y}$ lattice matched to InP in the presence of light waves in which the curves (**a**), (**b**) and (**c**) represent the perturbed three and two band models of Kane and that of the parabolic energy bands respectively

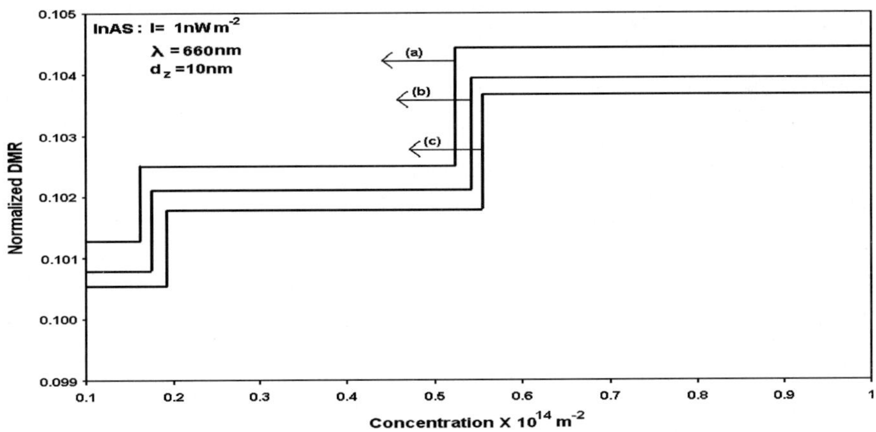

Fig. 10.52. Plot of the normalized 2D DMR as a function of electron concentration for ultra-thin films of n-InAs in the presence of light waves in which the curves (**a**), (**b**) and (**c**) represent the perturbed three and two band models of Kane and that of the parabolic energy bands respectively

10.12 The Formulation of the DMR in QWs of III–V, Ternary and Quaternary Materials Under External Photo-Excitation

In this section, we will study the DMR in QWs of III–V, ternary and quaternary materials under external photo-excitation. The DMR has been investigated numerically by taking the materials as stated in Sect. 10.1. Section 10.13 contains the result and discussion of this section.

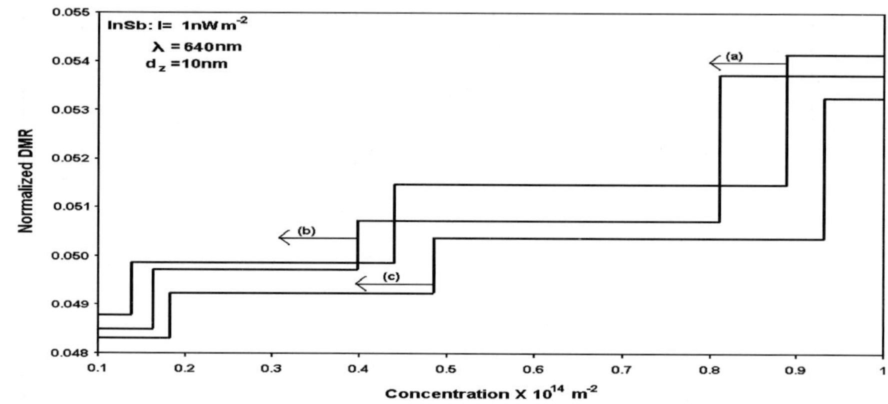

Fig. 10.53. Plot of the normalized 2D DMR as a function of electron concentration for ultra-thin films of n-InSb in the presence of light waves in which the curves (**a**), (**b**) and (**c**) represent the perturbed three and two band models of Kane and that of the parabolic energy bands respectively

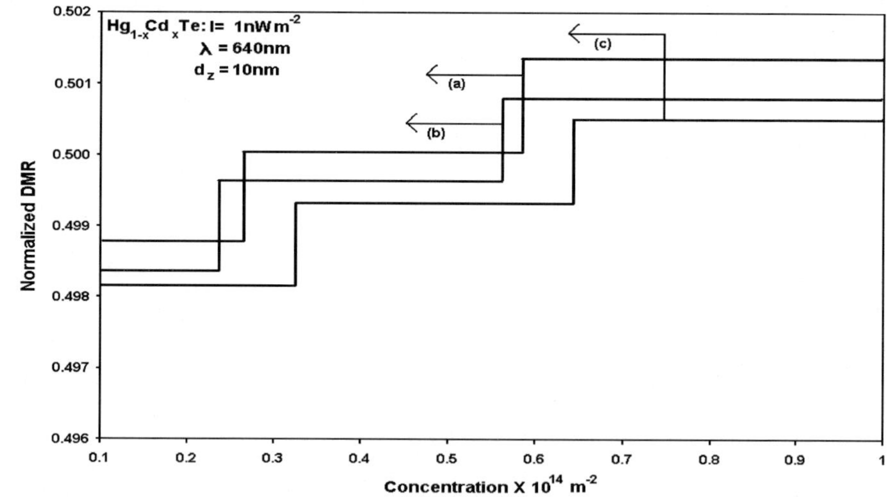

Fig. 10.54. Plot of the normalized 2D DMR as a function of electron concentration for ultra-thin films of n-$Hg_{1-x}Cd_xTe$ in the presence of light waves in which the curves (**a**), (**b**) and (**c**) represent the perturbed three and two band models of Kane and that of the parabolic energy bands respectively

The 1D electron energy spectrum in QWs of III–V, ternary and quaternary materials, whose unperturbed band structure is defined by the three band model of Kane, in the presence of light waves, can be expressed from (10.41) as

$$\frac{\hbar^2 k_x^2}{2m^*} = \beta_0 (E, \lambda) - \left[\frac{\hbar^2}{2m^*} \left(\frac{n_y \pi}{d_y} \right)^2 + \frac{\hbar^2}{2m^*} \left(\frac{n_z \pi}{d_z} \right)^2 \right], \tag{10.105}$$

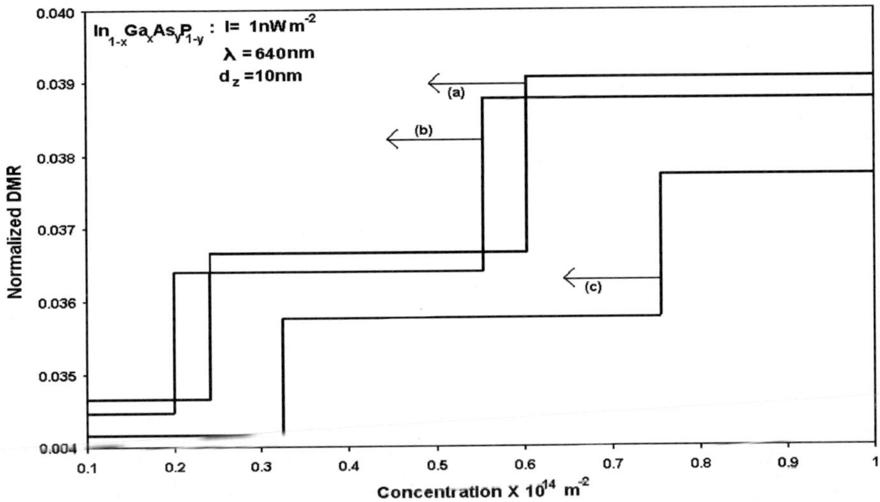

Fig. 10.55. Plot of the normalized 2D DMR as a function of electron concentration for ultra-thin films of n-$In_{1-x}Ga_xAs_yP_{1-y}$ lattice matched to InP in the presence of light waves in which the curves (**a**), (**b**) and (**c**) represent the perturbed three and two band models of Kane and that of the parabolic energy bands respectively

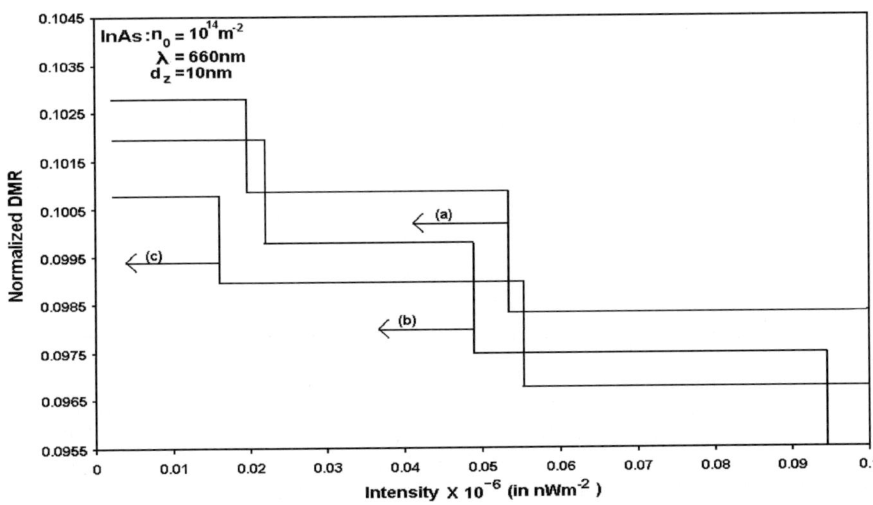

Fig. 10.56. Plot of the normalized 2D DMR as a function of light intensity for ultra-thin films of n-InAs in the presence of light waves in which the curves (**a**), (**b**) and (**c**) represent the perturbed three and two band models of Kane and that of the parabolic energy bands respectively

where $n_y\,(=1,2,3,\ldots)$ and $n_z\,(=1,2,3,\ldots)$ are the size quantum numbers along y and z directions while d_y and d_z are the nanothickness along the y and z directions respectively.

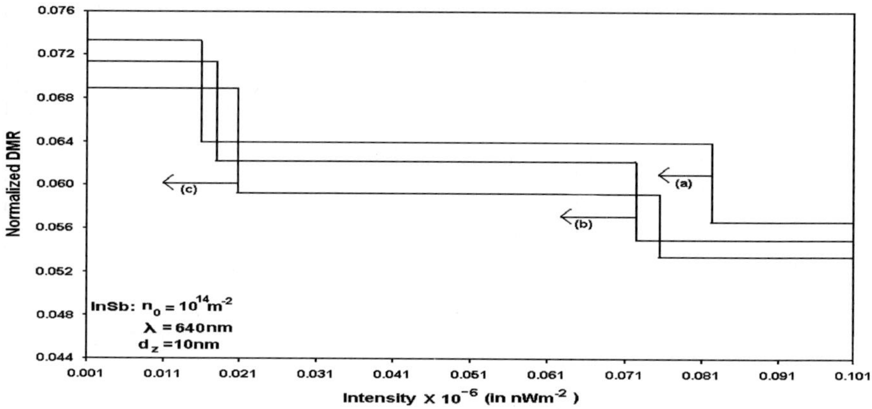

Fig. 10.57. Plot of the normalized 2D DMR as a function of light intensity for ultra-thin films of n-InSb in the presence of light waves in which the curves (**a**), (**b**) and (**c**) represent the perturbed three and two band models of Kane and that of the parabolic energy bands respectively

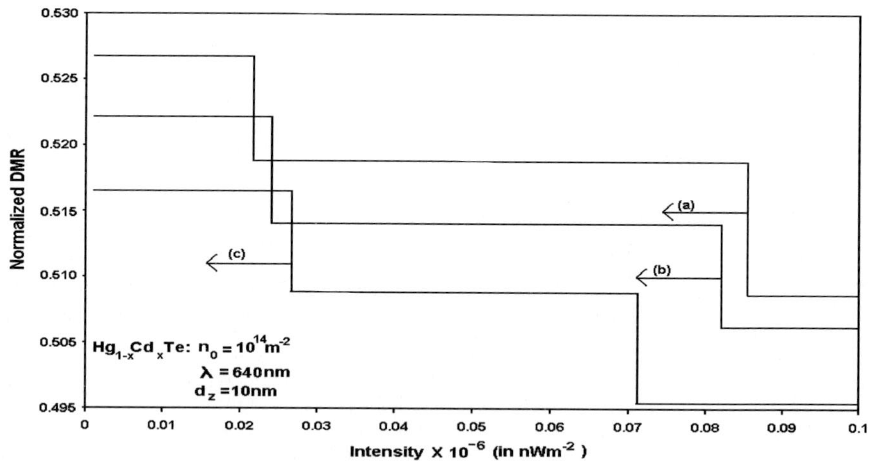

Fig. 10.58. Plot of the normalized 2D DMR as a function of light intensity for ultra-thin films of n-$Hg_{1-x}Cd_xTe$ in the presence of light waves in which the curves (**a**), (**b**) and (**c**) represent the perturbed three and two band models of Kane and that of the parabolic energy bands respectively

The sub-band energies ($E_{n_{l10}}$) can be expressed as

$$\beta_0(E_{n_{l10}},\lambda) = \left[\frac{\hbar^2}{2m^*}(n_y\pi/d_y)^2 + \frac{\hbar^2}{2m^*}(n_z\pi/d_z)^2\right]. \tag{10.106}$$

The EMM in this case can be written as

$$m^*(E_{\mathrm{F1DL}}, n_y, n_z, \lambda) = m^*\{\beta_0(E_{\mathrm{F1DL}},\lambda)\}', \tag{10.107}$$

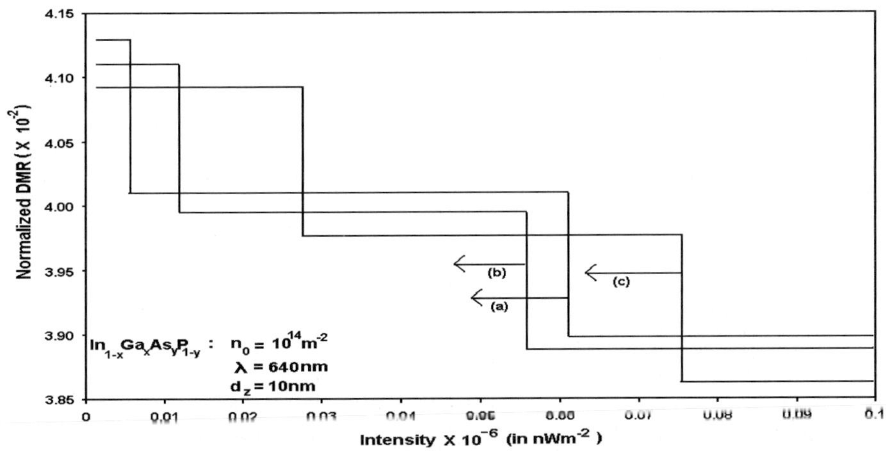

Fig. 10.59. Plot of the normalized 2D DMR as a function of light intensity for ultra-thin films of n-$In_{1-x}Ga_xAs_yP_{1-y}$ lattice matched to InP in the presence of light waves in which the curves (**a**), (**b**) and (**c**) represent the perturbed three and two band models of Kane and that of the parabolic energy bands respectively

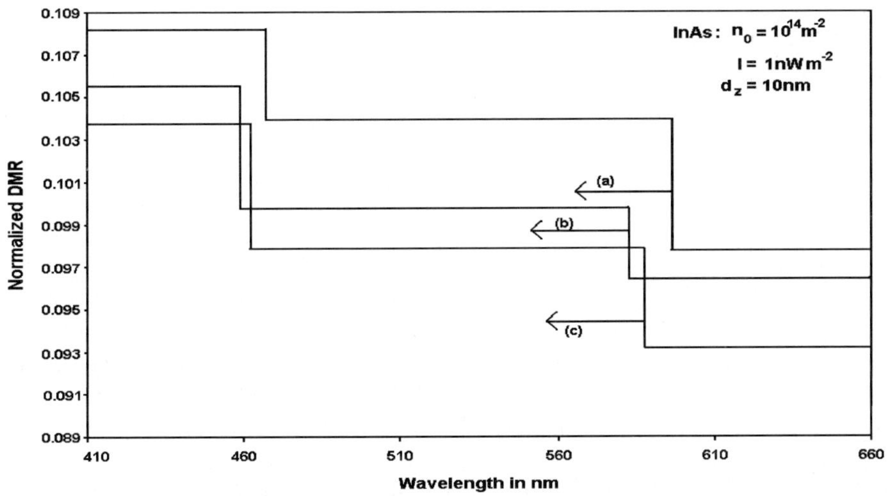

Fig. 10.60. Plot of the normalized 2D DMR as a function of wavelength for ultra-thin films of n-InAs in the presence of light waves in which the curves (**a**), (**b**) and (**c**) represent the perturbed three and two band models of Kane and that of the parabolic energy bands respectively

where E_{F1DL} is the Fermi energy in the present case, as measured from the edge of the conduction band in the vertically upward direction in absence of any quantization.

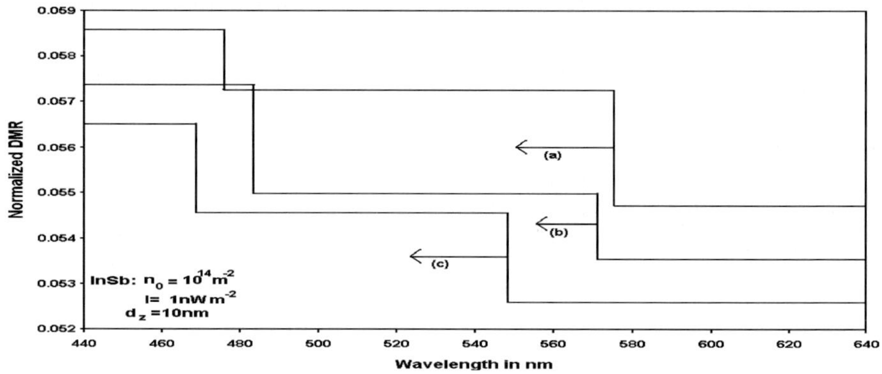

Fig. 10.61. Plot of the normalized 2D DMR as a function of wavelength for ultra-thin films of n-InSb in the presence of light waves in which the curves (**a**), (**b**) and (**c**) represent the perturbed three and two band models of Kane and that of the parabolic energy bands respectively

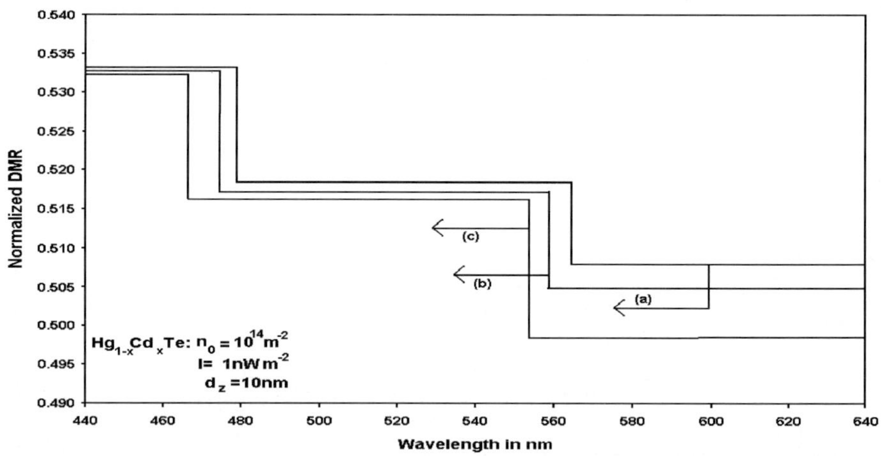

Fig. 10.62. Plot of the normalized 2D DMR as a function of wavelength for ultra-thin films of n-$Hg_{1-x}Cd_xTe$ in the presence of light waves in which the curves (**a**), (**b**) and (**c**) represent the perturbed three and two band models of Kane and that of the parabolic energy bands respectively

The one-dimensional density-of-states function ($N_{1D}(E,\lambda)$) is given by

$$N_{1D}(E,\lambda) = \left(\frac{g_v\sqrt{2m^*}}{\pi\hbar}\right)\sum_{n_y=1}^{n_{y\max}}\sum_{n_z=1}^{n_{z\max}} \{\beta_0(E,\lambda)\}'\left[\beta_0(E,\lambda) - (\hbar^2/2m^*)\phi(n_y,n_z)\right]^{-1/2} H(E - E_{n_{l10}}), \quad (10.108a)$$

where $\phi(n_y,n_z) = \dfrac{\hbar^2\pi^2}{2m^*}\left[\left(\dfrac{n_y}{d_y}\right)^2 + \left(\dfrac{n_z}{d_z}\right)^2\right]$.

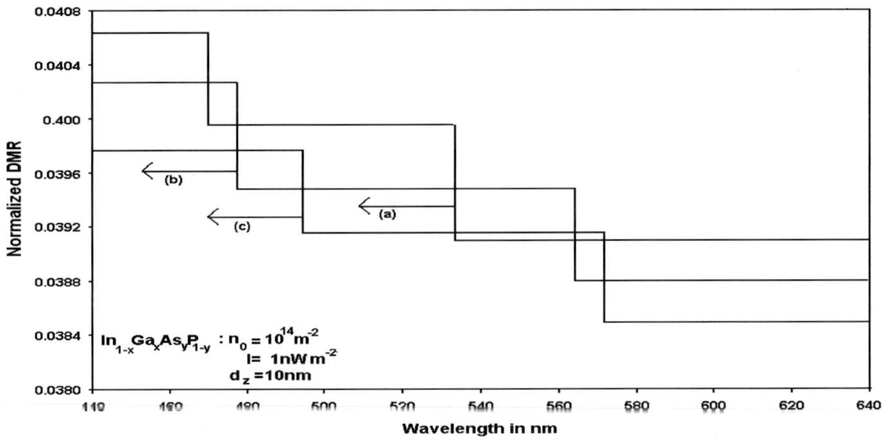

Fig. 10.63. Plot of the normalized 2D DMR as a function of wavelength for ultra-thin films of n-$In_{1-x}Ga_xAs_yP_{1-y}$ in the presence of light waves in which the curves (**a**), (**b**) and (**c**) represent the perturbed three and two band models of Kane and that of the parabolic energy bands respectively

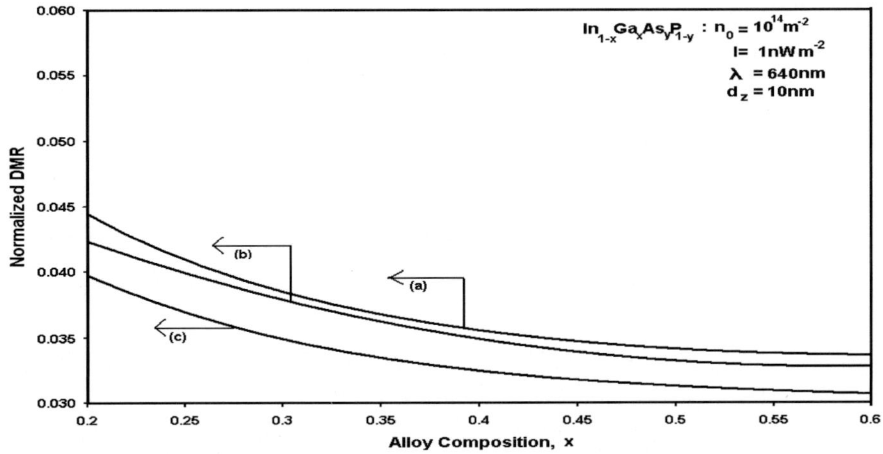

Fig. 10.64. Plot of the normalized 2D DMR as a function of alloy composition for ultra-thin films of n-$Hg_{1-x}Cd_xTe$ in the presence of light waves in which the curves (**a**), (**b**) and (**c**) represent the perturbed three and two band models of Kane and that of the parabolic energy bands respectively

Combining (10.108a), with the Fermi–Dirac occupation probability factor, the one dimensional electron concentration (n_{1D}) can thus be written as

$$n_{1D} = \frac{2g_v}{\pi}\frac{\sqrt{2m^*}}{\hbar}\sum_{n_y=1}^{n_{y\max}}\sum_{n_z=1}^{n_{z\max}}\left[M_{21}\left(n_y, n_z, E_{F1DL}, \lambda\right) + N_{21}\left(n_y, n_z, E_{F1DL}, \lambda\right)\right], \tag{10.108b}$$

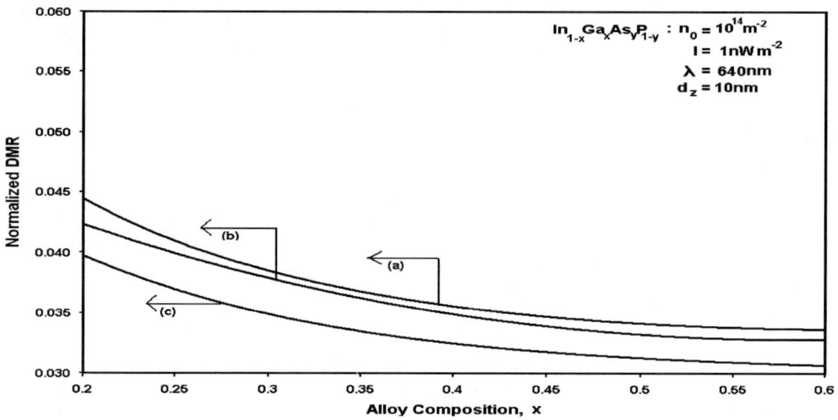

Fig. 10.65. Plot of the normalized 2D DMR as a function of alloy composition for ultra-thin films of n-$In_{1-x}Ga_xAs_yP_{1-y}$ in the presence of light waves in which the curves (**a**), (**b**) and (**c**) represent the perturbed three and two band models of Kane and that of the parabolic energy bands respectively

where $M_{21}(n_y, n_z, E_{\mathrm{F1DL}}, \lambda) \equiv \left[\beta_0(E_{\mathrm{F1DL}}, \lambda) - (\hbar^2/2m^*)\phi(n_y, n_z)\right]^{1/2}$ and $N_{21}(n_y, n_z, E_{\mathrm{F1DL}}, \lambda) \equiv \sum_{r=1}^{s} L(r)\, M_{21}(n_y, n_z, E_{\mathrm{F1DL}}, \lambda)$.

The use of (10.108b) and (1.11) leads to the expression of the 1D-DMR in this case as

$$\frac{D}{\mu} = \frac{1}{|e|}\frac{\sum_{n_y=1}^{n_{y\max}}\sum_{n_z=1}^{n_{z\max}}\left[M_{21}(n_y, n_z, E_{\mathrm{F1DL}}, \lambda) + N_{21}(n_y, n_z, E_{\mathrm{F1DL}}, \lambda)\right]}{\sum_{n_y=1}^{n_{y\max}}\sum_{n_z=1}^{n_{z\max}}\left[\{M_{21}(n_y, n_z, E_{\mathrm{F1DL}}, \lambda)\}' + \{N_{21}(n_y, n_z, E_{\mathrm{F1DL}}, \lambda)\}'\right]}. \tag{10.109}$$

In the absence of external photo-excitation, (10.108a) and (10.109) get simplified to (6.10) and (6.11) of Chap. 6, respectively.

Using (10.42), the expressions for the 1D dispersion relation, the sub-band energies, the EMM, the density-of-states function and the electron concentration for QWs of III–V, ternary and quaternary materials, whose unperturbed band structure is defined by the two band model of Kane, can, respectively, be written in the presence of photo-excitation as

$$\frac{\hbar^2 k_x^2}{2m^*} = \left\{\tau_0(E, \lambda) - (\hbar^2/2m^*)\phi(n_y, n_z)\right\}, \tag{10.110}$$

$$\tau_0(E_{n_{l11}}, \lambda) = \phi(n_y, n_z), \tag{10.111}$$

$$m^*(E_{\mathrm{F1DL}}, n_y, n_z, \lambda) = m^*\{\tau_0(E_{\mathrm{F1DL}}, \lambda)\}', \tag{10.112}$$

$$N_{1\mathrm{D}}(E,\lambda)=\left(\frac{g_{\mathrm{v}}\sqrt{2m^*}}{\pi\hbar}\right)\sum_{n_y=1}^{n_{y\max}}\sum_{n_z=1}^{n_{z\max}} \tag{10.113}$$
$$\{\tau_0(E,\lambda)\}'\left[\tau_0(E,\lambda)-(\hbar^2/2m^*)\phi(n_y,n_z)\right]^{-1/2}H(E-E_{n_{l11}}),$$

$$n_{1\mathrm{D}}=\frac{2g_{\mathrm{v}}}{\pi}\frac{\sqrt{2m^*}}{\hbar}\sum_{n_y=1}^{n_{y\max}}\sum_{n_z=1}^{n_{z\max}} \tag{10.114}$$
$$[M_{22}(n_y,n_z,E_{\mathrm{F1DL}},\lambda)+N_{22}(n_y,n_z,E_{\mathrm{F1DL}},\lambda)],$$

where $M_{22}(n_y,n_z,E_{\mathrm{F1DL}},\lambda)\equiv\left[\tau_0(E_{\mathrm{F1DL}},\lambda)-(\hbar^2/2m^*)\phi(n_y,n_z)\right]^{1/2}$ and $N_{22}(n_y,n_z,E_{\mathrm{F1DL}},\lambda)\equiv\sum_{r=1}^{s}L(r)\,M_{22}(n_y,n_z,E_{\mathrm{F1DL}},\lambda)$.

The use of (10.114) and (1.11) leads to the expression of the DMR in this case as

$$\frac{D}{\mu}=\frac{1}{|e|}\frac{\sum_{n_y=1}^{n_{y\max}}\sum_{n_z=1}^{n_{z\max}}[M_{22}(n_y,n_z,E_{\mathrm{F1DL}},\lambda)+N_{22}(n_y,n_z,E_{\mathrm{F1DL}},\lambda)]}{\sum_{n_y=1}^{n_{y\max}}\sum_{n_z=1}^{n_{z\max}}\left[\{M_{22}(n_y,n_z,E_{\mathrm{F1DL}},\lambda)\}'+\{N_{22}(n_y,n_z,E_{\mathrm{F1DL}},\lambda)\}'\right]}. \tag{10.115}$$

In the absence of external photo-excitation, (10.114) and (10.115) get simplified to (6.15) and (6.16) respectively.

Using (10.43), the expressions for the 1D dispersion relation, the sub-band energies, the EMM, the density of states function and the electron concentration for QWs of III–V, ternary and quaternary materials, whose unperturbed band structure is defined by the parabolic energy bands, can respectively be written in the presence of photo-excitation as

$$\frac{\hbar^2k_x^2}{2m^*}=\left[\rho_0(E,\lambda)-(\hbar^2/2m^*)\phi(n_y,n_z)\right], \tag{10.116}$$

$$\rho_0(E_{n_{l11}},\lambda)=\phi(n_y,n_z), \tag{10.117}$$

$$m^*(E_{\mathrm{F1DL}},n_y,n_z,\lambda)=m^*\{\rho_0(E_{\mathrm{F1DL}},\lambda)\}', \tag{10.118}$$

$$N_{1\mathrm{D}}(E,\lambda)=\left(\frac{g_{\mathrm{v}}\sqrt{2m^*}}{\pi\hbar}\right)\sum_{n_y=1}^{n_{y\max}}\sum_{n_z=1}^{n_{z\max}} \tag{10.119}$$
$$\{\rho_0(E,\lambda)\}'\left[\rho_0(E,\lambda)-(\hbar^2/2m^*)\phi(n_y,n_z)\right]^{-1/2}H(E-E_{n_{l11}}),$$

$$n_{1\mathrm{D}}=\frac{2g_{\mathrm{v}}\sqrt{2m^*}}{\pi\hbar}\sum_{n_y=1}^{n_{y\max}}\sum_{n_z=1}^{n_{z\max}} \tag{10.120}$$
$$[M_{23}(n_y,n_z,E_{\mathrm{F1DL}},\lambda)+N_{23}(n_y,n_z,E_{\mathrm{F1DL}},\lambda)],$$

where $M_{23}(n_y, n_z, E_{\mathrm{F1DL}}, \lambda) \equiv \left[\rho_0(E_{\mathrm{F1DL}}, \lambda) - (\hbar^2/2m^*)\phi(n_y, n_z)\right]^{1/2}$ and $N_{23}(n_y, n_z, E_{\mathrm{F1DL}}, \lambda) \equiv \sum_{r=1}^{s} L(r)\, M_{23}(n_y, n_z, E_{\mathrm{F1DL}}, \lambda)$.

The use of (10.120) and (1.11) leads to the expression of the DMR in this case as

$$\frac{D}{\mu} = \frac{1}{|e|} \frac{\sum_{n_y=1}^{n_{y\max}} \sum_{n_z=1}^{n_{z\max}} \left[M_{23}(n_y, n_z, E_{\mathrm{F1DL}}, \lambda) + N_{23}(n_y, n_z, E_{\mathrm{F1DL}}, \lambda)\right]}{\sum_{n_y=1}^{n_{y\max}} \sum_{n_z=1}^{n_{z\max}} \left[\{M_{23}(n_y, n_z, E_{\mathrm{F1DL}}, \lambda)\}' + \{N_{23}(n_y, n_z, E_{\mathrm{F1DL}}, \lambda)\}'\right]}. \tag{10.121}$$

In the absence of external photo-excitation, (10.120) and (10.121) get simplified to (6.19) and (6.20), respectively.

10.13 Result and Discussions

Using (10.108b), (10.109), (10.114), (10.115) and (10.120), (10.121), in Figs. 10.66–10.69, the DMR has been plotted as function of film thickness in QWs of n-InAs, n-InSb, n-$Hg_{1-x}Cd_xTe$ and n-$In_{1-x}Ga_xAs_yP_{1-y}$ lattice matched to InP in the presence of an external photo-excitation respectively. The curves (a)–(c) chronologically exhibit the DMR in QWs of the aforementioned materials whose unperturbed conduction electrons obey the three

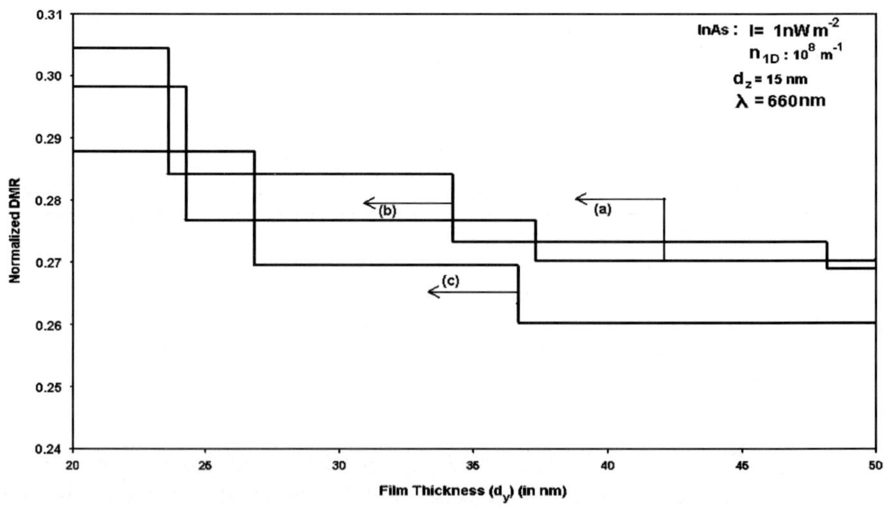

Fig. 10.66. Plot of the normalized 1D DMR as a function of film thickness for QWs of n-InAs in the presence of light waves in which the curves (**a**), (**b**) and (**c**) represent the three and two band models of Kane and that of the parabolic energy bands respectively

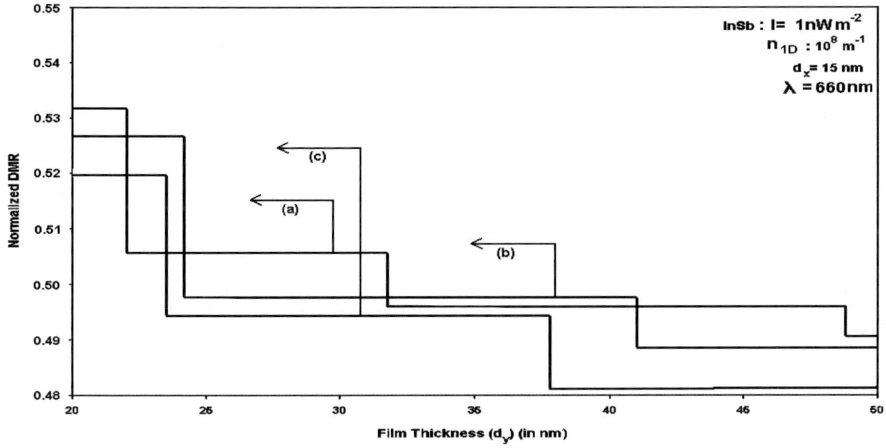

Fig. 10.67. Plot of the normalized 1D DMR as a function of film thickness for QWs of n-InSb in the presence of light waves in which the curves (**a**), (**b**) and (**c**) represent the perturbed three and two band models of Kane and that of the parabolic energy bands respectively

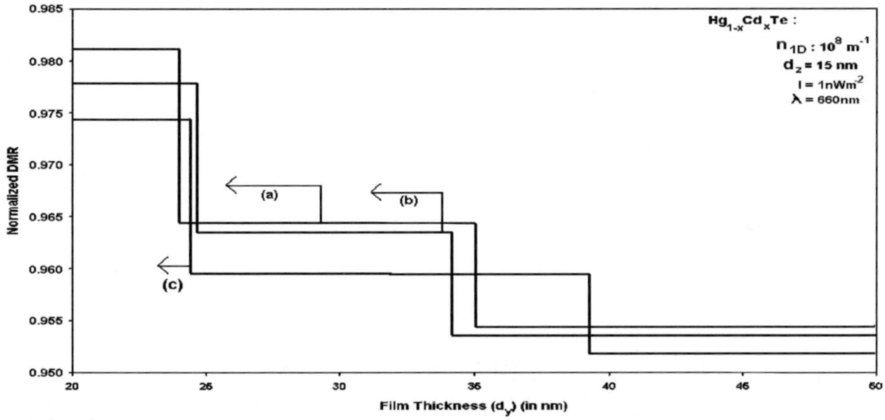

Fig. 10.68. Plot of the normalized 1D DMR as a function of film thickness for QWs of n-$Hg_{1-x}Cd_xTe$ in the presence of light waves in which the curves (**a**), (**b**) and (**c**) represent the perturbed three and two band models of Kane and that of the parabolic energy bands respectively

and the two band model of Kane together with the parabolic energy band. It can be observed from the said figures, that the DMR decreases as the film thickness is increased. Figures 10.70–10.73 exhibit the plot of the DMR vs. the electron concentration per unit length for the aforementioned cases. It should again be noted that the rate of change of the DMR against the respective variable functions totally depends on the energy spectrum constants of the

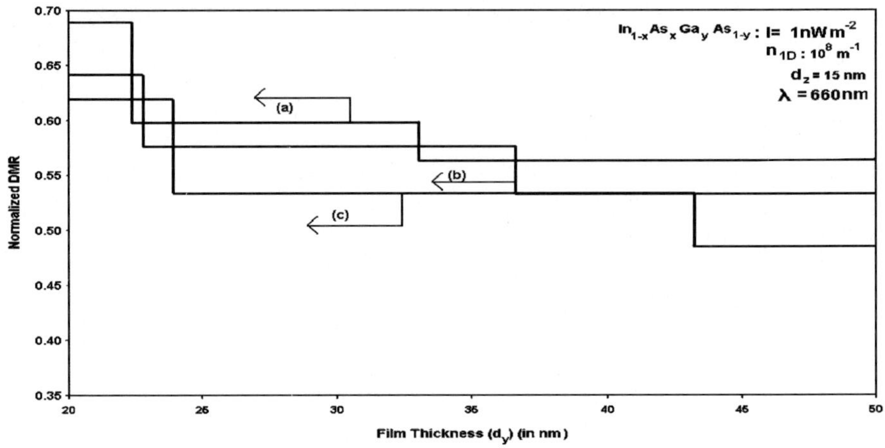

Fig. 10.69. Plot of the normalized 1D DMR as a functions of film thickness for QWs of n-$In_{1-x}Ga_xAs_yP_{1-y}$ in the presence of light waves in which the curves (**a**), (**b**) and (**c**) represent the perturbed three and two band models of Kane and that of the parabolic energy bands respectively

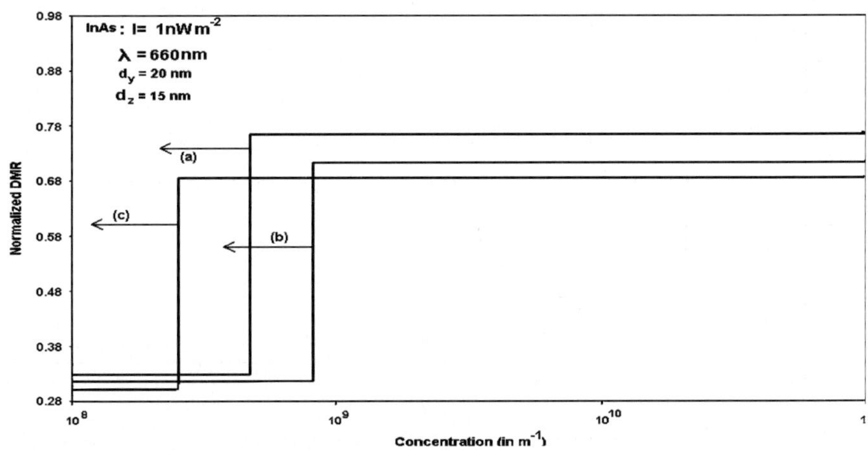

Fig. 10.70. Plot of the normalized 1D DMR as a function of electron concentration per unit length for QWs of n-InAs in the presence of light waves in which the curves (**a**), (**b**) and (**c**) represent the perturbed three and two band models of Kane and that of the parabolic energy bands respectively

respective materials. The variations of the DMR as functions of wavelength in the visible region and intensity have been plotted in Figs. 10.74–10.77 respectively. The Figs. 10.78 and 10.79 exibit the same for QWs of ternary and quaternary materials as function of alloy composition. The rest of the discussion of Sect. 6.3 is also applicable for Sect. 10.13.

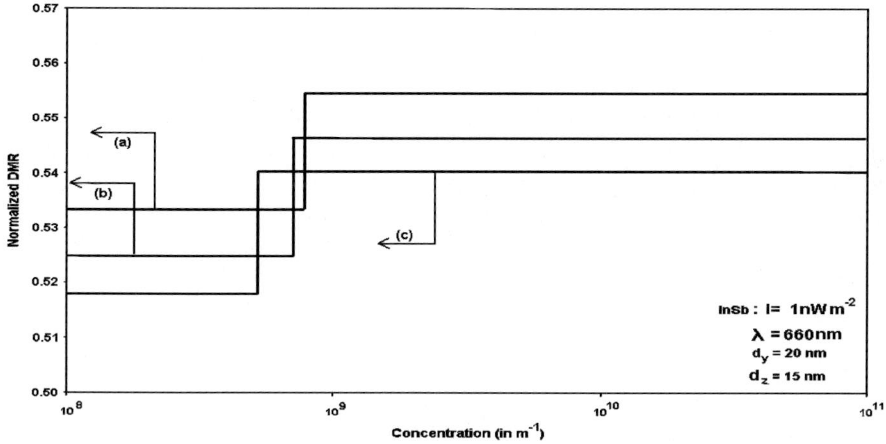

Fig. 10.71. Plot of the normalized 1D DMR as a function of electron concentration per unit length for QWs of n-InSb in the presence of light waves in which the curves (**a**), (**b**) and (**c**) represent the perturbed three and two band models of Kane and that of the parabolic energy bands respectively

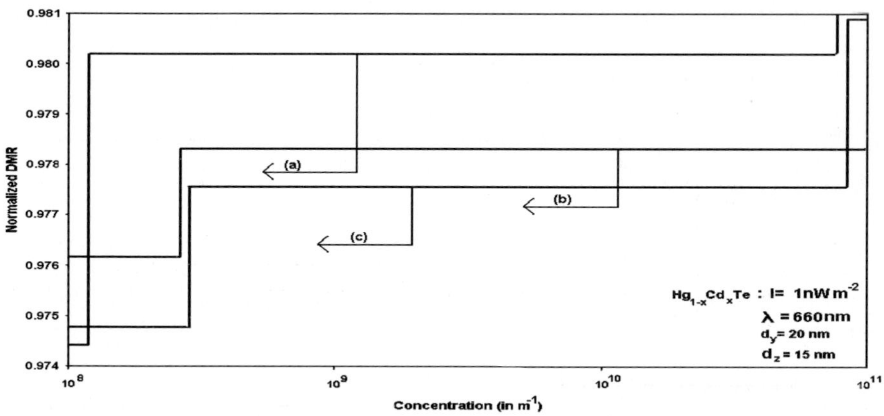

Fig. 10.72. Plot of the normalized 1D DMR as a function of electron concentration per unit length for QWs of n-$Hg_{1-x}Cd_xTe$ in the presence of light waves in which the curves (**a**), (**b**) and (**c**) represent the perturbed three and two band models of Kane and that of the parabolic energy bands respectively

10.14 Summary

In this chapter, we have presented the DMR in III–V, ternary and quaternary materials in the presence of light waves on the basis of newly formulated electron dispersion laws, whose unperturbed conduction electrons obey the three and two band model of Kane togetherwith the parabolic energy bands.

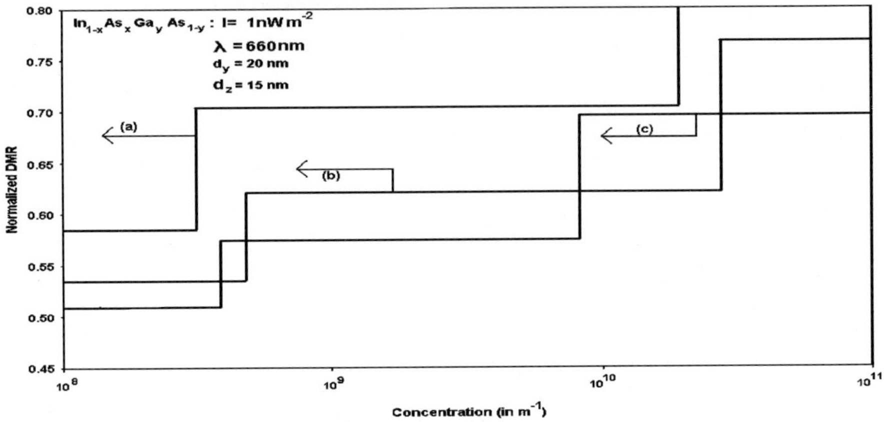

Fig. 10.73. Plot of the normalized 1D DMR as a function of electron concentration per unit length for QWs of n-$In_{1-x}Ga_xAs_yP_{1-y}$ lattice matched to InP in the presence of light waves in which the curves (**a**), (**b**) and (**c**) represent the perturbed three and two band models of Kane and that of the parabolic energy bands respectively

The influence of quantizing magnetic field and the cross-fields configuration on the DMR have been studied in the presence of an external photo-excitation. The influence of reduced dimensions (ultra-thin films and QWs) have further been investigated. In all the cases, the new electron energy spectra have been formulated for the purpose of investigation of the respective DMR. Under certain limiting conditions, all the results of the DMRs for all the materials leads to the well known expression of the Einstein relation for degenerate materials as investigated for the first time by Landsberg. For the purpose of condensed presentation, the specific electron statistics related to a particular energy dispersion law for a specific material and the corresponding DMR have been presented in Table 10.1.

10.15 Open Research Problem

R.10.1 Investigate all the appropriate systems and the corresponding open research problems from Chap. 2 up-to Chap. 9, in the presence of light waves.

Allied Research Problem

R.10.2 Investigate all allied research problems from Chap. 2 up-to Chap. 9, in the presence of light waves.

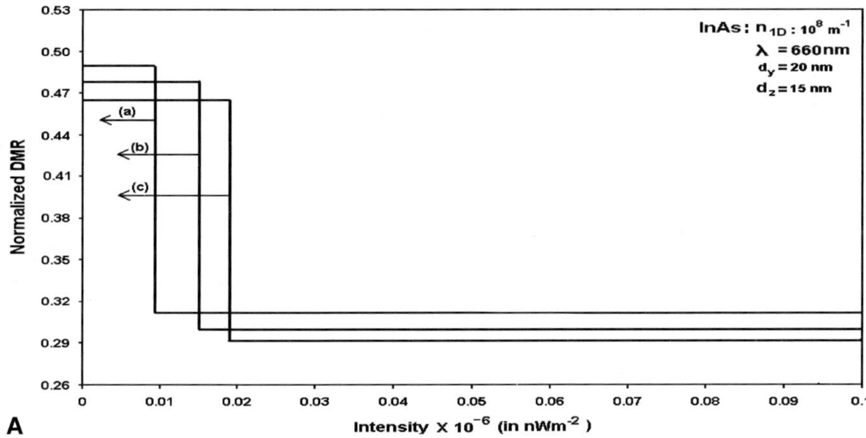

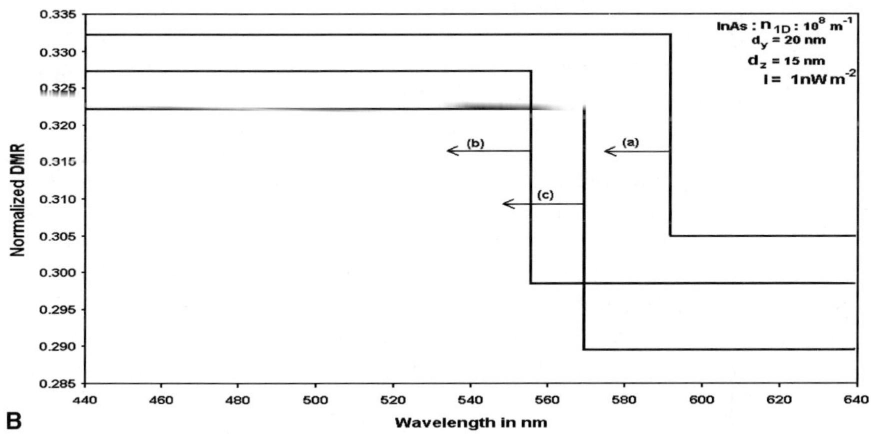

Fig. 10.74. Plot of the normalized 1D DMR as a function of light intensity and wavelength as shown in Figs. A and B for QWs of n-InAs in the presence of light waves in which the curves (**a**), (**b**) and (**c**) represent the perturbed three and two band models of Kane and that of the parabolic energy bands respectively

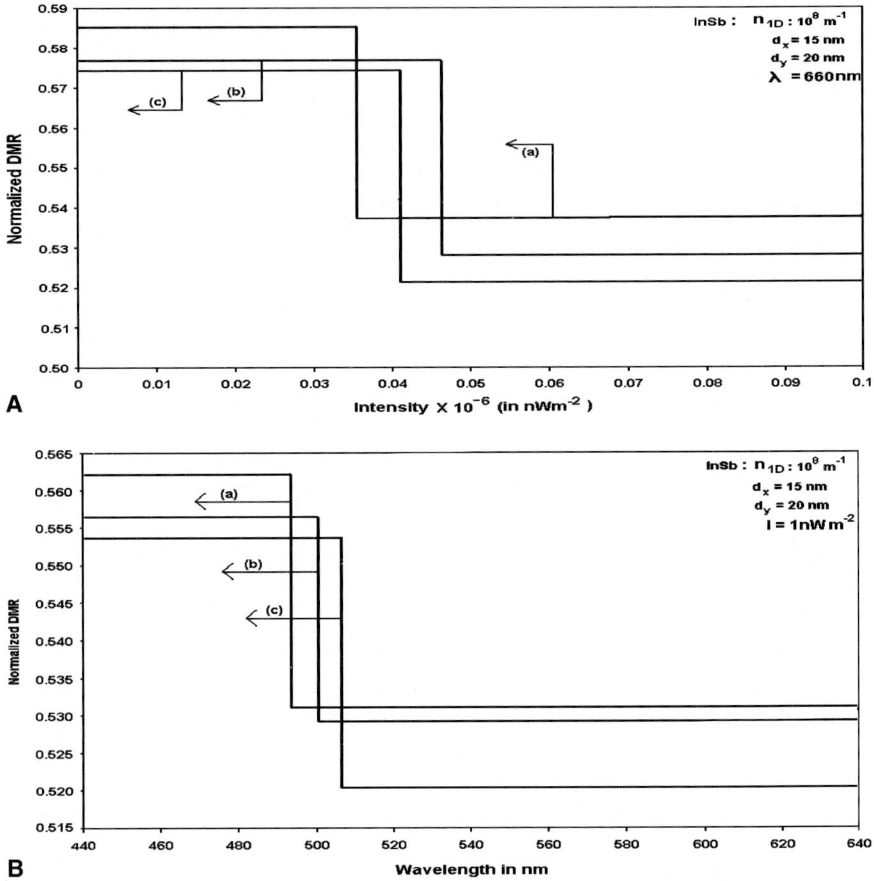

Fig. 10.75. Plot of the normalized 1D DMR as a function of light intensity and wavelength as shown in Figs. A and B for QWs of n-InSb in the presence of light waves in which the curves (**a**), (**b**) and (**c**) represent the perturbed three and two band models of Kane and that of the parabolic energy bands respectively

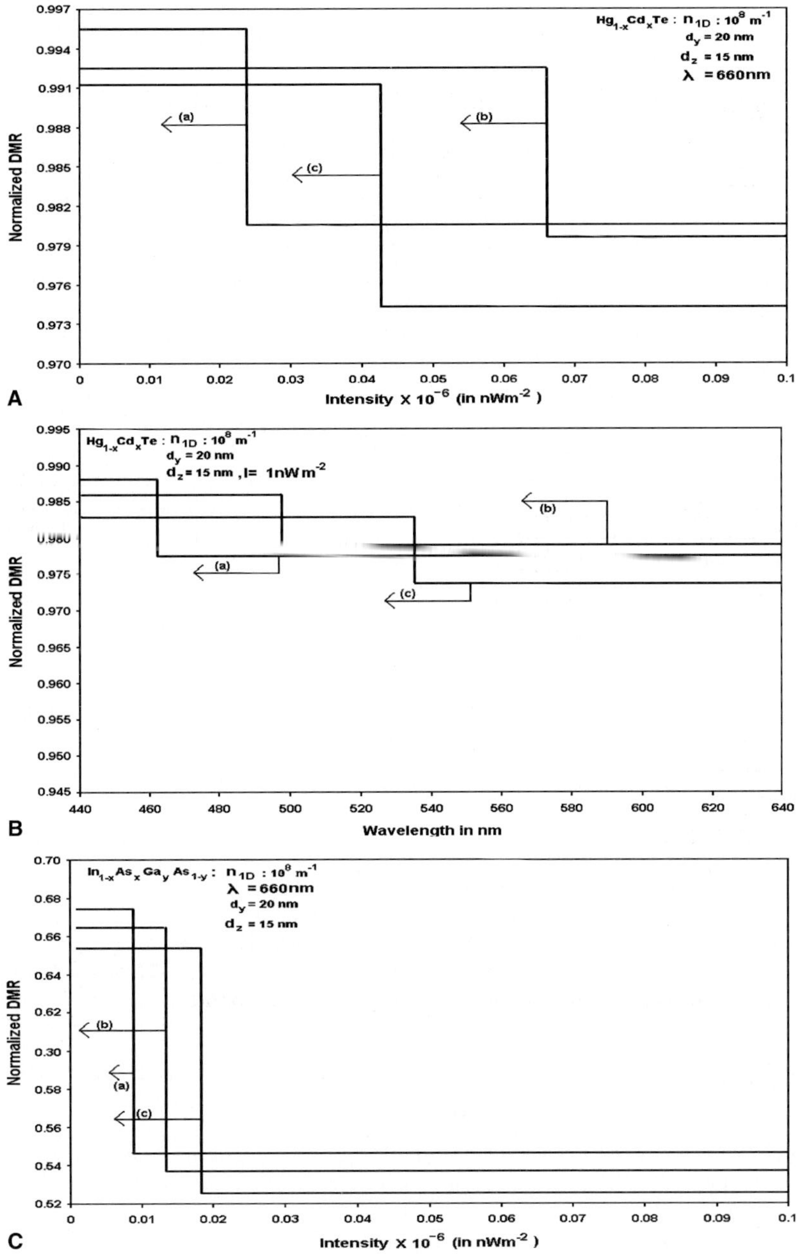

Fig. 10.76. Plot of the normalized 1D DMR as a function of light intensity and wavelength as shown in Figs. A and B respectively for QWs of n-$Hg_{1-x}Cd_xTe$ in the presence of light waves in which the curves (**a**), (**b**) and (**c**) represent the perturbed three and two band models of Kane and that of the parabolic energy bands respectively. The Fig. C exhibits the plot of the normalized 1D DMR as a function of light intensity for QWs of n-$In_{1-x}Ga_xAs_yP_{1-y}$ lattice matched to InP in which the curves (**a**), (**b**) and (**c**) represent the perturbed three and two band models of Kane and that of the parabolic energy bands respectively

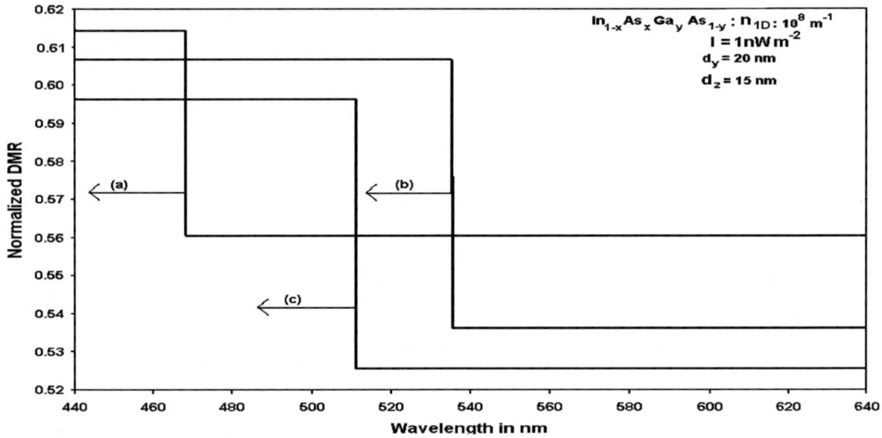

Fig. 10.77. Plot of the normalized 1D DMR as a function of wavelength for QWs of n-$In_{1-x}Ga_xAs_yP_{1-y}$ lattice matched to InP in the presence of light waves in which the curves (**a**), (**b**) and (**c**) represent the perturbed three and two band models of Kane and that of the parabolic energy bands respectively

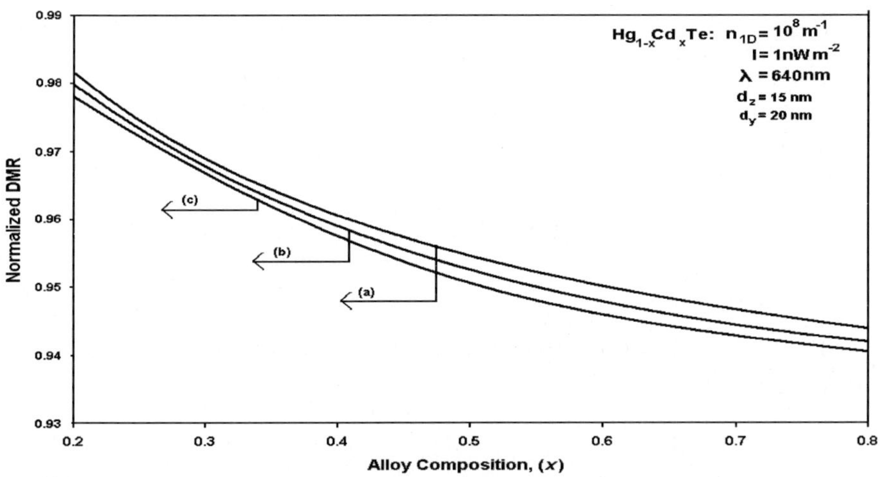

Fig. 10.78. Plot of the normalized 1D DMR as a function of alloy composition for QWs of n-$Hg_{1-x}Cd_xTe$ in the presence of light waves in which the curves (**a**), (**b**) and (**c**) represent the perturbed three and two band models of Kane and that of the parabolic energy bands respectively

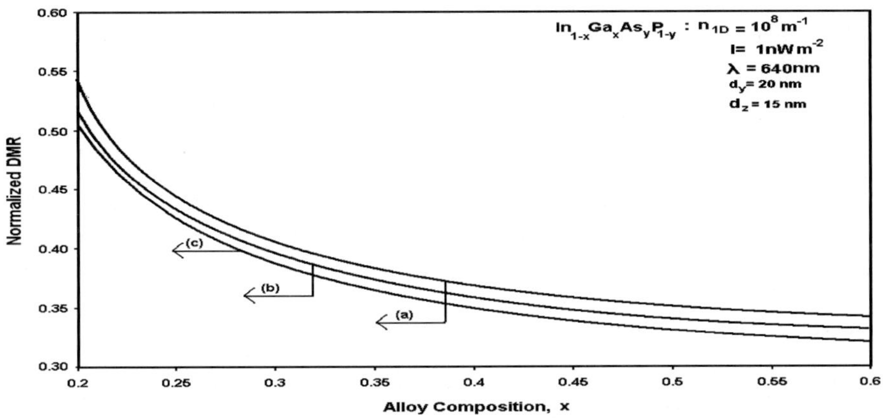

Fig. 10.79. Plot of the normalized 1D DMR as a function of alloy composition for QWs of n-$In_{1-x}Ga_xAs_yP_{1-y}$ lattice matched to InP in the presence of light waves in which the curves (**a**), (**b**) and (**c**) represent the perturbed three and two band models of Kane and that of the parabolic energy bands respectively

References

1. P.K. Basu, *Theory of Optical Processes in Semiconductors, Bulk and Microstructures* (Oxford University Press, Oxford, 1997)
2. K.P. Ghatak, S. Bhattacharya, J. Appl. Phys. **102**, 073704 (2007)
3. K.P. Ghatak, S. Bhattacharya, S.K. Biswas, A. De, A.K. Dasgupta, Phys. Scr. **75**, 820 (2007)
4. K.P. Ghatak, S. Bhattacharya, S. Bhowmik, R. Benedictus and S. Chowdhury, J. Appl. Phys. **103**, 094314 (2008)
5. K. Seeger, *Semiconductor Physics*, 6th edn. (Springer, Berlin, 1997)
6. B.R. Nag, *Electron Transport in Compound Semiconductors* (Springer, Berlin, 1980)
7. B.R. Nag, *Physics of Quantum Well Devices* (Kluwer, Netherlands, 2000)
8. R.K. Pathria, *Statistical Mechanics,* 2nd edn. (Butterworth-Heinmann, Oxford, 1996)

Table 10.1. The carrier statistics and the Einstein relation in presence of light waves in III–V, ternary and quaternary materials

External conditions	The carrier statistics	The Einstein relation
Bulk	In accordance with the dispersion relations whose unperturbed conduction electrons obey three and the two band models of Kane together with parabolic energy bands	The corresponding DMRs
	$n_0 = (3\pi^2)^{-1} g_v \left(\frac{2m^*}{\hbar^2}\right)^{3/2} [M_{10}(E_{F_L},\lambda) + N_{10}(E_{F_L},\lambda)]$ (10.46)	$\frac{D}{\mu} = \frac{1}{\vert e\vert} \frac{[M_{10}(E_{F_L},\lambda) + N_{10}(E_{F_L},\lambda)]}{[\{M_{10}(E_{F_L},\lambda)\}' + \{N_{10}(E_{F_L},\lambda)\}']}$ (10.47)
	$n_0 = (3\pi^2)^{-1} g_v \left(\frac{2m^*}{\hbar^2}\right)^{3/2} [M_{11}(E_{F_L},\lambda) + N_{11}(E_{F_L},\lambda)]$ (10.48)	$\frac{D}{\mu} = \frac{1}{\vert e\vert} \frac{[M_{11}(E_{F_L},\lambda) + N_{11}(E_{F_L},\lambda)]}{[\{M_{11}(E_{F_L},\lambda)\}' + \{N_{11}(E_{F_L},\lambda)\}']}$ (10.49)
	$n_0 = (3\pi^2)^{-1} g_v \left(\frac{2m^*}{\hbar^2}\right)^{3/2} [M_{12}(E_{F_L},\lambda) + N_{12}(E_{F_L},\lambda)]$ (10.50)	$\frac{D}{\mu} = \frac{1}{\vert e\vert} \frac{[M_{12}(E_{F_L},\lambda) + N_{12}(E_{F_L},\lambda)]}{[\{M_{12}(E_{F_L},\lambda)\}' + \{N_{12}(E_{F_L},\lambda)\}']}$ (10.51)
Magnetic field	$n_0 = \frac{g_v \vert e\vert B\sqrt{2m^*}}{\pi^2\hbar^2} \sum_{n=0}^{n_{max}} [M_{13}(E_{F_{BL}},B,\lambda) + N_{13}(E_{F_{BL}},B,\lambda)]$ (10.56)	$\frac{D}{\mu} = \frac{1}{\vert e\vert} \frac{\sum_{n=0}^{n_{max}} [M_{13}(E_{F_{BL}},B,\lambda) + N_{13}(E_{F_{BL}},B,\lambda)]}{\sum_{n=0}^{n_{max}} [\{M_{13}(E_{F_{BL}},B,\lambda)\}' + \{N_{13}(E_{F_{BL}},B,\lambda)\}']}$ (10.57)

	$n_0 = \frac{g_{\mathrm{v}} \lvert e \rvert B\sqrt{2m^*}}{\pi^2\hbar^2} \sum_{n=0}^{n_{\max}} [M_{14}(E_{\mathrm{F_{BL}}}, B, \lambda) + N_{14}(E_{\mathrm{F_{BL}}}, B, \lambda)]$ (10.62)	$\frac{D}{\mu} = \frac{1}{\lvert e \rvert} \frac{\sum_{n=0}^{n_{\max}} [M_{14}(E_{\mathrm{F_{BL}}}, B, \lambda) + N_{14}(E_{\mathrm{F_{BL}}}, B, \lambda)]}{\sum_{n=0}^{n_{\max}} [\{M_{14}(E_{\mathrm{F_{BL}}}, B, \lambda)\}' + \{N_{14}(E_{\mathrm{F_{BL}}}, B, \lambda)\}']}$ (10.63)
	$n_0 = \frac{g_{\mathrm{v}} \lvert e \rvert B\sqrt{2m^*}}{\pi^2\hbar^2} \sum_{n=0}^{n_{\max}} [M_{15}(E_{\mathrm{F_{BL}}}, B, \lambda) + N_{15}(E_{\mathrm{F_{BL}}}, B, \lambda)]$ (10.68)	$\frac{D}{\mu} = \frac{1}{\lvert e \rvert} \frac{\sum_{n=0}^{n_{\max}} [M_{15}(E_{\mathrm{F_{BL}}}, B, \lambda) + N_{15}(E_{\mathrm{F_{BL}}}, B, \lambda)]}{\sum_{n=0}^{n_{\max}} [\{M_{15}(E_{\mathrm{F_{BL}}}, B, \lambda)\}' + \{N_{15}(E_{\mathrm{F_{BL}}}, B, \lambda)\}']}$ (10.69)
Cross fields	$n_0 = \frac{2g_{\mathrm{v}} B\sqrt{2m^*}}{3L_x\pi^2\hbar^2 E_0} \sum_{n=0}^{n_{\max}} [M_{16}(n, E_{\mathrm{F_{BL}}}, \lambda) + N_{16}(n, E_{\mathrm{F_{BL}}}, \lambda)]$ (10.74)	$\frac{D}{\mu} = \frac{1}{\lvert e \rvert} \left[\frac{\sum_{n=0}^{n_{\max}} [M_{16}(n, E_{\mathrm{F_{BL}}}, \lambda) + N_{16}(n, E_{\mathrm{F_{BL}}}, \lambda)]}{\sum_{n=0}^{n_{\max}} [\{M_{16}(n, E_{\mathrm{F_{BL}}}, \lambda)\}' + \{N_{16}(n, E_{\mathrm{F_{BL}}}, \lambda)\}']} \right]$ (10.75)
	$n_0 = \frac{2g_{\mathrm{v}} B\sqrt{2m^*}}{3L_x\pi^2\hbar^2 E_0} \sum_{n=0}^{n_{\max}} [M_{17}(n, E_{\mathrm{F_{BL}}}, \lambda) + N_{17}(n, E_{\mathrm{F_{BL}}}, \lambda)]$ (10.80)	$\frac{D}{\mu} = \frac{1}{\lvert e \rvert} \left[\frac{\sum_{n=0}^{n_{\max}} [M_{17}(n, E_{\mathrm{F_{BL}}}, \lambda) + N_{17}(n, E_{\mathrm{F_{BL}}}, \lambda)]}{\sum_{n=0}^{n_{\max}} [\{M_{17}(n, E_{\mathrm{F_{BL}}}, \lambda)\}' + \{N_{17}(n, E_{\mathrm{F_{BL}}}, \lambda)\}']} \right]$ (10.81)

(Continued)

Table 10.1. *Continued*

External conditions	The carrier statistics	The Einstein relation
	$n_0 = \frac{2g_v B\sqrt{2m^*}}{3L_x \pi^2 \hbar^2 E_0} \sum_{n=0}^{n_{\max}} [\bar{M}_{17}(n, E_{F_{BL}}, \lambda) + \bar{N}_{17}(n, E_{F_{BL}}, \lambda)]$ (10.86)	$\frac{D}{\mu} = \frac{1}{\lvert e \rvert}\left[\frac{\sum_{n=0}^{n_{\max}} [\bar{M}_{17}(n, E_{F_{BL}}, \lambda) + \bar{N}_{17}(n, E_{F_{BL}}, \lambda)]}{\sum_{n=0}^{n_{\max}} \left[\{\bar{M}_{17}(n, E_{F_{BL}}, \lambda)\}' + \{\bar{N}_{17}(n, E_{F_{BL}}, \lambda)\}'\right]}\right]$ (10.87)
Ultrathin films	$n_{2D} = \frac{m^* g_v}{\pi \hbar^2} \sum_{n_z=1}^{n_{z\max}} [M_{18}(n_z, E_{F2DL}, \lambda) + N_{18}(n_z, E_{F2DL}, \lambda)]$ (10.92)	$\frac{D}{\mu} = \frac{1}{\lvert e \rvert} \frac{\sum_{n_z=1}^{n_{z\max}} [M_{18}(n_z, E_{F2DL}, \lambda) + N_{18}(n_z, E_{F2DL}, \lambda)]}{\sum_{n_z=1}^{n_{z\max}} [\{M_{18}(n_z, E_{F2DL}, \lambda)\}' + \{N_{18}(n_z, E_{F2DL}, \lambda)\}']}$ (10.93)
	$n_{2D} = \frac{m^* g_v}{\pi \hbar^2} \sum_{n_z=1}^{n_{z\max}} [M_{19}(n_z, E_{F2DL}, \lambda) + N_{19}(n_z, E_{F2DL}, \lambda)]$ (10.98)	$\frac{D}{\mu} = \frac{1}{\lvert e \rvert} \frac{\sum_{n_z=1}^{n_{z\max}} [M_{19}(n_z, E_{F2DL}, \lambda) + N_{19}(n_z, E_{F2DL}, \lambda)]}{\sum_{n_z=1}^{n_{z\max}} [\{M_{19}(n_z, E_{F2DL}, \lambda)\}' + \{N_{19}(n_z, E_{F2DL}, \lambda)\}']}$ (10.99)

	$$n_{2D} = \frac{m^* g_v}{\pi \hbar^2} \sum_{n_z=1}^{n_{z\max}} [M_{20}(n_z, E_{F2DL}, \lambda) + N_{20}(n_z, E_{F2DL}, \lambda)] \quad (10.103)$$	$$\frac{D}{\mu} = \frac{1}{\lvert e \rvert} \frac{\sum_{n_z=1}^{n_{z\max}} [M_{20}(n_z, E_{F2DL}, \lambda) + N_{20}(n_z, E_{F2DL}, \lambda)]}{\sum_{n_z=1}^{n_{z\max}} [\{M_{20}(n_z, E_{F2DL}, \lambda)\}' + \{N_{20}(n_z, E_{F2DL}, \lambda)\}']} \quad (10.104)$$
Quantum wires	$$n_{1D} = \frac{2g_v}{\pi} \frac{\sqrt{2m^*}}{\hbar} \sum_{n_y=1}^{n_{y\max}} \sum_{n_z=1}^{n_{z\max}} [M_{21}(n_y, n_z, E_{F1DL}, \lambda) + N_{21}(n_y, n_z, E_{F1DL}, \lambda)] \quad (10.108)$$	$$\frac{D}{\mu} = \frac{1}{\lvert e \rvert} \frac{\sum_{n_y=1}^{n_{y\max}} \sum_{n_z=1}^{n_{z\max}} [M_{21}(n_y, n_z, E_{F1DL}, \lambda) + N_{21}(n_y, n_z, E_{F1DL}, \lambda)]}{\sum_{n_y=1}^{n_{y\max}} \sum_{n_z=1}^{n_{z\max}} [\{M_{21}(n_y, n_z, E_{F1DL}, \lambda)\}' + \{N_{21}(n_y, n_z, E_{F1DL}, \lambda)\}']} \quad (10.109)$$

(*Continued*)

Table 10.1. *Continued*

External conditions	The carrier statistics	The Einstein relation
	$n_{1\mathrm{D}} = \frac{2g_{\mathrm{v}}}{\pi}\frac{\sqrt{2m^*}}{\hbar}\sum_{n_y=1}^{n_{y_{\max}}}\sum_{n_z=1}^{n_{z_{\max}}}[M_{22}(n_y,n_z,E_{\mathrm{F1DL}},\lambda)+N_{22}(n_y,n_z,E_{\mathrm{F1DL}},\lambda)]$ (10.114)	$\frac{D}{\mu} = \frac{1}{\lvert e\rvert}\frac{\sum_{n_y=1}^{n_{y_{\max}}}\sum_{n_z=1}^{n_{z_{\max}}}[M_{22}(n_y,n_z,E_{\mathrm{F1DL}},\lambda)+N_{22}(n_y,n_z,E_{\mathrm{F1DL}},\lambda)]}{\sum_{n_y=1}^{n_{y_{\max}}}\sum_{n_z=1}^{n_{z_{\max}}}[\{M_{22}(n_y,n_z,E_{\mathrm{F1DL}},\lambda)\}'+\{N_{22}(n_y,n_z,E_{\mathrm{F1DL}},\lambda)\}']}$ (10.115)
	$n_{1\mathrm{D}} = \frac{2g_{\mathrm{v}}\sqrt{2m^*}}{\pi\hbar}\sum_{n_y=1}^{n_{y_{\max}}}\sum_{n_z=1}^{n_{z_{\max}}}[M_{23}(n_y,n_z,E_{\mathrm{F1DL}},\lambda)+N_{23}(n_y,n_z,E_{\mathrm{F1DL}},\lambda)]$ (10.120)	$\frac{D}{\mu} = \frac{1}{\lvert e\rvert}\frac{\sum_{n_y=1}^{n_{y_{\max}}}\sum_{n_z=1}^{n_{z_{\max}}}[M_{23}(n_y,n_z,E_{\mathrm{F1DL}},\lambda)+N_{23}(n_y,n_z,E_{\mathrm{F1DL}},\lambda)]}{\sum_{n_y=1}^{n_{y_{\max}}}\sum_{n_z=1}^{n_{z_{\max}}}[\{M_{23}(n_y,n_z,E_{\mathrm{F1DL}},\lambda)\}'+\{N_{23}(n_y,n_z,E_{\mathrm{F1DL}},\lambda)\}']}$ (10.121)

11

The Einstein Relation in Heavily Doped Compound Semiconductors

11.1 Introduction

It is well known that the band tails are being formed in the forbidden zone of heavily doped semiconductors and can be explained by the overlapping of the impurity band with the conduction and valence bands [1]. Kane [2] and Bonch Bruevich [3] have independently derived the theory of band tailing for semiconductors having unperturbed parabolic energy bands. Kane's model [2] was used to explain the experimental results on tunneling [4] and the optical absorption edges [5, 6] in this context. Halperin and Lax [7] developed a model for band tailing applicable only to the deep tailing states. Although Kane's concept is often used in the literature for the investigation of band tailing [8, 9], it may be noted that this model [2, 10] suffers from serious assumptions in the sense that the local impurity potential is assumed to be small and slowly varying in space coordinates [9]. In this respect, the local impurity potential may be assumed to be a constant. In order to avoid these approximations, we have developed in this chapter the electron energy spectra for heavily doped semiconductors for studying the DMR based on the concept of the variation of the kinetic energy [1, 9] of the electron with the local point in space coordinates. This kinetic energy is then averaged over the entire region of variation using a Gaussian type potential energy. On the basis of the E–k dispersion relation, we have obtained the electron statistics for different heavily doped materials for the purpose of numerical computation of the respective DMRs. It may be noted that a more general treatment of many-body theory for the density-of-states of heavily doped semiconductor merges with one-electron theory under macroscopic conditions [1]. Also, the experimental results for the Fermi energy and others are the average effect of this macroscopic case. So, the present treatment of the one-electron system is more applicable to the experimental point of view and it is also easy to understand the overall effect in such a case [11]. In a heavily doped semiconductors, each impurity atom is surrounded by the electrons, assuming a regular distribution of atoms, and it is screened independently [8, 10, 12]. The interaction energy between electrons and impurities is known as the impurity screening potential. This energy is determined by the inter-impurity distance and the screening radius, which is known as the screening length. The screening radius changes with the electron concentration and the effective mass. Furthermore, these entities are important for heavily doped materials in characterizing the semiconductor properties [13, 14] and the devices [8, 15]. The works on Fermi energy and the screening length in an n-type GaAs have already been initiated in the literature [16–18], based on Kane's model. Incidentally, the limitations of Kane's model [9], as mentioned above, are also present in their studies.

At this point, it may be noted that many band tail models are proposed using the Gaussian distribution of the impurity potential variation [2, 9]. In this chapter, we have used the Gaussian band tails to obtain the exact E–k dispersion relations for heavily doped tetragonal, III–V, II–VI, IV–VI and stressed Kane type compounds. Our method is not at all related to the density-of-states (DOS) technique as used in the aforementioned works. From the electron energy spectrum, one can obtain the DOS but the DOS technique, as used in the literature cannot provide the E–k dispersion relation. Therefore, our study is more fundamental than those in the existing literature, because the Boltzmann transport equation, which controls the study of the charge transport properties of the semiconductor devices, can be solved if and only if the E–k dispersion relation is known. We wish to note that the Gaussian function for the impurity potential distribution has been used by many authors. It has been widely used since 1963 when Kane first proposed it. We will also use the Gaussian distribution for the present study.

In Sect. 11.2.1, on the theoretical background, the Einstein relation in heavily doped tetragonal materials has been investigated. Section 11.2.2 contains the results for heavily doped III–V, ternary and quaternary compounds whose undoped conduction electrons obey the three and the two band models of Kane together with parabolic energy bands and they form the special cases of Sect. 11.2.1. Sections 11.2.3–11.2.5 contain the study of the DMR for heavily doped II–VI, IV–VI and stressed Kane type semiconductors respectively. The last Sect 11.3 contains the results and discussion of this chapter.

11.2 Theoretical Background

11.2.1 Study of the Einstein Relation in Heavily Doped Tetragonal Materials Forming Gaussian Band Tails

The generalized unperturbed electron energy spectrum for the bulk specimens of the tetragonal materials in the absence of any doping can be expressed following (2.2) as

$$\frac{\hbar^2 k_z^2}{2m_{||}^*} + \left(\frac{b_{||}}{b_\perp}\frac{c_\perp}{c_{||}}\right)\frac{\hbar^2 k_s^2}{2m_\perp^*} = \left\{\frac{E(\alpha E+1)(b_{||}E+1)}{(c_{||}E+1)} + \frac{\alpha b_{||}}{c_{||}}\left[\delta E + \frac{2}{9}\left(\Delta_{||}^2 - \Delta_\perp^2\right)\right]\right.$$
$$\left. - \left(\frac{2}{9}\right)\frac{\alpha b_{||}}{c_{||}}\frac{\left(\Delta_{||}^2 - \Delta_\perp^2\right)}{(c_{||}E+1)}\right\}$$
$$- \left(\frac{\hbar^2 k_s^2}{2m_\perp^*}\right)\left\{\left(\frac{b_{||}}{b_\perp}\frac{c_\perp}{c_{||}}\right)\left[\left(\frac{\delta}{2} + \frac{\Delta_{||}^2 - \Delta_\perp^2}{6\Delta_{||}}\right)\frac{\alpha}{\alpha E+1}\right.\right.$$
$$\left.\left. + \left(\frac{\delta}{2} - \left\{\frac{\Delta_{||}^2 - \Delta_\perp^2}{6\Delta_{||}}\right\}\right)\frac{c_{||}}{c_{||}E+1}\right]\right\}, \tag{11.1}$$

where $b_{||} \equiv 1/\left(E_{\mathrm{g}} + \Delta_{||}\right)$, $c_\perp \equiv 1/\left(E_{\mathrm{g}} + \frac{2}{3}\Delta_\perp\right)$, $b_\perp \equiv 1/\left(E_{\mathrm{g}} + \Delta_\perp\right)$, $c_{||} \equiv 1/\left(E_{\mathrm{g}} + \frac{2}{3}\Delta_{||}\right)$ and $\alpha \equiv 1/E_{\mathrm{g}}$.

The Gaussian distribution $F(V)$ of the impurity potential is given by [2]

$$F(V) = \left(\pi\eta_{\mathrm{g}}^2\right)^{-1/2}\exp\left(-V^2/\eta_{\mathrm{g}}^2\right), \tag{11.2}$$

where η_{g} is the impurity scattering potential. It appears from (11.2) that the variance parameter η_{g} is not equal to zero, but the mean value is zero. Further, the impurities are assumed to be uncorrelated and the band mixing effect has been neglected in this simplified theoretical formalism.

We have to average the kinetic energy in the order to obtain the E–k dispersion relation in tetragonal materials including the band tailing effect. Using (11.1) and (11.2), we get

$$\left[\frac{\hbar^2 k_z^2}{2m_{||}^*}\int_{-\infty}^{E} F(V)\mathrm{d}V\right] + \left[\left(\frac{b_{||}}{b_\perp}\frac{c_\perp}{c_{||}}\right)\frac{\hbar^2 k_s^2}{2m_\perp^*}\int_{-\infty}^{E} F(V)\mathrm{d}V\right]$$
$$= \left\{\int_{-\infty}^{E}\frac{(E-V)\left[\alpha(E-V)+1\right]\left[b_{||}(E-V)+1\right]}{\left[c_{||}(E-V)+1\right]}F(V)\mathrm{d}V\right.$$
$$+ \frac{\alpha b_{||}}{c_{||}}\left[\delta\int_{-\infty}^{E}(E-V)F(V)\mathrm{d}V + \frac{2}{9}\left(\Delta_{||}^2 - \Delta_\perp^2\right)\int_{-\infty}^{E}F(V)\mathrm{d}V\right]$$
$$\left. - \left(\frac{2}{9}\right)\frac{\alpha b_{||}}{c_{||}}\left(\Delta_{||}^2 - \Delta_\perp^2\right)\int_{-\infty}^{E}\frac{F(V)\mathrm{d}V}{\left[c_{||}(E-V)+1\right]}\right\}$$
$$- \left(\frac{\hbar^2 k_s^2}{2m_\perp^*}\right)\left\{\left(\frac{b_{||}}{b_\perp}\frac{c_\perp}{c_{||}}\right)\left[\left(\frac{\delta}{2} + \frac{\Delta_{||}^2 - \Delta_\perp^2}{6\Delta_{||}}\right)\alpha\int_{-\infty}^{E}\frac{F(V)\mathrm{d}V}{\left[\alpha(E-V)+1\right]}\right.\right.$$
$$\left.\left. + \left(\frac{\delta}{2} - \frac{\Delta_{||}^2 - \Delta_\perp^2}{6\Delta_{||}}\right)c_{||}\int_{-\infty}^{E}\frac{F(V)\mathrm{d}V}{\left[c_{||}(E-V)+1\right]}\right]\right\}. \tag{11.3}$$

Equation (11.3) can be rewritten as [19]

$$\begin{aligned}
&\frac{\hbar^2 k_z^2}{2m_{||}^*} I(1) + \left(\frac{b_{||}}{b_\perp}\frac{c_\perp}{c_{||}}\right)\frac{\hbar^2 k_s^2}{2m_\perp^*} I(1) \\
&= \left\{ I_3(c_{||}) + \frac{\alpha b_{||}}{c_{||}}\left[\delta I(4) + \frac{2}{9}(\Delta_{||}^2 - \Delta_\perp^2) I(1)\right]\right. \\
&\quad \left. - \left(\frac{2}{9}\right)\frac{\alpha b_{||}}{c_{||}}\left(\Delta_{||}^2 - \Delta_\perp^2\right) I(c_{||})\right\} \\
&\quad - \left(\frac{\hbar^2 k_s^2}{2m_\perp^*}\right)\left\{\left(\frac{b_{||}}{b_\perp}\frac{c_\perp}{c_{||}}\right)\left[\left(\frac{\delta}{2} + \frac{\Delta_{||}^2 - \Delta_\perp^2}{6\Delta_{||}}\right)\alpha I(\alpha)\right.\right. \\
&\quad \left.\left. + \left(\frac{\delta}{2} - \left\{\frac{\Delta_{||}^2 - \Delta_\perp^2}{6\Delta_{||}}\right\}\right) c_{||} I(c_{||})\right]\right\},
\end{aligned} \tag{11.4}$$

where

$$I(1) \equiv \int_{-\infty}^{E} F(V)\mathrm{d}V, \tag{11.5}$$

$$I_3(c_{||}) \equiv \int_{-\infty}^{E} \frac{(E-V)\left[\alpha(E-V)+1\right]\left[b_{||}(E-V)+1\right]}{\left[c_{||}(E-V)+1\right]} F(V)\mathrm{d}V, \tag{11.6}$$

$$I(4) \equiv \int_{-\infty}^{E} (E-V)F(V)\mathrm{d}V, \tag{11.7}$$

$$I(\alpha) \equiv \int_{-\infty}^{E} \frac{F(V)\mathrm{d}V}{\left[\alpha(E-V)+1\right]}, \tag{11.8}$$

Let us substitute $E - V \equiv x$ and $x / \eta_{\mathrm{g}} \equiv t_0$, we get from (11.5)

$$I(1) = \left(\exp(-E^2/\eta_{\mathrm{g}}^2)/\sqrt{\pi}\right)\int_0^{\infty} \exp\left[-t_0^2 + (2Et_0/\eta_{\mathrm{g}})\right]\mathrm{d}t_0,$$

Thus,

$$I(1) = \left[\frac{1 + \mathrm{Erf}(E/\eta_{\mathrm{g}})}{2}\right]. \tag{11.9}$$

From (11.7), one can write

$$I(4) = \left(1/\eta_g\sqrt{\pi}\right)\int_{-\infty}^{E}(E-V)\exp(-V^2/\eta_g^2)\mathrm{d}V,$$

$$= \frac{E}{2}\left[1+\mathrm{Erf}(E/\eta_g)\right] - \left\{\frac{1}{\sqrt{\pi\eta_g^2}}\int_{-\infty}^{E} V\exp\left(-E^2/\eta_g^2\right)\mathrm{d}V\right\},$$

After computing this simple integration, one obtains. Thus,

$$I(4) = \eta_g\exp\left(-E^2/\eta_g^2\right)\left(2\sqrt{\pi}\right)^{-1} + \frac{E}{2}\left(1+\mathrm{Erf}(E/\eta_g)\right), = \gamma_0\left(E,\eta_g\right), \tag{11.10}$$

From (11.8), we can write

$$I(\alpha) = \frac{1}{\sqrt{\pi\eta_g^2}}\int_{-\infty}^{E}\frac{\exp\left(-V^2/\eta_g^2\right)\mathrm{d}V}{[\alpha(E-V)+1]}, \tag{11.11}$$

when, $V \to \pm\infty$, $\frac{1}{[\alpha(E-V)+1]} \to 0$ and $\exp\left(-V^2/\eta_g^2\right) \to 0$; therefore, using (11.11) one can write

$$I(\alpha) = \frac{1}{\sqrt{\pi\eta_g^2}}\int_{-\infty}^{+\infty}\frac{\exp\left(-V^2/\eta_g^2\right)\mathrm{d}V}{[\alpha E+1-\alpha V]}, \tag{11.12}$$

Equation (11.12) can be expressed as

$$I(\alpha) = \left(1/\alpha\eta_g\sqrt{\pi}\right)\int_{-\infty}^{\infty}\exp\left(-t^2\right)(u-t)^{-1}\,\mathrm{d}t, \tag{11.13}$$

where $\frac{V}{\eta_g} \equiv t$ and $u \equiv \left(\frac{1+\alpha E}{\alpha\eta_g}\right)$.

It is well known that [20]

$$W(Z) = (\mathrm{i}/\pi)\int_{-\infty}^{\infty}(Z-t)^{-1}\exp(-t^2)\mathrm{d}t, \tag{11.14}$$

in which $\mathrm{i} = \sqrt{-1}$ and Z is, in general, a complex number.

We also know that [20, 21],

$$W(Z) = \exp\left(-Z^2\right)\mathrm{Erfc}\left(-\mathrm{i}Z\right), \tag{11.15}$$

where $\mathrm{Erfc}(Z) \equiv 1 - \mathrm{Erf}(Z)$.

Thus, Erfc $(-iu) = 1 - \mathrm{Erf}\,(-iu)$,
Since, Erf $(-iu) = -\mathrm{Erf}\,(iu)$,
Therefore, Erfc $(-iu) = 1 + \mathrm{Erf}\,(iu)$.
Thus,

$$I(\alpha) = \left[-\mathrm{i}\sqrt{\pi}\,/\,\alpha\eta_g\right] \exp\left(-u^2\right)\left[1 + \mathrm{Erf}(\mathrm{i}u)\right], \tag{11.16}$$

We also know that [20]

$$\mathrm{Erf}\,(x + \mathrm{i}y) = \mathrm{Erf}\,(x) + \left(\frac{\mathrm{e}^{-x^2}}{2\pi x}\right)\left[(1 - \cos\,(2xy)) + \mathrm{i}\,\sin\,(2xy)\right.$$
$$\left. + \frac{2}{\pi}\mathrm{e}^{-x^2}\sum_{p=1}^{\infty}\frac{\exp\left(-p^2\,/\,4\right)}{\left(p^2 + 4x^2\right)}\right],$$
$$\left[f_{\mathrm{p}}\,(x,y) + \mathrm{i}g_{\mathrm{p}}\,(x,y) + \varepsilon\,(x,y)\right], \tag{11.17}$$

where $f_{\mathrm{p}}\,(x,y) \equiv [2x - 2x\,\cosh\,(py)\cos\,(2xy) + p\,\sinh\,(py)\sin\,(2xy)]$,
$g_{\mathrm{p}}\,(x\,,\,y) \equiv [2x\cosh\,(py)\sin\,(2xy) + p\sinh\,(py)\cos\,(2xy)]$,
and $|\varepsilon\,(x,y)| \approx 10^{-16}\,|\mathrm{Erf}\;\;(x + \mathrm{i}y)|$.

Substituting $x = 0$ and $y = u$ in (11.17), one obtains,

$$\mathrm{Erf}\,(\mathrm{i}u) = \left(\frac{2\mathrm{i}}{\pi}\right)\sum_{p=1}^{\infty}\left\{\frac{\exp\left(-p^2\,/\,4\right)}{p}\sinh\,(pu)\right\}, \tag{11.18}$$

Therefore, one can write

$$I\,(\alpha) = C_{21}\,(\alpha, E, \eta_g) - \mathrm{i}D_{21}\,(\alpha, E, \eta_g)\,, \tag{11.19}$$

where

$$C_{21}\,(\alpha, E, \eta_g) \equiv \left[\frac{2}{\alpha\eta_g\sqrt{\pi}}\right]\exp\left(-u^2\right)\left[\sum_{p=1}^{\infty}\left\{\frac{\exp\left(-p^2\,/\,4\right)}{p}\sinh\,(pu)\right\}\right] \text{ and}$$

$$D_{21}\,(\alpha, E, \eta_g) \equiv \left[\frac{\sqrt{\pi}}{\alpha\eta_g}\exp\left(-u^2\right)\right].$$

Equation (11.19) has both real and imaginary parts and therefore, $I(\alpha)$ is complex, which can also be proved by using the method of analytic continuation.

The integral $I_3\left(c_{||}\right)$ in (11.6) can be written as

$$I_3\left(c_{||}\right) = \left(\frac{\alpha b_{||}}{c_{||}}\right)I\,(5) + \left(\frac{\alpha c_{||} + b_{||}c_{||} - \alpha b_{||}}{c_{||}^2}\right)I(4)$$
$$+ \frac{1}{c_{||}}\left(1 - \frac{\alpha}{c_{||}}\right)\left(1 - \frac{b_{||}}{c_{||}}\right)I\,(1) - \left\{\frac{1}{c_{||}}\left(1 - \frac{\alpha}{c_{||}}\right)\left(1 - \frac{b_{||}}{c_{||}}\right)I\left(c_{||}\right)\right\}, \tag{11.20}$$

where

$$I\,(5) \equiv \int\limits_{-\infty}^{E} (E-V)^2 F(V)\mathrm{d}V, \tag{11.21}$$

From (11.21) one can write

$$I\,(5) = \frac{1}{\sqrt{\pi\eta_{\mathrm{g}}^2}}\left[E^2\int\limits_{-\infty}^{E}\exp\left(\frac{-V^2}{\eta_{\mathrm{g}}^2}\right)\mathrm{d}V - 2E\int\limits_{-\infty}^{E}V\exp\left(\frac{-V^2}{\eta_{\mathrm{g}}^2}\right)\mathrm{d}V\right.$$
$$\left.+\int\limits_{-\infty}^{E}V^2\exp\left(\frac{-V^2}{\eta_{\mathrm{g}}^2}\right)\mathrm{d}V\right],$$

The evaluations of the component integrals lead us to write

$$I\,(5) = \frac{\eta_{\mathrm{g}}E}{2\sqrt{\pi}}\exp\left(\frac{-E^2}{\eta_{\mathrm{g}}^2}\right) + \frac{1}{4}\left(\eta_{\mathrm{g}}^2 + 2E^2\right)\left[1 + \mathrm{Erf}\left(\frac{E}{\eta_{\mathrm{g}}}\right)\right] = \theta_0\,(E,\eta_{\mathrm{g}})\,, \tag{11.22}$$

Thus combining the aforementioned equations, $I_3\left(c_{||}\right)$ can be expressed as

$$I_3\left(c_{||}\right) = A_{21}\,(E,\eta_{\mathrm{g}}) + \mathrm{i}B_{21}\,(E,\eta_{\mathrm{g}})\,, \tag{11.23}$$

where $A_{21}\,(E,\eta) \equiv \left[\frac{\alpha b_{||}}{c_{||}}\left[\frac{\eta_{\mathrm{g}}E}{2\sqrt{\pi}}\exp\left(\frac{-E^2}{\eta_{\mathrm{g}}^2}\right) + \frac{1}{4}\left(\eta_{\mathrm{g}}^2 + 2E^2\right)\left\{1 + \mathrm{Erf}\left(\frac{E}{\eta_{\mathrm{g}}}\right)\right\}\right]\right.,$
$+\left[\frac{\alpha c_{||} + b_{||}c_{||} - \alpha b_{||}}{c_{||}^2}\right]\left\{\frac{E}{2}\left[1 + \mathrm{Erf}\left(E/\,\eta_{\mathrm{g}}\right)\right] + \frac{\eta_{\mathrm{g}}\exp(-E^2/\,\eta_{\mathrm{g}}^2)}{2\sqrt{\pi}}\right\}, + \frac{1}{c_{||}}\left(1 - \frac{\alpha}{c_{||}}\right)$
$\left(1 - \frac{b_{||}}{c_{||}}\right)\frac{1}{2}\left[1 + \mathrm{Erf}\left(E/\,\eta_{\mathrm{g}}\right)\right] - \left\{\frac{2}{c_{||}^2\eta_{\mathrm{g}}\sqrt{\pi}}\left(1 - \frac{\alpha}{c_{||}}\right)\left(1 - \frac{b_{||}}{c_{||}}\right)\exp(-u_1^2)\right\}$
$\left.\left[\sum\limits_{p=1}^{\infty}\left\{\frac{\exp(-p^2/\,4)}{p}\sinh\left(pu_1\right)\right\}\right]\right],$

$$u_1 \equiv \left[\frac{1 + c_{||}E}{c_{||}\eta_{\mathrm{g}}}\right] \text{ and } B_{21}\,(E,\eta_{\mathrm{g}}) \equiv \frac{\sqrt{\pi}}{c_{||}^2\eta_{\mathrm{g}}}\left(1 - \frac{\alpha}{c_{||}}\right)\left(1 - \frac{b_{||}}{c_{||}}\right)\exp\left(-u_1^2\right).$$

Therefore, the combination of all the appropriate equations together with the algebraic manipulations lead to the dispersion relation of the conduction electrons of heavily doped tetragonal materials forming Gaussian band tails as

$$\frac{\hbar^2 k_z^2}{2m_{||}^* T_{21}\,(E,\eta_{\mathrm{g}})} + \frac{\hbar^2 k_s^2}{2m_{\perp}^* T_{22}\,(E,\eta_{\mathrm{g}})} = 1, \tag{11.24}$$

where $T_{21}(E,\eta_g)$ and $T_{22}(E,\eta_g)$ have both real and complex parts and they are given by

$$T_{21}(E,\eta_g) \equiv [T_{27}(E,\eta_g) + iT_{28}(E,\eta_g)], T_{27}(E,\eta_g) \equiv \left[\frac{T_{23}(E,\eta_g)}{T_5(E,\eta_g)}\right],$$

$$T_{23}(E,\eta_g) \equiv \left[A_{21}(E,\eta_g) + \frac{\alpha b_{||}}{c_{||}}\left[\delta\gamma_0(E,\eta_g) + \frac{1}{9}\left(\Delta_{||}^2 - \Delta_{\perp}^2\right)[1+\mathrm{Erf}(E/\eta_g)]\right] - \left\{\frac{2}{9}\left(\frac{\alpha b_{||}}{c_{||}}\right)\left(\Delta_{||}^2 - \Delta_{\perp}^2\right)G_{21}(c_{||},E,\eta_g)\right\}\right],$$

$$G_{21}(c_{||},E,\eta_g) \equiv \frac{2}{c_{||}\eta_g\sqrt{\pi}}\exp\left(-u_1^2\right)\sum_{p=1}^{\infty}\left\{\frac{\exp(-p^2/4)}{p}\sinh(pu_1)\right\},$$

$$T_5(E,\eta_g) \equiv \frac{1}{2}[1+\mathrm{Erf}(E/\eta_g)],$$

$$T_{28}(E,\eta_g) \equiv \left[\frac{T_{24}(E,\eta_g)}{T_5(E,\eta_g)}\right],$$

$$T_{24}(E,\eta_g) \equiv \left[B_{21}(E,\eta_g) + \frac{2}{9}\frac{\alpha b_{||}}{c_{||}}\left(\Delta_{||}^2 - \Delta_{\perp}^2\right)H_{21}(c_{||},E,\eta_g)\right],$$

$$H_{21}(c_{||},E,\eta_g) \equiv \left[\frac{\sqrt{\pi}}{\eta_g c_{||}}\exp\left(-u_1^2\right)\right],$$

$$T_{22}(E,\eta_g) \equiv [T_{29}(E,\eta_g) + iT_{30}(E,\eta_g)],$$

$$T_{29}(E,\eta_g) \equiv \frac{T_{23}(E,\eta_g)T_{25}(E,\eta_g) - T_{24}(E,\eta_g)T_{26}(E,\eta_g)}{\left[(T_{25}(E,\eta_g))^2 + (T_{26}(E,\eta_g))^2\right]},$$

$$T_{25}(E,\eta_g) \equiv \left[\left(\frac{b_{||}}{b_{\perp}}\frac{c_{\perp}}{c_{||}}\right)\frac{1}{2}\left[1+Erf\left(\frac{E}{\eta_g}\right)\right] + \left(\frac{b_{||}}{b_{\perp}}\frac{c_{\perp}}{c_{||}}\right)\left(\frac{\delta}{2} + \left[\frac{\Delta_{||}^2 - \Delta_{\perp}^2}{6\Delta_{||}}\right]\right)c_{||}C_{21}(c_{||},E,\eta_g) + \left(\frac{b_{||}c_{\perp}}{b_{\perp}}\right)\left(\frac{\delta}{2} - \left[\frac{\Delta_{||}^2 - \Delta_{\perp}^2}{6\Delta_{||}}\right]\right)G_{21}(c_{||},E,\eta_g)\right],$$

$$T_{26}(E,\eta_g) \equiv \left(\frac{b_{||}}{b_{\perp}}\frac{c_{\perp}}{c_{||}}\right)\left(\frac{\delta}{2} - \frac{\Delta_{||}^2 - \Delta_{\perp}^2}{6\Delta_{||}}\right)\alpha D_{21}(\alpha,E,\eta_g) + \frac{b_{||}c_{\perp}}{b_{\perp}}\left(\frac{\delta}{2} - \frac{\Delta_{||}^2 - \Delta_{\perp}^2}{6\Delta_{||}}\right)H_{21}(c_{||},E,\eta_g), \text{ and}$$

$$T_{30}(E,\eta_g) \equiv \frac{T_{24}(E,\eta_g)\,T_{25}(E,\eta_g) + T_{23}(E,\eta_g)\,T_{26}(E,\eta_g)}{\left[(T_{25}(E,\eta_g))^2 + (T_{26}(E,\eta_g))^2\right]}.$$

From (11.24), it appears that the energy spectrum in heavily doped tetragonal semiconductors is complex. The complex nature of the electron dispersion law in heavily doped semiconductors occurs from the existence of the singularities in the corresponding undoped electron energy spectrum. It may be noted that the complex band structures have already been studied for bulk semiconductors and superlattices without heavy doping [22] and bear no relationship with the complex electron dispersion law as indicated by (11.24). The physical picture behind the formulation of the complex energy spectrum in heavily doped tetragonal semiconductors is the interaction of the impurity atoms in the tails with the splitting constants of the valance bands. More interaction causes more prominence of the complex part. When there is no heavy doping, $\eta_g \to 0$, and there is no interaction of the impurity atoms in the tails with the spin orbit constants. As a result, there exists no complex energy spectrum and (11.24) gets converted into (2.2) when $\eta_g \to 0$. Besides, the complex spectra are not related to the same evanescent modes in the band tails and the conduction bands.

The transverse and the longitudinal EMMs at the Fermi energy (E_{F_h}) of heavily doped tetragonal materials can be expressed respectively as

$$m_\perp^*(E_{F_h},\eta_g) = m_\perp^*\,\{T_{29}(E,\eta_g)\}'\big|_{E=E_{F_h}}, \tag{11.25}$$

and

$$m_{||}^*(E_{F_h},\eta_g) = m_{||}^*\,\{T_{27}(E,\eta_g)\}'\big|_{E=E_{F_h}}, \tag{11.26}$$

In the absence of band tailing effects $\eta_g \to 0$ and we get

$$m_\perp^*(E_F,0) = \frac{\hbar^2}{2}\left[\frac{\psi_2(E)\{\psi_1(E)\}' - \psi_1(E)\{\psi_2(E)\}'}{\{\psi_2(E)\}^2}\right]\Bigg|_{E=E_F}, \tag{11.27}$$

and

$$m_{||}^*(E_F,0) = \frac{\hbar^2}{2}\left[\frac{\psi_3(E)\{\psi_1(E)\}' - \{\psi_1(E)\}\{\psi_3(E)\}'}{\{\psi_3(E)\}^2}\right]\Bigg|_{E=E_F}. \tag{11.28}$$

Comparing these equations, one can infer that the effective masses exist in the forbidden zone, which is impossible without the effect of band tailing. For undoped semiconductors, the effective mass in the band gap is infinity.

The density-of-states function is given by

$$N_{HD}(E,\eta_g) = \frac{2g_v m_\perp^* \sqrt{2m_{||}^*}}{3\pi^2\hbar^3} R_{11}(E,\eta_g)\cos\left[\psi_{11}(E,\eta_g)\right], \tag{11.29}$$

where

$$R_{11}(E,\eta_g) \equiv \Bigg[\Bigg[\{T_{29}(E,\eta_g)\}'\sqrt{x(E,\eta_g)} + \frac{T_{29}(E,\eta_g)\{x(E,\eta_g)\}'}{2\sqrt{x(E,\eta_g)}} - \{T_{30}(E,\eta_g)\}'\sqrt{y(E,\eta_g)} - \frac{T_{30}(E,\eta_g)\{y(E,\eta_g)\}'}{2\sqrt{y(E,\eta_g)}}\Bigg]^2 + \Bigg[\{T_{29}(E,\eta_g)\}'\sqrt{y(E,\eta_g)} + \frac{T_{29}(E,\eta_g)\{y(E,\eta_g)\}'}{2\sqrt{y(E,\eta_g)}} + \{T_{30}(E,\eta_g)\}'\sqrt{x(E,\eta_g)} + \frac{T_{30}(E,\eta_g)\{x(E,\eta_g)\}'}{2\sqrt{x(E,\eta_g)}}\Bigg]^2\Bigg]^{1/2},$$

$$x(E,\eta_g) \equiv \frac{1}{2}\left[T_{27}(E,\eta_g) + \sqrt{\{T_{27}(E,\eta_g)\}^2 + \{T_{28}(E,\eta_g)\}^2}\right],$$

$$y(E,\eta_g) \equiv \frac{1}{2}\left[\sqrt{\{T_{27}(E,\eta_g)\}^2 + \{T_{28}(E,\eta_g)\}^2} - T_{27}(E,\eta_g)\right] \text{ and}$$

$$\psi_{11}(E,\eta_g) \equiv \tan^{-1}\Bigg[\Bigg[\{T_{29}(E,\eta_g)\}'\sqrt{y(E,\eta_g)} + \frac{T_{29}(E,\eta_g)}{2\sqrt{y(E,\eta_g)}}\cdot\{y(E,\eta_g)\}' + \{T_{30}(E,\eta_g)\}'\sqrt{x(E,\eta_g)} + \frac{T_{30}(E,\eta_g)\{x(E,\eta_g)\}'}{2\sqrt{x(E,\eta_g)}}\Bigg] \Bigg[\{T_{29}(E,\eta_g)\}'\sqrt{x(E,\eta_g)} + \frac{T_{29}(E,\eta_g)\{x(E,\eta_g)\}'}{2\sqrt{x(E,\eta_g)}} - \{T_{30}(E,\eta_g)\}'\sqrt{y(E,\eta_g)} - \frac{T_{30}(E,\eta_g)\{y(E,\eta_g)\}'}{2\sqrt{y(E,\eta_g)}}\Bigg]^{-1}\Bigg].$$

The oscillatory nature of the DOS for heavily doped tetragonal materials is apparent from (11.29). For, $\psi_{11}(E,\eta_g) \geq \pi$, the cosine function becomes negative leading to the negative values of the DOS. The electrons cannot exist for the negative values of the DOS and therefore, this reason is forbidden for electrons, which indicates that in the band tail, there appears a new forbidden zone in addition to the normal band gap of the semiconductor.

The use of (11.29) and the electron concentration at low temperatures can be expressed as

$$n_0 = \frac{2g_v m_\perp^* \sqrt{2m_\parallel^*}}{3\pi^2\hbar^3}\left[I_{11}(E_{F_h},\eta_g)\right], \tag{11.30}$$

where $I_{11}(E_{F_h},\eta_g) \equiv \left[T_{29}(E_{F_h},\eta_g)\sqrt{x(E_{F_h},\eta_g)} - T_{30}(E_{F_h},\eta_g)\sqrt{y(E_{F_h},\eta_g)}\right]$.

The DMR for heavily doped semiconductors is given by

$$\frac{D}{\mu}=\frac{n_0}{|e|}\frac{\partial n_0}{\partial\left(E_{\mathrm{F_h}}-\bar{E}_{\mathrm{hd}}\right)}, \tag{11.31}$$

where $\bar{E}_{\mathrm{hd}}$ is the electron energy within the band gap, as measured from $k=0$ and should be obtained from the dispersion relation of the heavily doped semiconductors under the conditions $E=\bar{E}_{\mathrm{hd}}$ when $k=0$.

For heavily doped tetragonal semiconductors, $\bar{E}_{\mathrm{hd}}$ is the smallest negative root of the equation

$$\left[T_{27}\left(\bar{E}_{\mathrm{hd}},\eta_{\mathrm{g}}\right)T_{29}\left(\bar{E}_{\mathrm{hd}},\eta_{\mathrm{g}}\right)-T_{28}\left(\bar{E}_{\mathrm{hd}},\eta_{\mathrm{g}}\right)T_{30}\left(\bar{E}_{\mathrm{hd}},\eta_{\mathrm{g}}\right)\right]=0, \tag{11.32}$$

For the purpose of condensed presentation, the numerical evaluation of the DMR by using (11.30)–(11.32) and the allied definitions have been presented in the section on the results and discussion of this chapter.

11.2.2 Study of the Einstein Relation in Heavily Doped III–V, Ternary and Quaternary Materials Forming Gaussian Band Tails

(a) Under the conditions, $\delta=0$, $m_{\|}^{*}=m_{\perp}^{*}=m^{*}$ and $\Delta_{\|}=\Delta_{\perp}=\Delta$, the electron dispersion law in this case assumes the form

$$\frac{\hbar^2k^2}{2m^*}=T_{31}\left(E,\eta_{\mathrm{g}}\right)+\mathrm{i}T_{32}\left(E,\eta_{\mathrm{g}}\right), \tag{11.33}$$

where

$$\begin{aligned}
T_{31}\left(E,\eta_{\mathrm{g}}\right)\equiv&\left(\frac{2}{1+\mathrm{Erf}\left(E/\eta_{\mathrm{g}}\right)}\right)\left[\frac{\alpha b}{c}\theta_0\left(E,\eta_{\mathrm{g}}\right)+\left[\frac{\alpha c+bc-\alpha b}{c^2}\right]\gamma_0\left(E,\eta_{\mathrm{g}}\right)\right.\\
&+\frac{1}{c}\left(1-\frac{\alpha}{c}\right)\left(1-\frac{b}{c}\right)\frac{1}{2}\left[1+\mathrm{Erf}\left(\frac{E}{\eta_{\mathrm{g}}}\right)\right]\\
&\left.-\frac{1}{c}\left(1-\frac{\alpha}{c}\right)\left(1-\frac{b}{c}\right)\frac{2}{c\eta_{\mathrm{g}}\sqrt{\pi}}\exp\left(-u_2^2\right)\left[\sum_{p=1}^{\infty}\frac{\exp\left(-p^2/4\right)}{p}\sinh\left(pu_2\right)\right]\right],
\end{aligned}$$

$$b\equiv\left(\frac{1}{E_{\mathrm{g}}+\Delta}\right),\ c\equiv\left(\frac{1}{E_{\mathrm{g}}+\frac{2}{3}\Delta}\right),$$

$$\begin{aligned}
u_2\equiv\frac{1+cE}{c\eta_{\mathrm{g}}}\ \text{and}\ T_{32}\left(E,\eta_{\mathrm{g}}\right)\equiv&\left(\frac{2}{1+\mathrm{Erf}\left(E/\eta_{\mathrm{g}}\right)}\right)\frac{1}{c}\left(1-\frac{\alpha}{c}\right)\left(1-\frac{b}{c}\right)\\
&\times\frac{\sqrt{\pi}}{c\eta_{\mathrm{g}}}\exp\left(-u_2^2\right).
\end{aligned}$$

Thus, the complex energy spectrum occurs due to the term $T_{32}\left(E,\eta_{\mathrm{g}}\right)$ and this imaginary band is quite different from the forbidden energy band.

The EMM at the Fermi level is given by

$$m^*\left(E_{F_h},\eta_g\right)=m^*\left\{T_{31}\left(E,\eta_g\right)\right\}'\big|_{E=E_{F_h}} \tag{11.34}$$

Thus, the EMM in heavily doped III–V, ternary and quaternary materials exists in the band gap, which is the new attribute of the theory of band tailing.

In the absence of band tailing, $\eta_g \to 0$ and the EMM assumes the form

$$m^*\left(E_F\right)=m^*\left\{I\left(E\right)\right\}'\big|_{E=E_F}, \tag{11.35}$$

The density-of-states function in this case assumes the form

$$N_{HD}\left(E,\eta_g\right)=\frac{g_v}{3\pi^2}\left(\frac{2m^*}{\hbar^2}\right)^{3/2}R_{21}\left(E,\eta_g\right)\cos\left[\vartheta_{21}\left(E,\eta_g\right)\right], \tag{11.36}$$

where

$$R_{21}\left(E,\eta_g\right)\equiv\left[\frac{\left[\left\{\alpha_{11}\left(E,\eta_g\right)\right\}'\right]^2}{4\alpha_{11}\left(E,\eta_g\right)}+\frac{\left[\left\{\beta_{11}\left(E,\eta_g\right)\right\}'\right]^2}{4\beta_{11}\left(E,\eta_g\right)}\right]^{1/2},$$

$$\alpha_{11}\left(E,\eta_g\right)\equiv\frac{1}{2}\left[T_{33}\left(E,\eta_g\right)+\sqrt{\left\{T_{33}\left(E,\eta_g\right)\right\}^2+\left\{T_{34}\left(E,\eta_g\right)\right\}^2}\right],$$

$$T_{33}\left(E,\eta_g\right)\equiv\left[\left\{T_{31}\left(E,\eta_g\right)\right\}^3-3T_{31}\left(E,\eta_g\right)\left\{T_{32}\left(E,\eta_g\right)\right\}^2\right],$$

$$T_{34}\left(E,\eta_g\right)\equiv\left[3T_{32}\left(E,\eta_g\right)\left\{T_{31}\left(E,\eta_g\right)\right\}^2-\left\{T_{32}\left(E,\eta_g\right)\right\}^3\right],$$

$$\beta_{11}\left(E,\eta_g\right)\equiv\frac{1}{2}\left[\sqrt{\left\{T_{33}\left(E,\eta_g\right)\right\}^2+\left\{T_{34}\left(E,\eta_g\right)\right\}^2}-T_{33}\left(E,\eta_g\right)\right]\text{ and}$$

$$\vartheta_{21}\left(E,\eta_g\right)\equiv\tan^{-1}\left[\frac{\left\{\beta_{11}\left(E,\eta_g\right)\right\}'}{\left\{\alpha_{11}\left(E,\eta_g\right)\right\}'}\sqrt{\frac{\alpha_{11}\left(E,\eta_g\right)}{\beta_{11}\left(E,\eta_g\right)}}\right].$$

Thus, the oscillatory density-of-states function becomes negative for $\vartheta_{21}\left(E,\eta_g\right)\geq\pi$ and a new forbidden zone will appear in addition to the normal band gap.

The electron concentration in the zone of low temperatures can be written as

$$n_0=\frac{g_v}{3\pi^2}\left(\frac{2m^*}{\hbar^2}\right)^{3/2}\frac{1}{\sqrt{2}}\left[T_{33}\left(E_{F_h},\eta_g\right)+\sqrt{\left\{T_{33}\left(E_{F_h},\eta_g\right)\right\}^2+\left\{T_{34}\left(E_{F_h},\eta_g\right)\right\}^2}\right]^{1/2}, \tag{11.37}$$

In this case, $\bar{E}_{hd}$ is given by

$$T_{31}\left(\bar{E}_{hd},\eta_g\right)=0, \tag{11.38}$$

One can numerically compute the DMR by using (11.37), (11.31), (11.38) and the allied definitions in this case.

(b) The dispersion relation in heavily doped III–V, ternary and quaternary materials whose undoped energy spectrum obeys the two band model of Kane is given by

$$\frac{\hbar^2 k^2}{2m^*} = \gamma_2 (E, \eta_g), \tag{11.39}$$

where $\gamma_2 (E, \eta_g) \equiv \left[\frac{2}{1+\text{Erf}(E/\eta_g)}\right] [\gamma_0 (E, \eta_g) + \alpha\theta_0 (E, \eta_g)]$.

As the original two band Kane model is an all zero and no pole function, the heavily doped counterpart with respect to dispersion law will be totally real and the complex band will vanish.

The EMM in this case can be written as

$$m^* (E_{F_h}, \eta_g) = m^* \left\{\gamma_2 (E, \eta_g)\right\}'\big|_{E=E_{F_h}}, \tag{11.40}$$

Thus, one again observes that the EMM in this case exists in the band gap.

In the absence of band tailing, $\eta_g \to 0$ and the EMM assumes the form

$$m^* (E_F) = m^* \left\{1 + 2\alpha E\right\}\big|_{E=E_F}, \tag{11.41}$$

The density-of-states function in this case can be written as

$$N_{HD} (E, \eta_g) = \frac{g_v}{2\pi^2} \left(\frac{2m^*}{\hbar^2}\right)^{3/2} \sqrt{\gamma_2 (E, \eta_g)} \left\{\gamma_2 (E, \eta_g)\right\}', \tag{11.42}$$

As the original two band Kane model is an all zero and no pole function, the heavily doped counterpart with respect to DOS will be totally real and the complex band will vanish.

The electron concentration at low temperatures is given by

$$n_0 = \frac{g_v}{3\pi^2} \left(\frac{2m^*}{\hbar^2}\right)^{3/2} \left\{\gamma_2 (E_{F_h}, \eta_g)\right\}^{3/2}, \tag{11.43}$$

In this case, $\bar{E}_{hd}$ is given by

$$\gamma_2 \left(\bar{E}_{hd}, \eta_g\right) = 0. \tag{11.44}$$

One can numerically compute the DMR by using (11.43), (11.31), (11.44) and the allied definitions in this case.

(c) The dispersion relation in heavily doped semiconductors whose unperturbed conduction electrons obey parabolic energy bands is given by

$$\frac{\hbar^2 k^2}{2m^*} = \gamma_3 (E, \eta_g), \tag{11.45}$$

where $\gamma_3 (E, \eta_g) \equiv \left[\frac{2}{(1+\text{Erf}(E/\eta_g))}\right] \gamma_0 (E, \eta_g)$.

As the original parabolic energy band is no pole function, therefore, the heavily doped counterpart will be totally real, which is also apparent from the expression (11.45).

The EMM in this case can be written as

$$m^*\left(E_{\mathrm{F_h}}, \eta_{\mathrm{g}}\right) = m^* \left\{\gamma_3\left(E, \eta_{\mathrm{g}}\right)\right\}'\big|_{E=E_{\mathrm{F_h}}}, \quad (11.46)$$

In the absence of band tailing, $\eta_{\mathrm{g}} \to 0$ and the EMM assumes the form

$$m^*\left(E_{\mathrm{F}}\right) = m^*. \quad (11.47)$$

It is well-known that the EMM in undoped parabolic energy bands is a constant quantity in general excluding cross-field configuration. But the same mass in the corresponding heavily doped bulk counterpart is a complicated function of Fermi energy and the impurity potential together with the fact that the EMM also exists in the band gap.

The density-of-states function in this case can be written as

$$N_{\mathrm{HD}}\left(E, \eta_{\mathrm{g}}\right) = \frac{g_{\mathrm{v}}}{2\pi^2}\left(\frac{2m^*}{\hbar^2}\right)^{3/2} \sqrt{\gamma_3\left(E, \eta_{\mathrm{g}}\right)}\left\{\gamma_3\left(E, \eta_{\mathrm{g}}\right)\right\}'. \quad (11.48)$$

As the original parabolic energy band model is a no pole function, the heavily doped counterpart will be totally real and the complex band will vanish.

The electron concentration at low temperatures is given by

$$n_0 = \frac{g_{\mathrm{v}}}{3\pi^2}\left(\frac{2m^*}{\hbar^2}\right)^{3/2} \left\{\gamma_3\left(E_{\mathrm{F_h}}, \eta_{\mathrm{g}}\right)\right\}^{3/2}. \quad (11.49)$$

In this case, $\bar{E}_{\mathrm{hd}}$ is given by

$$\gamma_3\left(\bar{E}_{\mathrm{hd}}, \eta_{\mathrm{g}}\right) = 0, \quad (11.50)$$

One can numerically compute the DMR by using (11.49), (11.31), (11.50) and the allied definitions in this case.

11.2.3 Study of the Einstein Relation in Heavily Doped II–VI Materials Forming Gaussian Band Tails

Using (2.27) and (11.2), the dispersion relation of the carriers in heavily doped II–VI materials in the presence of Gaussian band tails can be expressed as

$$\gamma_3\left(E, \eta_{\mathrm{g}}\right) = a_0' k_s^2 + b_0' k_z^2 \pm \bar{\lambda}_0 k_s, \quad (11.51)$$

Thus, the energy spectrum in this case is real as the corresponding undoped case as given by (2.27) is a no pole function.

The transverse and the longitudinal EMMs masses are respectively given by

$$m_{\perp}^{*}\left(E_{\mathrm{F_h}},\eta_{\mathrm{g}}\right)=m_{\perp}^{*}\left\{\gamma_3\left(E,\eta_{\mathrm{g}}\right)\right\}'\left[1-\left(\frac{\bar{\lambda}_0}{\sqrt{\left(\bar{\lambda}_0\right)^2+4a_0'\gamma_3\left(E,\eta_{\mathrm{g}}\right)}}\right)\right]\Bigg|_{E=E_{\mathrm{F_h}}}, \tag{11.52}$$

and

$$m_{||}^{*}\left(E_{\mathrm{F_h}},\eta_{\mathrm{g}}\right)=m_{||}^{*}\left\{\gamma_3\left(E,\eta_{\mathrm{g}}\right)\right\}'\big|_{E=E_{\mathrm{F_h}}}, \tag{11.53}$$

In the absence of band tailing effects $\eta_{\mathrm{g}}\to 0$, we get

$$m_{\perp}^{*}\left(E_{\mathrm{F}}\right)=m_{\perp}^{*}\left[1-\left(\frac{\bar{\lambda}_0}{\sqrt{\left(\bar{\lambda}_0\right)^2+4a_0'E}}\right)\right]\Bigg|_{E=E_{\mathrm{F}}}, \tag{11.54}$$

and

$$m_{||}^{*}\left(E_{\mathrm{F}}\right)=m_{||}^{*}. \tag{11.55}$$

Thus, in heavily doped II–VI materials, both the transverse and the longitudinal EMM exist in the band gap.

The volume in k-space enclosed by (11.51) can be expressed as

$$V\left(E,\eta_{\mathrm{g}}\right)=\frac{4\pi}{3a_0'\sqrt{b_0'}}\left[\left\{\gamma_3\left(E,\eta_{\mathrm{g}}\right)\right\}^{3/2}+\frac{3}{8}\frac{\left(\bar{\lambda}_0\right)^2\sqrt{\gamma_3\left(E,\eta_{\mathrm{g}}\right)}}{a_0'}-\left(\frac{3}{4}\frac{\bar{\lambda}_0}{\sqrt{a_0'}}\right)\right.$$
$$\left.\times\left(\gamma_3\left(E,\eta_{\mathrm{g}}\right)+\frac{\left(\bar{\lambda}_0\right)^2}{4a_0'}\right)\sin^{-1}\left[\frac{\sqrt{\gamma_3\left(E,\eta_{\mathrm{g}}\right)}}{\sqrt{\gamma_3\left(E,\eta_{\mathrm{g}}\right)+\frac{\left(\bar{\lambda}_0\right)^2}{4a_0'}}}\right]\right]. \tag{11.56}$$

Using (11.56), the density-of-states function in this case can be written as

$$N_{\mathrm{HD}}\left(E,\eta_{\mathrm{g}}\right)=\frac{g_{\mathrm{v}}}{2\pi^2a_0'\sqrt{b_0'}}\left[\left\{\gamma_3\left(E,\eta_{\mathrm{g}}\right)\right\}^{1/2}\left\{\gamma_3\left(E,\eta_{\mathrm{g}}\right)\right\}'+\frac{1}{8}\frac{\left(\bar{\lambda}_0\right)^2\left\{\gamma_3\left(E,\eta_{\mathrm{g}}\right)\right\}'}{4a_0'\sqrt{\gamma_3\left(E,\eta_{\mathrm{g}}\right)}}\right.$$
$$-\left(\frac{1}{2}\frac{\bar{\lambda}_0}{\sqrt{a_0'}}\right)\left\{\gamma_3\left(E,\eta_{\mathrm{g}}\right)\right\}'\sin^{-1}\left[\frac{\sqrt{\gamma_3\left(E,\eta_{\mathrm{g}}\right)}}{\sqrt{\gamma_3\left(E,\eta_{\mathrm{g}}\right)+\frac{\left(\bar{\lambda}_0\right)^2}{4a_0'}}}\right]$$
$$-\frac{\left\{\gamma_3\left(E,\eta_{\mathrm{g}}\right)\right\}'}{2}\left(\gamma_3\left(E,\eta_{\mathrm{g}}\right)+\frac{\left(\bar{\lambda}_0\right)^2}{4a_0'}\right)$$
$$\left.\times\left[\frac{1}{\sqrt{\gamma_3\left(E,\eta_{\mathrm{g}}\right)}}-\frac{\sqrt{\gamma_3\left(E,\eta_{\mathrm{g}}\right)}}{\left(\gamma_3\left(E,\eta_{\mathrm{g}}\right)+\frac{\left(\bar{\lambda}_0\right)^2}{4a_0'}\right)}\right]\right]. \tag{11.57}$$

Therefore, the electron concentration in the zone of low temperatures can be expressed as

$$n_0 = \frac{g_v}{3\pi^2 a_0' \sqrt{b_0'}} \left[\{\gamma_3 (E_{F_h}, \eta_g)\}^{3/2} + \frac{3}{8} \frac{\left(\bar{\lambda}_0\right)^2 \sqrt{\gamma_3 (E_{F_h}, \eta_g)}}{a_0'} \right.$$
$$\left. - \left(\frac{3}{4} \frac{\bar{\lambda}_0}{\sqrt{a_0'}}\right) \left(\gamma_3 (E_{F_h}, \eta_g) + \frac{\left(\bar{\lambda}_0\right)^2}{4a_0'}\right) \sin^{-1} \left[\frac{\sqrt{\gamma_3 (E_{F_h}, \eta_g)}}{\sqrt{\gamma_3 (E_{F_h}, \eta_g) + \frac{\left(\bar{\lambda}_0\right)^2}{4a_0'}}} \right] \right], \tag{11.58}$$

In this case, $\bar{E}_{\mathrm{hd}}$ is given by

$$\left\{\gamma_3 \left(\bar{E}_{\mathrm{hd}}, \eta_g\right)\right\} = 0. \tag{11.59}$$

Thus, one can numerically evaluate the DMR by using (11.58), (11.31), (11.59) and the allied definitions in this case.

11.2.4 Study of the Einstein Relation in Heavily Doped IV–VI Materials Forming Gaussian Band Tails

From (5.50), we can write

$$\frac{\alpha \hbar^4 k_s^4}{4m_t^+ m_t^-} + \hbar^2 k_s^2 \left[\left(\frac{1}{2m_t^*} - \frac{1}{2m_t^-}\right) + \alpha E \left(\frac{1}{2m_t^-} - \frac{1}{2m_t^+}\right) + \frac{\alpha \hbar^2 k_z^2}{4m_t^- m_l^+} \right]$$
$$+ \left[\left(\frac{\hbar^2 k_z^2}{2m_l^*} + \frac{\hbar^2 k_z^2}{2m_l^-}\right) + \frac{\alpha E}{2} \hbar^2 k_z^2 \left(\frac{1}{m_l^-} - \frac{1}{m_l^+}\right) + \frac{\alpha \hbar^4 k_z^4}{4m_l^+ m_t^-} - E(1 + \alpha E) \right] = 0, \tag{11.60}$$

Using (11.60) and (11.2), the dispersion relation of the conduction electrons in heavily doped IV–VI materials can be expressed as

$$\frac{\alpha \hbar^4 k_s^4}{4m_t^+ m_l^-} Z_0 (E, \eta_g) + \hbar^2 k_s^2 \left[\lambda_{71} (E, \eta_g) k_z^2 + \lambda_{72} (E, \eta_g)\right]$$
$$+ \left[\lambda_{73} (E, \eta_g) k_z^2 + \lambda_{74} (E, \eta_g) k_z^4 - \lambda_{75} (E, \eta_g)\right] = 0, \tag{11.61}$$

where

$$Z_0 (E, \eta_g) \equiv \frac{1}{2} \left[1 + Erf\left(\frac{E}{\eta_g}\right)\right], \quad \lambda_{70} (E, \eta_g) \equiv \frac{\alpha}{4m_t^+ m_t^-} Z_0 (E, \eta_g),$$

$$\lambda_{71} (E, \eta_g) \equiv \left[\frac{\alpha \hbar^2}{4m_t^- m_l^+} Z_0 (E, \eta_g) + \frac{\alpha \hbar^2}{4m_l^- m_t^+} Z_0 (E, \eta_g)\right],$$

$$\lambda_{72}(E,\eta_g) \equiv \left[\left(\frac{1}{2m_t^*}-\frac{1}{2m_t^-}\right)Z_0(E,\eta_g)+\alpha\left(\frac{1}{2m_t^-}-\frac{1}{2m_t^+}\right)\gamma_0(E,\eta_g)\right],$$

$$\lambda_{73}(E,\eta_g) \equiv \left[\left(\frac{\hbar^2}{2m_l^*}+\frac{\hbar^2}{2m_l^-}\right)Z_0(E,\eta_g)+\frac{\alpha\hbar^2}{2}\left(\frac{1}{m_l^-}-\frac{1}{2m_l^+}\right)\gamma_0(E,\eta_g)\right],$$

$$\lambda_{74}(E,\eta_g) \equiv \frac{\alpha\hbar^4 Z_0(E,\eta_g)}{4m_l^+ m_l^-} \text{ and } \lambda_{75}(E,\eta_g) \equiv \left[\gamma_0(E,\eta_g)+\alpha\theta_0(E,\eta_g)\right].$$

Thus, the energy spectrum in this case is real as the corresponding undoped material as given by (11.60) is a pole-less function.

The respective transverse and the longitudinal EMM in this case can be written as

$$m_\perp^*(E_{F_h},\eta_g) = \{2Z_0(E,\eta_g)\}^{-2}\left[Z_0(E,\eta_g)\left[-\{\lambda_{72}(E,\eta_g)\}'+\frac{\{\lambda_{78}(E,\eta_g)\}'}{2\sqrt{\lambda_{78}(E,\eta_g)}}\right] - \{Z_0(E,\eta_g)\}'\left[-\lambda_{72}(E,\eta_g)+\sqrt{\lambda_{78}(E,\eta_g)}\right]\right]\Bigg|_{E=E_{F_h}}, \quad (11.62a)$$

where $\lambda_{78}(E,\eta_g) \equiv [4\lambda_{70}(E,\eta_g)\lambda_{75}(E,\eta_g)]$,
and

$$m_{||}^*(E_{F_h},\eta_g) = \frac{\hbar^2}{4}\left[-\{\lambda_{84}(E,\eta_g)\}'+\frac{\{\lambda_{84}(E,\eta_g)\}'\lambda_{84}(E,\eta_g)+2\{\lambda_{85}(E,\eta_g)\}'}{\sqrt{(\lambda_{84}(E,\eta_g))^2+4\lambda_{85}(E,\eta_g)}}\right]\Bigg|_{E=E_{F_h}}, \quad (11.62b)$$

in which, $\lambda_{84}(E,\eta_g) \equiv \frac{\lambda_{73}(E,\eta_g)}{\lambda_{74}(E,\eta_g)}$ and $\lambda_{85}(E,\eta_g) \equiv \frac{\lambda_{75}(E,\eta_g)}{\lambda_{74}(E,\eta_g)}$.

Thus, we can see that both the EMMs in this case exist in the band gap. In the absence of band tailing effects $\eta_g \to 0$, we get

$$m_\perp^*(E_F) = \frac{\hbar^2}{2}\left[-\{\alpha_{11}(E)\}'+\frac{\alpha_{511}\{T_{311}(E)\}'}{2\sqrt{T_{311}(E)}}\right]\Bigg|_{E=E_F}, \quad (11.62c)$$

where $\alpha_{11}(E) \equiv \frac{2m_t^+ m_t^-}{\alpha\hbar^2}\alpha_{211}(E)$, $\alpha_{211}(E) \equiv \left[\frac{1}{2m_t^*}-\frac{\alpha E}{2m_t^+}+\frac{1+\alpha E}{2m_t^-}\right]$, $\alpha_{511} \equiv \frac{2m_t^+ m_t^-}{\alpha\hbar^2}\omega_{11}$, $(\omega_{11}) \equiv \left[\frac{\alpha^2}{16}\left[\frac{1}{m_t^- m_l^+}+\frac{1}{m_l^- m_t^+}\right]^2-\frac{\alpha^2}{4m_t^- m_t^+ m_l^- m_l^+}\right]^{1/2}$, $T_{311}(E) \equiv \frac{\omega_{311}(E)}{(\omega_{11})^2}$, $\omega_{311}(E) \equiv \left[\frac{\alpha E(1+\alpha E)}{m_t^+ m_t^-}+\left[\frac{1}{2m_t^*}-\left(\frac{\alpha E}{2m_t^+}\right)+\frac{(1+\alpha E)}{2m_t^-}\right]^2\right]$.

and

$$m_{||}^{*}(E_{\mathrm{F}}) = \left(\frac{m_{\mathrm{t}}^{+} m_{\mathrm{l}}^{-}}{\alpha}\right)\left[\left(\frac{\alpha}{2m_{\mathrm{l}}^{+}} - \frac{\alpha}{2m_{\mathrm{l}}^{-}}\right) + \frac{1}{2}\left\{\frac{2\left[\frac{1}{2m_{\mathrm{l}}^{*}} + \frac{1+\alpha E}{2m_{\mathrm{l}}^{-}} - \frac{\alpha E}{2m_{\mathrm{l}}^{+}}\right]\left(\frac{\alpha}{2m_{\mathrm{l}}^{-}} - \frac{\alpha}{2m_{\mathrm{l}}^{+}}\right) + \frac{\alpha(1+2\alpha E)}{m_{\mathrm{l}}^{-} m_{\mathrm{l}}^{+}}}{\left[\left[\frac{1}{2m_{\mathrm{l}}^{*}} + \frac{1+\alpha E}{2m_{\mathrm{l}}^{-}} - \frac{\alpha E}{2m_{\mathrm{l}}^{+}}\right]^{2} + \frac{\alpha E(1+\alpha E)}{m_{\mathrm{l}}^{-} m_{\mathrm{l}}^{+}}\right]^{1/2}}\right\}\right]\Bigg|_{E=E_{\mathrm{F}}}, \tag{11.63}$$

The volume in k-space enclosed by (11.61) can be written through the integral as

$$V(E,\eta_{\mathrm{g}}) = 2\pi \int_{0}^{\lambda_{86}(E,\eta_{\mathrm{g}})} \left[-\left[\lambda_{79}(E,\eta_{\mathrm{g}}) k_z^2 + \lambda_{80}(E,\eta_{\mathrm{g}})\right] + \sqrt{\lambda_{81}(E,\eta_{\mathrm{g}}) k_z^4 + \lambda_{82}(E,\eta_{\mathrm{g}}) k_z^2 + \lambda_{83}(E,\eta_{\mathrm{g}})}\right] \mathrm{d}k_z, \tag{11.64}$$

where

$$\lambda_{86}(E,\eta_{\mathrm{g}}) \equiv \left[\frac{\sqrt{\left[\lambda_{84}(E,\eta_{\mathrm{g}})\right]^{2} + 4\lambda_{85}(E,\eta_{\mathrm{g}})} - \lambda_{84}(E,\eta_{\mathrm{g}})}{2}\right]^{1/2},$$

$$\lambda_{79}(E,\eta_{\mathrm{g}}) \equiv \frac{\lambda_{71}(E,\eta_{\mathrm{g}})}{2\hbar^2 Z_0(E,\eta_{\mathrm{g}})},$$

$$\lambda_{80}(E,\eta_{\mathrm{g}}) \equiv \frac{\lambda_{72}(E,\eta_{\mathrm{g}})}{2\hbar^2 Z_0(E,\eta_{\mathrm{g}})}, \quad \lambda_{81}(E,\eta_{\mathrm{g}}) \equiv \frac{\lambda_{76}(E,\eta_{\mathrm{g}})}{4\hbar^4\left[Z_0(E,\eta_{\mathrm{g}})\right]^2},$$

$$\lambda_{76}(E,\eta_{\mathrm{g}}) \equiv \left[\lambda_{71}(E,\eta_{\mathrm{g}})\right]^2,$$

$$\lambda_{82}(E,\eta_{\mathrm{g}}) \equiv \frac{\lambda_{77}(E,\eta_{\mathrm{g}})}{9\hbar^4\left[Z_0(E,\eta_{\mathrm{g}})\right]^2},$$

$$\lambda_{77}(E,\eta_{\mathrm{g}}) \equiv \left[2\lambda_{71}(E,\eta_{\mathrm{g}})\lambda_{72}(E,\eta_{\mathrm{g}}) - 4\lambda_{70}(E,\eta_{\mathrm{g}})\lambda_{73}(E,\eta_{\mathrm{g}}) - 4\lambda_{70}(E,\eta_{\mathrm{g}})\lambda_{74}(E,\eta_{\mathrm{g}})\right],$$

$$\lambda_{83}(E,\eta_{\mathrm{g}}) \equiv \frac{\lambda_{78}(E,\eta_{\mathrm{g}})}{9\hbar^4\left[Z_0(E,\eta_{\mathrm{g}})\right]^2} \text{ and}$$

$$\lambda_{78}(E,\eta_{\mathrm{g}}) \equiv \left[4\lambda_{70}(E,\eta_{\mathrm{g}})\lambda_{75}(E,\eta_{\mathrm{g}})\right].$$

Thus,

$$V(E,\eta_{\mathrm{g}}) = \left[\left[\lambda_{87}(E,\eta_{\mathrm{g}})\right] \int_{0}^{\lambda_{86}(E,\eta_{\mathrm{g}})} \left[\sqrt{k_z^4 + \lambda_{88}(E,\eta_{\mathrm{g}}) k_z^2 + \lambda_{89}(E,\eta_{\mathrm{g}})}\right] \mathrm{d}k_z - \lambda_{90}(E,\eta_{\mathrm{g}})\right], \tag{11.65}$$

where $\lambda_{87}(E,\eta_g) \equiv 2\pi\sqrt{\lambda_{81}(E,\eta_g)}$, $\lambda_{88}(E,\eta_g) \equiv \frac{\lambda_{82}(E,\eta_g)}{\lambda_{81}(E,\eta_g)}$, $\lambda_{89}(E,\eta_g) \equiv \frac{\lambda_{83}(E,\eta_g)}{\lambda_{81}(E,\eta_g)}$ and $\lambda_{90}(E,\eta_g) \equiv 2\pi\left[\frac{\lambda_{79}(E,\eta_g)\{\lambda_{86}(E,\eta_g)\}^3}{3} + \lambda_{80}(E,\eta_g)\lambda_{89}(E,\eta_g)\right]$.

Using (11.21), (11.65) can be written as

$$V(E,\eta_g) = [\lambda_{87}(E,\eta_g)\lambda_{95}(E,\eta_g) - \lambda_{90}(E,\eta_g)], \tag{11.66}$$

in which,

$$\lambda_{95}(E,\eta_g) \equiv \left[\frac{\lambda_{91}(E,\eta_g)}{3}[-E_i[\lambda_{93}(E,\eta_g),\lambda_{94}(E,\eta_g)]\left[\{\lambda_{91}(E,\eta_g)\}^2 + 2\{\lambda_{92}(E,\eta_g)\}^2 F_i[\lambda_{93}(E,\eta_g),\lambda_{94}(E,\eta_g)]\right] + \left(\frac{\{\lambda_{86}(E,\eta_g)\}}{3}\right)\left[\{\lambda_{86}(E,\eta_g)\}^2 + \{\lambda_{91}(E,\eta_g)\}^2 + 2\{\lambda_{92}(E,\eta_g)\}^2\right]\left[\left[\{\lambda_{91}(E,\eta_g)\}^2 + \{\lambda_{86}(E,\eta_g)\}^2\right]^{1/2}\left[\{\lambda_{92}(E,\eta_g)\}^2 + \{\lambda_{86}(E,\eta_g)\}^2\right]^{-1/2}\right]\right],$$

$\{\lambda_{91}(E,\eta_g)\}^2 \equiv \frac{1}{2}\left[\sqrt{\{\lambda_{88}(E,\eta_g)\}^2 - 4\lambda_{89}(E,\eta_g)} + \lambda_{88}(E,\eta_g)\right]$, $E_i[\lambda_{93}(E,\eta_g),\lambda_{94}(E,\eta_g)]$ is the incomplete elliptic integral of the second kind and is given by [21],

$E_i[\lambda_{93}(E,\eta_g),\lambda_{94}(E,\eta_g)] \equiv \int_0^{\lambda_{93}(E,\eta_g)}\left[\left\{1 - \{\lambda_{94}(E,\eta_g)\}^2\sin^2\xi\right\}^{1/2}\right]d\xi$,

ξ is the variable of integration in this case, $\lambda_{93}(E,\eta_g) \equiv \tan^{-1}\left[\frac{\lambda_{86}(E,\eta_g)}{\lambda_{92}(E,\eta_g)}\right]$, $\{\lambda_{92}(E,\eta_g)\}^2 \equiv \frac{1}{2}\left[\lambda_{88}(E,\eta_g) - \sqrt{\{\lambda_{88}(E,\eta_g)\}^2 - 4\lambda_{89}(E,\eta_g)}\right]$, $\lambda_{94}(E,\eta_g) \equiv \frac{\sqrt{\{\lambda_{91}(E,\eta_g)\}^2 - \{\lambda_{92}(E,\eta_g)\}^2}}{\lambda_{91}(E,\eta_g)}$, $F_i[\lambda_{93}(E,\eta_g),\lambda_{94}(E,\eta_g)]$ is the incomplete elliptic integral of the second kind and is given by [21], $F_i[\lambda_{93}(E,\eta_g),\lambda_{94}(E,\eta_g)] \equiv \int_0^{\lambda_{93}(E,\eta_g)}\left[\left\{1 - \{\lambda_{94}(E,\eta_g)\}^2\sin^2\xi\right\}^{-1/2}\right]d\xi$.

Using (2.3a) and (7.66), the density-of-states function is given by

$$N_{HD}(E,\eta_g) = \frac{g_v}{4\pi^3}[\{\lambda_{87}(E,\eta_g)\}'\lambda_{95}(E,\eta_g) + \{\lambda_{95}(E,\eta_g)\}'\lambda_{87}(E,\eta_g) - \{\lambda_{90}(E,\eta_g)\}'], \tag{11.67}$$

Therefore the electron concentration at low temperature can be expressed as

$$n_0 = \frac{g_v}{4\pi^3}[\{\lambda_{87}(E_{F_h},\eta_g)\}\lambda_{95}(E_{F_h},\eta_g) - \{\lambda_{90}(E_{F_h},\eta_g)\}], \tag{11.68}$$

In this case, $\bar{E}_{hd}$ is given by

$$\left\{\lambda_{75}\left(\bar{E}_{hd},\eta_g\right)\right\} = 0. \tag{11.69}$$

Thus, one can numerically evaluate the DMR by using (11.68), (11.31), (11.69) and the allied definitions in this case.

11.2.5 Study of the Einstein Relation in Heavily Doped Stressed Materials Forming Gaussian Band Tails

Using (7.88) and (11.2), we can write,

$$\begin{aligned} &I(4)k^2 - T_{17}I(1)k_x^2 - T_{27}I(1)k_y^2 - T_{37}k_z^2I(1) \\ &= [q_{67}I(6) - R_{67}I(5) + V_{67}I(4) + \rho_{67}I(1)], \end{aligned} \quad (11.70)$$

where the symbols T_{17}, T_{27}, T_{37}, q_{67}, R_{67}, V_{67} and ρ_{67} have already been defined in the definitions of the symbols of (7.88) and

$$I(6) = \int_{-\infty}^{E} (E - V)^3 F(V)\,\mathrm{d}V. \quad (11.71)$$

Equation (11.71) can be written as

$$I(6) = E^3 I(1) - 3E^2 I(7) + 3EI(8) - I(9). \quad (11.72)$$

In which,

$$I(7) = \int_{-\infty}^{E} VF(V)\,\mathrm{d}V, \quad (11.73)$$

$$I(8) = \int_{-\infty}^{E} V^2 F(V)\,\mathrm{d}V, \quad (11.74)$$

$$I(9) = \int_{-\infty}^{E} V^3 F(V)\,\mathrm{d}V. \quad (11.75)$$

Using (11.2) and successively (11.73)–(11.75) together with simple algebraic manipulations, one obtains

$$I(7) = \frac{-\eta_{\mathrm{g}}}{2\sqrt{\pi}} \exp\left(\frac{-E^2}{\eta_{\mathrm{g}}^2}\right), \quad (11.76)$$

$$I(8) = \frac{\eta_{\mathrm{g}}^2}{4}\left[1 + \mathrm{Erf}\left(\frac{E}{\eta_{\mathrm{g}}}\right)\right], \quad (11.77)$$

and

$$I(9) = \frac{-\eta_{\mathrm{g}}^3}{2\sqrt{\pi}} \exp\left(\frac{-E^2}{\eta_{\mathrm{g}}^2}\right)\left[1 + \frac{E^2}{\eta_{\mathrm{g}}^2}\right]. \quad (11.78)$$

Thus, (11.72) can be written as

$$I(6) = \left[\frac{E}{2}\left[1 + \mathrm{Erf}\left(\frac{E}{\eta_g}\right)\right]\left[E^2 + \frac{3}{2}\eta_g^2\right] + \frac{\eta_g}{2\sqrt{\pi}} \exp\left(\frac{-E^2}{\eta_g^2}\right)\left[4E^2 + \eta_g^2\right]\right] \quad (11.79)$$

Thus, combining the appropriate equations, the dispersion relations of the conduction electrons in heavily doped stressed materials can be expressed as

$$P_{11}(E,\eta_g)\,k_x^2 + Q_{11}(E,\eta_g)\,k_y^2 + S_{11}(E,\eta_g)\,k_z^2 = 1, \tag{11.80}$$

where

$$P_{11}(E,\eta_g) \equiv \left[\frac{\gamma_0(E,\eta_g) - (T_{17}/2)\left[1+\mathrm{Erf}(E/\eta_g)\right]}{\Delta_{14}(E,\eta_g)}\right],$$

$$\Delta_{14}(E,\eta_g) \equiv \left[q_{67}\left\{\frac{E}{2}\left[1+\mathrm{Erf}\left(\frac{E}{\eta_g}\right)\right]\left[E^2+\frac{3}{2}\eta_g^2\right] + \frac{\eta_g}{2\sqrt{\pi}}\exp\left(\frac{-E^2}{\eta_g^2}\right)\left[4E^2+\eta_g^2\right]\right\} - R_{67}\theta_0(E,\eta_g) + V_{67}\gamma_0(E,\eta_g) + \frac{\rho_{67}}{2}\left[1+\mathrm{Erf}(E/\eta_g)\right]\right],$$

$$Q_{11}(E,\eta_g) \equiv \left[\frac{\gamma_0(E,\eta_g) - (T_{27}/2)\left[1+\mathrm{Erf}(E/\eta_g)\right]}{\Delta_{14}(E,\eta_g)}\right]$$

and $S_{11}(E,\eta_g) \equiv \left[\frac{\gamma_0(E,\eta_g) - (T_{37}/2)\left[1+\mathrm{Erf}(E/\eta_g)\right]}{\Delta_{14}(E,\eta_g)}\right]$.

Thus, the energy spectrum in this case is real.

The EMMs along x, y and z directions in this case can be written as

$$m_{xx}^*(E_{F_h},\eta_g) = \frac{\hbar^2}{2}\left[\left[\gamma_0(E_{F_h},\eta_g) - (T_{17}/2)\left[1+\mathrm{Erf}(E_{F_h}/\eta_g)\right]\right]^{-2} \left[\{\Delta_{14}(E_{F_h},\eta_g)\}'\left[\gamma_0(E_{F_h},\eta_g) - (T_{17}/2)\left[1+\mathrm{Erf}(E_{F_h}/\eta_g)\right]\right]\right] - \Delta_{14}(E_{F_h},\eta_g)\left[\frac{1}{2}\left[1+\mathrm{Erf}\left(\frac{E_{F_h}}{\eta_g}\right)\right] - \left\{\frac{T_{17}}{\eta_g\sqrt{\pi}}\exp\left(\frac{-E_{F_h}^2}{\eta_g^2}\right)\right\}\right]\right], \tag{11.81}$$

$$m_{yy}^*(E_{F_h},\eta_g) = \frac{\hbar^2}{2}\left[\left[\gamma_0(E_{F_h},\eta_g) - (T_{27}/2)\left[1+\mathrm{Erf}(E_{F_h}/\eta_g)\right]\right]^{-2} \left[\{\Delta_{14}(E_{F_h},\eta_g)\}'\left[\gamma_0(E_{F_h},\eta_g) - (T_{27}/2)\left[1+\mathrm{Erf}(E_{F_h}/\eta_g)\right]\right]\right] - \Delta_{14}(E_{F_h},\eta_g)\left[\frac{1}{2}\left[1+\mathrm{Erf}\left(\frac{E_{F_h}}{\eta_g}\right)\right] - \left\{\frac{T_{27}}{\eta_g\sqrt{\pi}}\exp\left(\frac{-E_{F_h}^2}{\eta_g^2}\right)\right\}\right]\right], \tag{11.82}$$

and

$$m_{zz}^*(E_{F_h},\eta_g)=\frac{\hbar^2}{2}\Bigg[[\gamma_0(E_{F_h},\eta_g)-(T_{37}/2)[1+\mathrm{Erf}(E_{F_h}/\eta_g)]]^{-2}$$
$$[\{\Delta_{14}(E_{F_h},\eta_g)\}'[\gamma_0(E_{F_h},\eta_g)-(T_{37}/2)[1+\mathrm{Erf}(E_{F_h}/\eta_g)]]]$$
$$-\Delta_{14}(E_{F_h},\eta_g)\left[\frac{1}{2}\left[1+\mathrm{Erf}\left(\frac{E_{F_h}}{\eta_g}\right)\right]-\left\{\frac{T_{37}}{\eta_g\sqrt{\pi}}\exp\left(\frac{-E_{F_h}^2}{\eta_g^2}\right)\right\}\right]\Bigg], \tag{11.83}$$

Thus, we can see that the EMMs in this case exist within the band gap. In the absence of band tailing effects $\eta_g \to 0$, we get

$$m_{xx}^*(E_F)=\hbar^2\bar{a}_0(E_F)\{\bar{a}_0(E_F)\}', \tag{11.84}$$

$$m_{xx}^*(E_F)=\hbar^2\bar{b}_0(E_F)\{\bar{b}_0(E_F)\}', \tag{11.85}$$

and

$$m_{xx}^*(E_F)=\hbar^2\bar{c}_0(E_F)\{\bar{c}_0(E_F)\}'. \tag{11.86}$$

The density-of-states function in this case can be written as

$$N_{HD}(E,\eta_g)=\frac{g_v}{3\pi^2}\{\Delta_{15}(E,\eta_g)\}^{-2}$$
$$\times\left[\frac{3}{2}\{\Delta_{15}(E,\eta_g)\}\sqrt{\Delta_{14}(E,\eta_g)}\{\Delta_{14}(E,\eta_g)\}' \right. \tag{11.87}$$
$$\left. -\{\Delta_{14}(E,\eta_g)\}^{3/2}\{\Delta_{15}(E,\eta_g)\}'\right],$$

where

$$\Delta_{15}(E,\eta_g)\equiv[[\gamma_0(E,\eta_g)-(T_{17}/2)[1+\mathrm{Erf}(E/\eta_g)]]$$
$$[\gamma_0(E,\eta_g)-(T_{27}/2)[1+\mathrm{Erf}(E/\eta_g)]]$$
$$[\gamma_0(E,\eta_g)-(T_{37}/2)[1+\mathrm{Erf}(E/\eta_g)]]]^{1/2}.$$

Using (11.87), the electron concentration at low temperatures can be written as

$$n_0=\frac{g_v}{3\pi^2}\left[\frac{\{\Delta_{14}(E_{F_h},\eta_g)\}^{3/2}}{\Delta_{15}(E_{F_h},\eta_g)}\right]. \tag{11.88}$$

In this case, $\bar{E}_{hd}$ is given by

$$\{\Delta_{14}(\bar{E}_{hd},\eta_g)\}=0. \tag{11.89}$$

Thus, one can numerically evaluate the DMR by using (11.88), (11.31), (11.89) and the allied definitions in this case.

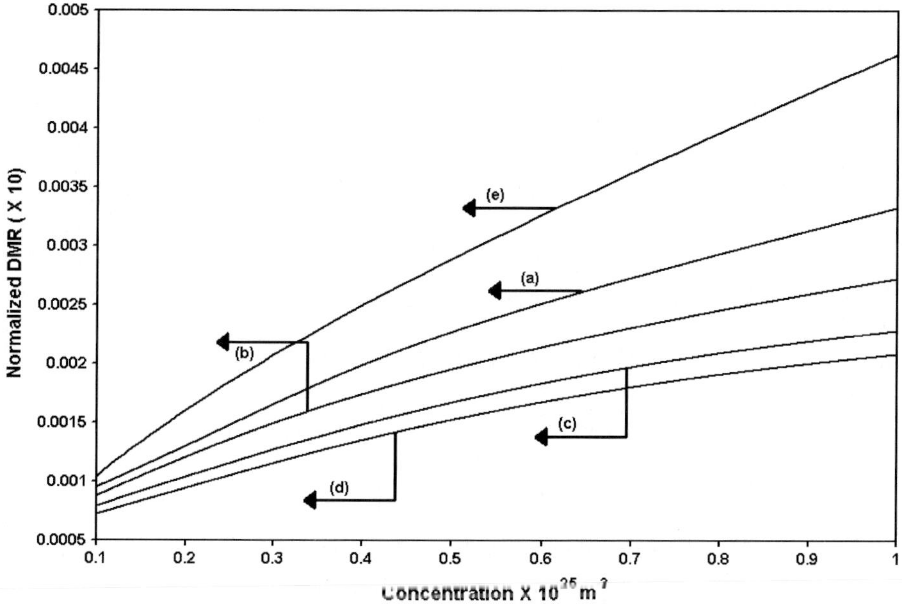

Fig. 11.1. The plot of the normalized DMR in the heavily doped n-Cd_3As_2 as function of electron concentration in accordance with (**a**) the generalized band model, (**b**) $\delta = 0$, (**c**) the three band model of Kane, (**d**) the two band model of Kane and (**e**) the parabolic energy bands

11.3 Result and Discussions

Using (11.30)–(11.32) together with the energy band constants as given in Table 2.1 of Chap. 2, the DMR has been numerically computed for heavily doped n-Cd_3As_2 as a function of electron concentration as shown in curve (a) of Fig. 11.1. The curve (b) corresponds to $\delta = 0$ and the curve (c) exhibits the dependence of the DMR on n_0 in accordance with the heavily doped three-band model of Kane, respectively. The plots (d) and (e) correspond to the heavily doped two-band model of Kane and that of parabolic energy bands respectively. By comparing the curves (a) and (b) of Fig. 11.1, one can assess the influence of crystal field splitting of the DMR in heavily doped tetragonal compounds. Figure 11.2 represents all cases of Fig. 11.1 for heavily n-$CdGeAs_2$. It appears from Figs. 11.1 and 11.2 that, the DMR in heavily doped tetragonal and nonlinear optical compounds increases with increasing carrier degeneracy as expected for degenerate materials without band tails.

One can numerically evaluate the DMR as a function of electron concentration in heavily doped III–V compounds by using n-InAs, and n-InSb as examples as shown in Figs. 11.3 and 11.4 by curves (a), (b) and (c) respectively, in accordance with the heavily doped three ((11.37), (11.31) and

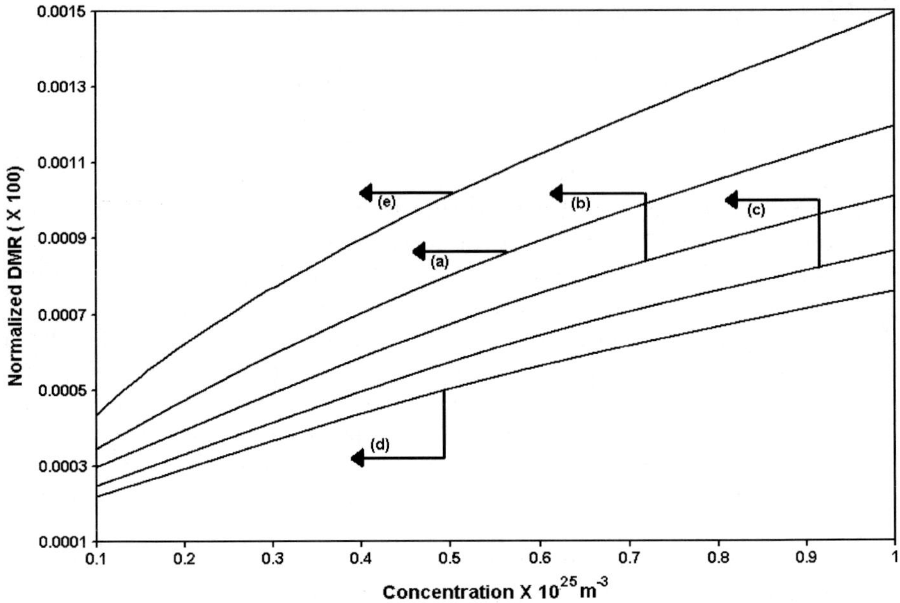

Fig. 11.2. The plot of the normalized DMR in the heavily doped n-$CdGeAs_2$ as function of electron concentration in accordance with (**a**) the generalized band model, (**b**) $\delta = 0$, (**c**) the three band model of Kane, (**d**) the two band model of Kane and (**e**) the parabolic energy bands

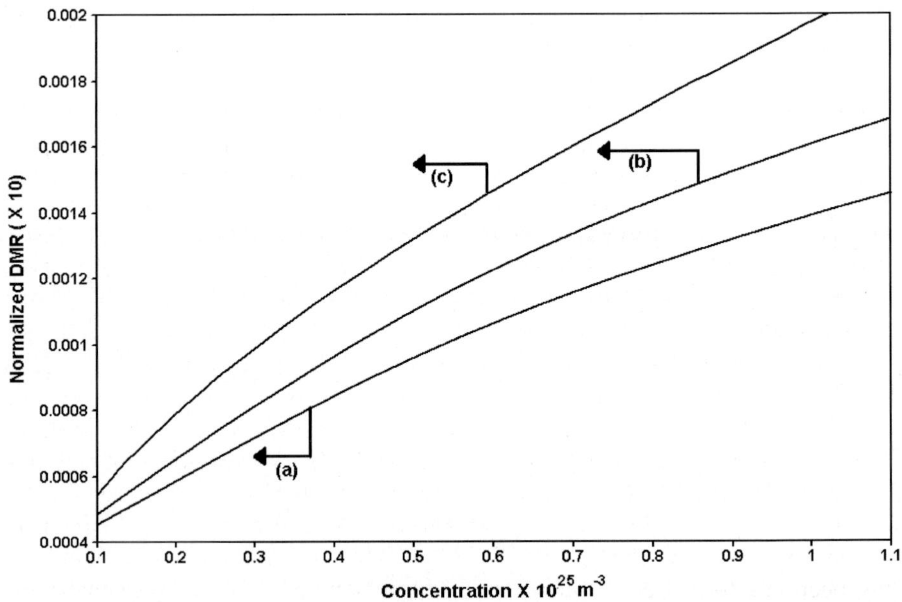

Fig. 11.3. The plot of the normalized DMR in heavily doped n-InAs as function of electron concentration in accordance with (**a**) the three band model of Kane, (**b**) the two band model of Kane and (**c**) the parabolic energy bands

(11.38)) and two band models of Kane ((11.43), (11.31) and (11.44)) together with the model of parabolic energy bands ((11.49), (11.31) and (11.50)).

Using n-$Hg_{1-x}Cd_xTe$ as an example of heavily doped ternary compounds, the DMR has been numerically plotted for all the band models as a function of electron concentration as shown in Fig. 11.5 for all cases of Fig. 11.3. It appears from Fig. 11.5 that the DMR in both cases of heavily doped ternary compounds increases with increasing electron concentration as usual for the degenerate compounds without band tails. Taking n-$In_{1-x}Ga_xAs_yP_{1-y}$ lattice matched to InP as an example of heavily doped quaternary compounds the DMR has been further been plotted as a function of electron concentration as shown in Fig. 11.6 in accordance with the three and two band models of Kane together with the isotropic parabolic energy band model for both the cases.

It appears that the DMR increases with increasing carrier degeneracy as usual. From Figs. 11.5 and 11.6, one can assess the influence of energy band constants on the DMR for heavily doped ternary and quaternary materials respectively.

Using (11.58), (11.31) and (11.59), the DMR has been plotted for the heavily doped p-CdS, as a function of hole concentration p_0 as shown by curves (a) and (b) in Fig. 11.7 for which $\bar{\lambda}_0 \neq 0$ and $\bar{\lambda}_0 = 0$, respectively. This has been presented for the purpose of assessing the influence of the splitting of the two spin states by the spin–orbit coupling and the crystalline field on the DMR.

Using (11.68), (11.31) and (11.69), in Fig. 11.8, the DMR has been plotted for the heavily doped (a) n-PbTe, (b) n-PbSnTe and (c) n-$Pb_{1-x}Sn_xSe$ DMR as a function of electron concentration in accordance with the heavily doped Dimmock model. For relatively low values of electron concentration, the values of the DMR for the three materials exhibit convergence behavior whereas for relatively large values of n_0, the numerical values differ widely from each other. Our present analysis is also valid for p-type IV–VI compounds with the proper change in the energy band constants.

Using (11.88), (11.31) and (11.89), in Fig. 11.9, the DMR has been plotted for the heavily doped stressed n-InSb as a function of electron concentration. For the purpose of assessing the influence of stress on the DMR in bulk specimens of stressed heavily doped n-InSb, plot (a) exhibits the DMR in the presence of the stress while plot (b) shows the same in the absence of the stress. In the presence of the stress, the magnitude of the DMR is being increased as compared with the same under stress free condition for both the cases. One important concept of this chapter is that the presence of singularities in the dispersion relation of the undoped materials creates the complex energy spectrum in the corresponding heavily doped samples. Besides, from the density-of-states function in this case, it appears that a new forbidden zone has been created in addition to the normal band gap of the semiconductor. If the basic undoped dispersion relation is pole-less, the corresponding heavily doped energy band spectrum will be real, although it may be the complicated

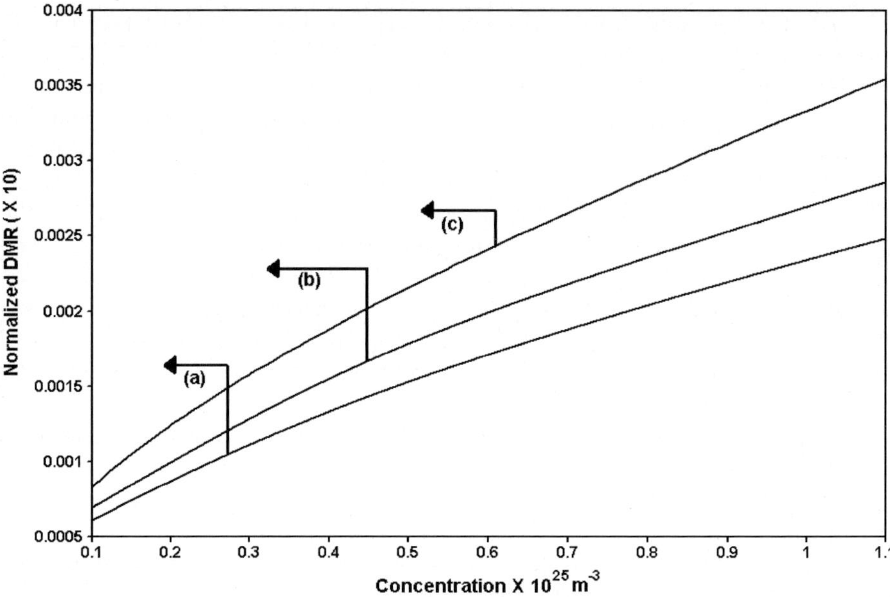

Fig. 11.4. The plot of the normalized DMR in heavily doped n-InSb as function of electron concentration in accordance with (**a**) the three band model of Kane, (**b**) the two band model of Kane and (**c**) the parabolic energy bands

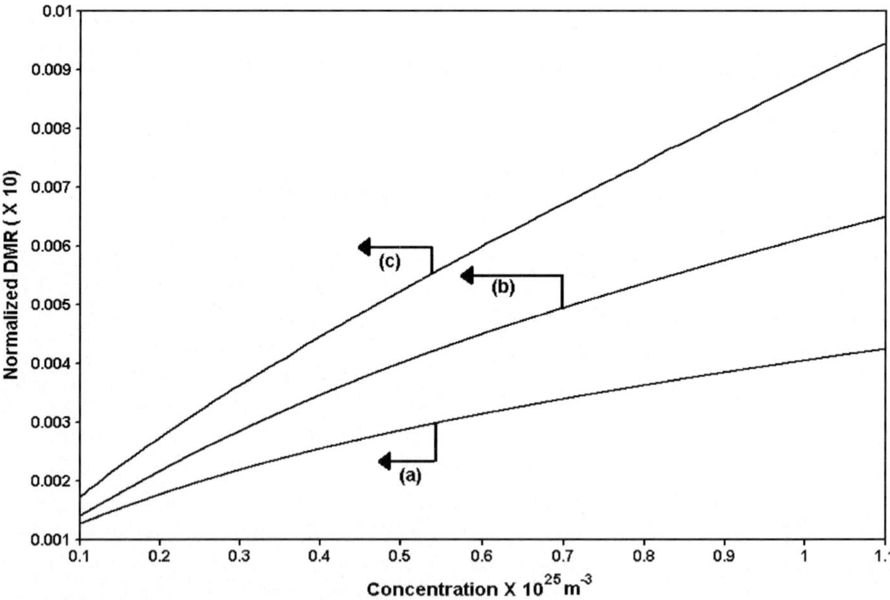

Fig. 11.5. The plot of the normalized DMR in heavily doped n-$Hg_{1-x}Cd_xTe$ as function of electron concentration in accordance with (**a**) the three band model of Kane, (**b**) the two band model of Kane and (**c**) the parabolic energy bands

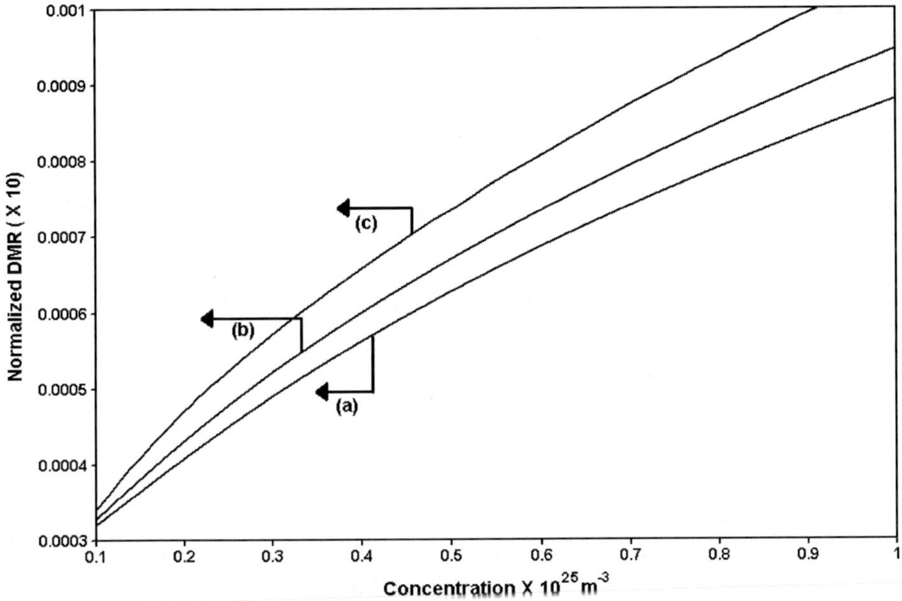

Fig. 11.6. The plot of the normalized DMR in heavily doped n-$In_{1-x}Ga_xAs_yP_{1-y}$ lattice matched to InP as function of electron concentration in accordance with (**a**) the three band model of Kane, (**b**) the two band model of Kane and (**c**) the parabolic energy bands

functions of exponential and error functions and deviate considerably from the undoped case.

Another important point in this context is the existence of the effective mass within the forbidden zone, which is impossible without the band tailing effect. It is an amazing fact that though heavily doped semiconductors have been extensively studied in the literature [1–22], the study of the carrier transport in heavily doped materials through proper formulation of the Boltzmann transport equation which needs in turn, the corresponding heavily doped carrier energy spectra is still one of the open research problems.

For the purpose of condensed presentation, the specific electron statistics related to a particular energy dispersion law for specific materials has been presented in Table 11.1.

11.4 Open Research Problems

11.1. Investigate the Einstein relation for all the materials as given in problems in R.2.1 of Chap. 2 in the presence of the Gaussian type band tails.

11.2. Investigate the Einstein relation in the presence of an arbitrarily oriented quantizing magnetic field in heavily doped tetragonal semiconductors

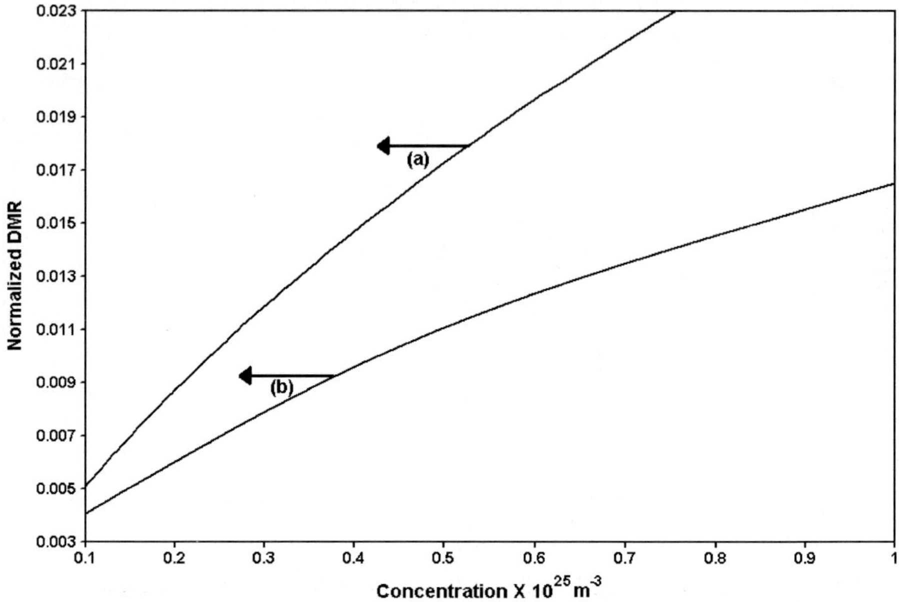

Fig. 11.7. The plot of the normalized DMR in heavily doped p-CdS as function of as function of hole concentration in accordance with (**a**) $\bar{\lambda}_0 \neq 0$ and (**b**) $\bar{\lambda}_0 = 0$

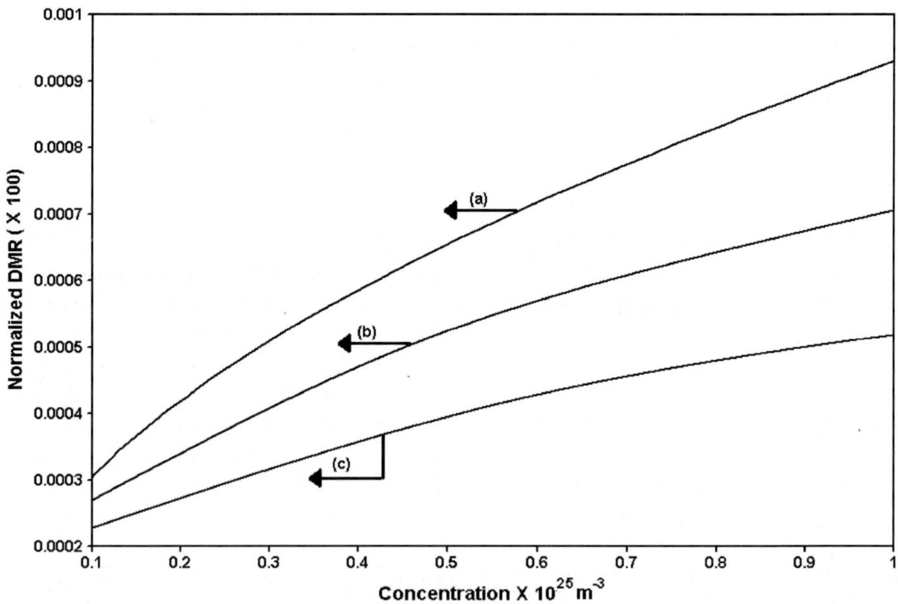

Fig. 11.8. The plot of the normalized DMR in heavily doped (**a**) n-PbTe, (**b**) PbSnTe and (**c**) $Pb_{1-x}Sn_xSe$ as function of electron concentration

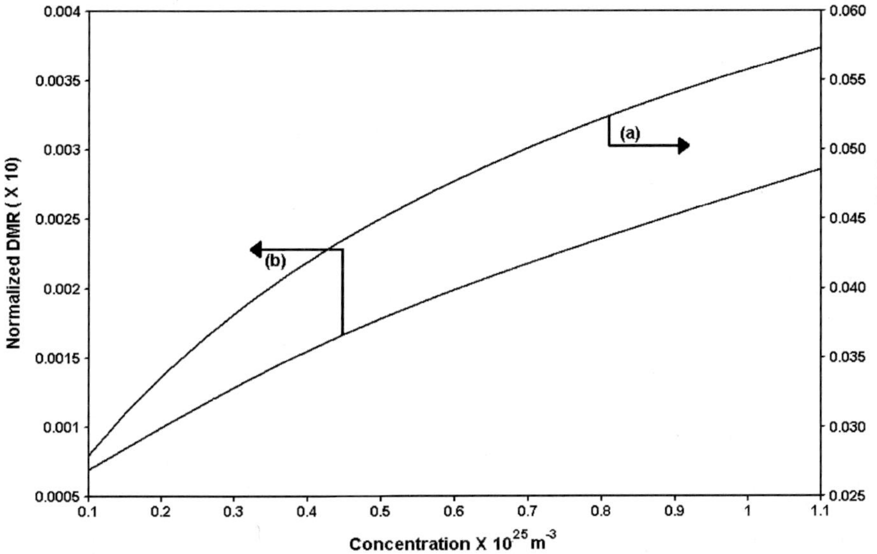

Fig. 11.9. The plot of the normalized DMR in heavily doped stressed n InSb as a function of electron concentration in which the curve (**a**) is in the presence of stress and curve (**b**) is under absence of stress

by including broadening and the electron spin. Study all the special cases for heavily doped III–V, ternary and quaternary materials in this context.

11.3. Investigate the Einstein relations for heavily doped IV–VI, II–VI and stressed Kane type compounds in the presence of an arbitrarily oriented quantizing magnetic field by including broadening and electron spin.

11.4. Investigate the Einstein relation for all the materials as stated in R.2.1 of Chap. 2 in the presence of an arbitrarily oriented quantizing alternating magnetic field by including broadening and electron spin under the condition of heavily doping.

11.5. Investigate the Einstein relation in the presence of an arbitrarily oriented quantizing magnetic field and crossed electric fields in heavily doped tetragonal semiconductors by including broadening and the electron spin. Study all the special cases for heavily doped III–V, ternary and quaternary materials in this context.

11.6. Investigate the Einstein relations for heavily doped IV–VI, II–VI and stressed Kane type compounds in the presence of an arbitrarily oriented quantizing magnetic field and crossed electric fields by including broadening and electron spin.

Table 11.1. The carrier statistics and the Einstein relation in the heavily doped tetragonal, III–V, ternary, quaternary, II–VI, IV–VI and stressed materials

Type of materials	The carrier statistics in quantum limit	The Einstein relation for the diffusivity mobility ratio in quantum limit
1. Tetragonal compounds	In accordance with the generalized dispersion relation as formulated in this chapter $n_0 = \frac{2m_{\perp}^{*} g_{\mathrm{v}} \sqrt{2m_{\|}^{*}}}{3\pi^2 \hbar^3} \left[I_{11}\left(E_{\mathrm{F_h}}, \eta_{\mathrm{g}}\right)\right]$ (11.30)	$\frac{D}{\mu} = \frac{n_0}{\lvert e\rvert} \frac{\partial n_0}{\partial\left(E_{\mathrm{F_h}} - \bar{E}_{\mathrm{hd}}\right)}$ (11.31) For heavily doped tetragonal semiconductors, $\bar{E}_{\mathrm{hd}}$ is the smallest negative root of $\left[T_{27}\left(\bar{E}_{\mathrm{hd}}, \eta_{\mathrm{g}}\right) T_{29}\left(\bar{E}_{\mathrm{hd}}, \eta_{\mathrm{g}}\right) - T_{28}\left(\bar{E}_{\mathrm{hd}}, \eta_{\mathrm{g}}\right) T_{30}\left(\bar{E}_{\mathrm{hd}}, \eta_{\mathrm{g}}\right)\right] = 0$ (11.32)
2. III–V, ternary and quaternary compounds	In accordance with the three band model of Kane which is a special case of our generalized analysis $n_0 = \frac{g_{\mathrm{v}}}{3\pi^2}\left(\frac{2m^*}{\hbar^2}\right)^{3/2} \frac{1}{\sqrt{2}} \left[T_{33}\left(E_{\mathrm{F_h}}, \eta_{\mathrm{g}}\right) + \sqrt{\left\{T_{33}\left(E_{\mathrm{F_h}}, \eta_{\mathrm{g}}\right)\right\}^2 + \left\{T_{34}\left(E_{\mathrm{F_h}}, \eta_{\mathrm{g}}\right)\right\}^2}\right]^{1/2}$ (11.37) (2.8). Equation (11.37) is a special case of (11.30). For the two band model of Kane $n_0 = \frac{g_{\mathrm{v}}}{3\pi^2}\left(\frac{2m^*}{\hbar^2}\right)^{3/2} \left\{\gamma_2\left(E_{\mathrm{F_h}}, \eta_{\mathrm{g}}\right)\right\}^{3/2}$ (11.43). Equation (11.43) is a special case of (11.37) For parabolic energy bands $n_0 = \frac{g_{\mathrm{v}}}{3\pi^2}\left(\frac{2m^*}{\hbar^2}\right)^{3/2} \left\{\gamma_3\left(E_{\mathrm{F_h}}, \eta_{\mathrm{g}}\right)\right\}^{3/2}$ (11.49). Equation (11.49) is a special case of (11.43)	$T_{31}\left(\bar{E}_{\mathrm{hd}}, \eta_{\mathrm{g}}\right) = 0$ (11.38) Equation (11.38) is a special case of (11.32). For the two band model of Kane, $\bar{E}_{\mathrm{hd}}$ is given by $\gamma_2\left(\bar{E}_{\mathrm{hd}}, \eta_{\mathrm{g}}\right) = 0$ (11.44) Equation (11.44) is a special case of (11.32). For parabolic energy bands, $\bar{E}_{\mathrm{hd}}$ is given by $\gamma_3\left(\bar{E}_{\mathrm{hd}}, \eta_{\mathrm{g}}\right) = 0$ (11.50) Equation (11.50) is a special case of (11.44)

3. II–VI compounds	$n_0 = \frac{g_v}{3\pi^2 a_0' \sqrt{b_0'}} \left[\{\gamma_3(E_{F_h}, \eta_g)\}^{3/2} + \frac{3}{8} \frac{(\bar{\lambda}_0)^2 \sqrt{\gamma_3(E_{F_h}, \eta_g)}}{a_0'} - \left(\frac{3}{4}\frac{\bar{\lambda}_0}{\sqrt{a_0'}}\right) \left(\gamma_3(E_{F_h}, \eta_g) + \frac{(\bar{\lambda}_0)^2}{4a_0'}\right) \sin^{-1}\left[\frac{\sqrt{\gamma_3(E_{F_h}, \eta_g)}}{\sqrt{\gamma_3(E_{F_h}, \eta_g) + \frac{(\bar{\lambda}_0)^2}{4a_0'}}}\right]\right]$ (11.58)	In this case, $\bar{E}_{hd}$ is given by $\{\gamma_3(\bar{E}_{hd}, \eta_g)\} = 0$ (11.59)
5. IV–VI compounds	$n_0 = \frac{g_v}{4\pi^3} [\{\lambda_{87}(E_{F_h}, \eta_g)\} \lambda_{95}(E_{F_h}, \eta_g) - \{\lambda_{90}(E_{F_h}, \eta_g)\}]$ (11.68)	In this case, $\bar{E}_{hd}$ is given by $\{\lambda_{75}(\bar{E}_{hd}, \eta_g)\} = 0$ (11.69)
6. Stressed compounds	$n_0 = \frac{g_v}{3\pi^2} \left[\frac{\{\Delta_{14}(E_{F_h}, \eta_g)\}^{3/2}}{\Delta_{15}(E_{F_h}, \eta_g)}\right]$ (11.88)	In this case, $\bar{E}_{hd}$ is given by $\{\Delta_{14}(\bar{E}_{hd}, \eta_g)\} = 0$ (11.89)

11.7. Investigate the Einstein relation for all the materials as stated in R.2.1 of Chap. 2 in the presence of an arbitrarily oriented quantizing magnetic field and crossed electric fields by including broadening and electron spin under the condition of heavy doping.

11.8. Investigate the Einstein relation in ultrathin films of heavily doped tetragonal, III–V, II–VI, IV–VI and stressed Kane type semiconductors.

11.9. Investigate the Einstein relation for heavily doped ultrathin films of all the materials as considered in problems R.2.1.

11.10. Investigate the Einstein relation in the presence of an arbitrarily oriented non-quantizing alternating magnetic field for the ultrathin films of heavily doped tetragonal semiconductors by including the electron spin. Study all the special cases for III–V, ternary and quaternary materials in this context.

11.11. Investigate the Einstein relations in ultrathin films of heavily doped IV–VI, II–VI and stressed Kane type compounds in the presence of an arbitrarily oriented non-quantizing alternating magnetic field by including the electron spin.

11.12. Investigate the Einstein relation for heavily doped ultrathin films of all the materials as stated in R.2.1 of Chap. 2 in the presence of an arbitrarily oriented alternating magnetic field by including electron spin and broadening.

11.13. Investigate the Einstein relation for all the problems from R.5.1 to R.5.2 under an additional arbitrarily oriented non-uniform electric field in the presence of heavy doping.

11.14. Investigate the Einstein relation for all the problems of R.5.3 under the arbitrarily oriented crossed electric and magnetic fields in the presence of heavy doping.

11.15. Investigate the Einstein relation for all the problems from R.5.1 to R.5.5 in the presence of finite potential well under the condition of heavy doping.

11.16. Investigate the Einstein relation for all the problems from R.5.1 to R.5.5 in the presence of parabolic potential well under the condition heavy doping.

11.17. Investigate the Einstein relation for all the problems from R.5.1 to R.5.5 in the presence of circular potential well under the condition of heavy doping.

11.18. Investigate the Einstein relation for accumulation layers of heavily doped tetragonal, III–V, IV–VI, II–VI and stressed Kane type semiconductors in the presence of an arbitrary electric quantization.

11.19. Investigate the Einstein relation in accumulation layers of all the materials as stated in R.2.1 of Chap. 2 under the condition of heavy doping and in the presence of electric quantization along arbitrary direction.

11.20. Investigate the Einstein relation in the presence of an arbitrarily oriented electric quantization for accumulation layers of heavily doped tetragonal semiconductors. Study all the special cases for III–V, ternary and quaternary materials in this context.

11.21. Investigate the Einstein relations in accumulation layers of heavily doped IV–VI, II–VI and stressed Kane type compounds in the presence of an arbitrarily oriented electric quantization.

11.22. Investigate the Einstein relation in accumulation layers of all the materials as stated in R.2.1 of Chap. 2 in the presence of an arbitrarily oriented quantizing electric field under the condition of heavy doping.

11.23. Investigate the Einstein relation in the presence of an arbitrarily oriented alternating magnetic field in accumulation layers of heavily doped tetragonal semiconductors by including the electron spin. Study all the special cases for heavily doped III–V, ternary and quaternary materials in this context.

11.24. Investigate the Einstein relations in accumulation layers of heavily doped IV–VI, II–VI and stressed Kane type compounds in the presence of an arbitrarily oriented alternating non-quantizing magnetic field by including the electron spin.

11.25. Investigate the Einstein relation in accumulation layers of all the materials as stated in R.2.1 of Chap. 2 in the presence of an arbitrarily oriented non-quantizing alternating magnetic field by including electron spin and heavy doping.

11.26. Investigate the Einstein relation in accumulation layers for all the problems from R.7.1 to R.7.6 in the presence of an additional arbitrarily oriented non-uniform electric field.

11.27. Investigate the Einstein relation in accumulation layers for all the problems from R.11.18 to R.11.20 in the presence of arbitrarily oriented crossed electric and magnetic fields.

11.28. Investigate the Einstein relation in accumulation layers for all the problems from R.11.18 to R.11.27 in the presence of surface states.

11.29. Investigate the Einstein relation in accumulation layers for all the problems from R.11.18 to R.11.27 in the presence of hot electron effects.

11.30. Investigate the Einstein relation in accumulation layers for all the problems from R.11.18 to R.11.23 by including the occupancy of the electrons in various electric subbands.

11.31. Investigate the Einstein relation in nipi structures of heavily doped tetragonal, III–V, II–VI, IV–VI and stressed Kane type materials.

11.32. Investigate the Einstein relation in nipi structures of all types of materials as discussed in problem R.2.1 as given in Chap. 2 under the condition of heavy doping.

11.33. Investigate the Einstein relation in the presence of an arbitrarily oriented non-quantizing alternating magnetic field for nipi structures of heavily doped tetragonal semiconductors by including the electron spin. Study all the special cases for heavily doped III–V, ternary and quaternary materials in this context.

11.34. Investigate the Einstein relations in nipi structures of heavily doped IV–VI, II–VI and stressed Kane type compounds in the presence of an arbitrarily oriented alternating non-quantizing magnetic field by including the electron spin.

11.35. Investigate the Einstein relation for nipi structures of all the materials as stated in R.2.1 of Chap. 2 in the presence of an arbitrarily oriented non-quantizing alternating magnetic field by including electron spin under the condition of heavy doping.

11.36. Investigate the Einstein relation for all the problems from R.11.31 to R.11.33 in the presence of an additional arbitrarily oriented alternating non-quantizing electric field.

11.37. Investigate the Einstein relation for all the problems from R.11.31 to R.11.33 in the presence of arbitrarily oriented crossed electric and magnetic fields.

11.38. Investigate the Einstein relation under magnetic quantization in heavily doped III–V, II–VI, IV–VI, HgTe/CdTe, strained layer, random, Fibonacci, polytype and sawtooth superlattices with graded interfaces.

11.39. Investigate the Einstein relation in heavily doped III–V, II–VI, IV–VI, HgTe/CdTe, strained layer, random, Fibonacci, polytype and sawtooth quantum well superlattices.

11.40. Investigate the Einstein relation in the presence of an arbitrarily oriented quantizing magnetic field by including electron spin in heavily doped III–V, II–VI, IV–VI, HgTe/CdTe, strained layer, random, Fibonacci, polytype and sawtooth Superlattices with graded interfaces and the corresponding effective mass superlattices.

11.41. Investigate all the problems from R.11.1 to R. 11.40, in the presence of light waves.

11.42. Investigate all the problems from R.11.1 up to R.11.41 in the presence of exponential, Kane, Halperin, Lax and Bonch-Bruevich band tails [12].

Allied Research Problems

11.43. Investigate the EMM for all the problems from R.11.1 to R.11.42.

11.44. Investigate the Debye screening length, the carrier contribution to the elastic constants, the heat capacity, the activity coefficient and the plasma frequency for all the materials covering all the cases of problems from R.11.1 to R.11.42.

11.45. Investigate the mobility for elastic and inelastic scattering mechanisms for all the materials covering all the cases of problems from R.11.1 to R.11.42.

11.46. Investigate the various transport coefficients in detail for all the materials covering all the cases of problems from R.11.1 to R.11.42.

11.47. Investigate the dia and paramagnetic susceptibilities in detail for all the materials covering all the appropriate research problems of this chapter.

11.48. Investigate all the problems from R.11.43 up to R.11.47 in the presence of exponential, Kane, Halperin, Lax and Bonch-Bruevich types band tails [12]

References

1. R.K. Willardson, A.C. Beer (eds.), *Semiconductors and Semimetals*, vol 1 (Academic, New York, 1966), p 102
2. E.O. Kane, Phys. Rev. **131**, 79 (1963); E.O. Kane, Phys. Rev. B. **139**, 343 (1965)
3. V.L. Bonch Bruevich, Sov. Phys.–Solid State **4**, 1953 (1963)
4. R.A. Logan, A.G. Chynoweth, Phys. Rev. **131**, 89 (1963)
5. C.J. Hwang, J. Appl. Phys. **40**, 3731 (1969)
6. J.I. Pankove, Phys. Rev. A **130**, 2059 (1965)
7. B.I. Halperin, M. Lax, Phys. Rev. **148**, 722 (1966)
8. R.A. Abram, G.J. Rees, B.L.H. Wilson, Adv. Phys. **27**, 799 (1978)
9. B.I. Shklovskii, A.L. Efros, *Electronics Properties of Doped Semiconductors*, vol 45 (Springer, Berlin, 1984)
10. E.O. Kane, Solid State Electron. **28**, 3 (1985)
11. P.K. Chakraborty, J.C. Biswas, J. Appl. Phys. **82**, 3328 (1997)
12. B.R. Nag, *Electron Transport in Compound Semiconductors* (Springer, New York, 1980)
13. P.E. Schmid, Phys. Rev. B **23**, 5531 (1981)
14. G.E. Jellison Jr, F.A. Modine, C.W. White, R.F. Wood, R.T. Young, Phys. Rev. Lett. **46**, 1414 (1981)
15. V.I. Fistul, *Heavily Doped Semiconductors*, ch 7 (Plenum, New York, 1969)
16. C.J. Hwang, J. Appl. Phys. **41**, 2668 (1970)
17. W. Sritrakool, H.R. Glyde, V. Sa Yakanit, Can. J. Phys. **60**, 373 (1982)
18. H. Ikoma, J. Phys. Soc. Jpn **27**, 514 (1969)

19. P.K. Chakraborty, A. Sinha, S. Bhattacharya, K.P. Ghatak, Physica B **390**, 325 (2007); P.K. Chakraborty, G.C. Datta, K.P. Ghatak, Physica Scripta **68**, 368 (2003); P.K. Chakraborty, K.P. Ghatak, J. Phys. Chem. Solids **62**, 1061 (2001); P.K. Chakraborty, K.P. Ghatak, Phys. Letts. A **288**, 335 (2001); P.K. Chakraborty, K.P. Ghatak, Phys. D Appl. Phys. **32**, 2438 (1999)
20. M. Abramowitz, I.A. Stegun, *Handbook of Mathematical Functions with Formulas, Graphs and Mathematical Tables* (Wiley, New York, 1964)
21. I.S. Gradshteyn, I.M. Ryzhik, *Tables of Integrals, Series and Products* (Academic, New York, 1965)
22. V. Heine, Proc. Phys. Soc. **81**, 300 (1963); J.N. Schulman, Y.C. Chang, Phys. Rev. B **24**, 4445 (1981)

12

Conclusion and Future Research

In this elementary book, we have investigated the simplified first order Einstein relation within the domain of the empirically well-known and well-adjusted Boltzmann transport equation. Such studies are extremely useful for practical analysis of the physical properties of semiconductor devices, in general.

Our analyses are valid under single electron approximation and invalid for totally quantized 3D wave vector space. The quantitative comparison between the theoretical formulations of the DMR for various materials under different physical conditions and the suggestion for the experimental determination of the same are not possible in many cases, as the experimental data of G are not available in the literature for all the compounds as considered here. Thus, the detailed experimental works are needed not only to uncover the phenomena, but also for an in-depth probing of the band structures of the different quantized materials which, in turn, control the key namely, the Boltzmann transport equation. In spite of such constraints, the new concepts, which have emerged from the present investigation, are really amazing in general and are discussed throughout the book.

We present to our readers the last set of challenging research (both open and allied) problems:

12.1 Investigate all the systems from Chap. 2 to Chap. 11 and the related research problems by removing all the mathematical approximations and establishing the appropriate uniqueness conditions.

12.2 Introducing new theoretical formalisms, investigate all the systems from Chap. 2 to Chap. 11 and the related research problems, in the presence of many body effects.

12.3 Introducing new theoretical formalisms, investigate all the systems of Chap. 2 and the related research problems for 3D quantization of the wave vector space of the charge carriers.

12.4 Investigate all systems as presented in Chap. 7 for magneto-accumulation layers in the presence of spin and broadening, considering the effects of surface states.
12.5 Investigate all the problems of Chap. 8 in the presence of an arbitrarily oriented alternating quantizing magnetic field, considering the influences of electron spin and broadening.
12.6 Investigate all the problems of Chap. 9 for quantum dot superlattices.
12.7 Investigate the Einstein relation for all types of quantum wire superlattices, as discussed in Chap. 9, in the presence of an arbitrarily oriented alternating quantizing magnetic field by including spin and broadening.
12.8 Investigate the Einstein relation for all types of quantum well superlattices in the presence of an arbitrarily oriented alternating quantizing magnetic field.
12.9 Investigate the higher-order Einstein relation and related research problems for all the systems from Chap. 2 to Chap. 11.
12.10 Investigate the Einstein relation for a nonlinear charge transport and related research problems for all the systems from Chap. 2 to Chap. 11.
12.11 Investigate the higher-order Einstein relation, after proper modifications, for all the systems from Chap. 2 to Chap. 11 and related research problems in the presence of magneto-size quantization.
12.12 Investigate the higher-order Einstein relation, after proper modifications, for all the systems from Chap. 2 to Chap. 11 and related research problems in the presence of an arbitrarily oriented high electric field.
12.13 Investigate the higher-order Einstein relation, after proper modifications, for all the systems from Chap. 2 to Chap. 11 and related research problems in the presence of hot electron effects.
12.14 Investigate the ambipolar Einstein relation, after proper modifications, for p-type materials for all the systems from Chap. 2 to Chap. 11.
12.15 Investigate the non-local Einstein relation and related research problems, after proper modifications, for all the systems from Chap. 2 to Chap. 11.
12.16 Investigate the Einstein relation for all the cases from Chap. 2 to Chap. 11 and related research problems introducing new theoretical formalisms for amorphous, functional, negative refractive index, macromolecular, organic, magnetic and other advanced materials.
12.17 Investigate the non-equilibrium Einstein relation and related research problems for all the cases from Chap. 2 to Chap. 11, introducing new theoretical formalisms.
12.18 Investigate problems 12.1 and 12.3–12.17 of this chapter, in the presence of many body effects.
12.19 Introducing new theoretical formalisms, investigate all appropriate problems of this chapter for quantum systems, where the Boltzmann transport equation is invalid.

Dear readers, we earnestly believe that you and only you can convert this cozy cottage of Einstein relation into a sky scraper, not only by solving the open and allied research problems, but also by extending the same in many other new directions, by inventing new research problems and new formalisms, both theoretical and experimental. The Einstein relation in nanostructures is the dual dance of the symmetry reduction of wave vector space of the charge carriers in semiconductors and quantum mechanics. Even after 30 years of continuous effort, we can see that the Einstein relation is really a sea.

Let us recall here the famous words of the Prince of Mathematics, Carl Friedrich Gauss (Prince of Mathematics-Carl Friedrich Gauss, by M. B. W. Tent, A. K. Peters Ltd. Wellesley, Massachusetts, 2006) "Nothing has been done, if something remains to be done". Make an honest academic commitment to delve deep into this topic in deference to this supreme mathematician.

Move with creativity and enjoy your work. Good luck!

Materials Index

Subject Index

Printing: Krips bv, Meppel, The Netherlands
Binding: Stürtz, Würzburg, Germany